About the Author

SIR MARTIN GILBERT is a leading historian of the modern world. A Fellow of Merton College, Oxford, since 1962 (an Honorary Fellow since 1994), he is the author of eighty books, including the multivolume official biography of Winston Churchill and the one-volume *Churchill: A Life*. He has written eight important books on the Holocaust, and published a pioneering series of historical atlases, of which the *Atlas of the Arab–Israeli Conflict* is now in its ninth edition. He is also the author of three books on Jerusalem, an acclaimed three-volume *A History of the Twentieth Century*, and comprehensive studies of both world wars, *The First World War* and *The Second World War*. Knighted in 1995, he lives in London. Details of all his books may be found on his Web site: www.martingilbert.com.

ISRAEL

A History

Other books by Martin Gilbert

Jerusalem Illustrated History Atlas
Sir Horace Rumbold: Portrait of a Diplomat
Jerusalem: Rebirth of a City
Jerusalem in the Twentieth Century
Exile and Return: The Struggle for Jewish Statehood
Israel: A History
Auschwitz and the Allies
The Jews of Hope: The Plight of Soviet Jewry Today
Shcharansky: Hero of Our Time
The Holocaust: The Jewish Tragedy
The Boys: Triumph Over Adversity
The First World War
The Second World War
D-Day
The Day the War Ended
In Search of Churchill
Empires in Conflict: A History of the Twentieth Century, 1900–1933
Descent into Barbarism: A History of the Twentieth Century, 1934–1951
Challenge to Civilization: A History of the Twentieth Century, 1952–1999
Never Again: A History of the Holocaust
The Jews in the Twentieth Century: An Illustrated History
*Letters to Auntie Fori: The 5,000-Year History of the Jewish People and
Their Faith*
The Righteous: The Unsung Heroes of the Holocaust
Churchill and America
Churchill and the Jews
Kristallnacht: Prelude to Destruction
The Somme: Heroism and Horror in the First World War
The Will of the People: Winston Churchill and Parliamentary Democracy

EDITIONS OF DOCUMENTS
Britain and Germany Between the Wars
Plough My Own Furrow: The Life of Lord Allen of Hurtwood
Servant of India: Diaries of the Viceroy's Private Secretary, 1905–1910
Surviving the Holocaust: The Kovno Ghetto Diary of Avraham Tory
Winston Churchill and Emery Reves: Correspondence 1937–1964

ISRAEL
A HISTORY

Revised and Updated Edition

MARTIN GILBERT

HARPER PERENNIAL

NEW YORK • LONDON • TORONTO • SYDNEY • NEW DELHI • AUCKLAND

HARPER ● PERENNIAL

First published in Great Britain in 1998 by Doubleday, a division of Transworld Publishers Ltd.

First U.S. hardcover edition published in 1998 by William Morrow and Co.

First Harper Perennial edition published 2008.

The Library of Congress catalogued the U.S. hardcover edition as follows:

Gilbert, Martin.
    Israel : a history / Martin Gilbert.
      p.   cm.
    Includes bibliographical references and index.
    ISBN 978-0-688-12362-8
    1. Israel—History. 2. Jews—Palestine—History—20th century. 3. Arab-Israel conflict.
I. Title.
    DS126.5.G53   1998
    956.94—dc21   97-32864
                                        CIP

ISBN 978-0-688-12363-5 (pbk.)

12 RRD 10 9 8 7

Dedicated to my friends in Israel,
who have introduced me, during thirty-five years,
to so many aspects of its history;
and to Shoshana Poznansky,
who is embarking on her own Israeli journey.

Contents

Contents

Preface to the 2008 edition

Ten years have passed since this book was completed. Those ten years have seen a surfeit of history in this land that, since creation of the Jewish State in 1948, has known – has been allowed – too little peace and quiet. Yet amid the conflict that sometimes seems almost to overwhelm the enthusiasms and hopes of Jewish statehood, progress is made, aspirations are fulfilled, the path to peace is still being taken. It is my hope that the two extra chapters, taking the story up almost to the sixtieth anniversary of statehood, will show not only the perils, fears and setbacks of the last ten years, but also the lines of a beneficent future.

Martin Gilbert
Merton College
Oxford
10 December 2007

Introduction

My aim in this book is to survey the history of the State of Israel during its first fifty years, and also to tell something of its founders and pioneers, going back to the second half of the nineteenth century. Ideology, politics, diplomacy and war each have their place in the narrative, as do the stories of many of the individuals – some famous, others not – who contributed to the building up and survival of the State, and whose aspirations and toil made it what it is today – a State of more than five million people, four million of them Jews.

Fifty years ago there were only half a million Jews in British Mandate Palestine. Their desire for statehood long preceded the declaration of the State. The pioneers and the ideologists, many of them from Russia, but deriving their inspiration and support from throughout the Jewish world, were beset by problems, both of material hardship and of the competing claims of rival movements. Zionism came to the fore as a national movement when Democratic Socialism and even Bolshevism were competing for many souls, including the soul of the Jewish people.

One feature of the evolution of Jewish statehood was the steady settlement of small groups of pioneers on the land, starting more than a hundred years ago. The farming settlements that the pioneers created from the last quarter of the nineteenth century onwards gradually transformed the landscape and, more importantly, formed the basis of the future structure of Israeli life, first in Turkish and British Mandate Palestine and then – for the process has been a continuous one – in Israel. These settlements, often named for Zionist leaders or biblical characters and places, reflect many aspects of the history and aspirations of Jewish immigration and endeavour. Long before the Holocaust destroyed the world of Jewish life and experience in Europe, and set its survivors on the road to Palestine, the building up of Jewish national institutions in Palestine, and of a Jewish population committed to statehood, had reached a high point of achievement, requiring only the decision of the

British to leave to set it on its national goals. It was not the Holocaust but the War of Independence, which began three years after the Holocaust ended, that enabled the Jewish State to come into being.

Over the past fifty years, Israeli society has faced a combination of pressures that are unusual in any nation: the pressures of continuous and massive immigration; five wars; the unpredictable cruelty of terrorist attacks (and, most recently, of suicide bombers); and a sense of the isolation and vulnerability of a small nation, each generation of which has lost loved ones in war and as a result of terrorist attacks. Israel is not only a nation that for the first three decades of its existence was surrounded by sworn enemies, but one that, following a victorious war in 1967, has had to share part of its own land with another people. This is not a novelty in history, but it is a painful situation that can only be resolved by supreme efforts of goodwill on both sides of the Palestinian–Israeli national divide.

This is an account in which history and current affairs merge: I have brought this first edition as up to date as possible. Every new day brings a new twist to the main themes of Israel's history since the foundation of the State fifty years ago: the aspirations of the old timers and the often conflicting desires of the newcomers, in a nation where immigrants make up the largest single sector of society; the tensions between the predominantly European Ashkenazi Jews and the Sephardi Jews, many of them of North African birth or origin; the tensions between the ultra-Orthodox Jews and those who adhere to a less extreme form of religious observance, or to no religious observance at all; the conflict between the Labour movement and ideology which ruled the State virtually unchallenged from 1948 to 1977 and since 1977 has competed for power and influence with the revisionist, Likud philosophy with its very different attitudes towards the nature of State power, the mood of State institutions and relationships with the Palestinian Arabs.

Every nation reaches watersheds that disturb the whole equilibrium of the society: Israel has already reached many in its first half-century of existence, among them the victory of June 1967 which resulted in the acquisition of a large Arab population, and the assassination in November 1995 of the Prime Minister, Yitzhak Rabin, by an Israeli Jew. The repercussions of this assassination will affect many aspects of life in Israel in the years to come.

Israel is a State based on the rule of law, and democratic values. Its institutions, among them the Knesset (Parliament) and the Supreme Court, reflect this philosophy of statehood. But the democratic nature of the institutions and the humanitarian basis of their responsibilities are sometimes questioned, even threatened, as seen most recently when death threats were uttered against the Chief Justice. Those within Israel who seek to maintain and enhance the democratic nature of the State face a complex struggle, as needy of vigilance today as it was at the time of the foundation of the State fifty years ago.

Although confronted with many problems, Israel possesses a strong will to succeed and prosper, to maintain its vigorous and fulfilling daily life, and to confound the critics who point to both external and self-inflicted problems as insoluble. This volume describes many instances when attempts have been made to resolve conflicts, and many where a harmonious resolution proved impossible. In Winston Churchill's words, 'The future, though imminent, is obscure'. But as Israel embarks upon its second half-century those who are responsible for its future have the task of nurturing its vibrant life, and of treasuring, guarding and enhancing its humane and democratic values.

Martin Gilbert
Merton College
Oxford
6 October 1997

Acknowledgements

During my visits to Israel in the 1970s I met some of those who were to become my friends and guides in years to come, among them the then President of the Hebrew University, Avraham Harman, the archaeologist Yigael Yadin, and the then Minister of Transport and Communications, Shimon Peres. Each of them, in different ways, influenced me to see the history of Israeli life and society as a dynamic, creative, often tormented, sometimes misguided, but essentially uplifting story. Shimon Peres also read my manuscript and made many valuable suggestions (without in any way being responsible for my portrayal of his own role and contribution to Israeli life).

My own conversations and correspondence over the past twenty-five years have been a source of many aspects of this book. I would like to thank those whose letters to me I quote, and whose archival material I have used, including David Harman, Gideon Raphael, Hannah Ruppin and Robbie Sabel, I thank. I am grateful to all those who have sent me material or responded to my historical queries: Clinton Bailey, Mark Faber, Jawdat Ibrahim, Rasie Ilan, Jack Kagan, Nicki Lyons, Yoram Mayorek (Central Zionist Archives), Stanley Medicks, David Neuman (Central Bureau of Statistics, Israel), Glenn Richter, Jay Shir, Sarah Zarfaty and Atara Rozik-Rosen, who also read the book in typescript. I am also grateful to Colette Avital, the Israeli Consul General in New York, and Carrie Riegelhaupt; and to the publisher's readers, Jennie West and Gillian Bromley. My particular thanks are due to my Publishing Director, Marianne Velmans, and to my Managing Editor, Katrina Whone. Kay Thomson made her usual substantial contribution to the work.

The book was also read in typescript by my friends, Walter Eytan and David Harman. Special thanks for her persistent and successful efforts in tracking down many sources for me are due to Enid Wurtman; and also to Yehuda Avner for generously giving me the full benefit of his long experience of Israeli public life.

List of photographs

List of maps

Photographic acknowledgements

Bildarchiv der Osterrichischen Nationalbibliothek, Vienna, 1; The Central Zionist Archives, 2, 6–8, 13, 17, 21, 42; S. Narinsky, 3; *Jewish Chronicle* photographic archive, 4, 40; Hulton Deutsch Collection Limited, 5, 38, 39, 43, 45, 52, 57; Government of Israel Press Office: 9, 16, 19, 22–26, 29, 30, 31, 32, 33, 36, 41, 44, 46, 47, 48, 49, 50, 52, 53, 54, 55, 56, 58, 59, 60, 61, 62, 63, 64; ASAP, Tel Aviv, 10, 11; Keren Hayesod Archive, 12; Imperial War Museum, 14, 15; Camera Press Ltd, 18; Stanley Medicks, 20; Hannah Safieh, 34; *Jerusalem Post* Picture Collection, 36 (Werner Braun), 37 (Bamahane); Associated Press, photographs 65 and 67 (Doug Mills), 66 (Nati Harnik), 68 (Zoom 77), 69, 70, 71, 72, 73 (Getty Images).

ISRAEL

A History

CHAPTER ONE

Ideals of statehood

Since the destruction of the Second Temple by the Romans in AD 70, the Jews, who were dispersed all over the Roman Empire, had prayed for a return to Zion. 'Next year in Jerusalem' was – and remains – the hope expressed at the end of every Passover meal commemorating the ancient exodus from Egypt. For two millennia the dream of such a return seemed a fantasy. Everywhere Jews learned to adapt to the nations within whose borders they lived. Frequent expulsion to other lands made a new adaptation necessary, and this was done. But Zion, which had been under Muslim rule almost without interruption since the seventh century, and under the rule of the Ottoman Turks since the early sixteenth century, was possible only for a few.

The perils of the journey could be severe. Of 1,500 Jews who travelled from Poland, Hungary and Moravia to Palestine in 1700, as many as 500 died on the way. But the imperative to return physically from exile never entirely died. In 1777 more than 300 Hasidic Jewish families, upholders of ultra-Orthodox Judaism, made the journey from Poland to Palestine. In 1812 some 400 followers of the Vilna Gaon – one of the enlightened Jewish sages of his generation – made the journey from Lithuania.

By the middle of the nineteenth century about 10,000 Jews lived in Palestine. More than 8,000 of them lived in Jerusalem. A few hundred lived in the holy city of Safed, in the north, where several Jewish sages were buried, in the mountain village of Peki'in (which had a tradition of continuous Jewish settlement since Roman times) and in nearby Tiberias on the Sea of Galilee. In the coastal town of Acre lived 140 Jews, mostly pedlars and artisans, but many without any means of support. There were several hundred Jews in Jaffa.

Most of the Jews in Palestine were immigrants from Poland and Lithuania. Many of them survived on charity, sending regular begging letters back to their original communities in Europe, and even dispatching special emissaries to raise funds. But the attraction of Palestine was growing. In 1862 a German Jew, Moses Hess, an advocate of the Jewish return to Palestine, wrote in his book *Rome and Jerusalem* of how his Jewish 'nationality' was connected 'inseparably' with the Holy Land and the Eternal City. Hess added, 'Without a soil a man sinks to the status of a parasite, feeding on others.'

Eight years later in 1870 a French educator, Charles Netter, with the approval of the Turkish authorities, founded an agricultural school at Mikveh Israel (Hope of Israel). The name was taken from a description of God in the book of Jeremiah. Located a few miles inland from Jaffa, Mikveh Israel was a settlement to which Jews living in countries where they experienced various educational disabilities and restrictions – initially Persia, Roumania and Serbia – could come to live and study.

It was in 1876 that the British Christian writer George Eliot (a pseudonym for Mary Ann Evans) completed her novel *Daniel Deronda*. It was to make its impact on many Jews, among them two Russian Jews, Eliezer Ben-Yehuda and I. L. Peretz, both of whom were to exert considerable influence, through their own writings, on Jewish national aspirations. One passage in the novel was often quoted:

> Revive the organic centre; let the unity of Israel which has made the growth and form of its religion be an outward reality. Looking towards a land and a polity, our dispersed people in all the ends of the earth may share the dignity of a national life which has a voice among the peoples of the East and the West – which will plant the wisdom and skill of our race so that it may be, as of old, a medium of transmission and understanding.

In 1878 a number of Jews from Jerusalem decided to establish a Jewish village in the Palestinian countryside. Their first effort was to buy some land near Jericho, but the Sultan refused to allow ownership to be transferred to Jews. They did manage to buy land from a Greek landowner in the coastal plain, and named their village Petah Tikvah (Gateway of Hope), but malaria, disappointing harvests, and quarrels among them led to failure. By 1882, when they abandoned the village, there were only ten houses and sixty-six inhabitants.

Also founded in 1878, by religious Jews from Safed who wanted to earn their own livelihood, and not be dependent on charity, was the village of Rosh Pinah. Lacking funds and experience, and frequently harassed by the Arabs from nearby villages, they gave up after two years, but Roumanian Jews, driven from Roumania by persecution and poverty, renewed the settlement in 1882, and obtained sufficient aid from the French Jewish philanthropist Baron Edmond de Rothschild to survive. Growing tobacco and planting mulberry trees for silkworms were two of their enterprises. Like so many of the Jewish settlements that were to be founded in Palestine, the name of Rosh Pinah was taken from a biblical phrase, in this case the 'head stone' from Psalm 118: 'The stone which the builders refused is become the head stone of the corner. This is the Lord's doing. It has become marvellous in our eyes.'

In Russia, following an upsurge of violent attacks against Jews – the pogroms – two movements were founded urging the emigration of Jews to Palestine to work as farmers on the Land of Israel (Eretz Yisrael) in order to 'redeem' it. One of the movements was known as the Bilu, from the Hebrew initials in the biblical phrase *Beth Jacob Lechu Venelcha*: 'O House of Jacob, come and let us go!' The young men and women of the Bilu, being secular and socialist in outlook, omitted the concluding words of the phrase: 'in the light of the Lord'.

The second Russian-born movement was Lovers of Zion (Hovevei Zion). It held its small founding conference in 1884, at Kattowitz, in German Upper Silesia, just across the border from Russian Poland. Its president, Odessa-born Judah Leib Pinsker (who had served as a physician in the Crimean War thirty years earlier, and been honoured by the Tsar) urged upon the thirty-six delegates present the importance of the 'return to the soil' of Palestine. But although both the Bilu and Hovevei Zion advocated emigration to Palestine and the building up of farms and villages there, the mass of dis-affected Russian Jews went to the United States, others to Britain, Western Europe and South Africa, in search of greater social equality and the chance of less poverty-stricken lives.

Only a small percentage of Russian Jewish emigrants (never more than 2 per cent) went each year to Palestine. But even this small percentage meant that 25,000 Jews reached Palestine between 1882 and 1903. Known as the First Aliyah, from the Hebrew word for ascent, in Palestine many of them lived by tilling the soil and by recourse to the financial support of the Rothschild family, which had for several years encouraged the work of Jews on the land and in the vineyards which the Rothschilds owned. Some of the Bilu pioneers worked as hired labourers at the Mikveh Israel agricultural school. One of them was Vladimir Dubnow, who took the Hebrew name Ze'ev (Wolf). On 21 September 1882 he wrote to his brother Simon – later a distinguished historian who believed that the greatest contribution of the Jews would be in the Diaspora, and who was murdered by the Nazis when in his eighties:

> Do you really think that my sole motivation in coming here is to better myself, with the implication that if all goes well I will have achieved this aim and that if it does not then I ought to be pitied? No. My ultimate aim, like that of many others, is greater, broader, incom-prehensible but not unattainable. The final goal is eventually to gain control of Palestine and to restore to the Jewish people the political

independence of which it has been deprived for two thousand years.

Don't laugh; this is no illusion. The means for realizing this goal is at hand: the founding of settlements in the country based on agriculture and crafts, the establishment and gradual expansion of all sorts of factories, in brief – to make an effort so that all the land, all the industry, will be in Jewish hands.

In addition, it is necessary to instruct young people and the future generation in the use of firearms (in free, wild Turkey anything can be done), and then – here I too am plunging into conjecture – then the glorious day will dawn of which Isaiah prophesied in his burning and poetic utterances. The Jews will proclaim in a loud voice and (if necessary) with arms in their hands that they are the masters of their ancient homeland. It doesn't matter whether that glorious day comes in another fifty years or more. Fifty years are but a moment for such an enterprise. Agree, my friend, that this is a sublime and magnificent idea.

Sixty-five and a half years were to pass before a Jewish State was established. At the time that Ze'ev Dubnow wrote his letter, there were scarcely 20,000 Jews in Palestine – one quarter of one per cent of world Jewry, and only one Jewish village. Dubnow himself was unable to settle in Palestine and returned to Russia after a few years, dying there more than forty years later. But slowly, with enormous difficulties, and yet with an incredible tenacity of purpose, the Jewish presence and Jewish enterprise in Palestine grew. In 1882 the town of Zichron Yaakov ('Memory of Jacob') was founded by Jewish immigrants from Roumania. It was financed by Baron Edmond de Rothschild, who named it after his father. That same year, 1882, a Gibraltar-born Jew, Hayyim Amzalak, who had emigrated to Palestine in 1830 – at the age of six – and was later British vice-consul for Jaffa, bought the land on which the first all-Jewish village in Palestine, Petah Tikvah, had been built four years earlier. He gave it to Bilu pioneers from Russia who were willing to try to make it work. As a British subject, Amzalak could buy land; Russian-born Jews were not allowed to do so.

Petah Tikvah was not only cursed by malaria, it suffered frequent attacks from the neighbouring Arab villages. But the pioneers persevered, and Baron Edmond de Rothschild – 'The Baron', as he was known – gave them the money needed to clear the malaria swamps which had defeated the first villagers.

Hayyim Amzalak also helped to finance the establishment of Rishon le-Zion (First to Zion), the first village to be built by settlers from outside Palestine – both Rosh Pinah and Petah Tikvah had been founded by local Jews from Safed and Jerusalem. Rishon le-Zion's settlers were ten pioneers from Russia. When funds began to run short and the water in the shallow wells had failed, an emissary from Rishon le-Zion travelled to Paris, where he persuaded Edmond de Rothschild to provide sufficient funds to dig a deep well. The Baron also sent out experts in agriculture and wine-growing.

The Carmel Oriental wine cellars, which he established, produce commercially successful wine to this day. The first Hebrew-language kindergarten and elementary school in Palestine were opened in Rishon le-Zion within a few years of its foundation. Once more it was the Bible which had provided the inspiration for the village's name, the phrase in the Book of Isaiah: 'The first shall say to Zion . . .'

In honour of the founding of the town, a Roumanian-Jewish poet, Naphtali Herz Imber, wrote a poem, *Hatikvah* (The Hope), which was to become the Zionist hymn, and later the State of Israel's national anthem. Imber read the poem to the farmers of Rishon le-Zion, one of whom, Samuel Cohen, who had emigrated from Moldavia four years earlier, set it to music. The poem read:

> As long as deep in the heart
> The soul of a Jew yearns,
> And towards the East
> An eye looks to Zion
>
> Our hope is not yet lost
> The age-old hope,
> To return to the land of our fathers
> To the city where David dwelt.

Within a few years, the second verse was changed to the version now in use:

> Our hope is not yet lost
> The hope of two thousand years,
> To be a free people in our land
> The land of Zion and Jerusalem.

In 1883 a Jewish immigrant from Russia, Reuben Lehrer, built a house in an Arab village, Wadi Hanin, in the coastal plain. Several other Jews, among them a fellow immigrant from Russia, Avraham Yalofsky, soon joined him. Until the War of Independence in 1948, Jews and Arabs lived peacefully side by side in the village. Lehrer called it Nes Ziona (Banner towards Zion). He and his friends planted citrus groves and engaged in bee-keeping. But an omen of the dangers that lay ahead for Jewish settlement in Palestine came only five years after the start of Jewish life in the village, when, in a nearby wadi, a group of Arabs – not from the village – ambushed Yalofsky and killed him.

Despite the hazards of life imposed by nature, and by man, each year saw some new effort at Jewish settlement. In 1884 a Russian-born Jew, Yehiel Michael Pines, the representative in Palestine of the newly established Moses

Montefiore Testimonial Fund (a Russian-based charitable fund, chaired by Judah Leib Pinsker, whose purpose was to support Jewish agricultural work in Palestine) bought, through the fund, the land needed for the Bilu pioneers to found another village, Gederah. The principal produce of their fields was to be grapes and grain.

Agriculture was also the chosen pursuit of the Lovers of Zion pioneers. Through it they believed they could 'redeem' the land and restore the Jews as a people. When a Russian-Jewish professor, Hermann Schapira, himself one of the founders of Lovers of Zion, proposed establishing a Jewish university in Jerusalem, he was challenged by Menachem Ussishkin, one of the young leaders of the movement, who declared that nobody but a visionary could think of a university in Palestine at that time: 'First and foremost we desire to create a farmer class in the Land of Israel, whose sons would not be "huge heads on chicken's feet" but ordinary people tilling the soil.'

The sense of nationality which was emerging among the Jews in Palestine received a boost in 1889, when a Russian-born Jew, Eliezer Ben-Yehuda, and a small number of like-minded friends, including Yehiel Pines, formed a group with the declared object of 'spreading the Hebrew language and speech among people in all walks of life'. A year later the group elected a committee which set about establishing Hebrew terms for modern words which were in daily use, and creating a uniform system of pronunciation – as every emigrant homeland had a different pronunciation, mostly derived from 2,000 years of evolving liturgical prayer, this was no easy task.

Ben-Yehuda drafted the committee's statement of purpose: 'To prepare the Hebrew language for use as a spoken language in all facets of life – in the home, school, public life, business, industry, fine arts, and in the sciences . . . To preserve the Oriental qualities of the language and its distinctive form, in the pronunciation of the consonants, in word structure and in style, and to add the flexibility necessary to enable it fully to express contemporary human thought.' This Sephardi pronunciation of Hebrew, which Ben-Yehuda favoured, was believed to be closer to the biblical tongue; the Ashkenazi pronunciation, with which he had been brought up in Russia, had changed over time, and by the impact of Yiddish, into something that he regarded as less mellifluous, and less 'genuine'.

There was a considerable element of excitement and ingenuity in the language committee's work. Ancient Hebrew roots were given modern forms. Arabic roots were also used. Non-Semitic words which had a Hebrew form were also incorporated. When, in 1890, a group of Lovers of Zion established a small farming settlement in Upper Galilee, on the west bank of the River Jordan, they gave it the Hebrew name Mishmar Ha-Yarden (Guard of the Jordan).

It was not only the Jews from Russia who were striving to introduce Hebrew as a spoken language. Several of the religious communities of Jews from Arab lands did likewise. Among them were several thousand Jews from the Yemen, one of whose leaders, Shalom ben Joseph Alsheikh, who emigrated to Jerusalem in 1891, was a leading educationalist in his community (and from 1908 until his death in 1944, at the age of eighty-five, chief rabbi of the Yemenite community of Jerusalem).

The Jewish population of Jerusalem grew considerably through immigration; Jews had been a majority there since the 1850s, and between 1864 and 1889 their numbers multiplied three-fold, reaching 25,000. There were then 14,000 Arabs in the city.

Among those who moved from Jerusalem to the growing number of Jewish villages and settlements was Abraham Shapira. He had been born in Russia in 1870 and was brought to Palestine by his parents when he was ten. When, in 1890, he moved to Petah Tikvah, he found that Arab farmers were pasturing their flocks on the Jewish fields, damaging the crops. Shapira set up a guardsmen's group which drove off the intruders and maintained the security of the village, enlisting the help of local Bedouin as well as of young Jewish settlers.

In 1890 another Jewish village was founded in the coastal plain. This was Rehovot (Wide Expanses; taken from the Book of Genesis). The land was bought from a wealthy Christian Arab landowner. The impetus of the founders was to establish a Jewish village that would not be dependent on the financial goodwill – and administrative supervision – of Edmond de Rothschild. After a decade of effort, and frequent attacks from the neighbouring Arab villages in which blows would be struck, property damaged and trees cut down, success came and the village grew, helped by the arrival of Jewish immigrants from Yemen.

Russian Jews, members of the Lovers of Zion who had emigrated from Vilna, Riga and Kovno, founded their own village in 1890. They called it by the Arabic name, Hadera (The Green), after the emerald green colour of the swamp vegetation around them. They could not have chosen a less hospitable site. It was not the local Arabs who tormented them, but the malarial mosquito. More than half of the inhabitants of Hadera died of malaria in the first twenty years of its existence. But with Egyptian drainage workers sent to them by Baron Edmond de Rothschild, and the planting of eucalyptus groves, they gradually drained the soil and were able to turn from field and vegetable-garden crops to citrus fruits.

The nineteenth century was coming to an end with considerable Jewish activity in Palestine. But among the sophisticated and assimilated Jews in Europe little was known of this. Palestinian Jewry was seen, wrongly, as an exclusively religious community, set in the ways of late medieval Judaism, black-garbed and clinging to piety and prayer in preference to manual labour and productive enterprise. One of those who had little knowledge of what had already been achieved in Palestine, especially by the Russian immigrants of the previous two decades, was a Hungarian-born Jew, Theodor Herzl. As a journalist in Paris since 1891, he was shocked by the anti-Semitism in France at the time of the Dreyfus case, when a Jewish officer was found guilty of treason, a charge later found to have been false. The harsh anti-Semitic tone of much of the criticism of Dreyfus appalled Herzl who also knew of the suffering of Jews in Russia. His calls for a return to Zion appealed above all to the Jews of Russia, for hundreds of thousands of whom the United States had hitherto been the main possible place of refuge and renewal.

The Dreyfus trial was a watershed in Jewish history. Jews everywhere asked themselves what had gone wrong with Jewish life. Why was there anti-Semitism? Three ways out of the trap seemed to present themselves: to become assimilated into the nation with whom one was living, to fight for a revolutionary socialism that would cure all the evils of the world including anti-Semitism, or to seek a 'normal' Jewish life in a Jewish land with a Jewish government. Herzl was drawn to the last option.

In Vienna, where Herzl had lived since he was eighteen, the election of Karl Lueger as mayor in 1893 was another catalyst. Lueger was head of a self-proclaimed anti-Semitic political party and won many of his votes by denouncing the Jews. In the month that Lueger triumphed electorally in Vienna, Herzl noted in his diary, 'The mood among the Jews is one of despair.' Herzl was convinced, however, that he could persuade both Jews and Gentiles to support the idea of a Jewish State in Palestine. He made contact with the Sultan of Turkey, sought to influence the German Kaiser and pressed his arguments vigorously with the leading Jews of Western Europe.

Some of those to whom Herzl expounded his ideas considered him insane. The first Rothschild whom he approached ignored his letter. But he found a much respected ally in the physician and philosopher, Max Nordau – at that time also a Jewish newspaper correspondent living in Paris – who told him, 'If you are insane, we are insane together. Count on me!' Nordau, like Herzl, had been born in Hungary. He was well known as a lifelong opponent of aristocratic corruption and monarchical oppression. In a much publicized book, *The Conventional Lies of our Civilization*, written in 1884 when he was thirty-five, he had condemned both contemporary society and hatred of Jews, which he portrayed as a symptom of national degeneration.

Herzl was in his mid-thirties when, without having been to Palestine, he espoused the idea of a Jewish nation there, and began to drive the idea forward. He launched a movement aimed at something more than the existing spasmodic settlement on the land of small groups of Jews, mostly from Tsarist Russia. His aim was the immediate return of the Jews to Palestine on a massive scale, from every one of the countries of the Diaspora, to land which would be theirs as a Jewish homeland, recognized as such by the Great Powers of the world.

Herzl's vehicle was an organization that he founded in 1896, the World

Zionist Organization. The term Zionism had been coined four years earlier. Like many of the 'isms' of the time, it appealed to those who sought a more ideal society, and one which they could direct themselves in the interest of their group. Nationalism was gaining in strength in many lands. In the Austro-Hungarian Empire, which Herzl knew well, Czechs, Slovaks, Ruthenians, Slovenes and Croats were among those minority groups hoping for national self-determination. Zionism proposed a national identity and a national home for the Jews. Herzl had the idea of opening negotiations with the Ottoman government to secure the land needed, and to build up a modern society in what he called the 'old-new' land.

Herzl's endeavours reached a culmination in February 1896 when he published a book entitled *The Jewish State*. In it he declared bluntly, 'Palestine is our ever-memorable historic home. The very name of Palestine would attract our people with a force of marvellous potency.'

The Jewish State was a major influence in the emergence of political Zionism. Herzl pointed to the part played by anti-Semitism in bringing the Jews to their existing desperate situation. 'We have honestly endeavoured', he wrote, 'to merge ourselves in the social life of surrounding communities and to preserve only the faith of our fathers. We are not permitted to do so. In vain we are loyal patriots.' It was the distress of the Jews, he believed, that bound them together. 'Thus united, we suddenly discovered our strength. Yes, we are strong enough to form a State, and, indeed, a model State. We possess all human and material resources necessary for the purpose.'

Herzl described some of the detailed work needed to set up a State in Palestine. Much of the book dealt with specific details of emigration, land purchase, house-building, labour laws, the proposed nature of manual work, the commerce, industry, education, welfare, and social life of the new State. It was to be a model society, ever developing and expanding, moulded in the image of European democracy but with Jewish values as its foundation. It was to be secular, not religious; bound by laws that derived from the European civil codes, guided by the European – and American – example of the separation of Church and State, looking forward to the twentieth century, not backward to the middle ages.

The Jews would not live in a ghetto in the new State, cut off from the modern, secular and scientific influences and challenges around them, but would, after so many centuries of isolation and even obscurantism, emerge into the world of modernity. The State would be a testimony to enlightenment and emancipation: it would witness the birth of the modern Jew in a Jewish land. Jews who lived in existing poor, backward societies, far from the modernity represented by Vienna or New York or London, would have the opportunity to put their dismal circumstances behind them.

Once the Jews were 'fixed in their own land', Herzl wrote, it would no longer be possible to scatter them all over the world. 'The Diaspora cannot take place again,' he added, 'unless the civilization of the whole earth shall collapse,' and he went on to proclaim, 'Here it is, fellow Jews! Neither fable nor deception! Every man may test its reality for himself, for every man will carry over with him a portion of the Promised Land – one in his head, another in his arms, another in his acquired possessions. But we must first bring enlightenment to men's minds. The idea must make its way into the most distant, miserable holes where our people dwell. They will awaken from gloomy brooding, for into their lives will come a new significance.'

Referring to those Jews, led by Judas Maccabeus, who had overthrown the rulers of Palestine in the pre-Roman era, Herzl wrote in his closing sentences, 'The Maccabeans will rise again. We shall live at last as free men on our own soil, and die peacefully in our own homes.'

Herzl did not stop his efforts with the publication of his book. Indeed, in the following months he pressed his point home in every quarter. In June 1896 he went by train to Constantinople. Stopping briefly in the Bulgarian capital, Sofia, he was welcomed by the local Jews with calls of 'Next Year in Jerusalem'. In Constantinople, however, the Turks would not give him any assurances about Palestine. In London that July, thousands of Jews came to hear him speak. Most of them were new immigrants from Russia, poor and eager for inspiration. In Paris six days later he was received by Baron Edmond de Rothschild, but the baron warned him against upsetting the Turks. Rebuffed, but not in any way defeated, Herzl wrote to a friend, 'There is only one reply to this situation. Let us organize our masses immediately.' The masses that mattered were the Jews of Russia – more than four million, the victims of political discrimination, economic pressures and physical attack. That November Herzl wrote to the Grand Duke Vladimir of Russia, an uncle of the Tsar, seeking to enlist imperial Russian support in what he described as 'the solution to a question as old as Christianity, a great and beautiful cause, designed to delight the noblest hearts. It is the return of the Jews to Palestine!' The Grand Duke was non-committal.

The ferment of anticipation stirred up among the Jews of Europe by Herzl's book elevated its contents from a mere theoretical tract to a call for debate and a blueprint for action. But it did not meet with universal Jewish approval; far from it. For most Orthodox Jews, Herzl's predominantly secular vision was disturbing; the return of the Jews to the Land of Israel (Eretz Yisrael) would only come after the arrival of the Messiah, and could not be brought forward by the efforts of mere mortals. Many secular Jews, for whom the peaceful future of their race lay in the peace and prosperity of the lands in which they already lived, were opposed to the idea of a separate Jewish State. On 18 March 1897 Herzl recorded in his diary a meeting with the Chief Rabbi of Vienna, Moritz Güdemann, who argued that the 'mission of Jewry' consisted, in future years as hitherto, 'in being dispersed throughout the world'.

Herzl was also opposed at first by those – mostly Russian – Jews for whom the fragmentary process of piecemeal settlement in Palestine was the key. They had looked for more than a decade to the slow, steady, farm by farm, settlement by settlement, building up of agricultural activity, with the aim of redeeming the Jewish soil by hard work and devoted individual effort. It was to this process that they had dedicated their energies and skills. Each step, each village was for them a triumph, as when, in 1896, Baron Edmond de Rothschild made it possible for a group of young settlers to establish what was to be the northernmost Jewish village in Palestine (and later the northernmost town in Israel), Metulla. The founders were chosen for their ability to defend the site as much as to farm it. They took the name from the local Arabic name for the site.

Herzl's call for Jewish statehood seemed too grandiose, too fraught with the complications of local Turkish and Arab opposition, too ambitious with regard to the accepted place of the Jew in the world, to be more than an extraordinary dream, an eccentricity. The work on settlements continued along its piecemeal, difficult but constructive path. One climax of these efforts was the finance provided in 1896 by the Jewish philanthropist Baron Maurice de Hirsch, through his Jewish Colonization Association (known as ICA). This enabled several colonies in Palestine to receive extra and essential financial help, among them Hadera and Mishmar Ha-Yarden.

De Hirsch's Jewish Colonization Association also financed the settlement of Russian and Roumanian Jews in Argentina, Brazil, the United States, Canada and Cyprus. This aspect of de Hirsch's philanthropy was not to the liking of Herzl and his followers, for whom Palestine alone was the intended focus of all Zionist efforts.

Herzl's ambitious proposal for full, and rapid, statehood was a vision that even many Russian-Jewish Zionists found difficult to accept. One of the leaders of the Lovers of Zion, Asher Ginsburg – known by the Hebrew name Ahad Ha'Am (One of the People) – was convinced that the real need was not for a Jewish political centre, but for a centre of spiritual regeneration. Ahad Ha'Am did not expect to see even the spiritual centre come into being for many years. On 29 July 1897 he wrote to a friend, 'I am already held by many to be a pessimist who sees and prophesies nothing but evil; so what good would come of what I have to say?'

Unlike Ahad Ha'Am, Herzl was an optimist, and he persevered with his efforts, making contact with Jews all over the world, and enlisting their support for his World Zionist Organization, committed to creating a Jewish State in Palestine. He was particularly energetic in persuading the Lovers of Zion, with their more modest aims of settlement in Palestine without sovereignty, not to reject his vision out of hand. On the morning of Sunday 29 August 1897, five months after the Chief Rabbi of Vienna had told him that the role of the Jew was as a catalyst in the lands of the Diaspora, Herzl opened the First Zionist Congress, in the Swiss city of Basle. This was the opening gathering of the World Zionist Organization, attended by more than 200 delegates, at least a quarter of whom were from Russia.

The scepticism of the Lovers of Zion towards the new movement had been overcome. Even Ahad Ha'Am, although not an official delegate, was present at the Congress and agreed to take part in the official photograph. Among the Jews who came to Basle that August were those from Palestine, from Arab lands, from Britain, and even from New York. American Zionism was represented by Rosa Sonnenschein, the editor of the *American Jewess*. In all, Jews from twenty-four different States and territories had gathered together.

Herzl, as President of the Congress, told the delegates, 'We have an important task before us. We have met here to lay the foundation-stone of the house that will some day shelter the Jewish people.' It was not to be a secretive or chance affair, he explained. And it was to be achieved by acting in the same manner as all aspiring national movements. 'We have to aim', Herzl insisted, 'at securing legal, international guarantees for our work.'

The delegates must not disperse once their deliberations were over, Herzl told them, but must continue to work without pause in search of their aim. 'At this Congress', he declared, 'we bring to the Jewish people an organization it did not possess before.' This organization, the Zionist movement, would enable an outcast people to act with dignity. The Jews were no longer, as they had in the previous decades, to 'steal into the land of their future'. Instead, they would negotiate their return openly, by a legal agreement with the Ottoman government, and with the formal approval of the Great Powers. That agreement, Herzl insisted, 'must be based on rights and not on toleration'. Once the negotiations were completed, he envisaged not the existing pace of piecemeal settlement – which he called derisively 'infiltration' – but 'the settlement of Jewish masses on a large scale'.

Herzl also spoke of the ethical aspect of a Jewish homeland. Zionism, he said, was 'a civilized, law-abiding, humane movement towards the ancient good of our people'. This theme was taken up by Max Nordau who stressed the material and moral misery of the Jews in the Diaspora. In eastern Europe, North Africa and Asia, where as many as nine-tenths of world Jewry were living, 'the misery of the Jews is to be understood literally. It is a daily distress of the body, anxious for every day that follows, a tortured fight for bare existence.'

Nordau drafted for the Zionist Congress a document called the Basle Programme. This began with the sentence, 'The task of Zionism is to secure for the Jewish people in Palestine a publicly recognized, legally secured homeland.' To fulfil this task, the Congress approved four specific tasks. The first was to 'encourage the systematic settlement of Palestine with Jewish agricultural workers, labourers and artisans'. The World Zionist Organization would 'organize and unite the Jewish people by the creation of groups in various countries whose object would be to foster the aims of the movement'. These groups were to be organized in accordance 'with the laws of their respective countries'. The Congress would also 'dedicate itself to strengthening Jewish consciousness and national feeling'. Finally, it would 'organize political efforts so as to obtain the support of the various Governments of the world for the aims of Zionism'.

This was a comprehensive and ambitious set of aims. Herzl was elated that they had all been endorsed. On 3 September 1897 he wrote in his diary, 'Were I to sum up the Basle Congress in a word – which I shall guard against pronouncing publicly – it would be this: At Basle I founded the Jewish State. If I said this out loud today, I would be answered by universal laughter. Perhaps in five years, and certainly in fifty, everyone will know it.'

CHAPTER TWO

Towards Zion

The programme of the Basle Congress, the idea of a determined effort to secure a Jewish homeland, caught the imagination of vast numbers of Jews throughout the world, but especially in Eastern Europe and Russia, where life under the rule of Christian sovereigns was fraught with problems. Herzl understood that a diplomatic initiative, negotiations, and an impressive programme were not enough. 'The foundation of a State', he wrote in his diary on 3 September 1897, 'lies in the will of a people for a State.'

Zionism, with its call for a Jewish revival in Palestine, made an immediate impact on Russian Jewry. But many Russian Jews were attracted to a quite different cause, that of revolutionary socialism. Not an escape from Tsarist tyranny to some distant land, but the destruction of Tsarism itself, was their ideal. On 7 October 1897, only five and a half weeks after the much publicized opening of the First Zionist Congress in Basle, a secret convention opened in Vilna of a new Jewish organization, the Bund – the Jewish Socialist Workers' Party – dedicated to the coming of a Russian socialist government. In March 1898 the Bund was formally admitted to the Russian Social Democratic Labour Party, as an autonomous body, and was to play a major part in the emergence of Russian revolutionary socialism. The Bund quickly became a mass movement, attracting many Jews by its demand for equal political and civil rights for Russian Jews, though Leon Trotsky, one of the leaders of revolutionary socialism in Russia (and a Jew) belittled the Bundist as 'Zionists who are afraid to become seasick'.

Henceforth the Zionists had to meet the challenge of the Bundists, whose appeal to a persecuted people was, like theirs, a strong one. Zionism was never a unanimous, or even a majority, movement among the Jews of Russia, the Jewish heartland; it made even less impact on the Jews of the United States, which was rapidly emerging as the second great centre of Jewish life. But Herzl was determined to make it work. When the Second Zionist

Congress opened on 28 August 1898, again at Basle, the number of delegates had more than doubled.

Amid the enthusiasm of the Second Congress, and the realization that the Zionist movement was gaining in size and respectability, little attention was paid to a report from one of the delegates, Leo Motzkin, who had just returned from Palestine. In his speech he stressed the 'established fact that the most fertile parts of our land are occupied by Arabs'. He even gave the figure of 650,000 for the number of Arabs, and warned of 'innumerable clashes between Jews and incited Arabs' that had taken place there. In his closing speech, Herzl made no reference to these uncomfortable, ominous facts.

Among those present at the Second Zionist Congress was a twenty-three-year-old Russian Jew, Chaim Weizmann, who had been studying chemistry at a Swiss university only 50 miles from Basle. That September he returned to Pinsk. He was not impressed by the state of Zionism there, writing to some friends, 'Propaganda is not yet being conducted as it should. A great shortage of speakers is felt. One ought to take advantage of these fiery times.' The Russian-Jewish world was indeed in ferment, with many Jews, especially students, turning not to the ideals of Zionism but to those of socialism and even revolution as the answer to discrimination and poverty.

A sense of impending success had seized Herzl's imagination. Visiting London, he told the 10,000 Jews who gathered to hear him speak in the East End on 3 October 1898, 'I fervently believe the time is very close when the Jewish people will very definitely be on the march.' Later that month Herzl left Vienna for the east, visiting Constantinople again, and making his first visit to Palestine. He timed his visit to coincide with that of the German Kaiser, William II. When the two men met by the roadside at the entrance to Mikveh Israel, the Kaiser commented that water was the key to building up the land. This, although true, was hardly the commitment of support for which Herzl had been looking.

On November 2, in Jerusalem, the Kaiser received a Jewish deputation including Herzl. The deputation presented the German monarch with an album of photographs of the existing Jewish settlements in Palestine. As a result of his two audiences with the Kaiser, Herzl felt that the World Zionist Organization, of which he was President, had been accorded serious recognition, giving Jewish national aspirations a status inconceivable a decade earlier. But while the Kaiser had expressed interest in the Jewish agricultural experiments in Palestine, he remarked in his talk with Herzl on November 2 that these experiments must be conducted 'in a spirit of absolute respect for the Sultan's sovereignty'.

The First and Second Zionist Congresses both advocated the establishment of a bank through which money could be directed to Palestine. It was incorporated in London on 20 March 1899 as the Jewish Colonial Trust (Juedische Colonialbank). The aims of the bank were set out in its statutes.

The first was 'to promote, develop, work and carry on industries, undertakings and colonization schemes . . . in particular of persons of the Jewish race into Palestine, Syria and other countries in the East'. Another aim was 'to acquire from any State or other authority in any part of the world any concessions, grants, decrees, rights and privileges whatsoever' for the employment of capital in Palestine, with a view 'to prospect, examine, explore, test and develop any mining, landed, agricultural and other properties, and to despatch and employ expeditions, agents and others'.

It was to take more than three years before the share capital was sufficiently subscribed by Jews all over the world for the bank to begin operation. To ensure the maintenance of political control, founder shares were held by the World Zionist Organization. Herzl headed the list of the bank's council members, who included Zionist luminaries from New York, Paris, Warsaw, Kiev, Moscow, Cologne and Sofia.

Herzl's efforts to alert European Jewry to the need for nationhood, and to the precarious nature of Jewish acceptance in otherwise civilized societies, were helped in 1899 by the publication of *The Foundations of the Nineteenth Century* by Houston Stewart Chamberlain. Born in Britain, Chamberlain had chosen to live in Germany where he married Richard Wagner's daughter. In his book he developed the theory of the 'blond' Nordic race, the originators of all that was noble in Western civilization. All that was bad, Chamberlain asserted, came from racial mixing, and particularly from any 'Nordic' inter-marriages with Jews. The Jews, he wrote, were a 'mongrel' people incapable of creativity; they were the universal corruptors. According to Chamberlain's so-called researches, not only King David and the biblical Prophets, but also Jesus, were not Jews at all but of ancient Germanic origin.

Houston Stewart Chamberlain's ideas won acclaim in Germany. The Kaiser was one of his supporters. His book was later to influence the most vicious anti-Semite of all, Adolf Hitler, with whom Houston Chamberlain developed a personal friendship. In the same year that *The Foundations of the Nineteenth Century* was published, a book was published in Cairo entitled *The Talmud Jew*. Its author, August Rohlings, likewise portrayed the Jew as a disruptive force in world history. Both books gave existing prejudices a false intellectual legitimacy, cloaking evil in spurious science.

For anti-Semites like Houston Stewart Chamberlain and August Rohlings, it was not only the wealth of the Rothschilds, but the cohesion of the Jews – who had often banded together in self-defence groups to ward off the attacks of pogroms in Russia – that gave proof of a worldwide Jewish 'conspiracy' which had to be crushed. For the Jews themselves, cohesion seemed their only chance of physical as well as spiritual survival. Zionism offered a cohesion that would lead to a national reality. In 1901 a Jew living in London, Ezekiel Wortsmann, wrote a pamphlet entitled *What Do the Zionists Want?*, in which he explained: 'We consider ourselves as strangers everywhere, even where we have been given complete civil rights, because we want to have a home of our own.' Wortsmann also argued that it was not enough to 'revive a people' in Palestine; it was also essential to 'revive its national tongue', the Hebrew language.

In Palestine itself, Hebrew was becoming more and more accepted each year as the common spoken language among the Jews. Eliezer Ben-Yehuda continued to compile a Hebrew-language dictionary which would both incorporate new words and enable the biblical language to be of use in the exigencies of modern life.

The Jews were a minority in Palestine at the turn of the century. But they formed compact and active communities, struggling, experimenting and growing. Despite the dull hand of Turkish bureaucracy, they lived their daily life with a sense of dignity and independence. Jews who arrived to join them were welcomed as if returning home.

On 29 December 1901 the Fifth Zionist Congress, meeting in Basle, set up a special fund, the Jewish National Fund, to buy land in Palestine. The Hebrew name, *Keren Kayemet Le'Yisrael*, came from a Talmudic dictum about good deeds, 'the fruits of which a man enjoys in this world, while the capital abides [*ha'keren kayemet*] for him in the world to come'. The money for the fund was to be collected from Jews in the Diaspora. 'The Jewish National Fund shall be the eternal possession of the Jewish people,' the Congress declared. 'Its funds shall not be used except for the purchase of lands in Palestine and Syria.' Palestine was then an integral part of Syria; some of the lands purchased in the coming few years were on the high ground just to the west of Damascus, known today as the Golan Heights.

The work of the Jewish National Fund was carried out by patient fund-raising throughout the Diaspora. It was particularly effective in Britain, Germany, the Austro-Hungarian Empire and the United States. Administered from Vienna, its central fund-raising symbol and practical means was the Blue Box, a small blue tin collecting box, placed in shops, offices, synagogue halls and private homes. It was from both wealthy Jews and from the small donations of hundreds of thousands of far from wealthy Jews, including many children, that the money for land purchase was to be raised. The fund also had a Golden Book in which the names of the larger donors were inscribed. Within three years of the fund being set up, enough money had been collected for the first land purchase – Kfar Hittim (Grain Village) in Galilee. The first settlers there were Jewish immigrants from the city of Lodz, in Russian Poland.

For the Jews of Russia, one cruel form of discrimination was the exclusion of Jewish children and students from the full rights of education. This exclusion spurred the development of a central feature of Zionism: the encouragement of educational opportunities. Herzl and the Zionists wanted Jews to be as free as anyone else to develop their own creative skills. On 8

June 1901, however, the Tsarist Ministry of Education directed that the existing 10 per cent quota for Jewish university students should not be calculated for the overall student body, but separately for each faculty. This greatly reduced the number of Jews in those faculties, like Law and Medicine, which Jewish students favoured.

In order to free Jewish education from its dependence upon hostile governments, the Zionists put forward the idea of a Jewish university in Palestine. Hermann Schapira, who had raised the idea in 1884 only to have it dismissed, and had then brought it to the First Zionist Congress in vain, continued to press for it and was supported by one of the younger Congress delegates, Chaim Weizmann. More than a thousand years earlier the academies of Sura and Pumbeditha had been the last fully autonomous Jewish intellectual centres. Yet even they had been subject to Mesopotamian overlordship. Since then, in Europe, North Africa and Asia every Jewish place of learning was at the mercy of local laws, often hostile to Jewish learning, while the numerical restriction of Jewish students in almost all places of learning effectively denied a higher education to many Jews who wished to pursue their studies.

As the restrictions on Jewish education in Russia grew more severe, the desire for specifically Jewish studies, whether religious, historical or literary, gave added power to the idea of an autonomous Jewish university in Palestine. On 3 May 1902 Herzl himself set out some of the arguments and some of the appeal of such a university in a letter to the Turkish Sultan, Abdul Hamid, to whom he wrote, 'The Jewish University should bring together all the scholarly qualities of the best universities, technical schools and schools of agriculture. The institution will offer nothing unless it is of the very first rank. Only then can it render real service to scholarship, to the students, and to the country.'

The idea of a Jewish university, and all that such a university implied, quickly became an integral part of Zionist thinking. During a conference of Russian Zionists which opened in Minsk on 4 September 1902, Ahad Ha'Am stressed the links between Zionism as a movement for national revival, and the cultural needs of the Jewish people. Both, he believed, could be given a new focus in Palestine. Among the resolutions which were passed in Minsk was one which expressed the sympathy of the conference with the idea of a Jewish university. At a Zionist conference held in Vienna at the end of October 1902, Chaim Weizmann joined with twenty-four-year-old Martin Buber in a motion aimed at facilitating 'a Jewish university in Palestine'. The mood of the conference was such that the phrase '*only* in Palestine' was insisted upon, following which Weizmann was appointed head of a small office in Geneva devoted to the task of advancing the university project.

Even as the Zionist movement was putting the idea of a Jewish university in Palestine at the forefront of its programme, Herzl was answering the questions of those who did not see why Palestine should be the destination of tens, or even hundreds of thousands of Jews. On 7 July 1902, while in London, he was asked about the settlement of Russian Jews in lands other than Palestine when he gave evidence to the Royal Commission on Alien Immigration. Why not continue to send the Jews to Argentina? Herzl was asked. After all, the great Jewish philanthropist, Baron de Hirsch, had provided substantial financial support for Jewish colonies across the Atlantic. That plan, Herzl replied, had been a failure 'because when you want a great settlement, you must have a flag and an idea. You cannot make those things only with money.'

Herzl then expressed the essence of Zionism when he told the Commissioners, 'With money you cannot make a general movement of a great mass of people. You must give them an ideal. You must put into them the belief in their future, and then you will be able to take out of them the devotion to the hardest labour imaginable.' He gave them Argentina as an example. 'Argentina has a very good soil,' he said, 'and the conditions for agricultural labour are much better than in Palestine, but in Palestine they work with enthusiasm and they succeed. I am not speaking of the artificially made colonies, but self-helping colonies, which have that great national idea.'

The Lovers of Zion were also looking for the means to secure land settlement in Palestine in more than a piecemeal way. In 1902 the Odessa Committee of the Lovers of Zion sent Ahad Ha'Am and the Russian-Jewish agronomist Jacob (Akiva) Ettinger to Palestine, to report on the state of the Jewish settlements there and on how they could be improved and expanded. Herzl, meanwhile, continued his diplomatic efforts to secure the Jews a legal place in Palestine. The Kishinev pogrom in 1903, in which more than fifty Russian Jews were killed by a mob rampaging through the town on Easter Sunday, seemed to give urgency to his efforts. A plan to settle the Jews at El Arish on the Mediterranean coast of the Sinai Desert so excited him that he decided against buying a family burial vault in the Vienna Jewish cemetery. In May the British-controlled Government of Egypt rejected the plan on the grounds that five times as much water would be needed for a Jewish settlement there as was available. Five days later Herzl bought the cemetery plot in Vienna after all. Four days after that, the British Colonial Secretary, Joseph Chamberlain, offered Herzl a Jewish homeland in Africa. To the outrage of those Zionists for whom Palestine was the one and only objective of all their efforts, Herzl accepted. He was determined to take up the first offer presented to the Jews by a Great Power, and to provide at least a place of temporary asylum for the Jews of Russia. In accepting, he split the Zionist movement. It was the first but not the last crisis in which the road already travelled seemed at risk.

The British offered Herzl a territory in Uganda, to be under the sovereignty of the British crown, into which a million Jews could immigrate and settle. The territory would be administered by the Jews and have a Jewish governor. When Nordau protested that Uganda was not Palestine, Herzl replied that, like Moses, he was leading the people to their goal via an apparent detour.

The Sixth Zionist Congress was soon to meet. Travelling to St Petersburg in August 1903, to show the Jews of Russia that he had not abandoned them, Herzl secured to their applause an end to the ban on Zionist organizations and fund-raising. Returning to Vienna through Vilna, he was greeted by a vast, admiring mass; so great was the throng at the railway station that gathered to see him off at the end of his visit that they were beaten back by the Tsarist police with considerable brutality.

The Sixth Congress was held in Basle. The idea of Uganda instead of Palestine, even as a temporary place of refuge, led to stormy arguments. The delegates from Kishinev were particularly vehement against any destination other than Palestine. Herzl worked busily behind the scenes to win over Nordau and to secure a majority. He succeeded: 295 for the Uganda scheme, 175 against, and 99 abstentions. Among those watching the proceedings from the gallery was Leon Trotsky who had chosen a very different path to salvation and who forecast the collapse of the Zionist movement.

The Zionist movement was certainly split. In Paris a Russian-Jewish student fired two pistol shots at Max Nordau with the words: 'Death to Nordau, the East African.' This was not the last time that a Jewish fanatic was to fire at a fellow Jew. Herzl continued on his travels in search of some form of land charter for Palestine, which remained his ultimate goal. Visiting Rome in January 1904, he was received by the Pope who said that if the Jews did settle in Palestine, the Catholic Church would be glad to convert them.

In April, at an emergency Zionist meeting in Vienna, Menachem Ussishkin – who in 1891 had astonished even his fellow Lovers of Zion by taking his bride to Palestine for their honeymoon – spoke with passion against Uganda. But Herzl refused to accept that Palestine alone must be the Zionist goal, and a state of 'armed peace' was declared.

Then, in July, at the early age of forty-four, Herzl died. He had been worn out by his frenetic, fevered, disputed labours and endless travels. 'Cut down in the flower of manhood,' the *Jewish Chronicle* wrote, 'and after efforts all too brief, he leaves the bulk of his people still in bondage and with the gates of their home relentlessly barred against them.'

The Uganda scheme was finished; Herzl's death effectively killed it, and even the British government had lost its enthusiasm. Inspired by the ideal of redeeming the Land of Israel through toil, a second wave of immigration – the Second Aliyah – began in 1904 and continued until the outbreak of the First World War. Mostly from Russia, Roumania and Eastern Europe, its members, numbering as many as 40,000 over the ten-year period, worked mostly as hired labourers on the farms established by the First Aliyah, or in the towns. They in turn established the first political Parties for Jewish workers in Palestine, a Hebrew-language press, and, in 1909, the first self-defence association, Ha-Shomer (The Watchman), and the first collective farm, the kibbutz.

One of those who inspired the Second Aliyah was Joseph Vitkin. Born in the Russian town of Mogilev, he had emigrated to Palestine in 1897 at the age of twenty-one. After working first as a labourer and then as a teacher he published a pamphlet in Hebrew – 'a call to the Youth of Israel whose Hearts are with their People and with Zion' – which influenced many young Russian Jews who were contemplating making their way to Palestine (known to them as the Land of Israel). Rejecting 'diplomatic' Zionism and the Zionism of Diaspora philanthropy, Vitkin insisted that Diaspora youth must emigrate to Palestine and make sacrifices for their homeland as other youth made personal sacrifices for their nations.

Another of those who inspired the Second Aliyah – of which he, like Vitkin, was a part – was Aharon David Gordon (best known as A. D. Gordon). Born in Russia in 1865, and working for three decades as a clerk, it was not until Gordon was almost fifty that he gave up his secure job and went to Palestine. There, he became a physical labourer. Working by day in the fields, at night he wrote about the dignity of toil, and about agricultural labour as the 'supreme act' of personal, national and universal redemption. It was by their return to the Land of Israel, Gordon insisted, that the Jews would return to the 'cosmic' well-springs of their distinctive Jewish creativity and spirituality.

Gordon's influence was enormous. All exploitation of Jew by Jew must, he insisted, be avoided. Both the land, and the means of production, should be collectively owned. There had also to be a personal transformation, the elimination of all desire for power or dominance. In their relations with non-Jews, the Jews in Palestine must avoid bettering themselves at the expense of others. The people of Israel (*Am-Yisrael*) must become a 'human people' (*Am-adam*). Its test would be with regard to the Arabs of Palestine. 'Our attitude towards them,' Gordon wrote, 'must be one of humanity, of moral courage which remains on the highest plane, even if the other side is not all that is desired. Indeed, their hostility is all the more reason for our humanity.'

Like the Zionist movement, the Jews of Palestine were far from united. Quite apart from the divisions between the predominantly secular immigrants from Russia and the existing largely Orthodox community already in Palestine, there were separate communities of Ashkenazi, Sephardi, Yemenite, Persian, Georgian, Bucharan and Moroccan Jews, many with their own languages and customs, whose independent institutions were often embattled with each other.

In an attempt to create a focus of unity, in 1903 Menachem Ussishkin journeyed from Russia and called a Great Convention at Zichron Yaakov. Among the points that were stressed was that Hebrew must become the

Jewish vernacular throughout Palestine. Although he would return to Russia, Ussishkin told the delegates (in the emotional language for which he was much admired), 'Even though we may be in the West, nevertheless our hearts always follow you in the East. May our right hand be forgotten and our tongues cleave to the roofs of our mouths if we do not devote our lives to the well-being of the Land of Israel! Depart, brethren, every man to his own home, and may the angels of peace accompany you upon your way. Go to the place from which you came, and where they wait expectantly to know all that you have done. Tell them that there are still men in the country who propose to revive the Jewish people. Go and bring life to the dry bones found so plentifully in our land.'

Before returning to Russia, Ussishkin established the Hebrew Teachers' Federation in Palestine. 'Whether the children in the village school learn more or less of the rudiments of elementary grammar,' he told the inaugural meeting, 'more or less of history, more or less of science, does not matter. What they have to learn, though, is this: to be strong and healthy villagers, to be villagers who love their surroundings and physical work, and most of all to be villagers who love the Hebrew tongue and the Jewish nation with all their hearts and souls.'

Among the immigrants of the Second Aliyah was Jacob Mintz, a Jew from a village near Minsk, who made his way to Palestine in 1904. Landing in Jaffa, he went with his wife and their two children to Petah Tikvah. Later he bought a plot of land at Nahalat Yehuda, near Rishon le-Zion, where he farmed. His brother, who might have gone with him to Palestine, decided instead to go to Britain, where he first worked as a pedlar in the East End of London; his descendants live in Britain today.

Such divided family objectives were common: the lure of Britain, France, Germany, and in particular the United States, was strong. Palestine, the Land of Israel in Jewish parlance and tradition, called for a particular sense of purpose and ideology. As well as being an ideological goal, Palestine could also be a place of refuge. In 1906, using money from the Jewish National Fund, a children's village was set up at Ben Shemen, halfway between Jaffa and Jerusalem, for orphans of the Kishinev pogrom.

On 7 September 1906 a twenty-year-old Russian Jew, David Gruen, landed at Jaffa. He had come from his home town of Plonsk, and made his way with thirteen others who had journeyed with him to Petah Tikvah. His aim was to till the soil of Palestine, and through labour to redeem the land, and the Jewish people. On the following morning he was at work in the orange groves hauling manure to put into the holes into which young orange trees would be planted. Like so many pioneers, he soon succumbed to malaria. A Jewish doctor told him he would be better off leaving the country. But he stayed, and, as David Ben-Gurion (son of Gruen) – the surname he was soon to choose in preference to his Russian one; it was also the surname of a

Jewish leader in Palestine in Roman times – was to play the leading part both in the establishment of a Jewish State forty-two years later, and in its growth and preservation.

Of Petah Tikvah, and also Rishon le-Zion, where most of the labourers were Arabs, and he himself was a hired hand, Ben-Gurion later wrote, 'we waged a struggle for Jewish labour. We regarded Arab labour in the Jewish villages as a grave danger, for we knew that the land would not be ours if we did not work it and develop it with our own hands. But there we never clashed with the Arabs or experienced Arab hatred.'

The Zionist imperative of Jewish labour was to lead over the years to growing Arab resentment as the area of land bought and farmed by Jews increased. Without Jewish land purchase, there could be no place for the growing number of immigrants; but with each such purchase, and the settlement of Jews on it, Arab hostility was exacerbated.

The proliferation of Jewish enterprise in Palestine in the decade before the First World War was considerable. In 1906, the year of Ben-Gurion's arrival in the country, the first Hebrew high school was founded, in Jaffa, by Russian-born Yehudah Metman-Cohen. The school began with four teachers and sixteen pupils. In 1908, also in Jaffa, the German-born Arthur Ruppin became, at the age of thirty-two, the head of the Palestine Office of the Zionist Executive.

Travelling widely throughout the land, Ruppin pressed for the establishment of Jewish farming settlements, arranged for the purchase of contiguous tracts of land, and set up an agricultural training farm at Kinneret, for Jewish labourers. Among the first workers there was Yitzhak Tabenkin, who had come to Palestine from Russia at the age of twenty-five, and who was to play a leading part in the evolution of Labour Zionism both in Palestine and later in the State of Israel (serving in the Knesset until 1969).

The money to buy the land at Kinneret came from the Jewish National Fund, as did the money needed that same year to buy farming land at Hulda, in the Judaean Hills: after Kfar Hittim and Ben Shemen, Kinneret and Hulda were the third and fourth purchases of the fund. Two years later, in 1910, the Roumanian-born Aaron Aaronsohn, who had been brought to Palestine at the age of six by his parents in 1882, set up the Jewish Agricultural Experiment Station at Athlit to test crops that would grow and flourish in the country; he had already discovered a wild species of wheat in Galilee. But the Zionist movement was not pleased when Aaronsohn advocated the employment of Arab labour on the new Jewish farms that were being set up. Most of the newly arrived Jewish labourers and teachers saw no future for the Jewish community in Palestine – the Yishuv (the Hebrew word for 'to settle', and 'settlement) – if it was to be dependent on Arab labour.

Aaronsohn also visited the United States, invited to do so by the Department of Agriculture in Washington. While there he acquired eight powerful American patrons for his work. Leading Jews from Chicago, New

York and Philadelphia were represented: the last city by the historian and communal leader Dr Cyrus Adler, and Samuel S. Fels – the inventor of the mothball.

Central to Zionist thinking was the Socialist concept of cooperative ventures, in which there would be neither owner nor manager, but in which the workers would share equally in the profits of their collective enterprise. The first cooperative society was founded in Petah Tikvah in 1900 for the marketing of citrus fruit. In 1906 the two main wine-growing towns, Rishon le-Zion and Zichron Yaakov, set up an Association of Wine Growers, becoming the collective owners and exploiters of the vineyards first set up by Baron Edmond de Rothschild.

It was also hoped that consumers would benefit from collective purchases and equal distribution: the first consumers' cooperative was established in Rehovot in 1906, but failed. Five years later a consumers' collective was set up in Jaffa, and four years after that one was established in Petah Tikvah. Ten years after the first experiment in Rehovot, under the pressure of wartime shortages, a national consumers' cooperative was established, its name, Hamashbir, deriving from the biblical word for ending famine. Hamashbir provided urban dwellers and farmers throughout Palestine with reasonably priced food, grain and corn. Fifty years after Hamashbir had been set up, it was a successful nationwide enterprise, supplying 800 cooperative stores throughout the country.

Collective farming was also an ideal for which the pre-First World War settlers strove, encouraged to do so by Arthur Ruppin, who in 1908 was put in charge of the Palestine Land Development Corporation, authorized to purchase land with the money raised by the Jewish National Fund. That year the training farm at Sejera, which had been in operation since 1901 – managed by the Russian-born agronomist Eliahu Krause – was handed over to the farm labourers. It then operated for a year without a manager or the imposition of any outside interests or directives. All the farm work was done by Jewish labourers who shared in the profits of their collective toil. Guard duty, which since the start of the farm in 1901 had been carried out by local Circassians – a tribe brought to the region from the Caucasus by the Turks a century earlier to serve as Ottoman imperial guards – was undertaken by the Jewish workers themselves.

The success of the collective farming experiment at Sejera led to a similar experiment later that year at the farm at Kinneret. When that also succeeded, Ruppin authorized the setting up of the cooperative settlement – or kibbutz – of Deganya (Cornflower), across the River Jordan from the Kinneret training farm. It was established in 1909 on land which was the fifth holding purchased by the Jewish National Fund. Deganya was initially worked by seven pioneers who had earlier been wage-earners at the Kinneret farm. Within five years, fourteen such kibbutz settlements had been established:

members were to own nothing beyond personal possessions; all profits made by the kibbutz were put into a common pot and this pot provided for the needs of all the members, including their food, and the education of their children, who slept and ate separately from their parents.

The use of exclusively Jewish labour at Sejera created tensions with the neighbouring Arab villages. At Passover 1909 Arabs from several of these villages murdered a number of Jews, 'simply', as Ben-Gurion, who was then working at Sejera, recalled, 'because they were Jews'. Self-defence became urgent. The kibbutzim, as well as the Jewish towns of Hadera, Rishon le-Zion and Rehovot, were protected from Arab attack by some forty members of Ha-Shomer. Blending in with local customs, the watchmen spoke Arabic, wore a mixture of Arab and Circassian dress, carried modern weapons and were, in many cases, expert horsemen. At harvest time they could call on a further 250 auxiliaries. The motto of Ha-Shomer was 'By blood and fire Judaea fell; by blood and fire Judaea shall rise.'

In the ten years following Herzl's death, the Turkish government was unwilling to grant the Jews any autonomous region in Palestine. But Jewish settlement there continued to grow, and was even accelerated, buoyed up by the Zionist vision, financed by the Jewish National Fund, and protected by Ha-Shomer. Jews from Russia, and also from Roumania – where persecution continued unabated – made their way to Palestine in their thousands by sea and overland. In Palestine, David Gruen joined the editorial board of the Labour Zionist journal *Ahdut* ('Unity'). It was then that he took the Hebrew surname by which he was to become best known: Ben-Gurion.

The primacy of the Hebrew language was central to Zionist activity in Palestine. Other languages fought for supremacy on the basis of overwhelming numbers: German and Yiddish were by far the most common languages spoken by the new immigrants, with Russian a close follower. Jews from Arab lands, of whom there were an increasing number, especially from North Africa, spoke Arabic and French. French was also the language of instruction of the worldwide Alliance Israelite Universelle school system – founded in Paris in 1860 – which had been operating in Palestine, largely on the basis of French-Jewish charity, since the 1870s. But mainstream Zionists insisted that Hebrew was the language of the future: among the Jews of Palestine it was rapidly becoming a vigorous contemporary language. In 1906 a Russian-born Jew, Boris Schatz, founded the Bezalel art school in Jerusalem. Hebrew was the language of instruction there from the first days.

In 1909 a Jewish town was established on the sand dunes just north of Jaffa – it was called Tel Aviv, the Hebrew for 'Hill of Spring'. It came to be known as 'the first all-Jewish city'. The Jewish population of nearby Jaffa, originally about 1,000 strong, had risen by immigration to more than 8,000. As a result, conditions of life in Jaffa had become as crowded and uncomfortable as in the Russian towns from which most of the immigrants had come – sometimes even more so. The new town provided welcome space. It also freed the immigrants from dependence on Arab landlords, who could raise rents at whim.

The land for Tel Aviv was bought from the Turks. The Jewish National Fund provided, through its head office in Cologne, the money needed to build the first sixty houses. The foundations were also laid for a Hebrew-language high school, the Herzliya Gymnasium, named after Herzl. Two of the first streets of the new town were named after Herzl and Ahad Ha'Am. When war broke out in 1914, there were more than 2,000 Jews living in Tel Aviv, in fewer than two hundred dwellings.

In 1911 the first Jewish hospital was opened in the predominantly Arab port of Haifa. Its founder was a German-Jewish doctor, Elias Auerbach, who had reached Palestine two years earlier. While still in Germany he had planned his own medical training in accordance with the needs of Palestine Jewry and, despite a shortage of both equipment and trained medical personnel, performed surgery in the new hospital. Twenty-two years later he was one of the founders of the Organization of German Immigrants, set up to receive the influx of refugees from Hitler's Germany (Auerbach himself had left Palestine for the duration of the First World War in order to serve in the German army on the Western Front). He died in Haifa in 1971, in his ninetieth year.

'You cannot make those things only with money' had been Herzl's proud assertion in 1902. And it was undoubtedly true that without the idealism of the desire for a national homeland, Zionism would not have succeeded. But money was an indispensable element in enabling that national home to be built up, especially the land purchased by the Jewish National Fund for the growing number of immigrants.

Many women, both as individuals and through material donations, contributed to the building up of Palestine and its Zionist institutions. In 1912 the Hadassah Women's Zionist Organization of America – American women who wanted to help medicine in Palestine – sent two of their number to establish a clinic in Jerusalem. In Palestine itself, women were active in seeking equality with men; in 1912 a girls' agricultural training farm was added to the Kinneret farm, and put under a woman director, Hannah Maisel-Shohat.

Another woman who reached Palestine in 1912 was Chassiya Feinsod. Born in the Polish city of Bialystok, and having trained for two years in Berlin as a kindergarten teacher, she was among the founders of the kindergarten system in Palestine, and later became the director of kindergarten teachers' seminaries. She had three sons who each made their contribution to the State of Israel: Yigael Yadin, a soldier and archaeologist; Yosef, an actor; and Mattatyahu, an air force pilot who was killed in action in the War of Independence.

*   *   *

In 1914 the supporters of the plan for a Hebrew University – one of the projects for which Chaim Weizmann had argued strongly at successive Zionist Congresses – sought to persuade an Englishman, Sir John Gray Hill, to sell them his house on the crest of Mount Scopus, overlooking both the Dome of the Rock and the Dead Sea. Much of the money was being raised in Russia by Lovers of Zion, at the instigation of Menachem Ussishkin who, thirty years earlier, had spoken so forcefully against such a project, but who had changed his mind to become a leading advocate. At a fund-raising gathering in Odessa in the winter of 1912, the main speaker, Joseph Klausner – who fifteen years later was to be professor of Hebrew Literature at the Hebrew University – appealed for funds for the university to an audience of 1,200 people, who, he recalled, 'listened with close attention which was not even interrupted when the electric light suddenly failed and the lecturer continued to speak in darkness for twelve minutes'.

The money was raised, and just in time. On 9 March 1914 – five months before the outbreak of the First World War – Arthur Ruppin wrote triumphantly in his diary, 'Today I succeeded in buying from Sir John Gray Hill his large and magnificently situated property on Mount Scopus, thus acquiring the first piece of ground for the Jewish University in Jerusalem.' What Ruppin had bought, for £2,100, was an option to purchase the site. It was to be almost four years before that option could be taken up.

Philanthropy was evident in every aspect of the building up of the Jewish community in Palestine. The Bezalel art school was funded by a German philanthropist, Otto Warburg, and a group of German Zionists, who helped to market the finished products of the school's craft workshop: today they are collector's items. An American Jew, Nathan Straus, provided the funds needed to set up a Jewish hospital in Jerusalem, with all the most modern medical facilities.

Jewish immigrants from Germany, encouraged by German diplomats and educators, were determined to try to make German, not Hebrew, the language of the Jewish community. Yiddish was also widely spoken, especially by the ultra-Orthodox Jews for whom the Hebrew language was a holy tongue and ought therefore to be used only in prayer. But the efforts of the Hebrew language committee set up in 1889 had borne fruit, and at the Eleventh Zionist Congress in 1913 it was authorized 'to serve as the centre of the renaissance and development of the Hebrew language'. In the Jewish schools in Palestine, Hebrew won the battle over German, while the French-speaking schools set up by the Paris-based Alliance Israelite Universelle agreed to introduce Hebrew into their curriculum.

No year passed without the establishment of a Jewish settlement of some sort. In 1914 a group of Lovers of Zion from Russia established Nahalat Yehuda (The Inheritance of Yehuda) just north of Rishon le-Zion, where several of them found employment as labourers, while establishing small farms of their own. They took their name in memory of Judah (Yehuda) Leib Pinsker, one of the leaders of their movement, who had died in 1891 at the age of seventy.

By 1914 there were 90,000 Jews living in Palestine, of whom 75,000 were immigrants. Since the setting up of the Jewish National Fund at the turn of the century, forty-three settlements had been established on the land, with a population of 12,000.

The majority of the Jewish immigrants in Palestine in 1914 were from Russia and Roumania. Those who were not tilling the soil, either as farmers in the settlements or as agricultural labourers, were working as shopkeepers, artisans and labourers in the towns. The number of Arabs was about half a million. The Arabs had also begun to find a political voice, and it had a strong anti-Zionist aspect to it. The two Jerusalem Arabs who were elected to the Ottoman Parliament in Constantinople in 1914 both stood on an anti-Zionist platform; that is, they called for a halt to all further Jewish immigration. Asked by the Jews why he was so strong in his denunciation of Zionism, one of them, Said Bey al-Husseini, replied that it was because of his desire for popularity and 'out of consideration for Arab public opinion'. The other successful candidate, Ragheb Bey al-Nashashibi (later mayor of Jerusalem under the British Mandate), told his would-be voters on the eve of the poll, 'If I am elected as a representative, I will dedicate all my energies, day and night, to remove the harm and danger awaiting us from Zionism and the Zionists.'

In the summer of 1914 the Turkish government imposed strict measures to prevent Jews who were not Ottoman subjects from settling in Palestine. This was not the first time that restrictions had been imposed, but these were the most severe. Then, in October, Turkey entered the First World War on the side of the Central Powers (Germany and Austria-Hungary). From that moment, Britain, France and Russia were Turkey's adversaries. The Jews of Palestine suffered immediately from the war conditions, as food supplies dwindled and the Turkish military authorities looked with suspicion upon the Jewish presence, so many Jews having come from Russia in the previous decade.

The Turkish military commander, Jemal Pasha, struck out at all manifestations of nationalist feeling, Jewish and Arab. Several Arab leaders were hanged in Beirut and Jerusalem. Eighteen thousand Jews were expelled from Palestine, or fled, 12,000 of them by ship from Jaffa to Alexandria. Jews known to be active in the Zionist movement were also expelled, among them Arthur Ruppin, who was sent into exile in Constantinople, and two of the leaders of Ha-Shomer – Manya and Israel Shohat – who were exiled to Anatolia.

With the coming of war a number of Jews set up a small military training unit, the Jaffa Group, to prepare for the defence of the Jewish settlements should the need arise. Two of the founders, Eliahu Golomb and Dov Hos, both immigrants from Russia, were later to hold high positions in the political life and defence of Palestine Jewry. At the request of his colleagues, Dov Hos volunteered for service in the Turkish army, becoming an officer. Because he focused his efforts on the defence of the Jewish settlements, he was accused by the Turks of a breach of military discipline. Escaping from Palestine, he was sentenced to death in absentia.

David Ben-Gurion and his Russian-born colleague, Yitzhak Ben-Zvi, were law students at the Turkish university in Constantinople in 1914. When war broke out they were on their way back to Palestine for the summer vacation. Jemal Pasha saw their names on a list of delegates to the forthcoming Zionist Congress and had them banished from the Ottoman Empire, 'never to return'. 'We were imprisoned until we could be deported,' Ben-Gurion later wrote, 'but since we were students at a Turkish university, we were treated with civility and during the day were permitted to stroll about in the compound of the administrative complex in which the jail was located. There I met an Arab student, Yihya Effendi, with whom I had been friendly back in Constantinople. When he asked me what I was doing there, I told him that I was imprisoned and that there was an order banishing me from the country. My friend Yihya Effendi responded with these words: "As your friend, I am sorry to hear it; as an Arab, I am glad." This was the first occasion on which I encountered an expression of political hostility on the part of an Arab.'

From the outbreak of the war, the idea had arisen in Zionist circles of the establishment of a Jewish Legion to fight alongside the Allies and against the Turks. The Legionnaires hoped to participate in the liberation of Palestine from the Turkish yoke and to find the Allies sympathetic to the idea of a Jewish homeland there. A Russian-born Jew, Vladimir Jabotinsky, who had travelled to Egypt in December 1914 as the war correspondent of a Moscow daily newspaper, called on the Zionists in exile there from Palestine to join in political and military activity alongside the British, French and Russians, and against the Germans and Turks.

Also in Egypt at the beginning of the war was Joseph Trumpeldor, a Jewish veteran of the Russo-Japanese war of 1904–5 – he had lost an arm fighting against the Japanese. Trumpeldor was a strong supporter of a Jewish military force. On 22 March 1915 a majority of the Palestine Refugees' Committee in Egypt passed a resolution 'to form a Jewish Legion and propose to England its utilization in Palestine'. Within a few days, 500 men had enlisted. Jabotinsky and Trumpeldor then prevailed upon the British government to allow the creation of a Zion Mule Corps, to serve on the Gallipoli peninsula, on which an Anglo-French force had just landed.

The Allied military enterprise at Gallipoli failed, but the work of the Zion Mule Corps was appreciated by the British commanders, and established the prospect of Jewish military participation in the conquest of Palestine. Of the 650 Jewish exiles from Palestine who enlisted in the Zion Mule Corps, 562 served on the Gallipoli peninsula; Trumpeldor took part in the 'V' Beach landing at the very outset of the campaign. When, in November, the evacuation of the peninsula began – the Allies having failed utterly in reaching their objectives – the Allied commander-in-chief wrote to Jabotinsky, 'The men have done extremely well, working their mules calmly under heavy shell and rifle fire, and thus showing a more difficult type of bravery than the men in the front line who had the excitement of combat to keep them going.'

In Palestine itself, the hardships of a country at war intensified, as Turkish troops struggled successfully to keep the British out of Palestine, but were being driven back in Mesopotamia almost to Baghdad. For the Jews, amid all the difficulties of war, life went on. On 4 May 1915 a child was born in kibbutz Deganya. His name was Moshe Dayan: he was named after nineteen-year-old Moshe Barsky, a member of Deganya, killed before the war by six Arabs who were trying to steal his mule. The young Dayan was the first child to be born on the kibbutz: in years to come he was to be among the State of Israel's foremost soldiers and statesmen.

Dayan's birth coincided with a plague of locusts. They arrived in a great swarm from across the Jordan. 'The members tried everything possible to destroy them or to protect the trees and crops,' Dayan later wrote. 'A sticky paste was smeared on the tree trunks, and the branches were wrapped in white sacking. Ditches were dug around the threshing floor and filled with water. But nothing helped. When the locusts moved on, their eggs hatched, and there were caterpillars everywhere. Little was salvaged.'

Later in the war, German air force pilots, who had been seconded to the Turkish army, found the homes at Deganya the perfect barracks in which to live and took them over. The members of the kibbutz had to move into the cowshed and storehouse for the rest of the war. During a cold and rainy winter all the children were ill; the young Dayan caught pneumonia.

Throughout the First World War, the defence of the outlying settlements in the north continued to be vigilantly maintained. In 1916 kibbutz Kfar Giladi was founded by members of Ha-Shomer on the north-west rim of the Huleh swamp. Its aim was both to guard the northern settlements against sporadic attacks by groups of neighbouring Arab villagers – a disturbing feature of the previous decade – and to increase the food supply available to the starving Jewish population throughout the country.

The existence and defensive energies of Kfar Giladi were decisive in ensuring that the future map of Israel included the 'finger' of Galilee – also known as the Galilee Panhandle.

In Tel Aviv a committee was set up to assist those suffering from the deprivation of war. The chairman of the committee was Meir Dizengoff, head of the local council, himself Russian-born and a former student revolutionary against Tsarist rule. Like many of the leaders of Palestinian Jewry, he was expelled by the Turks from Palestine, in his case to Damascus where he remained until liberated by the British in October 1918.

Jacob Mintz, who had settled in Palestine a decade earlier, was among those who were seized by the Turks and taken off for forced labour, surfacing the road to Damascus. That was the last his family heard of him. According to rumours, he contracted typhoid and died, as did many of those taken by the Turks at that time. His son, Aaron, who was then only twelve years old, left Nahalat Yehuda to look for his father and at least to find his body so that it could have a Jewish burial. In this he failed.

Among those who died during the war was Said Ben Shalom Levi. A Yemenite Jew, he had been one of the very first Yemenites to settle in Palestine, reaching Jaffa in 1888 and becoming a teacher. He became secretary of the Yemenite Workers' Union in Palestine, whose aim was to promote Jewish rather than Arab labour. His diary recorded the suffering of the Jews of Palestine during the war. Amid the growing number of deaths from hunger, especially children, a Russian-born Jewish doctor, Helena Kagan, who had reached Palestine shortly before the war, bought a cow and brought it into Jerusalem to be a source of precious milk for the sickly children under her care (the Turks had refused to recognize her Swiss medical doctorate, so she could practise only as a nurse).

Jewish children continued to be born even at this time of great hardship. On 17 March 1917 the kindergarten pioneer, Chassiya Feinsod, who had married Lipa Sukenik – a school algebra teacher – gave birth to her first son, Yigael. The name that she and her husband chose for their son expressed their faith that all would be well in the end: it is the Hebrew for 'He will be redeemed.' The suffering which all Jews in Palestine hoped would be redeemed reached a climax at Passover 1917 when the Turks expelled the Jews from both Jerusalem and Jaffa. One result of this was the decision of the agronomist Aaron Aaronsohn to try to rouse world opinion in support of the Jews of Palestine, and to raise large sums of money to relieve their distress.

The Aaronsohn family had set up a spy ring in Palestine working behind Turkish lines in support of the British, and hoping to see an end to Turkish rule. It was known as Nili. Its name consists of the initial letters of a Hebrew verse from the second book of Samuel, *Nezah Yisrael Lo Yeshakker* – The Strength of Israel will not Lie.

When Aaron Aaronsohn reached Britain, he heard that his sister Sarah, who was part of the Nili, had been captured and tortured for four days. She revealed nothing and then committed suicide. Spurred by grief and by a determination to rid Palestine of Turkish rule, Aaron Aaronsohn devoted himself to the task of persuading the British government to grant the Jews a national home in Palestine. His knowledge of the land of Palestine, its farmers and its pioneers, and his own family's brave contribution to the

Allied cause, stood him in good stead in the months to come. When the declaration of British support for a national home in Palestine was being drafted for the signature of the Foreign Secretary, A. J. Balfour, Aaronsohn's voice was a persuasive one.

It was Chaim Weizmann who, recognizing Aaronsohn's qualities, had brought him to Britain. Another of those whom Weizmann brought over during the discussions about the Jewish National Home was the Russian-born agronomist Akiva Ettinger, who had first visited the farming settlements in Palestine fifteen years earlier. In 1914 he had been appointed adviser and inspector of Jewish agricultural settlement in Palestine, but because of the outbreak of war had gone instead to The Hague. There, at the temporary office of the Jewish National Fund, he had prepared a series of detailed plans on the establishment, funding and working of new settlements.

It took more than a year of continuous negotiations between the Zionists and the British government before the Balfour Declaration was ready to be issued. As anti-war sentiment gained strength throughout Russia, fanned by Bolshevik propaganda, the British government became increasingly anxious to find a way to persuade the Jews of Russia to regard an Allied victory as an essential element in Jewish national aspirations. British hopes of defeating Turkey and supplanting her as the dominant power in the Middle East were in harmony with the wider Zionist hopes of a State in Palestine. But those Zionists who wanted the actual words 'Jewish State' to be included in the declaration had to be content with a phrase acceptable to the British, 'Jewish National Home'.

As finally sent the declaration itself took the form of a letter from A. J. Balfour to Lord Rothschild, dated 2 November 1917. It was to form the basis, once the First World War was over, of an upsurge in Jewish immigration to Palestine, and the 'close' settlement of the Jews on the land. It contained the emphatic assurance that 'His Majesty's Government view with favour the establishment in Palestine of a National Home for the Jewish People, and will use their best endeavours to facilitate the achievement of this object . . .'

The Balfour Declaration electrified Jews all over the world for whom a Jewish homeland in Palestine had been either a practical objective or a dream. Joseph Klausner recalled how in Odessa, during a meeting to welcome the Balfour Declaration, 'two hundred thousand Jews followed the motor-car of Ussishkin and his comrades, and the whole of Christian Odessa was astounded at the sight of the great Jewish procession, the like of which no Russian city had ever seen.' But in the United States, Ben-Gurion sounded a cautionary note. 'Britain has not given Palestine back to us,' he wrote. 'Even if the whole country were conquered by the British, it would not become ours through Great Britain giving her consent and other countries agreeing . . . Britain has made a magnificent gesture; she has recognized our existence as a nation and has acknowledged our right to the country. But only the Hebrew people can transform this right into tangible fact; only they, with body and soul, with their strength and capital, must build their National Home and bring about their national redemption.'

These were the proud words of a man who had already toiled in the fields and farms of Palestine for several years. But with the entry of the United States into the war, the Balfour Declaration raised the possibility of persuading the would-be Zionist pioneers in the United States to volunteer for service in Jewish military units being set up by Britain to fight in Palestine. Meanwhile, as recruiting got under way in the United States, on 7 December 1917, within a month of the Balfour Declaration, the British forces in Palestine commanded by General Allenby drove the Turks from Jerusalem. This left only the northern half of the country – including the Jewish settlements in the Jezreel Valley and Galilee – to be liberated.

With the British conquest of Jerusalem, the way was clear for the Zionists to work with the British to begin building up the Jewish National Home promised by the Balfour Declaration, and to create the Jewish national institutions through which that home could prosper. One of the earliest steps in this direction was the purchase on 31 January 1918 of the house and property of Sir John Gray Hill on Mount Scopus, on which the Zionists had acquired an option just before the outbreak of the war. For the sum of £6,500, the site was acquired that day for the future Hebrew University.

CHAPTER THREE

Beyond the Balfour Declaration
1918–1929

In the opening months of 1918, after Allenby's army had liberated Jerusalem, the war was still being fought in Europe, and its outcome was still uncertain. The Turks still controlled the north of Palestine, including the Jewish settlements in Galilee, and could not be dislodged. For nine months the front line lay only a few miles north of Jerusalem.

On 2 February 1918 a newly raised army battalion marched with fixed bayonets through the City of London. Onlookers noticed the unusual appearance of many of the soldiers; all 800 men wore a Star of David insignia on their uniforms. On the following day they set sail for Egypt, and further military training. This was the 38th battalion Royal Fusiliers, known as the Jewish Legion. The 38th was made up of Jews from many lands, but principally from Britain and Russia. A second battalion, the 39th, was being recruited in the United States.

David Ben-Gurion and Yitzhak Ben-Zvi had made their way to the United States as part of a recruiting drive for the Jewish Legion. The 'two Bens', as they were known, also set about organizing a group of American Jews, The Pioneer (Hehalutz), with the aim of returning to Palestine with large numbers of would-be immigrants when the war was over. They spoke in graphic tones of how, with the coming of war, the Turkish authorities in Palestine had turned against the Yishuv, the Jewish community there, expelling many, taking many others for forced labour, and making no efforts to prevent hunger and hardship among those who remained; 56,000 of the earlier Jewish population of 90,000.

One of those who responded to the appeal for volunteers to fight in the Jewish Legion was a Russian immigrant living in Milwaukee, Goldie Mabovitch, later Golda Meir. 'I had never met people like those Palestinians before nor heard stories like those they told about the Yishuv,' she later recalled, after meeting the Jewish Legion emissaries. 'This was my first clue about how terribly it was suffering from the brutality of the Turkish regime, which had already brought normal life in the country to a virtual standstill. They were in a fever of anxiety about the fate of the Jews of Palestine and convinced that an effective Jewish claim could be made to the Land of Israel after the war only if the Jewish people played a significant and visible military role, as Jews, in the fighting. In fact, they spoke about the Jewish Legion with such feeling that I immediately tried to volunteer for it – and was crushed when I learned that girls were not being accepted.'

In June 1918 the 38th battalion, and advance units of the 39th, were sent to the front line north of Jerusalem, and in the Jordan Valley. One company of the 38th, led by Jabotinsky, successfully forded the River Jordan. Two other companies, Americans of the 39th, drove the Turks from the town of al-Salt. 'By forcing the Jordan fords,' the commander of Allenby's right wing told the Jewish Legion when the war was over, 'you helped in no small measure to win the great victory gained at Damascus.' More than twenty Legionnaires were killed in action. A further thirty died of malaria.

The bulk of the 39th battalion Royal Fusiliers was still being recruited in the United States. Among the Russian-born Jews in America who wanted to live in Palestine, and who saw the battalion as a means to do so, was Nehemia Rubitzov, who enlisted as a private in the British army. Four years after he reached Palestine his son was born: the future Yitzhak Rabin.

Nehemia Rubitzov was one of 2,500 American Jews who joined the 39th battalion. A further 200 Jewish exiles from Palestine who were then living in the United States, among them Ben-Gurion and Ben-Zvi, also joined. But by the time the new recruits had completed their training – first in Canada, then in England and finally in Egypt – the war was over.

Even as the war continued, the British government was faithful to its Balfour Declaration pledge that it would use its 'best endeavours' to facilitate the establishment of a Jewish National Home in Palestine. To this end, it encouraged the establishment of a Zionist Commission to examine the future of Jewish settlement and institutions. Headed by Chaim Weizmann, the Zionist Commission reached Palestine while the war was still being fought. Among its members was Aaron Aaronsohn, who nine months later as a member of the Zionist delegation to the Paris Peace Conference was killed in an aeroplane accident over the English Channel while on his way from London to Paris. He was only forty-three years old.

Aaronsohn, like Ben-Gurion, understood that the future of the Jewish community in Palestine depended not only on immigration, but on the cultivation of the land. Less than 10 per cent of the land area of Palestine was under cultivation. The rest, whether stony or fertile, was uncultivated. No Arab cultivator need be dispossessed for the Zionists to make substantial land purchases. The potential of the land, on which fewer than a million people were living on both sides of the Jordan, was regarded as enormous.

On 29 January 1918, three months after the Balfour Declaration, Ben-Gurion, who was still in the United States drumming up support for emigration to Palestine, wrote in an article on the future potential of the land:

> According to an estimate of Professor Karl Ballod, the country's irrigable plains are capable of supporting a population of six million, to be sure under conditions of intensive cultivation and using proper modern irrigation methods. It is on these vacant lands that the Jewish people demands the right to establish its homeland.
>
> The demand of the Jewish people is based on the reality of un-exploited economic potentials, and of unbuilt-up stretches of land that require the productive force of a progressive, cultured people. The demand of the Jewish people is really nothing more than the demand of an entire nation for the right to work.
>
> However we must remember that such rights are also possessed by the inhabitants already living in the country – and these rights must not be infringed upon.
>
> Both the vision of social justice and the equality of all peoples that the Jewish people has cherished for three thousand years, and the vital interests of the Jewish people in the Diaspora and even more so in Palestine, require absolutely and unconditionally that the rights and interests of the non-Jewish inhabitants of the country be guarded and honoured punctiliously.
>
> In analysing the rights and interests of the Jews and non-Jews of Palestine, we note a characteristic difference between the two. The non-Jewish rights consist of *existing* assets, material or spiritual, which require legal guarantees for their preservation and integrity. The Jewish interests, which also include some existing assets, consist mainly of the *age-old* opportunities offered by the country, the economic and cultural potential of this semi-desolate land, the hidden wealth of natural resources and the soil which the Jewish people is destined to uncover and exploit through a creative effort and the investment of wealth and toil.
>
> The non-Jewish interests are conservative; the Jewish interests are revolutionary. The former are designed to preserve that which exists, the latter – to create something new, to change values, to reform and to build.

Ben-Gurion sought to combine the dynamic of Jewish settlement with the basically humane ideals of Judaism as it had evolved over the centuries. The rights of the inhabitants of land – not always respected in biblical times – were for him of great importance. Co-existence with the Arabs would, as he saw it, benefit the Arabs considerably, without in any way dispossessing them.

One symbolic step undertaken by the Zionist Commission while Britain and Turkey were still at war was the laying of the foundation-stones of the Hebrew University on Mount Scopus, overlooking both the Old City of Jerusalem to the west, and the Dead Sea and the mountains of Moab to the east. Weizmann, who had dreamed of such a university since the turn of the century, presided over the ceremony. Among those who wrote to congratulate him was Ahad Ha'Am, then living in London. 'I know that owing to present conditions the erection of the building will have to be postponed,' he wrote, 'so that for a long time – Heaven knows how long – the laying of the foundation-stones will remain an isolated episode without practical consequences. Nevertheless I consider it a great historical event.' Ahad Ha'Am's letter continued:

We Jews have been taught by our history to appreciate the real value of laying foundations for future developments. Our share, as a people, in the building up of the general culture of Humanity has been nothing else than the laying of its foundations long before the superstructures were built upon those foundations by other peoples.

When in time to come the Hebrew University stands proudly erect on the historic mountain, equipped with all ancient and modern instruments for the cultivation of mind and soul – what else will be its function but the laying of foundation-stones, on which our future national life will be rebuilt.

Since the beginning of our national movement in connection with the colonization of Palestine, we have always felt – many of us unconsciously – that the reconstruction of our national life is possible only upon spiritual foundations and that, therefore, the laying of those foundations must be taken in hand simultaneously with the colonization work itself.

In the first embryonic period, when the whole work in Palestine was still of very small dimensions and in a very precarious condition, the spiritual effort was concentrated in the then very popular Hebrew school at Jaffa, which was as poor and unstable as was the colonization itself. In the following period, the colonization work having been considerably enlarged and improved, the need for laying spiritual foundations made itself felt more vividly and found its expression in the creation of the 'Hebrew Gymnasium' at Jaffa – an institution incomparably superior to its predecessor.

Now we stand before a new period of our national work in Palestine, and soon we may be faced by problems and possibilities of overwhelming magnitude. We do not know what the future has in store for us, but this we do know: that the brighter the prospects for the re-establishment of our national home in Palestine, the more urgent is the need for laying the spiritual foundations of that home, on a

corresponding scale, which can only be conceived in the form of a
Hebrew University.

By a 'Hebrew University', Ahad Ha'Am continued, 'I mean – and, so, I am
sure, do you – not a mere imitation of a European university with Hebrew
as the dominant language, but a university which from the very beginning
will endeavour to become the true embodiment of the Hebrew Spirit of old
and to shake off the mental and moral servitude to which our people has
been so long subjected in the Diaspora. Only so can we be justified in our
ambitious hopes as to the future universal influence of the "Teaching" that
"will go forth out of Zion".'

These were high ideals, yet they reflected a strand in Zionism that was
already strong, and which Ahad Ha'Am himself had done much, through his
writings, to foster: that the Jewish State would not be merely another political entity in the world, but would have a deeper moral and ethical basis,
derived from the ethical legacy of the Old Testament. This attitude was
reflected three years later when Winston Churchill, then Secretary of State
for the Colonies, planted a tree on the still unfinished site of the university,
and told the leaders of Palestinian Jewry, 'We owe to the Jews a system of
ethics which, even if it were entirely separated from the supernatural, would
be the most precious possession of mankind, worth, in fact, the fruit of all
other wisdom and learning together.'

Churchill did not come to Palestine only with words: as part of Britain's
commitment to the Zionists, he granted a substantial economic concession
to the Russian-born engineer, Pinhas Rutenberg, and his Palestine Electric
Corporation, for the development of the water power of Palestine. Under
the Rutenberg Concession, the Zionists were able to expropriate land – even
Arab owned land – if needed for pumping stations and other facilities in
connection with damming the River Jordan. The Rutenberg works, just below
the Sea of Galilee, became a symbol of hope for the modernization of all
Palestine, Jewish and Arab alike. In the 1948 War of Independence they were
destroyed; today, following the Israel–Jordan peace treaty, the ruins are a
tourist attraction.

While on Mount Scopus, Churchill also spoke of what he called the
'blessing' that a Jewish National Home could be 'to all the inhabitants of this
country, without distinction of race and religion', and he went on to explain:

This last blessing depends greatly upon you. Our promise was a double
one. On the one hand, we promised to give our help to Zionism, and
on the other, we assured the non-Jewish inhabitants that they should
not suffer in consequence.

Every step you take should therefore be also for normal and material
benefit of all Palestinians. If you do this, Palestine will be happy and
prosperous, and peace and concord will always reign; it will turn into

a paradise, and will become, as is written in the scriptures you have just presented to me, a land flowing with milk and honey, in which sufferers of all races and religions will find a rest from their sufferings.

You Jews of Palestine have a very great responsibility; you are the representatives of the Jewish nation all over the world, and your conduct should provide an example for, and do honour to, Jews in all countries. The hope of your race for so many centuries will be gradually realized here, not only for your own good but for the good of all the world.

The status of Palestine with regard to the Jewish National Home depended entirely upon the fate of the Balfour Declaration. This had to be accepted by the victorious powers, and enshrined in the Mandates provisions of the League of Nations. To secure this, a Jewish delegation, headed by Chaim Weizmann, addressed the Paris Peace Conference on 27 February 1919. Of the four delegates, only Menachem Ussishkin spoke in Hebrew (the others spoke in English and French). By speaking in Hebrew he made an extraordinary impact on delegates who had never heard the biblical tongue used in ordinary speech. Ussishkin spoke, he said, on behalf of 'the largest Jewish community', the Jews of Russia – numbering more millions in those days than even the Jews of the United States.

In the immediate aftermath of the First World War there had been an upsurge in attacks on Jews in Russia, especially in the Ukraine, where, 250 years earlier, the Cossacks led by Bodgan Chmielnicki had wrought such havoc. Once more, Jews were singled out as targets for physical attack, looting and murder. After speaking to the Peace Conference of the historic attachment of the Jews to Palestine, of which they had been 'robbed' by the Romans, Ussishkin declaimed, 'restore that historic robbery to us!' and went on to tell the Allied statesmen of the years of exile and wandering:

Nowhere have we found rest for our weary spirit nor for our aching feet. Persecution, expulsion, cruel riots, unbroken distress – such have been our lot during all these generations in all the countries of the world, and in these very days – when the wielders of the world's destiny have proclaimed the liberation of the nations, the equality of the nations, and the self-determination of every separate nation – Russian Jewry, which I represent here, is undergoing fresh torrents of murder and rioting the like of which were never known even in the Middle Ages.

For us there is no way out save to receive, under your authority and subject to your supervision, one secure place in the world where we shall be able to renew our own lives and revive the national and cultural tradition which has come down to us from ancient times, and where can that secure spot be save in our historic country? Throughout all these generations we have not ceased to yearn for it, but have prayed the

God of Israel for our return thither. Not for a moment have we forgotten it, just as we have not forsaken our God, our tongue and our culture.

We let ourselves be slain for these possessions of ours rather than betray them. And on this very day I address you in our Hebrew tongue, the tongue of our kings and prophets which we have never forgotten. This tongue is bound up with all our national aspirations. At the beginning of the national revival in the Land of Israel, when we had barely begun our upbuilding work there, even before the war, we devoted our efforts to the revival of our language and our culture.

The Paris Peace Conference agreed to grant the Palestine Mandate to Britain, and to accept the promise of the Balfour Declaration to 'facilitate' the establishment of a Jewish National Home there. Herzl had confided in his diary in 1897 that he had founded the Jewish State; now, only twenty-two years later, the international community had made it possible, under Britain's aegis, for the Jews to build up the institutions and infrastructure of statehood. The Zionist Commission was already in Palestine, working to establish such a structure.

Reaching Palestine in November 1919, Ussishkin became Chairman of the Commission. It fell to him to fashion the administrative and structural form of the National Home: the Jewish National Institutions. A Jewish National Assembly was set up by the Jews of Palestine. From it, delegates elected a Jewish National Council (Va'ad Leumi). Ussishkin himself was elected to the Council by the vote of Sephardi Jews. He had always understood their needs, and recognized their unease at the Ashkenazi dominance of the Zionist movement, both in its early days, and through the Russian and Polish immigration that had constituted the main numerical influx of the previous three decades.

Ussishkin was very conscious of the fact that the Sephardi Jews had been the principal dwellers in the land for many centuries before the advent of political Zionism, and urged Sephardi participation in every institutional forum. He was determined that the Sephardi customs should not only be respected, but become an integral part of the new society. For him, Zionism was a 'Gathering of the Exiles' that would unite very diverse and hitherto isolated strands of Jewish life in the Diaspora into a single national unit.

In Upper Galilee, Jewish settlements were caught in the middle of a series of armed skirmishes between local Arabs and the French authorities who controlled the area immediately after the First World War. In 1919 the area was transferred to Britain. Arab attacks on the Jewish settlements continued. To protect them, Joseph Trumpeldor, the one-armed veteran of both the Russo-Japanese War and the Gallipoli campaign, who had just returned to Palestine from Russia, was asked to organize the defence of the northern settlements. On 1 January 1920 he reached Tel Hai, which he began to fortify.

On March 1, Tel Hai was attacked by a large number of armed Bedouin. The besieged Jews opened negotiations with their attackers. As talks were continuing, there was an exchange of rifle fire and Trumpeldor was wounded in the stomach. Fighting continued for the rest of the day. That evening, Trumpeldor died of his wound. His last words were said to have been, 'Never mind; it is worthwhile to die for our country.' Five other defenders were killed with him, among them twenty-three-year-old Sarah Chizik, who had come with her family from the Ukraine at the age of ten.

The death of Trumpeldor was to have a substantial impact on the Zionist movement. He was only forty when he was killed. Songs, poems and short stories were written about him. Children were named after him. His correspondence, diary and personal memoirs, which were published two years after his death, became a basic text for Zionist youth. Both socialist and right-wing Zionists were to find inspiration in his life's story. A youth movement, Brit Trumpeldor (Betar), founded in the Latvian city of Riga three years after Trumpeldor's death, became the standard bearer of militaristic and nationalistic ideology.

Among those who fought at Tel Hai was Shaul Meirov, who had come to Palestine from Russia ten years earlier at the age of thirteen, and was one of the first graduates of the Herzliya Gymnasium in Tel Aviv. In the 1930s he was to be the main organizer of illegal immigration, defying the British restrictions which had been imposed, and bringing in thousands of Jews clandestinely by sea and overland. After the death of his son Gur Meirov in the War of Independence, he took the surname Avigur (Gur's father).

The defence of Tel Hai, like that of Kfar Giladi during the First World War, ensured that the northern finger of Galilee was later to be a part of the State of Israel.

The wave of Jewish immigration to Palestine after the First World War was known as the Third Aliyah. It lasted for four years and saw the arrival of many Socialist-inspired pioneers. In all, 35,000 newcomers came during that short period. Some worked on road building. Others set up kibbutzim and founded the first moshavim: villages of smallholders in which elements of cooperative and private farming were combined. The first moshav was founded in September 1921 at Nahalal, in the Jezreel Valley. The name was taken from the Bible, where it refers to a town in the territory of the tribe of Zebulun. Two pre-First World War attempts to found a village there, first by Arabs and then by Germans, had failed, because of the malarial surroundings.

Wherever there was, as in the Jezreel Valley, abundant but often stagnant water the malaria mosquito flourished. Many new immigrants, especially those from countries like Britain and France to which they could easily return, were unable to take the privations which malaria imposed, and left Palestine for ever.

For fifteen years the Jewish settlers at Nahalal lived in wooden huts; the only concrete was used for their stables. A second moshav was founded three months later, east of Nahalal, at Kfar Yehezkel. It was named after Yehezkel (Ezekiel) Sassoon, an Iraqi Jew who served as Minister of Finance in five Iraqi governments between 1921 and 1925, and whose donation to the Jewish National Fund was a crucial element in the Jezreel valley purchase. Within the next decade, eight more moshavim had been set up in the valley. One of them, Kfar Gideon, was founded in 1923 by religious Jews from Roumania; they were to struggle against nature for twenty years, until sufficient water reserves were discovered. Herzliya was founded on the coast, as a moshav, in 1924; it later became a flourishing town

The inspiration for the moshav movement came from Eliezer Joffe who, born in Russian Bessarabia in 1882, had gone to the United States before the First World War, and had trained agricultural students there for life in Palestine. Joffe had emigrated to Palestine in 1910, and established an experimental farm at Ein Gannim (Garden Spring), near Petah Tikvah. Within a decade of the establishment of the first moshav, he founded a marketing cooperative, Tnuva (Produce), which was to become the largest of its kind in Israel, selling the produce of the settlements throughout the country. Another strong supporter of the moshav movement from its earliest days was Russian-born Yitzhak Elazari-Volcani (born Wilkansky), who had been a farm manager in Palestine before the First World War. In 1921, as the moshav movement began, he set up an experimental agricultural station, under the auspices of the Zionist Executive, and worked to improve the cultivation of the soil and the quality of the crops.

The kibbutz movement was also active in 1921, the Jewish National Fund having purchased an area of land at the foot of Mount Gilboa, in the Jezreel Valley. It was largely swamp, and cursed with malaria, but this did not deter the pioneers who set up two tented camps near Ein Harod (the Harod Spring). This was the site where, according to tradition, Gideon and his people had camped during their war against the Midianites; and perhaps also where Saul had camped against the Philistines – from Byzantine times the spring was believed to be the site of the battle between David and Goliath. Among its founders was Russian-born Yitzhak Tabenkin.

Ein Harod, and a nearby kibbutz, Tel Yosef, founded two years later and named after Joseph Trumpeldor, were part of a new kibbutz movement that advocated a single countrywide commune. Calling itself Gedud Ha-Avodah (The Labour Legion), it consisted of workers, including those contracted to build part of the Tiberias–Tabgha road in Galilee, and others laying the railway track between Ras al-Ain and Petah Tikvah. They had decided to form permanent communal settlements near their places of work, and defined their aim as 'the building of the land by the creation of a general commune of the workers of the Land of Israel'.

One of the main organizers of the Labour Legion was Shlomo Lavi (born

Levkovitz, in Russian Poland), who had been a labourer and later a watchman before the First World War. It was he who had conceived the idea of the 'large collective' which would combine agriculture, crafts and industry, and would be capable of absorbing new immigrants who had no previous agricultural training or experience. Another of the most active members, Yitzhak Landsberg – later Sadeh – who had been decorated for bravery while serving in the Russian army in the First World War, was to play a leading part in the War of Independence.

Not all the new settlements flourished. In 1921 the British authorities had allocated land in the Negev, near Tel Arad, for demobilized soldiers of the Jewish Legion. When no water was found, they were forced to move. It took them eleven years to find another site.

British military rule in Palestine continued for two years following the defeat of Germany. But the existence of the Balfour Declaration and the steady growth of immigration led to the establishment of new settlements even before the setting up of a civil administration. In 1920 kibbutz Kiryat Anavim (City of Grapes) was founded just outside Jerusalem. It was on a hillside opposite the predominantly Muslim Arab village of Abu Ghosh, with which it quickly established friendly relations; relations that were to remain good for the next eight decades, despite the turmoils and wars of future years.

The inspiration for founding Kiryat Anavim came from Akiva Ettinger, who had settled in Palestine as soon as the war was over as director of the agricultural settlement department of the World Zionist Organization. His aim was to institute, in this ancient stony landscape, the most modern methods of land reclamation, hill farming and afforestation. At first the hardships were such that there was a move to transfer the village to the more hospitable soil of the Jezreel Valley, but the settlers resisted this and determined to make their hillside flourish. Kiryat Anavim became a model for all hill settlements. The fruit orchards, dairy cattle and vineyards that they successfully set up were to be of inestimable value to Jerusalem when it was under siege in 1948.

Ettinger had a powerful ally in setting up Kiryat Anavim. This was Menachem Ussishkin who feared the isolation of the Jews of Jerusalem if there were not a series of Jewish settlements between the Holy City and the coast. The need for some form of Jewish settlement in the Jerusalem Corridor became all the more imperative as Jerusalem's own Jewish suburbs – on the garden city model – were established, and the city's Jewish majority, a demographic fact since the 1840s, felt cut off from the more populous Jewish towns and villages of the coastal plain. It was Ussishkin who used the money available to him as head of the Jewish National Fund to purchase the land on which Kiryat Anavim was then built.

The main work of the Jewish National Fund in 1920 was an ambitious scheme to which many Zionists were opposed. This was the proposed purchase of a substantial expanse of land in the Jezreel Valley (known in Hebrew as the Emek – the Valley). Much of the land was swamp, and infested with malaria, and the price being asked by an absentee Lebanese landowner was high. But it was for sale, and both Menachem Ussishkin and Arthur Ruppin – in a powerful alliance of the often divergent Russian and German Jewish outlooks – were emphatic that the land should be bought. There could be no such thing as overcharging for land, Ussishkin argued. 'The cost of land in Palestine would increase from year to year; while what was not redeemed today could quite possibly never again be redeemed by us.'

Ussishkin bought almost forty square kilometres of land. This was by far the largest Jewish land purchase in Palestine up to that time. After the purchase was made, he was astounded to learn that several of his colleagues on the Palestine Executive wanted to bring him before the court of the Zionist Congress for squandering money. He prepared his case, but during the Twelfth Zionist Congress, held in Carlsbad, Czechoslovakia between 1 and 14 September 1921, his action was approved.

Funds would still be needed, on a substantial scale, to enable the newly acquired land to be developed, to pay for the cost of irrigation, and also for the education of the children of the settlers. To this end, a special development fund, Keren Hayesod (Foundation Fund), had been set up the previous year, and was registered on 23 March 1921. Vladimir Jabotinsky was appointed its director of propaganda. The fund hoped to mobilize capital for its work in Palestine through donations from the 'Jewish masses' throughout the world. In 1921 Ussishkin set off for the United States to raise money for the fund. Among those who accompanied him was Albert Einstein, who had always shown sympathy for the Zionist ideal. Fifty years later, at the age of seventy-two, he was to decline the offer to become President of the State of Israel on Chaim Weizmann's death. The Ussishkin–Einstein mission was a success, the precursor of thousands of such missions in the years to come.

In December 1920 a major step forward was taken in unifying the different and competing Labour Zionist groups which had emerged since the First World War. This was the creation of the General Federation of Jewish Labour, the Histadrut, the main aim of which was the provision of work, cooperative supplies, vocational training, and education. As its secretary-general, the Histadrut elected David Ben-Gurion, who believed with a fierce conviction that its members were the 'army of labour' that would redeem the country. But when his sister Rivka, still in Russia, asked if he would help her come to Palestine, he would not do so. 'I do not believe that she will be able to work,' he explained, 'or to find work she is fitted for.'

For Ben-Gurion, as for many of those in the Histadrut, the task of building up the land was not one that could be shared by all Jews. Even within Palestine, the 4,000 members of the Histadrut were a minority of the 65,000 Jewish inhabitants; and many of them, as a result of the general economic depression at that time, were unemployed, or working in the harshest of conditions building roads for the Mandatory government.

The Balfour Declaration had included a phrase that nothing should be done with regard to establishing a Jewish National Home in Palestine that might be to the detriment of the other communities in Palestine. By far the largest of these communities, and indeed by far the largest group in the country, was the Arabs, some 500,000, as against 65,000 Jews. Arab hostility to Zionism had emerged before the First World War, and intensified after it. In 1920 there were violent Arab protests against further Jewish immigration. It was during the defence of the Jewish Quarter in Jerusalem against Arab attack that year that Yitzhak Rabin's father, Nehemia Rubitzov, who had volunteered to join the defenders, met his future wife, who was likewise a volunteer.

During the 1920 Arab riots an attempt was made, led by Jabotinsky, to protect the Jews from attack. For breaking the law by carrying arms, Jabotinsky was arrested by the British and imprisoned. The pre-war Ha-Shomer had been essentially small local groups set up to defend individual settlements, mostly in Galilee. What was needed now seemed to be a special defence organization, and in the Histadrut founding conference in December 1920 this was organized on a countrywide basis. The Histadrut would be responsible for all guard and defence duties, for which purpose a Haganah (Defence) organization was set up in March 1921. The Haganah was essentially a clandestine organization operating without the approval of the British authorities, but dedicated to maintaining the security of the settlements which the British could not do, through either lack of means or lack of will. The first tasks of the Haganah were the training of members and the purchase of arms. It was to defend Jewish property and protect Jewish life in Palestine until the establishment of the State twenty-seven years later.

The Haganah was set up not a moment too soon. In May 1921 Arab riots were renewed on an even more serious scale than the previous year. Many Jewish settlements were attacked, as well as the Jewish quarters of Jerusalem. When rioters tried to break into the village of Petah Tikvah, it was the guardsmen established thirty years earlier by Abraham Shapira who defended it, with Shapira himself still at their head. Shapira, a pioneer of Jewish self-defence in Palestine, lived to see the State of Israel established, and died in 1965 at the age of ninety-five. Chaim Weizmann wrote of him, 'He was a primitive person, spoke better Arabic than Hebrew, and seemed so much a part of the rocks and stony hillsides of the country that it was difficult to believe that he had been born in Russia. He was a man who in his own lifetime bridged a gap of thousands of years; who, once in Palestine, had shed his Galuth environment like an old coat.'

Galuth (Exile) is the Hebrew name for Diaspora. For Jews, it is regarded as a temporary existence, to be abandoned when the Messiah comes – or, in the case of modern Zionism, on reaching the land of Israel.

Among the successful defenders of the Neve Shalom Jewish quarter of Jaffa in the riots of May 1921 was Ephraim Chizhik, whose sister Sarah had been killed a year earlier at Tel Hai. It was not possible to defend all those who were attacked. During a week of rioting across the country, forty-seven Jews were killed. Among the Jewish settlements destroyed was Kfar Malal, which had been rebuilt after being destroyed four years earlier during the fighting between the British and Turkish armies. The recently appointed British High Commissioner, Sir Herbert Samuel – himself a Jew – sought to calm Arab anger by a temporary suspension of Jewish immigration. It was an act of appeasement that angered many Zionists.

Among those whose ships were unable to land their passengers as a result of the ban was the one on which Golda Meir was travelling. But it was only a temporary setback, and she was able to to make her way from Egypt to Palestine by rail a few weeks later. The Arabs had seen, however, that a show of violence could lead to political gains. The leader of the 1921 riots, Haj Amin al-Husseini, was appointed within a year Mufti of Jerusalem, in the hope that he would exercise a moderating influence. But he was to use his position as Mufti – the senior Muslim cleric in Palestine – to press the British to halt Jewish immigration altogether. Within a decade the Mufti had established himself as the predominant Arab political figure, challenging the traditional leadership of the Nashashibi and El Hadi families. The Mufti also soon eclipsed the influence of his uncle Musa Kazim Pasha al-Husseini whom the British had dismissed as mayor of Jerusalem following his encouragement of the anti-Jewish riots of April 1920 and his opposition to the recognition of Hebrew as one of the three official languages of Palestine (the others being English and Arabic).

Conditions for the new immigrants in Palestine continued to be hard, economically and emotionally. The Jewish community was not rich, much of the soil was not naturally fertile, housing in the towns was simple, even primitive, and the Arab majority was increasingly ill at ease with the influx of Jewish newcomers. Golda Meir recalled how, within a few hours of reaching Tel Aviv, one of her companions turned to her and said, 'Well Goldie, you wanted to come to Eretz Yisroel. Here we are. Now we can all go back – it's enough.' Six weeks later, Golda Meir wrote to a Zionist friend in the United States: 'Those who talk about returning are the recent arrivals. An old worker is full of inspiration and faith. I say that as long as those who created the little that is here are still here, I cannot leave and you must come.'

She added: 'I would not say this if I did not know that you are ready to work hard. True, even hard work is hard to find, but I have no doubt you will find something. Of course this is no America, and one may have to suffer a lot economically. There may even be riots again. But if one wants one's own land and if one wants it with one's whole heart, one must be ready for this. When you come I am sure we will be able to plan. There is nothing to wait for.'

Following the 1921 riots, the Haganah sent emissaries to Vienna to purchase revolvers and ammunition. To avoid detection by the British, this weaponry was smuggled into Palestine in various guises: beehives, refrigerators and steamrollers. A course of instruction in the use of arms was organized under the command of a former member of the Jewish Legion, Elimelekh Zelkovich (known as 'Avner'). When an Arab mob tried to attack the Jewish Quarter of the Old City of Jerusalem on 2 November 1921 – the fourth anniversary of the Balfour Declaration – an organized group of Jewish defenders drove them off, and four Jews were killed.

The rioting was a frightening experience for those caught up in it, often leaving vivid memories and scars. Four-year-old Yigael Sukenik, later the soldier-archaeologist Yigael Yadin, remembered for the rest of his life watching from his house in Jerusalem as a crowd of Arabs, wielding wooden clubs and shouting nationalist slogans, marched along the street outside.

In response to the continuing Arab riots, and in a further attempt to curb Arab hostility, the Mandate authorities, at Herbert Samuel's urging, introduced one permanent condition with regard to all future Jewish immigration. Such immigration would henceforth depend on the 'economic absorptive capacity' of Palestine at any given time. In fact, the Zionists themselves had begun to seek a limit to the number of immigrants, because of the economic burdens they imposed on the land and resources of the Zionist movement. The Zionist Executive went so far as to report to the Twelfth Zionist Congress that 'in the light of the economic conditions reigning in Palestine, and the World Zionist Organization's financial straits, the Executive considered sending penniless pioneers to Palestine undesirable and accordingly dispatched to all the principal migration offices telegraphic instructions to refrain from sending emigrants to Palestine for the time being.'

It was not the views of the Zionist Executive, however, but the realities of European politics, that determined whether immigrants would come to Palestine in large numbers or not. The anti-Jewish violence in the Ukraine, in which more than 100,000 Jews were murdered in the immediate aftermath of the First World War, was a powerful catalyst for immigration. Hardly had that violence ended than the rigours of Communist rule were imposed on the Ukraine, with a rapid elimination of private productive enterprise and hostility to all manifestations of Jewish cultural, educational and spiritual activity. Then, in 1924, the Polish government issued a series of economic decrees that discriminated against Jews, making it hard for many Jews to earn their livelihood at all.

The response of tens of thousands of Polish Jews was to sell whatever assets they had, take out what money they had in Polish banks and savings societies, and emigrate. For the first time since the mass emigration of Jews two decades earlier, the United States was imposing restrictions. But Palestine, the Jewish National Home, was open. It was also the scene of continual growth. Thus, in 1922, the year in which the Second Aliyah pioneer and ideologist A. D. Gordon died at Deganya, a new kibbutz was established seven miles south-east of Haifa. Given the name Yagur, after a biblical location mentioned in the Book of Joshua, it quickly became the largest kibbutz in the country, farming its own lands and providing dock labourers for Haifa port (Jewish dock labour was also a feature during those inter-war years of the port of Salonica, in Greece: the dockers were deported and murdered by the Nazis in 1943).

There were sustained Arab diplomatic attempts, including a high-level delegation to London, to persuade the British government to reduce Jewish immigration and even to halt it altogether. But in 1922 the Churchill White Paper, a British policy statement, emphasized that the Jews were in Palestine 'of right, and not on sufferance'. The means whereby the Zionists could take maximum advantage of this pledge came as a result of Article Four of the League of Nations Mandate for Palestine, which was ratified by the League Council in July 1922. This stated that 'an appropriate Jewish agency shall be recognized as a public body for the purpose of advising and co-operating with the administration of Palestine in such economic, social, and other matters as may affect the establishment of the Jewish National Home and the interests of the Jewish population in Palestine.' The article went on to recognize the Palestine representatives of the World Zionist Organization as such an agency (to be known as the Jewish Agency) 'so long as its organization and constitution are in the opinion of the Mandatory appropriate'.

Following the League Council's ratification, the Jewish Agency became the main organization through which Palestinian Jewry maintained its contacts both with world Jewry and with the Mandatory authorities and foreign governments. At the time of the ratification of the British Mandate, there were 83,790 Jews living in Palestine, just over 11 per cent of the population. The Arabs numbered three-quarters of a million. Through immigration, the Jews hoped gradually to reach and then surpass the Arab numbers. But while the Jewish population was to reach 30 per cent of the total by 1940, it came to face two obstacles: Arab immigration from countries throughout the Arab world, and British restrictions on Jewish immigration, imposed as a result of strong Arab opposition both from inside Palestine and from the Arab countries around.

Many Jews had difficulties in reaching Palestine which were not related to immigration procedures. Russian Jews, under a harsh Soviet regime, found it increasingly difficult to obtain permission to leave. One of those who wished to do so was the Hebrew writer Chaim Nahman Bialik, one of the most creative and inspiring figures in modern Hebrew literature, who became the national poet of the redemption of Israel. He had to intercede directly with a fellow-writer, Lenin's friend Maxim Gorky, before he was allowed to leave. Bialik went first to Berlin and then, after three years, to Palestine, making his home in Tel Aviv of which he wrote:

> I see a Hebrew creation such as Tel Aviv as decisive, as against the creations of hundreds of years in the Exile. I wonder whether there is a corner like this for Israel anywhere in the entire world. Such house-building, purely and solely by Jews, from top to bottom, this is a spectacular sight.
>
> This creation of an entire Hebrew city is sufficient to fill the hearts of the sceptics and doubters with faith that Israel's renaissance is an undeniable fact.
>
> Here in Tel Aviv, with all my senses I feel that I have not, nor could I have any other homeland but this place. And blessed be all the builders who have embarked on our eternal building in this corner, the only one in the world.

In the first years of the Mandate, Jewish immigration was virtually unimpeded. Between 1924 and 1927 more than 65,000 Polish Jews reached Palestine. A few of the new arrivals were farmers and labourers, like the previous waves of immigrants, but most were middle-class shopkeepers, businessmen, intellectuals and students. Many of these were prepared, and indeed eager, to turn to farming, as were the pioneers from Germany who set up a kibbutz called Beit Zera (House of Seed), just south of the Sea of Galilee, in 1927.

Also established in 1927 was the Ben Shemen youth village on the site of the children's village which had been destroyed in the First World War, the brainchild of Berlin-born Dr Siegfried Lehmann, who had taken on the responsibility for the rehabilitation of Jewish orphans in Lithuania after 1918. Lehmann brought many of these youngsters with him to Palestine. Lehmann's concept of work on the soil included the total creativity of rural living with folk music, folk dance and folk art, education for peace and tolerance – especially with the Arabs – a traditional Jewish, but non-Orthodox education, and a non-dogmatic outlook on life. He also encouraged student self-government as a preparation for future life on a kibbutz. The attraction of Ben Shemen for immigrant youth was considerable. But so substantial was the most recent immigration that the Fourteenth (1925) and Fifteenth (1927) Zionist Congresses decided that priority should be given to urban as opposed to rural settlements.

Urban settlements meant an increase in the demand for Jewish labour, thereby enhancing the strength of the Histadrut workers' organizations, among them the Solel Boneh (The Paver and the Builder) construction company and contracting firm, founded in 1924. One of the founders of Solel

Boneh was Dov Hos, a veteran of both the Turkish army and the Jewish Legion, who also represented the Jewish community of Palestine with the Mandate authorities. Other Histadrut organizations that grew in size and importance were the Hamashbir retail cooperative, the Tnuva marketing outlet and cooperative building societies through which residential workers' quarters were built in all the towns. Other workers' institutions which sought to provide a basis for economic security, knowledge and leisure were the Hasneh (Burning Bush) insurance company, the daily newspaper *Davar* (which was first published in June 1926), and the workers' sports organization, Hapoel (The Worker).

Among those who found work with Solel Boneh was Golda Meir's husband Morris Meyerson. His pay was in the form, not of cash, but of credit slips. The landlord, the milkman and their children's nursery school teacher would far rather have been paid in cash. 'On "pay day",' Golda later wrote, 'I used to dash to the little grocery store on the corner to try to talk the grocer into taking a chit worth one pound (100 piastres) for 80 piastres, which I knew was all he would give me for it. But even those 80 piastres were given not, God forbid, in cash but in a handful of more credit slips. With these I would then run to the woman who sold chickens, argue with her for twenty minutes or so and finally, on a good day, persuade her to take my slips (after she had deducted 10 or 15 per cent of their value) in exchange for a small piece of chicken with which I could make soup for the children.'

Golda Meir also remembered one of the stories circulating at that time about the economic hardship of life in Palestine. A Jew was heard to say that if he only had a good feather pillow to start off with, he could build himself a house. How? 'Very simple,' he said. 'Look, you can sell a good pillow for a pound. With this pound you can pay the membership fee of a loan society, which will entitle you to borrow up to ten pounds. With ten pounds in hand you can begin looking around for a nice little plot of land. Once you've found a plot, you can approach the owner with your ten pounds *in cash*, and naturally he will agree to take the rest in promissory notes. Now you are a landowner and you can look for a contractor. To him you say, "I have the land – now you build a house on it. All I want is a flat for myself and the family".'

In 1924 the Zionists were perturbed by an ultra-Orthodox leader, Israel de Haan, who was working to establish the Orthodox community as a separate entity distinct from the Zionists, and who seemed to be willing to enlist the support of non-Jews hostile to Zionism in order to advance the cause of ultra-Orthodoxy. He had travelled to Transjordan – British Mandated eastern Palestine – and held talks with the Emir Hussein, who was visiting his son, the Emir Abdullah, in Amman. That summer, de Haan prepared to visit Britain on behalf of the anti-Zionists, to try to persuade the British government that the Orthodox Jewish community in Palestine should not be under the authority of the secular Jewish institutions headed by the Jewish National Council. On the evening of June 24, three days before he was to depart on his mission, de Haan was assassinated on Haganah orders.

Many Zionist leaders and thinkers condemned de Haan's assassination. When the full facts were revealed fifty years later, a noted Zionist historian, A. J. Brawer, wrote that whether the murder 'was something necessitated by the times, or a tragic error, I, for one, am afraid that the stain of murder will not be eradicated from the judgement made by future historians'.

As the Haganah had continued to grow, arms caches were set up at several locations near Tel Aviv and in Galilee. In 1925 a Haganah officers' course was held on Mount Carmel, near Haifa.

Defence and education were two sides of the Zionist coin. Despite initial problems with funding, an Institute of Jewish Studies was set up on Mount Scopus in Jerusalem in 1924, on the site for a Hebrew University purchased by the Zionists just before the First World War. The money for the Institute came from an American philanthropist, Felix Warburg. Within a year, the Hebrew University was opened. Arthur Balfour came from London to be present. Weizmann watched with pride as his dream of a quarter of a century came into being. Jews from all over Palestine made the journey across the country, often by horse and cart, in order to be there. Christian and Muslim leaders graced the ceremony with their presence, and their goodwill. During his remarks at the opening, Chaim Nahman Bialik spoke with enthusiasm of the Zionist pioneers who were making their way to Palestine from Central Europe, and of how Jewish secular and religious values would be linked and enhanced:

> Thousands of our young sons, responding to the call of their heart, stream to this land from all corners of the earth to redeem it from its desolation and ruin. They are ready to spill out their longing and strength into the bosom of this dry land in order to bring it to life. They plough through rocks, drain swamps, pave roads, singing with joy.
>
> These youngsters elevate crude physical labour to the level of supreme holiness, to the status of a religion. We must now light this holy flame within the walls of the building which is now being opened on Mount Scopus.
>
> Let these youngsters build with fire the lower Jerusalem while we build the higher Jerusalem. Our existence will be recreated and made secure by means of both ways together.

The Hebrew University drew its teachers and scholars from all over the Jewish world. One of those who contributed substantially to an understanding of the topographical and historical roots of the National Home was the historian and geographer Samuel Klein. Born in Hungary, he had become a rabbi – first in Hungary, then Bosnia and finally Slovakia – and served in the First World War as a military chaplain with the Austro-Hungarian forces. He had visited Palestine in 1908 and become fascinated by its geography. For five years he both taught in Jerusalem and continued with his rabbinical duties in Slovakia, an early example of the commuting between Israel and Europe, and even the United States, which became common in academic circles half a century later. But Klein taught full time at the Hebrew University from 1929. His main contribution was his research into the Holy Books, especially the Talmud and Midrash, as primary sources for both the topography of the country and its early settlements. Before his death in 1940 he had trained a generation of scholars in the topography of Palestine and in the use of biblical and Hellenistic sources.

Another of the educational pioneers that year was Saul Aaron Adler, who at the age of five had been brought to England from Russia by his parents, and as a student had specialized in tropical medicine. During the First World War he served as a doctor and pathologist with the British and Indian troops on the Mesopotamia Front. In 1924 he left Britain for Palestine, joining the staff of the Medical School in Jerusalem, and becoming professor of the Parasitological Institute there four years later. He was best known to scholars as the translator of Darwin's *Origin of Species* into Hebrew. His chief scientific work was in the search for immunizations for malaria, cattle fever, leprosy and dysentery, all of which were features of the harshness of rural life in Palestine at that time.

Also in 1924, progress was made in the establishment of yet more agricultural settlements in the Jezreel Valley, where, on behalf of the Jewish National Fund, Akiva Ettinger was taking a leading part in both buying land and introducing mixed farming, mostly dairy farming and orchards. Ettinger was also a pioneer in afforestation. The variety of settlers spanned the whole spectrum of Jewish experience. That same year a moshav was established in the valley by a group of Polish Hasidic immigrants, led by their rabbis, the rabbis of Kozienice and Jablonow, who had persuaded their flocks to go with them to Palestine. The site was so poor, however, that after three years they moved to a site further west, less than 7 miles from Haifa, Kfar Hassidim. There, they had first to drain the malarial swamps; ten years later they established an agricultural school nearby.

As the number of settlements increased, the Jewish national institutions supporting them also grew, as did new projects designed to cater for their various needs. In 1925 a Jewish town was founded in the Jezreel Valley. Named Afula (The Town of Jezreel) its construction was made possible by American Zionists, members of the American Zion Commonwealth, who wanted the settlements in the valley to have an urban centre. It was located at what hitherto had been nothing more than a way station of the Haifa–Damascus narrow-gauge railway. But despite the hopes of the Americans, the kibbutzniks of the Jezreel Valley were not drawn to the town;

their way of life was far too self-contained and demanding, and when pleasures called, it was to the more bustling urban Haifa that most of them preferred to go. One institution in Afula whose facilities they did use, however, and welcome, was the regional hospital, the first to be set up in Palestine.

Among the moshavim founded in the Jezreel Valley in the 1920s was Kfar Baruch. Established in 1926, it was named after Baruch Kahana of Ploesti, in Roumania, who had given his wealth to the Jewish National Fund for the purpose of settlement in Palestine. Kfar Baruch was unusual in that its first farmers came from a variety of lands, not only from Poland, Germany and Roumania, but also from Kurdistan and Iraq. Even more unusual, some of the first settlers were 'Mountain Jews' from the Caucasus, a Jewish group whose ancestors had lived in the Caucasus for well over a thousand years, and who spoke an ancient Persian dialect, Judaeo-Tat.

Jews from Iraq were the main settlers at a kibbutz founded 14 miles south of Jerusalem in the last months of 1926. It was named Migdal Eder (The Tower of Eder), after a nearby site mentioned in the Bible. In 1929 it was one of several Jewish settlements that were attacked by Arab rioters, and had to be abandoned. Kfar Malal, which had been destroyed in the riots of 1921, and rebuilt a year later, successfully repelled several attacks.

Another kibbutz established in 1926 was Mishmar Ha-Emek (Guard of the Valley) on the western rim of the Jezreel Valley. Its first settlers were from Poland. Although Jewish immigration to Palestine came predominantly from Poland in the mid and late 1920s, immigrants came from every country in which Jews lived. In 1927, seventy-one-year-old Mullah Murad arrived from Persia. He had been for many years the secret leader of those Persian Jews from Meshed, in north-eastern Persia, who, almost a century earlier in 1839, had been forcibly converted to Islam. Under his Jewish name, Mordecai ben Raphael Aklar, he became the spiritual leader of the Meshed and Bucharan communities in Jerusalem.

As the number of immigrants rose, calling for a considerable increase in the tasks being carried out by Jewish workers, the Histadrut looked to the Jewish Agency to help finance its building and construction activities, and its social welfare institutions. But the Jewish Agency was itself entering a period of crisis. The cost of building up the Jewish National Home could not be met from its productive enterprises. Zionists overseas seemed to be losing their zeal. In 1927, as economic conditions worsened, twice as many Jews left Palestine as reached it. Seven thousand Jewish workers had no work: 5 per cent of the total Jewish population. That year there was a net gain in population, through natural increase, of only 289 people, while over the two-year period 1926–8 the ratio of Jews to Arabs actually fell, from 16.6 per cent to 16.2 per cent. This was disastrous from the perspective of those who looked towards an eventual Jewish majority in Palestine. 'With the means at our disposal,' Arthur Ruppin wrote in his diary on 1 January 1927, 'it may be possible to provide a livelihood for 2,000–3,000 immigrants a year, but we want to bring in 25,000–30,000! I cannot see how Zionism can make any progress with no more than its present financial resources.'

The diversity of the kibbutz movement during the Mandate period was considerable, and was to influence the political evolution of Jewish and later Israeli life. In 1927 Kibbutz Hameuhad (The United Kibbutz) movement was founded. Its members believed that every kibbutz must constitute an independent economic entity. Its founding fathers were Yitzhak Tabenkin and Shlomo Lavi. Tabenkin had worked as a farm labourer in Palestine before the First World War, when Lavi had also been a labourer (later he was a watchman). Both Lavi's sons were killed in the War of Independence. A writer and essayist, as well as a Labour political leader, he served after independence in both the first and second Knessets.

The Kibbutz Hameuhad movement favoured a large kibbutz; it was Marxist-oriented and advocated 'the wholeness of Israel'. By contrast, the Hever Hakvutzot (Collective Association) movement founded in 1927 considered itself non-Marxist and non-Socialist, and was in favour of a small, intimate kibbutz. Kibbutz Artzi (My Land), also founded that year, was entirely Marxist in thought; its members regarded Soviet Russia as the world of tomorrow. The religious Hakibbutz Hadati, founded eight years later, and also the moshavim, were cooperative rather than collective: unlike the kibbutz, members of the moshav lived with their children, did not eat collectively but in their own homes, and received a financial share of any profits made by the collective farming.

In search of funds with which to buy more land and build settlements and towns, and absorb those who came to live in them, in 1928 the Jewish Agency approached the British government for a substantial loan. It would be used, Chaim Weizmann explained in a note for the British Cabinet on 28 January 1928, 'for the sole purpose of promoting and expediting close settlement of Jews on the land, as contemplated by the Mandate'. When the request came before the Cabinet, it was supported by Lord Balfour, who informed his Cabinet colleagues on March 5 that he was 'not so sure' that Britain had in fact carried out its obligations to the Jews 'in a very generous spirit' since the Mandate was established. It was not Britain, he wrote, which had as yet provided either money or immigrants: 'The supply of money is due to Jewish idealism; the supply of immigrants is due to Jewish idealism combined with Jewish misery.'

Balfour pointed out to his colleagues that the Jews had already put money into Palestine to benefit all its inhabitants. He gave, as an example, the £100,000 a year spent on sanitation by the Jewish National Fund, 'though the whole community (and not the Jews only) benefit by it'. In other British colonial and dependent territories, such work was either not done, or was paid for out of taxes levied on all the inhabitants. Both Leopold Amery – the

Colonial Secretary – and Winston Churchill – then Chancellor of the Exchequer – supported the proposed Zionist loan. But, despite Balfour's advocacy, the Cabinet rejected it; and, in doing so, marked a step away from what had earlier been regarded as one of Britain's obligations under the National Home provisions of the Mandate.

In Palestine itself there was a serious increase in Arab hostility against the Jews during 1928. Under the influence of the Mufti it was being put about that the Jews had designs on the main mosque on the Temple Mount. On September 8 the Mufti submitted a memorandum to the Palestine government warning of 'the unlimited greedy aspirations of the Jews', whose aim, he declared, was 'to take possession of the Mosque al-Aksa gradually'. Three weeks later the Mufti presided over a Moslem Conference which urged restrictions on Jewish worship at the Wailing Wall. These included preventing the Jews 'from raising their voices or making speeches'.

The combination of economic problems, and a fall in immigration – news of the difficult economic situation in Palestine circulated widely in Poland – was hard enough for the Zionist institutions to bear. But when the Arab protests against any continuation of Jewish immigration broke out into physical violence, a period of grave danger ensued for the safety and morale of the Jews of Palestine. Other aspects of life in Palestine were potentially uplifting. Rich archaeological remains awaited excavation, and with them, the identification of ancient biblical sites with modern towns and settlements.

In 1926 the Hebrew University had created the post of Archaeologist in its faculty of Humanities. The man chosen was Lipa (who had changed his name to the Hebrew-sounding Eliezer) Sukenik, the former mathematics teacher from Poland whose wife was in charge of training kindergarten teachers. Sukenik had begun to identify the ruins of synagogues dating back 2,000 years, to the period of the last Jewish independence in Palestine. In July 1928 he wrote in his diary, 'The important work that lies before me is the creation of Jewish Archaeology.' Five months later the settlers at kibbutz Beit Alpha in the Jezreel Valley were digging a drainage ditch when they came across an extraordinary mosaic. On it they could make out the signs of the Zodiac and Hebrew letters. They hastened to contact Sukenik – who had lectured to them seven years earlier on the importance of archaeological finds – and sent a young kibbutznik to Jerusalem to inform him of their discovery. In January 1929 Sukenik travelled north, accompanied by his eleven-year-old son, Yigael, to excavate what he realized was an ancient synagogue.

As the members of the kibbutz helped Sukenik with the excavations, a new dimension was added to Jewish life in the National Home. 'Suddenly,' he later wrote, 'these people saw things that were never so tangible before. There was suddenly a feeling that this very parcel of land – for which they had suffered so much – wasn't just any piece of land but the place where their fathers and grandfathers had lived and died fifteen hundred or two thousand years before. All their work now had a different significance. Their history had been uncovered, and they could see it with their own eyes.'

The archaeologist Neil Asher Silberman, who half a century later worked for the Israel Department of Antiquities – and who studied under Sukenik's son – has written of the discovery of the Beit Alpha synagogue in those first months of 1929: 'For the settlers of the Jezreel Valley, it was validation of their physical presence; for the non-religious Jews of the country, it revealed a lively Judaism that freely borrowed motifs from other cultures; for the religious, it demonstrated the continuity of their religious rituals over millennia; and for the politically minded, it evoked a period in which the Jewish community of Palestine had maintained at least nominal autonomy.'

A single, dramatic, episode had linked the Jews of modern Palestine with those of antiquity.

CHAPTER FOUR

Threats and dangers
1929–1937

At the beginning of August 1929, at the Sixteenth Zionist Congress, meeting in Zurich, it was decided to enlarge the Jewish Agency in such a way as would make it 'coextensive with the Jewish people everywhere'. As a result of this decision, the Agency would no longer consist only of the leaders of the Zionist Organization, but would include leading Jews outside Palestine, in particular in Britain and the United States, who were in sympathy with the building up of a Jewish National Home in Palestine. Not only Zionist Jews, whose imperative was to live in Palestine, but non-Zionist Jews who nevertheless looked with sympathy on the idea of a Jewish home, would be represented at the highest level of policy-making.

The two most senior overseas non-Zionist members of the enlarged executive, Louis Marshall (who had been a leading force behind the enlarging of the Agency) from the United States, and Lord Melchett, from Britain, were appointed Chairman and associate Chairman respectively of the Jewish Agency Executive. Both men died shortly afterwards, however. This weakened the hoped for high-profile fund-raising campaigns in the United States and Britain. The worldwide economic depression which then struck the United States with particular ferocity was a further and prolonged setback to the enlarged Agency's fund-raising plans.

The immediate problem for the enlarged Executive came, however, not from the United States, but from within Palestine. The Arabs of Palestine were convinced that the enlarged Executive was intended to be (in the words of an official British report) 'a strong body of wealthy non-Zionists who were expected to provide funds for further Zionist activities in Palestine'. News that the Jewish Agency had been enlarged had, in the British view, 'spread quickly and was, in our opinion, a cause for increased apprehension and alarm among all classes of Arabs'.

There was a second strand of Arab disaffection, which had long preceded the enlargement of the Jewish Agency, and which was exacerbated
during the summer of 1929. Throughout the first six months of 1929, Jewish
prayers at the Wailing Wall had continued to be a focus of Arab protest. On
June 11, the High Commissioner wrote to the Mufti defending the right of
the Jews 'to conduct their worship' as in the past. But growing tension
between the two communities led, on August 23, to an attack by large
crowds of Arabs on individual, unarmed Jews in the Old City of Jerusalem.
According to the subsequent British commission of inquiry, 'large sections
of these crowds were bent on mischief, if not on murder'.

When news of the violence in Jerusalem reached Hebron, the Jewish
school there was attacked, and a Jew killed. On August 25 a large Arab
crowd made what the official British report described as 'a most ferocious
attack' on the Jewish Quarter. 'This savage attack,' the report continued, 'of
which no condemnation can be too severe, was accompanied by wanton
destruction and looting.' Within five hours, more than sixty Jews had been
killed, including many women and children.

The Arab violence spread rapidly. The kibbutz at Beit Alpha – where the
ancient synagogue mosaic had been excavated seven months earlier – was
among those that had to beat off Arab attacks. In the village of Motza, just
outside Jerusalem, six members of one family were killed, including two
children, and their bodies mutilated. On August 26, 'Arab mobs', as the
British report described them, killed and wounded forty-five Jews in
the northern town of Safed. In the suburbs of Jerusalem, 4,000 Jews were
forced to leave their homes, many of which were looted. When the attacks
ended at nightfall on August 29, the number of Jews killed throughout
Palestine was 133. Eighty-seven Arabs had also died, mostly shot by British
troops and police seeking to halt the violence. 'In a few instances', the official
report noted, 'Jews attacked Arabs and destroyed Arab property. These
attacks, though inexcusable, were in most cases in retaliation for wrongs
already committed by Arabs.'

The Haganah, set up less than nine years before, succeeded in protecting
a number of settlements. At Hulda, twenty-three Haganah members, led by
Ephraim Chizhik, one of the successful defenders of the Neve Shalom quarter
of Jaffa in 1921, held off more than 1,000 Arab attackers until, at the end of
their resources, they were rescued and evacuated by British forces. Chizhik
was killed in an Arab ambush during the retreat.

Many Jewish settlements had had to be abandoned during the riots. One
of them, Kfar Uriyyah in the Jerusalem Corridor, had been farmed since 1909,
though lack of water and poor access made life there difficult. Once abandoned it was not revived until 1943 when a group of Kurdish Jews, who had
worked in Jerusalem as stonecutters, made it their home. In 1949 it was
enlarged by an influx of new immigrants from Bulgaria.

*   *   *

The killing in Palestine came to an end, but the anti-Jewish propaganda continued. A Jerusalem Arab students' leaflet which was widely circulated on 11 September 1929 declared: 'O Arab! Remember that the Jew is your strongest enemy and the enemy of your ancestors since olden times. Do not be misled by his tricks, for it is he who tortured Christ (peace be upon him), and poisoned Mohammed (peace and worship be with him).' The leaflet urged an Arab boycott of all Jewish shops and trade in order 'to save yourself and your Fatherland from the grasp of the foreign intruder and greedy Jew'.

The British were well aware of the nature of Arab propaganda. As early as September 5 the officer commanding the British troops in Palestine telegraphed to the War Office in London with details of a manifesto 'full of falsehoods and inflammatory material which has been issued to Moslems in other countries'. Two weeks later, on September 29, the High Commissioner, Sir John Chancellor, telegraphed from Jerusalem to the Colonial Office in London: 'The latent deep-seated hatred of the Arabs for the Jews has now come to the surface in all parts of the country. Threats of renewed attacks upon the Jews are being freely made and are only being prevented by the visible presence of considerable military force.'

The High Commissioner went on to point out, 'Propaganda against immigration of Jews into Palestine has recently been conducted amongst Arabs in neighbouring countries on an extensive scale, and if there is any recrudescence of the disturbances in Palestine it is doubtful if incursions into Palestine by Arabs from beyond the frontier could again be prevented.'

On the very day of this warning of Arab pressures outside Palestine, the President of the Arab Executive in Palestine, Musa Kazim Pasha, warned a senior British official that unless the Jewish National Home policy was changed, 'there would be an armed uprising' among the Arabs. Musa Kazim Pasha added that such an uprising would involve, not only the Arabs of Palestine but 'participation of Moslems from Syria, Transjordan and perhaps Iraq'.

It looked as if the Arab hostility to the Jews would lead to full-scale war. In a further telegram to London on October 12, the High Commissioner pointed out that the Arabs of Palestine had recently obtained 'a considerable number of arms' from both Transjordan and the Hedjaz. Further arms were known to have entered the country from Syria. On October 26 a British police report warned that 'gangs of criminals, to attack Jews and British officials, have been formed, and will first function in areas at Haifa and Nablus'.

The Jews were shaken by the intensity of the Arab violence, but were determined not to surrender all that they had created in the past fifty years. On October 25, after a visit to one of the Jewish villages which had been attacked by the Arabs, Arthur Ruppin wrote in his diary, 'On Tuesday, I went from Tel Aviv to visit Hulda, most of which was destroyed and burnt to ashes during the disturbances. Many of the trees have also been burnt. There is nobody there. The place makes a terrible impression. I remembered what hopes we had when we built the first house there twenty years ago. But I was not depressed: we shall rebuild what has been destroyed. On the whole, it is strange that I am one of the few optimists. I have a profound mystical belief that our work in Palestine cannot be destroyed.'

In Ruppin's view, the Jewish community in Palestine would not only continue to exist, but would also 'animate Jewry in the Diaspora' to support it, and to build it up with funds and people. Ruppin also supported an organization set up to try to bridge the gap between Jews and Arabs by proposing a bi-national State in Palestine, one in which Jews and Arabs would have an equal share in the administration, regardless of the size of their respective populations, intermixed geographically, and with no borders between their various communities. The organization, set up in 1925, was called Brit Shalom (Covenant of Peace). One of its leading lights was the first Chancellor of the Hebrew University, San-Francisco-born Judah Magnes. But the idea of a bi-national State was not to the liking of the Zionist leaders.

In December 1929, at a meeting with Brit Shalom, Ben-Gurion explained his opposition: 'Our land is only a small district in the tremendous territory populated by Arabs – most sparsely populated, I might add. Only one fragment of the Arab people – perhaps 7 or 8 per cent, if we take into account only the Arabs of the Asian countries – lives in Palestine. However, this is not the case with respect to the Jewish people. For the entire Jewish nation this is the one and only country with which are connected its fate and future as a nation. Only in this land can it renew and maintain its independent life, its national economy and its special culture, only here can it establish its national sovereignty and freedom. And anyone who blurs this truth endangers the survival of the nation.'

The founding of new settlements had continued even in the year of riots. During 1929 veteran farm labourers founded Pardes Hannah in the coastal plain. 'Hannah's Citrus Grove' was named after a cousin of Baron Edmond de Rothschild. Also founded in 1929 was a coastal moshav, Netanya, named after the American–Jewish philanthropist, Nathan Straus. Within a decade it was a flourishing seaside town.

Two clear strands of thought were evolving within the Labour movement. The first was the Marxist, almost Communist orientation of Ha-Shomer Ha'tzair (The Young Watchman), which had been founded in Vienna during the First World War, and which favoured a bi-national Jewish and Arab State in Palestine, looking forward to a shared and intertwined Jewish–Arab destiny. The other strand was Mapai (*Mifleget Po'alei Eretz Yisrael* – the Party of the Workers of the Land of Israel), founded in 1930, which thought that the Socialism which it espoused was as important to Jewish life as the Bible. Mapai was soon to become the largest and most cohesive Zionist political

Party, its predominance lasting long after the establishment of the State of Israel. It was a Socialist-Zionist Party, committed to what it called 'constructive Socialism'. Its ideological leader was Berl Katznelson, who had come to Palestine from Russia in 1909. Its main activist was David Ben-Gurion. Both men were outspokenly pro-Bible and anti-Communist.

There was yet another division in the Zionist Labour ranks between those who wished to see the creation of a Jewish State immediately, even at the cost of a partition between Jewish and Arab areas, and those who were for the 'wholeness' of the country, even at the cost of postponing Jewish sovereignty for many years.

The diversity of Zionist political attitudes was remarkable in so small a community. The long hours spent in debate, and often in sharp disagreement, were also a manifestation of a vitality of spirit and aspiration that caused outsiders to marvel that the Jews had achieved so much, and to recognize their determination to build up their land with idealism as well as the sweat of their brow.

Among the leaders of Mapai was the Russian-born, German-educated Chaim Arlosoroff, who had come to Palestine from Berlin in 1924, and who had been convinced that the Zionist ideal could be realized in cooperation with the Arabs. The riots of 1929 had dented his confidence. He had also been distressed by the ultra-nationalist reaction to the riots by some Jews, warning that 'if one burdens a powerless people, lacking external means of power and fighting for its very survival, with a bloated programme of gestures and illusions, totally devoid of reality – then a terrible caricature is being created.'

The lessons of the 1929 riots were being discussed not only by settlers and political leaders, but by younger Jews who had lived through the disturbances and their aftermath. Twelve-year-old Yigael Sukenik (who later, as Yigael Yadin, was to marry Arthur Ruppin's daughter Carmella) wrote, in a school essay:

> When I remember the scenes at the Jerusalem vegetable market at the time of the disturbances, the image that comes to my mind is of a deserted market, empty of all vegetables. I have to ask myself: What would have happened if the disturbances had lasted for half a year, instead of just a month?
>
> Will we always be so dependent on those same Arabs, who at each and every disturbance withhold vegetables from us, and who will start to make fun of us for not being able to survive on our own vegetables, and for our dependence on them? Couldn't we fill the market with vegetables ourselves?
>
> Who can say if more disturbances won't break out in future, and maybe we'll be without vegetables again and the same Arabs will ridicule us for being so dependent on them.

In every settlement and in every neighbourhood we have to grow vegetables and send them to the markets of the cities, and then the Jews in the cities will not have to buy their vegetables from Arabs. And then the Arabs will no longer ridicule us, since we will not be dependent on them. And then it will seem to them that we are able to exist on our own. May this time come soon. Although the disturbances showed us the need for vegetables, we should learn from them other things as well.

As a result of the 1929 violence, the British government set up a Commission of Inquiry to examine the causes of Arab unrest. On 18 December 1929 the Commission examined a senior member of the newly enlarged Zionist Executive, Harry Sacher – a British journalist and lawyer who had spent nine years in Palestine. Sacher told them emphatically that the 'Jews have no intention of dominating or being dominated in respect of any other people in this country. They look upon their own right to create their own civilization as being neither greater nor less than the right of the Arabs to create their civilization.'

Among the other Zionists who gave evidence was Arthur Ruppin, who told the Commissioners – as he noted in his diary on December 31 – 'that enough land would become available for Jewish settlement' as the Arab farmers would change over to intensive farming, a change-over which they would be able to afford 'with the money they would acquire through selling part of their land to the Jews'.

Published in March 1930, the official report stressed that 'the claims and demands which from the Zionist side have been advanced in regard to the future of Jewish immigration into Palestine have been such as to arouse among Arabs the apprehension that they will in time be deprived of their livelihood and pass under the political domination of the Jews.' The report continued, 'There is incontestable evidence that in the matter of immigration there has been a serious departure by the Jewish authorities from the doctrine accepted by the Zionist Organization in 1922 that immigration should be regulated by the economic capacity of Palestine to absorb new arrivals.'

This was a serious criticism. The report went on to recommend an early declaration of the British government as to the policy which was to be pursued 'in regard to the regulation and control of future Jewish immigration into Palestine'. But mere control or limitation of Jewish immigration was not what the Arabs required. On 1 May 1930 the Colonial Secretary in the British Labour government, Lord Passfield, met with an Arab delegation of senior members of the recently established Arab Executive for Palestine. The delegation, headed by Musa Kazim Pasha and his nephew the Mufti, told the Cabinet's Palestine Committee that the Arab position could be summed up as: 'The Arabs did not accept the Mandate, wished for the abandonment of the Balfour Declaration, demanded the establishment of democratic institutions and the prohibition of the sale of land to the Jews.'

A second official British report was issued in October 1930. Instructed to examine the previous report's conclusions in detail, it stated that, given existing Arab farming methods, there was insufficient land in Palestine to meet the needs of Jewish immigrants. On the basis of this assertion, the British government intimated that future Jewish immigration might have to be curtailed even more rigorously than in the past. The Zionists were dismayed. Deeply angered, Weizmann resigned as President of the Zionist Organization and the Jewish Agency. At a meeting with two senior British ministers, the Colonial Secretary, Lord Passfield, and the Foreign Secretary, Arthur Henderson, Weizmann declared, on November 17, 'We feel our work in Palestine has been during the last few years constantly subjected to scrutiny and inquiry. It is like a plant which is being constantly taken out of the soil to look at its roots; it is not very good for the plant.'

At a second meeting with Passfield and Henderson on the following day, Weizmann warned that two-thirds of the Jewish people – mostly in Eastern Europe – 'live under conditions which, without any attempt on my part to harass your feelings, are such as to destroy slowly but surely the vital elements of a great race'. The mass of Jews in those lands were eager to emigrate. Both Poland and Roumania had instituted anti-Semitic policies. Both had declared that they had 'too many' Jews. In Turkey and in Arab countries the Jews were likewise 'seeking an outlet'. And yet, Weizmann continued, 'while they are seeking an outlet, every door of those countries into which the Jews emigrated in the past is gradually being closed before them: America, South Africa, Canada, Mexico, each used to be a country of immigration; they are closed now.'

A few moments later, Weizmann was asked whether it was 'anti-racial', or economic, causes that barred the Jews' way. It was both, he replied, and in a powerful, anguished statement, set out the Jewish dilemma as he saw it, and the place of Palestine in the Jewish perspective:

> We have, throughout our history, not disappeared among the nations with whom we have lived. We have remained different; different in religion, different in outlook, and we have therefore been placed in a minority among the people with whom we have lived.
>
> They being in the majority, we were placed in a position which rendered us economically inferior to that majority, so a complex situation has been produced which, as I said, in its external effects moves the Jews to go from one country to another, and today the world is closed to our emigration; with the possible exception of Brazil and Argentina, and with the possible exception of a small infiltration into any other country, there is no room in the world for the mass of Jews who can find no room in their respective countries.

The moral effect of this situation has combined with an age-long tradition of attachment to Palestine, a tradition which forms an integral part of our religion, a tradition which, I think, has made us what we are; a stubborn, stiffnecked attachment to a country which many of us have not seen, but which has been the central part of our history, which has made us look upon Palestine as the country where we wanted to find the realization of an age long dream.

Following this appeal, the British government agreed to abandon the immigration clauses of the 1930 White Paper. Even so, in the letter which he wrote to Weizmann, and which he read out in the House of Commons on 13 February 1931, the Prime Minister himself, Ramsay MacDonald, warned that if, in consequence of the existing Zionist policy of using only Jewish labour, Arab labour were to be replaced, 'that is a factor in the situation for which the Mandatory is bound to have regard'.

The situation inside Palestine was as worrying to the Zionists as Britain's new-found coolness. On 3 December 1931 Arthur Ruppin wrote to a friend, 'Undoubtedly the Arabs have greatly strengthened their political position during the past few years and are much less ready to make concessions to the Jews than they were ten years ago.' Ruppin no longer saw any hope in reliance upon Arab goodwill. His letter continued:

At most, the Arabs would agree to grant national rights to the Jews in an Arab state, on the pattern of the national rights in Eastern Europe. But we know only too well from conditions in Eastern Europe how little a majority with executive power can be moved to grant real and complete national equality to a minority. The fate of the Jewish minority in Palestine would always be dependent upon the goodwill of the Arab majority, which would steer the State.

To the Jews of Eastern Europe, who form the overwhelming majority of all the Zionists, such a settlement would be completely unsatisfactory, and it would kill their enthusiasm for the Zionist cause and for Palestine. A movement which would agree to such a compromise with the Arabs would not be supported by the East European Jews and would very soon become a *Zionism without Zionists*.

On the last day of 1931 Ruppin wrote, pessimistically, that it was in his view 'doubtful whether a Jewish minority will be able to preserve its national individuality and independence against an Arab majority if the latter controls the machinery of state'. His pessimism was justified. Three months later, on 9 April 1932, a new British High Commissioner in Palestine, Sir Arthur Wauchope, telegraphed to the Colonial Office in London: 'I have learnt from my private conversations that the Jews' objection to any Legislative Council is due to belief that the Arab leaders chiefly desire a Legislative Council in order to check the advancement of a National Home for the Jews. They are not altogether incorrect in this belief.'

The reiterated Arab demand for a Legislative Council, coupled with Arab hostility to Jewish immigration, led the British to adopt a policy of cautious verbal restraint. On April 12, at a meeting of the Cabinet's Committee on Palestine, the official minutes recorded general agreement with the view 'that considerable embarrassment had been caused by past pronouncements of a too specific and definite nature, and that it would be desirable to publish as little as possible with regard to the Government's intentions'.

The future of Jewish immigration into Palestine had become subjected to the policy of a distant government that was afraid to offend either side, but particularly unwilling to aggravate Arab opinion, given the Arab States with which Britain had relations, and the sixty million Muslims in British India, whose loyalty to the imperial crown was an essential feature of the pacification of the Indian subcontinent. One of the reiterated questions asked behind the scenes in British official discussions in London was whether the Jewish needs in Palestine could really be allowed to endanger the wider British need for the support of the Muslim world. One of the greatest attractions of Zionism for Jews worldwide was that the Jews in Palestine would be masters of their own destiny, as the Jews elsewhere were not. Yet the pledges of the Mandate and the activities of the Jewish Agency were both under growing threat from the British willingness to defer to Arab pressure.

In many ways Jewish and Arab nationalism in Palestine were pursuing parallel paths. The Arabic-speaking schools – Muslim and Christian – in Palestine were supported and administered by the Mandate government. H. E. Bowman, the Director of Education, wrote that the Arab leaders in Palestine 'did their utmost to make the schools the nucleus of nationalist inspiration. Their aim was to embarrass the Government by giving the schools a nuisance value; and, at the same time, to inculcate in the Arab youth a passionate nationalism which would show itself in overt acts whenever an opportunity served.' In quoting this, the historian Elie Kedourie commented, 'The same, of course, holds true, *mutatis mutandis*, of the Zionist schools.'

The 1929 riots had not halted Jewish immigration, though they were intended to do so. Nor did they deter Jews emigrating to Palestine from countries where the standard of living, as well as the general security, was much higher. Among the immigrants in 1930 was a twenty-four-year-old American from Louisville, Kentucky, Shimon Agranat. Settling in Haifa, he practised law, and served as a local magistrate. Two years after the establishment of the State, he was appointed a justice of the Supreme Court, of which he became President in 1965.

One result of the 1929 riots was a sharpening of the divide between the

Zionists who still sought compromise with the Arabs, and those who believed that military confrontation was the only option. The Betar youth movement, founded in 1923, was emerging as the principal advocate of a militant Zionism, and as an uncompromising opponent of the Zionist-socialist youth groups. With its red-brown uniform, Betar was criticized by mainstream Zionists as having a 'fascist character', reminiscent of Mussolini's Blackshirts. In 1931 the first world conference of Betar was held in Danzig. Jabotinsky was elected its head.

Training in defence was proclaimed the prime duty of every Betar member. In Palestine, Betar work brigades – known from 1934 as 'mobilized groups' – undertook training in self-defence, drilling, street-fighting, the use of rifles and pistols, and military tactics. Membership in 1931 was 22,300. It was to reach 90,000 in the next seven years.

Other youth groups also flourished, among them the Socialist-oriented Scout movement under the leadership of a high school teacher, Zion Ha-Shimshoni, whose members admired most of all the Maccabees, the upholders of Jewish sovereignty in Palestine 2,000 years earlier. Hiking, learning about the land, and exploring its villages and remote places, were the weekly excitement. Annual summer camps were held at Kiryat Anavim, in the hills just to the west of Jerusalem. A high point of the Scouting year was the hike to the Dead Sea and climbing the fortress of Masada: among the youngsters who made the climb was Yigael Sukenik, who, three decades later (as Yigael Yadin), was to excavate the mountaintop with extraordinary results.

Immigration from Russia, which had been so important in the first decade of the Mandate, dried up almost completely in the second decade. As Stalin's control over every aspect of Soviet life intensified emigration virtually ceased. In 1932 one of the very last groups of Russian pioneers to reach Palestine founded a kibbutz at Afikim, just south of the Sea of Galilee, near the confluence of the Jordan and Yarmuk Rivers (the name Afikim means 'stream courses'). At first they raised poultry and cattle, and grew bananas, dates and grapefruit. Thirty years later they were to set up a plywood factory, largely for export, and became a partner in a nearby factory producing synthetic materials. By 1967, with 1,290 members, Afikim was one of the largest communal settlements in Israel.

Also founded in 1932, by veteran agricultural labourers of the Second and Third Aliyah, was Kfar Azar, only 6 miles east of Tel Aviv. A moshav, it raised vegetables, dairy cattle and poultry. Another moshav, Kfar Bilu, was also founded that year, in the coastal plain not far from Rehovot. It was part of a 'Thousand Families Settlement Scheme', designed to reverse the trend of living in towns, and to augment the agricultural productivity on which the towns increasingly depended. Another settlement founded in 1932 was Kfar Yonah, in the coastal plain five miles from the sea. Funding came from the estate of a Belgian Zionist, Jean Fischer, who had died in 1929, and whose Hebrew name (Yonah) the settlement commemorates. Fischer's son Maurice, one of the settlement's founders, was later Israeli ambassador to Paris.

The Jewish Legion veterans, whose attempt to found a settlement in 1921 had failed through lack of water, were more fortunate in 1932: on July 19 they founded a moshav, Avihail (Father of Strength), on a waste stretch of sand dunes halfway between Haifa and Tel Aviv. It managed to grow citrus fruits, and survived. On that same long stretch of sand dunes, Barukh Chizhik, whose sister Sarah and brother Ephraim had both been killed in Arab attacks in 1920 and 1929 respectively, acquired a farm near Herzliya in 1932. An expert on citrus cultivation, he worked to improve strains of fruit trees; he had already published (in 1930) a collection of popular legends on the flora of Palestine.

It was in 1932 that Gershon Agron founded the *Palestine Post*. Ukrainian-born, Agron (his original name was Agronsky) had lived in the United States as a child, coming to Palestine with the Jewish Legion in 1918, at the age of twenty-four. The idea behind a daily English-language newspaper was twofold: to enable the British officials and residents in Palestine to see the full range of Zionist activities and aspirations, and to give the English-speaking Jews in Palestine news from the wider world.

The expansion of the Jewish national institutions was a continual process, building up the structure on which statehood could eventually be based. In 1932 the Jewish Agency moved its agricultural research station from Tel Aviv to Rehovot. Here, two years later, Chaim Weizmann founded a scientific research institute in which Jewish scientists from all over the world were encouraged to participate in the most modern scientific experiments. In October 1932 twelve young German Jews reached the youth village of Ben Shemen. They were the first of five thousand youngsters brought from Germany by Youth Aliyah, set up in Berlin by Recha Freier, and later entrusted by the Jewish Agency to Henrietta Szold, the founder of Hadassah.

New villages, a growing urban life, institutions that provided a range of social and research services: the Jewish community in Palestine had much to be proud of by 1932, only a decade since the League of Nations had confirmed the Jewish National Home aspect of the Palestine Mandate. Yet the growing hostility of so many of the Arab leaders to the Zionist endeavour was a permanent cloud over the whole enterprise. There were then, according to the precise British census figures, 192,137 Jews in Palestine, and 1,073,827 Arabs. The Arab population had reached a million the previous year. As the economic situation inside Palestine improved, partly as a result of British road building and social welfare activities, partly as a result of Jewish constructive enterprise, the Arab population grew, both by immigration from the surrounding Arab countries and by natural increase.

The director of the Political Department of the Jewish Agency, Chaim

Arlosoroff, was pessimistic about the prospects of Arab–Jewish under-standing. He had in mind some revolutionary action whereby the Jews would come to power in Palestine, although still a minority, and use the period of their power to push for mass immigration and the emergence of a Jewish majority. It was not clear how he envisaged this being realized, but he wrote to Weizmann in a private letter on 30 June 1932, 'Zionism cannot, in the given circumstances, be turned into a reality without a transition period of the organized revolutionary rule of the Jewish minority; that there is no way to a Jewish majority, or even to an equilibrium between the two races (or else a settlement sufficient to provide a basis for a cultural centre) to be established by systematic immigration and colonization, without a period of a nationalist minority government which would usurp the State machinery, the administration and the military power in order to forestall the danger of our being swamped by numbers and endangered by a rising (which we could not face without having the State machinery and military power at our disposal).' During this 'period of transition', Arlosoroff added, 'a systematic policy of development, immigration and settlement would be carried out.'

In 1932 a Polish Jew, Yitzhak Persky, entered Palestine as a 'capitalist' immi-grant, one of the categories permitted by the British (entering with £1,000 or more). He left his family in their Polish village – Vishneva – until such time as he could establish himself and they could join him. His son Shimon Peres later recalled:

> For the next two years our contact with him was by letter. He wrote in a bold, confident hand, his letters full of the sweeping optimism that characterized him. Sometimes he enclosed snapshots of himself and his partner, Kabak, both decked out in light summer suits, my father tall and sun-tanned, the other man shorter, but also relaxed and confident-looking.
>
> Eventually, we followed: by train to Istanbul and then aboard a Polish steamship to Jaffa, where our vessel was instantly besieged by a bustling maelstrom of small boats and barges, with longshoremen in red tar-bushes and broad pantaloons offering everything from palm branches to green-coloured iced lemonade. But Father was at hand, in a boat of his own, and he smoothly shepherded us through customs and immi-gration and on to our new home in what was then the centre of Tel Aviv.
>
> Rehov Ha'Avoda, or Work Street – our apartment was at No. 8 – was a turning off King George Street, one of the main thoroughfares of the new Jewish city. The two street names seemed to symbolise our new condition: the main street named in honour of the British colonial ruler, and the side street articulating the Zionist vision of the future Jewish State.

The pressure of immigration from Poland continued to rise, as Polish anti-Semitism impinged on the lives of so many of Poland's three million Jews. Zionist youth groups were also active in every Polish town. One of these movements, Gordonia, named after the Zionist pioneer A. D. Gordon, provided the motivation and settlers for a kibbutz founded in 1933 in Lower Galilee, west of Nazareth. Named Kfar Ha-Horesh (Woodland Village), its members were employed in its first years planting forests on the surrounding and hitherto barren hills. Their largest plantation was the King George V Forest.

Gordonia, which had a large following in Poland, and sent many young-sters to Palestine, was declaratively non-Socialist. Its members refused to celebrate the First of May or to carry the Red Flag.

On 30 January 1933 Hitler came to power in Germany. The anti-Jewish frenzy that had accompanied his political campaigns and path to power was imme-diately translated into action. Physical attacks on German Jews, their exclusion from the professions and a campaign to drive them out altogether from thousands of villages and even small towns led to a sudden upsurge in emigration. German Jews began to reach Palestine in ever-growing numbers. The pattern of immigration, dominated for the previous ten years by Jews from Poland, changed dramatically. Not only was emigration from Germany encouraged by the new German authorities, but the restrictions on immigration to Palestine were at that time minimal. The resulting influx of refugees from Germany altered the demographic balance of Palestine Jewry.

Few of the German Jewish newcomers were Zionists. Their motivation was to escape persecution. But many of them were sympathetic to the aims of Zionism and entered into the spirit of the Zionist experiment. Indeed, they were soon to be found in every branch of Jewish national life, above all in the professions as doctors, teachers, scientists, architects, photographers, painters and poets. In Palestine the Mapai leader Chaim Arlosoroff, who had begun to doubt Britain's commitment to continuing Jewish immigration, focused his energies on organizing the rapid, mass emigration of Jews from Germany. In the spring of 1933 he travelled to Germany to open negotia-tions with the German Nazi authorities for the transfer of German-Jewish property to Palestine. That summer, however, while walking on the beach with his wife Sima in Tel Aviv, he was shot and murdered.

The leader of a clandestine group of Jewish underground 'activists', Abba Ahimeir, was charged by the Palestine police with plotting Arlosoroff's murder. Born Shaul Heisinovitch in Russia, thirty-five years earlier, Ahimeir had, as a young man, been attracted by both Communism and Socialism. In 1928 he had joined Jabotinsky's Revisionists, advocating active opposition to the Mandatory government. In the interests of a Jewish State, he believed that illegal action was justified.

Ahimeir had become the leader of an extremist Revisionist faction, whose newspaper, *Hazit Ha-Am* (The People's Front) violently attacked the Labour movement and the official Zionist leadership, including Arlosoroff. Two rank-and-file Revisionists, Avraham Stavsky and Zevi Rosenblatt, who were also charged with the murder, were identified by Arlosoroff's widow. They denied the accusation. The district court acquitted Ahimeir and Rosenblatt, but convicted Stavsky. He was later acquitted, however, by the Supreme Court for lack of corroborating evidence. The defence accused the police of manipulating the widow's testimony and other evidence for political reasons, and argued that the murder was connected with an intended sexual attack on Mrs Arlosoroff by two young Arabs. One of these Arabs, who was in prison on another murder charge, twice confessed to having been involved in Arlosoroff's murder, but twice retracted his confession, accusing Stavsky and Rosenblatt of having bribed him to confess.

In the aftermath of Arlosoroff's murder members of the Labour movement, with few exceptions, regarded the widow's testimony as proof of the existence of criminal fascist tendencies among Revisionists. For their part, the Revisionists, and many other non-Labour circles including the Ashkenazi Chief Rabbi, Abraham Isaac Kook, maintained Stavsky's innocence and denounced the affair as a blood libel perpetrated by Jews against Jews.

The Arlosoroff murder continued to exacerbate relations between the Labour Zionists and the Revisionists for many years. At the time of Yitzhak Rabin's assassination in 1995, his widow Leah remembered how it took place 'within days of my first setting foot in Palestine as a little girl'. She and her husband had often discussed it in later years. 'No matter who killed Arlosoroff,' she wrote, 'the Revisionists had created a climate that provoked his death. They spread vicious rumours and promoted articles contending he was a Nazi collaborator.'

In August 1933 the Eighteenth Zionist Congress, held that year in Prague, called for the Jewish National Home to be built up 'as speedily as possible and on the largest scale'. Within a month of this declaration, Musa Kazim Pasha, the President of the Arab Executive, again called for an immediate halt to all further Jewish immigration. That October, as immigration from Germany climbed, Arab protesters attacked British public buildings and police posts in Nablus, Jaffa and Jerusalem. In the resulting violence, one policeman and twenty-six Arab rioters were killed. Bloodshed was becoming a cruel by-product of the political debate. But the building up of the Jewish National Home, though often overshadowed by violence, was never halted. In 1933 Polish and German immigrants combined to establish Kfar Pines, in the northern part of the coastal plain. They took their name from Yehiel Michael Pines, the Russian-born Hebrew scholar who, in the 1880s, had been one of the early advocates of Hebrew as the spoken language of the Jews of Palestine, and the patron of the pioneering Gederah settlement in 1884. Thus, in the names of the new villages, the founders and visionaries were remembered.

The impetus to found new settlements was not restricted to new immigrants. In 1933 a group of veteran farmers founded Kfar Vitkin, in the Hefer plain north of Tel Aviv. The land for Kfar Vitkin had taken three years to prepare for farming. It was to become the largest moshav in the country, with more than 1,100 inhabitants ten years later (after which numbers fell). Orange groves, dairy cattle and poultry were its main produce. It was named after Josef Vitkin, a Russian-born pioneer who had emigrated to Palestine in 1897.

Another settlement founded in 1933, in the coastal plain, was Mishmar Ha-Sharon (Guard of the Sharon). Not only was it a pioneer in intensive, irrigated farming, it also pioneered the commercial raising of flowers – including the biblical Lily of Sharon; biblical experts say that this lily was almost certainly a rose: the kibbutz grew both.

Also founded in 1933, by a group of veteran pioneers who had come from Russia with the Third Aliyah a decade earlier and had been at the centre of the defence of Hulda during the 1929 riots, was Kfar Chaim, named after Chaim Arlosoroff. Kfar Chaim flourished, its prosperity based on citrus groves and dairy cattle.

As the number of immigrants from Germany rose, a group of them banded together to form a farming settlement of their own in 1934. The place which they chose was on the coast only six miles from the border with Lebanon. It was called Nahariya, from the stream – *nahar* – on which it was located. The first villagers were middle-class immigrants, and it proved hard for them to change from their former professional and commercial lives to that of farmers. Soon, however, they found a new occupation as Nahariya developed into a small seaside resort. But a few miles to the east were a number of Arab villages and in those early days a sense of isolation from the main areas of Jewish settlement was always present.

It was not only the sand dunes of the coast and the emptiness of northern Galilee that attracted Zionist endeavour. In the spring of 1934, David Ben-Gurion made a journey through the Negev Desert to the Gulf of Akaba. Twenty-five years later he would tell his colleagues in government what he found there: a few mud huts 'at the end of nowhere' which served as a 'primitive post' for the British Mandate police. It was called Um Rashrash, the Arabic for 'Mother of the Rustling Sound'. Shimon Peres, one of those to whom Ben-Gurion described his journey, later wrote:

> Away to the south lay the clear blue waters of the Gulf, sparkling in the sun, and set against the crimson mountains of Edom. Apart from the two mud huts, the beach was empty, and behind it, stretching to the north, rolled the parchment-like desert. This spot was more evocative of the last vestige of a romantic mood conjured by Lawrence of Arabia than of the future gateway to the new world.

Ben-Gurion had a vision of this site, at the southernmost tip of what was then Palestine and reachable at that time only after an arduous trek, as a bridge between a future Jewish State and the continents of Africa and Asia. And he kept coming back to this idea.

He was firmly convinced that the wilderness of the Negev, empty of people, would one day become a vital centre of Israel development. The exploitation of potash, phosphates and other minerals would provide the basis of large industrial plants whose products would be carried over the desert to Eilat, and this route would serve as an 'overland Suez Canal'.

'The bare Negev and the open Eilat,' Peres added, 'seemed to him to offer the most fruitful development prospects for Israel.'

The efforts of the Zionist leaders to come to some agreement with the Arabs of Palestine in the early 1930s were continuous. The most important of these efforts came on 18 July 1934, when Ben-Gurion and Dr Magnes met Auni Abdul Hadi, the leader of the movement devoted to Palestinian Arab independence. Ben-Gurion asked Abdul Hadi bluntly, 'Is it possible to reconcile the ultimate goals of the Jewish people and the Arab people?' and went on to tell his Arab interlocutor:

> Our ultimate goal is the independence of the Jewish people in Palestine, on both sides of the Jordan, not as a minority but as a community of several millions. In my opinion, it is possible to create over a period of forty years, if Transjordan was included, a community of four million Jews in addition to an Arab community of two million. The goal of the Arab people was independence, and the unity of all Arab countries. If the Arabs agreed to our return to our land, we would help them with our political, financial and moral support to bring about the rebirth and unity of the Arab people.

Abdul Hadi became 'enthusiastic when he heard this', Ben-Gurion noted in his account of their talk, 'and said that if with our help the Arabs would achieve unity he would agree not to four million, but to five or six million, Jews in Palestine. He would go and shout in the streets, he would tell everyone he knew, in Palestine, in Syria, in Iraq, in Damascus and Baghdad: Let's give the Jews as many as they want, as long as we achieve our unity.' Then, Ben-Gurion recorded, Abdul Hadi's enthusiasm abated, and he 'reverted to his mocking and sceptical tone and asked what guarantees the Arabs would obtain. The Jews in Palestine would increase in number to four million, while the Arabs in the other countries would be left with the English, the French, and the promise given by the Jews. Did we think the Arabs could rely on our promises and declarations?' Ben-Gurion's account continued:

I told him that if we should reach agreement on the main point we would seek together practical means whereby each side could insure the interests of the other. Even we had not yet attained four million in the country, the realization of Zionism was a long process, and the rebirth of the Arab people would also not come about overnight.

Auni then asked Ben-Gurion whether the Jews would help the Arabs get rid of France and England. 'I answered that I had to speak frankly on this matter too,' Ben-Gurion wrote. 'We would not fight against the English. We, too, had grievances against the Mandatory Government, perhaps no less than those held by the Arabs. But the English had helped us, and we wanted them to continue to do so. And we were faithful to our friends. The building up of the Arab economy, the raising of the level of culture, public education, the development of the various Arab countries – all these preceded and conditioned political liberation. In that positive task we were prepared to render all possible assistance to the Arab people. The only question was whether the Arabs were prepared to let us work peacefully and undisturbed in Palestine.'

When Dr Magnes asked Auni whether the Arabs were willing to 'sacrifice Palestine in order to attain the broader goal in the other Arab countries', Ben-Gurion interjected to say that 'we did not wish the Arabs to "sacrifice" Palestine. The Palestinian Arabs would not be sacrificed so that Zionism might be realized. According to our conception of Zionism, we were neither desirous nor capable of building our future in Palestine at the expense of the Arabs.'

Ben-Gurion then told Auni:

> The Arabs of Palestine would remain where they were, their lot would improve, and even politically they would not be dependent on us, even after we came to constitute the vast majority of the population, for there was a basic difference between our relation to Palestine and that of the Arabs. For us, this Land was everything and there was nothing else. For the Arabs, Palestine was only a small portion of the large and numerous Arab countries.
>
> Even when the Arabs became a minority in Palestine they would not be a minority in their territory, which extended from the Mediterranean coast to the Persian Gulf, and from the Taurus Mountains to the Atlantic Ocean.

For the Jewish people, Ben-Gurion told Auni, 'it was essential that they be the majority here, as otherwise they would not be independent.' But the Arabs, with the vast Arab hinterland and neighbouring Arab regions, 'could not turn into a minority'. Auni seemed to agree with this, for when Magnes asked him whether the Arabs in the various countries 'really felt their unity' he replied that 'while this might not yet be true of the masses, the Arab intelligentsia in all countries – Syria, Iraq, Saudi Arabia, Tunis, Morocco – did feel that they belonged to one culture, one past, one nation'.

The conflict of national interest centred, as Ben-Gurion had seen, upon a basic inequality that whereas many Arab nations already existed, and the Arab world extended from the Atlantic coast of Morocco to the waters of the Persian Gulf (and the Muslim world far beyond that), the Jews had no existing country in which they were sovereign, or a majority. This inequality was heightened in the perception of Palestinian Jewry by the events in Germany. There was no Jewish sovereign state, however small, to which the Jews of Germany could go as of right and settle in their own land. Jewish nationalism, as formulated by the Zionists for the past third of a century, had not yet secured such a safe and sovereign haven.

In 1934 as the persecution of Jews in Germany continued and it became clear that the Nazi racial policies were not – as some people hoped – going to abate, 42,000 Jews entered Palestine with the permission of the British. This was a record number of immigrants for any year in the country's history. One of the first moshavim to be established by Jews from Germany was Kfar Bialik, founded just inland from Haifa Bay in 1934, and named after the Russian-born Hebrew writer Hayyim Nahman Bialik who had just died and who had lived briefly in Berlin before emigrating to Palestine a decade earlier. Another new settlement, Kfar Shmaryahu, named after the early Russian-born Zionist leader Shmaryahu Levin, was founded three years later. Its immigrants were mostly educated middle class professionals; its main activity, from the outset, was intensive poultry breeding.

It was not only German Jews who were on the move after 1933. Immigration from Poland and Roumania also remained high, as fears spread of Nazi and anti-Jewish movements gaining the ascendancy elsewhere. In 1934 a group of religious Jews from Eastern Europe founded Kfar Ha-Ro'eh, a moshav which quickly became the spiritual centre for the religious moshav movement. As well as farming, it gave a central place to religious studies in its yeshiva (religious academy). The name of the moshav was made up of the Hebrew initials of one of the spiritual mentors of religious Jews in Palestine, Rabbi Abraham Isaac Kook.

Suspicion of British policy, and a fear that the British would in the end call a halt to Jewish immigration, led the Revisionist movement to press for even more rapid Jewish immigration and for immediate Jewish settlement, not only in Palestine, but also in Transjordan. In 1934 Jabotinsky introduced for his youth movement followers the Betar Oath: 'I devote my life to the rebirth of the Jewish State, with a Jewish majority, on both sides of the Jordan.'

The Revisionists carried out policies which led to an exacerbation of British antagonism towards Jewish activities. In 1934 they hired a ship that brought 117 'illegal' Jewish immigrants from Europe to Palestine: Jews who did not have the necessary permission to immigrate. This was the first Revisionist venture in a move which was to see fifteen more ships set off that same year from the ports of the Black Sea towards Palestine. Most of them were intercepted by the British navy and their human cargo either interned or deported.

Both the aims and the methods of the Revisionists roused hostility and criticism in British circles. In a 'very secret' memorandum dated 28 March 1934, the then Colonial Secretary, Sir Philip Cunliffe-Lister, warned his Cabinet colleagues that 'illicit immigration' had assumed 'alarming proportions', and would be strictly combated. But there was 'no use blinking the fact', his memorandum continued, 'that today Arabs and Jews are diametrically opposed on the whole subject of immigration. The Arabs claim that nothing will satisfy them except a complete embargo on all further immigration; and Jewish extremists do not make matters easier by their claims to unlimited immigration, and by their avowed determination to make Palestine not merely a National Home, but a Jewish State.'

The Revisionists had made it clear that for them Jewish statehood was the aim of all immigration and settlement activity. The Jewish Agency avoided any open mention of a State. But the Colonial Secretary believed that 'even the moderate Jewish leader is appealing to his constituents, I have no doubt, that the picture of the Jewish National Home often expands into one of the Jewish State'. British fears of Jewish statehood went side by side with the constructive efforts of the Zionists to create national institutions at every level. In 1934 Hannah Chizik, sister of the agriculturist Barukh and of Sarah and Ephraim who had been killed during Arab riots, founded a boarding school in Tel Aviv where girls from the kibbutzim could learn agriculture. Behind the school – which was a typical example of Tel Aviv Bauhaus architecture, was a field where cows could graze and a cowshed. The girls sold the cows' milk and the fruit and vegetables which they grew in a small shop at the school entrance. From these sales they financed their education.*

Anxious to build a Jewish entity, and to increase the possibilities for Jewish immigration, the Jewish Agency employed only Jews in the major Zionist enterprises, a policy that was resented by Arabs in search of work. Fearing a further outbreak of Arab violence, the Haganah secretly purchased rifles and ammunition abroad and smuggled them into Palestine by sea. In October 1935 part of a consignment from Belgium broke open while being unloaded at Jaffa. In protest, the Arabs declared a general strike and held demonstrations in Jaffa, Haifa and Jerusalem. That autumn, hoping to improve relations with the Arabs, the Jewish Agency added 50 per cent to the land hitherto reserved for Arabs in the Huleh basin. This was land that had been allocated to a Jewish group by the Mandate administration.

---

* The school closed before the end of the Mandate and the building – known as Beit Hannah after Hannah Chizik – fell into disuse. In 1996 it was restored, and the ground floor opened as a café – the Café Apropo. It was here on 22 March 1997 that a Palestinian Arab killed himself and three Israeli women.

In 1935 an even higher number of Jews entered Palestine than in the previous year: 61,000 as against 42,000 (among them was two-year-old Igael Tumarkin, brought by his parents from Dresden, who was to be one of the leading painters and sculptors of the State of Israel). The British government, committed since 1921 to consider the 'economic absorptive capacity' of the country in its immigration policy, began to look with alarm at more Jews entering Palestine than the land could economically support. Here was a clear case, so the Arabs argued, for imposing strict immigration restrictions. Two years later, however, a British Royal Commission, headed by Lord Peel, commented on this upsurge in Jewish immigration: 'Far from reducing economic absorptive capacity, immigration increased it. The more immigrants came in, the more work they created for local industries to meet their needs, especially in building: and more work meant more room for immigrants under the "labour schedule". Unless, therefore, the Government adopted a more restrictive policy, or unless there were some economic or financial set-back, there seemed no reason why the rate of immigration should not go on climbing up and up.'

This was a powerful endorsement of Jewish immigration. Among the settlements founded in 1935 was Beit Ha-Shittah (House of the Acacia Tree), in the Jezreel Valley. It was named after the biblical site of the defeat of the Midianites by Gideon. The founders were a combination of Jews from Germany and Palestinian-born Jewish youth, the latter known as Sabras, from the cactus fruit that is tough and prickly outside, sweet and succulent inside. Also founded by immigrants from Germany that year was Yedidyah (Friend of God), in the coastal plain. It was named after the Jewish philosopher Philo of Alexandria, whose influence on early Christianity was considerable; his name had been Hebraized as Yedidyah in the nineteenth century. Love of God, as opposed to fear of God, was the virtue he stressed.

Also founded in 1935 was Sha'ar Ha-Amakim (Gateway of the Valleys), 8 miles inland from Haifa, on a ridge overlooking the Jezreel Valley to the north and the Zebulun Valley to the south. Its founders were immigrants from Yugoslavia.

Events in Germany in 1935 made Palestine seem more than ever an essential focus of Jewish hopes. On September 15 a Nazi Party convention held in Nuremberg accepted two special statutes, known subsequently as the Nuremberg Laws, whereby no Jew could be a German citizen. This put even greater pressure on German Jews to go to Palestine, where the Jewish National Home though not a Jewish State certainly felt like one to all the intents and purposes of daily life there. Two-thirds of the population was Arab, but most of the Jewish population lived in compact areas that were contiguous with each other. Tel Aviv was an entirely Jewish city. Other Jewish towns had flourished, or were being built up. The Yishuv, the Jewish community, prided itself on the modernity of its life and culture.

Every aspect of Jewish life was catered for in Palestine. In 1936, the year of Hitler's much-vaunted Berlin Olympics, the Maccabi Sports Organization held its own international games (often called the Jewish Olympics) near Tel Aviv. Jews came to compete from almost every country of the Diaspora, including Germany. After the games were over many of the competitors, not only from Germany but from Austria and Czechoslovakia, decided to remain in Palestine. They did so 'illegally', having been granted permission by the Mandate authorities to enter the country only in order to take part in the games. They were fortunate to have somewhere to go, where their presence would be accepted without demur, for the Maccabi Organization decided that year to found and to finance Kfar Ha-Maccabi, a kibbutz just inland from Haifa.

Also in 1936, the Palestine Orchestra was established. Its founder was the Polish-born child prodigy, Bronislaw Huberman, who in 1892, at the age of ten, had played the violin before the Emperor Francis Joseph in Vienna. Huberman gathered together a group of mostly German-Jewish refugee musicians, and persuaded Arturo Toscanini to conduct the opening concerts that December in Tel Aviv and Jerusalem. The repertoire included music by the German-Jewish composer Felix Mendelssohn, whose works had been banned by the Nazis. In their first Jerusalem concert in 1937, which was attended by the Peel Commissioners, all the works were by Beethoven.

The literary life of Jewish Palestine also flourished. The influx of Jewish writers, poets, composers, artists and actors from many different lands combined with the working face of Zionism to create a strong sense of nationality. In 1936 the Hebrew Writers' Association chose as its chairman Ukrainian-born Yosef Aharonovitch, who had lived in Palestine for thirty years. A leader of the Labour movement, and director since 1922 of the Bank Ha-Poalim (Workers' Bank) in Tel Aviv, he wanted all agricultural and industrial work on Jewish settlements to be carried out by Jews. He was a strong advocate of integrity in public life and efficiency in the national Jewish institutions and wrote many articles to this end. He also believed that preparations should be made for mass immigration.

Academic life benefited as much as literary and artistic life from immigration. In 1936 a leading teacher of rabbinical studies, Chanokh Albeck, reached Palestine from Berlin, where he had been teaching for the previous decade. He was immediately made Professor of Talmud at the Hebrew University, where he taught for the next twenty years, and was the author of many scholarly works of biblical scholarship in both German and Hebrew.

Kibbutz Ha-Zorea (The Sower) was established in 1936 by 150 recent immigrants, German-Jewish youth of the Werkleute (Men of Action) youth movement, who had come to Palestine two years earlier. The kibbutz was on the south-west rim of the Jezreel Valley. Its founders took part in the planting of a substantial forest in the nearby hills. One of those who visited Ha-Zorea several times in its early days was a patron of the Werkleute,

Wilfred Israel, the son of a wealthy Berlin Jewish department store owner, who had become active in helping Jewish emigration from Germany. He was killed in an air crash in the Second World War while travelling from Lisbon to London. A museum in kibbutz Ha-Zorea was named after him and it contains many works of Far Eastern art which he donated to them in his will.

Between 1933 and 1936 the Jewish population of Palestine increased from 234,967 to 384,078: from just over 20 per cent to just under 30 per cent. In protest against any further Jewish immigration, the Arabs began a general strike on 15 April 1936. That day three Jews travelling between the Arab towns of Tulkarm and Nablus were attacked and killed. On the following day, a small dissident Jewish group, calling itself the Irgun Zvai Leumi (National Military Organization, also known in short as Irgun, and, from its Hebrew initials, as Etzel), which held Jabotinsky as its inspiration, carried out a reprisal killing against Arab workers near Tel Aviv. Within forty-eight hours, Arab gangs were searching for Jews throughout Jaffa, attacking them and setting fire to Jewish-owned shops.

On May 7 the Arab leaders met in Jerusalem and demanded an immediate end to all Jewish immigration, a ban on any further Jewish land purchase, and an Arab majority government. Jewish farms were attacked all over Palestine: Jewish houses were burnt, shops looted and whole orchards destroyed. Attacks on individual Jews led, within a month, to the deaths of twenty-one Jews, several of them women and children.

The British responded to this Arab violence by announcing on May 11 that they intended 'to suppress all outbreaks of lawlessness'. Six Arab rioters had been killed by the police by the middle of May. Denouncing the reprisals carried out by the Irgun, the Jewish Agency urged the Jews to exercise restraint. Jews continued to be killed throughout Palestine – leading to a total death toll of eighty by October. In trying to keep the peace, British troops killed more than 140 Arabs, while thirty-three British soldiers were killed in armed clashes with Arab bands.

For the Jews, it was galling to see what little effect the British protection could have. During the summer of 1936 thousands of Jewish-farmed acres were destroyed and fruit orchards cut down in deliberate acts of Arab vandalism. Jews were killed while travelling in buses, or even sitting in their homes. Whole Jewish communities fled, among them the ninety-four Jews, mostly from Kurdistan and other Muslim countries, whose families had lived in the predominantly Bedouin town of Beisan since the beginning of the century; the 350 Jews living in the predominantly Arab town of Acre; and all but one Jewish family of the ten families who lived in the Arab village of Peki'in, where, according to tradition, their ancestors had lived since Roman times.

The Haganah reacted to the Arab attacks by establishing mobile patrols trained specifically to counter-attack, and to do so at night. One of these patrols, led by a former officer in the Red Army, Yitzhak Landsberg (later Sadeh), and based on Kiryat Anavim just outside Jerusalem, set the pattern by using the threatened settlement as a base for counter-attacks on the nearby Arab villages from which the raiders had come. 'We learned to move in the darkness,' one of its members later recalled, 'what to wear, how to communicate with the other patrol members, how to whisper orders, how to fall and how to rise, how to shoot and how to hide, how to listen and what to listen to, how to establish a front-line position and how to construct it – and above all how to anticipate the intentions of an armed attacker before he could mount his attack.'

On 18 May 1937 it was announced in the House of Commons that a Royal Commission would be set up to investigate the causes of unrest in Palestine. Lord Peel was appointed Chairman, with Sir Horace Rumbold, a former British Ambassador to Berlin, as his deputy. The Commissioners made an extended tour of Palestine, visiting the main Arab centres, and several Jewish settlements, as well as Tel Aviv. John Martin, the Commission's Secretary, later recalled an incident which made a considerable emotional impact on Lord Peel and his colleagues. Visiting a Jewish agricultural settlement the Commissioners saw a man living in a rough hut, but with a piano and musical scores. Sir Horace Rumbold was certain that he had met the man before. On asking his name, it appeared that he was a well-known German musician from Leipzig who had once played at the British Embassy in Berlin. 'We all felt uncomfortable about his plight,' Martin recalled. Rumbold began to commiserate with the man. 'This is a terrible change for you,' he said, condoling. But the musician replied, to Rumbold's surprise, 'It is a change, from Hell to Heaven.'

CHAPTER FIVE

Hopes . . . and blows
1937–1939

The year 1937 was to be a decisive one for the Jews of Palestine. It was believed that the Peel Commission would suggest the creation of two independent States in Palestine, one Jewish and the other Arab. The concept of partition put the enticing prospect of statehood before the Jews. At the same time they feared the danger – from their perspective – of a truncated area of control. Even as the Peel Commissioners, having returned to Britain, continued to take evidence, the Zionists persevered with the establishment of new settlements. On April 9 a new type of settlement was set up. During the night its founders built a watchtower and a wooden perimeter fence and the settlement was in existence by dawn. The first tower-and-stockade village was established just north of the Arab village of Beisan, less than a mile from the River Jordan. It was settled by Jews from Kurdistan and was named Beit Yosef, after the labour Zionist leader Yosef Aharonovitch who had just died.

Within a month, the first tower-and-stockade village east of Beisan was founded, between the town and the River Jordan. It was named Maoz (Stronghold). An Englishman who was making his first visit to Palestine, Maurice Pearlman, was present on the day the settlement was established and heard from the settlers about the initial plans whereby a group of Palestinian-born Jewish youth would venture to this remote region, notorious for the depredations of roving Bedouin:

> These boys and girls – most of them were still in their teens – had taken a pledge, while at secondary school, to found a communal settlement of their own. As Palestinian-born Jews they were well acquainted with the country and its inhabitants; all spoke both Arabic and Hebrew; they were thus well fitted to colonize the most difficult and dangerous areas. A piece of land had been set aside for them by the responsible Jewish

colonizing body, but the last eighteen months was considered too dangerous a period for settlement, so that their requests to settle had been continuously refused.

The group, however, chafed under these postponements and decided that it was idle to wait any longer. A few of them visited the site, Maoz, in the desolate plains south-east of Beisan. They found it covered with very tall weeds, and abounding with Bedouin encampments. Returning like their Biblical forbears, the 'spies', they reported, in the Joshua and Caleb tradition, that it was not as dangerous as was suggested, and that they must buy a tractor, no matter how heavily it would involve them in debt, and prepare the land for settlement.

Mass persuasion, aided by the *fait accompli* – purchase of the tractor – finally influenced the responsible Jewish leaders to their view, and plans were made for their establishment in their new settlement. Ein Harod, at the eastern end of the Valley of Esdraelon, the largest settlement near Maoz, became the centre of feverish activity, barricades, bungalows, and a watchtower being built ready to be transported to the new site.

The day came when the settlement was to be established. Maurice Pearlman was an eyewitness of how it was done:

Zero hour was fixed at 3.30 on a certain Tuesday morning, and the new group, accompanied by members of the neighbouring settlements, set out with their portable structures in a procession of lorries, and nosed their way eastwards.

The road extends to the eastern end of the Valley and then turns northwards to the Syrian border. Maoz, however, lies at the southern end of the Beisan area, and so, leaving the road at the point where it turns, the lorries prepared to crush out their laborious way across the Beisan wilderness, covered with tall rambling weeds. But they had gone no more than a few yards when they came upon a path, newly beaten out. They travelled along it until they were within a short distance of Maoz. Just ahead, they perceived the dull glow of a lighted cigarette in the darkness. As they came closer, they made out the silhouette of a young boy seated on a tractor.

The figure detached itself and came towards them.

'Shalom, Shalom, Hacol Beseder! Hallo, hallo. Everything's OK!' he called, quite calmly.

'Shalom,' they replied.

He was a youth of eighteen, a member of the new group. When the date of settlement had been decided, he had volunteered to go out some time before and crush out a path with the tractor. But he had a job in town during the day, and besides it was felt advisable not to attract too much attention to the impending settlement of this area. The tractor had therefore been brought out four nights before and hidden midst the weeds, and he had decided to work it during the night.

For three nights before the 'great day', he had come out to this spot and carefully and quietly had worked, methodically burning the weeds, then crushing them with the tractor. He was alone and completely unarmed. There was the ever-present danger that he would be surprised by Bedouin, in which case only his wits, his Arabic and perhaps a cigarette may have saved him.

Yet he showed no outward sign of relief when he met the great settling party from Ein Harod. 'Hacol Beseder – everything's OK.' Nothing more. And they followed him to the newly-cleared site, and hastily prepared to build up their homestead.

By late afternoon, a new Jewish village had sprung up. Bungalows had been erected, barricades run up round the boundaries, and the wooden watchtower, complete with lamp and dynamo, put in position. When night fell the lamp was switched on mid dead silence. It threw a powerful beam across the surrounding waste. Its symbolism was apparent to all, and it was the only moment when they permitted themselves to be sentimental. It was the light of a new life and a new era. An area from which civilization had departed 2,000 years before was being reclaimed.

A new settlement had been established.

Not long after the establishment of Maoz, a veteran Zionist leader, Hayyim Sturman, who had come to Palestine from the Ukraine in 1906, and had been a pioneer of Jewish self-defence both before and after the First World War, was killed, together with two friends, when their car went over a mine laid by Arab insurrectionists. A founding member of Ein Harod, he had been a familiar and encouraging figure to the young men and women who had just set up Maoz. In his memory they renamed their kibbutz Maoz Hayyim (Hayyim's Stronghold). Ten years later Hayyim Sturman's son Moshe was killed in the War of Independence; and eleven years after that in 1969 his grandson Hayyim was killed during an Israeli Army raid on Green Island at the mouth of the Suez Canal.

On 30 June 1937 a tower-and-stockade kibbutz was established at Tirat Zevi (Zevi's Castle), 6 miles south-east of Beisan, and less than a mile from the Jordan border. Its founders were religious pioneers from Central Europe. It too was established in a single night. From the outset its members were active in religious study and education, while at the same time farming. They took their name from Rabbi Zevi Hirsh Kalischer who, after the revolutions of 1848, pointed to the many national movements then active in Europe, and chastised his fellow-Jews for being the only people without such an aspiration. His book urging a return to Palestine and the carrying out of wide-spread and intensive agricultural labour as an essential prelude to spiritual redemption was published in 1862.

Four years before his death, Kalischer had hoped to settle at Mikveh Israel and to supervise the carrying out of religious duties connected with the Land of Israel, but at the age of eighty he was too old to leave Germany. Agriculture in Palestine on a large scale had been his main theme, both for the Jews already there and for those persecuted in Europe. The pioneers of Tirat Zevi could not have chosen a more apposite religious champion.

Also in 1937, the Zionist leader and historian Nahum Sokolow was commemorated in the name of the tower-and-stockade outpost of Sde Nahum (Nahum's Field) established by recent immigrants from Poland and Austria. It was in the low-lying hills two miles to the north of Beisan.

It was not only defended settlements that were being erected by the Jews in Palestine. With the enthusiastic financial support of a British Jew, Miriam Sacher – whose husband Harry had been one of the draftsmen of the Balfour Declaration and an early advocate of the Hebrew University – a hotel was built at Kaliya, at the northern end of the Dead Sea, to which the hard-working civil servants, businessmen and politicians of the Jewish community could descend, especially in winter, for rest and relaxation. A soak in the waters of the Dead Sea was highly prized by those with aches and pains. Miriam Sacher also hoped that the hotel could become a meeting place between Arabs from Transjordan and Jews from Palestine, a haven where discussion could take place in a calm and friendly atmosphere, more than 1,000 feet below sea level.

The Zionist imperative in 1937 was defence as well as discussion. That spring a senior Haganah officer, Yosef Avidar, set up a series of countrywide courses to train officers in the techniques that had been learned during the previous year's Arab revolt. Night patrols and mock attacks on Arab villages were central to this training. In the summer of 1937 a three-day course was held at Kiryat Anavim, the kibbutz near Jerusalem from which some of the first night patrols had set out. The British authorities also looked to the Jews for help as Arab attacks on British soldiers intensified. Working with the Jewish Agency, the British established the Jewish Settlement Police, equip-ping them with trucks and armoured cars and allowing them to act immediately against any Arab threat.

Yitzhak Landsberg, a pioneer of the Haganah night patrols – who had been one of the founders of the Labour Legion commune in 1921 – was also working out a new system of defence through attack. These were the field companies: permanent mobile patrols that were set up throughout the areas of Jewish settlement in Palestine. They were known by the Hebrew acronym Fosh – from *plugot sadeh* (field companies). Landsberg himself changed his name to Sadeh (Field), the name by which he was to lead many acts of Jewish resistance to Arab attacks in the years ahead. More than fifty Jewish settlements were protected by the Haganah in 1937, including the recently established stockade-and-tower settlements in the Beisan Valley. That year, two sustained Arab attacks on Tirat Zevi and Maoz were driven off.

Charged with advising on the future of both the Jewish and Arab presence in Palestine, the Peel Commissioners cross-questioned two former British Cabinet Ministers about what they had envisaged the future to be when the Mandate was secured. David Lloyd George, the Prime Minister at the time of the Balfour Declaration and the Mandate, said that the idea was 'that a Jewish State was not to be set up immediately by the Peace Treaty without reference to the wishes of the majority of the inhabitants. On the other hand, it was contemplated that when the time arrived for according representative institutions to Palestine, if the Jews had meanwhile responded to the opportunity afforded them by the idea of a National Home and had become a definite majority of the inhabitants, then Palestine would become a Jewish Commonwealth.'

Winston Churchill was likewise called before the Commissioners. In answer to a question from Lord Peel, he declared that the Jewish right to immigration ought not to be curtailed by the 'economic absorptive capacity' of Palestine (a phrase used in his own White Paper of 1922). He also spoke of 'the good faith of England to the Jews'. The British government had certainly committed itself, Churchill said, 'to the idea that some day, somehow, far off in the future, subject to justice and economic convenience, there might well be a great Jewish State there, numbered by millions, far exceeding the present inhabitants of the country . . . We never committed ourselves to making Palestine a Jewish State . . . but if more and more Jews gather to that Home and all is worked from age to age, from generation to generation, with justice and fair consideration to those displaced and so forth, certainly it was contemplated and intended that they might in the course of time become an overwhelmingly Jewish State.'

Sir Horace Rumbold, who took up the questioning, asked Churchill whether there was not 'harsh injustice' to the Arabs if Palestine attracted too many Jews from outside? Churchill replied that even when the Jewish National Home became 'all Palestine' there was no injustice. 'Why is there harsh injustice done,' Churchill asked his questioner, 'if people come in and make a livelihood for more, and make the desert into palm groves and orange groves? Why is it injustice because there is more work and wealth for everybody? There is no injustice. The injustice is when those who live in the country leave it to be desert for thousands of years.'

Later in the session, Rumbold asked Churchill 'at what point' he would consider the Jewish National Home to be established, and Britain's undertaking to the Jews fulfilled. Churchill replied, 'When it was quite clear the

Jewish preponderance in Palestine was very marked, decisive, and when we were satisfied that we had no further duties to discharge to the Arab population, the Arab minority.'

As the Peel Commission continued its deliberations, it became clear that one solution it was likely to propose was the two-State solution: a State for the Jews and a State for the Arabs, each to govern the area in which it was in a majority. There were rumours inside Palestine that the Peel Commission would, among its other geographic decisions, limit the Jewish area around the Sea of Galilee to the western shore. But to place the small Jewish area on the eastern shore within the Arab State would effectively cut off the Jewish State from the main water resources of the area, and from control of the sea itself. The Zionists decided to act at once, and to set up a settlement on the eastern shore, at Ein Gev (Water Hole). One young Jew, Teddy Kollek, an immigrant from Austria, has described in his memoirs how:

> We summoned up our energies and carried out a lightning-quick settlement of the 'tower-and-stockade' type on the eastern side of the lake. It was completed within one day in June 1937.
>
> Hundreds of people came in dozens of trucks and, within a few hours, put up a stockade around the area, a few huts, and a watchtower with a searchlight.
>
> By evening we were settled, and a few hours later, when we heard the Peel Commission Report broadcast over the radio, one of our members started chanting, 'Now we belong to Emir Abdullah' (the Emir of Transjordan).
>
> We all laughed and danced. Spirits were high. We had succeeded. We were right on to the water and ready to fight for it.

In presenting the Peel Commission Report to the Cabinet on 25 June 1937 the new British Colonial Secretary, William Ormsby-Gore, had strongly advised his colleagues to accept the partition of Palestine into a Jewish and an Arab State, calling the report 'a lucid and penetrating analysis' which had led him to accept 'without hesitation the Commission's diagnosis of the root of the trouble as a conflict of irreconcilable national aspirations'. He also accepted that partition, although a 'drastic and difficult operation', had 'the best hope of a permanent solution, just to both parties and consonant with our obligations both to Jews and to Arabs'.

Another assessment of the effect of partition reached the Cabinet from the Commander of the British Forces in Palestine, Lieutenant-General Sir John Dill, who telegraphed to the Chief of the Imperial General Staff on June 29 that the report 'so cuts across Zionists' aspirations that Jewish resistance to it will be strenuous. Long-sighted Dr Weizmann may be willing to accept and bide his time, but if he does he is unlikely to carry world Jewry with him. Although Jewish opposition to the Report may be frantic, consider it unlikely to take form of armed resistance. Nevertheless Jewish restraint on reprisals is likely to weaken.'

General Dill also gave his view of the Arab reaction. The larger proportion of Arabs, he felt, were in favour of acceptance, while the Mufti and what he called 'dangerous elements' were against. The report was likely, he believed, to 'split Jewry, larger proportion being against and very few in favour'. The Jews, Dill concluded, were likely 'to turn every political stone to undermine the Report but unlikely to use force'.

When the Royal Commission published its report on July 7 all of Weizmann's geographical worries were confirmed. The Jewish State was to be a small one. Not only the Negev, and the eastern half of the Sea of Galilee, but even the Rutenberg concession on the River Jordan, granted to the Jews by Britain in 1922 as a means of developing hydroelectric power, were to be excluded from it.

According to the report, the Arabs were afraid that they would be 'overwhelmed and therefore dominated by Jewish immigrants'. The report went on to point out that even if Jewish immigration were restricted to 30,000 a year – the number for 1936 – Jews would outnumber the Arabs by 1960, while if immigration remained at the 1935 figure of 60,000 then the Jews would outnumber the Arabs by 1947. To avoid an immediate exacerbation of Arab fears, the Commissioners recommended an annual limit to Jewish immigration of 12,000, for a period of five years. But they warned that this would be only a palliative, not a solution. The Arabs would continue to want the Jews out of Palestine, in order to have for themselves 'the same national status as that attained, or soon to be attained, by all the other Arabs of Asia'.

As it was impossible to devise any form of government for Palestine which would satisfy both Arabs and Jews, the report concluded that partition into two separate States, with Jerusalem remaining under British control, 'seems to offer at least a chance of ultimate peace'. Partition would mean 'that the Arabs must acquiesce in the exclusion from their sovereignty of a piece of territory, long occupied and once ruled by them'. For their part, the Commissioners concluded, 'the Jews must be content with less than the Land of Israel they once ruled and have hoped to rule again'.

On July 16 Jabotinsky wrote privately to Churchill to protest against the Peel partition plan. Although, he wrote, the Jews themselves had not 'fully realized' the full implications of the report, 'they surely want, above all, room for colonization and a Holy Land that is a Holy Land'. Partition, Jabotinsky was convinced, 'kills all their hopes'.

The Peel Commission report was debated by the Twentieth Zionist Congress held in Zurich in August 1937. The principle of partition was accepted. It meant, after all, the establishment of a Jewish State far sooner than had been conceivable even two years earlier. Jewish statehood fulfilled Herzl's vision. But the actual area proposed by Peel was voted to be unacceptable. The Arabs rejected partition altogether. In September 1937, in Damascus, 400 Arabs, representing all the Arab States as well as Palestine itself, resolved that Palestine was 'an integral part of the Arabian homeland', and insisted that Britain had to choose 'between our friendship and the Jews'.

By the end of 1937 violence had broken out again. The Arabs did not want even the small Jewish areas of Palestine to be ruled by Jews. The plan collapsed, Arab attacks on the British were renewed, Arab attacks on Jews intensified, and Jewish reprisals on Arabs were instituted – against the wishes of the Jewish Agency – by extremists who were convinced that violence, and even terror, were the route to statehood.

On September 26 Arab terrorists murdered Lewis Andrews, the District Commissioner for Galilee, who had been responsible for arranging the Peel Commissioners' travels through Palestine. In government circles in London the pressure of the various Arab States in the Middle East against any possible Jewish State, however small, was continuous and effective. A senior civil servant, George Rendel, prepared a memorandum to be signed by Anthony Eden, the Foreign Secretary, and circulated to the British Cabinet, which stated, as if by Eden himself: 'It has been suggested to me that there is only one way in which we can now make our peace with the Arabs, and avoid the dangers I have indicated above, that is, by giving the Arabs some assurance that the Jews will neither become a majority in Palestine, nor be given any Palestinian territory in full sovereignty.'

Such a solution, the memorandum asserted, 'would be welcomed by King Ibn Saud'. Not to adopt such a policy was likely not only to involve the British government 'in continuing military commitments of a far-reaching character in Palestine itself, but also to bring on them the permanent hostility of all the Arab and Moslem Powers in the Middle East'. This Foreign Office Cabinet paper marked the beginning of Britain's new policy. Its aim, Eden himself explained in a personal message to the British Ambassador in Washington on November 26, was to seek some alternative to the Peel Commission, an alternative 'which would not give Jews any territory exclusively for their own use'.

Dr Weizmann, who had returned to his home at Rehovot in Palestine, understood exactly what was happening in London. On the last day of 1937 he wrote to the Permanent Under-Secretary of State at the Colonial Office, Sir John Shuckburgh, that he had been told 'that under the pressure of Indian Moslems, Arab Kings, Italian intrigue and, last but not least, anti-Zionist Jews' the notion was gaining ground 'in high quarters' that there should be in Palestine a single Arab State, with 'the reduction of the Jews to permanent minority status'. Weizmann went on to tell Shuckburgh:

> Jews are not going to Palestine to become in their ancient home 'Arabs of the Mosaic Faith', or to exchange their German or Polish ghetti for an Arab one.

Whoever knows what Arab government looks like, what 'minority status' signifies nowadays, and what a Jewish Ghetto in an Arab State means – there are quite a number of precedents – will be able to form his own conclusions as to what would be in store for us if we accepted the position allotted to us in these 'solutions'.

Weizmann's letter continued, angrily but also anxiously:

It is not for the purpose of subjecting the Jewish people, which still stands in the front rank of civilization, to the rule of a set of unscrupulous Levantine politicians that this supreme effort is being made in Palestine.

All the labours and sacrifices here owe their inspiration to one thing alone: to the belief that this at last is going to mean freedom and the end of the ghetto.

Could there be a more appalling fraud of the hopes of a martyred people than to reduce it to ghetto status in the very land where it was promised national freedom?'

Neither the Arab States' pressures on the British government, nor British political pressures on the Zionist leaders, deterred the Jewish enterprise in Palestine. Each year saw new settlements established by new groups of immigrants who undertook intensive agricultural training at an existing settlement before embarking on a new one. In 1937 a group from several countries – including the United States, Germany and Poland – established a kibbutz in the southern coastal plain at Kfar Menachem, named after Menachem Ussishkin. A moshav set up on the same site a year earlier had found it impossible to farm, partly as a result of the poor soil and partly because of Arab attacks, but the new settlers persevered, driving off a number of renewed Arab attacks, and their enterprise survived.

Jewish immigrants from Galicia, one of the heartlands of pre-First World War Jewish life in the Austro-Hungarian Empire, founded a kibbutz in 1937, 3 miles inland from Haifa Bay. They called it Usha, the name of the town in which many of the scholars who survived the crushing of Bar Kochba's revolt by the Romans had lived, studied and legislated. Among the second-century Laws of Usha is one that lays down 'that a man must support his son until he is twelve years old' – and thereafter help him in his trade.

One stimulus for the creation of settlements was acts of terror. On 9 November 1937 five young Jews were killed by Arabs in the hills 8 miles west of Jerusalem. Early the following year, members of the Gordonia youth group, immigrants from Lodz in Poland, set up overnight a tower-and-stockade kibbutz in their memory. It was called Ma'ale Ha-Hamisha (Ascent of the Five). The settlers planted fruit orchards and vineyards.

*   *   *

In 1938 Jews from Czechoslovakia, a country threatened by Hitler's territorial demands, established a kibbutz in the Haifa Bay area only a few miles from several Arab villages of Western Galilee; their priority was to secure their homes within a stockade. They named it Kfar Masaryk, after the first President of Czechoslovakia, Thomas G. Masaryk, who had died the previous year. The location was a brackish swamp near the mouth of a stream flowing (or failing to flow) into the Mediterranean, so they had to carry out much drainage work and land reclamation before cultivation could begin.

In the far north of Palestine, in 1938, Polish immigrants founded a tower-and-stockade kibbutz at Eilon, in the hill country only two miles from the Lebanese border. The ground was exceptionally rocky, even for a rocky land. But when it was cleared, fruit orchards were planted and the kibbutz thrived. Its name was a mixture of two local trees, the oak (allon) and the pistachio (elah). Another tower-and-stockade settlement set up in 1938 was Kfar Ruppin – named after the founder of the kibbutz movement, Arthur Ruppin. It was situated on the River Jordan south of the Sea of Galilee. Jews from Germany, Czechoslovakia and Hungary were among its pioneers. Because of the hot climate, date palms and pomegranates could be grown.

Jews from the Polish province of Galicia founded their second kibbutz in two years in 1938. They called it Tel Yitzhak, in memory of Yitzhak (Isaac) Steiger, the founder of the Zionist youth movement in Galicia. Also founded in 1938 was Allonim (Oaks), just off the Haifa–Nazareth road. Its pioneers were graduates from the first group of Youth Aliyah to have come from Germany six years earlier.

As immigration grew, the educational needs of the Jews of Palestine increased. Among those who began teaching at the teachers' seminary in Jerusalem in 1938 was Zvi Adar, who had been born in Palestine during the First World War. For the next thirty years he was to teach about Jewish identity in past ages, and to seek to elucidate, and then to transmit, the educational and humanistic values to be found in the Bible. The link of the Jews to their land was also being stressed by the followers of Chief Rabbi Abraham Isaac Kook, who had died in 1935, and whose teaching – that it was the divine plan that led the people of Israel to make the Land of Israel its country – was to have a considerable impact on successive generations.

The Jewish world was being divided by its attitude to the future of Palestine. As the situation in Germany worsened, and pressure on immigration continued to grow, Weizmann still wanted to work with the British, and to devise an operable compromise. Within Palestine, the Jewish Agency continued to denounce Jewish reprisals and all forms of Jewish terror action against the Arabs. But opposition to the standard bearers of the Zionist cause was growing. On 1 February 1938 the Revisionists, led by Jabotinsky, held a World Congress in Prague. Six days later they passed a series of resolutions, the principal one of which was their opposition 'to any plan whatsoever which would deprive the Jewish people of their right to establish a Jewish majority on both sides of the Jordan'. The Jewish Agency and the 'old Zionist organization' were described as 'traitors' who had abandoned 'the ideals of Zionism as propagated by Herzl'. In Palestine the Revisionists decided to strike back at the repeated Arab attacks on Jewish settlements and travellers, attacks which the Jewish Agency had insisted must be met by restraint. Henceforth, most acts of Arab terror were met, often within a few hours, by equally savage acts of reprisal by the Revisionists' military arm, the Irgun.

In the two years 1936 and 1937, 60,000 Jews had entered Palestine as legal immigrants. But on 10 March 1938, as Arab protests continued, the Colonial Office in London informed the new High Commissioner in Palestine, Sir Harold MacMichael, that no more than 2,000 immigrants of independent means – with a personal capital of £1,000 – and not more than 1,000 Jewish workers, should be allowed to enter Palestine in the six months from April to September 1938. This not only reduced drastically the number of Jews entering Palestine, but effectively held the Jewish–Arab ratio at just below 30 per cent.

On this basis, there could never be a Jewish majority; and in 1938–9 the Jewish population (of just over 400,000) increased by 4,235, while the Arab population rose by 65,413. Since the first weeks of 1938 the Jews were also confronted in Palestine with an increase in Arab violence, including the murder of Jewish farmers (some shot, some stabbed), and shots being fired at Jewish buses and trucks as they were driving along the highways. One route by which Arab bands infiltrated into Palestine was from the north, across the Lebanese border. To prevent this, land was bought at Hanita, on the border itself, and in March 1938 the Jewish Agency's defence force, the Haganah, decided to fortify the site, clear the fields and open an access road.

Four hundred men were chosen to take part in the operation, which was to be completed in a single day in order to present the British authorities and the local Arab inhabitants with a *fait accompli*. The work was under the command of the leader of the Haganah field companies, Yitzhak Sadeh, with two deputies, Yigal Allon and Moshe Dayan. In his memoirs Dayan recalled the pre-dawn operation:

> We moved out of our assembly point before dawn and headed northward for Hanita. We had to leave the vehicles on the road and laboriously climb the rocky slopes. While one group started hacking out a smooth track, the rest of us carried heavy loads of fortification equipment and materials by hand.
>
> On the hilltop site we began erecting a wooden watchtower and the standard perimeter fence, a double wall of wood filled with earth and boulders. We hoped to do all this during the day so that the tented

compound would be defended by nightfall, when we expected the first attack. But night came and we had not completed the fortifications. There had been too much to do, and we were also hampered by a strong wind. We could not even put up the tents.

At midnight we were attacked.

The attack was driven off, and the watchtower and stockade erected. A few weeks after Hanita had been established, a captain in the British army, Orde Wingate (later killed in action behind Japanese lines in Burma in the Second World War), was to join these Jewish soldiers and help them to protect their fields and settlements from Arab attack by a method already pioneered by Yitzhak Sadeh – the night ambush of the would-be attacking forces. Wingate was not a Jew but, like many of his generation in Britain, he was deeply versed in the Bible and, in his case, regarded the various biblical prophecies about the return of the Jews to their land as something to be welcomed, and in which he could participate.

The Haganah mobile patrols gained invaluable experience with the establishment by Wingate of Special Night Squads commanded by Yitzhak Sadeh. 'There were times,' Dayan wrote of Wingate, 'when he would march on, driven by an iron will. He had an unshakeable belief in the Bible. Before going on an action he would read the passage in the Bible relating to the places where we would be operating and finding testimony to our victory – the victory of God and the Jews.'

For the Jews of Europe, already the victims of hatreds beyond their control, March 1938 saw a further deterioration in their circumstances. On March 15 Hitler entered Vienna and direct Nazi rule was at once extended to Austria. Within only a few days, the whole apparatus of anti-Jewish persecution – from the imposition of the Nuremberg Laws to the beatings, torture and deportations to Dachau concentration camp – was instituted in its most vicious form. Overnight more than 180,000 Jews were added to the German Reich. At the same time a further 40,000 people of Jewish descent – many of them baptized Christians – were declared to be, racially, Jews.

In Palestine, the tense situation was exacerbated by the death sentence passed by a British court on a young Revisionist, Shlomo Ben Josef, who had opened fire on an Arab bus. No one had been killed in the incident, and in England the *Manchester Guardian* called, on June 6, for clemency. On June 29 the British Consul General in Riga telegraphed to the Foreign Office: 'Last night a stone was thrown through my study window wrapped in a paper bearing the message, "The Jewish people will never forget the blood of their brother Ben Joseph".' But the Mandate government decided that an example should be set, and Ben Josef was executed, the first Jew to be hanged in Palestine by the British.

During the summer of 1938 the situation in Palestine deteriorated. The hopes of so many tens of thousands of Jews for a quiet and productive life were being challenged in the most cruel way. On June 24 two Arabs working in a Jewish-owned stone quarry near Haifa were wounded by Arab raiders. The wounded men were taken to hospital, but two of the raiders entered the hospital in search of them, killing by mistake another Arab, a patient from Nablus. On June 29 an Arab terrorist threw a bomb at a Jewish wedding party at Tiberias: seven Jews were wounded, including three children. Parallel with these events, the Irgun struck back in violent acts of reprisal after nine Jews had been killed during June in Arab attacks on individual Jews throughout Palestine. On July 6 a single Jewish terrorist bomb killed twenty-five Arabs in Haifa. On July 11, Arabs killed two Jews in Haifa, and on July 12, also in Haifa, an elderly Jew was stoned to death.

The killings and counter-killings continued. On July 21 four Jewish workers were killed at the Dead Sea. Four days later, in Haifa, a Jewish terrorist bomb killed thirty-nine Arabs in the melon market.

In Palestine, the Jewish newspaper *Davar* denounced the Irgun reprisals as both 'shameful and calamitous', while the newspaper *Ha-Aretz* declared them to be 'a criminal gamble with the fate of the Jewish community'. The Zionist leaders were equally strong in their condemnation. The terrorists, declared Ben-Gurion, were 'miserable cowards'. Yitzhak Ben-Zvi warned that the terrorists 'were stabbing the community in the back'. At the end of July, the Histadrut – the General Federation of Jewish Workers – issued an official manifesto stating that the bloodshed of reprisals was 'a disgrace and a madness'.

At the youth agricultural village of Ben Shemen, which had been estab-lished a decade earlier by Dr Siegfried Lehmann, one of the new boys in 1938 was fifteen-year-old Shimon Persky (later Shimon Peres). 'Despite the atmosphere of Arab hostility which pervaded every corner of the village and every facet of our lives,' he recalled, 'Lehmann was a committed political dove who believed that Zionism should and must offer far-reaching conces-sions to the Arabs in order to achieve an understanding with them.' To this end, Lehmann was active in the Brit Shalom movement, which had Albert Einstein's support.

Peres later wrote of the life of a teenager at Ben Shemen:

> On Saturdays, we went on hikes through the surrounding countryside, in the foothills of the Jerusalem range. The same Arabs who took pot-shots at us by night would welcome us during these outings, in the best tradition of Semitic hospitality. Perhaps Dr Lehmann's reputation as a peace activist, which was well known throughout the area, secured at least the daytime truce.
>
> At night, however, there could be no such carefree relaxation, and the older pupils took an active part in guarding the village. Soon after I arrived at Ben Shemen, I swore my oath of allegiance to the Haganah,

by the light of a candle, with a Bible and a revolver on the table before me. Soon after that, I started weapons training, and within a short time I had been appointed commander of one of the guard-posts.

This was not, in fact, an especially high-ranking assignment, but for me it was fateful. My guard-post was just beyond the home of Gelman, the carpentry teacher, from which emerged a barefoot young girl with long-brown plaits and a face of Grecian grace. I was instantly totally smitten. Her name was Sonia, and she was eventually to become my wife. I sought to impress her by reading to her, sometimes by the light of the moon, selected passages from Marx's *Das Kapital*.

In Germany and Austria, the situation of the Jews was becoming desperate. In Poland and Roumania, anti-Semitism had again become open, vociferous and violent. Throughout Europe the possibility of Nazi rule, or pro-Nazi regimes, encouraged and stimulated anti-Jewish feeling. As the number of Jewish refugees grew, and the number of would-be refugees intensified, the States which were taking in refugees began to seek ways of regulating the flow. At an international conference held in the French town of Evian, on the shore of Lake Geneva, the British government's representative explained that only very limited numbers could henceforth enter Palestine. Britain itself, which had already taken in more than 40,000 German Jewish refugees since 1933, was also imposing restrictions following a growth in public unease at the large numbers of new immigrants. The United States' delegate announced that his country would adhere to its existing quotas: 27,370 refugees a year from Germany and Austria. Given such insistence on the letter of the law, it would have taken twenty years – until 1958 – before all the Jews, even of the German Reich alone, could reach asylum in the United States.

The United States' decision to maintain its strict quotas served as a guide-line for other States; the delegate from Peru described the 'wisdom and caution' of the United States as being a 'shining example that guided the immigration policies' of Peru itself. Only the Dutch, the Danes and the Dominican Republic agreed to let in Jewish refugees without restrictions. On the last day of the Evian conference, July 15, a resolution was passed which stated that 'the countries of asylum' – such as they were – 'are not willing to undertake any obligation towards financing involuntary emigration'. Other resolutions expressed the Conference's 'sympathy' towards the plight of the refugees – Jewish and non-Jewish alike. On the following day a journalist from a Swiss newspaper asked one of the Palestinian Jews present, Golda Meir, for her comments. 'There is one ideal I have in mind', she replied, 'one thing I want to see before I die – that my people should not need expressions of sympathy any more.'

The Haganah decided to reactivate 'illegal' immigration. A small ship, the *Poseidon*, was chartered in Europe, and made its way to Palestine without being intercepted. Late in 1938 the Haganah set up a special organization, Mossad le-Aliyah Bet (Organization for Immigration B), to find the ships, to hire the crews, to collect the would-be immigrants, and to arrange for them to be spirited ashore once they had reached the coast of Palestine. The head of the new organization was Shaul Meirov, one of the founders of the Haganah. Ten years later, during the War of Independence, he would be in charge of Israel's arms purchases in Europe.

In 1939 a group of graduates from the Mikveh Israel agricultural school founded a moshav in the coastal plain north of Tel Aviv. They called it Kfar Netter, after the founder of Mikveh Israel, Charles Netter. That same year, recent immigrants from Poland and Lithuania were at the forefront of the founding of kibbutz Amir (named after the Hebrew word for the top of a tree). On the bank of the Upper Jordan and the edge of the Huleh swamp, the settlement was in constant danger of flooding during the winter rains and the subsequent melting of the snow on Mount Hermon, whose snow-covered peaks, inside French Syria, provided a picturesque backdrop to the harsh life of daily toil.

In the hills above the Sea of Galilee, a group of religious immigrants founded a small settlement at Alummot (Sheaves) in 1939. The high and somewhat isolated spot had earlier served as a fruit tree nursery for the Palestine Jewish Colonization Association. Further tower-and-stockade settlements were also being established in exposed and isolated regions. On May 3 immigrants from Lithuania and Poland founded Dafna (Laurel Tree). It was located less than a mile to the south of the Lebanese border, and only three miles west of the Syrian border. On the very next day, pioneers from Roumania established kibbutz Dan on the Syrian border itself. It was named after the biblical city that was the chief town of the tribe of Dan, and the northernmost place shown to Moses before his death ('And Moses went up from the plains of Moab unto the mountain of Nebo . . . And the Lord shewed him all the land of Gilead, unto Dan').

A third settlement was established that month less than a mile from Dan. Together with Dan and Dafna it was known as a Ussishkin Fortress, after the pioneer Zionist, Menachem Ussishkin, who had died two years earlier. Its founders came from Eastern Europe, and it was called She'ar Yashuv (A Remnant Shall Return), from the verse in Isaiah: 'For though thy people Israel be as the sand of the sea, yet a remnant of them shall return.' Fifty-seven years later, above She'ar Yashuv, two helicopters crashed and seventy-three Israeli soldiers were killed: one of the helicopters fell on the village.

An important step forward in overseas Jewish help for the Jews of Palestine came in 1939, with the founding of the America Palestine Fund. This amalgamated a number of committees already in existence that were giving financial support to educational, cultural and social service institutions in

Palestine. Hitherto, these committees had been competing for funds from American Jewry. Now they could act in unison. The flagship of American support which went back to before the First World War was the Hadassah Organization of America, whose new hospital on Mount Scopus had just been opened.

The situation inside Palestine had become critical. As a result of continual hostility to Jewish immigration from Egypt, Iraq and Saudi Arabia, Syria and Yemen, Britain decided to turn its back on the basic premise of the 1922 Churchill White Paper, that Jewish immigration could continue until such a time as there was a Jewish majority. This was at the very moment when Nazi racial laws, already in force in Germany and Austria, were being extended to the Sudetenland region of Czechoslovakia. A new White Paper, named after the Colonial Secretary Malcolm MacDonald (the son of Britain's first Labour Prime Minister) and known to the Jews of Palestine as the Black Paper, set out new and, for the Jews, draconian rules which made the attainment of a Jewish majority impossible. Over the coming five years (that is, until May 1944) a total of 75,000 Jews would be allowed to enter Palestine, plus a further 25,000 emergency cases. After that, majority rule would be instituted: that is to say, the Arab majority would be given legislative powers.

The Arabs made it clear that they would use their legislative powers to halt any further Jewish immigration. The Arab majority would therefore be fixed for all time and Jewish statehood would be an impossibility. There were 445,457 Jews in Palestine in 1939, just under a third of the total population. That number would be allowed, under the 1939 White Paper, to rise to 545,457. The Arab population in 1939 stood at 1,501,698 (and with an annual natural increase in excess of 30,000). The Jews would therefore be in a permanent minority. They would also, under the White Paper, be unable to buy land outside the relatively restricted area of their existing settlements.

The Jews of Palestine were always capable of facing adversity with imagination. In the immediate aftermath of the 1939 White Paper the Jewish Agency, and the organizations grouped around it, the Jewish National Institutions, established a number of new settlements. One of these was Mahanayim (Two Camps), at a biblical location ten miles north of the Sea of Galilee. Three earlier efforts had been made to put a settlement there. The first was thirty-seven years earlier in 1902, when settlers hoped to establish tobacco growing in the region, but failed. The second attempt was a few years later, when Jews from the Caucasus were sent there to raise cattle. That experiment failed. The third attempt was in 1918, to establish a workers' farm, but that also failed. This fourth attempt succeeded, with crop farming providing a livelihood.

Also established in May 1939, as an immediate reaction to the White Paper, was Shadmot Devorah (Fields of Deborah). The founders were Jews from Germany. It lay just to the east of Mount Tabor, and was named after one of the most tenacious British benefactors of Palestine – and later Israeli –

Jewry, Dorothy (Devorah) de Rothschild. Two months later the country's most southerly tower-and-stockade settlement was founded, kibbutz Negba (Towards the Negev). Its founders were pioneers from Poland. There were Arab villages all around them in the open, rolling countryside.

Religious Jewish immigrants from Germany set up a small farming community in 1939 in the fertile area along the River Jordan 5 miles south of the Arab town of Beisan. They called it Sde Eliyahu (Elijah's Field) in memory of Rabbi Elijah Gutmacher, a nineteenth-century Hasidic rabbi and mystic who rejected the current Orthodox thinking that the Jewish people should wait passively for the coming of the Messiah. Instead, he argued, they should do all in their power to hasten redemption by engaging in constructive work in the Land of Israel. 'Settling in the Holy land,' he had written in the mid-nineteenth century, 'making a beginning, redeeming the sleeping land from the Arabs, observing the commandments that can be observed in our day – making the land bear fruit, purchasing land in the Land of Israel to settle the poor of our people there – that is an indispensable foundation-stone for complete redemption.'

It was not only the Jews of Palestine who saw the 1939 White Paper as an injustice. All the surviving members of the Peel Commission wrote a letter to *The Times*, stating that the 'cessation of immigration' that would result from the White Paper policy did not 'eliminate the fear of domination; it only transfers it from Arab minds to Jewish'. Addressing himself to the British Prime Minister, Neville Chamberlain, Winston Churchill told the House of Commons: 'It is twenty years since my right hon. Friend used these stirring words: "A great responsibility will rest upon the Zionists, who, before long, will be proceeding, with joy in their hearts, to the ancient seat of their people. Theirs will be the task to build up a new prosperity and a new civilization in old Palestine, so long neglected and misruled." Well, they have answered his call. They have fulfilled his hopes. How can he find it in his heart to strike them this mortal blow?'

Henceforth, at the very moment when Hitler annexed the Czech provinces of Bohemia and Moravia and extended Nazi racial laws to the large Jewish population of Prague, British policy was directed towards stopping Jewish immigrants reaching Palestine in excess of the restrictions. Pressure was put on Roumania, Bulgaria, Yugoslavia and Greece to prevent Jews making their way along the Danube or through the Adriatic and Aegean Seas towards Palestine. During a debate in the House of Commons on 30 July 1939, Malcolm MacDonald revealed that, to show its displeasure at the continuation of 'illegal' immigration, the British government had decided 'a short while ago' to suspend all legal immigration. This suspension, he told the House, would remain in operation for a further six months. At the same time, 'illegal' immigrant ships would still be intercepted and turned back.

When Alfred Duff Cooper, a Conservative minister who had resigned from the Cabinet after Munich, spoke in the debate, it was the first time he had ever spoken on Palestine. 'It seems to me,' he said, 'that the latest announcement that because illegal immigration is succeeding, legal immigration has to be stopped, is another lamentable proof of failure. It is like a petulant schoolmaster who, because some boys play truant, keeps in those who come to school.' The Jews and Arabs, Duff Cooper went on, were both 'old friends' of Britain. If two old friends come to you for help, you help the one who has 'the greatest need', clearly, in this instance, the Jews. And he added, 'Before these islands began their history, a thousand years before the Prophet Mohammed was born, the Jew, already exiled, sitting by the waters of Babylon, was singing: "If I forget thee, O Jerusalem, may my right hand forget its cunning."'

During his speech Duff Cooper spoke of what a 'hateful experience' it was for the Arab to see his land 'passing out of his hands into those of another race', but went on to ask, 'What hateful experiences are other races going through at the present time? Compare it, for a moment, with the long torture that is being inflicted on the Jews.' His speech ended with an appeal to the government not to close Palestine to the Jews, and he declared:

> In the course of their long persecution, they have begun once again to see a hope of return. It is us, it is the British people, British statesmen, the fore-runners of right hon. Gentlemen on the Front Bench, who have raised that hope in their hearts.
>
> It is the strong arm of the British Empire that has opened that door to them when all other doors are shut. Shall we now replace that hope that we have revived by despair, and shall we slam the door in the face of the long-wandering Jew?

On 23 August 1939 a pact was signed between Hitler's Germany and Stalin's Soviet Union. In arranging for the eventual partition of Poland between them, the pact sealed the fate not only of Poland, but of Poland's three million Jews. The news of the signing of the pact was announced in the middle of one of the sessions of the Twenty-First Zionist Congress, being held in Geneva. On the following day Arthur Ruppin, one of the delegates from Palestine, noted in his diary, 'The news exploded like a bomb.'

In his final words to the Congress, on the evening of August 24, Weizmann told the delegates: 'If, as I hope, we are spared in life and our work continues, who knows – perhaps a new light will shine upon us from the thick black gloom,' and he ended, 'The remnant shall work on, live on, until the dawn of better days. Towards that dawn I greet you. May we meet again in peace.' The official protocol of the Congress recorded how, at this point: 'Deep emotion grips the Congress. Dr Weizmann embraces his colleagues on the platform. There are tears in many eyes. Hundreds of hands are stretched out towards Dr Weizmann as he leaves the hall.'

CHAPTER SIX

The Second World War
1939–1945

On 1 September 1939 the Germans invaded Poland. The Second World War had begun. At 9.15 that night, far from the European spotlight, the *Tiger Hill*, a cargo ship with 1,400 illegal refugees on board, ran ashore off Tel Aviv. As it did so, in the glare of searchlights, the passengers sang *Hatikvah*. The Mandate authorities boarded the ship and took the 'illegals' to the Sarafand Detention Camp, on the Tel Aviv to Jerusalem road. They also took off the bodies of two refugees who had been killed during the previous night when Marine Police had opened fire as the freighter approached the coast. A third refugee, twenty-seven-year-old Mrs Yona Shimshelevitz, who had been wounded during the shooting, was taken to the Hadassah Hospital in Tel Aviv, where she died on the afternoon of September 2.

For two days German troops pushed deep into Poland, and German aircraft bombed Polish troops and civilian refugees. Hundreds of Jews were among those killed while fleeing from their homes. Then, on September 3, Britain and France declared war on Germany. Despite their new and formidable preoccupations, the British government continued to intercept all boats trying to bring Jews 'illegally' to Palestine.

On September 25 a Cabinet Committee decided that Britain could not assist in the emigration in wartime of 'Reich nationals' – that is, German and Austrian Jews. That same day the British Foreign Office rejected a suggestion by the Italian government to help facilitate the passage of German Jewish refugees through Italian ports. 'Illegal' immigrants – by now the vast majority of all immigrants – who did manage to reach Palestine were being interned at Sarafand, and at another former army camp at Athlit, just south of Haifa. The head of the Middle East Department of the Colonial Office suggested to Malcolm MacDonald on October 10 that it should be publicly announced that when the war was over these internees would be 'transferred to the country from which they came'. MacDonald agreed, noting four days later that, in his view, the British government should adopt the policy of 'sending them back to mainland Europe after the war'.

Further illegal immigrants who were caught were also detained in Athlit. Many others managed to avoid capture and, helped by the Jewish settlements along the coast to make their way inland, became part of an estimated 25,000 'illegals' who were living and working as if they were citizens.

As 1939 came to an end, British pressure on foreign governments continued; so much so that on December 20 the Colonial Office wrote to the Foreign Office of Turkey's continuing laxity in allowing refugee boats to pass through the Bosphorus on their way from the Black Sea: 'We therefore suggest that the Turkish government might be invited to enact legislation prohibiting their merchant marine from engaging in this traffic, on the lines of the law enacted by the Greek government at our request in the earlier part of this year.' Of the refugee ship whose passage had prompted this protest, the Colonial Office wrote, 'We trust that, even if the *Sakaria* cannot be turned back, she will at least be refused all facilities at Turkish ports on her passage through.'

On December 29 illegal immigration was discussed at the Foreign Office where one official minuted, 'The only hope is that all the German Jews will be stuck at the mouth of the Danube for lack of ships to take them.'

For the Jews in Palestine, war had brought a new spirit of determination. They rallied to Ben-Gurion's striking declaration, 'We will fight with the British against Hitler as if there were no White Paper; we will fight the White Paper as if there were no war.' Recruitment into the Allied forces, and a desire to serve the Allied cause, went parallel with the hope of bringing in, legally or 'illegally', as many European Jews as could escape the torment of Nazi rule. Recruitment was on a large scale.

With the coming of war in Europe, the Haganah also made efforts to improve its organization. A new post, that of Chief of Staff, was given to Yaakov Dostrovsky, Russian-born, who had been brought to Palestine as a child. In recent years he had been the commander of the Haganah in Haifa. From his first days as Chief of Staff he worked to turn the Haganah from an effective but diffuse self-defence organization into a model army. As his personal assistant he chose Yigael Sukenik, then twenty-three, who later recalled their first encounter: 'He said to me, "Listen, I'm receiving endless amounts of material from regional commanders, instructors, and course commanders. As part of your job," he told me, "go over these things. Whatever you think you can respond to, answer in my name – I give you the authority."'

To maintain secrecy, and avoid British scrutiny and searches, the Haganah had two headquarters in Tel Aviv – two small apartments five minutes' walk from each other – and switched frequently between them. Names had also to be hidden. Dostrovsky took the surname Dori, and Sukenik was given the name Yadin.

The Haganah also developed a small air force, consisting mostly of crop-dusting aircraft. One of its founders was the Labour leader Dov Hos, a pilot and pioneer aviator. From the first days of the war he worked for the formation of Jewish units in the British army.

Within the borders of Mandate Palestine Zionist work and expansion continued. On 9 October 1939 a group of young immigrants from Germany and Central Europe founded a kibbutz at Beit Ha-Aravah (House of the Steppe), 1,235 feet below sea level on the northern shore of the Dead Sea. By using the fresh waters of the nearby River Jordan to sweep away the extremely saline soil, they made the land – which had been desolate since biblical times – capable of producing vegetables, fruit and fodder. As the weather in winter was hot they could grow summer fruits in winter. The new kibbutz maintained good relations with its Arab neighbours in Jericho, 5 miles to the north-west.

On the coastal plain, the small farming settlement of Beit Yitzhak (Isaac's House) was founded in 1940 by immigrants from Germany. Many of them had been university teachers and professors until Hitler's legislation deprived them of their livelihood. It was named after a German Zionist, Yitzhak (Isaac) Feuerring, whose financial contribution had enabled the settlement to be established. Citrus orchards and a fruit preserves factory became its main livelihood.

Also founded in 1940, on the ruins of ancient Caesarea, was the coastal kibbutz of Sdot Yam (Fields of the Sea). The first settlers were Palestinian-born Jews and Youth Aliyah graduates from Germany and Hungary. At first fishing was the main activity. Among the settlers was Hungarian-born Hannah Szenes, who was soon to volunteer to serve as a parachutist, and to be dropped in Royal Air Force uniform – and with the code name 'Minnie' – into German-occupied Yugoslavia.

Immigrants from Holland and Czechoslovakia founded Sde Nehemyah in 1940. It was in the far north of the country, on the edge of the Huleh swamp. The name commemorated Nehemiah de Lieme, a Dutch Zionist leader, and former managing director of the Jewish National Fund, who had just died, at the age of fifty-eight, shortly after the German invasion of Holland in May 1940.

As the Haganah built up its weaponry looking to a period when it might be necessary for the Jews of Palestine to defend themselves against the Arabs on a greater scale than hitherto, the British conducted frequent searches to uncover these weapons. Even Dr Lehmann's youth village at Ben Shemen was not exempt from a search. One day in 1940,

Shimon Peres has written, 'we awoke to find the entire village sur-rounded by British soldiers, with plain-clothes CID men looking down at us from perches in the trees around the central courtyard.' His account continues:

> The rumour soon went round that they were looking for illegal Haganah weapons. We pupils were quickly instructed to assemble in a building called 'the store', where we were to pretend to be deeply engrossed in our studies. In fact, our role was to conceal an arms 'slick' hidden beneath the floor of the building. The tactic worked: the soldiers did not interrupt our 'lesson' to search the place, and the arms there were not found.
>
> But that was the only 'slick' that remained undiscovered. The British unearthed substantial stores of rifles, machine-guns and hand-grenades, and carted the whole lot off with them, along with Dr Lehmann and two or three of the senior staff.
>
> We were both shocked and proud. On the one hand, the 'siege' and search had left us somewhat traumatized, but on the other, we were pleased that so much weaponry should have been hidden in our little village, and that our Dr Lehmann wasn't quite so naïve a peacenik as we had thought. Clearly he took his responsibility for the security of the village very seriously indeed.

The Haganah work was continually disrupted, but never halted. Its head-quarters staff operated under a number of disguises. Yigael Yadin gained regular access to the British military base at Sarafand in his disguise as a milk and margarine delivery man. His wife Carmella, then a sergeant in the Auxiliary Territorial Service (ATS) of the British army, would hear the guard call out to her when Yadin drove up to the gate with his milk float: 'Sarge, the marge!'

At a meeting of the War Cabinet on 12 February 1940, Churchill, who had entered the government as First Lord of the Admiralty on the outbreak of war, raised the question, which both Weizmann and Jabotinsky had raised with him personally, of allowing the Jews of Palestine to be armed and trained in their own defence. Churchill told his former political opponents, now his colleagues, that, as the official minutes recorded, 'It might have been thought a matter for satisfaction that the Jews in Palestine should possess arms, and be capable of providing for their own defence. They were the only trustworthy friends we had in that country, and they were much more under our control than the scattered Arab population.' Churchill added that 'he would have thought that the sound policy for Great Britain at the begin-ning of the war would have been to build up, as soon as possible, a strong Jewish armed force in Palestine':

In this way we should have been able to use elsewhere the large and costly British cavalry force, which was now to replace the eleven infantry battalions hitherto locked up in Palestine.

It was an extraordinary position that at a time when the war was probably entering its most dangerous phase, we should station in Palestine a garrison one-quarter of the size of our garrison in India – and this for the purpose of forcing through a policy which – in his judgement – was unpopular in Palestine and in Great Britain alike.

Churchill was not successful. Lord Halifax, the Foreign Secretary, called his proposal 'discreditable', on the grounds that the problem was 'a much wider one' than Palestine alone, 'and had repercussions throughout the Muslim world'. Neville Chamberlain supported Halifax, saying, 'We certainly could not give the Jews alone freedom to arm.'

This same War Cabinet of 12 February 1940 discussed the 1939 White Paper decision to limit Jewish land purchase in Palestine. Once more, Churchill put forward the Zionist arguments. It was, he said, 'a short-sighted policy', which would 'put a stop to agricultural progress in Palestine'. Such a policy would not only be bad for agriculture, but would 'cause a great outcry in American Jewry'. The Secretary of State for War, Oliver Stanley, defended the restrictions on Jewish land purchase, telling his colleagues that it would be 'a good thing' for Palestine to slow down the intensive land cultivation. 'There was a great danger', Stanley asserted, 'of over-production of citrus fruits.'

Malcolm MacDonald supported Stanley. 'It would be an error of judgement', he believed, 'to exaggerate the influence of the Jewish element in the United States.' Churchill made one further appeal on behalf of maintaining Jewish land purchase rights from the Mediterranean to the Jordan, as laid down in the Mandate, telling his colleagues, as the minutes of the meeting recorded, 'that he, personally, ascribed to government encouragement very little of the credit for the great agricultural improvements which had taken place in Palestine. Broadly speaking, they were all the result of private Jewish efforts. So far as the Government of Palestine had played any part in agricultural development, it was with Jewish money wrung from the settlers by taxation.' Nor, the minutes continued, was Churchill impressed 'by the political grounds on which it was attempted to justify our action in bringing this great agricultural experiment to an end. The political argument, in a word, was that we should not be able to win the war without the help of the Arabs; he did not in the least admit the validity of that argument.'

Churchill's protest was unsuccessful, and the land purchase regulations were put into force. In a letter to the Secretary-General of the League of Nations on 28 February 1940, the Colonial Office in London wrote, 'it cannot be too often repeated that somehow and at some time the Jews and Arabs in Palestine will have to learn to live together in peace'. In the view of the

British government 'the continuation of wholly unregulated transfers of land from Arabs to Jews is bound to exacerbate the present differences between the two communities, and thus postpone indefinitely the harmony which alone can bring contentment in Palestine.' 'The effect of these regulations', Ben-Gurion declared that same day, 'is that no Jew may acquire in Palestine a plot of land, a building, or a tree, or any right in water except in towns and a very small part of the countryside . . . They not only violate the terms of the Mandate but completely nullify its purpose.'

On September 9 the European war reached Tel Aviv when Italian aircraft bombed the city killing 107 Jews. Yitzhak Rabin was returning home from the sea when the bombs exploded less than half a mile in front of him. He was horrified by the carnage, which he was to recall vividly while on a visit to London more than fifty years later.

Early in November 1940 two ships, the *Milos* and the *Pacific*, reached Haifa port with 1,771 'illegal' immigrants on board. The High Commissioner, Sir Harold MacMichael, refused to allow them to land, and they were transferred to a French ocean liner, the *Patria*, which the British had specially chartered in order to deport them to the Indian Ocean island of Mauritius. On November 20, while the immigrants were still being trans-ferred to the *Patria*, MacMichael broadcast a blunt communiqué setting out the new deportation policy. The British government, he declared, 'can only regard a revival of illegal Jewish immigration at the present juncture as likely to affect the local situation most adversely, and to prove a serious menace to British interests in the Middle East. They have accordingly decided that the passengers shall not be permitted to land in Palestine but shall be deported to a British colony and shall be detained there for the duration of the war.' The 'ultimate disposal' of the deportees, MacMichael added, 'will be a matter for consideration at the end of the war, but it is not proposed that they shall remain in the colony to which they are sent or that they should go to Palestine. Similar action will be taken in the case of any further parties who may succeed in reaching Palestine with a view to illegal entry'.

On November 24, while the *Patria* was still at anchor off Haifa, another immigration ship, the *Atlantic*, with 1,783 refugees on board, was escorted into Haifa Bay by the Royal Navy. On the following morning as the first 200 of those 'illegals' were being transferred to the *Patria*, explosives, planted by the Haganah to immobilize the ship and halt the deportation, blew up more forcefully than intended, and within fifteen minutes the *Patria* had sunk, drowning more than 250 refugees.

Not knowing what their fate would be, the survivors of the *Patria* explo-sion, and the illegals on board the *Atlantic*, were taken ashore. Karl Lenk, one of the *Atlantic* passengers, later recalled what proved for him to be a very brief moment of freedom (he was shortly to be deported by the British to a camp in Mauritius where he died before the end of the war). Recalling

the moment that he and his fellow illegals were brought ashore in Haifa, he
wrote:

How marvellous to stand on dry land once again, to breathe fresh air
and not to be sandwiched between bodies! The harbour was new and
the customs hall, bustling with activity, was spotlessly clean. We were
closely searched. All papers were impounded and receipts were issued.

We were loaded into buses and Jewish drivers were taking us to disin-
fection stations. Our clothes were treated separately. Some people's
heads were crawling with lice. All of us had our hair doused with a
pungent liquid. We had a hot shower and our own doctors looked us
over. It was only many months later that I learned how worried they
had been about my emaciated state and how delighted they were that
I had pulled through. I had been in such poor condition that the doctors
were not prepared to risk giving me typhoid and smallpox injections at
the time. Dr Kummermann thought that with my cough I would surely
soon be released anyway.

Back on the buses, we were driven over well surfaced suburban
roads. New buildings were rising everywhere. The street signs were in
English, Hebrew and Arabic. We passed modest sized houses
surrounded by colourful flowering shrubs. Heavily laden asses trudged
along the pavements led by Arab boys. A wrecked house here and there
bore silent witness to the prevailing state of civil unrest. Buses full of
children passed by.

A large Jewish school was being built. We saw the modern buildings
of a rural kibbutz set in obviously lovingly tended fields which
contrasted with the barren land further away, where the monotony
of the landscape was further enhanced by the scrub and the
occasional corrugated iron roofed shack. We passed through ancient
Akko, whose narrow streets were surrounded by a medieval wall. The
town stands on a tongue of land which is dominated by the green and
golden roof of the mosque and by the minarets which gleamed in the
sunshine. The bus skirted the town and we were relieved not to be
taken to its infamous prison. Eventually the bus took a left turn and a
barrack camp came into view, surrounded by a barbed wire fence.

We were taken to the 'Office for Refugees', which seemed reassuring,
inasmuch as there was another office which dealt with political prisoners. We were told that we would have to do our own cleaning,
cooking and baking. Police supervised the issue of rations and the
preparation of the food. The wooden barracks had corrugated iron roofs
and accommodated thirty men each.

The scale of the *Patria* tragedy led the British government to announce
that, whereas the remaining 1,600 refugees on board the *Atlantic* would still

be deported to Mauritius, the 1,900 refugees who had been on the *Patria* when it sank would be allowed to remain in Palestine. This decision led to an immediate protest from the British Commander-in-Chief in the Middle East, General Wavell, who telegraphed to the Secretary of State for War, Anthony Eden, on November 30: 'Have just heard the decision re *Patria* immigrants. Most sincerely trust you will use all possible influence to have decision reversed. From military point of view it is disastrous. It will be spread all over Arab world that Jews have again successfully challenged decision of British government and that policy of White Paper is being reversed. This will gravely increase prospect of widespread disorders in Palestine necessitating increased military commitments, will greatly enhance influence of Mufti, will rouse mistrust of us in Syria and increase anti-British propaganda and fifth column activities in Egypt. It will again be spread abroad that only violence pays in dealing with British.'

The reply to General Wavell came, not from the War Office or the Colonial Office, but from Churchill himself, who telegraphed to the General on December 2: 'Secretary of State has shown me your telegram about *Patria*. Cabinet feel that, in view of the suffering of these immigrants, and perils to which they had been subjected through the sinking of their ship, it would be necessary on compassionate grounds not to subject them again immediately to the hazards of the sea. Personally, I hold it would be an act of inhumanity unworthy of British name to force them to re-embark. On the other hand Cabinet agreed that future consignments of illegal immigrants should be sent to Mauritius provided that tolerable conditions can be arranged for them there.'

Churchill's telegram continued, 'I wonder whether the effect on the Arab world will be as bad as you suggest. If their attachment to our cause is so slender as to be determined by a mere act of charity of this kind it is clear that our policy of conciliating them has not borne much fruit so far. What I think would influence them much more would be any kind of British military success.'

Churchill's telegram was decisive, and Wavell's protest was overruled. The *Patria* deportees were allowed to remain in Palestine, first in the internment camp at Athlit then, within a year, at liberty. Nor was Churchill's judgement at fault, for on December 14 a military intelligence report on the effect of the *Patria* decision on the Arabs concluded that it had been 'remarkably small'.

The aftermath of this episode proved, however, a blow to the Zionists, for on December 26 the British government suspended the quota for legal immigration for three months, thus halting all immigration until March 1941. This decision was reached despite Churchill's insistence only two days before that the government, as he minuted, 'have also to consider their promises to the Zionists, and to be guided by general considerations of humanity towards those fleeing from the cruellest forms of persecution'.

Having received this clear indication of Churchill's attitude, the Permanent Under Secretary of State, Sir John Shuckburgh, minuted, that same day, in deciding not to inform Churchill of the suspension of the quota, 'Our object is to keep the business as far as possible on the normal administrative plane and outside the realms of Cabinet policy and so forth.' Subsequently, the quota for April to September 1941 was also suspended, and no immigration certificates issued for that period either.

The deportation of those who had reached Palestine on board the *Atlantic* proceeded. Karl Lenk was once again an eyewitness. 'At 2 p.m. on December 8,' he recalled, 'we were officially told that only those who had been on the *Patria* would be permitted to stay. The rest of us would be deported the following day. After all that we had been through we were determined to make a desperate gesture of defiance. Some of us were over eighty years old and it was more than doubtful that they would ever see their relatives if they were now carried off to a prison on a faraway tropical island.' Lenk's account continued:

> We were to get up at 5 a.m., ready for departure at six, with one piece of hand luggage and two blankets each. All protests were useless. It was the decision of the Mandatory government. They were trying to spirit us away befor the Yishuv could react to the *fait accompli* and before it could rally to our assistance. It was essential to gain time. We might be safe if we could manage to hold out until noon.
>
> The few Jews among the police had been taken off duty. Troops with rifles and bayonets were drawn up outside the barbed wire perimeter. We decided on passive resistance. Word was passed from barrack to barrack and also to the women's compound. We went to bed naked, left our luggage unpacked and refused to get up in the morning. We were making our protest all right, but there was little doubt about the final outcome – we just weren't in a shape to take on the Palestine police.
>
> Some older people urged compliance and argued that things would be worse if we made trouble (this kind of argument has long been the subject of many a wry Jewish joke!), but the youngsters were determined to resist and to show the British that we would not go like sheep to the slaughter after suffering unspeakable hardship to get to our ancient homeland. We tried to show that we meant to stay. The British showed us in turn that they were going to have their orders obeyed.

A British police officer appeared and again demanded that the intended deportees got dressed.

> The order was ignored by all except one man who had been most vociferous in urging resistance the night before. An hour had been gained, but there was still a long way to go until noon.

The police returned and repeated the order which was ignored yet again. They pulled off the blankets and began to wield their truncheons. I suffered a bruised elbow which presently became very sore. A second unit arrived carrying truncheons and batons. The futility of further resistance became apparent – our hut held mostly elderly people.

We dressed and started packing, but before we could finish another gang of thugs was upon us and drove us outside. I grabbed my attaché case and a couple of blankets; the rest stayed behind, but I could not have carried any more anyway. Those who were not out by then were thrown out by the scruff of the neck. Outside, baton wielding policemen chased several youngsters up the street and rained blows down upon them.

The youngsters were covered in blood as they were driven back screaming. They had occupied an adjoining hut and had refused to dress; they had crept out and run all over the camp stark naked urging continued defiance. They were caught one by one and beaten up until they collapsed bleeding. Finally they were thrown back inside. Now they did want to cover themselves, but the clothes and luggage had been taken away. Barefoot, wrapped in a blanket and still bleeding they passed the women's compound and the compound of the *Patria* survivors on the way to the bus. The latter had been kept indoors throughout.

The women were sobbing, but the police were having fun. They taunted us, laughed, jeered and shouted: 'Look at the bloody Jews!' One man had an epileptic fit and was thrown about like a sack of potatoes.

Both Churchill and the Jewish Agency had expressed the hope that the conditions of detention of the deported immigrants in Mauritius should not be onerous. As early as 13 November 1940, Churchill's Private Secretary, John Martin, had insisted, on Churchill's instructions, that the deportees should be 'decently treated in Mauritius'. But the government department responsible held a different view, which was expressed succinctly on 11 January 1941 in a Colonial Office minute which laid down that the conditions of the detention of the deportees 'should be sufficiently punitive to continue to act as a deterrent to other Jews in Eastern Europe'. The Jews deported to Mauritius were held in camps. Until Zionist pressure in London was able to effect a change (in 1944) conditions in the camps were harsh. Detainees were also prevented from volunteering for service with the Allied forces until late in the war.

As the war continued, the Zionists were appalled by the mounting campaign of terror against the Jews of Poland, and by the continuing refusal of the British government to modify in any way the immigration or land purchase restrictions of the 1939 White Paper. In February 1940, in a book entitled *The Jewish War Front*, Jabotinsky argued that the Jews were also an integral part of the Allied war effort and, with copious and disturbing quotations, showed that the terrible Jewish sufferings in Poland, although played down in the British press, were well known to all those who read the news agency telegrams, and thus to all those responsible for government policy. The Jews, Jabotinsky argued, must now work towards full statehood; and he added, 'The Jewish State is a true and proper war aim. Without it, the ulcer that poisons Europe's trouble cannot be healed: for without it there can be no adequate emigration of the millions whose old homes are irretrievably condemned; without it there can be no equality; and without this no peace.'

Jabotinsky died a few months later, in the United States, at the age of sixty. Not since the death of Herzl thirty-six years earlier was there such a sense of loss in the Jewish world, even among Zionists for whom Jabotinsky's Revisionism had been too extreme. His fund-raising activities in the United States in 1921, his prewar plan to evacuate one and a half million Jews from Poland to Palestine, his support for Jewish military units in both world wars, his writing and his oratory, marked him out as one of the giants of the Zionist movement. But the bitterness created by the Revisionist split and policies meant that when the State of Israel was established, eight years after Jabotinsky's death, Ben-Gurion refused to allow his body to be reinterred in Israel. It was not until 1964 after Ben-Gurion's retirement from politics that this decision was reversed. Today Jabotinsky's remains are buried in Jerusalem, in the military cemetery on Mount Herzl.

Jewish soldiers from Palestine had a part to play in the Middle East in the British war effort. For the military operation against the Vichy French forces in Syria, the British made use of specially trained Haganah 'shock troops' (Plugot Mahatz), known by their Hebrew initials as Palmach. One of those recruited by the Haganah to serve in the Palmach was Yitzhak Rabin, who was then with a group of Labour youth at a kibbutz north of Haifa, training to establish a new kibbutz. 'At the end of May 1941,' he wrote in his memoirs, 'there were rumours that German units had reached Lebanon, with the knowledge and consent of the Vichy government, and the long-awaited order arrived. By dusk the next day, I was in kibbutz Hanita, on the Lebanese border, together with about twenty equally puzzled but eager young men. In the kibbutz reading room we were met by a group of top-echelon Haganah leaders, including Moshe Dayan, Yitzhak Sadeh and Yaakov Dori.'

Rabin's account continued:

> Dori was the first to address us and told of the forthcoming British invasion of Greater Syria, including Lebanon, to prevent Axis forces from using the area as a spring-board for invading Palestine simultaneously from the north and south. In response to a British request, the Haganah had decided to co-operate in the campaign, and that is why we had

been brought to the border area. I was elated: at last I was about to take part in a battle on a global scale.

In truth, that fantasy was a gross exaggeration. We were divided up into two- and three-man sections and began foot patrols along the border until early June. Then my unit was informed that our task was to cross the border in advance of the Australian forces and cut the telephone lines to prevent the Vichy French from rushing reinforcements to the area . . .

At nightfall we crossed into Lebanon. The route to our objective and back was about fifty kilometres – to be covered on foot, of course.

As I was the youngest, I was given the job of climbing up the telephone pole. We had only received our climbing irons that day and hadn't had time to practise. Unable to use the irons, I took off my boots (which was the way I usually climbed), and shinned up the pole and cut the first wire, only to find that the pole was held upright by the tension of the wires. The pole swayed, and I found myself on the ground. But for lack of choice, up again I climbed, cut the wire, made my way down and repeated the operation on the second pole. Our mission completed, we buried the pieces of wire and made our way back to Hanita by a short cut, covering the distance quickly.

The story of the Haganah's participation in the invasion of Syria, Rabin reflected, 'might never have been remembered, even as a footnote to history, had it not been for the fact that on that same night, in a clash with a Vichy French force, Moshe Dayan lost his eye'. There was also a tragedy for the new Jewish force, when a commando unit of twenty-three men was sent by sea to attack the oil refineries in the Lebanese port of Tripoli. They set off in high spirits, and were never seen again.

As war came even to the borders of Palestine, the Jewish Agency continued to establish new settlements. In 1941 a group of youngsters from Germany set up kibbutz Yavneh, equidistant from the Arab towns of Yibnah and Isdud. A religious group, their kibbutz was to become the centre of the religious kibbutz movement. That same year the Haganah conducted a two-month training course for officers in the Carmel mountains.

In opposition to the Haganah, the Irgun believed that it must continue to fight the British in Palestine, and try to seize power. Avraham Stern, who had formed a breakaway 'Irgun in Israel' movement (also known as the Stern Gang), tried to make contact with Fascist Italy in the hope that, if Mussolini were to conquer the Middle East, he would allow a Jewish State to be set up in Palestine. When Mussolini's troops were defeated in North Africa, Stern tried to make contact with Nazi Germany, hoping to sign a pact with Hitler which would lead to a Jewish State once Hitler had defeated Britain. After two members of Stern's group had killed the Tel Aviv police chief and two of his officers, Stern himself was caught and killed. His followers continued on their path of terror.

David Ben-Gurion, returning to Palestine having spent a year in Britain and the United States, asked Arthur Ruppin to prepare material for a postwar Allied peace conference on Palestine. 'I hope that you will show that there is a way of bringing five million Jews to Palestine,' Ben-Gurion told him.

During 1941, as the German army was advancing through the Western Desert towards Egypt, prewar immigrants from Germany founded a kibbutz in the Negev, ten miles east of Gaza. The idea was to have a number of such settlements that might serve as a Jewish presence, should the British be forced to withdraw from Palestine. The name of the settlement, Dorot (Generations), was composed from the initials of the Labour leader and pioneer aviator Dov Hos, his wife Rivkah and their daughter Tirzah, who had all been killed the previous year in a road accident.

The mass murder of the Jews of Europe had begun with the German invasion of the Soviet Union on 21 June 1941. By the spring of 1942 as many as one million Russian Jews and many hundreds of thousands of Polish Jews had been murdered, and death camps had just begun operation, or were about to start, at four remote locations in German-occupied Poland: Chelmno, Belzec, Sobibor and Treblinka. To the last of these as many as three-quarters of a million Jews were to be deported and killed, including almost all the Jews of Warsaw, the largest prewar Jewish city population outside New York.

On 6 May 1942, as the first indications of the scale of these killings reached Britain and the United States, and as the British War Cabinet decided formally that 'all practicable steps should be taken to discourage illegal immigration into Palestine', an Extraordinary Zionist Conference was held at the Biltmore Hotel in New York. Its theme, expressed most forcefully by the Chairman of the Jewish Agency Executive, David Ben-Gurion, was that the Jews could no longer depend upon Britain to establish a Jewish National Home in Palestine, and that to secure this goal the Jewish Agency should replace the British Mandate as the Government of Palestine. When the conference ended on May 11 a majority of the delegates were committed to the establishment of Palestine as a 'Jewish Commonwealth', as well as an end to all immigration restrictions. The Biltmore Programme, as it was known, established for the first time the call for a Jewish State as the official policy of the Zionist movement.

On 18 May 1942, after another British War Cabinet discussion on illegal immigration, it was officially decided that 'no steps whatsoever should be taken to facilitate the arrival of "refugees" into Palestine'. It was the War Cabinet secretariat itself that put the word 'refugees' in inverted commas. Three weeks later, in a debate in the House of Lords, Lord Moyne, who had left the Colonial Office at the end of February 1942, argued that the Jews were not really a 'race' at all. 'It is very often loosely said that the Jews are Semites,'

Moyne told the House on June 9, 'but anthropologists tell us that, pure as they have kept their culture, the Jewish race has been much mixed with Gentiles since the beginning of the Diaspora,' and he added, 'during the Babylonian captivity they acquired a strong Hittite admixture, and it is obvious that the Armenoid features which are still to be found among the Sephardim have been bred out of the Ashkenazim by an admixture of Slav blood.'

To the Arabs, however, Moyne asserted, the Jews were 'not only alien in culture but also in blood'. Palestine could not absorb three million dispossessed Jews of Europe. 'Immigration on this scale would be a disastrous mistake and indeed an impracticable dream.' The Arabs, he said, 'who have lived and buried their dead for fifty generations in Palestine, will not willingly surrender their land and self-government to the Jews.'

Even as the war between the Allies and the Axis continued on land, at sea and in the air, and even as Hitler's war against the Jews gained in savage intensity, the relationship between the Zionists and the British was on a collision course. On 25 May 1942, at the annual dinner of the Anglo-American Palestine Committee, Dr Weizmann again declared – as he had done at the Biltmore Hotel less than three weeks earlier – that 'Palestine alone could absorb and provide for the homeless and the stateless Jews uprooted by the war'. Weizmann added, 'It is to canalize all the sympathy of the world for the martyrdom of the Jews that the Zionists reject all schemes to resettle these victims elsewhere – in Germany, or Poland, or in sparsely populated regions such as Madagascar.' It was Hitler who, in 1940, had first suggested Madagascar as a place where all the Jews of Europe might be sent, before the policy turned to physical extermination.

In Palestine, the Jewish Agency's work in settling the land continued. The war in Europe was not allowed to halt the creation of new kibbutzim. Many of those who were sent to build them had arrived from Europe before the war and had been trained at existing settlements by their respective kibbutz movements for the task of opening up a new corner of the country to Jewish cultivation.

In an attempt to populate the south, a kibbutz was set up in 1942 on the very edge of the Judaean hills, in an area surrounded by Arab villages, by East European immigrants who had arrived before the war. It was called Gat, after several Canaanite cities of ancient times (four such cities are listed in the conquests of Pharaoh Thutmose III). In the coastal plain, close to the Arab villages of the Samarian hills, Jews who had emigrated from Czechoslovakia just before the war set up a farming kibbutz. They named it Narbata, after the nearby ruins of the city in which Jews from Caesarea had sought refuge after the war against Rome in AD 66. They later renamed it Ma'anit (Furrow).

In the Huleh Valley, on the edge of malarial swamps, a kibbutz was established in 1943. It was named after the French Jewish Socialist and prewar

Prime Minister, Léon Blum, who was even then being held by the Germans in various concentration camps (he survived the war to become Prime Minister again). The founders of Kfar Blum were of two kinds: pioneers from the Baltic States and English youngsters from the Habonim movement. This somewhat unusual mixture led people to call Kfar Blum the 'Anglo-Baltic' kibbutz. Funds for its establishment had come largely from the United States, from the American Labour Zionist Order. It was they who wanted the kibbutz to bear the name of Léon Blum, who in 1936 had proposed a Popular Front of all political parties in France (including the Communists) to fight the growing Fascist tendencies in French political life.

Also founded in 1943 was Kfar Etzion, on the site of the Iraqi-Jewish kibbutz that had been abandoned in the Arab riots of 1929. An attempt had been made to reconstitute the kibbutz in 1935, by a Jewish citrus grove owner, S. Z. Holzmann, who wanted to turn it into a mountain village and holiday resort. Etzion was the Hebrew rendering of his name. But the 1936 rebellion had brought his project to a halt. The new founders of 1943, Orthodox Jews from Poland, were thus making a third attempt to create a Jewish presence in the Hebron Hills. Much of their work was the planting of forests on the stony hillsides. Three other villages – Massuot Yitzhak (Isaac's Beacon), Ein Zurim (Spring of Rock) and Revadim (Layers) – were soon added.

Another kibbutz founded in 1943, by a group of immigrants from Poland, was Yad Mordechai. Only eight miles north of the Arab city of Gaza, it was placed there in order to push Jewish settlement as far south as possible along the coastal plain (today it is two and a half miles from the border of the Gaza Strip at the Erez crossing point). It was named after the leader of the Warsaw ghetto uprising, Mordechai Anielewicz, who had just been killed in battle against the Germans, having led a group of armed Jews against the far superior German military regime that was deporting the last Jews from Warsaw to their deaths. The story of the heroism of the Warsaw Ghetto revolt gave courage to Jews all over the world that, despite the hopelessness of revolt inside Nazi Germany, the spirit of revolt was not dead, and the determination of men and women not to go down without fighting was as strong as at any time in Jewish history.

Just south of the Gaza–Beersheba road, a group of religious pioneers from Germany founded Be'erot Yitzhak (Isaac's Wells) on 9 August 1943. The Isaac from whom they took their name was Isaac Nissenbaum, a leading figure among religious Zionists in Poland, Lithuania and Latvia before the First World War, and one of the founders of the Jewish National Fund. He had been murdered in the Warsaw Ghetto in the previous year, at the age of seventy-four.

Also in 1943, to the south of Beersheba, fifteen young men and women from the Scout movement founded kibbutz Revivim. It was originally called Tel Zofim (Hill of Scouts). The later name, Revivim (Dew Drops) was taken from Psalm 65: 'Thou crownest the year with Thy goodness; and Thy paths drop fatness.' Revivim was the southernmost Jewish settlement at that time. Its founders hoped to master the harsh conditions of the Negev, and to create working relations with the Bedouin of the region. A severe shortage of water led them to work out a system of diverting the flash-flood waters from a nearby wadi to irrigate their small plantations of date palms, pomegranates and olives.

It was a chief of the Azazma tribe, Salama Ibn Said, who had sold the land for Revivim to the Jews. A poet from the same tribe, Ayyad Awwad Ibn Adesan, travelled from encampment to encampment reading a poem he had written opposing the sale of land and criticizing both Ibn Said and another chief who sold land to the Jews, Id Ibn Rabia. Adesan warned of the danger of the Bedouin being disinherited from their own space and freedom, and mocked the use to which the chiefs put their new found wealth:

> Look at Ibn Said and Rabia, Oh my!
>> They've built houses of stone, painted red and so high!
> Their wives stand around in a thin chemise gown,
>> Fried foods and soft bread are their only renown.

Adesan's poem also contained a warning:

> The land that was spacious, yet narrow will be,
>> You'll find nowhere to rest 'tween the hills and the sea.

*   *   *

In the late summer of 1943 there was a change in British policy, at Churchill's urging, which enabled several thousand Jews to enter Palestine from Nazi-dominated Europe. Any refugee who could get, by rail or sea, out of the Balkans to Istanbul (an escape route just opened up) could proceed to Palestine irrespective of existing quotas. Ehud Avriel, one of the Jewish Agency representatives in Istanbul, later recalled that at the end of August 1943, while he was in Istanbul, he received an official message through the British Embassy in Ankara to the effect that 'from now on every Jew who is able to reach the shores of Turkey under his own steam – repeat: under his own steam – is to be issued a visa to Palestine and to be sent on his way by British authorities'.

The first boat to benefit from Britain's new policy was the *Milka*, which had come from the Roumanian port of Constanta with 240 refugees from the Bukovina. To greet them at the Haidarpasha railway station, the terminus for Asia Minor and the Middle East, were four railway cars with a notice on them saying: 'Reserved for Passengers to Palestine'. Before the end of the year a second boat, the *Maritza*, also reached Turkish waters, bringing with it 224 survivors from the forced marches and concentrations camps of Transnistria. They too continued unmolested from Turkey to Palestine.

Despite his intense preoccupations with the daily conduct of the war, and with the mounting problems of landing a massive Allied force in northern Europe, Churchill had continued to follow the Jewish and Zionist dimension of the war, and to support the idea – which the Peel Commission had put forward in 1937, and which many Jews had also accepted, albeit reluctantly – of two separate States in Palestine, one Arab and one Jewish. On 12 January 1944 Churchill wrote to his senior War Cabinet colleagues, 'Some form of partition is the only solution,' and thirteen days later he informed the Chiefs of Staff Committee, 'Obviously we shall not proceed with any form of partition which the Jews to do not support.'

In 1944 the Jews of Palestine reeled from the impact of the terrible fragments of news that were reaching them almost daily about the Holocaust. That year the Warsaw-born poet Nathan Alterman, who had settled in Palestine in 1925 as a boy of fifteen, and was to become one of the outstanding modern Israeli poets, wrote a poem, 'Plagues of Egypt', in which he turned the ten plagues of biblical times – which are so central to the Jewish collective memory and Passover service – into a prototype of the history of mankind with its unending cycle of war and renewal, writing, in reflection of the mood of the time, 'Howling and disaster, Father, go endless on and on.'

When the few survivors who did reach Palestine during the war told their stories, there was unprecedented anguish. There was also a renewed determination that, when the war ended, a Jewish State would make it possible for Jews in distress wherever they might be to have a haven and a home. What the borders of that home should be was the subject of much discussion. Ben-Gurion was especially keen on an extension of Jewish settlement southward, even as far as the northernmost tip of the Red Sea, on the Gulf of Akaba (from which Solomon had sailed to the lands of the Queen of Sheba). Ten years had passed since his visit to Um Rashrash, at the head of the Gulf of Akaba. In 1944 he authorized the funds and Yitzhak Sadeh, head of the Palmach, outlined the plans for an expedition to that distant region, where the borders of the Palestine Mandate, the Hashemite Kingdom of Transjordan and Egypt converged.

A group of teenagers agreed to make the journey. They were joined by the senior scout of the Palmach, Chaim Ron, the zoologist Dr Mendelssohn, and the archaeologist Shmarya Gutman. The trek was to take three weeks, with map-making and exploration combined. One of the group was Shimon Peres, who later recalled:

> We had twelve camels, on which we piled our gear, and took turns to ride them. We also had revolvers and hand-grenades concealed under the false bottoms of our water canisters. We headed south towards the former Turkish border point, Um Rashrash (today's Eilat), but before we got there we began to be followed by a Bedouin camel patrol,

commanded by a British officer named Lord Asquith (the son of the former Prime Minister).

Since the Negev was a closed military area in which trekking was forbidden without a formal permit from the Mandatory Government – which, of course, we did not have – we naturally aroused their suspicions. In the end, we were arrested, piled into an army truck and taken back to Beersheba – then a wholly Arab town – where we found ourselves incarcerated in the local jail.

In Beersheba, Peres recalled, 'we were arraigned before a magistrate and sentenced to two weeks in prison. I, as leader of the illicit expedition, was additionally sentenced to pay a substantial fine. The escapade made newspaper headlines at the time as a daredevil sort of prank. But it was to prove its value during the War of Independence, when Ben-Gurion asked me for the maps we had drawn up – and asked Sadeh and me to plan a route for the army units sent southwards to take Um Rashrash.'

In January 1944 the Irgun called upon the Jews of Palestine to revolt. Their demand was 'immediate transfer of power in Eretz Yisrael to a Provisional Hebrew government'. Their method, expressed succinctly by their leader, Menachem Begin, was 'a prolonged campaign of destruction'. Yet Begin, like Ben-Gurion, realized that the Allied winning of the war was of paramount importance, and, whilst attacking several British police stations in August 1944 in search of arms, the Irgun decided 'not to attack military installations so long as war with Nazi Germany was in progress'. This decision, Begin wrote, 'was punctiliously obeyed until May 1945'.

On 28 September 1944 Churchill announced in the House of Commons that a Jewish Brigade Group would be set up, to be trained and armed by Britain as a front-line military unit. Five days later, on October 3, the Mufti of Jerusalem, who was then in Berlin, wrote to the head of the SS, Heinrich Himmler, proposing the establishment of 'an Arab-Islamic army in Germany'. The German government, Haj Amin added, 'should declare its readiness to train and arm such an army. Thus it would level a severe blow against the British plan and increase the number of fighters for a greater Germany.'

The Mufti was convinced that the setting up of an Arab army under German auspices 'would have the most favourable repercussions in the Arab-Islamic countries' and proposed November 2, the anniversary of 'the infamous Balfour Declaration', as the date of the public announcement. A senior SS officer had already reported to Himmler that in a conversation on September 28 the Mufti 'noted happily that the day is nearing when he will head an army to conquer Palestine'. But no such announcement was made, and the Mufti's army remained a figment of his own anti-Zionist imagination. The Mufti himself remained in Berlin, supervising Nazi propaganda broadcasts to the Middle East, and organizing parachute drops in

British-controlled areas. A month after the Mufti's letter to Himmler, four Arab parachutists were dropped into northern Iraq. Their aim was to carry out sabotage work against British oil installations. But the local village headman informed the police and the saboteurs were caught.

On 4 November 1944 Weizmann and Churchill met to discuss the future of Palestine. If the Jews could 'get the whole of Palestine', Churchill told the Zionist leader, 'it would be a good thing, but if it came to a choice between the White Paper and partition, then they should take partition'. Churchill also told Weizmann that 'he too was for the inclusion of the Negev' in the future Jewish State.

During their meeting, Churchill criticized the Jewish terrorism which had broken out against the British in Palestine. He went on to ask Weizmann to see the new British Minister-Resident in Cairo, Lord Moyne, who would, Churchill believed, be sympathetic. 'Lord Moyne', Churchill said, 'has changed and developed in the past two years.'

On 6 November 1944, before Weizmann could take Churchill's advice, Lord Moyne was assassinated in Cairo. His killers were two young Palestinian Jews, Eliahu Hakim and Eliahu Bet Tsouri, members of the Stern Gang. Accusing Moyne of a catalogue of crimes, including his speech of February 1942 in which he opposed large-scale Jewish immigration, and the expulsion of Irgun and Stern Gang members to Ethiopia, they committed what Churchill described in the House of Commons eleven days later as 'a shameful crime'. Churchill went on to say, 'If our dreams for Zionism are to end in the smoke of assassins' pistols, and our labours for its future to produce only a new set of gangsters worthy of Nazi Germany, many like myself would have to reconsider the position we have maintained so consistently and so long in the past.'

Weizmann shared Churchill's sense of outrage, as did the Jewish Agency Executive, which issued an immediate statement calling upon the Jewish community in Palestine 'to cast out the members of this destructive band, deprive them of all refuge and shelter, to resist their threats, and render all necessary assistance to the authorities in the prevention of terrorist acts and in the eradication of the terrorist organization'. There then began what was known as the 'saison': the period during which the Jewish Agency handed over 700 Irgun names to the British authorities. Most of these were then arrested. On November 11 the executive of the Histadrut, on which Golda Meir served, denounced both the Stern Gang and the Irgun as 'fascist'. Eight days later, at a meeting of the Zionist movement's highest parliamentary forum, the Small Zionist Actions Committee, Ben-Gurion produced a plan for 'disgorging' terror by helping the authorities to break the Irgun and Stern Gang entirely. Eliahu Golomb, the head of the Haganah underground, argued that the struggle was between 'Zionist democracy and Jewish Nazism'.

On December 16 Golomb held a secret meeting with Nathan Friedman-Yellin, one of the Stern Gang triumvirate. The most that Friedman-Yellin would promise was that the Stern Gang would not carry out any further terror actions while the two assassins were on trial (they were executed on 22 March 1945), and would not make any attempts on Churchill's life.

While British and Jewish leaders expressed their horror at Lord Moyne's assassination, the deliberate murder of Europe's Jews was reaching its terrible climax. For many British policy-makers, however, the overriding concern was still to stop Jewish refugees from reaching Palestine. On 24 December 1944 the new High Commissioner in Palestine, Lord Gort, telegraphed to the Foreign Office from Jerusalem to ask that the Soviet government – whose troops had entered the Balkans – be asked to close both the Roumanian and Bulgarian frontiers on the grounds that 'Jewish migration from South East Europe is getting out of hand'.

At the moment of Gort's vexation, Palestinian Jewry was making its hardest sacrifices for the British and Allied cause. Several dozen Palestinian Jews had volunteered to be parachuted deep inside German-controlled Europe. They undertook a double task: to carry out intelligence work for the British, and to make contact with the surviving Jewish communities. One of these parachutists, the Hungarian-born Hannah Szenes, was captured by the Germans on the Yugoslav–Hungarian border, and later tortured and then executed in Budapest on 4 November 1944. Another parachutist, Enzo Sereni, was murdered in Dachau two weeks later.

The Jews of Palestine were proud of what they had contributed to the Allied cause. More than 30,000 Palestinian Jews had served in the British army, and fought in Greece, Crete, North Africa, Italy and northern Europe. 'This record', a Zionist memorandum pointed out two years later, 'may be compared with the total of 9,000 Arabs who enlisted in Palestine, but who hailed partly from Transjordan, Syria and the Lebanon; long before the end of the war, this total was reduced by at least one-half through desertions and discharges.' Often, when the Zionists came up against British hostility, they would refer in similar tones to the disparity between the Jewish and Arab contribution to the Allied war effort.

A particular worry for those Palestinian Jews who were hoping to survive Arab hostility was the creation, on 22 March 1945, of an Arab League committed to the establishment of a 'United Islam' throughout the Arab lands of North Africa and the Middle East. The substantial differences between the members of the League were to find a focus of unity in the years to come in their denunciation of the establishment of a Jewish State. For a quarter of a century Zionists had quoted Lord Balfour's speech at the Royal Albert Hall on 12 July 1920 when he said, 'So far as the Arabs are concerned, I hope they will remember that it is we who have established an independent Arab sovereignty of the Hedjaz. I hope they will remember it is we who desire in

Mesopotamia to prepare the way for the future of a self-governing, autonomous Arab State, and I hope that, remembering all that, they will not grudge that small notch – for it is no more than that geographically, whatever it may be historically – that small notch in what are now Arab territories being given to the people who for all these hundreds of years have been separated from it.'

By 1945 there were five Arab countries in the Middle East: Syria (including Lebanon), Iraq, Saudi Arabia, Yemen and Transjordan. Between them, they had a total land area of 1,184,000 square miles and a total population of more than fourteen million. Compared with this, Palestine from the Mediterranean to the River Jordan was 10,500 square miles (equal to less than one per cent of the Arab territories that had been liberated from the Turks in 1918), with a population of 1,140,000 Arabs and 517,000 Jews. Even in Palestine the Jews were in a minority – twenty-eight years after the Balfour Declaration. Even if British restrictions on Jewish immigration were lifted, and the quarter of a million survivors of the Holocaust and other displaced persons then refused admittance were allowed in, the Arabs would still have been in a majority. It was an ill-omen for Jewish statehood.

The Jews played their part in all the Allied armies, and their losses were often heavy. Of the 500,000 Soviet Jewish soldiers who fought in the ranks of the Red Army, as many as 200,000 were killed in action. In addition, more than 750 Palestine Jews were killed fighting in the British ranks. Weizmann's younger son Michael was one of 2,500 British Jews who were killed: he had been reported missing while on a flying mission in February 1942. When the war ended on 8 May 1945, the 'victory' had a bitter, tragic taste for all Jews, and for the shattered remnant of Eastern European Jewry there seemed little hope of a decent future in the lands so drenched with the blood of their loved ones. The Jewish death toll will never be fully known. Among the million and a half children who were murdered were many whose names are unknown. Hundreds of thousands of individuals, whose names are found before the war in lists of doctors, nurses or teachers, students or lawyers, writers or poets disappeared without trace. The contribution they would have made to their countries of birth, and to Israel, is incalculable.

CHAPTER SEVEN

The British Mandate continues
1945–1946

On 8 May 1945 the German armies surrendered unconditionally. Germany and Austria were occupied by the Allied armies and divided into four Zones of Occupation: American, Russian, British and French. Dr Weizmann was full of hope that, as soon as the Middle East was discussed by the victorious Allies, a Jewish State would be on the international agenda, and that, just as in 1922 the League of Nations had confirmed the wartime British pledge to the Jews for a National Home, so in 1945 the victorious powers would move that pledge forward one more step to statehood. But whereas Churchill was sympathetic to the idea of Jewish statehood, this was true neither of the Conservative Party which he led, nor, in fact, of the Labour Party which was hoping to come to power, even though it explicitly supported Jewish statehood at its Party Conference in the autumn of 1944. A British General Election had been called for July 5. It would be the first in ten years.

On June 29 Churchill dictated a letter to Weizmann in which he warned the Zionist leader that one solution to 'your difficulties' might be the transfer of the Mandate from Britain to the United States which, he wrote, 'with her great wealth and strength and strong Jewish elements, might be able to do more for the Zionist cause than Great Britain. I need scarcely say I shall continue to do my best for it. But, as you will know, it has few supporters, and even the Labour Party now seems to have lost its zeal.'

Sensing this new mood, on June 26 Ben-Gurion declared, at a Press Conference in New York, 'If the British government really intends now to maintain and enforce the White Paper, it will have to use constant and brutal force to do so.' To attempt to maintain the White Paper policy would, he warned, 'create a most dangerous situation' in Palestine. Irgun activities were stepped up, the earlier Irgun pledge not to attack military targets until the end of the war having been regarded as fulfilled. On July 13 a British army truck carrying gelignite fuses was ambushed, and the constable on escort duty killed in the attack. On July 23, in a combined action, the Irgun and the Stern Gang blew up a railway bridge near Yibnah, on the British railway supply line from Egypt to Palestine.

On August 24 the officer administering the government of Palestine, J. V. W. Shaw, wrote to the new Colonial Secretary, Arthur Creech Jones – the Labour Party having come to power in Britain at the beginning of July as a result of the general election – quoting Ben-Gurion's New York speech and also the words of the American Zionist leader, Abba Hillel Silver, to the effect that 'Jews will fight for their rights with whatever weapons are at their disposal'. The picture, Shaw warned, was 'a sombre one', and he went on to explain:

> The young Jewish extremists, the product of a vicious education system, know neither toleration nor compromise: they regard themselves as morally justified in violence directed against any individual or institution that impedes the complete fulfilment of their demands.
>
> In a similar spirit their ancestors in the second century BC laid waste Palestine until a ravaged countryside and ruined cities marked the zenith of Hasmonaean power. The prototypes of the Stern Group and National Military Organization are the Zealots and Assassins according to whose creed even Jews married to Gentiles were worthy of death in Roman times.
>
> These Zealots of today, from Poland, Russia and the Balkans have yet to learn toleration and recognition of the rights of others. As the Foreign Secretary said recently of the Balkans, these people do not understand what we understand by the meaning of the word democracy.
>
> The Jewish Agency may deplore terrorism; but every immoderate speech, such as those quoted at the beginning of this letter, the flagrant disregard on the one hand for the authority of Government in maintaining law and order and on the other for the Arab case, the chauvinism and intolerance of their educational system all contribute to an atmosphere in which the fanatic and the terrorist flourish.

It was not the extremists, however, but the pioneer youth of the various Labour movements, with their deep love of the land and of democratic values, who were at the forefront of Jewish effort in Palestine as the war came to an end, and as hopes of statehood were revived. Labour Zionism had long been the main thrust of creative effort throughout the Land of Israel. During 1945 a group of such youngsters established Misgav Am (Stronghold of the People) at the very north of Upper Galilee, on the Lebanese border. It was 2,770 feet above sea level, and could be reached only by steep footpaths.

Even if attempts in the past to reach good relations with the Arabs had failed, there was still a strong desire among the Labour Zionists to live together with the Arabs, and not, as many of the extremists hoped, to make them subordinate to Jewish nationalist needs, or even to drive them out of Palestine altogether.

Following the German surrender, Jews in Europe who had been liberated from the concentration camps faced an uneasy future. Few had any family left alive, or homes to go to; throughout Eastern Europe, the homes which had been confiscated from the Jews by the Germans were now lived in by Poles, Czechs, Hungarians, Ukrainians, Lithuanians and Latvians. The Jewish survivors joined an estimated eight million people who had been displaced by war. Known as Displaced Persons (DPs) they found refuge in camps set up specially for them by the United Nations Relief and Rehabilitation Agency, UNRRA.

The Jewish DPs began to organize themselves and to make demands of their own. They wanted the Jewish people to be recognized as a people, not, as hitherto in the DP camps, as Poles, Hungarians, Czechs etc. They also demanded the establishment of a Jewish State, and their own emigration to it. Calling themselves She'erit Ha-Peletah (the biblical 'surviving remnant', or 'the saving remnant'), they elected camp committees, published newspapers, and set up branches of the Zionist Organization inside the camps. In July 1945 they held a conference at the St Ottilien DP camp near Munich. Ninety-four delegates came, from camps in the American, British and French Zones of Occupation in Germany and Austria. At the conclusion of the conference, the participants called for the immediate establishment of a Jewish State in Palestine, the recognition of the Jewish people as an equal member of the Allied nations, and Jewish participation in the peace negotiations.

On 1 August 1945, a date that was to prove a turning point in the path to Jewish statehood, Earl G. Harrison, the dean of the Faculty of Laws of the University of Pittsburgh, produced, at the urgent request of President Truman, a report on the state of the DPs in Europe. The Jews did not want to be repatriated to Poland, Harrison explained, for anti-Semitism had already led, since the end of hostilities in May 1945, to the murder of several hundred Jews who had been making their way back to their former homes. Physical attacks on Jews in Poland were on the increase.

In his report, Harrison pointed out the harsh conditions in the DP camps, and the overwhelming desire of the Jewish DPs to go to Palestine. But British regulations, which were being strictly enforced, made this impossible. The Jewish Agency had demanded 100,000 Palestine certificates for the DPs. Britain had refused. 'The only real solution of the problem,' Harrison wrote, 'lies in the evacuation of all non-repatriatable Jews in Germany and Austria, who wish it, to Palestine.'

Truman accepted Harrison's report and pressed the British to issue the required Palestine certificates. The British Labour government again refused. But the plight of the DPs had become a feature of the international agenda. In October 1945, two months after the publication of the Harrison report,

Ben-Gurion, as Chairman of the Jewish Agency Executive, visited the DP camps, accompanied by General Eisenhower. Ben-Gurion was forcibly struck by the determination of the DPs to reach Palestine. It was this desire that had led many of them to make their way from Eastern Europe to the DP camps in Germany, hoping that this would prove a way-station on the road to Palestine.

At Ben-Gurion's suggestion, and with Eisenhower's approval, the American military commanders in Germany agreed to set up a 'temporary haven' for the DPs in the American Zone of Occupation. At the very moment when the British were refusing to allow any further Jews to cross into their zone from the east, the Americans provided a lifeline. When UNRRA conducted a poll among the Jewish DPs, it emerged that 96.8 per cent wanted to go to Palestine.

As 1945 came to an end, the Jewish Agency continued to press for an end to the immigration restrictions of the 1939 White Paper. In London, a special Palestine Committee of the Cabinet reported on October 19 that Britain would support the appointment of an Anglo-American Committee to examine 'the situation of the Jews in Europe' and to make recommendations 'for its alleviation'. However, the Chairman of this committee, Herbert Morrison, the Deputy Prime Minister, and a Labour Party stalwart since before the First World War, warned in his report of Muslim opposition to any relaxation of the immigration restrictions. President Truman's wish that the survivors of the concentration camps should be allowed into Palestine had, Morrison pointed out, already led to protests, not only from the Arab States but also from the Muslims of British India.

On October 10 Weizmann went to see the Foreign Secretary, Ernest Bevin, in the hope of changing the British government's mind, but his effort was in vain. A month later, on November 13, Bevin announced that the 1939 White Paper policy restrictions would continue.

In the first weeks of 1946 the British government increased its efforts to prevent Jews leaving Europe for Palestine. One main route led through the British Zone of Austria into Italy. On 10 January 1946, in an attempt to block this route, Sir Noel Charles, the British Ambassador in Rome, proposed to the Foreign Office in London 'that I be authorized officially to inform the Italian government that we are concerned at the growing illegal emigration traffic and would welcome any measures they can take to check it. I could then suggest that the Italian government should institute an exit visa regulation which would give them a basis for taking action against illegal emigrants.' Sir Noel added, 'I should also like to be able to assure the Italian government that we are doing our utmost to check clandestine influx of refugees from Austria.'

The head of the Refugee Department at the Foreign Office, Douglas Mackillop, had a rare insight into the reasons for the exodus of Polish Jewry, noting on January 12 that it arose 'partly for racial and economic reasons, readily understandable since the new Poland does not offer them the same opportunities as the old, and there is a spontaneous general wish on the part of European Jewry to go to Palestine'. Mackillop added, 'Though it is magnified and artificially fostered by Zionist propaganda, it is a real aspiration.'

Despite this unusual Foreign Office understanding of the 'real aspiration' of Polish Jewry, the British Cabinet continued to exert every possible pressure to keep the Polish Jews inside Poland. On 25 January 1946 the Cabinet's Overseas Reconstruction Committee met in London, with Ernest Bevin as chairman. The purpose of the meeting was to discuss what Bevin described as 'the serious problem' caused by the exodus of Jews from Eastern Europe into the British and American Zones of Germany and Austria, and from there into Italy. The Chancellor of the Duchy of Lancaster, John Burns Hynd, told his colleagues that during his own recent visit to Germany 'he had spoken with Jews at Belsen Camp, and these were all in favour of going to Palestine'. Hynd gave as his personal advice that it would be 'undesirable to let it be thought that there would be an early opportunity for large numbers of Jews to go to Palestine, as this would not only encourage the present movement, but would also encourage Polish Jews forming part of the large mass of Poles who were now being repatriated from our Zone in Germany to Poland, to refuse to go home'.

To prevent Jews reaching Palestine had become a Labour government priority, some said obsession. The Foreign Office had considered sending 'an appeal' from the British to the Soviet government 'to suspend the movement of Polish Jews from Poland', Bevin told the Cabinet Committee, but felt that such an approach would be 'unprofitable'. It had also considered asking the Polish government to create 'more tolerable conditions' for Jews already in Poland, but feared that if Britain did this the Poles 'would resent the imputation' of anti-Semitism. Bevin, however, did not believe that Polish anti-Semitism was the principal cause of the Jewish desire to leave. According to his information, he said, 'most of the Jewish migrants from Poland were influenced by political rather than racial motives in their efforts to reach Palestine'.

A struggle now began, in Palestine and in Europe, between the British government and the Jews. It was a struggle marked by increasing bitterness and extremism on both sides. The British, determined to halt the now swelling tide of 'illegal' immigration from liberated Europe to Mandatory Palestine, went so far as to return captured immigrants from the waters of the eastern Mediterranean to which they had sailed, to the DP camps in Germany from which they had fled. In Europe itself, at the frontier crossings between Austria and Italy, British troops halted concentration camp survivors who were on their way to the Adriatic and to Palestine and held them in former prisoner-of-war camps.

Inside Palestine the Irgun reacted by attacking British military installations. As the subsequent death sentences on those caught, and the resulting reprisals instituted by the Irgun, increased the tension, acts of even greater violence marked their efforts. Force alone, the Irgun believed, could bring an end to the White Paper, open Palestine to unimpeded Jewish immigration, and make possible a Jewish majority, and Jewish statehood. When Bevin himself publicly warned the Jews not to 'push to the head of the queue', it was not the homeless and displaced alone who saw this remark as a crude anti-Semitic quip, which grossly belittled the reality of the refugee plight.

In Palestine even teenagers caught pasting up posters faced the full rigour of the mounting bitterness and hatred. One was Asher Tratner, a pupil at Haifa High School, who was shot and wounded in the hip while, unarmed, he was putting anti-British posters on walls. Menachem Begin, the Irgun commander, recalled in his memoirs how Tratner, although wounded, was neither seen by a doctor nor sent to hospital. Instead, Begin wrote:

> He was dispatched, his wound open and bleeding, to the Acre Jail. The wound festered. His jailers tied him to the bed. The boy had to wipe the blood and pus from his wound with strips torn from his shirt. The guards continued to maltreat him.
>
> I was told by Rabbi Blum, whom the authorities had appointed prison chaplain, that he had drawn the attention of the British to the critical condition of the young prisoner. The reply was characteristic: 'Rabbis should concern themselves with the souls of the prisoners, not with their bodies. Mind your own business.'
>
> Asher's spirit was not broken, but his body was destroyed. When at last the prison doctor was brought he was diagnosed with severe blood poisoning and the boy was removed to hospital. But it was too late. Even amputation of his leg did not save Asher. After weeks of suffering in the Acre Jail and the Haifa hospital, he died.

*   *   *

The Jewish response to the continuing immigration restrictions was to accelerate the mass movement of 'illegals' across the borders and mountain passes of Germany, Austria and Italy, and over the seas, from ports in the Adriatic and the Aegean, in the hope of being able to land, secretly and unnoticed, on the coast of Palestine. The poet Nathan Alterman caught the mood of those who helped in this task in a poem which he dedicated to 'The Italian captain' of one of the ships:

> The work of our boys is wrapped in the secrecy of night.
> Let us bless it as we do our daily bread.
> See now from ship to shore

They carry their people on strong shoulders.
Here's to this cold, staunch night!
To this dangerous, backbreaking life!

Here's to the little boats, captain!
And to the others on the way!
To the boys who got the command
To steer their craft through the fog
At the right time, to the right place
Without compass or map.

About them a certain story will be told
By the waves and the open sky.
How they fought their people's Battle of Trafalgar
On a lonely boat cutting through the water.
The clouds float above. The wind holds firm.
The work is done, as the sky's overhead!
Let's drink to it, captain,
To our next meeting on the seas.

A day will come, and you'll sit in a corner of a tavern.
You'll sit, Granddad, with a bottle of Chianti,
And you'll smile, chewing your tobacco,
And say: Well, fellows, I'm an old man now.

Between August 1945 and May 1948 as many as 40,000 Jews made their way to Palestine clandestinely. Zionist emissaries from Palestine helped to organize the crossing of borders, false documents, reception areas, and holding camps along the way. Between August and December 1945 eight small boats reached Palestine from ports in Italy and Greece, with 1,040 immigrants in total. With few exceptions, they managed to enter Palestine without being caught by the British. The inhabitants of the Jewish towns and villages near the coast where they were put ashore quickly spirited them away, as they had done their predecessors before and during the war.

From January to July 1946 a further eleven boats brought 10,500 immigrants. Intercepted by the British, they were interned in a detention camp at Athlit, but were allowed to enter Palestine soon afterwards. In the DP camps in Europe, tens of thousands more Jews waited for the day when they would be allowed to go to Palestine without restriction. Helped financially by the American Jewish Joint Distribution Committee (The Joint), Hebrew schools were established in the DP camps, which gave education to 10,000 children, and kindergarten activities to 2,000 more. There were newspapers, theatrical performances, orchestral music, sports, and even agricultural training. But the conditions in the camps were not good, the prolonged confinement and uncertainty after so many terrible experiences breeding dissension and bitterness.

The Anglo-American Committee of Inquiry was continuing its researches, visiting Europe and Palestine. Its most harrowing time was in the DP camps. One member of the committee was a British Labour Member of Parliament, Richard Crossman. 'Did the DPs really want to go to Palestine?' Crossman had asked himself before setting out. 'Or was this idea the result of Zionist propaganda?' he wondered. As the committee members visited the camps, they realized that Zionist emissaries had already been there, and had taken charge of much of the educational and social work in them. 'These Zionists had already shown themselves to be a powerful force,' Crossman commented. 'They had even successfully prevented a United Nations plan to send unaccompanied Jewish orphans to care in Switzerland. This action had greatly angered the policymakers in London, and even Washington.' But, Crossman recalled:

> Even if there had not been a single foreign Zionist or a trace of Zionist propaganda in the camps these people would have opted for Palestine. Nine months had passed since VE-Day and their British and American liberators had made no move to accept them in their own countries. They had gathered them into centres in Germany, fed them and clothed them, and then apparently believed that their Christian duty had been accomplished.
>
> For nine months, huddled together, these Jews had had nothing to do but to discuss the future. They knew that they were not wanted by the western democracies, and they had heard Mr Attlee's plan that they should help to rebuild their countries. This sounded to them pure hypocrisy. They were not Poles any more; but, as Hitler had taught them, members of the Jewish nation, despised and rejected by 'civilized Europe'. They knew that far away in Palestine there was a National Home willing and eager to receive them and to give them a chance of rebuilding their lives, not as aliens in a foreign state but as Hebrews in their own country. How absurd to attribute their longing for Palestine to organized propaganda! Judged by sober realities, their only hope of an early release was Palestine.

The Anglo-American Committee visited the camps amid much hostility. This was not surprising, for, Crossman recalled, the inmates 'regarded the British officials as prison warders', and had no appreciation for a statesman, Bevin, 'who forbade them to go to Palestine and offered them no other homeland'. At a camp near Villach, in southern Austria, the committee came across several hundred Polish Jews – men, women and children – who had been prevented by the British from crossing the nearby border into Italy, and had been held for seven months in a disused prisoner-of-war camp, guarded and confined by British troops. Crossman recalled:

> The camp policeman, elected by the inmates, was a Polish boy of sixteen who had spent six years of his life in concentration camps. His had been the survival of the fittest, of the personality able to cajole, outwit or bribe.
>
> I asked him if he had relatives in America and he replied that his mother was there. I asked him whether he wrote to her and his handsome face contorted with passion: 'I have cut her off, root and branch. She has betrayed the destiny of my nation.' I asked him what he meant, and he replied: 'She has sold out to the Goy. She has run away to America. It is the destiny of my nation to be the lords of Palestine.' I asked him how he knew this was the Jewish destiny and he replied: 'It is written in the Balfour Declaration.'
>
> It was useless to argue with that boy or to tell him that the Balfour Declaration had promised no such thing. He was the product of his environment, of six years under the SS in concentration camps and nine months under the British at Villach.

The Anglo-American Committee proceeded from Europe to Palestine, where it continued to take evidence. Among the Zionist leaders who appeared before it was Golda Meir, who told the Committee that the 160,000 members of the Histadrut were willing to receive 'large masses of Jewish immigrants, with no limitations and no conditions whatsoever'. She went on to try to explain to the Committee 'what it means to be the member of a people whose every right to exist is constantly being questioned'. The Jewish workers in Palestine, she said, 'have decided to do away with this helplessness and dependence upon others within our generation. We want only that which is given naturally to all people of the world, to be masters of our own fate – *only* of our fate, not of the destiny of others; to live as of right and not on sufferance, to have the chance to bring the surviving Jewish children, of whom not so many are left in the world now, to this country so that they may grow up like our youngsters who were born here, free of fear, with heads high.'

Golda Meir added: 'Our children here don't understand why the very existence of the Jewish people as such is threatened. For them at last, it is natural to be a Jew.'

While the Anglo-American Committee was preparing its report, which it was believed would include the recommendation of an immediate grant of 100,000 immigration certificates to the Jews of the DP camps, British pressure on the Italian government grew. In April 1946, as had been suggested by the British Ambassador in Rome, a diplomatic protest from London led to the Italians refusing to allow two 'illegal' immigrant ships, called by the Jews the *Dov Hos* and the *Eliahu Golomb*, to leave the port of La Spezia, on the Italian Riviera. For their part, the 1,014 refugees, most of whom were from Poland and Central Europe, refused to leave the ships. Instead, they called a hunger strike. If force were used against them, they said, they would commit mass suicide, and sink the ships.

In Palestine, Golda Meir proposed a hunger strike by fifteen Zionist leaders, as a gesture of support with the refugees, and as a means of forcing the British to allow the ships to sail. Before starting the strike, the Zionist leaders went to see the Chief Secretary of the Palestine government, Henry Gurney. In her memoirs, Golda Meir recalled: 'He listened, then he turned to me and said, "Mrs Meyerson, do you think for a moment that His Majesty's Government will change its policy because *you* are not going to eat?" I said, "No, I have no such illusions. If the death of six million didn't change government policy, I don't expect that my not eating will do so. But it will at least be a mark of solidarity."'

The hunger strike was, in fact, successful. On May 8 the *Dov Hos* and the *Eliahu Golomb* sailed for Palestine, their passengers having been granted immigration certificates for the next month's quota. But even as this small success had been notched up by the Zionists, the Haganah received information that the British intended to deport 200 'illegal' immigrants – almost all of them survivors of the Holocaust – who were being held in the detention camp of Athlit, on the coast. The Palmach was instructed to force a way into the camp and to release the immigrants. They would then be taken to a nearby kibbutz and dispersed throughout the country.

Yitzhak Rabin was chosen as deputy commander of the operation. He later recalled how the Palmach 'took advantage of the fact that the British permitted welfare workers and teachers to enter the camp by infiltrating a group of Haganah physical training instructors. Their mission was to organize the immigrants and overpower the Arab auxiliaries guarding the perimeter so that the raiding force could break in.' His account continued:

I commanded the assault force that set out on a moonless night and halted about ninety metres from the fence of the brightly lit camp. We cut the wire and before reaching the second, inner fence we ran into our 'teachers', who reported that they had managed to break the firing pins in the Arab auxiliaries' rifles.

The Arabs cocked their guns, pressed the triggers, and nothing happened. Quickly we forced our way in and hurried past the immigrants' quarters to the British billets. There was no sign of an alert. The plan had succeeded beyond expectations: the British were fast asleep.

Once the immigrants had assembled, the pull-out commenced in the incomprehensible and menacing silence. Our battalion commander ordered me to remain in the camp for about half an hour until the immigrants could reach the trucks. It was a bizarre sensation: the camp was brightly lit and silent, and a Jewish auxiliary policeman walked right past us, determined to see nothing.

At a quarter to two we withdrew, running as fast as we could to catch the trucks. (In fact we reached them before the immigrants did.) But our detailed planning had overlooked two difficulties: one psychological, the other physical. The immigrants refused to be parted from their bundles, the only possessions they had left; and the infants and toddlers, who had to be carried, hampered their parents' movements.

The battalion commander decided to leave with those immigrants who had reached the assembly point and ordered me to wait for the rest and bring them to Kibbutz Beit Oren, on the Carmel ridge. Then a passing British truck opened fire and was silenced, resulting in the death of a British sergeant (the only casualty of the whole operation). To mislead the British, the commander sent the trucks in one direction and led his batch of immigrants off in another.

I began to muster the hundred or so survivors whose fate was now in my hands and moved off with about sixty soldiers from various platoons. We made slow progress scaling the Carmel, and I ordered the troops to carry the children on their shoulders. I picked a child up myself. It was an odd feeling, to carry a terrified Jewish child – a child of the Holocaust – who was now paralysed with fear. As my shoulders bore the hopes of the Jewish people, I suddenly felt a warm, damp sensation down my back. Under the circumstances, I could hardly halt.

Dawn broke as we crawled along, and it would be full light before we could reach Beit Oren. We prepared to hide in the woods throughout the day, and I sent two of my people to reconnoitre the vicinity of the kibbutz. They reported having found a gap in the British encirclement of the settlement and thought we could manage to get through. I decided to try.

We filtered through as quickly as possible, then dispersed the immigrants and hid our weapons in previously prepared caches. The British brought up reinforcements and tried to break through the kibbutz gate, but in the meanwhile thousands of citizens from Haifa streamed out to Beit Oren and the kibbutz to which the battalion commander had led his group. The British put up road blocks, but they were reluctant to open fire on such a multitude. By afternoon the whole area was teeming with people, and the immigrants were swallowed up in a human sea. The British conceded defeat.

*   *   *

In searching for places other than Palestine into which Jewish survivors could be sent, the British government ruled out Britain itself. During a Cabinet meeting on 30 October 1945 the Home Secretary, Chuter Ede, had made it clear that, on the basis of the Labour Party's declared immigration policy, it would not be 'right' for Britain to contemplate 'any large-scale addition to our foreign population at the present time'. Some liberated youngsters had already been allowed in and were in special hostels run by Jewish voluntary organizations.* But the Home Secretary stressed that 'any large-scale extension of schemes of this kind would involve unjustifiable demands on available housing accommodation.'

In addition to considerations of the availability of housing and clothing, the Home Secretary added, 'there is the consideration that at a time when demobilization is in process, the admission of a large number of foreigners would create apprehension as to competition in the labour market'. The survivors would have to look elsewhere than either Palestine or Britain. Even the United States maintained its quotas. Madagascar was suggested, not for the first time, as a possible destination for the Jews; Hitler had toyed with the idea in 1940. On 6 April 1946, however, the British Consulate-General in Madagascar reported in confidence to the Foreign Office in London that while Madagascar might be suitable for 200 colonists 'of the peasant class', stress should be laid by Britain 'on providing the right type of colonist in the first instance, and not city-bred Jews who were worn and emaciated through long confinement in concentration camps.'

On April 20, exactly two weeks after this discouraging communication, the Anglo-American Committee of Inquiry completed its report, urging the British to end the land-purchase restrictions imposed on the Jews as part of the 1939 White Paper, and to grant 100,000 Palestine certificates immediately. The British again refused to allow immigration on anything approaching that scale.

One result of this deeply frustrating impasse was an experiment in unity: the establishment of a Jewish Resistance Movement, under Ben-Gurion's authority, whereby the Irgun and Stern Gang agreed to work hand in hand with the Haganah, with a coordinated strategy. This policy reached its peak of effectiveness on the night of June 16–17, when the Palmach isolated Palestine from the neighbouring States by destroying ten road and rail bridges, and damaging the Haifa railway workshops. Twelve days later, on June 28 – later known as Black Saturday – the British retaliated by sealing the Jewish Agency buildings and arresting 3,000 Jews throughout Palestine, including most of the senior members of the Zionist Executive.

Two days after the arrests, Richard Crossman – whose work on the Anglo-American Committee had ended – told a hostile House of Commons, 'I must say frankly to the House that what we are trying to impose on the Jewish community is a reimposition of the White Paper, something which no Jew in Palestine accepts as either law or order. This affects not only the extremists of the Left but also of the Centre. No Jew anywhere, least of all Dr

---

* I have written about these youngsters, 732 in all, in *The Boys, Triumph over Adversity*, Weidenfeld and Nicolson, London (and Holt, New York), 1996.

Weizmann or the Haganah, can be won over to support the government by the arrests of thousands of their brothers.' It was 'impossible to crush a resistance movement', Crossman added, 'which has the passive toleration of the mass of the population'.

The British did succeed, that June 29, in discovering a series of underground storerooms in kibbutz Yagur, where 600 rifles, light machine-guns, pistols and small mortars had been hidden. 'It was difficult to tell,' wrote a British major who took part in the raid, 'who were more surprised – the Jews whose ingenuity had not been quite good enough, or the troops who were awed by the sight of a find of such unexpected proportions. As it was the first really successful arms search, troops of other formations, including the 6th Airborne Division, visited the scene in order to see for themselves the various devices by which weapons had been concealed so that in the event of further searches, they would have some idea of how to set about the task.'

The Jews intensified their efforts at concealment, while at the same time starting clandestine arms manufacture, making their own Sten guns, handgrenades and two-inch mortars.

Among those who had reason to remember Black Saturday was Yitzhak Rabin. A few weeks earlier, on the eve of a Palmach attack on the headquarters of the Police Mobile Force – one of the most active searchers-out of clandestine arms – his motorcycle had crashed and his leg was badly hurt. After three weeks in hospital in Haifa he was sent home to Tel Aviv on crutches. That Saturday, he later wrote:

> I was awakened at dawn by the roar of vehicles in the street. Soon there was a sharp knock at the door and a British paratroop captain (we called them 'anemones') asked, 'Rabin?'
>
> My father nodded in the affirmative, whereupon three squads of paratroopers burst into the flat, armed with Brens and sub-machine guns. Another platoon had surrounded the house with barbed-wire barricades. This was an imposing military operation!
>
> Together with my father and a visitor who was staying with us, I was dragged out to a British army truck and barely managed to clamber aboard. We were taken to a nearby school, where I saw Moshe Sharett, the head of the Jewish Agency's political department. He, too, had been arrested. In fact, as only later became clear to us, so had most of the members of the Jewish Agency Executive (fortunately Ben-Gurion was abroad at the time) and thousands of people suspected of belonging to the Palmach. The operation became known in the annals of our modern history as 'Black Saturday'.
>
> At first we were taken to a tent camp at Latrun, halfway between Tel Aviv and Jerusalem, and two days later we were driven to large warehouses in the Rafa area, at the southern end of what is now the Gaza

Strip. During the next few days some 1,600–2,000 Jewish detainees were brought to the camp from all over the country.

I was particularly troubled by my father's detention, for which I was to blame. He was in a state of distress because he had been hauled off before he could get to his false teeth, and consequently he could scarcely eat a thing. Fortunately, two weeks later he was released, and after a month our visitor was also freed.

In the camp, we began to get ourselves organized. Dr Chaim Sheba, who had been sent by the Jewish Agency to take care of us, examined me and had me sent to the camp hospital. Later he tried to procure my release, but the head of British intelligence retorted, 'He'll remain in detention even if he breaks both legs!'

Of the Jewish Agency leaders, only Ben-Gurion escaped arrest on Black Saturday, as he was in Paris. While he was there, depressed at what had happened to his colleagues, he met a very elated Ho Chi Minh, who believed (wrongly) that he was about to achieve independence for the people of Vietnam. When Ben-Gurion explained about the arrests, and the struggle between Britain and the Jews of Palestine, Ho Chi Minh offered him another, apparently more immediately obtainable, homeland: the highlands of central Vietnam. Ben-Gurion politely declined.

On 4 July 1946, forty-two Jews were murdered by Poles in the town of Kielce. Most of those murdered had made their way to Kielce, not to live there, but as a brief staging post on their way to Palestine. The 'Kielce pogrom', as it quickly became known – in an echo of the violent mob attacks on Jews in Tsarist Russia – led to a massive upsurge in the number of Jews – survivors of the Holocaust – trying to get out of Poland and Eastern Europe. Within twenty-four hours of the Kielce murders, 5,000 Jews were on the move from their homes in Poland towards the Czechoslovak border. These new refugees also were on their way to Palestine. But their route through Austria was barred at the entrance to the British Zone of Occupation, which lay astride the road and rail route to Italy. By the end of 1946 the 100,000 Jewish DPs in Germany had been augmented by a further 130,000.

In Palestine, the arrest of the Zionist leaders had created a ferment of discontent. Weizmann alone, of the leaders then in Palestine, had not been seized. On July 9, as President of the Jewish Agency, he issued a statement on behalf of those who had been imprisoned. The excuse for the arrests, he said, for the seizure of the Jewish Agency archives and for the countryside searches and arrests, had been the 'deplorable and tragic' acts of Jewish terrorism of recent months. Yet these acts 'have sprung from despair of ever securing, through peaceful means, justice for the Jewish people'.

Within three weeks the violence reached a climax. On July 22 the Irgun blew up a wing of the King David Hotel in Jerusalem. Ninety-one people were killed, including British administrators working in the hotel. Many of these administrators were Arabs and Jews. There was a shock of horror among Jews and Arabs alike in Palestine, and among critics of Jewish terrorism throughout the Diaspora, and in the non-Jewish world.

The British dead included the Postmaster-General of Palestine, G. D. Kennedy, a veteran of the retreat from Mons in 1914. One of the Arabs killed, Jules Gress, a senior assistant accountant with the Secretariat, and a Catholic, had been an officer in the Turkish army in the First World War, when he was taken prisoner by the British. Among the Jews killed were Julius Jacobs, a senior Mandate official and secretary of the Jerusalem Music Society, who had served in the 2nd Jewish Battalion under Allenby; Dr Wilhelm Goldschmidt, a refugee from Hitler's Germany in 1933 who had risen to be an assistant legal draftsman to the Government of Palestine; and Claire Rousso, a nineteen-year-old telephone operator in the Secretariat. One of several drivers killed was an Armenian, Garabed Paraghanian. Twenty British soldiers were also killed.

The Jewish Agency denounced what it called 'the dastardly crime' perpetrated by a 'gang of desperadoes', and called on the Jews of Palestine 'to rise up against these abominable outrages'. The Sephardi Chief Rabbi, Ben Zion Uziel, spoke of his 'loathing and abhorrence' at the crime. The Jewish Community Council warned of the 'abyss opening before our feet by irresponsible men' who had carried out a 'loathsome act'. Watching the Jewish mourners at one of the many funeral processions, the *Palestine Post* commented: 'The faces were set and expressionless, as though they were burying not only our countrymen, but cherished hopes.'

At three o'clock on the afternoon of July 23 all work and traffic stopped in Jewish Jerusalem at three o'clock, in mourning for the dead. As a result of the King David bomb, the recent agreement between the Haganah and the Irgun broke down, the Jewish Resistance Movement was at an end, and the Irgun and Stern Gang began to operate once more separately from, and often in defiance of, the Haganah.

On July 29, a week after the King David explosion, the British Cabinet discussed an American government request that the movement of 100,000 Jews from Europe to Palestine should begin within a month. During the discussion the Prime Minister, Clement Attlee, argued that the movement should not begin until plans for Jewish and Arab 'provincial autonomy' – Britain's latest idea – had been further developed. On the following day, despite Attlee's doubts as to the wisdom of the move – like Churchill before him, Attlee found himself outnumbered by his colleagues on Zionist issues – it was agreed to deport to Cyprus all 'illegal' refugees who were caught trying to enter Palestine. At the same time, it was suggested that Weizmann be asked to use his influence 'to persuade illegal immigrants not to resist transfer to staging camps in Cyprus', on the understanding that they would soon be transferred to Palestine as part of the 100,000. But Jewish fears were now too roused to trust such a proposal. For both Weizmann and the Jewish Agency immediate and unimpeded immigration to Palestine itself was the dominant need, on which no compromise was possible.

A serious blow to Zionist hopes of Labour government support had come on July 24, with the publication of a plan drawn up by yet another Cabinet Committee on Palestine, and announced in the House of Commons on July 30 by Herbert Morrison. Under this new plan, Palestine would be divided into three areas. The first area, constituting 43 per cent of the Mandate area, and including Jerusalem, was to remain under British control. The second area, 40 per cent of the country, was to become an area of Arab provincial autonomy. The third area, the remaining 17 per cent of Palestine, was to become the area of Jewish provincial autonomy. All three areas would be under British rule.

Only if the Jews accepted this plan, Morrison announced, would the 100,000 refugees of the Anglo-American Committee's recommendation be allowed into Palestine. Even so, those allowed in would have to be refugees primarily from Germany, Austria and Italy. Adult refugees from Poland, Roumania and Eastern Europe would not be included in the total, only orphan children from these areas. Nor would Britain agree to receive these refugees in Palestine, even under these restrictions, unless the United States government agreed in advance to undertake 'sole responsibility for their sea transport', as well as providing them with food 'for the first two months after their arrival'.

The Jews rejected the Morrison plan as providing too small an area, with many Jewish settlements excluded from it.

From August 1946 to December 1947 the 51,700 immigrants who arrived illegally off the shores of Palestine – on thirty-five ships – were taken to Cyprus, put in camps, surrounded by barbed wire, and guarded by armed British soldiers. Many of the soldiers disliked the task assigned them; they were from the same army, and wearing the same British uniform, that had liberated so many of the DPs from Bergen-Belsen a year and a half earlier.

Throughout the second half of 1946, the British government refused to relax the Palestine immigration quota system which had been in force since 1939. The Jewish Agency, its leaders still in detention, continued to prepare for statehood and, unable to dislodge the British from their policy towards the DPs, decided to establish inside Palestine as many settlements as possible in one of the remotest and potentially most vulnerable regions of the country, the Negev.

On 6 October 1946 – the Day of Atonement – eleven kibbutz settlements were established during a single night. Their inhabitants, and their simple infrastructure, were rushed southward by truck, and primitive but effective facilities were prepared by dawn. One of these settlements was Kfar Darom, not far from Gaza. Two years later it was to be captured by the Egyptian army after a prolonged siege. It was reconstituted only in 1970, three years after Israeli forces overran the Gaza Strip. Today it is one of the Jewish enclaves in the Gaza Strip, excluded from the Palestinian Authority. Further east, Mishmar Ha-Negev (Guard of the Negev) was also founded that night, its members coming from as far afield as Latin America and Bulgaria. In the War of Independence it was to serve as a base for the capture of the Arab town of Beersheba, 10 miles to the south.

Also founded in the Negev on October 6, sixteen miles north of Beersheba, was Shoval, named after a biblical town believed to have been in the area. Its founders were Palestinian Jews and Jews from South Africa. They pioneered contour-ploughing methods to prevent erosion, and established friendly relations with the Bedouin tribe that camped nearby. The kibbutz also built a reservoir to capture the water that came throughout that dry region in sudden, short but destructive flash floods.

One of the most remote settlements founded on October 6 was Gal-On (Monument to Strength). Its founding members were from Poland. Some of them had survived in the wartime ghettos or had fought as partisans against the Germans. They chose their name to commemorate those who had been killed in the wartime ghetto revolts.

The settlements set up on 6 October 1946 were not the only ones to be established that year. On rocky ground on the Tiberias–Rosh Pina road, five miles south of Rosh Pina, kibbutz Ammiad (My Nation for Ever) was established in 1946 by a group of Palestinian-born youngsters, whose first and heavy task was to reclaim the land for farming. Two other new settlements were founded that year by demobilized Jewish soldiers who had served in the British army, in the Royal Engineers. One was a moshav in Lower Galilee, within sight of Mount Tabor. It was named Kfar Kisch after a distinguished British soldier, Brigadier Frederick Kisch, under whom many of them had served. Kisch had been killed by a mine during the North African campaign. The second settlement was Kfar Monash, also a moshav, located in the Hefer Plain and named after a distinguished Australian First World War military leader and Jew, General Sir John Monash. The land on which Kfar Monash was built had been acquired through the donations of Australian Jewry.

Another settlement established in 1946 was Yehiam, in northern Palestine, only six miles – across rugged terrain – from the Lebanese border, and just below the imposing ruins of the twelfth-century crusader fort of Judin. Its establishment in such a remote area, so near to the international border, was an important political statement by the Jews of Palestine. They wanted to be present throughout the land, even at its extremities, and in its most remote regions. The settlement was named after Yehiam Weitz, who had been killed less than four months earlier during the night the bridges were blown up by the Haganah.

*   *   *

Some Jewish survivors sought refuge in places other than Palestine. More than 15,000 survivors were allowed to enter Latin America in 1946, most of them choosing Argentina. Nearly 13,000 were given permission to settle in the United States, 7,000 in Canada, and 6,000 in Australia. But still the magnet of Palestine was by far the strongest.

So desperate were the Jews of Palestine for some form of sovereign entity, in which they could be their own rulers, and into which survivors could come without any numerical restrictions, that both Weizmann and Ben-Gurion were prepared to accept – as the Jewish Agency had been ten years earlier – some form of partition. On 28 October 1946, on the eve of the release of the Jewish leaders from detention, Ben-Gurion wrote to Weizmann, 'We should, in my opinion, be ready for an enlightened compromise even if it gives us less in practice than we have a right to in theory, but only so long as what is granted to us is really in our hands. That is why I was in favour of the principle of the Peel Report in 1937 and would even now accept a Jewish State in an adequate part of the country.' Ben-Gurion added, 'We are the generation which came after you, and which has been tried, perhaps by crueller and greater sufferings; and we sometimes, for this reason, see things differently. But fundamentally we draw from the same reservoir of inspiration – that of sorely tried Russian Jewry – the qualities of tenacity, faith, and persistent striving which yields to no adversity or foe.'

The Twenty-Second Zionist Congress opened in Basle on 9 December 1946, in the same hall that Herzl had used for the first congress in 1897. The newly established settlements in Palestine, Weizmann told the delegates, 'have, in my deepest conviction, a far greater weight than a hundred speeches about resistance – especially when the speeches are made in New York, while the proposed resistance is to be made in Tel Aviv and Jerusalem'. Weizmann did not minimize the cause of Jewish bitterness. Indeed, his speech contained a powerful reproach of Britain's White Paper policy, for, as he told the delegates:

> Whenever a new country was about to come under Gestapo rule, we asked that the gates of the National Home be opened for saving as many as possible of our people from the gas-chambers. Our entreaties fell on deaf ears; it seemed that the White Paper was more sacred for some people than life itself.
>
> Sometimes we were told that our exclusion from Palestine was necessary in order to do justice to a nation endowed with seven independent territories, covering a million square miles; at other times we were informed that the admission of our refugees might endanger military security during the war.
>
> It was easier to doom the Jews of Europe to a certain death than to evolve a technique for overcoming such difficulties.

For many of the delegates in Basle, the situation described by Weizmann justified armed resistance against the British in Palestine. But in his speech he urged a halt to all anti-British acts of terror and violence. What was needed, he said, was 'the courage of endurance and the heroism of super-human restraint'. Terrorism was 'a cancer in the body politic' of Palestinian Jewry. Accused by some of those in the hall of being a demagogue, Weizmann cried out in anger, 'If you think of bringing the redemption nearer by un-Jewish methods, if you lose faith in hard work and better days, then you commit idolatry and endanger what we have built.'

Weizmann ended his appeal for moderation with a final, anguished outburst. 'Would that I had a tongue of flame, the strength of prophets,' he said, 'to warn you against the paths of Babylon and Egypt. "Zion shall be redeemed in judgement", and not by any other means.'

Following Weizmann's appeal for moderation there was a stormy debate. Among those who witnessed it were thirty-one-year-old Moshe Dayan and twenty-three-year-old Shimon Peres, who thirty years later were to be among Israel's political leaders. Ben-Gurion was angered to the point of despera-tion, returning to his hotel room (the very one Herzl had stayed in) and packing his bags. Shimon Peres recalled the sequel, as he and a mutual friend, Arieh Bahir – from the Jordan Valley kibbutz Afikim – entered Ben-Gurion's room:

> Ben-Gurion wheeled round and stared at us. 'Are you coming with me?'
>
> 'Where are you going?' we asked.
>
> 'To form a new Zionist movement. I have no more confidence in this Congress. It's full of small-time politicians, pathetic defeatists. They won't have the courage to make the decisions that are needed at this time. Only the Jewish youth, all over the world, will provide the courage needed to face the historic challenges facing Zionism. After one-third of our nation has been wiped out – among them some of our finest young people – the survivors have no hope other than to rebuild their lives in the historic homeland, the only land that can and must open its gates wide to welcome them.'
>
> Ben-Gurion was in a fighting mood. Bahir looked at me, and I motioned to him to tell Ben-Gurion that we would go with him. This calmed his fury somewhat, and soon we plucked up the courage to suggest that, before slamming the door on the Congress, he try one more time to win over the Mapai faction. 'If there's a majority there, we'll all stay; and if not, we won't be the only ones to leave with you; a great many more will come too.' Ben-Gurion agreed to make this last effort.
>
> Meanwhile, word spread through the Congress corridors that a profound crisis had erupted. Mapai moderates like Sprinzak, Kaplan and Moshe Sharett, who tended towards Weizmann, nevertheless were loath

to lose Ben-Gurion. As so often in the past and in the future, even those of his colleagues who differed with him and bitterly criticized his leadership retained a deep respect for his vision and for his iron will. They knew that, ultimately, there was no one like him, and no substitute for him.

At seven that evening, the crucial meeting of the Mapai faction began. As chairman the faction chose Golda Meir, who was considered a Ben-Gurionite even though she opposed partition. She was second to none in her ability to run a stormy and emotion-laden meeting.

This meeting began with a vehement attack by Ben-Gurion on 'Mr Engineer Kaplan', whom he accused of living in yesterday's world and of failing to understand that only a sovereign Jewish state, no matter how small, could provide salvation for the Holocaust refugees.

Kaplan, despite his usually calm and moderate demeanour, hit back as hard as he could. He asserted that 'Mr Advocate Ben-Gurion' was living in a dream world and was proposing fantasies instead of pragmatic policies.

The meeting raged on through the night. Only at dawn was the vote called, and Ben-Gurion won by a majority – a slender majority, but a majority none the less. A wave of relief swept over us. In our hearts, we knew that at that moment the Jewish State had been born. Nothing now could deter Ben-Gurion from his goal.

By 171 votes to 154, Weizmann's appeal for restraint was defeated. 'Perhaps it was in the nature of things that the Congress should be what it was,' he reflected in his memoirs, 'for not only were the old giants of the movement gone . . . but the in-between generation had been simply wiped out; the great fountains of European Jewry had been dried up. We seemed to be standing at the nadir of our fortunes.'

Weizmann's commanding authority, which had been virtually unchallenged for almost thirty years, was over. From that moment, Ben-Gurion was the effective head of the Zionist movement.

In Palestine, the British began a series of ever more thorough searches for arms caches hidden in the Jewish settlements. Jews caught in possession of arms were arrested, imprisoned, and even flogged. Following one such flogging, on 29 December 1946 the Irgun seized a British major and three sergeants, and flogged them in retaliation. The bitterness between the British authorities and the Jewish community was intensifying to the point of hatred.

CHAPTER EIGHT

Year of decision
1947

With the opening weeks of 1947 the violence increased: on January 1 an Irgun group, attacking a British police post, killed a policeman. The British government was coming close to the end of both its patience and its self-confidence. That day, at a meeting of the Cabinet's Defence Committee, it was agreed 'that to continue this policy in Palestine in present circumstances placed the Armed Forces in an impossible position'. Three days later the Secretary-General of the Arab League, Azzam Pasha, announced that the Arabs would vote against any partition scheme that might be put forward, and would continue to oppose any further Jewish immigration.

Such Arab hostility was well known. But on the morning of January 7 a new factor was introduced into the Middle East discussion, which made Arab goodwill even more essential. On that day a 'Top Secret' memorandum, written four days before, was circulated to the British Cabinet, entitled 'Middle East Oil'. Its authors were Ernest Bevin, the Foreign Secretary, and Emanuel Shinwell, the Minister of Fuel and Power (and a Jew). The memorandum was illustrated with facts and charts to show 'the vital importance for Great Britain and the British Empire of the oil resources of this area'. The Middle East, the memorandum stressed, was likely to provide 'a greater proportion of the total world increase of production than any other oil-bearing region'. By 1950, the 'centre of gravity' would shift from Persia 'to the Arab lands', with Saudi Arabia, Bahrain, Kuwait and Iraq being the main oil producers.

Bevin and Shinwell went on to warn of the grave risks involved in offending the Arabs 'by appearing to encourage Jewish settlement and to endorse the Jewish aspiration for a separate State'. Bevin favoured a plan put forward by the Arab States for a 'unitary' State in Palestine. 'The certainty of Arab hostility to partition is so clear,' he wrote, 'and the consequences of permanently alienating the Arabs would be so serious,' he told his Cabinet colleagues on January 14, 'that partition on this ground alone must be regarded as a desperate remedy.' Such a decision would 'contribute to the elimination of British influence from the vast Moslem area between Greece and India' and would 'jeopardize the security of our interests in the increasingly important oil production in the Middle East'.

Bevin told his colleagues that he favoured 'an independent unitary State' in Palestine, with special rights for the Jewish minority, but incorporating 'as much as possible of the Arab plan'. He went on to explain that he did not accept Arab demands to halt Jewish immigration altogether, although, he wrote, 'steps must be taken to prevent a real flooding of the country by Jewish immigrants'.

Bevin argued that 'a Jewish government' in Palestine would never accept the partition lines as final, but would eventually seek to expand its frontiers. 'If Jewish irredentism is likely to develop after an interval,' he commented, 'Arab irredentism is certain from the outset. Thus the existence of a Jewish State might prove a constant factor of unrest in the Middle East.'

For one more week the British Cabinet debated Palestine and the possible solutions under British rule. Then, on 22 January 1947, it decided (but did not make its decision public for another three weeks) that if it could not get 'an agreed solution' between the Jews and the Arabs, it would turn the problem over to the United Nations – the post-1945 successor to the League of Nations, which had authorized the Palestine Mandate twenty-six years earlier.

Britain's will-to-rule had reached its end. The constant combating of violence inside Palestine, the embarrassment of guarding Holocaust survivors in camps in Cyprus – not to speak of the ferocious winter in Britain during which food and fuel shortages and hardship were hitting every home – made the decision easier. Five months later Britain was to give up by far the brightest jewel of the Empire, India. By contrast, the troubles in Palestine had made the Mandate a lump of base metal.

On 15 February 1947 the British government made its public announcement that the problem of Palestine would be handed over to the United Nations. Both Jews and Arabs prepared to put their respective cases to the United Nations. If Jewish statehood were to come to pass here was yet another challenge to meet, another obstacle to overcome. But the prospect of an end to British rule did not halt Jewish terrorist activity. On March 1 the Irgun blew up the British Officers' Club in Jerusalem, killing fourteen officers. The Jewish Agency leaders were appalled at this act of terror. Ben-Gurion, who had taken over responsibilities for defence in addition to his chairmanship of the Jewish Agency Executive, feared that the actions of the Irgun and the Stern Gang would undermine all the Jewish Agency's efforts to create a unified, disciplined and effective Jewish central command.

Amid violence, the instinct for normal, constructive, cultural life could not be extinguished. At the beginning of 1947 the Palestine Orchestra invited the American Jewish composer Leonard Bernstein to conduct a series of concerts. Two days after his arrival in Palestine he wrote to a friend, 'There is a strength and devotion in these people that is formidable. They will never let the land be taken from them, they will die first. And the country is beautiful beyond description.'

But the daily turmoil impinged even upon music. 'The situation is tense and unpredictable,' Bernstein wrote a few days later, 'the orchestra fine and screaming with enthusiasm (first rehearsal this morning). I gave one down-beat today to the accompaniment of a shattering explosion outside the hall. We calmly resumed our work. That's the method here. An Englishman was kidnapped at our hotel last night, the police station was blown up today, a truck demolished in the square – and life goes on; we dance, play boogie-woogie, walk by the Mediterranean (which is out of a fairy book) and we hope for the best.'

'Life goes on': but the kidnappings and violence led to swift reprisals. On April 16 the British executed four members of the Irgun – Dov Gruner, Mordechai Alkahi, Yehiel Dresner and Eliezer Kashani – in Acre Prison. A week later, two members of the Stern Gang – Meir Feinstein and Moshe Barzani – committed suicide in the Central Prison in Jerusalem just before their execution.

'Life goes on': on May 1, in the Edison Cinema in Jerusalem, Leonard Bernstein introduced his *Jeremiah* symphony. The audience was enraptured. Not since Toscanini's concerts in Jerusalem in 1936, wrote the *New York Times*, 'has a conductor been recalled so many times and given a similar ovation'. For Jewish Jerusalemites, music always had the power to lift their spirits amid the tensions. Bernstein was deeply moved. 'I was in tears,' he wrote. 'I have never seen anything like it, that hysterical, screaming audience'.

Travelling northward at the end of his two-week visit, Bernstein conducted in the open air at Ein Harod, one of the kibbutzim in the Jezreel Valley. An estimated 3,500 people came to watch and listen. The *Boston Morning Globe* reporter, Arthur Holzman, wrote, 'They had come by truck or wagon and on foot, they lay atop cars, stood in the aisles and spilled over onto the platform. Because of Government road restrictions, many of them would have to spend the rest of the night until dawn lying in trucks or in the open fields. It was as if Heaven had sent them this genius to help them forget their troubles.'

Those troubles remained the daily concern of Ben-Gurion, who, having taken over the Jewish Agency's defence portfolio, brought with him what Yitzhak Rabin described as 'a new spirit'. In his memoirs, Rabin recalled how Ben-Gurion summoned all the Haganah officers from battalion commanders upward and, 'for the first time, urged us to prepare for war on a scale never before envisioned. Yigal Allon came back from one briefing elated.' Ben-Gurion had asked Allon whether the Jews were capable of standing up to an invasion by the Arab armies. 'We have the basis,' Allon replied. 'If the present nucleus grows by substantial numbers and if we get suitable equipment, we could indeed face the Arab armies.'

Those two 'ifs', Rabin reflected, 'were cardinal problems'. The reality was harsh. In April 1947, Ben-Gurion totalled up the arms in the Haganah's possession: these amounted to 10,073 rifles, 1,900 sub-machine-guns, 186 machine-guns and 444 light machine-guns. Heavy equipment consisted of 672 two-inch mortars and 96 three-inch mortars. 'It goes without saying,' Rabin wrote, 'that there was not one single cannon, heavy machine-gun or anti-tank or anti-aircraft weapon, not to mention armour, air or naval force. Nor were there any means of land transportation. In view of this stock of arms, talk of facing up to the regular Arab armies sounded like lunacy.'

In May 1947 Shimon Peres was sent from his kibbutz to join work at the Haganah High Command, located at the Red House on the Tel Aviv seafront. His task was to mobilize new manpower. Later he became responsible for arms procurement overseas. Among those with whom he worked was the head of the Haganah mission in the United States, Teddy Kollek. The kibbutz youth, trained as farmers, were being mobilized by the Haganah for all the demands of warmaking.

A special session of the United Nations was opened at the beginning of May 1947 to hear the arguments about the future of Palestine. A leading American Zionist, Abba Hillel Silver, addressed it on May 8 – two years to the day after the end of the Second World War. 'The Jews were your allies in the war,' he said, 'and joined their sacrifices to yours to achieve a common victory. The representatives of the Jewish people of Palestine should sit in your midst. We hope that the representatives of the people which gave to mankind spiritual and ethical values, inspiring human personalities and sacred texts which are your treasured possessions, and which is now rebuilding its national life in its ancient homeland, will be welcomed before long by you to this noble fellowship of the United Nations.'

The Arabs also put their case. That same day the representative of Iraq, Dr Fadhil Jamail, insisted that the Arabs of Palestine should not, as he put it, 'suffer for the crimes of Hitler'. But on May 12, Ben-Gurion pointed out, in answer to this line of argument, that the homeless Jewish refugees were being brought to 'our own country' – to Jewish towns and villages, not to Arab inhabited areas. The Jews who had returned, he said, 'are settled in Petah Tikvah, Rishon le-Zion, Tel Aviv, Haifa, Jerusalem, Deganya, the region of the Negev, and other Jewish towns and villages built by us'.

On May 15 the United Nations set up a Special Committee on Palestine, known popularly by its initials as UNSCOP. That same day, in Palestine itself, two British officers were killed dismantling a mine on the railway. Two weeks later, 400 Jews from French North Africa reached Palestine: the first

Jews to come from Arab lands as 'illegals'. As with all those caught by the Royal Navy, they were taken to Cyprus and interned: the *New York Times* of June 1 called it the Haganah's 'first non-European rescue bid'.

On June 2 UNSCOP elected its Chairman, Emil Sandström, a Swedish Supreme Court Judge. As UNSCOP set off for Palestine, the Arab Higher Committee declared a complete boycott throughout Palestine of all its proceedings, giving as its reason that its terms of reference included a study of the problem of the Jewish Displaced Persons who, leaving Europe in a series of 'illegal' ships, were still being intercepted by the British and deported to Cyprus.

Even before UNSCOP began its work, it was clear that the fate of the DPs would be a predominant issue. On April 14, 2,552 'illegal' immigrants had reached Haifa on board the *Guardian*. Three of them had been killed while unsuccessfully resisting a Royal Navy boarding party which was in the process of transporting them to Cyprus. On April 22 a further 769 'illegals', arriving on board the *Galata*, had likewise been trans-shipped to Cyprus, followed by another 1,442 who arrived on board the *Trade Winds* on May 17, then a further 1,457 on board the *Orletta* on May 24, and then a further 399 on board the *Anal* on May 31. Churchill, who was then leader of the Conservative Opposition, called this process a 'squalid war'.

The UNSCOP members were already in Palestine when in July yet another 'illegal' ship, named by the Jews *Exodus 1947*, reached the Palestine coast from Genoa. On board were many German and Polish survivors. After three of them had been killed in a skirmish with the Royal Navy, the ship was brought into Haifa under British naval escort. But its 4,500 refugee passengers were ordered by Bevin to go, not to Cyprus – as had happened to so many previous 'illegals' – but back to Europe. To this end, they were at once forcibly transferred to another ship, the *Empire Rival*.

Among those who recalled the impact of the transfer of the refugees on the members of UNSCOP was one of the two Jewish liaison officers to the Committee, Aubrey (later Abba) Eban. It was he who persuaded four of the eleven committee members, including the chairman, to go to Haifa to see for themselves what was happening. There, Eban wrote in his autobiography, the four members watched 'a gruesome operation'. The Jewish refugees had decided 'not to accept banishment with docility. If anyone had wanted to know what Churchill meant by a "squalid war", he would have found out by watching British soldiers using rifle butts, hose pipes and tear gas against the survivors of the death camps. Men, women and children were forcibly taken off to prison ships, locked in cages below decks and sent out of Palestine waters.'

When the four members of UNSCOP came back to Jerusalem, Eban recalled, 'they were pale with shock. I could see that they were preoccupied with one point alone: if this was the only way that the British Mandate could continue, it would be better not to continue it at all.' A few days before the refugees were sent back to Europe, Golda Meir declared, in a speech in Tel Aviv, 'To Britain we must say: it is a great illusion to believe us weak. Let Great Britain with her mighty fleet and her many guns and planes know that this people is not so weak, and that its strength will yet stand it in good stead.'

On Bevin's orders the *Empire Rival* was sent first to Port de Bouc, in southern France, where the refugees refused to land, and then to Hamburg. Forced by British troops to disembark on the hated German soil from which so many of them had fled, these 'illegals' were sent to a DP camp at Pöppendorf. According to the Yugoslav member of UNSCOP, the *Exodus 1947* saga 'is the best possible evidence we have' for allowing the Jews into Palestine.

On July 4 Ben-Gurion appeared before UNSCOP. His evidence opened with a survey of Jewish history since biblical times. 'With an indomitable obstinacy,' he said, 'we always preserved our identity. Our entire history is a history of continuous resistance to superior physical forces which tried to wipe out our Jewish image and to uproot our connections with our country and with the teaching of our prophets.' At the same time, Ben-Gurion added, 'an unbroken tie between our people and our land has persisted through all these centuries in full force'. This tie had been strengthened in each generation through 'Jewish homelessness and insecurity' in the Diaspora:

> The homelessness and minority position make the Jews always dependent on the mercy of others. The 'others' may be good and may be bad, and the Jews may sometimes be treated more or less decently, but they are never masters of their own destiny; they are entirely defenceless when the majority of people turn against them.
>
> What happened to our people in this war is merely a climax to the uninterrupted persecution to which we have been subjected for centuries by almost all the Christian and Moslem peoples in the old world.
>
> There were and there are many Jews who could not stand it, and they deserted us. They could not stand the massacres and expulsions, the humiliation and discrimination, and they gave it up in despair. But the Jewish people as a whole did not give way, did not despair or renounce its hope and faith in a better future, national as well as universal.

In his evidence to UNSCOP, Ben-Gurion also paid tribute to Britain: 'It will be to the everlasting credit of the British people,' he said, 'that it was the first in modern times to undertake the restoration of Palestine to the Jewish people.' At the same time, in Britain itself, Jews 'were and are treated as equals. A British Jew can be and has been a member of the Cabinet, a Chief Justice, a Viceroy.'

Speaking of the fate of European Jewry, Ben-Gurion noted that in a recent

Gallup Poll taken in the American Zone of Germany, 14 per cent of the Germans questioned had condemned Hitler's massacre of the Jews, 26 per cent had been 'neutral', and 60 per cent had approved the killings. 'The Jews do not want to stay where they are,' he said. 'They want to regain their human dignity, their homeland, they want a reunion with their kin in Palestine after having lost their dearest relations. To them, the countries of their birth are a graveyard of their people. They do not wish to return there and they cannot. They want to go back to their National Home, and they use Dunkirk boats.' Were a Jewish State to be established, it would mean the settlement 'of the first million Jews' in the 'shortest possible' time.

On July 8 Dr Weizmann gave evidence before UNSCOP. The Jews, he said, had 'an abnormal position' in the world: one that had beset them throughout their history. It was characterized by one thing above all, 'the homelessness of the Jewish people'. There were, of course, many individual Jews who had 'very comfortable homes', particularly in America, Western Europe, Scandinavia, and 'formerly in Germany', but, as a 'collectivity' they were homeless. Weizmann continued:

> They are a people, and they lack the props of a people. They are a disembodied ghost. There they are with a great many typical characteristics, many strong characteristics which have not disappeared throughout centuries, thousands of years of martyrdom and wandering, and at the same time they lack the props which characterize every nation.
>
> We ask today: 'What are the Poles? What are the French? What are the Swiss?' When that is asked, everyone points to a country, to certain institutions, to parliamentary institutions, and the man in the street will know exactly what it is. He has a passport.
>
> If you ask what a Jew is – well, he is a man who has to offer a long explanation for his existence, and any person who has to offer an explanation as to what he is, is always suspect – and from suspicion there is only one step to hatred or contempt.

It was his duty, Weizmann continued, although in the past he had never thought it would be necessary, 'to try to explain: "Why Palestine?"' and he went on to ask:

> Why not Kamchatka, Alaska, Mexico, or Texas? There are a great many empty countries. Why should the Jews choose a country which has a population that does not want to receive them in a particularly friendly way; a small country; a country which has been neglected and derelict for centuries? It seems unusual on the part of a practical and shrewd people like the Jews to sink their effort, their sweat and blood, their substance, into the sands, rocks and marshes of Palestine.

Well, I could, if I wished to be facetious, say it was not our responsibility – not the responsibility of the Jews who sit here – it was the responsibility of Moses, who acted from divine inspiration. He might have brought us to the United States, and instead of the Jordan we might have had the Mississippi. It would have been an easier task. But he chose to stop here. We are an ancient people with an old history, and you cannot deny your history and begin afresh.

For the establishment of the Jewish National Home, Weizmann told the Committee, the Jews owed the British 'a sincere tribute', but the time had come for the 'Home' to evolve. Partition would be the best evolution in the current circumstances. 'If it is final, the Arabs will know, and the Jews will know, that they cannot encroach upon each other's domain.' Partition would represent 'a new and great sacrifice on the part of the Jewish people', but its advantage would be that it could not be further 'whittled down' or 'bargained down'.

Yet the area of Palestine which the Jews would have after partition, Weizmann insisted, 'must be something in which Jews could live and into which we could bring a million and a half people in a comparatively short time. It must not be a place for graves only, or graveyards, or, as you sometimes see on very full trains, "standing room only".'

Whatever was done, Weizmann told the Committee, it should be done quickly. 'Do not let it drag on. Do not prolong our agony. It has lasted long enough and has caused a great deal of blood and sorrow on many sides.'

On July 13 Emil Sandström and two of his UNSCOP colleagues held a secret meeting with the leaders of the Haganah, in the Jerusalem suburb of Talpiot. The three men wanted to know if the Haganah had the means, and the will, to protect the Jewish areas against Arab attack, in the event of the establishment of a Jewish State. The six Haganah representatives made a strong impression. One of them, Yigael Yadin, spoke emphatically of how, if the Arab States around Palestine were to launch an attack across the borders of the Jewish State, the Haganah would attack the enemy's naval and air bases far behind the front line.

Enquiring as to how many artillery pieces the Haganah possessed, the three UNSCOP members were told that the number could not be revealed 'for security reasons'. There were in fact none, though efforts were soon to be made to purchase some abroad.

The UNSCOP deliberations continued against a background of violence. At the beginning of July, the Irgun took two British sergeants as hostages, hoping to prevent the execution of three of their members who had been sentenced to death. After searching in vain for the two sergeants, the British authorities executed the three men in Acre Prison: their names were Avshalom Haviv, Yaakov Weiss and Meir Nakar. On July 31, as a reprisal against these executions, the Irgun killed the two British sergeants whom they were holding captive, and booby-trapped the ground underneath their bodies.

Following the death of the two sergeants and the publicity surrounding it, the British public demanded that the troops be brought home. In Palestine several Jews were murdered by British soldiers as a counter-reprisal.

UNSCOP held its last meeting on August 31. Its majority report – signed by the representatives of Canada, Czechoslovakia, Guatemala, the Netherlands, Peru, Sweden and Uruguay – was issued on September 1. It proposed the creation of two separate and independent States, one Arab and one Jewish, with the city of Jerusalem as a *corpus separatum* under international trusteeship. Under these proposals the Jewish State would contain 498,000 Jews and 407,000 Arabs. The Arab State would contain 725,000 Arabs and 10,000 Jews. The city of Jerusalem and its environs, including the Arab towns of Bethlehem and Beit Jalla, would contain 105,000 Arabs and 100,000 Jews. The Negev would be part of the Jewish State, Western Galilee part of the Arab State.

Just as the Palestinian Arab leaders had rejected the Peel Commission's similar proposal ten years before, their political leadership, the Arab Higher Committee, rejected these proposals for an Arab and a Jewish State. The Jewish Agency, however, accepted the UNSCOP proposals, subject to further discussions on the actual boundary lines, even though the new plan would keep a quarter of the Jews of Palestine outside the area of Jewish statehood. Acceptance would involve a sacrifice, Abba Hillel Silver declared on October 2, but this sacrifice 'will be the Jewish contribution to a painful problem and will bear witness to my people's international spirit and desire for peace'.

Ben-Gurion realized that as a result of the UNSCOP decision to establish an Arab as well as a Jewish State in Palestine, the Jews were likely to have to fight for whatever territory they had been allotted. He also contemplated the possibility of fighting to extend the area allotted to the Jews. In his instructions to the General Staff of the Haganah he stressed that the 'full capacity' of Palestinian Jewry – the Yishuv – was to be mobilized, in order 'to safeguard the entire Yishuv and the settlements (wherever they may be), to conquer the whole country or most of it, and to maintain its occupation until the attainment of an authoritative political settlement.' Men and women were equally conscripted. Asked by his staff officers for operational clarification in advance of the United Nations vote on partition, Ben-Gurion told them, 'If the decision is favourable we shall defend every settlement, resist every attack, and maintain services to the Jewish Yishuv and to all the Arabs who so desire; we shall not restrict ourselves territorially.'

Ben-Gurion made a serious effort, shortly before the United Nations vote on the partition proposals, to seek the neutrality of King Abdullah of Transjordan, whose British trained and officered army, the Arab Legion, was the strongest fighting force in the Middle East. The king had long been at loggerheads with Haj Amin al-Husseini, the Mufti of Jerusalem, for the moral leadership of the Arabs of the whole region, and for spiritual control of the Muslim Holy Places in Jerusalem.

Abdullah's secret interlocutor was to be Golda Meir. 'He had agreed to meet me – in my capacity as head of the Political Department of the Jewish Agency – in a house at Naharayim (on the Jordan), where the Palestine Electric Corporation ran a hydro-electric power station,' she later recalled. 'I came to Naharayim with one of our Arab experts – Eliahu Sasson. We drank the usual ceremonial cups of coffee and then we began to talk. Abdullah was a small, very poised man with great charm. He soon made the heart of the matter clear: he would not join in any Arab attack on us. He would always remain our friend, he said, and, like us, he wanted peace more than anything else. After all, we had a common foe, the Mufti of Jerusalem, Haj Amin al-Husseini. Not only that, but he suggested that we meet again, after the United Nations vote.'

*   *   *

On 29 November 1947 in New York the General Assembly of the United Nations debated the UNSCOP proposals. During the debate, the Soviet representative, Andrei Gromyko (later Foreign Minister), astonished the Zionist representatives by his warm endorsement of their desire for statehood. 'The Jewish people had been closely linked with Palestine for a considerable period in history,' he said. 'As a result of the war, the Jews as a people have suffered more than any other people. The total number of the Jewish population who perished at the hands of the Nazi executioners is estimated at approximately six million. The Jewish people were therefore striving to create a State of their own, and it would be unjust to deny them that right.'

When the vote was taken, the partition proposal was accepted by thirty-three votes to thirteen, with ten abstentions. Britain was among those States which abstained. All six independent Arab States voted against the plan, as did four Muslim States – Afghanistan, Iran, Pakistan and Turkey – and three other States, Cuba, Greece and India. Among those in favour of partition were the United States, the Soviet Union, Australia, Canada, France, the Netherlands, New Zealand, Poland and Sweden.

For the Jews of the Diaspora, the news that there was to be a Jewish State in Palestine represented, as the American Zionist Emergency Council declared, 'a milestone in the history of the world', which had 'ended 2,000 years of homelessness for the Jewish people'.

There was also a voice of caution. On the day after the United Nations vote, Weizmann expressed in his diary his fear of the future role of the religious Parties in the Jewish State. 'We must clearly differentiate between legitimate religious aspirations and the State's obligation to defend them, and the excess of power sometimes revealed by supposedly religious groups,' he wrote. 'We must be firm if we lust for life.'

The United Nations partition resolution contained a recommendation that a Free Port for Jewish immigration should be opened – in Haifa – on 1 February 1948. At last the Holocaust survivors would be able to go to Palestine. They would also have someone to defend them against Arab attacks on the settlements and inter-urban transport, for in the autumn of 1947, in anticipation of renewed and intensified conflict with the Arabs, the Haganah was reorganized under the political control of Ben-Gurion, and a National Command established, representing all the political parties of Palestinian Jewry.

Reformed in this way while it was still regarded by the British as an illegal organization, the Haganah operated through clandestine methods and behind a heavy veil of secrecy. Its newly appointed Operations Officer was Yigael Yadin. Four brigades were set up, in the north, centre and south of the country, and in Jerusalem. Supply and manpower branches were established. Men who had Second World War battle experience with the British army's Jewish Brigade were recruited.

Special hideouts were prepared for Haganah arms and operational documents. All members had code-names. The Haganah itself was known both as 'the aunt' and 'the organization'. Bullets were 'cherries' or 'plums'. Rifles were 'pipes'. Pistols were 'sprinklers'. At the centre of the military preparedness of the Haganah was the Palmach – the strike force set up in 1941 to fight alongside the British in Syria. In the autumn of 1947 it had 2,100 men and women under arms, and a thousand more who had completed their training and returned to civilian life, but could be mobilized at any time.

In preparation for the coming conflict, the Haganah was divided into two sections, the field and garrison armies (HISH and HIM). The field armies were made up of men between the ages of eighteen and twenty-five, the garrison forces of women nurses and signallers, and of men over twenty-five. There were also youth battalions, the Gadna, which carried out pre-military training for those under eighteen. Medical services were drawn from the existing Palestinian Jewish health organizations, the Magen David Adom (Red Shield of David – the equivalent of the Red Cross and Red Crescent), the Workers' Sick Fund, and the Hadassah hospitals and clinics.

Even in the DP camps, and in the detention camps in Cyprus, the Haganah was active in training those who would eventually be able to enter Palestine. To enable rudimentary rifle training to be carried out in the camps, local carpenters were persuaded to make wooden rifles. How the battle would come, and the manner in which it would be fought, was a matter of considerable speculation in the Haganah. But there was general acceptance that war with the surrounding Arab countries was inevitable. A month before the partition resolution, Israel Galili, the chief of the Haganah High Command, expressed the following view, which proved in all respects to be an accurate one. 'The trouble will not begin,' he said, 'with someone blowing the big horn and announcing throughout the country, "Let's start rioting." The beginnings are normally "not serious", sporadic, insignificant. A few shots here and there, some isolated attacks; these accumulate, ignite and spread gradually, step by step.'

Israel Galili added, 'As far as we know, it is the Mufti's belief that there is no better way to "start things off" than by means of terror, isolated bombs thrown into crowds leaving movie theatres on Saturday nights. That will start the ball rolling. For no doubt the Jews will react, and as a reaction to a reaction there will be an outbreak in another place. Some rowdies will toss hand-grenades into Jewish taxis while they are passing through areas populated by Arabs; once more the Jews will retaliate in a violent way, and thus the whole country will be stirred up, trouble will be incited, and the neighbouring Arab countries will be compelled to start a "holy war" to assist the Palestinian Arabs.'

The Mufti himself was not in Palestine. Having helped to raise an SS division in Bosnia for Hitler during the Second World War, he had been imprisoned in France awaiting trial as a war criminal, but had escaped by the skin of his teeth (much as he had escaped from the British in Palestine in 1937). He made his way to Egypt where, until his death twenty-five years later, he exhorted the Palestinian Arabs to challenge every aspect of Jewish settlement and demanded the total expulsion of all Jews from Palestine.

CHAPTER NINE

Undeclared war
November 1947–April 1948

In New York, the United Nations General Assembly had voted for the establishment of a Jewish and an Arab State in Palestine. But Britain would remain in Palestine until the moment of transfer, a date not yet decided upon. The United Nations resolution had stated, with regard to the timing of the transfer, somewhat imprecisely, that it should be 'as soon as possible, but in any case not later than 1 August 1948'. Transfer of power might possibly have to wait another nine months, until the summer of 1948.

Hard though the wait might be for the Jews of Palestine, when the news of the United Nations vote reached them there was rejoicing in the streets. A mini-State was better than no State; even one from which Jerusalem was excluded (it was to remain under United Nations control). The dream of sovereignty was about to come to pass. But all around Palestine were Arab States who had obtained their independence earlier, who had built up national armies, and who were linked in the Arab League, founded in 1945 with its headquarters in Egypt, that voiced implacable enmity to the possibility of Jewish statehood.

Ben-Gurion was at the Dead Sea when news of the partition vote reached Palestine. Aged sixty-one and suffering from chronic backache, he would spend as many weekends as possible at the Kaliya Hotel, bathing in the medicinal waters. He was asleep when the news of the vote reached the hotel and was woken up to be told it. Returning by car to Jerusalem, he saw the crowds dancing in the streets, but he himself, he wrote in his diary, felt as if he were a 'man in mourning among the celebrators'.

Among those who were celebrating that night was the Palestinian-born soldier, Moshe Dayan, who later recalled in his memoirs:

> I felt in my bones the victory of Judaism, which for two thousand years
> of exile from the Land of Israel had withstood persecutions, the Spanish

Inquisition, pogroms, anti-Jewish decrees, restrictions, and the mass slaughter by the Nazis in our own generation, and had reached the fulfilment of its age-old yearning – the return to a free and independent Zion.

We were happy that night, and we danced, and our hearts went out to every nation whose UN representative had voted in favour of the resolution. We had heard them utter the magic word 'yes' as we followed their voices over the airwaves from thousands of miles away.

We danced – but we knew that ahead of us lay the battlefield.

'My father's strong hands lifted me from my bed and sleep,' Dayan's daughter Yaël later wrote. 'Mother woke up Udi, and we dressed in a hurry. I half knew what was happening, and the excitement was contagious. We hurried – leaving Assi, who was still a baby, to sleep – to the community hall in the centre of the village. Everybody was there, old and young, dancing to the music, kissing and embracing and crying with joy, until the rising sun painted the sky pink and it was time to milk the cows. If there was a vein of sadness in the faces of the dancers, it was a small indication of the price that would be paid. How many of these young people would be dancing here next year? How many of the children would be orphaned? How many mothers would be without sons?'

Among those who were to die on that battlefield in the months ahead was Moshe Dayan's own younger brother Zorik; for the Arabs of Palestine had turned violently against the United Nations decision. Even the 'mini' Arab State was rejected by the Mufti from his place of exile in Egypt, and by their other leaders in Palestine who focused their hatred on Jewish statehood. From the moment of the United Nations vote, Arab terrorists and armed bands attacked Jewish men, women and children all over the country, killing eighty Jews in the twelve days following the vote, looting Jewish shops, and attacking Jewish civilian buses on all the highways.

For the Arabs outside Palestine, a similar wave of anti-Jewish hatred led to violence against Jews in almost every Arab city: in British-ruled Aden, scene of a savage attack on Jewish life and property, eighty-two Jews were killed on December 9. In Beirut, Cairo, Alexandria and Aleppo, Jewish houses were looted, and synagogues attacked. In Tripolitania more than 130 Jews were murdered by Arab mobs.

There followed, in Palestine, five and a half months of terrorism and violence. 'Jews will take all measures to protect themselves,' the Jewish National Council declared on December 3. This was essentially a call for a vigilant defence. On December 13 the Jewish Agency, representing a majority of Palestinian Jewry, denounced what was becoming a rising tide of Irgun reprisals, calling them 'spectacular acts to gratify popular feeling.'

On December 11 the British Secretary of State for the Colonies, Arthur Creech Jones, told the House of Commons that Britain was 'responsible for law and order' in Palestine. But although the British were still in charge of Palestine, they could not control it. Groups of armed Syrians were active in the north of the country. A self-styled Arab Liberation Army, founded in Syria by the Arab League, and armed by Lebanon and Transjordan, was likewise in action in many areas. It was equipped not only with rifles and automatic weapons, but with light tanks. Local Arab forces in Jaffa were commanded by an Iraqi officer, one of many volunteers from Arab countries who had reached the city.

As Israel Galili had forecast a month earlier, the beginning of the conflict was marked by frequent but isolated acts of terror. On the day after the partition resolution, Arab riflemen opened fire on an ambulance making its way to the Hadassah Hospital on Mount Scopus. No one was hurt. That same day, a bus that was taking Jewish civilians from Netanya to Jerusalem was attacked by three Arabs with a machine-gun and hand-grenades. Five Jews were killed; one of them, twenty-two-year-old Shoshana Mizrachi Farhi, had been on her way to Jerusalem for her wedding.

This attack on the bus came to be the symbol of the beginning of war. The other incidents that day were hardly remembered. Yet half an hour after the first bus attack, and at almost the same spot, a second bus was attacked and one passenger was killed. She was Nehama Cohen, a pathologist at the Hadassah Hospital, who had been returning to duty. Later that day, twenty-five-year-old Moshe Goldman was shot dead at the Jaffa–Tel Aviv boundary. These were the first seven victims in a struggle that was to take 6,000 Jewish lives.

Thus began a war in which, by the time it ended, one per cent of the Jewish population of Palestine had been killed. There was to be no respite. On December 1 the Arab Higher Committee declared a three-day general strike. On the second day of the strike, 200 Arabs, many of them youngsters, broke into the commercial centre in Jerusalem, looting and burning Jewish-owned shops. The British troops nearby made no effort to intervene. When the Haganah sent a platoon of men to defend the centre, the British forces, still regarding the Haganah as an illegal organization, formed a barrier across the road and prevented them from approaching the scene of the Arab attack, and then forced the Jews back up the road.

In southern Tel Aviv, Arabs from Jaffa attacked the Hatikvah Quarter. They were driven off by the Haganah only after a prolonged and bloody battle, in which almost seventy Arabs were killed. Fighting between Jewish and Arab units became a daily fact of life throughout the country. On December 10 a detachment of Palmach soldiers was attacked while patrolling the water pipeline near the Arab village of Shu'ut in the Negev. The commander of the detachment, Assaf Shechnai, had told his men that the headman of the village would never attack them: 'He is my good friend.' Such friendship was real enough in pre-partition days. It meant nothing in war. Shechnai and his men were killed.

*   *   *

The greatest danger to the Jews would come when the Arabs felt strong enough to isolate their settlements and cut the roads. Confronted with the prospect of many Jewish settlements being attacked by neighbouring Arab villages, Ben-Gurion told two senior Haganah officers, 'We should adopt a system of aggressive defence. With every Arab attack we should be ready to reply with a decisive blow, destroy the site, or expel the inhabitants and take their place.'

That December, kibbutz Revivim, 12 miles south of Beersheba, was surrounded by armed Bedouin and could make no contact except by radio with the outside world. There was a call within the Haganah for the settlement to be evacuated, and for other isolated settlements in the Negev likewise to be abandoned, as the ability to defend them did not exist. Ben-Gurion decided, however, that no attempt would be made to 'shorten lines' in the Negev by evacuating the more distant settlements. For him the Negev was an integral part of Jewish Palestine, and the centre of future settlement and growth. Not one isolated settlement would be abandoned. Every settlement would have to make plans for its own defence, and if necessary to withstand a siege.

The Negev settlements would not be left entirely to their own devices, however. Such weapons as could be spared would be provided. Reinforcements would be sent when they were available. A special Palmach brigade was formed to keep the Negev roads open, and the first armoured cars produced locally by a newly established Haganah Armour Service were sent to the Negev to protect the water pipelines on which the settlements depended. But Revivim, the most southerly settlement, lay beyond the reach even of the special Palmach brigade. It had to hold out unaided.

On December 11 there was a concerted Arab attack on the Jewish Quarter of the Old City of Jerusalem, where 2,500 Jews were living. For six hours Haganah members fought them off. Three Arabs were killed, among them Issa Mohamed Iraqash, believed to be one of the leading inciters of violence in the Old City. A Red Shield ambulance which went to the Old City for the Haganah wounded was fired on, but succeeded in taking the men to the Hadassah Hospital. Also on December 11, ten Jews were killed when their convoy, carrying food and water to the Etzion bloc settlements, was ambushed just south of Bethlehem. A rapid-firing Bren gun was believed to have been used in the attack. Red Shield ambulances brought the dead and wounded back to Jerusalem.

It was increasingly dangerous for Jews to travel to Jerusalem from the coast. On December 12 a British Overseas Airways Corporation truck being driven by an Arab from Lydda Airport to the city was stopped by an Arab gang who ordered the Arabs on board to scatter. Three Jews on the truck were then taken down and shot. Two were members of the BOAC staff at the airport, Yitzhak Jian and David Ben Ovadia. The third was a cook at the airport restaurant, Joseph Litvak.

An extra element in the fighting that December was the dramatic increase in activity of the Irgun and the Stern Gang. Both groups still regarded terrorism as the most effective method to be used against both the outgoing British rule and the Arabs. They were also still repeatedly denounced by the Jewish Agency, and opposed by the Haganah. Small in numbers and se-cretive in method, they carried out a series of actions which exacerbated the tensions between the two communities. On December 13 several Irgun members driving in two cars near the Damascus Gate bus station hurled two bombs into the crowd and opened fire with automatic weapons. Five Arabs were killed including a fourteen-year-old boy, Ahmed Amin Hamma.

As part of the British forces in Palestine, the Arab Legion, commanded by a British officer, Glubb Pasha, took no part in the first two weeks of fighting. But in mid-December a group of legionnaires attacked a Haganah motor convoy that was on its way to the isolated youth village of Ben Shemen. Fourteen Haganah men were killed.

As the road from the coast to Jerusalem made its way up the narrow defile of the Bab al-Wad (The Gate to the Valley), travellers were particularly vulnerable to Arab sniping. Among seven Jews killed while making the journey on December 27 was the German-born Hans Beyth, who had just completed arrangements for the care of 20,000 young survivors of the Holocaust and other youngsters from Europe. He was murdered while on his way home from Haifa and Athlit, where he had been welcoming chil-dren on their arrival from a youth village in Cyprus.

The daily killing grew in scale and viciousness. On the day after the December 27 ambush on the road to Jerusalem, ten people were killed in the city itself, five Arabs and five Jews. The five Arabs were killed when members of the Stern Gang forced their way into an Arab home and opened fire on those inside. One of the five Jews who were killed was stabbed to death near the Damascus Gate on his way to a funeral. Another of the Jews, Miriam Meir, the mother of six children, was hanging up her washing on the roof of her house in the Jewish Quarter of the Old City when she was shot dead by an Arab sniper. Another Jew, Dr Hugo Lehrs, a British government medical officer at the Beit Safafa Government Contagious Diseases Hospital, was walking with an Arab doctor and an Arab nurse from one hospital building to another when three armed Arabs approached. 'Which is the Jew?' they asked. The Arab doctor and nurse then stood aside. Shots were fired and Dr Lehrs was killed.

On December 29 three members of the Irgun left a bomb at the Damascus Gate bus station. Fifteen Arabs were killed, among them a ten-year-old boy, Suaal Amashe, and an eleven-year-old girl, Namal Shamaa. In retaliation, two Jews – a father and daughter – and two British policemen, were shot dead by Arab gunmen. Palestine was descending into the abyss.

*   *   *

Under the partition resolution, thirty-three Jewish settlements were to be excluded from the area of Jewish sovereignty. Many of those were in isolated areas. As Arab attacks on isolated areas increased, the military experts of the Haganah decided, despite Ben-Gurion's wish to the contrary, that the defence of these isolated settlements was too great a burden for the forces at their disposal to carry. Orders were given for the settlements of the Etzion bloc, the southern Negev, western Galilee and the Jerusalem area to be evacuated. But the settlers themselves insisted that they remain put and defend themselves. They regarded themselves, not as military burdens on the restricted Haganah forces, but as defensive barriers and potential offensive bases of the State-to-be.

At the end of 1947 the Haganah decided to take reprisals against those individual Arabs who had personally and actively participated in attacks on Jews. These retaliatory actions in which the Arabs were killed were accompanied by the distribution of leaflets and radio broadcasts (the latter illegal under the existing British law) which warned the Arabs that if they decided on hostilities they would be the losers, but that if they refrained from attacking Jews, the Jews would not initiate action against them. The Jewish National Council issued a proclamation to the Arabs: 'Expel those among you who want blood to be shed, and accept the hand which is outstretched to you in brotherhood and peace.'

The two dissident Jewish groups, the Irgun and the Stern Gang, were opposed to the Haganah policy of strictly limited reprisals and the offer of an outstretched hand. After an Arab attack on the Commercial Centre in Jerusalem immediately following the partition resolution, the Irgun had set fire to the Rex cinema in Jerusalem, a cinema frequented by Jews and Arabs. A few days later, an Irgun member threw a hand-grenade into the crowded Arab vegetable market opposite the Damascus Gate in Jerusalem. A dozen Arabs were killed. Further Irgun attacks on Arabs – mostly on Arab civilians – took place in Jaffa, Yehudiah and in several Arab villages near Haifa.

As Jews and Arabs battled, the British were in the main bystanders. Their one thought was to leave Palestine altogether; to hear news of the date on which they would depart. Still, the relics of the old policy died hard and the date of open immigration, 1 February 1948, was still over a month off. On 29 December 1947 – exactly a month after the United Nations resolution – learning that two large ships, *Pan York* and *Pan Crescent*, with 7,000 immigrants on board each, were about to enter the Mediterranean from the Black Sea, the Commander of the British forces in the Middle East issued an 'operations order' for the arrest and boarding of the two ships before they could reach the coast of Palestine.

The Jewish Agency leaders did not want a confrontation with the British. They knew that immigration would be possible in two months' time. Ben-

Gurion gave orders that there was to be no resistance. The British boarded the ships, and they were taken to Cyprus, where their 14,000 passengers disembarked on the morning of 1 January 1948.

On 5 January 1948 a truck filled with oranges, under which explosives were hidden, was driven into Jaffa by two Stern Gang members, who, knowing exactly when the explosives were timed to go off, entered a nearby café and calmly drank coffee before leaving on foot for Tel Aviv. In the explosion that followed, more than twenty Arabs were killed, many of them in the coffee houses near the city's clock tower. That same morning the Haganah carried out a reprisal action against a military target, in this case the Semiramis Hotel in Jerusalem, which was being used as military headquarters by an Arab para-military organization (the Najada). It later emerged that the Spanish consular representative, who lived in the hotel, had been among the dead, as were eleven Christian Arabs who had not been involved in the conflict in any way. The Najada youth had not planned to meet there until the evening.

Two days later the Irgun planted a bomb at the Arab National Guard outpost at the Jaffa Gate, through which the Arabs had been refusing to allow food to pass to the Jewish Quarter in the Old City. Fourteen Arabs were killed. Three of the Irgun men were shot dead by British police. British statistics issued two days later gave the death toll in violent incidents in Jerusalem over the previous six weeks as 1,069 Arabs, 769 Jews and 123 Britons.

On January 10 the Arab Liberation Army, based in Syria, made its first attack across the border into Mandate Palestine. The objective of this Syrian-sponsored attack was the Jewish border settlement of Kfar Szold, which lay only 200 yards from the Syrian border. Nine hundred Arab soldiers joined the assault, but the defenders, who numbered less than a hundred, were helped by a British armoured unit that was stationed nearby, and which opened fire with its machine-guns and 2-pounder artillery pieces against the invaders. At the same time, Royal Air Force planes attacked an Arab military stronghold just across the border.

When the British Ambassador in Damascus protested to the Syrian government about the attack on Kfar Szold, the Syrian Prime Minister is said to have replied, 'Pretty soon the Arab armies will teach the Jews a lesson they will never forget.'

Within the Jewish Agency there were those who saw the fighting as opening up opportunities for extending Jewish settlement. On January 11, Yosef Weitz, the Director of the Lands Department of the Jewish National Fund, wrote in his diary, with regard to Arab tenant farmers in two villages in a predominantly Jewish area in the north of Palestine, 'Is it not now the time to be rid of them? Why continue to keep in our midst these thorns at a time when they pose a danger to us?'

In the next four months, while the British Mandate was still in place, the question of the expulsion of Arabs from their villages, and the subsequent blowing up of Arab houses, was to be much discussed, and acted upon. In February, when Bedouin who grazed their flocks on mostly Jewish-owned land in the Beisan Valley began to cross the River Jordan into Transjordan, Weitz wrote in his diary, 'Is it possible that now is the time to implement our original plan: to transfer them there.' But when Weitz asked Ben-Gurion to expel the Arabs from all the areas within the proposed Jewish area of statehood, Ben-Gurion refused to agree. Weitz then took it upon himself to encourage local Haganah units in several areas to evict Arab tenant farmers on Jewish land and then to burn the farmers' homes to the ground.

During February the Arabs focused their attentions on the more remote Jewish settlements. One of these was Nevatim, in the Negev, which was completely cut off by Arab attackers. The nearest Jewish settlement was Beit Eshel, five miles away. From there, the pilot of a small Piper Cub aeroplane, which had just arrived from Tel Aviv, decided to fly on to the beleaguered settlement. The pilot, Pinhas Ben-Porat, took with him a machine-gun and a volunteer. Later he described the sequel:

> I took as many hand-grenades as they could spare and prepared to take off. The settlers helped by pushing the plane, putting its wings on, and opening the gate. (The wings had been taken off soon after landing in order to enable the Piper Cub to be pushed into the narrow courtyard.)
>
> I removed the two side doors from the plane, so that it would be possible to operate through the opening, and fastened the gunner to his chair, with the machine-gun placed so that it would not hit the propeller or the wheels. The magazines were arranged next to me, so that I could help the gunner load while piloting the plane. I took off along the mud track leading to the settlement.
>
> A few minutes later I saw Nevatim. The terrain around the settlement was criss-crossed by valleys and creeks. Near the settlement were the water tower and some vegetable plantations. I observed large-scale Arab movements but could not determine the number of attackers, because the terrain provided good cover.
>
> Since it was impossible to fire in the direction of flight, but only side-ways and downwards, I would instruct my companion when to pull the trigger, while shifting the plane toward the target. We fired wherever we saw people, or wherever we thought they might be; we fired, threw our hand-grenades – and soon the rout set in. Presumably this was the first time in their lives that the Bedouin had tasted bullets from above.
>
> I flew low once or twice, in order to fire under the two bridges near the settlement, assuming that the attackers would try and find shelter there. Then I circled again, searching for a convenient landing spot and

came down near the settlement fence. I left the engine running, so as to be ready to take off again any minute. I went up to the fence to find someone, but the courtyard was empty. No one could be seen. Everyone was in position and did not dare to leave the trenches. I shouted into the courtyard for someone to come out, because I was unable to climb over the barbed wire.

An excited young man came toward me; from all sides they shouted to him: 'Bend down! Lie down!' They were not certain the attack had passed. I said to him: 'I came to fetch the wounded.' Meanwhile the men emerged from their trenches. Their faces showed surprise at my arrival and that of my plane. Two settlers had been wounded, one of them seriously.

It was not easy to evacuate the seriously wounded man through the barbed wire. I cut two strands in the fence and with the gunner's help lifted the stretcher over the fence. The wounded man suffered greatly but controlled himself. I removed the seats and arranged a place for him to lie near the pilot seat. I was compelled to leave the gunner, a member of Beit Eshel, behind in Nevatim, because the take-off area was so small that I doubted I would be able to take off at all.

A few minutes later the pilot landed at Beit Eshel. There, without removing the wounded man from the plane, a doctor rendered emergency aid, and once more the pilot took off to evacuate the wounded man to Tel Aviv. He was back in Tel Aviv at eleven in the morning, three hours after he had left the city. After cleaning his plane, and removing the cartridges, he took off again for Lydda airport, where his unsuspected Piper Cub, clearly owned by a private company and piloted by a commercial pilot, came down innocently to refuel.

The following day, the Arabs informed the British authorities in Beersheba that a Red Army plane (half of the Piper Cub was painted red) had been used by the Zionists in Nevatim. The British ascribed the story to 'fertile oriental imagination'.

Also besieged in the first and second week of January 1948 were the four isolated Jewish settlements in the Hebron Hills: Kfar Etzion, Massuot Yitzhak, Ein Zurim and Revadim (known as the Etzion bloc). A thousand Arabs, drawn from the villages in the area, had surrounded the bloc, under personal command of Abdel-Kader al-Husseini, the Mufti's cousin. There were fewer than thirty armed defenders. On the eve of the attack, hundreds of Arab men, women and even children arrived from their villages carrying empty bags in which to put the loot they anticipated would be theirs within a few hours. But the attack was driven off, and 150 Arabs killed.

The bloc, still closely besieged, appealed for arms, ammunition and reinforcements. On January 16 a relief column set off to try to get through to them. They carried explosives with which it was hoped that the bloc's defenders would blow up the road bridges between them and Hebron, along which Arab reinforcements were being brought both to the bloc and to Jerusalem.

Thirty-five men left Hartuv on foot, without a radio, making their way across the hills. Not one of them arrived at the Etzion bloc. It is believed that an Arab farmer whom they met on the way gave them false directions and then alerted nearby Arab villagers to their whereabouts. The Arab desire to revenge the humiliating defeat a few days earlier was strong. Before the thirty-five could reach the besieged settlements they were ambushed. None survived. Two days later the bodies were found by a search party and brought into the bloc. 'After their funeral', wrote the bloc's commander, 'the isolation was emphasized and clarified in all its starkness. With lowered heads and heavy hearts, we felt that we had lost live contact with Jerusalem, and our loneliness was underlined in all its cruelty.' The siege continued.

The Arabs had been able to enlist the support of a number of deserters from the British army, and also of a few Yugoslav and Polish soldiers stationed in Palestine. A number of German prisoners-of-war, who had only recently been released from the camps where they had been held in Egypt since 1942, also volunteered their services in the Arab cause. The first use of foreign volunteers was on January 14 in Haifa when a postal delivery truck was commandeered by volunteers, and exploded in the centre of the Jewish part of town. Fifty Jews were injured. Two weeks later, on the night of February 1, a British armoured car, likewise commandeered by British army deserters, was loaded with explosives and driven to the offices of the *Palestine Post* in the centre of Jewish Jerusalem. The building was demolished.

Two isolated Jewish suburbs in Jerusalem itself were attacked at the beginning of February – Yemin Moshe below the Jaffa Gate and Mekor Hayim in the south of the city. Both attacks were driven off. The Jews were determined not to let Jerusalem fall. In the north an unexpected problem confronted the Palmach during an assault on a village in Upper Galilee from which Arab attacks had come. A Jewish girl leading her section was ordered to reach a building, throw in a grenade and then enter the building. As she was about to throw her grenade she heard a baby crying. She decided not to throw the grenade. When questioned by her superiors – who were understandably concerned that the building had not been taken – she quoted her explicit instructions, 'You are to hit only gang leaders and agitators.'

As the Arab attacks intensified, the Haganah felt obliged to rethink its policy. In January, after the platoon that had tried to reach the besieged Etzion bloc was destroyed, a new policy was decided on. There was to be no more 'individual distinction'. The new instruction read, 'If Salameh is the base from which Tel Aviv is attacked, then Salameh will be the target for a counter-attack; we can no longer differentiate between those who actively participate in the attacks on Tel Aviv, and those who do not. From now on only a differentiation in location is possible: we will continue to draw a line between village and city, between areas dominated by the Mufti and those under control of his opponents.'

It was thus decided to engage in counter-attacks as bitter and as effective as possible wherever Jews were the victims of attack. Only areas that had remained quiet hitherto would not be affected. It was also decided to undertake counter-attacks that would echo the type engaged in by Arabs: when they harassed Jewish communications, their communications would be harassed; if they besieged Jewish cities, Arab cities would be similarly besieged.

What had begun as spasmodic actions and responses had turned into a full-scale war, even though the British were still nominally the rulers of Palestine. In Jerusalem, the Arabs living in the western parts of the city had mostly left by the end of January. In the suburb of Romema at the entrance to the city the Jewish inhabitants, a minority, had wanted to leave, but the Haganah had persuaded them to stay. It was the Arabs who left. 'The eviction of Arab Romema had eased the traffic situation', the Haganah commander in Jerusalem, Israel Zablodovsky, reported to Ben-Gurion on January 20. As to one of the most prosperous Arab residential suburbs in the city, Zablodovsky added laconically, 'Talbiyeh is also increasingly becoming Jewish, though a few Arabs remain.'

On February 5 Ben-Gurion instructed the Haganah to allow Jewish civilians to move into abandoned and conquered Arab districts in Jerusalem. Two days later he gave an account of the situation in the city to the leaders of Mapai, many of whom had come from Tel Aviv and the coast. 'From your entry into Jerusalem through Lifta-Romema,' he told them, 'through Mahane Yehuda, King George V Street and Mea Shearim – there are no strangers left. One hundred per cent Jews. Since Jerusalem's destruction in the days of the Romans – it hasn't been so Jewish as it is now. In many Arab districts in the West – one sees not one Arab.' Ben-Gurion added, 'I do not assume this will change.'

What had happened in Jerusalem, Ben-Gurion told his party leaders, 'could well happen in great parts of the country – if we hold on.' In the coming 'six or eight or ten months of the war' there would be 'great changes in the composition of the population of the country'.

On February 15 the Haganah captured the Arab village of Caesarea. The Arab homes in the village were on land leased from the Jews. Those inhabitants who had not already fled were ordered to leave. The twenty villagers who would not go were then expelled, and thirty houses destroyed. Six were left intact for lack of explosives. Yitzhak Rabin, the Palmach operations officer, had opposed the blowing up of the houses (which were owned by Jews), but had been overruled by his superiors.

On February 16 the Arab Liberation Army, having been repulsed at Kfar Szold as a result of British military intervention, attacked again, in an area where there were no British forces. This time the Arabs were commanded by one of their most charismatic leaders, Fawzi al-Kaukji. Their objective was the settlement of Tirat Zvi, in the southern Beisan Valley. But the settlers were prepared and had carefully organized themselves in their separate military tasks with mutual support between them, and the attacking army was driven off having suffered heavy casualties.

On the night of February 22 three British military vehicles, again commandeered by British army deserters, were exploded on Ben Yehuda Street in Jerusalem. Fifty-two Jews were killed: most of them were asleep when the explosives went off. Two of the drivers of the vehicles were the same two British deserters who had earlier been involved in the *Palestine Post* explosion. In reprisal, the Stern Gang killed ten British soldiers. That night the British authorities agreed that their troops would no longer enter the Jewish parts of the city. Zipporah Porath, a young American student who had come to Palestine four months earlier, and who found herself acting as a nurse on Ben Yehuda Street, wrote to her parents in the United States, 'Are there words to describe senseless human tragedy? Will I, can I, ever forget this day? I am becoming like the Jews who live here: every shock and sorrow nurtures them to grim restraint and fierce dedication.'

On March 1 at a plenary session of the Jewish National Council the Chairman, David Remez – who for thirteen years had been Secretary-General of the Histadrut – proposed the establishment in its place of a Provisional Council of State. His proposal was accepted. The Provisional Council became the legislative body of the Jews of Palestine, who, with its creation, took a clear and essential step towards statehood. Ten days later, on March 11, a car belonging to the United States Consulate General in Jerusalem flying the Stars and Stripes was driven by an Arab driver into the courtyard of the Jewish Agency building where the Provisional Council was located. The Agency staff, recognizing both the car and the driver, thought no more of it. The driver walked off and the car blew up, killing thirteen people.

Among the dead were a thirteen-year-old messenger boy, Chaim Polotov (his first name means 'Life' in Hebrew) and seventy-one-year-old Dr Leib Jaffe, one of only five men still alive who had been a delegate to the First Zionist Congress in 1897.

Several members of the staff of the United Nations Commission established after the partition resolution the previous November were in Jerusalem at the time of the Jewish Agency bomb. Escorting them were two Jewish Agency officials, Walter Eytan and Chaim Herzog (Eytan had worked in British Signals Intelligence during the Second World War; Herzog had seen active service with the British army). Herzog, whose wife Aura was injured in the explosion, went straight from the Agency building to the office of one of the staff of the commission, a Norwegian, Colonel Ragnvald Alfred Roscher-Lund. 'Only then,' Herzog later wrote, 'did I realize that I was covered with dust and my shirt was soaked with blood. I took Roscher-Lund to the Agency and showed him the ruins. Workers were already laying bricks and repairing the internal damage to the building. As secretaries cleaned up and workers moved silently carrying bricks, his eyes filled with tears. "Such a nation will never be defeated," he said.'

The United Nations Security Council continued to debate the future of Palestine. On March 19 – eight days after the Jewish Agency forecourt bomb – the United States representative, Warren Austin, announced that in view of the continuing bloodshed, partition was no longer possible. It was the view of the United States that Palestine ought to be put under the control of the United Nations. This was a blow to Jewish hopes of statehood. Behind it lay deep United States fears, felt most keenly by the new Secretary of State, George C. Marshall, the most powerful member of the Truman Cabinet, that the Soviet Union would be able to take advantage of chaos in the Middle East by throwing its support behind one or other of the warring parties. Marshall warned the Jews against declaring a State of their own. Seeking to gain advantage from the Americans' reticence the Soviet Union was showing considerable verbal support for the Jews of Palestine, and, through the new Communist government in Czechoslovakia – which had come to power in February – was allowing arms to be bought by emissaries of the Haganah.

The American proposal for a United Nations Trusteeship over Palestine was challenged within twenty-four hours by Ben-Gurion, who told a specially summoned press conference in Tel Aviv, 'It is we who will decide the fate of Palestine. We cannot agree to any sort of Trusteeship, permanent or temporary. The Jewish State exists because we defend it.' Behind these tough words lay an even tougher decision: to plan for the establishment, if necessary without international support, of a provisional government.

There were 50,000 Jews being held behind barbed wire in the British detention camps in Cyprus. All of them had been caught while trying to enter Palestine by ship, and had been detained. Many of them were able-bodied; some had previous experience of fighting in the Second World War. The Haganah was desperate to train as many as possible for service in Palestine as soon as the British would let them leave. On March 20 the first course was started on the island for weapons instructors. Seventy mock rifles, made out of wood, and wooden grenades, were the principal weapons. The Hebrew language was taught as part of the weapons drill.

In Palestine, the Haganah was forced to abandon the coastal highway south of Tel Aviv–Jaffa on March 26. The Jewish settlements and units in the Negev were cut off. Only two small Piper Cub aircraft maintained contact between south and north. Isolated settlements throughout the country relied upon the Piper Cubs to bring them minimum supplies – and also newspapers from which they could follow the course of the war elsewhere. The Ot Hagvurah (Medal for Bravery) – the highest decoration for bravery – was given to Zvi Siebel, who landed at Ben Shemen when the settlers were pinned down in their positions by Arab gunfire. This was to be the sole Ot Hagvurah awarded to a pilot in the War of Independence. Only twelve such decorations were awarded in the whole war.

Slowly the British troops began to withdraw from their outlying camps and fortified strongholds. The Operations Officer of the Haganah, Yigael Yadin, was uneasy at the prospects for a prolonged struggle. In summing up the events of February, he informed the Haganah High Command, 'The enemy is freely choosing the place and intensity of action. Our loss of strategic in-itiative at this stage of development will cause an extremely dangerous situation in the stages to come.'

Following Yadin's warning, the Haganah prepared an operational plan – Plan D – to be put into effect when the British evacuation had reached such a point that no British intervention would be possible. 'The objective of this plan,' Yadin explained in its preamble, 'is to gain control of the territory of the Hebrew State and defend its borders.' It also aimed, he explained, at gaining control 'of the areas of Jewish settlement and population' outside the area allocated to the Jewish State.

In Plan D, the Haganah addressed, for the first time, the question of how to deal with Arab towns and villages. The strategic basis of the plan was that if an Arab town or village was close to a Jewish settlement or town, and prevented continuity between the Jewish cities and neighbouring settle-ments, or disrupted the essential military lines of communication, it should be occupied. The man who had established the original strategy of the Haganah, Eliahu Golomb, had laid down clearly several years earlier: 'Haganah does not in any way intend to create a force to dominate others. We cannot imagine a situation in which Nablus, an Arab city surrounded by Arab villages, would be captured by a Jewish force and turned into a Jewish city.'

Whether Plan D contradicted this basic principle was much debated. In defence of those who said that the principle had not been altered, only the circumstances and the emphasis, the rest of Golomb's original policy state-ment was cited: 'Our function is to defend by force of arms, wherever and whenever and against whomever such defence will be required, our freedom of settlement, of immigration, of development and of self-determination. We do not envisage a force which will dominate others, which will terrorize or subject others, but a force which will defend the right of the Jews to come to their country, to settle it and to lead in it a free and sovereign existence.'

What was at stake in March 1948 was, in the view of those who devised Plan D, the very 'sovereign existence' of the Jews in their land. Hence the instruction in the plan that if Arab towns or villages occupied strategic points, were located on essential routes of communication, or were used as enemy bases, the Haganah must undertake 'destruction of the armed force and the expulsion of the population outside the borders of the State'.

How vulnerable the Jews were with regard to their 'sovereign existence' was made devastatingly clear when a Haganah convoy sent to relieve the isolated kibbutz of Yehiam at the end of March was attacked by Arab forces and destroyed; all forty-six Jewish soldiers lost their lives.

On March 22 Arab forces at the Bab al-Wad and in the hills leading from there to Jerusalem cut the city off from the coast for all Jewish traffic. For more than a week Haganah companies tried in vain to break the siege. Then on March 31 Ben-Gurion summoned the High Command of the Haganah to his home in Tel Aviv. 'There is only one burning question,' he told them, 'and that is the war for the road to Jerusalem.'

Hitherto, Haganah military activity had been carried out at company level. But in order to try to relieve Jerusalem a force of 1,500 men – three times as large as any previous Haganah operation – was considered essential to open the road and safeguard the convoys.

The Jerusalem road operation was given the code-name Nachshon, after the former Hebrew slave (Nachshon, the son of Aminadav) who, at the time of the Exodus, had been the first to jump into the parting waters of the Red Sea, and to make his way forward on dry land. To carry out Operation Nachshon, arms were requisitioned from Haganah units throughout the country, and operations elsewhere were brought almost entirely to a halt. But most of the arms being brought from the north were held up because of a curfew imposed by the British on the Haifa–Tel Aviv road – a curfew imposed at that very moment in reaction to an independent Irgun operation. Only a few of the arms could be brought down by sea, and Operation Nachshon was almost called off. It was saved by the arrival, on the night of April 1–2, of the first shipment of arms to reach Palestine by air from Czechoslovakia. Fortunately for the Haganah, there was an airfield in the south, Beit Darass, which had just been evacuated by the British.

The airfield at Beit Darass was made operational for a single night. As the DC-4 cargo plane approached the landing strip was illuminated for a few minutes by a series of electrical flashes. The plane landed, the arms were unloaded, the plane refuelled and took off. The arms – 200 rifles, 40 machine-guns, and 150,000 rounds of rifle and machine-gun ammunition – were dispersed to the neighbouring settlements. The following morning, as rumours spread that a plane had landed, British officers returned to inspect the field. They found nothing. On the next day, April 3, a Yugoslav cargo ship docked at Tel Aviv. Without the British seeing what was happening, Jewish dock workers unloaded a further arms cache – which had been hidden under the ship's cargo of potatoes and onions – and it too was spirited away. With this shipload, a further 500 rifles, 200 machine-guns, and more than 5,000,000 rounds of ammunition reached the Haganah forces.

This arms shipment was a further instalment of the weaponry bought from the Czechoslovak government by a small purchasing mission sent there by Ben-Gurion, and headed by the Austrian-born Ehud Avriel. These arms gave Operation Nachshon a chance of success. In preparation two preliminary operations were launched, both planned with the utmost care. In the first a small, specially trained Haganah detachment struck at the base of the Mufti's area commander, Hassan Salameh, who had set up his headquarters in a two-storey building near the Arab town of Ramle, hidden by citrus groves. The building had earlier been a British officers' training base. The Haganah attack was a success. Most of the building was blown up, and several of Salameh's senior personnel were killed. In a second preliminary operation, the Haganah captured the Arab village of Kastel, overlooking the road to Jerusalem only five miles west of the city.

Operation Nachshon was launched on the evening of April 6. Despite the success of the two preliminary attacks, the main attack did not go according to plan. Two Arab villages which were to have been occupied, Saris and Beit Machzir, drove the attackers away, while Arab forces based in the Arab village of Kolonia, on the outskirts of Jerusalem, managed to cross the main road and attack the Jewish children's farm at Motza. The farm, from which the children had been evacuated, was defended by a small Haganah detachment which was overrun. Further west, however, several Arab villages south of the road were captured.

A convoy of trucks which left the coast that night managed to get through to the city without loss. Its journey took ten hours. It was the first convoy to reach Jewish Jerusalem in two weeks. The city itself was under continuous bombardment from the Arab Legion artillery located on the high ground north of the Damascus Gate. The besieged Jewish Quarter of the Old City was being specially targeted. On April 13 a kindergarten was hit and twenty children injured.

Throughout the second and third weeks of April the Arabs strove to regain control of the Tel Aviv–Jerusalem highway. For six days they attacked the Jewish force that had captured the Arab village of Kastel, dominating the road on its eastern end. The Arab commander, Abdel Kader al-Husseini, a relative of the Mufti, who had been visiting Damascus in search of reinforcements of his own, hurried back to the Jerusalem sector and took personal charge of the attack on Kastel. On April 9 the Jewish defenders of Kastel could continue no more. Their supplies were exhausted. The last order given to them by their platoon commander Shimon Alfasi was: 'All privates will retreat – all commanders will cover their withdrawal.' Alfasi was killed in the battle, covering the retreat. His order became a watchword for many future actions. Reinforcements, arriving too late, could do nothing more than provide covering fire for the retreating soldiers.

The recapture of Kastel was a moment of triumph for the Arab forces, and could have marked a turning point in their military fortunes. But in the very last stage of the battle, Abdel Kader was killed. He had believed the battle to be over, and had walked up to a Haganah machine-gunner who was still at his post, and whom he believed to be an Arab already in possession of the Jewish position.

For twenty-four hours the Arabs who had captured Kastel held the Jerusalem road, and the siege returned. But the death of Abdel Kader was a psychological blow to them from which they did not recover. The leaders of the various units, unwilling to adopt a static defence – rather than sweeping on towards Jerusalem – and having no strong leader to guide them, quarrelled among themselves, and set off northward, back to their villages. When Haganah soldiers prepared to attack Kastel again, they found that it had been abandoned.

The day of the recapture of Kastel was also the day on which Irgun and Stern Gang forces attacked the Arab village of Deir Yassin, on a hill even closer to Jerusalem. It was the last village on the western side of Jerusalem whose Arab inhabitants had not largely or totally fled. The attack, in which 245 Arabs were killed – many of them women and children – generated a controversy and a bitterness that remain to this day a contentious issue in Israeli life. The official account written in 1961 by Lieutenant-Colonel Netanel Lorch, who had fought in the war, and was later head of the Military History Division of the Israeli General Staff, describes how Irgun and Stern Gang forces 'massacred hundreds of villagers, took the rest prisoner and paraded them proudly through the streets of Jerusalem'. The Jewish Agency and the Haganah High Command both 'immediately expressed their deep disgust and regret'. A Jewish Jerusalemite, Harry Levin, wrote in his diary, 'None of the barbarities the Arabs have committed in the past months can excuse this foul thing done by Jews. Most Jews I have spoken to are horrified.'

CHAPTER TEN

The conflict intensifies
April–May 1948

April 15 was a black day in the history of the Arab–Jewish conflict. Although Operation Nachshon had succeeded in opening the road to Jerusalem, and was disbanded that day, its triumph had already been overshadowed by the killings at Deir Yassin, and was to be eclipsed in the minds of the Jewish public by the counter-killings which followed swiftly in their wake. The most savage of these was the Arab attack on April 15 in which seventy-seven Jewish doctors, nurses and patients, on their way in armoured buses to the Hadassah Hospital enclave on Mount Scopus, were ambushed and killed.

Among the dead in the Hadassah convoy was Chaim Yassky, an ophthalmologist and Director of the Hadassah Medical Organization, whose pioneering work on the scourge of trachoma had saved the eyesight of tens of thousands of Arabs. As a young man in Odessa before the First World War, he had taken part in the successful Jewish self-defence there against the Russian pogromists. That night Zipporah Porath wrote in her diary, 'All Jerusalem is walking round asking itself, Is there no end to it?'

Of the twenty-eight people in the Hadassah convoy who were saved, only eight had not been wounded.

That evening, amid the violence and anguish of fighting, the National Opera held its first performance in Tel Aviv. The opera was the creation of Edis de Philippe from Brooklyn and Mordechai Galinkin from Leningrad. The first performance was Massenet's romantic opera *Thais*, sung in Hebrew. It was attended by many leading figures including Ben-Gurion. 'Noisy accompaniment,' wrote Joan Comay, 'was supplied by the gunfire from nearby skirmishes between Tel Aviv and Jaffa.'

During the final phase of the battle for the Jerusalem road, the commander of the Arab Liberation Army, Fawzi al-Kaukji, had tried to take advantage of the Haganah's preoccupation with capturing the settlement of Mishmar

Ha-Emek, using field guns which he had been given by the Syrian army. The battle lasted for eight days. At one point much-needed ammunition was dropped to the defenders from one of the Piper Cubs. At another point, a British officer managed to secure a twenty-four-hour cease-fire, to enable the evacuation of all the children and wounded soldiers in the settlement. The Haganah and Kaukji's forces battled for control of several Arab villages in the hills above and behind Mishmar Ha-Emek. One of these villages was overrun by the Arabs eleven times, but retaken each time by the Haganah. Then on April 12 the Haganah took the offensive, isolating Kaukji's forces, who were forced to retreat. Kaukji was unlucky as, unknown to him, a single, isolated building not far from the settlement had been used for several years as the advance training base for all Haganah commanders from platoon leader upwards, so the area was very familiar to them.

Kaukji had secured the help of Druse soldiers – with whom Syria's relations were historically uneven – from the Jebel Druse mountain region of Syria in return for payment. As the battle for Mishmar Ha-Emek was reaching its climax, he had appealed to them to enter the battle. 'I am in trouble,' he signalled. 'If you do not help me, my complaint will be to God.' In his written instructions to his men to go forward the Druse commander told them (in what one historian has described as 'probably a unique document in modern military history') that they had no specifically Druse mission or interest in the war; that they were mercenaries who received pay; and that they were therefore obliged to do what they were told.

The Druse went into action, fighting bravely against Haganah positions just to the east of Ramat Yohanan. After suffering heavy losses the Druse were driven off. It was during this battle that Moshe Dayan's brother Zorik was killed.

The British forces were still in Palestine, but were withdrawing rapidly. No one yet knew the date of the final withdrawal, which under the United Nations partition resolution could have been as late as September, but could also be as early as May or June.

In Tiberias local Jewish and Arab leaders worked to maintain friendly relations between the two communities. But at the beginning of April the Arab Liberation Army decided to take the initiative, isolated the Jewish Quarter in the lower town, and effectively blocked all movement between Upper Galilee and the Jewish settlements elsewhere in Palestine. On April 18 the Haganah took retaliatory action, using Plan D as its strategic justification, and, helped by a company of Palmach soldiers, cut Arab Tiberias in two. There was no need to attack the two halves. The Arab population decided to leave, and, assisted by British army units in their southward trek, crossed into Transjordan.

The Haganah commander issued a declaration of victory: 'The Arab citizens have left this city. I herewith declare an autonomous Jewish regime in this city.' The commander of the Arab Liberation Army appealed to King Abdullah to send him reinforcements 'which would save the city and capture it from the Jews'. Abdullah replied curtly that he had already sent Arab Legion vehicles to evacuate non-combatants.

On April 21, three days after the Haganah capture of Tiberias, the general commanding the British forces in Haifa informed the Jewish and Arab leaders in the city that he was planning to withdraw all British troops to the port. For twenty-four hours, Arab and Jewish forces battled for control of the city. The British general wagered a friend a bottle of whisky that no one side could control Haifa in under a week. He lost his bet. Haganah forces, using a three-pronged attack (Operation Scissors) from the Jewish-held into the Arab-held areas, secured the whole city in forty-eight hours.

The Arabs were unfortunate that the city's chief magistrate, Ahmad Bey Khalil, and the Arab militia commander, Amin Bey Azzadin – a Lebanese Druse – both left the city on the first night of the battle, by sea for Beirut in a French steamer, ostensibly to seek military reinforcements. They never returned. After they sailed, Amin Bey's deputy, Yunis Nafa'a, a former sanitary inspector with the Haifa Municipality, assumed command of the Arab forces, but he too left the city by sea within twenty-four hours of assuming command.

At the invitation of the British general, Arab and Jewish municipal leaders came together, and the general asked if the Arabs were willing to surrender. They requested time to consult. When they returned, they said they were not prepared to sign any surrender document, but would leave the city. Abba Hushy, the Jewish Mayor (Haifa had a Jewish majority), pleaded with them to remain in the city where they had lived for so long, and had worked side by side with the Jews. They refused to stay. With the departure of Ahmad Bey, Amin Bey and Yunis Nafa'a they felt deserted. A British Military Intelligence Report regarded Amin Bey's flight as 'probably the greatest single factor' in the demoralization of the Haifa Arab community. 'The desertion of their leaders and the sight of so much cowardice in high places completely unnerved the inhabitants,' the British assessment concluded.

Abandoned by their leaders, the Arabs of Haifa began an exodus by sea to Acre and Lebanon, and by land to Nazareth. The Jews of Haifa made efforts to persuade the Arabs to stay. Visiting the port on April 22, Yosef Weitz, the Director of the Lands Department of the Jewish National Fund, wrote in his diary that the state of mind among the Arabs who were fleeing 'should be exploited'. The inhabitants of Haifa who had not left should be pressed to leave. 'We must establish our State', Weitz noted on April 22. But this was not the policy of the local Haganah and it was not followed up. On April 25 the correspondent of *The Times* reported to London: 'The Jews wish the Arabs to settle down again to normal routine, but evacuation continues.' On May 5 a British military observer noted, 'The Jews have been making extensive efforts to prevent wholesale evacuation, but their propaganda appears to have had very little effect.'

While the Haganah troops distributed leaflets calling on the Arabs to remain calm and return to work (and forbidding looting by Jewish soldiers), on the Arab side, the Arab Higher Committee (then based in Damascus) was urging the Arabs to leave; one reason it gave was that Arab air forces intended to bomb Haifa and hit at the Jews that way. Another influence was the Mufti of Jerusalem who was hoping to set up a provisional Palestine Arab government in the north of the country, and who was insisting that the invasion by the Arab armies was close and that the whole country would then fall into Arab hands. Until then, it was safest to be far away from likely areas of conflict.

Only a few thousand Arabs remained in Haifa. They and their descendants live in Haifa to this day. In several villages outside Haifa, the hope that Weitz had of an enforced Arab exodus was satisfied. When he went to see the adjutant of the local Haganah commander on April 24 he was told that two nearby Arab villages were being evacuated. 'I was happy to hear from him that this line was being adopted by the command,' Weitz wrote in his diary, 'to frighten the Arabs so long as flight-inducing fear was upon them.'

On April 20, with Kastel in the hands of the Haganah, a convoy set off for Jerusalem from Tel Aviv. Among those travelling with it were David Ben-Gurion and Yitzhak Sadeh, the Chief of Staff of the Haganah. The commander of the Palmach's Harel Brigade, Yitzhak Rabin, who was responsible for the road, protested that the journey was still not entirely safe. The Arabs could still attack in the Latrun area. His caution was overruled. 'The convoy set out at dawn, headed by the escorting force,' he later recalled, 'and, directly behind it, an armoured bus carrying Ben-Gurion. The lead vehicles got through without any opposition, but then the Arabs began their attack – perhaps the heaviest onslaught ever launched on a convoy to Jerusalem. Hundreds of Arab irregulars had deployed along the road from Latrun to Bab al-Wad and their fire managed to bring many vehicles to a halt. I was at the rear of the convoy and, when I received word of the attack, I ordered our reserve armoured company forward to take up positions and prevent the attackers from reaching the road. The fire was particularly heavy in the Latrun section.'

Owing to the gravity of the situation, Rabin recalled, 'I even ordered two stolen British armoured cars to be brought out of concealment and sent into action to rescue the convoy. After a prolonged battle we managed to extricate most of the vehicles. The toll was over twenty dead, many wounded and the loss of twenty vehicles. Our troops had fought valiantly but this was cold comfort in the face of the heavy loss of life. And, of course, the road was again closed behind us.'

Ben-Gurion had wanted to spend Passover with the troops in Jerusalem, and he did so. He also supervised the plans to extend Jewish control in the city. Several Arab villages in the Jerusalem corridor had already been blown up, their deserted buildings dynamited one by one. On April 24 an attempt was made to drive the Arab forces from their fortified position on the high ground of Nebi Samwil (the tomb of the Prophet Samuel) to the north of the city, but it was driven off and twenty-five of the Jewish attackers were killed. Four days later, Yitzhak Rabin occupied the Arab suburb of Sheikh Jarrah, hoping to link up the Mount Scopus enclave with the rest of Jewish Jerusalem, but following an ultimatum from the British, one of whose army camps was nearby, he withdrew.

Throughout Palestine Arabs were fleeing from the scene of battle, or from imminent battle, and their villages being destroyed. On April 21 Weitz wrote in his diary, 'Our army is steadily conquering Arab villages and their inhabitants are afraid and flee like mice. You have no idea what happened in the Arab villages. It is enough that during the night several shells will whistle over them and they flee for their lives. Villages are steadily emptying, and if we continue on this course – and we shall certainly do so as our strength increases – then villages will empty of their inhabitants.'

In northern Galilee, the Haganah's Plan D required control of the overwhelmingly Arab town of Safed, which had a small Jewish population, mostly of religious Jews, and a Jewish presence dating back to antiquity. Plan D also required control of the Arab villages in the high ground around Safed, from which the many Jewish settlements in the 'Finger of Galilee' could be threatened at will, and often were. There were 10,000 Arabs living in Safed, and 1,500 Jews. The Jews were mostly elderly religious living in the orbit of the synagogues and shrines of the many ancient rabbis who had made Safed their home. Since the Arabs had attacked a Jewish bus making its way to Safed in February, the Jewish quarter of the town had been under siege, despite the presence of British forces.

As the siege continued, food supplies ran short. Vegetables, fruit, milk and eggs became unobtainable. Even water and flour were in desperately short supply. Each day, the Arab attackers drew closer to the heart of the Jewish quarter, systematically blowing up Jewish houses as they pressed in on the central area.

Then, on the evening of April 16, a few hours before the British were to leave Safed to its fate, a platoon of Palmach soldiers, based on the heights of nearby Mount Canaan, made its way secretly into the centre of the Jewish quarter of Safed and took control of the defences. The arrival of the platoon raised the morale of the besieged, who began to strengthen their fortifications under Palmach guidance. With the approval of the rabbis, whose writ ran large among the predominantly Orthodox Jews of Safed, permission was given to continue with this defensive work over the Passover.

The departure of the British from Safed nearly spelt doom for the besieged Jews for in his last official act the British commander handed over the keys and control of the British police fortress on the western slope of Mount Canaan to the Arabs, as well as three fortified positions inside the city: the Shalva House, the ancient citadel and the municipal police station. Fifteen miles north of Safed, another British police fort was handed over to the Arabs that day, at Nebi Yusha, on the main road leading from Lebanon into the Huleh valley. A Palmach attack on the fort was driven off, and twenty-eight Palmach soldiers killed.

A mile to the south was the Jewish settlement of Ramot Naftali, named after both the biblical tribe that lived in that area and Orde Wingate. It had been founded by soldiers who had served in the British Army in the Second World War: they called themselves the Wingate Group, in honour of Orde Wingate, the British soldier who had helped the Jews in Palestine before the war, and had been killed behind Japanese lines in Burma in 1944.

Ramot Naftali was attacked by Arab forces and besieged. Help from ground troops was impossible; but some supplies and ammunition were dropped to the defenders by Piper Cubs. Flying in one of them, Orde Wingate's widow Lorna threw down the Bible which he had always carried with him on his military campaigns in Palestine and Burma. At night, using steep mountain paths, the children of the settlement were evacuated to safety by a company of soldiers, each one of whom carried a child on his shoulders down the 2,000-foot precipitous slopes to the safety of Ayelet Ha-Shahar in the plain below. Today part of that evacuation route is on the Israel Track, the recently established cross-Israel hiking path.

To examine the situation in northern Galilee at first hand, the Palmach commander, Yigal Allon, flew north from Tel Aviv in a Piper Cub on April 21. When he returned on the following day he advocated an offensive strategy to seize control of the main Arab-held positions and all main roads. Allon recommended 'the harassment of Beisan in order to increase the flight from it' and 'the harassment of Arab Safed in order to speed up its evacuation'. This was Operation Yiftach (named after Jephtah, a judge in biblical times whose home had been across the River Jordan, overlooking Upper Galilee). It was to start as soon as the British completed their withdrawal from the area. On April 28 British troops pulled out of the last police fortress in their control in Upper Galilee, at Rosh Pinah in the valley immediately below Mount Canaan. The fort was then occupied by the Haganah.

On April 25 the first Jewish initiative against the Arabs of Jaffa took place. It was carried out by an Irgun detachment. But the Arabs beat the attackers off, and did so again on the following day. The Irgun High Command made contact with the Haganah and asked for a plan of joint operations. An agreement was quickly reached, coming into force on April 26, the day of the second Irgun setback. That day all Haganah commanders were sent an instruction which read, 'For the time being, this is to inform you that wherever there are Irgun positions, they will from now on come under the command of Haganah area commanders, through the Irgun position commanders. The Irgun will not undertake any action unless agreed upon beforehand with Haganah. This includes the acquisition of arms from the British army. The Irgun will be prepared to undertake operations on request from Haganah High Command, which it is able to carry out.'

Haganah commanders were ordered to avoid any course of action that might lead to conflict with the Irgun. A common strategy was being worked out under Plan D with regard to contiguity and security of Jewish settlements, in this case along the Mediterranean coast.

The British had not yet left Palestine and were smarting under the Haganah success in taking over Haifa, which it had been assumed the Arabs would hold. The British government was under pressure from the Arab States for having let Haifa 'go to the Jews'. To protect the Arab positions in Jaffa, British forces – which elsewhere were withdrawing from Palestine on a daily basis – were rushed from Cyprus. A tank battalion and an artillery regiment were landed on April 27. Ernest Bevin informed the British commanders that they must prevent the capture of Jaffa by the Jews 'at all costs'. Haganah positions in Tel Aviv were shelled by the newly arrived artillery regiment, and a British air attack took place on the isolated Jewish suburb of Bat Yam, to the south of Jaffa port.

A frontal assault on Jaffa seemed impossible. On April 29 the Haganah captured the two Arab villages just east of Bat Yam, from which attacks on Jewish road traffic into Tel Aviv had frequently been launched. Jaffa was effectively surrounded. But Kaukji, responding to the appeals of its defenders, sent substantial reinforcements, including artillery, and Tel Aviv was again shelled, this time by Arab gunners.

While Jaffa awaited a combined Haganah and Irgun assault, and as the date of the complete British withdrawal from Palestine drew nearer, both Jewish and Arab troops tried to secure the bases elsewhere in the country which would become crucial in the event of continuing conflict. On April 29, following the British evacuation of Zemach – a small town at the southern end of the Sea of Galilee, on the Haifa–Damascus railway – the Haganah secured the police fortress there. That same day, as British troops left the police fortress guarding the Haifa–Damascus road at Gesher, where it crossed the River Jordan, Haganah troops occupied it.

The Arab Legion, crossing the road bridge over the Jordan (there were three bridges there, the railway bridge, the modern road bridge and an ancient Romano-Turkish bridge), attacked the police fort and nearby settlement of Gesher. So confident were the Arabs of success that high-ranking Arab Legion officers, senior officials of the Transjordan government, and the heir to the throne of Transjordan, Crown Prince Talal, came to witness the Arab Legion victory. The Jewish settlers were told to evacuate their settlement within an hour and to hand it over to the Arab Legion. They refused to do so, and awaited the attack. When it came, it failed; the Arab Legion forces were unable to break into the police fort, and withdrew across the river.

The first phase of the Palmach plan to capture Safed was to secure a corridor through the mountains north of the city by capturing the Arab village of Birya. This was successful. But when the main offensive began on May 6, the Arab Liberation Army brought up artillery pieces that were able to shell the Jewish Quarter. British Army Headquarters in Haifa offered to mediate and to seek a truce during which the Jewish women and children could be evacuated. The Palmach rejected this offer and on the evening of May 10 renewed its assault on the Arab fortified positions. In the battle for the police station the company commander, Yitzhak Hochmann, was killed. In the battle for the Shalva House the platoon commander in charge, Abraham Licht, was also killed. But by morning, all three positions were in Palmach hands.

The Arabs of Safed began to leave, including the commander of the Arab forces, Adib Shishakli (later Prime Minister of Syria). With the police fort on Mount Canaan isolated, its defenders withdrew without fighting. The fall of Safed was a blow to Arab morale throughout the north. The city was by far the most important Arab centre in the region. As news of its fall spread the Arab villagers in the Huleh Valley abandoned their homes and fled north-wards into Lebanon. On May 4 the Palmach launched Operation Broom. Its aim was to clear out the Arab population from the area south of Rosh Pina between the north–south road and the Jordan River. The operation was helped by the Arab psychology at that time. With the invasion of Palestine by regular Arab armies believed to be imminent – once the British had finally left in eleven or twelve days' time – many Arabs felt that prudence dictated their departure until the Jews had been defeated and they could return to their homes.

The objective of Operation Broom was to destroy the bases from which Arab attacks were being made on Jewish positions, and 'to join lower and upper Galilee with a relatively wide and safe strip' – of Jewish territory. The attack began with mortar shelling. As many as 2,000 Arabs fled eastward into Syria. On May 10 Safed fell to the Palmach forces. 'The echo of the fall of Arab Safed carried far,' Yigal Allon wrote in the official history of the Palmach. 'The confidence of thousands of Arabs of the Huleh was shaken. We had only five days left, until 15 May. We regarded it as imperative to cleanse the interior of Galilee and create Jewish territorial continuity in the whole of Upper Galilee.' Allon's account continued:

> The protracted battles reduced our forces, and we faced major tasks in
> blocking the invasion routes. We, therefore, looked for a means that
> would not oblige us to use force to drive out the tens of thousands of
> hostile Arabs left in Galilee and who, in the event of an invasion, could

strike at us from behind. We tried to utilize a stratagem that exploited the defeats in Safed and in the area cleared by Broom – a stratagem that worked wonderfully.

I gathered the Jewish *mukhtars*, who had ties with the different Arab villages, and I asked them to whisper in the ears of several Arabs that giant Jewish reinforcements had reached Galilee and were about to clean out the villages of the Huleh, to advise them, as friends, to flee.

The flight encompassed tens of thousands. The stratagem fully achieved its objective, and we were able to deploy ourselves in face of the invaders along the borders, without fear for our rear.

Isolated for many months, the settlers of the Etzion bloc – on the road between Bethlehem and Hebron – held out against repeated Arab attacks. The children, and some women, had been evacuated towards the end of 1947. As the battle intensified on 4 May 1948 there were 110 settlers at Kfar Etzion, about 100 in Massuot Yitzhak, forty in Revadim, and thirty in Ein Zurim. A number of soldiers were also there: a company of Haganah and a platoon of Palmach, the escorts of a convoy that had failed to reach Jerusalem and had been forced to return to the bloc. Supplies were flown in by Piper Cub to a landing strip that had somehow been levelled in the hilly, stony terrain.

A renewed and sustained Arab attack on the Etzion bloc was launched on May 4. The main attackers were soldiers of the Arab Legion, assisted by infantrymen recruited from the nearby Arab villages. Twelve of the defenders were killed in that assault, but the Arab losses were higher and the attack was called off.

The British had stood aside from much of the conflict but their forces were not entirely free from danger. In a Stern Gang attack on a British army post near Tel Mond, seven British soldiers had been killed. On May 7 the chief of British Military Intelligence in Palestine summoned Chaim Herzog, who had become the deputy head of the Jewish Agency's Security Department, to warn him, as Herzog later wrote, 'that the British were now prepared to take violent action against the Jews. As a first step, every armoured vehicle – the only thing that ensured what little freedom of travel we had – would be confiscated and destroyed. I protested that the entire community could not be held responsible for the actions of a small group, but he was not sympathetic. Later that evening Roscher-Lund told me that the British were looking forward to additional operations by the Stern Group so they could settle accounts with the Jews once and for all.'

On May 10, in an attempt to see if war with Transjordan could be averted, the Jewish Agency sent one of its most formidable negotiators, Golda Meir, on a second secret mission to King Abdullah of Transjordan. She was authorized to seek a territorial agreement with him, along the lines of the United Nations partition plan. Accompanied by Ezra Danin, one of the Jewish Agency's Arab experts, Golda Meir travelled in disguise. She later recalled:

> I would travel in the traditional dark and voluminous robes of an Arab woman. I spoke no Arabic at all, but as a Moslem wife accompanying her husband, it was most unlikely that I would be called upon to say anything to anyone. The Arab dress and veils I needed had already been ordered, and Ezra explained the route to me. We would change cars several times, he said, in order to be sure that we were not followed; and at a given point that night someone would turn up not far from the king's palace to lead us to Abdullah.
>
> The major problem was to avoid arousing the suspicions of the Arab Legionnaires at the various check posts we had to pass before we got to the place where our guide was to meet us.
>
> It was a long, long series of rides through the night. First into one car, then out of it, into another for a few more miles and then, at Naharayim, into a third car. We didn't talk to each other at all during the journey. I had perfect faith in Ezra's ability to get us through the enemy lines safely, and I was much too concerned with the outcome of our mission to think about what would happen if, God forbid, we were caught.
>
> Luckily, though we had to identify ourselves several times, we got to our appointed meeting place on time and undetected. The man who was to take us to Abdullah was one of his most trusted associates, a Bedouin whom the king had adopted and raised since childhood and who was used to running perilous errands for his master.
>
> In his car, its windows covered with heavy black material, he drove Ezra and myself to his house. While we waited for Abdullah to appear, I talked to our guide's attractive and intelligent wife, who came from a well-to-do Turkish family and complained to me bitterly about the terrible monotony of her life in Transjordan. I remember thinking that I could have done with some monotony myself at that point, but I only nodded my head sympathetically.

The talks for which Golda Meir's journey had been undertaken were about to begin:

> Then Abdullah entered the room. He was very pale and seemed under great strain. Ezra interpreted for us and we talked for about an hour. I started the conversation by coming to the point at once. 'Have you broken your promise to me, after all?' I asked him.
>
> He didn't answer my question directly. Instead he said: 'When I made that promise, I thought I was in control of my own destiny and could do what I thought right. But since then I have learned otherwise.' Then

he went on to say that before he had been alone, but now, 'I am one of five,' the other four, we gathered, being Egypt, Syria, Lebanon and Iraq. Still, he thought war could be averted.

'Why are you in such a hurry to proclaim your State?' he asked me. 'What is the rush? You are so impatient!' I told him that I didn't think that a people who had waited 2,000 years should be described as being 'in a hurry', and he seemed to accept that.

'Don't you understand,' I said, 'that we are your only allies in this region? The others are all your enemies.' 'Yes,' he said, 'I know that. But what can I do? It is not up to me.' So then I said to him: 'You must know that if war is forced upon us, we will fight and we will win.' He sighed and again said 'Yes. I know that. It is your duty to fight. But why don't you wait a few years? Drop your demands for free immigration. I will take over the whole country and you will be represented in my parliament. I will treat you very well and there will be no war.'

I tried to explain to him that his plan was impossible. 'You know all that we have done and how hard we have worked,' I said. 'Do you think we did all that just to be represented in a foreign parliament? You know what we want and to what we aspire. If you can offer us nothing more than you have just done, then there will be a war and we will win it. But perhaps we can meet again – after the war and after there is a Jewish State.'

The king was not prepared to recognize Jewish statehood in any form, only Jewish autonomy under his own sovereignty. Golda Meir and Ezra Danin returned empty handed to Tel Aviv. Negotiations between a senior Arab Legion officer, Colonel Goldie – an Englishman – and Haganah officers, aimed (from the Transjordanian perspective) at coordinating plans with the Jews in such a way as 'to avoid clashes without appearing to betray the Arab cause', were likewise unsuccessful.

Ben-Gurion had to face a fateful decision: whether or not to declare statehood. It would also be a decision to give the Haganah the responsibility for survival or extinction. The United States was calling for a three-month ceasefire. The implication was that the United Nations would then take control of the country until a political solution acceptable to both sides could be worked out. From the Jewish point of view to accede to the American request would postpone statehood indefinitely.

On May 12 Ben-Gurion summoned his Provisional Council to an emergency meeting at the Jewish National Fund building in northern Tel Aviv. He had asked Golda Meir to be there, to report on whether Abdullah would agree to keep the peace along the River Jordan. 'When I entered the room,' she later recalled, 'he lifted his head, looked at me and said, "*Nu?*" I sat down and scribbled a note. "It did not work," I wrote. "There will be war. From Mafrak, Ezra and I saw the troop concentrations and the lights."'

Golda Meir added, 'I could hardly bear to watch Ben-Gurion's face as he read the note, but, thank God, he didn't change his mind – or ours.' There should be no delay in declaring statehood; that was Ben-Gurion's view. But it was far from unanimous. Declaring statehood meant risking the triumph of Arab armies over the new State. Half the council preferred to postpone statehood and accept a cease-fire.

Ben-Gurion summoned Yigael Yadin and Israel Galili from Haganah head-quarters in central Tel Aviv. Their view as to whether an Arab invasion could be resisted would be crucial. That morning at dawn the Arabs had renewed their attack on the isolated Etzion bloc, south of Jerusalem. The last attack had been repelled eight days earlier. This time the Arabs renewed it in force. As well as local Arab forces, the British-trained Arab Legion, the most profes-sional of Transjordan's forces, was also taking part in the attack.

Before setting out from Haganah headquarters, Yadin sent the defenders of the Etzion bloc a message urging them to hold out. On reaching the Jewish National Fund building, he and Galili were confronted by a very anxious group of people. Ben-Gurion began by telling Yadin, 'To you, there should be no political consideration. The members want to know what will happen if we establish the State and the Arabs invade. Do we have any possibility of maintaining our position? You, Yadin, must give a purely military answer. Don't tell us what you want, or what you think might be desirable, but just if you think that there is a possibility for our forces to stand firm.'

After so many months of planning, this was the moment of decision. Yadin began with a report on the Arab attack that was even then taking place on the Etzion bloc, describing how terrifying it was for the defenders to cope with the relentless artillery shelling and the arrival of Arab Legion armoured cars with heavy guns. Yet the same situation, he warned, could be expected on four other fronts if Haganah forces had to face the armies of Lebanon, Syria, Egypt and Iraq. 'The conventional forces of the neighbouring states with their equipment and weapons have an advantage,' Yadin confided. 'But one must not make a purely military determination of weapon against weapon and unit against unit.' Although the Haganah had no heavy arma-ment, he said, newly purchased artillery was expected to reach the country soon. Meanwhile, the Haganah had undergone training to knock out armoured vehicles with makeshift bombs.

Yadin continued: 'The question is, to what extent our people will be able to prevail against that force, considering the morale and ability of the enemy and our own tactics plan. It's been proven many times that the numbers and the formations are not always decisive.' The Haganah forces had a high level of training and determination. 'If I wanted to sum it all up and be cautious, I'd say that at this moment, our chances are about even. If I wanted to be more honest, I'd say that the other side has a significant edge.'

This was a dangerous moment. The Provisional Council was confronted with a grave dilemma. A 'significant edge' for the Arab armies could mean the destruction of the would-be State. Better perhaps to call in the United Nations and hold off the declaration of statehood. Ben-Gurion wrote in his memoirs, 'Yadin spoke with very great caution. He did not say that we would be able to stand.'

Ben-Gurion then told the meeting that there would be terrible dangers in suddenly calling a halt to the current military preparations and depending on the mercy of others, especially when heavy arms were already acquired and the mass conscription of Jewish youth was under way. 'If we can increase our forces, widen training, and increase our weapons, we can resist and even win.'

Yadin and Galili returned to Haganah headquarters. Yadin's immediate task was to amend Plan D – which related only to the conflict with the Arabs inside Palestine – in order to include in it the means of combating the invasion by five Arab armies across the Lebanese, Syrian, Transjordanian and Egyptian borders. He had only three days to amend the long and carefully considered plan.

With Yadin and Galili on their way back to Haganah headquarters, Ben-Gurion, as chairman of the Provisional Council, called for a decision by a show of hands on the United States proposal for a cease-fire. By six votes to four, the proposal was rejected. It was then agreed that statehood would be established on the following Friday afternoon, in two days' time.

Midnight had come and gone. A historic decision had been made. A Jewish State – the first for 2,000 years – would come into being, and would seek to defend itself against considerable odds. There remained one further point to be resolved, the name of the new State. By a vote of seven to two, the Council decided that it would be 'Israel'.

Throughout the deliberations in Tel Aviv about whether or not the Jewish forces could stand up against their Arab adversaries, the Arab attack on the Etzion bloc had continued. The defenders' only contact with the outside world had been by radio. The messages sent by the defenders to Haganah headquarters in Jerusalem told of the death of Mosh, the commander, and contained a desperate appeal for planes (birds):

> 12.15 p.m: We are heavily shelled. Our situation is very bad. Their armoured cars are 300 yards from the fence. Every minute counts. Hurry the dispatch of planes.

> 12.53 p.m: Hundreds of Arabs advance on Kfar Etzion, about 300 yards away. The situation is desperate. We have many killed and wounded. Mosh among those killed. Without immediate aerial support we will be lost.

1.00 p.m: Kfar Etzion reports it has no knowledge about developments in the other settlements of the bloc. Send birds.

1.45 p.m: Heavy fire of artillery, mortars and machine-guns. The birds have not yet appeared. We have about 100 killed and many wounded. Establish contact with the Red Cross or help in any other way.

A platoon commander, Zvi Ben Yosef, was said to have died humming a song that he had composed:

> And if I die in battle,
> you will take up my arms
> in revenge.

A Piper Cub on its way to the bloc was forced to turn back because of a technical fault. By midnight on May 12 there were insufficient men unwounded for a counter-attack to be launched. By the morning of May 13 each of the settlements was cut off from the others. Meanwhile, in the Jordan Valley, the Arab town of Beisan was occupied by the Haganah on May 12. Its occupation established, as Plan D laid down, continuity of Jewish-controlled areas between the settlements north and south of it. The Jewish troops found Beisan deserted; its 3,000 inhabitants had fled. The Zionist imperative of settling the land was possible where hitherto it had been impossible; with the capture of Beisan a new Jewish settlement, Sheluhot (Shoots or Sprouts), was founded just two miles to the south of the town by immigrants from Germany and Hungary.

On May 13, the day after the capture of Beisan, following twenty-four hours of negotiation between the Arab Emergency Committee in Jaffa and the Haganah High Command, the terms of the Arab surrender of Jaffa were signed. The last British troops had just left Jaffa, committed to their timetable of withdrawal. Immediately after the surrender was signed, the Haganah entered Jaffa and established a military government. Of the city's 70,000 Arab inhabitants, 67,000 left, some by boat, some by car and truck, most on foot, carrying what little they could gather together and take with them of their possessions. On the following day, with Arab forces about to take over the airport at Lydda, Jewish forces occupied the British police fortress at Beit Dagan on the road from Tel Aviv to Jerusalem.

Throughout the second week of May, more and more Arab suburbs of Jerusalem were taken over by the Haganah. Among those wounded in the battle for Katamon was David Elazar, who had come to Palestine from Sarajevo at the age of fifteen, in 1940, and was to be Chief of Staff of the Israeli army in the war of 1973.

The battle for the Tel Aviv–Jerusalem road intensified. The ancient hill fort of Gezer, which overlooked the road from the south, was captured by the Haganah on the night of May 13–14. It had once been given by the Egyptian Pharaoh as a wedding present to King Solomon.

In the Negev, Haganah operations were being carried out in stages. The main Arab village dominating the north–south road was that of Breir. It had been decided in mid-April to establish a Jewish settlement, Bror Khayil, named after a Jewish settlement known to have been in the area in ancient times. Bror Khayil was to be closer to the road than the Arab village. It had been erected on the night of April 19–20. Fences and barracks were all in place by dawn, and guarded – to the astonishment of the Arabs of Breir a mere thousand yards away.

But still, under Plan D, the Haganah had to secure the road, which the Arabs, using ditches and barbed wire barriers, had cut to traffic. To secure the road it was felt that Breir must be taken. There was little opposition from the inhabitants, whose village fell on May 12. Haganah bulldozers then filled in the ditches and pushed aside the barbed wire barriers, opening the road and giving access to the Jewish settlements to the south. Two other villages that had harassed traffic on the roads, Huleikat and Kaukaba, were also overrun. Their inhabitants fled eastward into the Hebron Hills.

But the danger to the Jewish settlements was not over. On May 13, the day after the capture of Breir, volunteer troops from Egypt, members of the Muslim Brotherhood, attacked Kfar Darom, to the south. Although no regular Egyptian forces had crossed the Palestine–Egypt border, the troops attacking Kfar Darom were led by a regular Egyptian officer, with support from guns and tanks of the Egyptian army. For the Jewish defenders in the south, it was an ominous development, so close to Britain's final withdrawal.

As Arab Legion armoured cars penetrated Kfar Etzion on May 13, the defenders fought to the last. 'A desperate Masada battle was waged in the village,' was the last radio message received in Jerusalem. When the battle ended there were only four Jewish survivors. One of them reported that when the area commander of the forces at the centre of Kfar Etzion went forward with some of his men under the white flag of surrender the Arab Legion forces ceased firing, but when the sound of battle was heard elsewhere, they reopened fire.

There then occurred an incident that has scarred the memory of that time until today. When an Arab Legion officer called for a cease-fire, fifteen of the defenders piled their arms and, as ordered to do so by the officer who wished to take a photograph of the surrender, lined up in a row. At that moment another Arab, armed with a sub-machine-gun, came up and opened fire. All fifteen Jews were killed.

The battle was followed by a massacre of the surviving Jewish civilians. 'You kill your enemy, OK', commented Uzi Narkiss, the young commander of the Palmach group that had been the last to reach the Etzion bloc, 'but why do you have to dismember him?'

Also killed by the triumphant victors was an Arab family who had maintained friendly relations with the Jews throughout the troubled times and had found safety in Kfar Etzion when the fighting began.

From the Etzion bloc settlements of Revadim, Massuot Yitzhak and Ein Zurim, the surviving Jews were led into captivity. It was 9.30 on the morning of May 14, the day that Britain left Palestine for ever.

There was one part of Palestine whose future still depended, not on the Jews or the Arabs, but on the United Nations. This was Jerusalem. During debates on May 14 a series of resolutions were put forward at the General Assembly in New York that would have put Jerusalem under United Nations rule. Three such resolutions were proposed: first by Guatemala, then by Australia and finally by the United States. Each resolution was rejected, the Arab nations being emphatic that Jerusalem must be not an international, but an Arab city (despite its Jewish majority). 'It was not a passive default,' one of the Jewish Agency's participants in the scene, Abba Eban, has written of the United Nations role, 'but an active relinquishing of responsibility in a critical hour. Israel would never forget the lesson. If the United Nations would not take responsibility in time of peril, by what right could it claim authority when the danger was passed? At six o'clock New York time, when the Mandate was formally ended, the representative of Iraq arose exultantly to cry, "The game is up!" The General Assembly had lost its right of succession.' There was to be no United Nations administration in Palestine, only a struggle for power between the Jews and Arabs in direct confrontation. The Iraqi representative found this much to be welcomed. In war or peace, Arabs and Jews would have to work out the destiny of their land alone.

CHAPTER ELEVEN

The War of Independence
May 1948 to the first truce

At five o'clock on the afternoon of 14 May 1948, in the main hall of the Tel Aviv Museum, a ceremony took place that inaugurated the State of Israel. The ceremony began with the singing of the Jewish anthem 'Hatikvah'. A few moments later David Ben-Gurion, as Prime Minister and Minister of Defence of the newly created provisional government, put his signature to Israel's Declaration of Independence. As the other signatories completed their work, 'Hatikvah' was played again, by the Palestine Symphony Orchestra. Palestine was no more. Although the last of the British troops and administrators were not to leave the country until the following day, May 15, the day of their departure was a Saturday – the Jewish Sabbath – a day on which those religious Jews who would have to sign the Declaration could not wield a pen. Indeed, according to Jewish tradition, the Sabbath began the previous evening – that is, at sunset on May 14. Hence the Friday afternoon declaration.

'For two thousand years the revival of the Jewish State in Palestine had been the passion and dream of a scattered people,' Walter Eytan wrote ten years later. The decision to declare independence, he commented, 'was Israel's alone. The courage to take it grew from the stored-up anguish of those two thousand years. The debate which was still going on in the United Nations might have been a metaphysical disputation for all the effect it had on the events in Tel Aviv.'

The declaration opened by describing the Land of Israel as the birthplace of the Jewish people, and looking back at the land's distant past. 'Here their spiritual, religious and political identity was shaped. Here they first attained to statehood, created cultural values of national and universal significance, and gave to the world the eternal Book of Books.' They had 'kept faith' with the land during all the years of dispersal, 'and never ceased to pray and hope for their return to it, for the restoration in it of their political freedom'. The declaration continued:

Impelled by this historic and traditional attachment, Jews strove in every successive generation to re-establish themselves in their ancient homeland. In recent decades they returned in their masses. Pioneers, ma'pilim – immigrants coming in defiance of restrictive legislation – and defenders, they made deserts bloom, revived the Hebrew language, built villages and towns, and created a thriving community, controlling its own economy and culture, loving peace but knowing how to defend itself, bringing the blessings of progress to all the country's inhabitants, and aspiring towards independent nationhood.

The declaration went on to recount the historic stages, starting with the First Zionist Congress in 1897, the Balfour Declaration of 1917, and the League of Nations Mandate, which gave 'international recognition to the historic connection between the Jewish people and Eretz Yisrael, and to the right of the Jewish people to rebuild its National Home.' The 'catastrophe which recently befell the Jewish people – the massacre of millions of Jews in Europe' was, the declaration stated, 'another clear demonstration of the urgency of solving its homelessness by re-establishing in the Land of Israel the Jewish State, which would open the gates of the homeland wide to every Jew, and confer upon the Jewish people the status of a fully privileged member of the comity of nations'. Through the partition resolution of November 1947, the recognition by the United Nations 'of the right of the Jewish people to establish their own State is irrevocable'.

Such was the preamble. The document then went on to 'declare the establishment of a Jewish State in the Land of Israel, to be known as the State of Israel'. Until that moment, very few indeed of those listening to the broadcasting of the declaration knew what the name of the State was going to be. Since the partition resolution half a year earlier, many different names had been proposed, among them Zion, Ziona, Judaea, Ivriya and Herzliya. Postage stamps printed in advance in anticipation of the declaration of statehood were marked 'Doar Ivri' (Hebrew mail) since no one knew what the name would be.

The new State had its name. One of its earliest civil servants, Walter Eytan, has commented:

> The moment the name was proclaimed, everyone realized instinctively that it could in fact have no other.
>
> The children of Israel, the people of Israel, the land of Israel, the heritage of Israel – all these had existed, in reality and metaphysically, for so many thousands of years, they had exercised such influence on the evolution of mankind that the State of Israel was their logical consequence and culmination.
>
> Ben-Gurion's choice of name was hailed as a stroke of genius; in fact it arose out of the innermost historic or tribal consciousness of us all.

There could have been no more effective introduction of the new State to the world. 'Israel' on its visiting-card was as eloquent as could be: it was so evocative and so immanently true that nothing more needed to be said. It made obvious to the world not only who we were but that we were what we had always been, and that if the State of Israel as such was a newcomer on the international scene, it was in fact but the natural outward form, in modern terms, of a mystery and a people whose roots went back to the earliest ages of man.

The Declaration of Independence continued with an assurance that Israel would be open for Jewish immigration, and for 'the ingathering of the exiles'. It would foster the development of the country 'for the benefit of all inhabitants'. It would be based on 'freedom, justice and peace, as envisaged by the prophets of Israel'. It would ensure 'complete equality of social and political rights to all its inhabitants, irrespective of religion, race or sex', and would guarantee freedom of religion, conscience, language, education and culture. It would be faithful to the Charter of the United Nations (of which it was not yet, of course, a member).

The declaration went on to appeal to the Arabs of Israel (the borders of which were nowhere defined) to 'preserve peace and participate in the upbuilding of the State on the basis of full and equal citizenship' and 'due representation in all its provisional and permanent institutions'. As to Israel's Arab neighbours, the new State extended to them and their peoples 'an offer of peace and good neighbourliness'. The State of Israel was 'prepared to do its share in common effort for the advancement of the entire Middle East'.

The declaration ended with an appeal to the Jewish people 'throughout the Diaspora to rally round the Jews of the Land of Israel in the task of immigration and upbuilding, and to stand by them in the great struggle for the realization of the age-old dream – the redemption of Israel'.

As the thirty-eight signatories – including a future President of the State, Yitzhak Ben-Zvi, and two future Prime Ministers, Moshe Sharett and Golda Meir, as well as two Revisionist representatives – appended their signatures to the Declaration of Independence, with its high ideals, a new nation was born. Yitzhak Rabin, who was then commanding the Palmach's Harel Brigade in the Jerusalem corridor, recalled:

> An ancient radio in Kibbutz Ma'ale Ha-Hamisha, a few miles outside Jerusalem, conveyed Ben-Gurion's voice to us as he proclaimed the establishment of the State of Israel. Our weary troops strained to catch the portent of his words.
> One soldier, who was curled up in a corner in a state of complete exhaustion, opened a bleary eye and pleaded, 'Hey men, turn it off. I'm dying for some sleep. We can hear the fine words tomorrow.'

Someone got up and turned the knob, leaving a leaden silence in the room. I was mute, stifling my own mixture of emotions. None of us had ever dreamed that this was how we would greet the birth of our State, but we were filled with an even stronger sense of determination now the State existed.

One signature appears further down the declaration than the others. It is that of the Communist Party leader, Meir Vilner. He had felt the need to obtain permission to sign from the Soviet Union. This reached him on the day before the ceremony, when he was in Jerusalem. He had signed at once, a day early, and left the space above for the others to sign.

Chaim Weizmann was in New York when the State of Israel was declared. 'The demoralizing illusions of trusteeship and truce were behind us,' he wrote later that year in his memoirs. 'We were now face to face with basic realities, and this is what we had asked for. If the State of Israel could defend itself, survive, and remain effective, it would do so largely on its own; and the issue would be decided, as we were willing it should be, by the basic strength and solidity of the organism which we had created in the last fifty years.'

Weizmann had already written to President Truman seeking recognition for the new State, should it be declared; and in an immediate and direct response American recognition was issued from Washington only eleven minutes after the State had been proclaimed in Tel Aviv. A special messenger had to be sent to the White House to say what the name of the new State would be.

On the day that independence was declared, Egyptian aircraft bombed Tel Aviv. They were the first bombs dropped on the city from the air since 1940, when Mussolini's air force had attacked it. Then, more than a hundred Jews had been killed. There were no deaths in this Egyptian raid.

John Barrard, an overseas volunteer who had served for almost four years in the South African air force in the Second World War, later recalled how during the raid, 'three Egyptian Spitfires virtually demolished the aircraft clustered in and on the apron of the single hangar at Sde-Dov, the Tel Aviv airfield. In any other air force those aeroplanes would have been written-off. But there was no choice. They were painstakingly restored, parts made or cannibalized and they flew again.'

Barrard added, 'Fortunately a number of clapped-out Austers, the little aeroplanes used by flying clubs, had been purchased from the RAF in January 1948, the officer in charge having been persuaded that they were virtually useless. They were dispersed and the 21 "scrap" Austers were reassembled into a smaller number of usable aircraft. But they were slow and had no night-flying instruments. So in the first hectic weeks after the Declaration of the State, I worked almost continuously round the clock at

Sde Dov, fitting those little planes with Venturi Tubes and basic night-flying instruments, so that they could be used as bombers, with crude hand-made bombs chucked over the side.'

The role of these little planes, the Austers, Pipers, Fairchilds, 'in the first crucial weeks before the Czechoslovakian airlift brought Messerschmitts,' Barrard wrote, tends 'to be overlooked. But they were vital in checking the better equipped enemy's rapid advances, flying in mail and urgently needed medical supplies to isolated areas. One of those little Austers stood mounted on a plinth outside Sde Dov, in recognition of the role they played. Sadly, it was later removed. A great pity, as it put the inequality of resources into perspective and served as a reminder of the grave danger the new little State was in, in May 1948.'

During the first day of independence a notice was posted up throughout Tel Aviv by the Haganah staff. Its tone echoed that of the notices circulated to senior civil servants in London during the autumn of 1940:

> The enemy threatens invasion. We must not ignore the danger. It may be near. The security forces are taking all necessary measures. The entire public must give its full help.
> 1. Shelters must be dug in all residential areas and the orders of Air Raids Precautions officers must be obeyed.
> 2. Mass gatherings in open areas and streets must be avoided.
> 3. Every assistance must be given to the commanders of the security forces in erecting barriers, fortifications, etc.
>   No panic. No complacency.
>   Be alert and disciplined.

Ze'ev Sharef, the secretary to the new government, later recalled that when the first issue of the Provisional Government's *Official Gazette* was published, containing the Declaration of the Establishment of the State of Israel, 'I waited for the first copies and took two, one of them for David Ben-Gurion. When I brought it to him, he asked:

'"What's new in the city?"

'"Tel Aviv is rejoicing and gay," I answered.

'He returned soberly, "I feel no gaiety in me, only deep anxiety as on November 29, when I was like a mourner at the feast."'

In New York, Abba Eban was monitoring the first hours of statehood. 'News came rolling in thick and fast,' he wrote. The Soviet Union and Guatemala had recognized Israel. Egyptian planes had bombed Tel Aviv. Ben-Gurion had made his first broadcast as Prime Minister. 'And yet,' Eban added, 'as the tumult of Jewry's greatest day in modern history swept through the streets of New York, Weizmann lay silent in the darkened hotel room, with a few of us around him. Cables came from Tel Aviv telling of familiar

Zionist leaders bearing new and glamorous ministerial titles. But there was no news or greeting for Weizmann. A sense of abandonment and ingratitude invaded his mood. Suddenly a bellboy appeared with flowers, fruit and – a telegram from the Zionist Labour leaders in Tel Aviv.'

The telegram, from those members of the Labour movement with whom Weizmann had worked most closely in the years leading up to statehood, read: 'On the occasion of the establishment of the Jewish State we send our greetings to you who have done more than any living man towards its creation. Your help and stand have strengthened all of us. We look forward to the day when we shall see you at the head of the State established in peace.' The signatories were Ben-Gurion and the four senior members of his Provisional Council of State, Eliezer Kaplan, Golda Meir, David Remez and Moshe Sharett.

'I went out into Park Avenue,' Eban wrote. 'It was dark and late. Back in our hotel . . . I waited until midnight when the *New York Times* with its banner headlines gave us the news: victory in Washington and the United Nations, but danger in the Middle East.'

During the day, Justice Felix Frankfurter, one of the senior figures of American Jewry, wrote to Weizmann, 'Mine eyes have seen the coming of the glory of the Lord; happily you can now say that, and can say what Moses could not.' Two days later, the Provisional Council of State elected Weizmann as Israel's first President. Travelling to Washington, he was received by Truman with all the solemnity to be accorded to a Head of State (the first Israeli President to be invited to Britain on a State Visit was Weizmann's nephew Ezer, forty-nine years later). Weizmann used his visit to Washington, as he had used so much of his time in the past, on behalf of his people, obtaining a promise from the American President of a loan of $100 million for the economic development of the new State.

The morning of 15 May 1948 saw the British gone (except for a small garrison at Haifa port to supervise last-minute evacuations) and Israel independent. It also saw the start of a new war: the military intervention of five independent Arab States. Not always working in harmony and seldom if ever coordinating their military strategies, the attacking forces were nevertheless a formidable threat as they crossed into Israel from every direction. Israel was not unprepared. Its army, which had numbered 4,500 at the time of the partition resolution six months earlier, exceeded 36,500. Twenty per cent of Mandate Palestine was under its control. But 1,200 Jews had already been killed in battle.

The British opened the detention camps in Cyprus on May 15. Several hundred new arrivals in Palestine who had been trained by the Haganah while in the camps were hurried to the battlefield.

Iraqi troops also crossed the River Jordan on May 15, the first full day of Israeli independence. Crossing the river at Gesher – whose settlers had driven off an Arab Legion attack two weeks earlier – on the following morning they attacked the British police fort and the village. The villagers held their fire until the attacking troops were almost upon them, and then drove them back. But the settlement was besieged for seven days. On one occasion an Iraqi armoured vehicle reached the gate of the police fort and blew it up, but was itself hit by an Israeli 'Molotov cocktail' and went up in flames.

There were many other clashes in the north in the opening days of the war. On the night of May 15 a company of Israeli soldiers crossed the Lebanese border and marched over rough, hilly terrain seven miles into Lebanon, where they blew up a road bridge over the River Litani. This seriously impeded the ability of the Lebanese to send armoured vehicles and troop transports southward. That same night, Syrian troops came down from the Syrian Heights (now known as the Golan Heights) and, supported by thirty armoured vehicles, by far the largest military force in the area, overran two Jewish settlements in the plain below, Ma'agan on the shore of the Sea of Galilee, and Sha'ar Ha-Golan on the River Yarmuk. At Zemach Jewish defenders who had taken up positions in the former British police fort held up the Syrians for two days – until all forty-two of them were killed. The Syrians then moved westward to the River Jordan. In Tel Aviv Ben-Gurion gave the order to Moshe Dayan, the commander of the Haganah forces in the area: 'Hold the Jordan Valley'.

While Dayan was on his way to Deganya, two of its senior settlers, Yosef Baratz and Ben-Zion Yisraeli, were on their way – through the night of May 19 – to Tel Aviv, to beg Ben-Gurion to send reinforcements to the north, as a matter of the gravest urgency. His advice to his two distraught visitors was, 'We don't have enough artillery, enough aeroplanes. Every front needs reinforcements. The situation is extremely grave in the Negev, in the Jerusalem area, and in Upper Galilee.' In his diary, Ben-Gurion recalled how Baratz broke down and cried at the thought of the destruction of his family and of Deganya.

Ben-Gurion sent the two men to see Yigael Yadin. All he could tell them was that Dayan was already on his way north, and that they should let the Syrian tanks approach to within a few yards of the gates and then disable them with hand-thrown Molotov cocktails. The two men were furious that no reinforcements could be sent north to help them. Yadin later recalled, 'I was suddenly shocked at that moment, when I realized that the fall of Deganya would mean that the whole north of the country might be lost. In the south, the Egyptian army was advancing on Tel Aviv. Jerusalem was cut off, and the Iraqis were putting pressure on the middle of the country. This was a moment that I suddenly felt that the dream of generations was about to disintegrate.'

Yadin went to see Ben-Gurion, to ask that four 65-millimetre field artillery pieces – without aiming sights – that had been unloaded a few days earlier at Tel Aviv and which Ben-Gurion wished to send to the Jerusalem road, be sent instead to the north. 'The situation there was critical,' Yadin continued. 'Ben-Gurion was adamant. We sat together for three hours. Never have two men leading a war for the entire country sat for so long over so little. I pounded Ben-Gurion's glass-topped desk with my fist and the glass shattered.'

Ben-Gurion agreed to send two of the artillery pieces to the north. 'Since it would be impossible to send them to Jerusalem before tomorrow night,' he wrote in his diary, 'they can in the meantime serve an important function in the Jordan Valley.' Dayan had already reached Deganya.

'Father was surrounded by close friends,' Yaël Dayan later wrote, 'some of whom had no business being there but followed him faithfully. Nahalal fighters, his brother-in-law Israel, friends from the settlement of Yavniel – all hand-picked and capable.'

Crossing the River Jordan, at 4.30 on the morning of May 20, the Syrians attacked Deganya, advancing with tanks and flame-throwers. A bloody battle ensued. The Syrians were on the verge of overrunning the kibbutz when, at about noon, the two 65-millimetre guns arrived on the scene. Opening fire on the Syrian tanks and infantry, the two guns wrought havoc, psychological as well as physical, and the Syrians withdrew. A thrill of relief spread through the new State: a well-armed enemy could be repelled. One of the Syrian tanks, which had become stuck in the settlement's perimeter during the attack, has remained there as a memorial to this day.

The siege of Gesher ended when the two artillery pieces that had saved Deganya from the Syrians were rushed southwards, and opened fire on the Iraqi forces still besieging the fort from the hills above. The Iraqis fled.

Throughout the second full day of Israel's independence, Lebanese troops were advancing from the north; Syrian, Iraqi and Transjordanian troops from the east; Egyptian troops from the south. Even Saudi Arabia sent a small contingent, that fought under Egyptian command. Israel had its back to the sea. And even there it was vulnerable, as Egyptian troops were landed from the sea at Majdal.

The first Egyptian attack was on the small, isolated settlement of Kfar Darom, a religious kibbutz seven miles south of Gaza. The thirty able-bodied defenders had already driven off an attack by units of the Egyptian Muslim Brotherhood before the declaration of statehood. When their stock of hand-grenades ran out, the settlers put explosives in the small bags that held their phylacteries, and threw them at their attackers. When the Egyptian army advanced with tanks, the defenders used the single PIAT (its name derives from its official designation Projectile Infantry Anti-Tank) anti-tank weapon that had been rushed up to them during the night to good effect, knocking out the lead tanks, whereupon the other tanks turned and withdrew.

Kfar Darom was left to its own devices, its perimeter guarded, the occasional shell lobbed into it, but with no further attempt to take it. The Egyptian army moved on, five miles south-east, to kibbutz Nirim. There, half the forty defenders were killed or wounded, but the perimeter held. An air attack on the following day did not break the will of the defenders. As at Kfar Darom, the Egyptians then moved on, leaving Nirim surrounded but unconquered, shelled from time to time, but never surrendering.

Neither Kfar Darom nor Nirim offered any hindrance to the Egyptian advance into Israel. A third kibbutz, Yad Mordechai, which had been established five years earlier, was regarded as a more serious obstacle lying as it did on the main road from Gaza to Majdal, and preventing the link-up of these two bases. A larger force was assembled to overrun this third kibbutz, though it too had relatively few defenders, farmers and army reinforcements constituting no more than a single infantry company. Against them were assembled two infantry battalions, one armoured battalion and one artillery regiment. The assault began before dawn on May 24.

The battle for Yad Mordechai lasted five days. When the last machine-gun was out of action, and the last rifle ammunition spent, the defenders withdrew, creeping through the Egyptian lines under cover of night, carrying their wounded with them.

The five days during which Yad Mordechai held out were as crucial for Tel Aviv as the defence of Antwerp had been in 1914 for the preservation of the Channel Ports. Antwerp held out long enough for the Germans to be cheated of their entry into Calais. Yad Mordechai's resistance saved Tel Aviv. For in those five days, not only were Tel Aviv's defences strengthened, but reinforcements were found which, sent to the south, could hold the line halfway between Tel Aviv and Gaza.

Another settlement attacked at the start of the second week of the War of Independence was Negba, which had been founded by pioneers from Poland in the summer of 1939. When the Egyptian army opened its attack on Negba with an artillery barrage on May 23, the 145 defenders – seventy-five members of the settlement and seventy full-time Haganah soldiers – had no artillery with which to reply. Against them were ranged 2,000 Egyptian regular soldiers. After nine days of continuous bombardment, during one of which 6,000 shells fell on the settlement, the Egyptians launched a tank and infantry assault. It began at dawn on June 2. The defenders possessed a single PIAT anti-tank weapon. In his book *The Army of Israel*, Lieutenant-Colonel Moshe (formerly Maurice) Pearlman described the assault:

> When the tanks reached to within 200 yards of the settlement, they began to seek out weak points in the fence through which they might break. With the leading tank sixty yards from the settlement perimeter, the Jews opened up with their lone PIAT. They scored a direct hit with their first round. The tank retired under cover of a smoke screen.
>
> The next two tanks were also allowed to come within sixty yards of

the defence post when the PIAT again went into action. The second and third rounds failed to explode, but two direct hits were scored with the fourth and fifth.

The fourth tank made for another point in the settlement fence, approaching a Jewish position that was without any armour-piercing weapon. It succeeded in crashing its way through the perimeter and advanced to within five yards of the defence post. The Jews flung two Molotov Cocktails on the tank, but they did not explode. They followed it up with hand-grenades, the splinters of which must have entered the turret through the slit, for the tank promptly retired. Two other tanks were disabled by mines. The seventh failed to advance.

The Egyptian infantry platoons approached to within 200 yards of the settlement and opened up a blistering fire, under cover of which one platoon advanced another hundred yards. But it received concentrated fire from the settlement and retired. During this time, the Egyptian air force carried out two further attacks on Negba.

Fighting continued for four hours. At 1100 hours, a two-inch mortar smoke screen was thrown round the disabled tanks and when it cleared the tanks were already out of range of the settlement weapons, being dragged towards the Police Station by the armoured cars. The infantry too had retired. But the shelling was maintained.

At 1400 hours, the Egyptians renewed their attack. Planes again came in to assault the settlement from the air. Tanks and cars – fourteen in all – again led the way, followed by infantry. They came more warily this time. But this time, the settlement of defenders opened fire when the enemy came within a range of 300 yards. The Jews fired with all they had. Two Egyptian Bren carriers were hit by three-inch mortars. Shortly thereafter, the Egyptians retired from the field.

The Egyptian attackers had lost six tanks, two Bren carriers and 100 killed and wounded. They had also lost a Spitfire, brought down by one of Negba's machine-gunners. The Jews lost eight men killed and twelve wounded.

Just to the north of Tel Aviv, at Ramat Gan, the Israeli General Staff Headquarters monitored every phase of the battle, and, in the early stage of the fighting, came under bombardment from long-range Egyptian artillery fire. One of those working there at the time, Shimon Peres, later wrote of how, one morning, shells dropped in the courtyard of the little house where Ben-Gurion was working, and one of the sentries at the door was badly injured. Peres hurried to Ben-Gurion's office with Nehemiah Argov, Ben-Gurion's *aide de camp*, 'and we begged him to go down to the bomb-shelter. He contented himself with putting on a steel helmet, then looking at us with an expression of childish astonishment he said, "Look, I'm in the middle of writing something, how can I break off now?" But when he was told that the sentry had been wounded he left his papers and came running out to help us with the stretcher. He insisted on being one of the bearers and would not leave the injured man until he was safely in the hospital. He saw no reason why bombs should determine his way of life – he determined it! And the fact that they were falling on the roof of his house was quite irrelevant in comparison with the urgency of his work.'

With independence, a Provisional Council of State had been created, to govern the country until such time as elections could be held. The council's first legislative act was a Law and Administration Ordinance which abolished all restrictions on Jewish immigration. The ordinance also retroactively validated the immigration of every Jew who had at any time entered the country, including those tens of thousands who had done so in contravention to the British Mandate regulations (the 'illegals').

Even as the war continued, new settlements were being founded. One of the first was Shoresh (Root), established in the Jerusalem corridor 12 miles west of Jerusalem, by survivors of the Holocaust from Roumania and Czechoslovakia. The nearby Arab village of Saris, which had been overrun by the Haganah a month earlier in April 1948, was levelled to the ground. Also established in the early phase of the war, on May 23, was Allonei Abba, in the western Lower Galilee. Like the founders of Shoresh, many of its founding settlers were survivors of the Holocaust. They came from Czechoslovakia, Roumania and Germany. The name of their settlement derived from the two words allon (oak) – for the natural Tabor oaks that grew nearby – and Abba, for the wartime Jewish hero, Abba Berdichev, who had been captured and killed after being parachuted from Palestine into Slovakia in 1943 on a mission to help the British war effort and to make contact with the Jews.

To the south of Jerusalem, a violent battle was taking place at kibbutz Ramat Rahel, from which the walls of the Old City could be seen to the north. Twice the Egyptians overran the kibbutz. Twice the Israelis retook it. For a while it looked as if the Egyptian army would be able to advance up the Bethlehem Road to the very heart of the Jewish city. But when reinforcements arrived from the Harel Brigade, which had just returned from the Jerusalem corridor to the city, the Egyptians were finally pushed back southward.

In Tel Aviv, a bitter dispute was raging between Ben-Gurion and Yadin about Ben-Gurion's desire to attack the Arab Legion at Latrun and to break the Arab hold on the road to Jerusalem. Ben-Gurion was emphatic that the attempt should be made. Yadin was equally emphatic that the troops and weaponry needed were not available. Ben-Gurion wanted the operation carried out by the newly formed Seventh Brigade, commanded by a Haganah veteran, Shlomo Rabinowitz. 'Shlomo will carry out the operation, Yigael will plan,' Ben-Gurion noted in his diary. But Yadin pointed out that several hundred of the brigade's 2,000 men were survivors of the Holocaust who had just arrived in Israel from British internment camps in Cyprus. Many had never fired a rifle. Many spoke no word of Hebrew. The brigade had not a single field radio.

On May 22 Ben-Gurion ordered Yadin to launch the attack on the British police fort at Latrun 'without any delay'. Yadin argued that the Seventh Brigade was not yet ready, and that in any case, the danger to Jerusalem was not as great as Ben-Gurion believed. As the Prime Minister expostulated on the importance of Jerusalem, Yadin told him, 'You don't have to explain the significance of Jerusalem to me. I was born there, and my whole family is there in the siege.' Ben-Gurion would not give up his demand for an attack. Yadin later recalled, 'I told him it would be murder, but that didn't help me.'

The battle for Latrun was to begin on May 24. That morning Yadin went to inspect the troops. Many did not even have water bottles. Others, who had some previous experience of fighting, had been given rifles of a type they had never seen before. Yadin asked Ben-Gurion for a postponement, but Ben-Gurion was adamant that Shlomo Rabinowitz must lead his men against Latrun that night. A message from Yitzhak Rabin, in Jerusalem, clinched the matter. 'Shlomo must break through with reinforcements tonight,' Rabin wrote. An Arab Legion armoured unit had tried to break through the Haganah positions opposite the New Gate. To the south of the city, the Egyptians had again attacked Ramat Rahel.

That night the battle for Latrun began. The arrival of the Israeli troops in the region of the fort was hampered by confusion and poor organization. It was four in the morning before they were ready to attack. The Arab Legion was waiting, well dug in, well armed and well prepared. One of the first companies to advance was led by Ariel Sheinermann. He and his men were assailed by a punishing barrage of fire. As most of them were cut down and killed, Rabinowitz, not yet aware of the disaster on the battlefield, signalled to headquarters in Tel Aviv: 'The festivities have begun.'

The fighting continued throughout the morning. As the heat of the day intensified, many of the men – without water canteens – became desperate from thirst. Arab Legion artillery pounded them as they lay in the grass. The Arab Legion defenders were greatly helped by the unusual heat wave. Many of the Jewish attackers were recent immigrants who were quite unused to such heat. Cries for water, and of fear and desperation, often in Yiddish, were heard all over the battlefield. Many collapsed from the heat and thirst. Snipers picked them off when they rose. The evacuation of the wounded and the dead took most of the day. When the withdrawal was completed eighty men lay dead.

Among those killed at Latrun was Reuven (Ruvka) Oppenheim, a recent immigrant. His body was never found. He had arrived in Israel eight months earlier, with his father, brother and sister. What made his story unusual was that he was not only a survivor of the mass murder of Jews in his home town, Nowogrodek, during the Second World War, but had survived with his family. Together with his mother and four-year-old sister, Mariaske, Ruvka had been smuggled out of the ghetto by his brother and father, who were living in the forests of White Russia in a 'family camp' established there in 1943 by the Bielski brothers. Until liberation, Ruvka Oppenheim fought with the partisans against the Germans. He and his family then emigrated to Palestine. After joining the army in November 1947 he came home on leave only once, at Passover.

Sheinermann, who was wounded at Latrun, survived. Later, as Ariel Sharon, he was to become one of Israel's most controversial commanders and politicians. Among those seriously wounded in the battle for the Jerusalem road was Avigdor Arikha, a survivor of the Holocaust (who had brought with him to Israel the drawings he had made of the conditions in his slave labour camp in Roumania). He was to become one of Israel's finest painters.

As the battle was being fought at Latrun, to the south of Jerusalem the Israeli forces drove the Egyptians out of Ramat Rahel for the final time. Nor did they stop there. Between Ramat Rahel and Bethlehem, visible from both, was an Arab Legion stronghold, located in the Monastery of Mar Elias. On May 26, the day after the recapture of Ramat Rahel, Israeli troops attacked the stronghold. On entering it, they found the Arab soldiers gone. The only occupant left in the monastery was a lone Russian Orthodox nun, who had refused to abandon her room even at the height of the battle raging all around.

Another battle was in its last stages that day, the battle for the Jewish Quarter of the Old City. For two more days the Arab Legion pounded the quarter with artillery, systematically destroying its buildings and driving the defenders, whose ammunition was running out, into an ever-decreasing area. On May 28 Yitzhak Rabin went up to Mount Zion in Jerusalem, where, he later wrote, 'I witnessed a shattering scene. A delegation was emerging from the Jewish Quarter bearing white flags. I was horrified to learn that it consisted of rabbis and other residents on their way to hear the Legion's terms for their capitulation. That same night the Jewish Quarter surrendered to the Arab Legion.'

The loss of the quarter was a bitter low to the Israelis, both militarily and morally. Together with the Wailing Wall just below it, the quarter represented the spiritual centre of Jewish Jerusalem, with its many synagogues and houses of prayer. As its inhabitants were led into captivity in Transjordan, there was a sense of humiliation among the Jews elsewhere in the country. But this Arab victory, the first success of the Arab Legion in Jerusalem, was also its last. The tide of battle was turning in the city, with the boundaries of Jewish Jerusalem elsewhere being strengthened, and extended.

Among the defenders of the Jewish Quarter who had been badly wounded was a twenty-two-year-old English girl, Esther Cailingold. She was taken by the Arab Legion to a hospital in the nearby Armenian Quarter. Offered a cigarette, she declined, because it was the Sabbath. Not long afterwards she died of her wounds. After her death a letter was found under her pillow, addressed to her parents. 'We had a difficult fight,' she wrote. 'I have tasted hell, but it has been worthwhile, because I am convinced the end will see a Jewish State and all our longings. I have lived my life fully, and very sweet it has been to be in our land.'

Ben-Gurion saw Jerusalem as the centre to the battle. Yadin was certain that the Negev was the place where it would be decided. The British commander of the Arab Legion, Glubb Pasha, was looking for victory on the West Bank of the Jordan. But whichever part of the Jewish areas of settlement was under attack had to be defended. In the north, Arab control of Acre had cut off the rest of the northern coastline from contact with the Haganah. Most in danger was the small seaside resort and farming village of Nahariya with 1,400 inhabitants. Its only contact with Jewish Palestine was by small motor boats that plied from time to time to and from Haifa. On May 29, immediately after the capture of Acre, Nahariya was reunited with the rest of the country, and was included in the Jewish State from which it had been excluded under the United Nations partition plan.

Also on May 29 a kibbutz was established two miles north-east of Acre. It was named Shomrat, after the Hebrew word *shomer* (watchman). Its settlers were men and women from Hungary, Czechoslovakia and Roumania, who had been training in another settlement specifically for this task.

May 29 was also the day on which the first Messerschmitts of the Israeli air force were ready to go into action. They had been flown from Czechoslovakia, where Ehud Avriel, one of the wartime Jewish Agency representatives in Istanbul, had been active in the search for arms of all sorts. That week he reached agreement with the Czechs to buy another thirty Messerschmitts, thirty Spitfires and nine Mosquitoes. The Mosquitoes were capable of flying direct to Israel without refuelling.

Henceforth one or two Czech aircraft reached Israel each day. They were not a large enough force to challenge the Arab air forces in aerial combat, but were able to serve as spotter aircraft for the ground forces, to fly in military and food supplies to isolated outposts, and even to bombard Arab troop concentrations in both Amman and Damascus. When Egyptian troops were sent by sea towards Gaza, the troop transports were attacked from the air. But the planes' first action, on that Saturday, May 29 – war-making could not be suspended for the Sabbath – was a crucial attack on the Egyptian forces assembled at the Arab village of Isdud (later Ashdod), less than 20 miles south of Tel Aviv, which were about to launch their attack on the city.

The bridge at Isdud was blown up by Israeli sappers, in a daring raid. But it could be repaired at any time. An Egyptian armoured column was ready to march north without delay. The four newly arrived Messerschmitts had not been checked or tested; indeed, they had not even taken off since being reassembled after their arrival in Israel by sea. But there was no time for such essentials except, once they had reached the air strip at Ekron, for the painting of the Star of David insignia on the side of the planes. The maintenance mechanic who painted the first star, Benjamin (Benny) Peled, was later to learn to fly, and after a distinguished career in the Israeli air force, to become its Commander-in-Chief (in May 1973).

One of the pilots at Ekron, Ezer Weizman (Chaim Weizmann's nephew), later recalled the Messerschmitts' first flight after the decision was taken to attack the Egyptian armoured column in order at least to delay it:

> Our agitation was enormous. We looked over the maps, selecting our route and picking out our bombing runs. Then we set off at dusk: four planes, four pilots. It was a very short distance from Tel Nof to the battlefield near the Ashdod bridge. No sooner had we taken off than we were swooping down on the Egyptian column.
>
> Anti-aircraft fire pursued me as I dived toward my target. Hurtling downward, I was astounded and even somewhat frightened as I caught a glimpse of the Egyptian force, whose size exceeded all my expectations: thousands of soldiers and hundreds of military vehicles lined the highway to Tel Aviv.
>
> I dropped two bombs and began the steep climb. With the anti-aircraft fire still at my back, I dived once more, heading toward the row of armoured vehicles. As I dropped more bombs, I saw Egyptian soldiers fleeing in all directions.
>
> When I pulled out of my dive, I caught a glimpse of my 'number two,' Eddie Cohen, a volunteer from South Africa. As I watched, Eddie's plane plunged downward, dipping lower and lower until it crashed. Eddie's first mission for the Israeli air force was also his last.
>
> Our attack had been successful; it did delay the Egyptian column. The blow we had inflicted was a heavy one, according to the notions then current. However, our triumph was mingled with despondency. The mission had ended with the loss of a fellow pilot as well as one-quarter of the air force's combat planes.

More aircraft arrived in Israel from overseas, some crated on board ship, others flown in by their pilots. From the United States, Flying Fortresses, the B-17s, the stalwart bombing aircraft of the Second World War, were flown to Israel illegally from an American base in the Atlantic by a group of volunteers headed by Al Schwimmer. On May 18 Ben-Gurion had written in his diary, 'Al Schwimmer, who is very capable, has gone into action.' The Flying Fortresses that Schwimmer acquired were both a practical and a morale boost. At the same time the boats which a month earlier had brought in 'illegal' immigrants were repaired and fitted with such small – and even First World War – artillery pieces as were available. One of these ships, the *Eilat*, in coordination with the Messerschmitts, attacked Egyptian troop carriers off Gaza.

Boris Bressloff, a British Jew with no previous military experience, served as a navigator during these air attacks. 'We were able to airlift over 3,000 pounds of bombs,' he recalled, 'but the difficult thing was to get rid of them in the right place. Quite often, when the bombardier said "bombs away", the armourer, who was hoping to observe their departure from the bomb bay, would shout over the intercom, "we've still got the ****ng things! They've stuck." We would then have to make another run, much to the delight of the enemy gunners, who by this time had got our range. On the second run the armourer would then have the unenviable task of assisting the bombs to leave by hanging out into the bomb bay with one hand, and triggering the releases with a screwdriver.'

A stream of volunteers, like Boris Bressloff, had begun to arrive in Israel from abroad, men and women who left their work and families in order to be of service on the battlefield. Many of them had experience of Second World War activities: as engineers, signalmen, artillerymen, armoured corps experts, doctors and nurses, infantrymen, seamen and pilots. There were a number of Christians among them, who did not want to see the Jewish State destroyed. These volunteers went by the name of Mahal, the Hebrew acronym for volunteers from abroad (*Mitnavdei Hutz la-Aretz*).

An estimated 5,000 volunteers came to serve with the Israel Defence Forces, the largest national groups being 1,500 from the United States and Canada, 1,000 from Britain, 500 from South Africa. Even Finland made its contribution of thirty volunteers. More than 150 Mahal volunteers were killed in action. About 300 of those who came stayed on in Israel after the war, or returned to settle there soon afterwards. But most returned home, their service done, their contribution made to the survival of the Jewish State. Yitzhak Rabin was among those Israeli soldiers who never forgot the contribution of the volunteers. On his visits to Britain in later years, especially as Prime Minister, he always set time aside to meet and talk with those who had once served under him.

Forty-four years after the War of Independence, in dedicating the Mahal memorial in Israel, Rabin told the former soldiers, and the families of those who had died fighting, 'You came over here, not knowing the conditions; many of you didn't know even the language, the Hebrew language. I remember then, in the air force, the language was English, because most of the pilots of the first squadron of fighters were Mahal people from English-speaking countries – from the United States, Canada, Britain, South Africa and other places. My generation will never forget what you have done, how much you made it possible for us to achieve what has been achieved.'

Mahal volunteers with experience of aerial warfare were often assigned to take part in air attacks on the day they arrived in the country. Pilots were in such short supply that there was no time for formalities or practice. On one occasion a Mahal crew that was arriving across the Mediterranean in a Flying Fortress was directed by the Israeli air force radio controller to bomb Cairo while still on its way to Israel. The crew thus carried out their first bombing mission without having seen the country for which they were fighting.

The influx of former detainees from the British camps in Cyprus had created a further source of manpower. The immigrant recruits who arrived after May 15 were given the name Gahal (from the Hebrew acronym of *Gyus Chutz la-Aretz* – recruits from abroad). They included Jews from the DP camps, from North Africa and from Eastern Europe. But then the Cyprus influx was halted by the British government after a Security Council resolution on May 29, with regard to the provisions of the truce that the United Nations was calling for that there should be no immigration of persons of military age. Military age was defined as between eighteen and forty-five. The detention camps were closed once more.

On May 31, on Ben-Gurion's orders (as Minister of Defence), the Haganah formally gave up its underground and clandestine status, and was transformed into the regular army of the new State. Its name was incorporated into the name of the new army: *Zeva Haganah le-Israel* (Israel Defence Forces) – known by its Hebrew and English initials as both Zahal and the IDF. The forces of the Irgun and Stern Gang were invited to join the new army, and to fight as equals for the defence of the land. They promised to do so.

It was not a moment too soon for a single, unified Israeli army. In the fighting in the south Egyptian forces were engaged in a fierce battle with a much smaller Israeli force, renewing their hopes of pushing northward to Tel Aviv. In the early hours of June 3 Yadin was asked by Ben-Gurion to explain the military situation in the south to the foreign journalists who had gathered in a Tel Aviv hotel from all over the world. 'The reports we were receiving were not encouraging,' he later recalled. 'I had no idea what to say.'

Yadin, like Ben-Gurion, understood the power of morale. In 1940 Winston Churchill had spoken with confidence of victory at a time when no logical mind could conceive of how victory could come, only the spectre of defeat. Yadin told the journalists that although the fighting around Isdud was bitter, he had just received a dispatch from the Israeli commander that indicated that the bulk of the Egyptian invasion force – under continuous Israeli attack from the air and from artillery barrages – had been completely encircled by Israeli troops. 'This was a complete bluff,' Yadin later recalled. 'There was not a single word of truth to this at the time I gave this statement.' But he understood the enormous power of the international news media. 'The next morning,' he recalled, 'we intercepted a radio transmission between the High Command in Cairo and the field commander. The High Command was raging, "What is your situation? Why are you not reporting?" And the commander responded, "We're being attacked fiercely but we're holding our ground!" So they told him from Cairo, "You don't even know your own situation! You're completely surrounded!" And that threw the Egyptian troops in the field into a panic.'

'It wasn't *me* that they believed,' Yadin reflected. 'The foreign journalists reported it as a fact.'

On the morning of June 4 three Egyptian ships – a troop carrier, a large landing craft and a warship – were spotted sailing northward along the coast towards the city. When they were opposite Jaffa, and within sight of the houses along Tel Aviv's Hayarkon Street where Yadin had held his press conference, three Israeli aircraft attacked them. For just over two hours the air–sea battle continued. It ended when the Egyptian ships turned southward and steamed away. One of the Israeli aircraft was shot down, and its two crew members killed. One was the pilot David Sprinzak, son of the Secretary of the Histadrut. The other was the bombardier Mattatyahu Sukenik, Yadin's youngest brother.

Yadin's father, Professor Sukenik, was in Jerusalem when his son was killed. The news was brought to him in the form of a letter written by Yadin and sent from Tel Aviv by a light aircraft which landed on the makeshift airstrip in the Valley of the Cross. The letter read:

> Don't weep, don't grieve. When Matti took off yesterday to chase away the enemy ships from the shores of Tel Aviv, he knew very well what awaited him. He was not frightened by the risk. He was not terrified or deterred. He fulfilled his duty to his people and his country with bravery.
>
> So don't cry, because the danger did not shake him. With bravery he fulfilled his duty to the people and homeland.
>
> So don't cry. Just be proud that he was one of us. And let us all cherish his memory with the love and admiration he deserves.
>
> Yigael

Another family to be bereaved by war was that of Moshe Dayan. 'We children played war games, talked in military jargon, and bragged about our fathers and uncles,' his daughter Yaël later wrote, and she added:

> The realization of what it was really like came with the shock of the first death in the village, in the family, someone we knew – and the first grave. The agony of this war shattered me with the cry of my grandmother Dvorah when she was told that her youngest son, Zorik, had

been killed. I had never seen her cry before, or I didn't remember. I had never thought of her as old before, or even vulnerable. But after Zorik's death my wise and all-knowing grandmother, discplined and brave, was a broken person.

Indeed, we were all helpless in our anguish. Zorik was special to each of us. He was not a typical Dayan, with his optimism, his jolly, unmalicious humour, his vitality and sensitivity. Zorik kidded me and made me laugh; he teased me and told me funny stories. With Zorik I rode a horse for the first time. Zorik was Dvorah's baby, my mother's young brother, a friend to Reuma her sister, the village's naughty darling.

Zorik's body was left for three days in the fields of Ramat Yohanan, unapproachable because of snipers. It was a warm May and the last of the spring flowers, poppies and wild chrysanthemum, covered the fields where my father stayed with us, and Mimi insisted in her sobbing that Zorik was still alive. I knew the baby would never know his father, and I panicked. It could happen to me, and to my brothers. I had no control over the things that mattered, and there, in the coffin draped in black, could be my own father.

The proximity of death was frightening. The candles burning in my grandparents' house, the sobbing turning into screams by the grave, the disbelief on Shmuel's pained face, my father's inability to ease his mother's anguish – memories were told and retold, as if talking about him, or reading his poems, or remembering things he said would resurrect him.

I was never lied to, and reality was never bent in order to accommodate my age. Storks did not bring children into the world, and the dead did not disappear into the clouds in the sky to dwell with angels. Zorik would never return. He was buried in Shimron, and we laid fresh flowers on his grave. Mimi was a widow, my beloved grandparents were bereaved parents, and I found little if any comfort in my father reassuring me, time and time again, that he was not going to be hurt. He had got his bullet already, he said, and he survived it. Now he knew how to avoid them and he had no intention of being hit again. There was no god in our lives to pray to, so I had to take his word and try hard to believe what he said.

*   *   *

In the first week of June, Israeli forces entered the Arab town of Jenin (on what was later to be called the West Bank). But Jenin was recaptured by Iraqi soldiers after a fierce battle on June 4. Because of the speed of the withdrawal, many of the Israeli dead had to be left behind on the battlefield. It was to be several years before an Israeli army chaplain was allowed, by the eventual Jordanian rulers, to gather the remains and bury them in accordance with Jewish religious tradition.

By the end of the first week in June, more than 200,000 Arabs had fled from the area under Israeli control, leaving 155 villages deserted. A further 73,000 had fled from Jaffa and Acre, which had been allocated to the Arab State by the United Nations, and 40,000 more had fled from Jerusalem. On June 5 Yosef Weitz went to see Ben-Gurion, to set out for him the extent of this exodus. Weitz suggested that Ben-Gurion begin negotiations with the Arab governments about settling these Arab refugees in the Arab States. Ben-Gurion noted in his diary, 'I don't think this is possible in the midst of a war. At the same time, we must make immediate preparations for settlement of the abandoned villages with the assistance of the Jewish National Fund.'

In the second week of June, the Israeli army secured the region around Mount Tabor, in Galilee. A few days later, on June 16, a kibbutz was established there. Its founders were native-born Israelis, and also immigrants from the United States, Bulgaria, Turkey, Germany and South America. They called it Ein Dor, from the city in the Bible famous for the witch visited by Saul ('the woman that divineth by a ghost').

The Arab Legion's hold on the British police fort at Latrun was still disrupting all Israeli efforts to keep open the road to Jerusalem, and to link what from the Israeli perspective were the head and heart of the new State (Jerusalem and Tel Aviv, or vice versa). After the disastrous attack on May 25, Latrun had remained under Arab control. With the main road to Jerusalem still cut, pressure on the Israeli outposts closer to the city increased. The highest ground to the north of the road, Radar Hill – site of a British Second World War radar installation – was seized by the Arabs and held by them.*

The continued blocking of the road to Jewish Jerusalem was creating extreme hardship inside the city. The food ration, for soldiers and civilians alike, was reduced to two pieces of bread, thin soup, tinned vegetables, and a gallon of water – for drinking, cooking and washing – per person per day. The water shortage was particularly acute, with long queues waiting patiently at water distribution points, subjected to spasmodic but frightening shell fire. A second attack on Latrun seemed imperative. Ben-Gurion was confident that if he unified the military command of the whole area under a first-class commander, all would be well. His choice fell on a graduate of West Point, Colonel David Marcus, who had served on Eisenhower's staff in Europe at the end of the Second World War, and had volunteered for service with the Haganah at the start of hostilities. On May 28 Ben-Gurion appointed Marcus to be Supreme Commander of the Jerusalem Front. He was given the rank of Colonel, and a *nom de guerre*, Mickey Stone, to hide his identity.

---

* This high ground was known as Ha-Radar (The Radar). In recent years an Israeli town was built there (it lies just across the Green Line). Almost imperceptibly the name changed, and it is now Har Adar (Adar Hill, Adar being one of the months in the Hebrew calendar).

Under Stone's command the second Israeli attack on Latrun took place five days after the first on May 30. During the attack on the police fort several Israeli armoured vehicles, part of the Armoured Brigade, broke into the courtyard and reached the main gate in the walls. But the sappers whose task was to place explosive charges against the walls were hit by Arab artillery fire from outside the fort and without them the armoured vehicles had to pull back. 'I can still recall with emotion,' Chaim Herzog (then second in command of the Armoured Brigade) recalled half a century later, 'the death of one radio operator, a young girl called Hadassah Lipshitz. She was in one of the halftracks, a light armoured truck, left over from World War II and primitive even at the time. She was on the radio to us at headquarters when she was shot – and she talked to us, getting weaker and weaker, until the moment she died.' Herzog added, 'As a staff officer, one who plans the battles, you don't see the deaths as they happen. You hear noises and you see flames. You only see the death afterward when the bodies are collected. I've never been sure which is worse for commanders, to see those around them go down in the midst of action or to see the cold bodies in the quiet of the aftermath, knowing their heroic sacrifices were made at the commander's direction. I suppose the answer is that both choices are equally terrible and wasteful.'

The infantry soldiers who would have entered the fort at Latrun once the gate was blown open fell back. The explosion of land-mines had led to panic in their ranks. There was nothing that Colonel Stone could do but order a retreat, signalling to Ben-Gurion as the battle ended: 'Plan good. Artillery good. Armour excellent. Infantry disgraceful.' The second battle of Latrun was over. The road to Jerusalem remained closed.

On June 8, knowing that the truce being called by the United Nations was imminent, Ben-Gurion demanded a third attempt to capture Latrun. His order was that the ridge must be taken on the following night 'whatever the cost'. But he was opposed by the Palmach commander, Yigal Allon, and by Allon's brigade commanders. Their reason was not only that a third attack might fail, but that it was no longer needed. In the previous few days a possible alternative route to Jerusalem, just beyond the range of the Arab Legion artillery at Latrun, had been discovered by chance by two Israeli soldiers making their way down from Jerusalem on leave.

Using this route, a formation of 150 soldiers, most of them new recruits who had reached Tel Aviv direct from the British detention camps in Cyprus before the camps had been sealed off again, were sent up to Jerusalem and reached the city without incident. Two Arab villages that lay astride the route, Beit Jiz and Beit Sussin, were occupied by the Seventh Brigade. Yadin then gave the order for a bulldozer to clear a track that could be used by vehicles. Four jeeps also managed to negotiate the track, to a point where an Arab orchard came to an end at a precipitous hillside 400 feet high. To negotiate this bluff was the main obstacle to opening a passable road.

Armed with information about this alternative route, Allon and Stone sent Rabin to the coast to urge Ben-Gurion to change his mind about attacking Latrun a third time. Rabin later recalled the Prime Minister's reaction to his mission:

> Ben-Gurion was furious and it seemed to me that Latrun had become something of an obsession for him.
>
> 'Is it or is it not possible to attack Latrun tonight?' he barked.
>
> 'It's possible,' I said, 'but the chances of success are close to zero.'
>
> 'Why wasn't I notified yesterday that you had no intention of attacking Latrun?' he flared at me.
>
> 'I don't know. I have been charged with submitting this proposal on behalf of Allon and Stone. Latrun is not sacred. The purpose of taking it is to safeguard our link with Jerusalem. If that purpose can be gained by other means, why must we shed our blood over Latrun?'
>
> Ben-Gurion's anger did not subside. 'Why didn't Allon tell me that he doesn't intend to attack Latrun?'
>
> His rage reached its peak as he shouted, 'Yigal Allon should be shot!' I was astounded, barely able to mumble, 'Ben-Gurion, what are you saying?' But he did not withdraw his remark. 'Yes, you heard me correctly!'

For two hours Ben-Gurion continued to argue in favour of a third attack on Latrun. Then he conceded. 'You have left me with no choice,' he told Rabin. 'I approve.' It was June 10. The truce was due to come into effect on the following morning. The attack on Latrun was cancelled. The new road became Jerusalem's life-line. Known at first as Route Seven (after the Seventh Brigade), it was soon renamed the Burma Road after the Second World War supply route from Burma to China, which had enabled the Chinese to be supplied over the mountains in their struggle against Japan. It was Chaim Herzog who went specially to Tel Aviv to put before Ben-Gurion a list of the heavy earthmoving equipment and an estimate of the number of workers needed to create a road that could be used by lorries bearing food supplies and other equipment. The first problem was how to carry supplies from the orchard to the valley below. Herzog later wrote:

> I suggested that we put hundreds of porters at the bluff to take supplies from trucks, which could approach through the orchard. Porters would then carry the supplies to the valley below, where trucks from Jerusalem would be waiting to take the supplies back to the city. Ben-Gurion immediately gave orders to supply everything we needed.
>
> This momentous development opened the supply route to Jerusalem even before the first truce was negotiated. Using bulldozers, tractors, and manual labour, the engineers began the nearly impossible task of

creating a passable road to the bluff at the head of the orchard and a road to the valley below.

At night, against the background of Jordanian shelling, the scene was almost unreal: hundreds of porters silently carrying food and supplies down the hill to waiting trucks and jeeps and even mules. Even herds of cows were led along this route because we desperately needed to ship beef into the city. Hollywood later made a movie of this extraordinary operation, *Cast a Giant Shadow*, with Kirk Douglas as Mickey Stone.

On June 10, the day on which a third Israeli assault on Latrun was abandoned, and the day before the cease-fire was to come into effect, Syrian forces in the north attacked the settlement of Mishmar Ha-Yarden, on the western bank of the River Jordan, near the B'not Yaakov bridge. Their aim was to overrun the settlement, and then push westward to the main north–south road, which was the only road into the finger of Galilee.

In attacking Mishmar Ha-Yarden, the Syrians had air mastery, as well as eight tanks and ample artillery. Women and children had been evacuated; the men fought alone. By midday the settlement was overrun, and those defenders who were captured taken to prison in Damascus. It was not until the end of the day, when a survivor of the action reached Mahanayim, five miles to the west, that the fate of the settlement was known. So elated were the Syrians at their victory that they called it Fatih-Allah – the Capture of God.

After the war, Mishmar Ha-Yarden was rebuilt a mile further from the river; the ruins of the original settlement remain ruins to this day.

Following a series of appeals from the United Nations General Assembly, which since May 28 had been represented in Israel by a Swedish diplomat, Count Folke Bernadotte, the truce came into being on June 11, after nearly four weeks of war. Both sides agreed that the truce would also last for four weeks. As the fighting died down, it became clear that Israel was in control of much more territory than she could have hoped for a month earlier. Both eastern and western Galilee were under complete Israeli control, as were the Jezreel Valley from Haifa to the River Jordan, the coastal strip down to a point just north of Isdud, a large pocket in the central Negev, though not the main Negev town of Beersheba, and, kept open at such heavy cost, a corridor from the coast to Jerusalem.

CHAPTER TWELVE

From the first truce to the second truce

On the first night of the truce, 11 June 1948, the commander of the Jerusalem Front, Mickey Stone (Colonel Marcus), who was staying that night with a Palmach battalion near Abu Ghosh, left his tent and went outside the camp perimeter to relieve himself. Hearing footsteps, a sentry called out for the password. Marcus did not speak Hebrew, and had wrapped himself in a white sheet, which the sentry thought was Arab dress. The sentry fired a warning shot and repeated his demand. Then two more shots were fired, and Marcus was killed. 'His death hit me very hard,' Chaim Herzog has written, 'for we had talked so much about our dreams for the future. One night I was standing on the Burma Road near the serpentine track that mounted the bluff. People were moving supplies, and the area was covered in a heavy cloud of dust. Suddenly a hush fell as a jeep with a coffin strapped to it slowly made its way up the bluff, moving toward the coast. It was Mickey. His body was on its way to the airport, where it would be flown to West Point for burial. Watching the jeep disappear, we felt orphaned – and in a sense we were.'

Marcus's body was flown back to New York, escorted by Moshe Dayan, then a senior army officer, and Joseph Hamburger, the former captain of the *Exodus*.

Israel's losses between the declaration of statehood and the first truce were 876 soldiers and 300 civilians. During the truce, Israel made every effort to improve its military position. The Burma Road by-passing the main Jerusalem highway was widened and paved. Weapons were brought in, largely from Czechoslovakia. The immigrants from the Cyprus detention camps were integrated into the existing, battle-tested units. There was also an episode during the first truce that was to cause considerable soul-searching in later years. A former member of the Haganah, Meir Tubiansky, who had been an officer in the British army during the Second World War, and had been serving in the last months of the Mandate as a liaison officer with the electricity generating company (run by British personnel) disappeared. So suddenly did he leave Intelligence headquarters that his beret and pipe were still in his desk drawer. It later emerged that he had been accused by the head of Israel's Military Intelligence, Isser Be'eri, of having given vital intelligence to the British, who had passed it on to the Arabs.

During the first truce, Tubiansky was taken to the captured Arab village of Beit Jiz, near Latrun, court martialled, found guilty, and shot within hours of his trial. When, a year later, Ben-Gurion heard about the episode, he ordered an inquiry. Be'eri was found guilty of having falsified documents and was dismissed from the army in disgrace. Tubiansky was posthumously rehabilitated, and reburied with full military honours.

On July 2 a new kibbutz was founded, overlooking the River Jordan and facing the Syrian Heights. It was only five miles from the area held by the Syrian forces who had seized Mishmar Ha-Yarden. The kibbutz was named Kfar Ha-Nasi (President's Village) in honour of Chaim Weizmann. It was made up of members of the Habonim youth movement who had been preparing for months to set up a kibbutz in this area. Most of them came from England and Australia. A week after their kibbutz was established, its members found themselves in the centre of a battle as Syrian troops broke the truce in an attempt to seize the western side of the River Jordan, that ran half a mile beyond the kibbutz boundary. But the Syrian forces were driven off, and the slow, hard, patient work of construction began, within sight of the Syrian machine-gun posts on the facing hills to the east.

During the first truce, the United Nations mediator, Count Bernadotte, produced a plan of partition of his own, hoping to secure a peaceful political solution acceptable to both sides. Under this plan, the Negev would go to the Arabs and Galilee to the Jews. Jerusalem would come under Arab Transjordanian rule, with its Jewish inhabitants granted urban autonomy. Given Jerusalem's substantial Jewish majority, this was a serious blow to Jewish hopes of control of at least the Jewish sections of the city. When Bernadotte's proposals were rejected by both Israel and the Arabs (who did not want the Jews of the city to have any autonomy at all), a negotiated political solution was ruled out, and a renewal of hostilities became inevitable.

There was, however, an internal Jewish conflict before the Arab–Israel conflict broke out anew. On June 19 the ship *Altalena*, which had sailed from southern France, approached the coast of Israel. On board, as well as 900 immigrants, were arms destined for the Irgun – including 5,000 rifles and 270 light machine-guns – and many Irgun volunteers, among them a large group from Britain. Ben-Gurion wanted these weapons sent to the Jerusalem Front, and had the authority to order this under the unified command structure.

The Irgun rejected Ben-Gurion's appeal that the arms be handed over to the Israeli forces, and that the men on board become an integral part of those forces. Ben-Gurion was outraged. The Irgun had been officially incorporated into the Israel Defence Forces three weeks earlier. He wanted no separate or factional armies, and was determined that the Israel Defence Forces would alone carry arms, and defend the State. He also feared that Menachem Begin, the Irgun commander, might want to set up some form of Irgun control in the areas allocated by the United Nations to the Jewish State.

On the evening of June 20 the *Altalena* reached the coast off Kfar Vitkin. Ben-Gurion asked his senior military advisers what troops were available in the vicinity who could be brought into action, and what action to take. Yigael Yadin said he had 600 men nearby, and could assemble more. He then asked Ben-Gurion, 'What are our orders? Should the brigade commander be told to concentrate his forces and simply threaten to use them, or should he concentrate his forces and go into action if the threat does not work?' To act meant to shoot, Ben-Gurion replied, telling Yadin, 'A threat is meaningful only if it is backed by a willingness of the government to carry it out.'

Ben-Gurion ordered the Irgun to hand over the ship to the Israeli army. The Irgun refused, and firing ensued between Irgunists who had arrived to unload the ship and the Israeli soldiers sent to prevent any landing. Two Israeli soldiers and six Irgunists were killed before the Irgunists on shore surrendered. They then renewed their earlier oath to obey the instructions of the government. But the boat sailed off.

On the following day, June 21, the Government Press Office issued an announcement with Ben-Gurion's full approval:

> The Provisional Government and the High Command of the Defence Forces wish to make clear that they are determined to stamp out immediately this traitorous attempt to deny the authority of the State of Israel and of its representatives.
>
> The Provisional Government and the High Command will not permit the enormous efforts made by the Jewish people in this country to secure their independence and sovereignty, while fighting a bloody conflict forced on them by external enemies, to be undermined by an underhanded attack from within.
>
> Jewish independence will not endure if every individual group is free to establish its own military force and to determine political facts affecting the future of the State.
>
> The Provisional Government and the High Command call on all citizens and soldiers to unite in the defence of national unity and the authority of the people.

That night, Ben-Gurion and Levi Eshkol, his deputy Minister of Defence, slept at army headquarters in Ramat Gan, just inland from Tel Aviv. Shimon

Peres was with them, 'a rifle in hand, in case army headquarters were stormed by protesters.'

While Ben-Gurion slept, the *Altalena* sailed southward to Tel Aviv. There, at dawn, it was visible to all the citizens. Joining those on board was Menachem Begin. Ben-Gurion repeated his demand for the surrender of the arms. Again it was refused, with Begin broadcasting from the ship by radio that army snipers had been ordered to assassinate him, which the army insisted was untrue. An Irgun statement published that morning in the news-papers declared: 'We warn those agitating for a civil war, and their supporters, that if such a war begins, it will not be limited to a single area.'

At a meeting of the Provisional Council of State on June 21, Ben-Gurion set out the issues at stake from his – and as he saw it the nation's – perspec-tive. What the Irgun was doing, he said, was 'an attempt to destroy the Army and to kill the State. If the Army and the State surrender to an independent force the Government might just as well pack up and go home.' There were those in his government who wanted to open negotiations with the Irgun. 'I am as much a compromiser as the rest of you,' he told them. 'But there are things on which there can be no compromise, for the very soul of the State is at stake. They must agree to turn over the ship, to accept the authority of the Government. Once they agree, we will be generous; we will not harm anyone. At the most, there will be a few arrests. All of us want to avoid bloodshed. But there is no room for negotiations. They must turn over the ship to the Government and accept the authority of the Army. It is our Army. If they do so, there will be no battle. If the affair is really over, there will be an amnesty. But there is no room for compromise or negotiations. The future of the war effort is at stake.'

Ben-Gurion called for additional troops. Some came from the army base at Sarafand, others, members of the Negev Brigade, had just reached Tel Aviv for a short period of leave, having infiltrated through the Egyptian lines. The Irgunists on the *Altalena* were raked from bow to stern by machine-gun fire from the Israeli forces that were assembling by the beach, and the ship was set ablaze. Ben-Gurion in reporting the event to the Provisional Council declared, 'Blessed be the gun which set the ship on fire; that gun will have its place in Israel's War Museum.'

Jew had fought against Jew, in full sight of the Jewish metropolis. Among the forty Irgun dead was a leader of the Irgun's pre-State illegal immigration movement, Avraham Stavsky, who had been one of those charged with the murder of Chaim Arlosoroff in 1933, and a Jewish volunteer from Cuba. Stavsky, who had successfully run the British blockade of Palestine thirteen times without incident, had been wounded on board, and was killed by machine-gun bullets when swimming towards the shore.

John Altmann, who had been born in Stettin and reached Britain as a refugee shortly before the outbreak of the Second World War, was also among those on board the *Altalena*. 'There were soon internal explosions and a lot of smoke', he recalled. 'At the bottom of the main companionway the Captain had had the crates of detonators put separately from the shells. He sent us below to bring them up to dump them overboard. We did not understand his logic but with the benefit of hindsight I now realize that the citizens of Tel Aviv have much to thank him for. A statue would not be over-doing it. Mr Begin got hysterical stating that he wished to go down with the ship. It was explained to him that the ship could not go down, only up, and the Captain who kept his cool, literally had Mr Begin dumped overboard.' After that, Altmann wrote, 'we all abandoned ship, the deck which was of metal was burning the soles of our feet. We all jumped overboard and swam to the shore. The small arms fire from the hotel balcony continued and a few people were hit in the water. The ship's propeller was threshing furiously as the captain was trying to free the keel and get out to sea. A young volunteer from Morocco with whom I had been friendly on board, was hit in the water and was sucked into the path of the propeller. He did not survive. The rest of us made it to the beach.'

As the immigrants and the Irgunists jumped into the sea to escape the burning ship, Israel Defence Forces soldiers were instructed to go into the water and help to rescue them. Many of the Irgun newcomers went straight to the battle fronts: Joe Kohn from Philadelphia was killed in Jerusalem during the last attack on the Jewish Quarter of the Old City.

On June 22, Ben-Gurion summoned another meeting of the Provisional Council of State. He was concerned about the many public expressions of regret that fire had been opened against the Irgun, and that Irgunists had been killed. Rabbi Meir Berlin, a member of the Council, warned that 'the killing of Jews by other Jews' was bound to affect the war effort adversely: 'It will be said that the State is weak and its future is threatened.' Ben-Gurion took the opposite view, telling the Council that the matter went much further than a mere military confrontation:

> An armed uprising against a State or an army can be repressed by mili-tary force, but military force is not sufficient to eliminate the danger. The chicanery of the dissidents was in large measure the result of the support given to them in various quarters.
>
> At one time it might have been possible to explain, though difficult to justify, this support. Now it is difficult even to explain. We are in the midst of a life-and-death struggle.
>
> At the moment there is a truce, but the war has not ended. Enemy armies are entrenched inside the country. Jerusalem is surrounded by the Arab Legion and its artillery. The Negev road is in the hands of a large Egyptian force. Mishmar Ha-Yarden is held by the Syrians. Moreover, additional Arab forces are poised on the borders.
>
> The arrogant action of armed gangs inside the country gravely

endangers our ability to defend our own future and the future of the entire Jewish people.

This danger will not end until the citizens of this country and Jews everywhere understand the tragic significance of the very existence of such organizations. Not only the army, but the entire people, must help to uproot this evil.

Ben-Gurion faced another internal challenge during the first truce: a challenge by senior army officers against his nominees for the command of four 'fronts' that were being established, within which to group the existing twelve brigades. The frequent illness of the Chief of Staff, Yaakov Dori, added to the growing unease. Ben-Gurion's authority in combining the Prime Minister's office and that of the Minister of Defence (something which Churchill had done in Britain in the Second World War) was also being challenged.

A committee of five members of the Provisional Council of State examined the complaints. Among its conclusions was severe criticism of his 'interference' in the battles for Latrun and the Jerusalem road. Ben-Gurion was so angered that he threatened to resign. He was dissuaded from doing so by the Operations Officer, Yigael Yadin, who prevailed upon him to accept a compromise, whereby decision-making power would reside in a War Cabinet of five members, under Ben-Gurion's chairmanship. This was also an echo of the British wartime system.

With Dori chronically ill, the functions of Chief of Staff were taken over by Yadin, who had given up a career in archaeology to become a professional soldier. He was thirty-four years old. Shimon Peres, who often saw Ben-Gurion and Yadin together, later wrote:

Often Ben-Gurion and Yadin would get into raging arguments, in the heat of a battle, over the interpretation of some obscure phrase in the Bible. Ben-Gurion cherished these sparring matches.

Yadin was a charismatic figure who would make his points during strategy sessions with drama and force, his burning eyes surveying each participant in turn. He would prepare meticulously for every such meeting, mastering the issue down to the smallest details and compiling a cogent statement of his position.

But the quality in Yadin which won Ben-Gurion's admiration most of all was his totally impartial devotion to the State – and complete aloofness from all party-political considerations. He made a powerful impact on the entire General Staff, including Ben-Gurion himself, when he ordered the arrest of one of the top officials at the Ministry of Finance who had failed to appear for reserve duty. 'There are no special privileges in this army,' Yadin announced. 'Everyone is equal before the law.'

Ben-Gurion, who fought a running battle against that common

Jewish trait of 'doing people favours' which quickly grows into tendentiousness and discrimination, especially appreciated Yadin's uncompromising stand in this affair.

A feature of Yadin's personality which I myself found particularly admirable was his custom never to criticise Ben-Gurion behind his back – or to withhold criticism of the Premier to his face.

*   *   *

The isolated settlement of Kfar Darom, south of Gaza, had continued to hold out against the Egyptians since May. On the last day of June it was reached by a convoy from the commando battalion, which intended to evacuate the wounded and the women. But the Egyptians were able to prevent the convoy from leaving, so that the meagre food supplies that were by then all that remained in Kfar Darom had to be shared by the defenders with the commandos.

Several attempts were made by the Israeli air force to drop food by parachute. But Kfar Darom was only 40 yards square. Egyptian anti-aircraft fire meant that the drops had to be at night, and, inevitably, most of the food came down in Egyptian-held territory. Where a parachute drop landed in Kfar Darom itself, or in the no man's land around the kibbutz, the Egyptians opened fire on it with their artillery to prevent the supplies being collected.

After eight days, the commando unit managed to escape from Kfar Darom, taking some of the less seriously wounded, and some of the girls of the settlement with them. Then, on the night of July 7–8, the settlers, for whom no further help could be sent, agreed to be evacuated. On the morning of July 8, not realizing that the kibbutz was deserted, the Egyptians opened fire with a heavy-artillery barrage, and then sent a combined armoured and infantry force against the settlement. They found it empty.

The four weeks' truce was due to end at midnight on July 9. In an attempt to prevent fighting breaking out again, Count Bernadotte asked both the Jews and the Arabs to agree to an unconditional cease-fire for the next ten days. Israel accepted this proposal. The Arab governments rejected it. 'What legitimate objectives can anybody have,' Abba Eban asked the United Nations Security Council four days later, 'which can be threatened by the preservation of peace for ten days? What are the ambitions which rest upon so flimsy a moral foundation that they cannot endure ten days and ten nights of peace?'

In the Jerusalem area, the ending of the truce was the moment Ben-Gurion had been waiting for to extend the area of the city under Israeli control. As fighting flared up again, a company of boys of sixteen and seventeen, commanded by twenty-one-year-old Oded Chai (who had seen active service in the Jewish Brigade in Europe in 1945), set off to capture the high ground to the west of the city, beyond which lay the three Arab villages of Beit Masmil, Ein Karim and Malha. Hardly had the group started when Oded Chai was killed by a sniper's bullet. Asking for a ten-minute postponement, the new commander, Eliahu Lichtenstein, reorganized the company and, with a new battle cry, 'For Oded!' they attacked and captured the hill. Now known as Mount Herzl, it was the hill seen by tens of millions of television viewers worldwide during the funeral of Yitzhak Rabin in 1995.

On the following night, using Mount Herzl as their starting point, the Israeli forces drove the Arabs from Beit Masmil (now the western Jerusalem suburb of Kiryat Ha-Yovel) and Ein Karim. To secure the railway line, an Irgun unit attacked Malha on the night of July 13–14. It suffered heavy casualties: seventeen Irgunists were killed, among them Nathan Cashman, from London, who had arrived in Israel on board the *Altalena*.

That night Haganah troops attacked the Arab village of Beit Safafa and seized the houses of the village that lay on either side of the railway line, from which, since the start of hostilities in May, the isolated southern Jewish suburb of Mekor Hayim (Source of Life) had been under frequent attack.

In the north, where the Syrians had already broken the truce to seize a bridgehead over the River Jordan, there was a fierce battle, starting on July 10, in which the Israeli forces struggled to push the Syrians back across the Jordan to the former international border. The Syrians had air superiority. On the ground, the Israeli artillery pieces were outranged by the Syrian guns. Despite the taking and retaking of Syrian positions during the course of ten days of intense fighting, the bridgehead remained intact.

With the ending of the first truce, fighting was resumed on all the fronts. It lasted for ten days. The principal Israeli effort was in the centre of the country, where Arab control of the two Arab cities of Lydda and Ramle provided a permanent threat both to all north–south and to east–west (Tel Aviv–Jerusalem) communications. The Israeli High Command also hoped to seize the Arab town of Ramallah, north of Jerusalem, to widen the corridor to the city, and make it less vulnerable to attack through its northern suburbs. Lydda and Ramle were captured. Ramallah was not.

The capture of Lydda and Ramle was a central part of Plan D, these two large Arab towns being astride the main north–south road and railway, and hostile. A commando unit was ordered to break into them, to create the confusion needed to allow infantry forces to capture the towns later. The commander of the unit, Moshe Dayan, later recalled the raid:

> The unit advanced in a column, headed by an Arab Legion armoured
> vehicle with turret and a two-pounder captured in yesterday's operations.
> A gunner had somehow been found, a wireless operator, and a name,
> 'The Terrible Tiger,' and immediately it was put to service in our ranks.
> Following 'The Tiger' came the half-track and jeep company. When

the column was about one kilometre from the city, it encountered heavy fire. The Terrible Tiger would halt from time to time, and return fire to the fortified positions, with the convoy continuing on its way. With every shot of The Tiger – sandbags would be observed crumbling into Arab positions, and smoke columns would go into the air. Arabs were seen escaping from their positions, which had been hit.

That was the moment. Follow in the footsteps of the escapees. The main road was blocked, and the unit moved along a side track, which, fortunately, had not been mined.

Soon afterward we penetrated through the line of positions. The column moved slowly, in line, along the narrow track. Barricades were breached from time to time. Firing went on at full speed. The jeeps, whose only armour was their weapons, their light machine-guns, shot at windows, at fences, at sandbag positions. The cactus fence was cut by machine-guns, as if with a sickle. We increased our speed and left the positions behind us.

From the junction at the entrance of Lydda (known to the Israelis as Lod), Dayan's column continued south toward Ramle, which it entered under a hail of Arab gunfire. His account continued:

Only The Terrible Tiger turned north and entered Lod. He roamed its street firing in all directions, and then opened up a duel with the police fortress. It was there that the convoy found it, on its way back from Ramle, with the wireless operator bleeding profusely.

The visit to Ramle was exceedingly brief. The unit had been badly hit. There were four dead in the jeeps. The half-tracks had a number of wounded, some of them seriously. A jeep belonging to the Scouts remained burning near the police station. Most of the tyres were punctured, radiators had fallen off. One jeep had lost its hood and its machine-gun; in another, bullets had severed the brakes. Some men were missing, wounded who had fallen off jeeps near the police station. Morale was dropping. The men were bandaging their wounds, and replacing the punctured tyres.

Time was precious. The longer the unit delayed, the more difficult its withdrawal would be. Reports were now being received of Arab Legion armoured vehicles on their way to the battle. The Arab Legion unit in the police fortress halfway between Lod and Ramle, which had been taken by complete surprise when the commando battalion made its first appearance, had indeed mistaken it for reinforcements it was expecting. Meanwhile, it had recovered from its shock, realized its mistake and manned its positions.

The only withdrawal route ran right outside the police fortress, unless an attempt could be made east of Ramle along the Latrun road.

The order to retreat was given. The wounded were loaded onto the half-tracks. The convoy took off. It passed by the police fortress, under cover of The Tiger. The battalion returned to Ben Shemen. Its casualties: nine dead and seventeen wounded. When the column left Lod on its way back to Ben Shemen, it was found that one of the wounded was missing. Its commander asked for permission to go back into the city and look for his comrade. The permission was granted. The entire raid lasted no more than forty-seven minutes.

That night, Israeli infantrymen attacked Lydda in force and seized part of it. On the following morning, Arab Legion forces attacked the Israeli positions but were driven off. In the fighting, thousands of Arab civilians panicked, left their houses, and were caught in the crossfire: 250 were killed. The Israelis lost no more than four soldiers. As the fighting continued, Yigal Allon, at military headquarters, asked Ben-Gurion with regard to Lydda, 'What shall we do with the Arabs?' to which Ben-Gurion replied, 'Expel them.'

Orders were issued at once, and began: 'The inhabitants of Lydda must be expelled quickly without attention to age. They should be directed towards Beit Nabala.' The orders were signed by the Chief of Operations of the fighting in the central area, Yitzhak Rabin.

Ramle also fell to a sustained Israeli attack on July 12. Only the former British police fort that stood between the two towns was still unconquered. But when Israeli forces finally moved against it that day, it was found to be empty. Its defenders had fled. Within twenty-four hours, 50,000 Arabs were moving eastward from Ramle and Lydda, seeking the safety of the Samarian hills. Some were taken by the Palmach in trucks and buses to the border of the area controlled by the Arab Legion.

On the outskirts of Lydda, Shmarya Gutman, a Jewish archaeologist, was an eyewitness of the Arab exodus. 'A multitude of the inhabitants walked one after another,' he recalled four months later. 'Women walked burdened with packages and sacks on their heads. Mothers dragged children after them. Occasionally, warning shots were heard. Occasionally, you encountered a piercing look from one of the youngsters in the column, and the look said, "We have not yet surrendered. We shall return to fight you."'

On the eastward march into the hills, and as far as Ramallah, in the intense heat of July, an estimated 335 refugees died from exhaustion and dehydration. 'Nobody will ever know how many children died,' Glubb Pasha commented. In a Palmach report written soon afterwards, possibly by Allon, the exodus from Ramle and Lydda was said to have averted a long-term Arab threat to Tel Aviv, and in addition to have 'clogged the routes of advance of the Legion', and imposed on the Arab economy the burden of 'maintaining another 45,000 souls'.

This strategic argument led to a fierce debate inside the Israeli political establishment. The co-leader of Mapam, Meir Ya'ari, noted how 'easily' did the military planners 'speak of how it is possible and permissible to take women, children and old men and to fill the roads with them because such is the imperative of strategy', and he added: 'I am appalled.'

On July 12, as the battles for Lydda and Ramle were at their height, Israeli troops drove the Iraqi soldiers from the Arab village of Ras al-Ain (in Hebrew Rosh Ha-Ayin – Head of the Spring). It was from here, by the headwaters of the River Yarkon, that Jerusalem had pumped up much of its water during the British Mandate. Control of the village by the Arabs since November 1947 had been the main cause of Jerusalem's severe water shortage. An Iraqi counter-attack was driven off.

The northern Negev was dominated by Egyptian forces, intent on pushing northward to Tel Aviv. On July 12 they attacked the settlement of Negba, launching the assault with an artillery barrage and air attack. The battle lasted for more than seven hours. At one moment the Egyptians reached the inner fence. But by nightfall they had failed to penetrate into the settlement, and they withdrew, leaving a tank and several Bren gun carriers behind.

An Egyptian army officer who was taken prisoner a few days after the battle for Negba spoke of the astonishment among the Egyptian officers of his command. 'How did Negba hold out? There were 150 defenders in the place, 4,000 shells were spent on it. According to our calculations 50 per cent were to have been hit and a further 25 per cent to be busy treating casualties. If so, how were you able to defend Negba?'

The settlement newspaper answered the Egyptian's question: 'Quantitatively, the balance of forces was in your favour; but qualitatively, it was in ours. There was one item which the Egyptians were lacking – a clear idea about the purpose of the battle. But this is precisely what we possessed in full measure: fearless spirit, the knowledge and conviction about the purpose of the battle, the realization that with our own bodies we barricaded the way north.'

After failing to capture Negba, the Egyptian forces turned, on July 14, against the isolated settlement of Gal-On, 15 miles to the east. Here too they were halted, partly by the minefield surrounding the settlement, and partly by the accuracy of the defenders' fire. On July 15 it was Be'erot Yitzhak that was attacked. At one point in the battle the Egyptians were so close to the perimeter that the defenders used hand-grenades to drive them off. A detachment of the commando battalion of the Negev Brigade, and an artillery battery, arrived just as the settlement was about to be overrun. As the Egyptians withdrew, seventeen of the settlement's defenders lay dead. Their buildings had been completely destroyed.

On July 15 the United Nations Security Council resolved that the Arab rejection of the ten-day extension to the cease-fire proposed by Bernadotte constituted a breach of the peace, and ordered a permanent cease-fire under penalty of sanctions. The Arabs rejected this, and the fighting continued. On July 18 the Israeli forces in the Negev took the initiative, attacking the Egyptian troops who were encamped at the village of Kharatya. The village was overrun, but then the Egyptians mounted a counter-attack. Two of their tanks were about to break into the village when an Israeli soldier, Ron Feller, who was carrying a hand-held PIAT anti-tank gun, with only two rounds of ammunition, crawled along a cactus fence until he was only 25 yards from the tanks, separated the cactus with his hands, and fired. The first tank was hit and disabled. The second tank turned around and left the battlefield. The rest of the Egyptian armoured force followed it. In recognition of what he had done, Feller was one of twelve soldiers made Heroes of Israel.

The Israeli hopes of using the ten days of fighting between the truces to drive the Egyptians from the northern Negev were disappointed. Attacks on the Egyptian positions in the villages of Iraq el-Manshiya and Faluja were themselves driven off.

During the ten days' fighting, Israeli forces in Galilee captured the Arab town of Nazareth. Some 300 Iraqi troops who were holding the police fort in the city fled before the battle reached them. Tens of thousands of the city's Arab inhabitants fled as the battle drew near. Also escaping from Nazareth was Fawzi al-Kaukji, the commander of the Arab Liberation Army, though his headquarters in the fort was found intact, and all his orders and reports fell into Israeli hands. These included the papers of a German sabotage expert whose activities had been evident to the Israelis since the beginning of the year.

During the battle for Nazareth there was a certain amount of looting by the Israelis. In a few instances 'greed for loot' (as Lieutenant-Colonel Netanel Lorch described it) prevented units from adequately organizing the defence of a locality which they had just captured, so that the Arabs were able to counter-attack successfully. But looting had been specifically condemned on the eve of the battle. 'No soldier will indulge in looting in the city,' warned the last operational order before the assault on the city. 'The operation commander has received strict instructions concerning the measures to be taken against anyone ignoring or contravening this order. Our soldiers are civilized, and they act with respect for the religious feelings of others. Yet if any one of them be found wanting, he will be prosecuted immediately, without hesitation, and will be heavily punished.' This order was carried out.

On July 16 the Israeli flag flew over Nazareth. Sixteen members of the Arab Liberation Army had been killed in the battle, including one Briton, one German and one Iraqi. One Israeli soldier had been wounded.

Following the capture of Nazareth, many Arab villages in the neighbourhood surrendered. One village brought all its arms and military equipment to a nearby Jewish settlement on a horse-drawn cart. When an Israeli jeep with two soldiers on it approached the Arab village of Daburiya, the villagers ran up white flags and surrendered. Ten years earlier, Daburiya had caused Wingate and his Night Squads heavy casualties. By the time the second truce came into effect on July 19, the road from Haifa to Nazareth was under Israeli control. The road from Tel Aviv to Haifa was likewise cleared of pockets of Arab resistance, including the two villages of Ein-Ghazal and Jaba, from which the Arabs had fled four months earlier. With their conquest, the road was opened in its entirety to Israeli traffic for the first time in eight months.

The fighting between the two truces was also used by the Israeli forces to widen their area of control at the entrance to the Jerusalem Corridor. On July 13 the Arab village of Tsora – the birthplace of the biblical Samson – was captured from the Egyptians. This gave the Israeli forces control of a stretch of the Tel Aviv to Jerusalem railway. A few days later, the former Jewish settlement of Hartuv, which had been founded in 1895 and twice abandoned – first in the Arab riots of 1929 and again in May 1948 – was recaptured, as was the nearby police fort, which had been under the control of the Egyptian army.

With Latrun and the foothills on the road to Jerusalem still in Arab Legion hands, a third attempt to capture Latrun was begun on the night of July 14–15. Instead of attacking the police fort, which had held off the two previous assaults, this third attempt struck at the ridge behind the fort. But the Arabs were able to bring reinforcements along a 'Burma Road' of their own, linking Latrun with the high ground of Radar Hill, north-west of Jerusalem, and the Israeli attack was driven off. The appearance of Arab armoured vehicles on a rocky hillside where no road or path was known to exist was a blow to the Israeli attackers, who had no anti-tank weapons with them to stop the armour in its tracks.

The Israeli troops withdrew. They were unable to take back with them three gravely wounded men, and had to leave them behind. A medical orderly, David Miller, refused to let the wounded men be left alone and, against specific orders, insisted on remaining with them and looking after them. All four were killed by the Arab Legion.

In Jerusalem, the imminence of the truce, and earlier failures to dislodge the Arab Legion from the area north of the Damascus Gate, led to a last-minute attempt to gain some definite advantage before the truce came into effect. The United Nations Security Council had voted for the cease-fire in Jerusalem to come into effect on July 17, forty-eight hours before the cease-fire else-where. A two-pronged attack was planned towards the Jewish Quarter, through the New Gate to the north and through the Zion Gate to the south. Both attacks failed. The Irgun, which undertook the New Gate operation, broke through the gate but failed to capture the building just inside it, the Collège des Frères, which dominated that corner of the Old City. The attack on the Zion Gate was intended to be started by an explosive device that would punch a hole in the wall large enough to enable the troops to rush in through it. But the device was faulty, and the wall remained intact.

Outside Jerusalem, the second truce was to begin on July 19. Two days earlier, in the southern entrance to the Jerusalem Corridor, the Israelis seized the Arab village of Deir Rafat, overlooking the Tel Aviv to Jerusalem railway. Also on July 19, north of the Jerusalem Corridor, Israeli forces hoped to reach the Latrun–Ramallah road, which was in the hands of the Arab Legion and constituted a permanent threat to the Jerusalem Corridor from the north. An attack was planned on the road on the last night before the truce. At the opening of the attack, the Arab village of Shilta was overrun, but not the village of Midya – the birthplace of the Maccabeans. The fiercest battle was for the high ground known as Hill 318, where the Arab Legion forced the Israeli soldiers to withdraw. More than forty Israeli soldiers were killed during the battle, most of them observant Jews from the specially created 'Orthodox' company of the Palmach.

On July 18, as the hour of the truce drew near, a fourth and final attempt was made by the Israeli army to capture Latrun. In preparation for the assault, tanks were transferred from the northern front. During the attack, one of these tanks opened fire on the police fort and hit it with the second of its two shells. A third shell jammed, however. The shell-casing had stuck in the barrel. The tank turned back in order to find a place beyond Arab range where it could repair the gun. Seeing the tank turning, the other tanks crews thought that this was the signal to retreat. As the tanks withdrew, the commander of the attack was forced to pull the infantry back as well, to prevent them being exposed to Arab Legion fire without the support of the tanks. The battle was over, and Latrun remained in Arab hands (as it was to do until 1967).

The work of the Mahal volunteers was continuous. Within thirty-six hours on July 18–19, a South African surgeon, Dr Stanley Levin – one of 500 South Africans who had volunteered to help out for the duration of the war – performed twenty-eight successive operations on severely wounded soldiers, without leaving the operating theatre.

1. Theodor Herzl, founder of
political Zionism.

2. The young David Ben-Gurion,
future Prime Minister.

3. Aaron Aaronsohn, a leading agronomist
in Palestine before 1914.

4. Vladimir Jabotinsky, founder of the
Revisionist movement.

6. Jewish farm pioneers near Kiev, preparing for immigration to Palestine, 1923.

5. Menachem Ussishkin and Chaim Weizmann visiting Palestine with the Zionist Commission after the First World War.

7. A group of Jewish fishermen in their boat, Uzkov, Ukraine, preparing for immigration to Palestine, 1924/5.

8. Jewish stonecutters in a quarry near Jerusalem, 1925.

9. German Jewish youngsters, members of Youth Aliyah, reach Ein Harod, 19 February 1934.

10. A watchtower under construction in the Jordan Valley settlement of Sha'ar Ha-Golan, 21 August 1937.

11. Erecting a water drill, Kfar Menachem, 1 October 1937.

12. Young German Jewish immigrants rejoicing on the quayside, Tel Aviv, 1939.

13. Hearing the news of the Nazi–Soviet Pact at the Zionist Congress, Geneva, August 1939. In the front row, left to right, Moshe Shertok (later Sharett), David Ben-Gurion, Chaim Weizmann, Eliezer Kaplan.

14. Wartime barbed wire, Jerusalem: British military vehicles drive along the Jaffa road.

15. Postwar violence: the south wing of the King David Hotel blown up, July 1946.

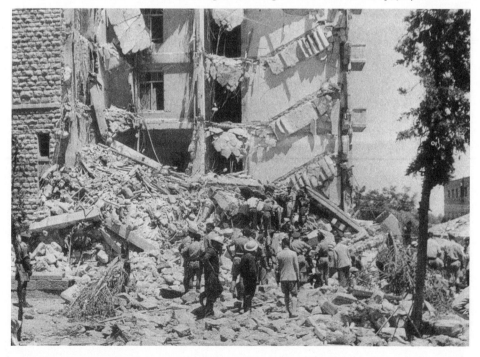

16. Tel Avivians rejoicing at the partition resolution, November 1947.

17. Underneath a portrait of Theodor Herzl, and flanked by the flag of the new State, David Ben-Gurion reads the Israeli declaration of independence, 14 May 1948.

18. Jerusalemites celebrate statehood, Ben Yehuda Street, 14 May 1948.

19. A wounded man evacuated after the bombing of Tel Aviv by Egyptian aircraft, 15 May 1948.

20. Overseas volunteers (members of Mahal) form a Star of David with their rifles, Galilee, November 1948; in the front, kneeling, a British volunteer, Stanley Medicks.

21. Egyptian soldiers at the Iraq Suwedan Police Fort surrender, November 1948.

22. Armistice negotiators at Rhodes, January 1949. Left to right: Yigael Yadin, Yitzhak Rabin and Yehoshafat Harkabi.

23. Independence Day, May 1949: a Tel Aviv street scene.

24. President Harry S. Truman presented with a gift by David Ben-Gurion, Washington, May 1951. Abba Ebon looks on.

25. A new immigrants' camp, November 1952.

26. Immigrant tented housing, with new housing beyond, December 1952.

27. Israeli troops advancing into the Sinai Desert: November 1956.

28. Moshe Dayan and soldiers at Sharm el-Sheikh: November 1956. On Dayan's left, in peaked cap, Yitzhak Rabin.

29. North African immigrants at their ma'abara in Athlit, 8 April 1954.

30. Tel Aviv street scene, July 1957.

31. Yigael Yadin (with stick) and David Ben-Gurion on Masada, 10 November 1963. Paula Ben-Gurion is next to her husband; Miriam Sacher far right.

32. An Israeli reprisal raid into the West Bank village of Samua, 13 November 1966.

33. The Six Day War, 1967: ruined Egyptian war equipment in the Mitla Pass.

34. The Six Day War: an Israeli tank and Israeli troops reach the YMCA building
in East Jerusalem.

35. The Six Day War: Colonel Motta Gur (profile to the camera) and his troops, overlooking the Dome of the Rock and the Old City, a few moments before launching the attack on the Old City.

36. The Six Day War: Jordanian soldiers surrender on the Temple Mount, just below the Dome of the Rock.

37. The Six Day War: Israeli troops overlooking the Wailing Wall.

38. The Six Day War: Israeli soldiers of the armoured tank division during the advance into Syria.

39. The Six Day War: Syrian tanks captured during the Israeli advance on the Golan Heights.

CHAPTER THIRTEEN

From the second truce
to the armistice agreements

The second truce began on 19 July 1948. Two months had passed since the start of the War of Independence. The State of Israel was still not a single geographic unit; the Egyptian army lay astride the road leading to the settlements in the Negev. In the north, several Jewish settlements were in the area controlled by the Arab Liberation Army. The Israeli-controlled Jerusalem Corridor was still a narrow strip of land, the main road of which was in part in the hands of the Arab Legion.

Israel used the time of the second truce for consolidation of what had been gained. A narrow-bore water pipeline was laid parallel to the improvised Burma Road to supply Jewish Jerusalem, which had suffered from a desperate water shortage for several months, the Arabs having cut the water pipeline from the coastal plain.

The Government of Israel set up, during the second truce, a new organization, known from its Hebrew initials as Nahal (Pioneer Fighting Youth). The aim was to provide a combination of military and agricultural training for members of the youth movements, and for members of the Youth Aliyah (young immigrants) villages. During their term of military service the members of Nahal also underwent training for the establishment of new settlements; to this end, each of their groups (*garinim*) would be mobilized together and kept together in training, on military service and in the army outposts that would then become fully fledged agricultural settlements.

Also during the second truce, some Israeli soldiers (mostly members of the Palmach) were released from the army to undergo special training in existing settlements, before being sent to establish settlements in remote or strategic areas that had come under Israeli control.

On August 10 there was a concert in Tel Aviv. Zipporah Porath, the young American student who had come to Palestine in 1947 for a year and had quickly become committed to the new State, wrote to her parents in the United States:

> The hall was full with too many people, sitting, waiting, sweating, fanning the summer heat into each other's faces. Yehudah pointed knowingly at the majestic crop of white hair two rows in front of us – a great painter (Reuven Rubin); at the lady in black with a half smile on her face – our first Minister to Russia (Golda Meir); at the grey suit and the black moustache – our first Foreign Minister (Moshe Sharett); and when our Prime Minister, B-G, entered, everyone clapped in recognition.
>
> The conductor (Izler Solomon) decided it was time to put in an appearance, and the concert was on. The Prime Minister opened his tie and the first two buttons of his shirt. The Foreign Minister removed his grey jacket.

Zipporah Porath had been in Jerusalem during the siege. As the concert began, her mind drifted back to those harsh days:

> So this was the Holy Land and this its Holy City. I heard the thud, the passing whizz, the whistling, whining sounds, and then everything exploded. The music had stopped abruptly. I raised my head and saw a body floating in blood, a little child with deep black shiny eyes; I found myself staring at a woman in a red dress with shiny black buttons.
>
> The Prime Minister mopped his forehead and the sides of his face with a ball of white handkerchief. The lady in black leaned back contentedly. The conductor's arms were spread out like the wings of an airplane and the music soared again. It banged against the windows and the pillars, it shattered the ceiling, collapsed and fell into a heap of rubble and flying plaster . . . one great big heap of crushed stone and people in the middle of Jerusalem's Ben Yehuda's Street.

Many people were to have similar nightmares, both by day and night, as a result of what they had been through in the struggle for independence.

The second truce continued. During it, supplies were dropped by air to two of the isolated Negev settlements east of Beersheba, Beit Eshel and Nevatim. It was also used to intensify recruiting and improve training. A special organization was set up to treat disabled soldiers and to help with their rehabilitation after they had left hospital. It was named Elisha, after the disciple of Elijah famed in the Bible for his healing powers.

Since the first days of statehood, arms purchased by Israel's first emissaries abroad had been flown in by air from Czechoslovakia. The second truce was a period of intensified air flights. In the first week of the truce, eleven flights arrived. Among the cargoes were Messerschmitt planes (which had been crated up and had to be reassembled), spare propellers, raw materials for

Israel's fledgling military industry, and large quantities of arms. But on August 12 the Czech government – Communist for the previous six months and deferring to a hardening of Soviet policy toward Israel – ordered a halt to the shipments within forty-eight hours. The final, frantic flights brought out not only the regular shipments, but the planes, crews and equipment that had been needed at the air staging-post inside Czechoslovakia.

In Iraq a leading Jewish businessman, Shafiq Adas, was accused of exporting arms to Israel, via Italy, disguised as junk. Found guilty of treason, despite there being no evidence that he had ever shipped arms, he was hanged on August 15 in front of his newly built mansion in Basra. After his body was taken down from the scaffold and a doctor confirmed that he was dead, he was hanged again, and his body then mutilated by the vast crowd of Iraqi onlookers.

Following the ending of arms shipments from Czechoslovakia, Israeli emissaries abroad sought other sources of supply. Switzerland and Italy sold artillery and anti-aircraft guns. In Europe some United States surplus tanks were purchased. As the barrels of the tanks' guns were deliberately spiked – to prevent them being re-used as weapons – before they were sold, they had to be redesigned once they reached Israel. Three shipments of tanks, with ten tanks in each shipment, reached Haifa port during September.

From Mexico a quantity of filed guns was purchased. Thirty-two of them reached Israel in September, but were found unsuited to the battlefield. High-octane aircraft fuel from Mexico was of considerable value, however, when the second truce ended. In addition to a dozen Flying Fortresses, which arrived in Israel just as the second truce was about to begin – and which made a few bombing sorties over military targets in Cairo, Damascus and Amman – 300 automatic weapons were purchased from United States war surplus in Hawaii – the inventory of the Pacific war – and almost 800 automatic weapons from the continental United States.

At the beginning of October a ship left Antwerp with a mass of weapons and other equipment purchased in Europe. The cargo included signalling equipment, high-octane aircraft fuel, batteries, motorcycles, cars, and – most immediately useful of all – fifty jeeps, which on arrival were equipped with light machine-guns and infiltrated into the Negev on the eve of the renewal of hostilities.

The weapons brought in to Israel from Europe and the United States in August, September and October were to ensure that when fighting was renewed, Israel would have a certain edge, though not an overwhelming one, over its enemies.

Starting on August 22 a new air base was established in the Negev, near Ruhama. To reach it the planes had to fly across the region controlled by the Egyptian army. The base could not be reached by road without being subject to Egyptian ground attack. An average of eight two-way flights were made every night for more than fifty nights: a sign of how determined Israel was to drive the Egyptians out of the soil of former Mandate Palestine. Two thousand tons of equipment were flown in and distributed to the isolated (and hitherto seemingly abandoned) Negev settlements. Almost 2,000 passengers were also flown in, many of them recruits for the Negev Brigade, and soldiers of the brigade who had not seen their families for eight months were flown back north.

Israeli diplomatic representation was an essential feature of statehood: among its immediate needs were the purchase of arms, support in the United Nations (of which Israel was not then a member) and support with regard to Jewish immigration. On 2 September 1948 Golda Meir arrived in Moscow as Israel's first diplomatic representative to the Soviet Union. Her office and living quarters were a suite of rooms in the Hotel Metropole. There were more than 2.5 million Jews living in the Soviet Union, but the Soviet authorities would not allow them to emigrate: emigration was not an option open to any Soviet citizen, and Jews suffered from widespread anti-Semitism. 'From the beginning,' Golda Meir recalled, 'we had "open house" in my rooms on Friday evenings. I had hoped that local people would drop in – as people do in Israel – for a piece of cake and a cup of tea with us, but it was a very naïve hope – though the Friday night tradition was to continue long after I left Moscow. Newspapermen came, Jews and non-Jews from other embassies came, visiting Jewish businessmen (such as furriers from the States) came, but never any Russians, and never any Russian Jews.'

The apparent passivity of the Jews of the Soviet Union changed quite unexpectedly on the Jewish New Year. Two days earlier the Soviet Jewish writer and propagandist Ilya Ehrenburg had written an article in *Pravda* stressing that the State of Israel had 'nothing to do' with the Jews of the Soviet Union. In the Soviet Union, he added, there was 'no Jewish problem'. Golda Meir, having read that admonition, assumed that the Jews of Moscow had been warned off trying to greet her. When she had gone to the synagogue a week earlier it had been far from crowded; presumably after Ehrenburg's warning article it would be almost empty. When the Israel legation staff went on Jewish New Year to the Moscow synagogue Meir's shock at what she saw was overwhelming. She later wrote:

> The street in front of the synagogue had changed. Now, it was filled with people, packed together like sardines, hundreds and hundreds of them, of all ages, including Red Army officers, soldiers, teenagers and babies carried in their parents' arms. Instead of the 2,000-odd Jews who usually came to synagogue on High Holidays, a crowd of close to 50,000 people was waiting for us.
>
> For a minute, I couldn't grasp what had happened – or even who they were. And then it dawned on me. They had come – those good, brave Jews – in order to be with us, to demonstrate their sense of

kinship and to celebrate the establishment of the State of Israel.

Within seconds, they had surrounded me, almost lifting me bodily, almost crushing me, saying my name over and over again. Eventually, they parted ranks and let me enter the synagogue; but there, too, the demonstration went on. Every now and then, in the women's gallery, someone would come to me, touch my hand, stroke or even kiss my dress.

Without speeches or parades, without any words at all really, the Jews of Moscow were proving their profound desire – and their need – to participate in the miracle of the establishment of the Jewish State, and I was the symbol of the State for them.

I couldn't talk, or smile, or wave my hand. I sat in that gallery like a stone, without moving, with those thousands of eyes fixed on me. Now we were together again, and as I watched them, I knew that no threat, however awful, could possibly have stopped the ecstatic people I saw in the synagogue that day from telling us, in their own way, what Israel meant to them.

The service ended, and Golda Meir got up to leave:

But I could hardly walk. I felt as though I had been caught up in a torrent of love so strong that it had literally taken my breath away and slowed down my heart. I was on the verge of fainting, I think. But the crowd still surged around me, stretching out its hands and saying 'Nasha Golda' (our Golda) and 'Shalom, shalom', and crying.

Out of that ocean of people, I can still see two figures clearly. A little man who kept popping up in front of me and saying '*Goldele, leben zolst du. Shana Tova!*' (Goldele, a long life to you and a Happy New Year) and a woman who just kept repeating: 'Goldele! Goldele!' and smiling and blowing kisses at me.

It was impossible for me to walk back to the hotel, so although there is an injunction against riding on the Sabbath or on Jewish holidays, someone pushed me into a cab. But the cab couldn't move either because the crowd of cheering, laughing, weeping Jews had engulfed it.

I wanted to say something, anything, to those people, to let them know that I begged their forgiveness for not having wanted to come to Moscow and for not having known the strength of their ties to us. For having wondered, in fact, whether there was still a link between them and us. But I couldn't find the words.

All I could say, clumsily, and in a voice that didn't even sound like my own, was one sentence in Yiddish. I stuck my head out of the window of the cab and said: '*A dank eich vos ihr seit geblieben Yidden*' (Thank you for having remained Jews), and I heard that miserable inadequate sentence being passed on through the enormous crowd as though it was some wonderful, prophetic saying.

Ten days later, Golda Meir attended the Day of Atonement service. Another vast concourse gathered. She later recalled how, when the rabbi recited the closing words of the service – 'Next year in Jerusalem' – 'a tremor went through the entire synagogue'.

In the years ahead, the Jews of the Soviet Union were to suffer an increase in persecution, and were to be isolated from the outside Jewish world. But the State of Israel worked out means of maintaining contact with them and, in due course, a method of campaigning to enable them to leave. The eventual mass emigration of Soviet Jews to Israel – more than 750,000 in the twenty-five years 1971–1996, was to be one of the largest migrations in Jewish history.

On 17 September 1948 Count Bernadotte, the United Nations mediator, was on his way by car to have tea with the Military Governor of Jerusalem, Dov Joseph, when his car was ambushed and he was killed. Also killed with him was a French officer, Colonel Serot, who during the Second World War had helped to save Jews on the eve of deportation. The fatal shots were fired by Yehoshua Cohen, a member of the Stern Gang. His orders had come from Tel Aviv, where one of the triumvirate which had decided upon the assassination was Yitzhak Ysernitsky (later, as Yitzhak Shamir, Prime Minister of Israel). Cohen claimed that the deed was done because Bernadotte had proposed an Arab administration for Jerusalem.

Golda Meir was in Moscow when she learned of Bernadotte's assassination. 'I thought the end of the world had come,' she later wrote, 'and I would have given anything to have been able to fly home and be there during the ensuing crisis.' Bernadotte's murder was forcefully denounced by the Israeli government. Ben-Gurion ordered, 'Arrest all Stern Gang leaders. Surround all Stern bases. Confiscate all arms. Kill any who resist.' The killing of Bernadotte marked an end to whatever tolerance of Jewish underground activities had survived the shelling of the *Altalena*. The government would allow no separate armed activity, and from that moment, clamped down rigorously on all dissidents. Two leading members of the Stern Gang, Nathan Friedman-Yellin and Mattityahu Shmulevitz, were later sentenced to eight and five years imprisonment respectively for belonging to an illegal organization. Ysernitsky managed to evade arrest, and was one of those who issued a manifesto that declared: 'The war of liberation goes on!'

As well as arresting more than 250 Stern Gang members, Ben-Gurion issued a twenty-four-hour ultimatum to the Irgun to hand over their weapons and dissolve themselves 'unconditionally'. The ultimatum was signed by the Chief of Staff, Yigael Yadin. It stated that if the arms were handed over, the Irgun disbanded, and its members joined the Israel Defence Forces, 'none of you will suffer for the infringements you have hitherto committed against the laws of Israel, and you will be treated like every other Jew.' But if they failed to comply, the ultimatum concluded, 'the army will act with all the means at its disposal'.

A South-African-born member of the Irgun, Doris Katz, recalled how many of the 'hot blooded' Irgun members, 'jealous of the prestige of their beloved organization, insisted the Government's challenge be taken up'. The Irgun felt confident that in Jerusalem they could still maintain their independence. The ultimatum, reported the correspondent of *The Times*, 'was received with considerable resentment both by its members and by the majority of the population of the city'. At a press conference in the Eden Hotel, in the centre of Jewish Jerusalem, Begin's spokesman, Shmuel Katz (the husband of Doris Katz) told journalists that the Irgun strength in the city 'is sufficient to ensure that any attack would involve the attackers in heavy losses'. Katz added, 'We are not prepared to shed the blood of Haganah soldiers whose lives the Government of Israel is so lightly prepared to throw away.' With this reasoning, Menachem Begin accepted the ultimatum. On September 20 the Irgun ceased to exist as an independent military organization. The Irgun base in Katamon was dismantled, and the Irgun wounded were moved to army hospitals.

The handing over of arms was carried out 'without incident', Doris Katz wrote angrily, 'in spite of the characteristic gaucherie and tactlessness, sending a unit of Palmach (who had carried out the brutal attack on the *Altalena* from the Tel Aviv shore) to receive the arms.' And she added (writing in 1953), 'Perhaps one day, when our Government reaches maturity, they will learn how to be gallant and gracious, and how to recognize and acknowledge worth even in their political opponents.' The bitterness between the Irgun and Palmach, between the Revisionists and Labour, between Ben-Gurion and Begin, and their respective supporters, was to survive the first fifty years of statehood.

Ben-Gurion had not only sought to prevent a separate Irgun army, he was intent on ensuring that no independent or semi-independent military forces would exist within the new State. To this end, he pressed for the abolition of the Palmach National Command, which had operated since May within the Israel Defence Forces yet maintaining a separate existence. Ben-Gurion insisted that the Palmach headquarters and command structure should be dissolved. On September 14 he met at Na'an with all sixty-four Palmach commanding officers. 'The Palmach differed from other bodies up until the war,' he told them, 'because it was constantly under arms. Its men were better trained than other members of the Haganah and it was composed of pioneering youth. But how does it differ from all the other units now that there is general mobilization, and all men are being constantly trained and learning skills that could not have been learned during the underground period?'

In defending the continuation of the Palmach National Command, Yigal Allon argued that it might be needed again, as it had been during the attack on the *Altalena*, to combat dissident activity, and even during 'a possible civil war'. Perhaps, Allon added, the very existence of the Palmach 'will prevent a civil war'. But Ben-Gurion was insistent that there could only be one Israel Defence Force, and that it was through that unity that the unity of the State would be preserved. The Palmach National Command was thereupon deprived of the ability to give operational instructions to Palmach brigades. Nor could it mobilize soldiers any more, train them, or carry out 'ideological education'.

Allon accepted Ben-Gurion's order. The Palmach would not seek to be an army within an army.

On October 1 a National Palestinian Council meeting in Gaza elected the Mufti, Haj Amin al-Husseini, as its president (he was then still in exile in Egypt) and declared itself to be the provisional government of 'All-Palestine'. Auni Abdul Hadi, who in 1921 had been one of the Palestinian Arab leaders urging Winston Churchill to call a halt to all further Jewish immigration, was appointed Minister for Social Affairs.

King Abdullah of Transjordan immediately denounced the All-Palestine government, which would not, he said, be allowed jurisdiction in any of the areas under control of the Arab Legion. Abdullah had no intention of allowing the followers of the Mufti to gain the political or territorial rewards of Transjordan's military efforts.

Despite Abdullah's opposition, on October 12 the Egyptian, Syrian and Lebanese governments gave their formal recognition to the All-Palestine government in Gaza. Five days later, Iraq did likewise. To counter the implications of this, Abdullah worked with great energy to extend his control of the Arab-held areas of Palestine. After considerable efforts to persuade the Arabs of Palestine that Transjordan was their natural and best protector, on December 1 the King convened a conference in Jericho which adopted a resolution that 'Palestine and Transjordan constitute one indivisible unit.' The resolution went on to say that the unity of Palestine and Transjordan constituted 'a prior condition for any final settlement'. It was also resolved that Transjordan would 'unite' with Palestine (something that Abdullah had requested of the British, in vain, twenty-five years earlier). The name of the united region was declared as the 'Arab Hashemite Kingdom'. Palestinian members, the King announced, would be elected to the Parliament in Amman.

Egypt, Syria and Iraq refused to accept these changes: Abdullah was isolated and the Arab Hashemite Kingdom never came into being. However, Abdullah's determination to annex as much of Palestine as he could was clear, and was not relinquished. The All-Palestine government likewise failed to survive; Egypt did not want any such independent authority in Gaza.

*　　　*　　　*

In the hill country east of Nahariya, a group of settlers from Hungary founded kibbutz Ga'aton in October 1948. Even as they set up their first buildings and border fence, they were fired on by Arabs from the surrounding hills. They took the name of their kibbutz from a town mentioned in historical accounts of the return of the Jewish exiles from Babylon, believing it to be the nearby ruin known to the Arabs as Ja'tun.

Since the coming into force of the second truce, Egyptian forces were still in place along the Mediterranean coast, to a point twenty miles south of Tel Aviv. Despite the Ruhama airlift, the Israeli settlements in the Negev were effectively cut off from the rest of Israel, with Egypt in control of Beersheba and the roads running out of it to the south. A military plan was devised – code-named Operation Yoav – to split the Egyptian forces into three segments, and then to attack each one separately. The final objective was the capture of two Arab towns, Beersheba and Gaza. 'Having taken full advantage of the respite,' Yitzhak Rabin recalled, 'we were ready and eager for action. But there was one catch. To avoid the political handicap of taking the blame for breaking the truce, we had to find some pretext for renewing the fighting.'

Under the terms of the second truce, Egypt had agreed to allow Israeli convoys to drive through the Egyptian lines to these settlements. But these convoys had not, in practice, been allowed through. It was decided, in Rabin's words, 'to send a supply convoy through as a deliberate act of provocation. When the Egyptians opened fire on it they would provide us with an adequate pretext to renew the fighting. Our only fear was that the Egyptians would change their tactics and actually allow the convoy through! With the plan set to commence on the evening of October 15, that is precisely what happened. Every additional mile the convoy covered unmolested, the nearer our nerves stretched to breaking point. The excuse for our attack was slipping out of our grasp. In the end, with the aid of a random shot here, another there, we had our pretext. The air force launched its bombing raids and the operation went forward.'

The fighting against the Egyptian army began on October 15, and continued for seven days. On the first day Israeli planes carried out a bombing raid of considerable intensity on the Egyptian air base at El Arish, on the Sinai Mediterranean coast. The railway line from El Arish to Rafa was attacked from the air and blown up. But an Israeli attack on the Egyptian forces cutting off the rest of Israel from the Negev was a failure. An attack on the Egyptian stronghold at Iraq el-Manshiya was driven off at heavy cost to the attackers. During the battle, a recent immigrant, Shalom Zimbel, helped carry wounded men from the battlefield. He made the journey, under Egyptian machine-gun fire, twelve times during the day, taking six wounded men to safety. He was unhurt. That evening, when returning to base, he was killed by an Egyptian artillery shell.

By the morning of October 17, after a night of hard fighting, the bulk of the Egyptian forces began to withdraw. But 5,000 men remained near Faluja, and refused to pull back to the south. This 'Faluja pocket' held out tenaciously against all Israeli attacks, and refused to surrender.

The fighting in the northern Negev created yet more chaos for the Arab villagers, who had been about to harvest their crops, and who saw their fields becoming battlegrounds and their villages, held by the Egyptians as strongholds, pulverized by Israeli artillery fire. Many of them fled eastward to the Hebron Hills. On October 19 the United Nations Security Council was called together in emergency session at Lake Success, just outside New York. At the request of Bernadotte's American successor as mediator, Dr Ralph Bunche – the grandson of a black slave – it adopted a resolution calling for 'both parties' to withdraw to the positions they had held at the end of the truce.

The Security Council resolution recognized both sides of the coin: 'the problems of the convoys to the Jewish settlements' to the south, in the Negev, and 'the problems of the dislocation of large numbers of Arabs and their inability to harvest their crops' in the area where fighting was taking place. It called for immediate negotiations, and the 'permanent stationing' of United Nations observers throughout the area.

Israel agreed to comply with the Security Council resolution, and Ben-Gurion instructed the Israeli commanders that they had only two more nights (October 19 and 20) in which to open a way to the Negev. One main Egyptian point of resistance remained in the road south: six fortified hills around the Arab village of Huleikat. Assisting the Egyptian troops there was a company of Saudi Arabian soldiers.

For the Israeli troops, the battle for Huleikat was one of the fiercest of the War of Independence. In the attack on one of the six hills, the section leader, Moshe Lichter, lost both his legs while attacking a machine-gun position. The hill was captured, but he died soon afterwards. Lichter's machine-gunner, a man known as Emsi, was gravely wounded. To the medical orderlies who came to evacuate him, he apologized for having ceased fire, but, he explained to them, his ammunition had run out. He, too, died soon afterwards of his wounds.

The Egyptians were driven from Huleikat, and the road to the Negev was opened. From midnight on October 19 Israeli vehicles were using it for troop movements and supplies. But the Israelis failed to eliminate the Faluja pocket, or to capture the former police fort at Iraq Suwedan, despite repeated attacks on it throughout the evening of October 20 and during the night.

Having surrounded and isolated the Egyptian troops in the Faluja pocket, Israeli forces used the final twelve hours before the cease-fire to try to drive the Egyptians from Beersheba. A plan to drive them from Gaza was given up because of the strength of the Egyptian forces there. Beersheba was less well defended. It was also within driving and marching distance of several Israeli settlements which had survived their own long isolation. Both the infantry and armour of the attacking forces were concentrated at Mishmar Ha-Negev, thirteen miles north of Beersheba. Artillery and mortar support was concentrated at Hatzerim, five miles west of the town. The attack began at four in the morning on October 21. It was spearheaded by troops of the Mahal (foreign volunteers) and Gahal (immigrant soldiers), many of them from France and North Africa.

Machine-gun fire from the Beersheba police fort prevented the infantry from breaking into the town. A single Israeli anti-tank gun was brought up at eight o'clock. It fired four rounds, and the Egyptians inside the fort raised the white flag. The Egyptian commander had not been told that the road to the Negev had been opened by the Israelis, and had been quite unprepared for the attack. An hour after the surrender of the fort, the whole town was in Israeli hands. 'Maybe the Egyptians lost their spirit to fight,' Uzi Narkiss later reflected. 'They did not imagine that we would dare to capture the capital of the Negev.'

The renewal of fighting in the south on October 15 also saw fighting renewed in Jerusalem. That day a Muslim Brotherhood unit – Egyptian fundamentalists who had long struggled to dominate the Egyptian political scene – blew up the end house in the Arnona suburb. On the following day the Arab Legion blew up an Israeli outpost on Mount Zion. The Israelis retaliated by blowing up an ancient mausoleum on the same slope, in which the Legion had its local headquarters; all the Legionnaires inside were killed. On the night of October 18–19 an Israeli raid to blow up an Egyptian fortified artillery point overlooking the Bethlehem road was successful, but one of the Israeli sappers, who was wounded in the explosion he had helped create, was captured and tortured, and his mutilated body carried in triumph through the streets of Bethlehem.

Despite these raids and counter-raids, the cease-fire lines in Jerusalem did not change, nor could the Israeli army use the renewal of hostilities to try to break into the Jewish Quarter which, since its capture five months earlier, had been systematically looted, and its many synagogues destroyed.

Jerusalem was divided, and was to remain so until 1967. Concrete walls to keep out snipers' bullets, and barbed wire barriers, marked the line of division, which no one crossed except at their peril. On the Israeli side, buildings were found in which university students, no longer able to study in the isolated enclave on Mount Scopus, could resume their studies. The Burma Road was opened to civilian traffic to and from the coast. A new settlement was established on the Jerusalem to Tel Aviv railway, Tal Shahar (Morning Dew). Its name was in honour of the United States Secretary of the Treasury, Henry Morgenthau Jnr, an American Jew whose sympathies with Israel were well known. The first settlers were immigrants from Roumania. Another settlement was also founded near the cease-fire line that year, east of Caesarea. A youth village, Allonei Yitzhak (Oaks of Yitzhak), it trained immigrant children, mostly in agriculture. Among its financial sponsors was the Hadassah Organization of America. Its name was in memory of Yitzhak Gruenbaum, head of the Rescue Committee of the Jewish Agency during the Second World War.

As the cease-fire drew near, Israeli forces in the Jerusalem Corridor drove the Arabs from Beit Shemesh, whose Jewish youth training farm and cement factory had been opened shortly before the start of the war, but abandoned when war broke out. This gave Israel control of another section of the Tel Aviv to Jerusalem railway, at the entrance to the Sorek gorge.

South of Jerusalem, the Arab village of Beit Jibrin, at the base of the Hebron Hills, was overrun, together with the former British police fort there, just before the cease-fire came into effect. Another Arab village captured in the final hours of the fighting was Beit Natif. It was here that the thirty-five members of the relief column had been killed on the eve of independence during their attempt to relieve the settlers in the Etzion bloc. From the captured village the Israeli troops could make out in the distance the burned-out ruins of the settlements of the bloc, but retaking the site lay beyond their military capacity.

On October 21 the Government of Israel took a decision that was to have a lasting and divisive effect on the rights and status of those Arabs who lived within its borders: the official establishment of military government in the areas where most of the inhabitants were Arabs. Regulations promulgated for these areas established security zones and prohibited permanent residents from leaving them without a permit. Entry into the zones was also not allowed to those who were not permanent residents unless they were in possession of a permit. In addition, the regulations permitted the military administration to remove permanent residents from the security zones and transfer them elsewhere: this led in the summer of 1949 to the expulsion of the Arab residents of three villages – Khisas, Qeitiya and J'auna – who were not subsequently permitted to return.

The third cease-fire came into effect at three in the afternoon of October 22. In the south, where most of the fighting had taken place, Israel regained possession of the ruins of Nitzanim and Yad Mordechai. In the north, the truce was broken within a few hours by Kaukji, who did not consider that his Arab Liberation Army was responsible in any way to the United Nations. Surrounding the remote settlement of Manara, and threatening Misgav Am, he posed a danger to the whole of Upper Galilee, raising fears in Israel that the Syrians might take the opportunity to attack from the east.

A military plan was devised to secure Upper Galilee as far as the northern boundary of the Palestine Mandate. It was given the code-name Operation Hiram, after the biblical King Hiram of Tyre, the ally of David and Solomon, who had sent cedars of Lebanon to build the Temple in Jerusalem. The operation began on the night of October 28–29, and took sixty hours. Advancing eastward from Nahariya, northward from the Nazareth–Tiberias road, and westward from Safed, Israeli forces attacked Kaukji throughout the areas within his control.

From Safed Israeli units, striking towards Kaukji's heartland, captured twelve villages, and also Meron – one of the oldest Jewish communities in Galilee. Among the volunteers fighting in Galilee was Captain Tom Derek Bowden, known as Captain Appell (his *nom de guerre*). A British army officer, not a Jew, he had been badly wounded in the British attack on Syria in 1941 during which Moshe Dayan, his guide in that campaign, had lost his eye. In 1944 Captain Bowden was parachuted into Arnhem, and captured. As he was carrying some letters from his girl friend in Haifa, a Jewish girl, Hannah Appell, he was sent to Belsen where he was later liberated by the British. He had already fought at Latrun before he was sent up to Galilee as a company commander.

During the battle for Meron another overseas volunteer, Sam Freiman, a Polish-born survivor of the concentration camps who had been taken to Britain after the war, served as Captain Appell's runner. 'We started an attack at night and by early morning we had taken Meron,' Freiman recalled. 'Then I fell asleep near a tree. I slept so well (never have I slept so well in my life). When I woke up, we started walking up the mountain, we sat there for a few hours. Then the Syrians started to shell us and for a while it was very frightening until the shelling stopped. Captain Appell then sent me down the mountain to bring up other soldiers who were taking over our position. I remember the soldiers were older men who had come from the German camps. They were carrying mattresses and tins of food and while I was taking them up they were very noisy. I was scared that the Syrians would hear us and start shelling again. We got up to the trenches and as I was sitting in the trench there was a Canadian soldier sitting next to me, who suddenly collapsed. A sniper got him in the head, he was wearing a helmet but it did not help, the man was dead.'

They then attacked Kaukji's stronghold in the Arab village of Jish – the ancient Gush Halav in which, 2,000 years earlier, the Jews had held out against the Romans. A Syrian battalion took part in the defence of Jish, but was unable to help Kaukji drive off his attackers. The Israeli deputy commander recalled the course of the battle:

> Opposition was fierce. It seemed that in Jish there were considerably larger forces than had been anticipated according to the available information. The sun rose and shed light on the chaos in Jish. In the village square, near the stone building which in mandatory times had served

as a health clinic, and which was now occupied by the commander and staff of the second Yarmuk Brigade, a column of Lebanese buses was standing, painted with bright oil colours and covered with inscriptions.

Those were the buses which during the previous night, just a few hours before our assault on Jish, had brought a Syrian battalion consisting of four hundred men from Marj-Ayun, where it had previously been deployed.

The soldiers themselves, well dressed and equipped, ran about between the houses and in the narrow lanes of the village and among the neighbouring fig orchards, individually or in groups, and tried to resist our attack in an area with which they were completely unfamiliar.

Kaukji's men fled toward the mountains surrounding Jish, firing sporadically and ineffectively against the advancing armour. A truck loaded with ammunition burned fiercely adding to the confusion and the noise of battle. A concentrated fire attack, followed by the assault of an infantry company into the village, wiping up isolated nests of opposition, terminated the battle.

In addition to the buses, we captured two of our own homemade (sandwich) armoured cars, taken from us in the Yehiam convoy and now marked with the symbol of the Liberation Army, a crooked dagger, dripping blood, piercing a Shield of David.

A short 37-mm anti-tank gun fully loaded and prepared to fire was found in the morning in an abandoned position. Arms and ammunition of French make were found in large quantities, and considerable amounts of supply. From the brigade HQ we collected bags full of documents.

Later on prisoners began to come in. Kaukji's men were broken, and officers and soldiers of the Syrian army huddled close together, observed us with fearful eyes. One young Syrian officer murmured incessantly, in English, 'I spent two years at the military academy.'

The Syrians had fought the hardest of all the Arab units during the battle of Jish, and had suffered the heaviest casualties, 200 killed. That afternoon Kaukji evacuated his forces from the highest of the mountains of northern Palestine, Mount Meron, 1,208 metres above sea level.

The Israeli forces turned next to the Arab town of Tarshiha, the most heavily fortified town within Kaukji's domain. His German mining expert had sought to defend it by a skilfully placed barrier of mines. On the eve of the battle, an Israeli sapper, Zigi Shapira, working at night, discovered and dismantled forty-seven mines. With the dawn, he was forced to stop his work, and the mine barrier proved effective. But on the following night Israeli artillery fire on Tarshiha was followed, at dawn, by the sight of white flags flying from the rooftops. Kaukji's men had gone, and the town surrendered.

Inside Tarshiha, the Israelis found most of Kaukji's heavy artillery. He and his men had fled northward, taking a little-known track that led to the crusader castle at Monfort. From there they made their way across the mountains to Lebanon. The isolated Jewish settlement of Eilon, which lay just to the west of the route of Kaukji's retreat, was liberated from the daily fear of attack. Its defenders could hardly believe that the troops coming up the steep and winding road towards it were not yet another attempt by the Arab Liberation Army to break its resistance, but Israeli troops.

In the Galilee panhandle, the virtual siege of Manara was broken, and Israeli forces, crossing into Lebanon, advanced as far as the River Litani. Israel was in control of the whole of the Mandate region of Upper Galilee.

Despite the truce, the Israeli army was concerned to get rid of the threat of the Egyptian forces in the Faluja pocket, and at the former police fort of Iraq-Suwedan (known as the Monster on the Hill), which could still threaten the road to the Negev. To avoid the setback of the previous failed and costly attack, the Israeli forces practised their tactics on the recently captured police fort at Bet Gubrin. The attack began on November 9 and, after a greater barrage of artillery fire than the Israeli forces had used before, the defenders of the fort surrendered. A diversionary attack on Iraq-Suwedan was also successful. One of the Israeli attackers, Gana Simantov, lost both his legs in the assault. Despite his terrible injuries he continued to fight. For his courage, he received the prestigious Hero of Israel award.

Only the Faluja pocket resisted all Israeli assaults and bombardments. A thousand Egyptian soldiers fought on, their food supplies dwindling, their resources further strained when they were joined by several hundred soldiers fleeing from Iraq-Suwedan. The morale of the men was also affected by a severe shortage of cigarettes.

The commander of Israel's Southern Front, the Palmach leader Yigal Allon, was so impressed by the Egyptian fortitude in defending the Faluja pocket that he offered the commander, Sayid Taha (known as the Sudanese Tiger) repatriation to Egypt with all his men if they would surrender. The two men met at Gat after the fall of Iraq-Suwedan. Among those with Taha was Major Gamal Abdel Nasser, later, as President of Egypt, Israel's implacable enemy.

Allon promised Taha that he and his men would not have to suffer the indignity of becoming prisoners-of-war, but could go home at once. According to the record of their discussion, Allon, after praising the courage of the Egyptian commander and his men, told him:

'You must understand, Colonel, that the position along this front has been decided. Your brigade is now besieged on all sides, and has no hope of breaking out. It is my duty to avail myself of any opportunity to destroy your formation. What purpose would be served by your desperate stand? The fate of the second half of the pocket will not differ from that of the first which fell into our hands the day before yesterday.'

Taha was quiet for a few seconds, then replied: 'Yes, sir, but as long as I have soldiers and ammunition, there is no reason why I should stop fighting.'

Allon: 'I respect your courage, Colonel, but do you not agree that lives of men are precious, and that there is no logic in sacrificing your men fighting against a people whom, in fact, you do not consider your enemy, and who have only good intentions toward you? I do not offer you a degrading surrender, but an honourable one with full military honours and full respect, with a possibility of immediate return home. Consider this. Let us save the blood of our soldiers and cease fighting. I cannot desist from that as long as foreign armies are on our soil. But you do not fight on your soil. You must realize that I am right.'

Taha: 'Sir, there is no doubt that your position is better than mine. The planning of the operation and the courage of your soldiers are admirable. You have broken through our strongest lines, and the face of the victorious Egyptian army was covered with shame. I do not delude myself to believe that by continuing my stand I will be able to change the balance of forces and save the situation on our front. However, there is one thing which I can save and that is the honour of the Egyptian army. And therefore I will continue to fight here to the last, or to the last man, unless I receive different orders from my government.'

Allon: 'As a soldier I well understand your sentiments. But surely you must know that under extraordinary circumstances a local commander must take responsibility for decisions, as was done by Von Paulus, for instance. Believe me, Colonel, your government is not worthy of commanding as brave an officer as yourself. Consider your soldiers and the Egyptians and lay down your arms.'

Taha: 'No, sir, I have no alternative but to continue the fighting. Surely you know that I am saving the honour of the Egyptian army.'

Allon: 'May I ask you to give your government, or your Supreme Command, a description of the real situation, and to request their consent to your surrender?'

Taha: 'I will report the content of our conversation.'

Sayid Taha fought on. He was encouraged to do so by the fact that the siege was not complete. On October 30, six Egyptian aircraft flew over the pocket and dropped supplies, 'accompanied', Taha noted in his diary, 'by the triumphant cries of residents and soldiers'. He added, 'The visit of the planes did more than anything else to lift our morale. The Jews were taken by surprise.' Much-needed medical supplies were among the packages dropped, as well as ammunition and cigarettes. 'The soldiers were more interested in the cigarettes than in the ammunition,' Taha added.

On October 31 a number of Egyptian soldiers managed to enter the pocket. They were returning from leave in Egypt and were able, by giving Beersheba a wide berth, to avoid detection by the Israelis. After that the siege lines were tightened. But small groups of soldiers still managed to arrive on foot from Gaza, and even, on one occasion, in a camel train from the Hebron Hills. Israeli bombing raids had begun to replace the Egyptian supply drops. Colonel Taha's diary testifies to the efficacy of the raids. On November 17 he wrote, 'Enemy planes attack daily. The sight of the three Flying Fortresses has become a regular feature every day at 5.00 p.m.' On the following day his entry read, 'The night before last was the hardest and cruelest Faluja has witnessed. The enemy's air activity continued for twelve hours without interruption.' And again, a few days later, 'Last night was even more difficult and cruel than its predecessor. The number of bombs dropped on Faluja reached about a thousand. Enemy planes have destroyed the water pump.'

An attempt was made to prepare a breakout from the Faluja pocket. It was the idea of a British officer serving under Glubb Pasha, Major Lockett, who entered the pocket from Transjordanian-held territory on November 20 and proposed to Taha a scheme whereby the forces in Faluja would destroy all their heavy equipment and then break out, using only their rifles and pistols. 'You've got to take the risk,' were Lockett's words to Taha when he expressed his unease at the plan. When Taha informed the Egyptian High Command of what Lockett proposed, the response was blunt. 'It is impossible to rely on a plan whose initiator is Glubb Pasha', the Egyptian Commander-in-Chief in Gaza replied, 'and it is impossible to keep the details of the plan from the Jews if it originated in Amman. The evacuation of the troops by foot through areas held by Jews means a massacre for these troops. Reject the plan and expel the drunken Lockett from your stronghold. It will not bring honour on our army but will be a grave disaster. Defend your positions to the last bullet and the last man. Only that is proper for Egyptian soldiers.'

The siege continued. No attempt was made to break out. On November 24, Taha noted in his diary in triumph, 'Today a convoy arrived from Bethlehem with a load of ammunition and medical supplies and a quantity of tinned food. The convoy, consisting of forty-five camels, arrived in Faluja in spite of the Jews, an event which greatly gladdened us and relieved our distress.'

Israel granted the Red Cross access to the besieged men, so that further medical supplies reached them.

CHAPTER FOURTEEN

The last four months of the war

On 4 November 1948 the United Nations Security Council, after consulting with the acting mediator for Palestine, Dr Ralph Bunche, called for the withdrawal of all forces to the positions they had held on October 14. It also called for negotiations 'conducted directly between the parties, or failing that, through the intermediaries in the service of the United Nations, permanent truce lines and such neutral or demilitarized zones as may appear advantageous, in order to ensure henceforth the full observance of the truce in that area.' Failing an agreement, the permanent lines and neutral zones were to be established 'by decision of the Acting Mediator'.

Israel declined to withdraw from the areas of the Negev which it had conquered from the Egyptians, or to allow Egyptian forces back into those areas, all of which lay within the borders of Mandate Palestine.

On November 23, while the Faluja pocket remained besieged, Israeli forces made their way through the Negev Desert to the isolated outpost at Sdom (the biblical Sodom) on the Dead Sea, which had been cut off from any contact overland for more than six months. Some supplies had been flown in to a small, improvised landing strip, but there was no road access at all. The Israeli convoy had to make its way on tracks and defiles along which no four-wheeled vehicles had passed before. Their success in reaching Sdom extended the boundaries of the new State 20 miles further east and south.

On November 29, there was time for a moment of ceremonial to mark the first anniversary of the United Nations partition resolution. In that year a State had been created and defended, and had survived. Zipporah Porath was in Haifa, working as a nurse, on that first anniversary. 'From the roof of the hospital,' she wrote to her parents in the United States, 'I watched this morning's parade, a parade of soldiers of the Jewish State. Not partisans or underground fighters. Soldiers standing erect and proud, in rain puddles six inches deep, wearing shabby outfits – winter uniforms still haven't reached us – listening to lofty words of accomplishment and tribute.'

On the Jerusalem front, the Israeli and Arab Legion commanders, Lieutenant-Colonel Dayan and Lieutenant-Colonel Abdullah al-Tall, came to an agreement on November 30 for a new 'sincere' cease-fire. Not only did the fighting stop between the two opposing forces, but an agreement was negotiated to allow Israeli convoys to go up to Mount Scopus on a regular basis to relieve the Israeli policemen guarding the enclave. A 'City Line' was established, running through Jerusalem at the point where the armies confronted each other, and separating them. Chaim Herzog, who later commanded the Jerusalem District, wrote in his memoirs:

> In the heart of the city, this separation was emphasized by a stone wall designed to prevent sniping. In the no man's lands it was a series of barbed-wire entanglements, minefields, and makeshift obstacles.
>
> These no man's lands were created because Dayan and Tall had used a thick china marker to delineate the City Line on a map. The line was two or three millimetres wide on the map – which translated to forty to sixty metres on the ground. Their line covered houses and even whole streets indiscriminately.
>
> In the heat of the summer, the marker tends to melt and had covered certain areas, thus widening the City Line even further. Neither side could agree on what belonged to whom, and people were killed because of the thickness of a pencil.
>
> Guests at the reconstructed King David Hotel sat on the terrace or on their room's balcony overlooking Jordanian sentries perched on the city wall within small-arms distance. The intervening no man's land turned into a mass of concertina wire, garbage and putrefying animals that had been blown up by the mines. Nothing could be rebuilt along the front line – it was simply too dangerous.

*　　*　　*

The Arab world was divided over the future of Arab Palestine. King Abdullah of Transjordan refused to recognize the Mufti's All-Palestine government in Gaza, and on December 1 convened a conference in Jericho at which representatives of the main West Bank towns expressed their desire for union with Transjordan. The main resolution to this effect had been agreed upon the previous day at King Abdullah's villa at Shuneh between Abdullah and the Mayor of Hebron, Sheikh Muhammad Ali al-Jaabari. When one of the resolutions of the Jericho Conference welcomed Abdullah as the ruler of 'all Palestine', the British government warned him that he would lose British support unless he limited his territorial aspirations to Arab Palestine, as defined by the United Nations partition resolution of November 1947.

The Egyptian government opposed Abdullah's attempt to assert his influence over all the Arab-controlled areas of the former Mandate. At the same time, the Egyptian government was under public pressure as a result of the military setbacks in the Negev, and was criticized for its weakness by the Muslim Brotherhood. As Muslim Brotherhood protests grew, the authorities in Cairo took a tough stand, dispersing demonstrators whose slogans declared 'Palestine for the Arabs'. During the course of the demonstration, the commander of the Cairo police was killed. In retaliation, the Egyptian Prime Minister, Nokrashy Pasha, declared the Muslim Brotherhood illegal. On December 28 he was murdered by a Brotherhood member.

Israel had one more objective before it felt able to call a halt to the fighting against Egypt. This was to drive the Egyptian army out of the remaining areas of Mandatory Palestine south-west of Beersheba, along the western edge of the Negev. In December, the isolated kibbutz of Revivim, 15 miles south of Beersheba, which had held out, first against Bedouin forces and then against the Egyptian army for a whole year, was liberated. It had to be entirely rebuilt.

A new offensive plan, Operation Horev, was devised. Its objective, set out by Yigael Yadin on December 10, was succinct: 'Defeat of the invading enemy force and its expulsion from the boundaries of the country.' The operation was to be launched on December 20. In planning their advance along the one surfaced road from Beersheba through Bir Asluj to El Auja, on the old Mandate border with Egypt, the Israelis were confronted by a series of Egyptian strongholds. But there was another route, built by the Romans 2,000 years earlier, connecting the two towns, which went past the ruins of two ancient Nabatean cities, Ruheiba (Rehovot in the Bible) and Halsa (Halutza). Palmach scouts had explored this route several years earlier.

The Roman road (known as the Ruheiba trail) was explored again to see if it was usable by tanks. The Operations Officer of the attacking force, Yitzhak Rabin, concluded, 'It is difficult, but possible, after a certain amount of repairs to the track.' Those repairs were put in hand. But torrential rain at the very moment when the attack should have begun on the night of December 20–21 made it impossible to start, as the beds of the Negev wadis overflowed and the tracks became eddies of mud.

The attack was postponed to the night of December 22–23. In his order of the day on the eve of battle, the commander of the Southern Front, Yigal Allon, while noting that the Egyptians – fearful 'of our victorious army' – were confined to two narrow defensive lines, one between Gaza and Rafa and the other between Bir Asluj and the Sinai Desert, far from their rear bases and reinforcement centres, went on to tell his men:

> However, the task has not yet been finished. The truce imposed on us
> by the United Nations has slowed the tempo of the war and has held

up the disintegration of the enemy. The campaign and its momentous results cannot be summed up yet, because it is not yet completed.

The enemy braces himself, and his aim is to strike a blow at us in order to wipe out his disgrace. We will, therefore, nullify his plot with a crushing blow. We will expel his forces from the boundaries of the State of Israel and liberate the remaining areas of the Negev still in his hands.

This operation is to complete what has begun in previous operations. Just as we prevailed over the enemy then by virtue of the strength and the spirit of our soldiers, of the infantry and other corps and services, merged together into one army and following one order, so shall we prevail now.

The house of Israel places its trust in our army. The people believe in the fighters of the Southern Front and look forward to a final and decisive victory over the invaders. For the final defeat of the Egyptian invaders! To bring about peace speedily! To victory! We shall assault the enemy!

The operation began that night, with a diversionary attack against Hill 86, an Egyptian position overlooking the Gaza–Rafa road. It was an uneven battle, unlucky for the attackers. Field telephones had not arrived, and the Israeli army radios were out of order. Continuing rain meant that many of the attackers' weapons were jammed by mud. Then, from the Egyptian defences, four half-tracks emerged, spitting fire. These were flamethrowers, a weapon that the Egyptians had not used before. The clothes of some of the Israeli soldiers furthest forward caught fire. The attackers pulled back, then advanced again, but had to be called off as more Egyptian armour was brought up. 'Of all the brigade's battles,' wrote the commander of the action, Nahum Golan, 'I do not remember one where in so short a time so many acts of heroism and devotion by soldiers were carried out.'

Amid these struggles on the battlefield, the Canadian government announced its recognition of the State of Israel. On December 24 the Canadian Minister for External Affairs, Lester Pearson, informed Moshe Sharett, his Israeli opposite number, 'The State of Israel was proclaimed on May 15, 1948. During the seven months that have elapsed, the State of Israel has, in the opinion of the Canadian government, given satisfactory proof that it complies with the essential conditions of statehood. These essential conditions are generally recognized to be external independence and effective internal government within a reasonably well-defined territory.'

The main military success, even amid failure on the battlefield, was to convince the Egyptian High Command that the cutting of the Gaza–Rafa road was the principal Israeli objective. Even when the attack was beaten off, the Egyptians considered the coast road to be the crucial sector, and the isolation of Gaza the chief Israeli objective. This deception was reinforced by an attack south-east of Khan Yunis, on the village of Abasan, by the aerial bombing of Rafa and El Arish (along which reinforcements to Gaza would have to pass) and by the bombardment of Gaza itself from the sea.

Meanwhile, the main attack, along the Roman road from Halsa to Auja, was ready to be launched. It was set to begin on the morning of December 24, but once again heavy rains and floods forced a twenty-four-hour postponement. The attack was finally launched on the morning of December 25. Among the troops were many from Mahal and Gahal who had led the attack on Beersheba two months earlier.

The Egyptians, though taken by surprise, fought tenaciously. The Israeli plan to make the Roman road passable was not completed in time. In the fighting to reach El Auja, one of the Israelis who was killed was a former Stern Gang fighter, 'Jinji' (Ginger), whom, in Mandate days, the British had long sought in vain.

After three days of hard fighting, and many setbacks, the Israelis reached El Auja just after midday on December 28. They did not stop there but, pushing past the former frontier post, crossed the Egyptian border into the Sinai. The move across the Egyptian border was carried out without Ben-Gurion's approval. Once inside Egypt, the Israeli troops were attacked by mistake by Israeli aircraft who did not realize the border had been crossed. One soldier was killed. Another, who carried with him, as a lucky charm, the flag of the Jewish Brigade Group from the Second World War, was able to raise it and show that the troops were indeed Israeli.

The Egyptian defences had collapsed. On December 29, Israeli troops, pushing deep into Sinai, established a base at Abu Ageila, 20 miles west of the border. From there, Israeli hit-and-run raids attacked the road between El Arish and Rafa, and towards El Arish itself. They also continued to drive westward another 20 miles into Sinai, to Bir Hama and Bir Hasuna.

One more Israeli hope remained: to force the Egyptians out of the Gaza Strip. It was hoped by the High Command in Tel Aviv that the raids on El Arish, and deep into Sinai, would encourage the Egyptians – even panic them – into retreating from Gaza. But the Egyptians refused to be lured away from Gaza, even under threat of Israeli units far behind their lines. Even the Egyptians in the Faluja pocket had stayed stubbornly on the defensive, refusing to admit the helplessness of their position and surrender. Indeed, they succeeded in mounting a counter-attack on the Israeli forces surrounding them, and forcing a tactical Israeli withdrawal.

With the capture of Abu Ageila on December 29, the United Nations Security Council called on Egypt and Israel to institute a cease-fire. Egypt, desperate to retain Gaza and push the Israelis out of Sinai, called on the Arab Legion and Iraq to open attacks against Israel elsewhere. They declined to do so. The Transjordanians went further, urging the Egyptians to pull back their troops from the Hebron Hills, which they had controlled since the previous

May, and to allow Transjordanian forces to take over there, releasing the Egyptian troops for service elsewhere. But the Egyptians did not want to see Abdullah extend his control to the Hebron Hills, which would make him master of all Arab-controlled Palestine except for the Gaza Strip.

Only Britain came to Egypt's aid. On December 30 the British government issued an ultimatum to Israel that unless Israeli troops obeyed the Security Council resolution of November 4, it would employ force against Israel in accordance with the Anglo-Egyptian Treaty of 1936. Because there was no British diplomatic representative in Israel, the ultimatum was submitted through the State Department in Washington, which transmitted it to the United States Ambassador to Israel, James MacDonald.

Ben-Gurion was in Tiberias when MacDonald asked to see him. Driving to Galilee, MacDonald presented the British ultimatum. Ben-Gurion rejected it. Britain had no right to act on behalf of the Security Council, he said, and Israel had every right to act in self-defence. The Israeli forces had not invaded Egypt, but had entered it as part of a tactical manoeuvre; and they had been ordered to withdraw. This was indeed so. Ben-Gurion despatched the order to evacuate the Sinai to the Negev Brigade headquarters on the morning of December 31.

The hit-and-run raid on El Arish had, however, developed into something more substantial. After the airfield several miles to the south of the town had been captured on December 30, Israeli armour and infantry moved forward to the outskirts of the town itself. By dawn on the morning of December 31 the commander of the Israeli forces, Yigal Allon, and his Chief of Operations, Yitzhak Rabin, felt confident of capturing El Arish, reaching the sea and cutting off the Egyptian forces in the Gaza Strip from all overland contact with Egypt. But even as the two men were looking out over the sand dunes at the prize that they were confident was within their grasp, a runner reached them from Negev Brigade headquarters with a coded message from Israeli General Headquarters in Tel Aviv. They must evacuate the whole Sinai peninsula within twenty-four hours.

Allon did not want to give up the capture of El Arish. Instructing Rabin to take charge of the operations in the south, he flew to Tel Aviv to put the case for the town's capture to Ben-Gurion. At General Headquarters Allon spoke to Yigael Yadin, who told him, 'You needn't try to persuade me; I agree with you already. Try to persuade BG.' Ben-Gurion was still in Tiberias. Allon went to see Moshe Sharett, Israel's Foreign Minister, at his home in Tel Binyamin, and asked to be allowed to take El Arish, destroy the base, and then evacuate immediately. But Sharett was convinced that the capture of the town would bring the British into the war, and declined to authorize Allon's plan.

Allon then asked if Israeli troops could remain for four days in the Sinai Desert, during which he would make preparations as if he was going to attack El Arish. The Egyptians would bring troops back from the Gaza Strip, and he would then attack the strip opposite Gaza and Rafa, well within Israeli territory. Allon was convinced that the United States would accept an Israeli explanation that four days were needed to withdraw the forces from Sinai.

Sharett saw the logic of Allon's new plan, and did not dismiss it out of hand. But when they both telephoned Ben-Gurion, the Prime Minister turned it down. He did, however, authorize Allon to use the El Auja–Rafa road (which lay mainly inside Egyptian territory) in order to mount an attack on Rafa.

Britain was not the only opponent of any further Israeli advance into Egypt. On December 31, a few hours after Ben-Gurion had ordered the withdrawal from Sinai, President Truman telegraphed to Ben-Gurion demanding the 'immediate withdrawal' of all Israeli forces from Sinai, and warning that unless Israel complied, the United States 'would be compelled to reconsider its connection with Israel'. These were stern words from the first, and by far the most powerful, nation to have recognized Israel seven months earlier.

As Ben-Gurion had already promised Britain, on 1 January 1949 the Israeli withdrawal from Sinai began, with a strategic bridge over the Wadi El Arish and the buildings at Abu Ageila being demolished on the way. On 2 January 1949 the last Israeli troops left Sinai. That day two Israeli Spitfires attacked an Egyptian train on the El Arish to Gaza railway. That same day a single Egyptian plane flew over Jerusalem, dropping its bombs near the Sha'are Zedek Hospital. Seven people were injured.

The war with Egypt was not over. Israel still hoped to drive the Egyptians out of the Gaza Strip. For four days, as a result of Ben-Gurion's compromise with Allon, Israeli forces struggled to capture Rafa, with the aim of cutting off and surrounding Gaza. The fighting was intense, but, as the hour for the truce drew nearer, Rafa remained in Egyptian hands. On January 5 a sandstorm made the assault all the more difficult as it enabled an Egyptian armoured column to counter-attack without being seen soon enough to be driven off.

The sandstorm continued throughout January 6, bringing all fighting to a halt. That day the Egyptian government informed the United Nations of its willingness to enter into immediate negotiations for an armistice agreement in compliance with the Security Council resolution of November 16, something it had hitherto set its heart against.

With Egypt willing to negotiate an armistice, Israel had no option but to accept the cease-fire, which came into effect at two in the afternoon of January 7. Later that same afternoon there was a final and indecisive clash between Israeli and Egyptian tanks and half-tracks on the Rafa–El Arish road. By seven o'clock the battle was over. Egypt was in control of the Gaza Strip. Israel had driven the Egyptian forces out of the Negev. The only Egyptian troops on Israeli soil were those still trapped in the Faluja pocket. They too, together with the Israelis surrounding them, were subject to the cease-fire.

*   *   *

A few hours before the cease-fire came into effect, four British fighters from their base in the Canal Zone flew over the Israeli–Egyptian border to make sure that the Israelis had indeed withdrawn from Sinai. Flying above Rafa, and then inland, they were thought by Israeli observers on the ground to be Egyptian warplanes. The Israeli air force sent up its Spitfires to intercept them. One of the Israeli pilots was Ezer Weizman, who had been in the very first bombing raid against an Egyptian armoured column south of Tel Aviv at the end of May. Weizman's opponent in the air was Flight Lieutenant Brian Spragg. 'I got behind him and let loose on his tail,' Spragg later recalled. In his log, Spragg noted at the time: 'Had a tussle with a Yiddish Spit.'

Three of the four British Spitfires were shot down near Nirim, twelve miles inside the Negev. One British pilot was killed. 'I saw his aircraft spiralling down,' another of the British pilots, Douglas Liquorish, later recalled, 'and a second later I felt the bullets tearing into my plane with one ending up buried in the seat armour right behind my head.'

The British government demanded an explanation, and at the same time reinforced the British military garrison at Akaba, the Transjordanian port at the head of the Red Sea, with a battalion of soldiers. In a surprise gesture of support for Israel, President Truman criticized the flight of the British planes over the battle area, and also the reinforcing of Akaba. The British flights had revealed, however, that there were still some Israeli detachments – as authorized to Allon by Ben-Gurion – on the Egyptian side of the road running from Rafa to El Auja. The Israeli High Command had to give an immediate order for these to be withdrawn. When the units themselves protested (for they felt that their presence on Egyptian soil could have been used as a bargaining counter in the coming armistice negotiations) they were informed by Allon, 'There is no room for any appeal or delay in carrying out the order, however harsh it may be.'

On January 18, in an attempt to defuse the crisis, the British Foreign Secretary, Ernest Bevin – whom many Israelis blamed for the harshness of British rule in Palestine in its final phase – ordered the immediate release of the remaining Jewish detainees in Cyprus. Within four weeks, the last of the detainees, many of them survivors of the Holocaust, had reached Haifa. An Israeli ship, the *Azma'ut* (Independence) brought them across the eastern Mediterranean.

Inside Israel a struggle was taking place between the secular and religious members of the government; it came to a climax on 20 January 1949, when the Cabinet debated the question of whether or not non-kosher meat could be imported into Israel. An official Israeli government bulletin later that day recalled that 'the three members of the Cabinet who belong to the religious parties – Mr Moshe Shapiro, Minister of Immigration, Rabbi Y. M. Levin, Minister of Social Welfare, and Rabbi J. L. Fishman, Minister for Religion and War Victims – had during three months, abstained from the deliberations of the Cabinet because of their disagreement with the Government's views on this question.' That day, by a majority of six votes to three, 'the Cabinet has now adopted the suggestion of the religious bloc, according to which the importation of meat will in future be jointly controlled by the Ministry of Commerce and Industry and the Ministry for Religion.'

'In plain words' wrote Arthur Koestler, the Hungarian-born journalist and author who was then visiting Israel, 'the Cabinet has capitulated to the rabbis and no non-kosher meat will be imported into Israel.' Slowly, the Orthodox rabbis were finding their political power, which within five decades they were using to secure considerable financial advantages as well as spiritual benefits for their constituents: the group within the State with the highest birthrate, often as many as ten children per family.

On 24 February 1949, under the auspices of the United Nations mediator Dr Ralph Bunche, an armistice was signed between Egypt and Israel. It was the first such agreement between Israel and any of its warring neighbours. The aim of the armistice was not merely to end the fighting but, as its terms stated, to 'facilitate the transition . . . to permanent peace'. The phrase was taken from the United Nations Security Council resolution of November 16. A second such armistice agreement was signed with Lebanon on March 23. But with regard to Jordan, Israel still had one further military objective: the former Turkish (and British) police post at Um Rashrash on the Gulf of Akaba. As early as 1944 Ben-Gurion had expressed his determination that Israel would take Eilat, and his conviction that it could do so without bloodshed.

Israeli and Transjordanian officials were already on the island of Rhodes, discussing armistice terms, when Ben-Gurion gave the order for the attack on Eilat to be launched. On March 9 two Israeli columns set off, competing to be the first to reach the warm waters of the Red Sea. Several Arab Legion units lay in their path, but, on orders from Amman, they were withdrawn across the Transjordanian border, before they could be drawn into battle. On March 11 the first Israeli troops reached the Gulf of Akaba. A white sheet was found, a Star of David was drawn on it in blue ink, and Um Rashrash was claimed for Israel. That same day, at Rhodes, the Israeli delegation signed a cease-fire. The armistice negotiations could continue.

Israel was in control of all the Negev, from the Mediterranean shore north of Gaza to the former Turkish police post at Um Rashrash, on the Gulf of Akaba, and northwards along the Arava Valley to the southern shore of the Dead Sea. To the south of Beersheba the Negev was a desert wasteland, but it was Israel's sovereign territory.

The armistice with Transjordan was concluded in Rhodes on 3 April 1949. Among its provisions was that Israel would have access to the Hadassah

Hospital and Hebrew University buildings on Mount Scopus, which remained Israeli sovereign territory. But this access, running through Transjordanian-held East Jerusalem – and past the site of the Hadassah convoy massacre of 1947 – was never implemented. The most that Transjordan would allow was a small guard to be shuttled back and forth at fortnightly intervals. Medical services, and academic studies, could not be resumed. Israel had to find new sites for the Hadassah Hospital and the Hebrew University.

Several dozen settlements had been destroyed during the War of Independence. One of them, Beit Ha-Arava, on the northern shore of the Dead Sea, had held out, though completely isolated, for six months. Then its settlers – who had founded their kibbutz nine years earlier – were evacuated by boat to Sdom, at the southern end of the sea. Later they set up two new settlements in Galilee, Kabri and Gesher Ha-Ziv. The hotel at Kaliya, where Ben-Gurion had been when the United Nations voted on partition, was destroyed in the fighting. Later, the site became a Jordanian army base.

Immigration never ceased entirely even at the height of the battles, or indeed in the final year of British rule. In the twelve months from March 1947 to March 1948 the British had allowed 6,000 of the Jewish DPs in their zone of Germany to go to Palestine. A further 1,460 Palestine certificates were issued during that period to DPs in the French and American Zones. When the State of Israel was established in 1948, two-thirds of the remaining DPs went to Israel. The first of them had begun to arrive from 1 February 1948, the date laid down in the United Nations partition resolution for the opening of a free port for immigrants. With the ending of the fighting, the numbers became a flood. Then the British closed the Cyprus camps again, and did not open them until February 1949. Once reopened, they were emptied with great speed and several tens of thousands of new immigrants reached Israel. The challenge to Israel over the next few years was to find them places to live, and a livelihood.

Twelve years after the end of the War of Independence, Yigael Yadin reflected on the question of where the victory had come from. 'If we are to condense all the various factors, and they are many, which brought about victory,' he wrote, 'I would not hesitate to credit the extraordinary qualities of Israel's youth, during the War of Independence, with that victory. It appears as if that youth had absorbed into itself the full measure of Israel's yearning, during thousands of years of exile, to return to its soil and to live in liberty and independence, and like a giant spring which had been compressed and held down for a long time to the utmost measure of its compressibility, when suddenly released – it liberated.'

CHAPTER FIFTEEN

The ingathering of the exiles

The first national elections of the new State were held on 25 January 1949, before the signing of the armistice agreements. The left-wing Mapai (the Party of Eretz Yisrael Workers), which had been the predominant political grouping since 1930, gained forty-six seats out of the 120 which made up the Constituent Assembly (later the first Parliament or Knesset). This was the largest single grouping of seats. The more extreme left-wing Mapam (United Workers' Party) gained nineteen seats. The United Religious Front – a newly formed grouping of all the religious Parties – gained sixteen seats. The right-wing Herut List, headed by Menachem Begin, secured fourteen seats. Two Arab members represented the Democratic List of Nazareth.

Under proportional representation, several small and sectional Parties gained a few seats. The Sephardi List won four seats, and the Yemenite Association one seat. Maki (the Communist Party) won four seats. A Fighters' List, followers of the Stern Gang, gained one seat, which was taken by Nathan Friedman-Yellin.

Ben-Gurion, as leader of Mapai, remained Prime Minister. An implacable enemy of Menachem Begin's Herut, and of his Revisionist philosophy, he refused to contemplate any Herut participation in his administration. He was equally adamant at not having any Communist representation. He was convinced that both these Parties would serve only to destroy democracy, the Communists in favour of the dictatorship of the Left, Herut in favour of a dictatorship of the Right.

When the election was over, Ben-Gurion announced his four-word guiding principle in coalition forming: 'Without Herut and Maki'. He was never to deviate from this. Unable to obtain sufficient votes to form a coalition, Herut remained in the political wilderness for almost twenty years; a humiliating situation for those of its members who felt that their activities in the pre-State era – though denounced as terrorism by the bulk of Palestinian

Jewry – had been heroic, and a contributory factor in Britain's departure. Ben-Gurion was insistent that Herut was not one of the 'constructive forces' in the State, and during his long political ascendancy they were excluded from the principal positions of patronage and power. It was not until shortly before his death in 1973 that Ben-Gurion was to mellow in his implacable distaste towards Begin.

As Ben-Gurion awaited the day when he would be asked to form a government – and this would have to wait until Israel had elected its first President – he set out in a broadcast four principles which he intended to be his guide:

1. The acceptance of collective responsibility for the policies of the Government by all parties participating in it.
2. A foreign policy aimed at achieving friendship and cooperation with the United States and the Soviet Union; a pact between Jews and Arabs; loyal support for the United Nations; and a strengthening of world peace.
3. A Labour majority in the Government and cooperation with constructive forces.
4. Complete civil equality for women – Jewish, Christian and Muslim – and the abolition of all existing discrimination against women as embodied in Turkish and Mandatory laws that are still on the books. These laws deny women full equality and civil rights. On the day we established the State we proclaimed that it would accord equal rights to all its citizens without regard to religion, race or sex. Let us be faithful to that declaration.

Elaborating on these principles at a public meeting six days later, on February 4, Ben-Gurion stressed the need for a 'partnership' between the State of Israel and the Jews of the Diaspora. There must also, he said, be a partnership inside the country, 'between the workers and the other constructive forces among the people'. His policy would be 'to mobilize the assistance of all the Jews in the Diaspora and all the constructive forces in Israel'. He was proud that the principles he had enunciated derived from the experiences and needs of the Jewish people. 'We have not borrowed values from others,' he said. 'We did not adopt the approach of the German Social Democrats when they were still the leading force in the international Labour movement before the First World War. We did not copy the ideas of the British Labour Party when it grew in importance and influence. We did not follow the footsteps of Soviet Communism. We paved our own path; it is along this path that we have moved forward and will continue to do so. We are the sons of a people whose fate differs from that of all other peoples, and we are faced with a task that was not imposed upon the workers of any other country.'

*   *   *

Ben-Gurion believed that the armistice agreement with Egypt would be a precursor to a peace agreement. On 14 February 1949, ten days before the armistice was signed, the Knesset held its first session. It was convened in Jerusalem, many members coming up from the coast for the occasion. As there was as yet no parliament building, the meeting was held in the Jewish Agency building. It was a very special occasion in Jewish history, the first Jewish parliament in a Jewish sovereign State. An onlooker noted in his diary, 'The main streets, which have known pain and suffering, want and siege, took on a festive appearance.'

The main item of business was to elect a President. The Mapai candidate was Dr Chaim Weizmann. The Revisionists put forward Professor Joseph Klausner. When a Herut Member of the Knesset, Aryeh Ben-Eliezer, spoke on behalf of Klausner – taunting supporters of the government with the words 'we have never argued with collaborators, and we don't intend to begin now' – there were cries of 'Boo! Fascists! Followers of Mussolini! Collaborators!' The bitterness of the decade-long conflict was not to be resolved during the first fifty years of statehood.

Weizmann was elected President by 83 votes to 15. At the end of the meeting, the Speaker declared: 'I hereby close our session in Jerusalem, where we have fulfilled the dream of generations, a dream that did not die despite suffering, expulsion, torture and *autos-da-fé*.' The provisional stage in the life of the Knesset had been completed. The representatives of the people of Israel had come together 'to establish and shape the independent State of Israel, to strengthen and safeguard it so that all our far-flung exiles can be gathered together in this land. At this session we have chosen a President. In the sessions to follow we will choose a permanent Government. We have taken the first step toward building our lives. Israel's salvation is at hand.'

On February 24, Weizmann, as President, entrusted Ben-Gurion with the task of forming a government. Ben-Gurion thereupon turned to those whom he called the 'constructive' political Parties. The coalition was made up of forty-six members of Mapai, two Arab members representing the Democratic List of Nazareth, sixteen members of the United Religious Front, five members of the Progressive Party, and four Sephardi List members. Mapam, on the left of Mapai, remained outside the coalition and the government of its own volition: it did not want to be in any government with the religious parties.

The Cabinet consisted of twelve Ministers, chosen from the coalition members. Ben-Gurion became Prime Minister and Minister of Defence. Dov Joseph – who had been governor of Jerusalem during the siege – became Minister of Rationing and Supply, forced to deal with the many continuing shortages of basic foodstuffs. Golda Meir became Minister of Labour and Social Insurance, the area of responsibility which she had undertaken within the Jewish Agency during the British Mandate. The Minister of Justice, Pinhas

Rosen (born Felix Rosenbluth, in Berlin in 1887), a representative of the Progressive Party of which he was a founder, laid the basis for the legal system of Israel. Rosen's particular achievement was the establishment of a judiciary that would be independent of both the Knesset and the Cabinet.

Moshe Sharett became Minister for Foreign Affairs, the Jewish Agency portfolio he had held before independence. Israel's relations with the world beyond the ring of hostile Arab States was clearly going to be crucial. Hoping to avoid being buffeted or exploited by the two opposing power blocs that had emerged in the aftermath of the Second World War – led by the Soviet Union and the United States – the Knesset approved, on March 9, that among the basic principles of the government's programme was 'friendship with all freedom-loving States, and in particular with the United States and the Soviet Union.'

On 10 March 1949 the Provisional Government gave way to the first Cabinet of the new State.

On March 23 an armistice agreement was signed between Israel and Lebanon, and Israel withdrew from the villages in southern Lebanon which it had occupied during the war. Negotiations for the armistice with Transjordan were taking place on the island of Rhodes. But the delineation of the border between Israel and Transjordan seemed an intractable problem, and, in an attempt to break the deadlock, after nightfall on March 23 two Israeli negotiators, Yigael Yadin and Walter Eytan – accompanied by Moshe Dayan and Yehoshafat Harkabi (a future Director of Military Intelligence), who had just returned from Rhodes – were taken to King Abdullah at his villa at Shuneh.

Despite the initial setback when the Israeli gift of a Bible was opened by the King at the page where a coloured map showed Israel at the time of Solomon (including Amman), Yadin entranced the King by a flawless recital in Arabic of an ancient Arabic poem, and the negotiations proceeded well. By two in the morning agreement had been reached and signed. The King gave the Israelis roses from his garden as a farewell gift. They returned to Jerusalem to report their success to Ben-Gurion.

A week later, the Transjordanian Prime Minister, Tawfiq Abdul Huda, who had been in Beirut at the time of the Israeli meeting with the King, raised objections, and on March 30 Yadin and Eytan returned to Shuneh. The issue was the line of the border in the Hebron Hills. The Prime Minister wanted adjustments in the line. 'Both Yadin and I were acutely conscious of the Transjordanians' right to take up the position they did,' Eytan wrote to Moshe Sharett a few days later. 'We were, after all, discussing the future of villages that were wholly Arab in population and situated in territory under Arab control' – Iraqi troops were still in occupation in the Hebron Hills, awaiting the details of the armistice terms. 'In spite of all guarantees and fine phrases,' Eytan added, 'it was as clear to the Transjordanians as to us that the people of these villages were likely to become refugees as soon as the Iraqis withdrew, and possibly even before.'

Abdullah did not want any further delay, and after failing to persuade Yadin to allow him to take Beit Jibrin 'as a birthday present' (it was the King's birthday on the following day) he agreed to abide by the original line of a week earlier. Three days later, on April 3, in the Hotel des Roses in Rhodes, the Transjordanian and Israeli delegates signed the agreement that had in fact been made in Shuneh, rewritten by Dr Ralph Bunche in the somewhat more legalistic language favoured by the United Nations. In the Knesset, Menachem Begin called for a no-confidence vote for the 'abandonment' of so much of the land of Israel. His call was brushed aside.

Allan Burke, who had served in the Royal Navy throughout the Second World War, including the Atlantic and Russian convoys and the Sicily and Normandy landings, was asked in March 1949 to bring to Israel the Royal Canadian Navy frigate, *Strathadam*, which had been built in 1944 and seen one year's war service in the Atlantic. Purchased by the new State as Israel's first major warship, the *Strathadam* was re-registered under the Panamanian flag and taken to Marseille. There Jewish volunteers under Burke's command prepared to sail her to Israel. During the weeks that the frigate was at Marseille, Burke recalled, 'intelligence reports said that Egyptian frogmen aboard an Egyptian merchantman berthed nearby might try to sabotage it. A heavy guard was put on the ship round the clock: volunteers recruited from a camp at St Jérome packed with refugees waiting to be taken to Israel. Further reports said the Egyptians were going to attack on a certain night.'

Although the engine and boiler repairs had not been completed, Burke obtained clearance from the harbour authorities to go out on 'engine trials'. Taking on board just enough fuel to reach Haifa, he ordered one harbour pilot to take the vessel out, and another to take it in, and requested fuelling facilities for the following day. Leaving harbour in thick fog, he dropped the first pilot, navigated the approaches to the port, and steamed for Haifa. On arrival, he was made officer-in-command of the Israel Flotilla.

On 11 May 1949 – almost a year after Israel had declared its independence – the United Nations voted to accept Israel as a member State. She became the 59th State of the world organization. This brought to fruition those aspirations of Jews in the Diaspora which had animated the Basle Programme in 1897 and the Balfour Declaration twenty years later. But the States nearest to Israel geographically – Lebanon, Syria, Transjordan, Saudi Arabia and Egypt – had without exception cast their votes against recognition, and had declined to turn their armistice agreements into peace treaties. In a debate in the Knesset four months later, on June 15, the Foreign Minister, Moshe Sharett, warned the assembled parliamentarians, 'The storm which has been raging around us will not soon be stilled. Nor do we hold the certainty in our hearts that it will not break out anew, with greater violence. Our vital interest is in a comprehensive peace soon, we are duty-bound to lend our best efforts to its achievement. But, with all our striving for peace, we must not lose patience if it tarries in coming. If destiny has so decreed it – we are strong enough to wait with composure.'

A fourth and final armistice agreement was signed with Syria on July 17. The War of Independence had lasted a year and seven months. In recognition of his work as a mediator and facilitator of the armistice agreements, Ralph Bunche was awarded the Nobel Peace Prize. Like the other armistice agreements, the one between Israel and Syria was intended to lead to an eventual peace treaty, but did not do so. Damascus, like Cairo, Amman and Beirut, remained the implacable adversary of the very concept of a Jewish State.

There was distress in Israel when it was learned that among the books that were translated into Arabic and widely distributed in the Arab capitals was the *Protocols of the Learned Elders of Zion*, a pre-First World War Russian forgery (that had been exposed in the law courts between the wars) claiming to be the blueprint of a Jewish world conspiracy to subvert national governments and to rule the world.

At a conference held in Lausanne in 1949, the United States pressed Israel to take in between 200,000 and 300,000 of the estimated half million Palestinian refugees (the Israeli estimate was 520,000, the United Nations' estimate 726,000, and the British government's estimate 810,000). Israel resisted the Lausanne proposals. Some Israelis feared that among those returning might be fifth columnists dedicated to terror. Others feared that if Israel offered to take back 100,000 refugees, as part of a compromise proposal, there would be considerable international pressure to take more. There was also resentment in Israel at American pressure on the refugee issue, and fear that there would be public discontent at the return of large numbers of refugees, particularly among those Jews who had begun to arrive in large numbers from Arab lands.

On December 8 the General Assembly of the United Nations created UNRWA – the United Nations Refugee and Works Administration for Palestinians. It began operations five months later. Israel undertook responsibility for those Arabs who had fled to those parts of the country that ended up inside Israel's borders, the so-called 'internal refugees', between 25,000 and 32,000, who could not return to their own homes and villages because they had been either razed to the ground or occupied by Jews. About 35,000 Arab refugees had also found their way back across the border since the end of hostilities. In camps set up in all the States bordering on Israel, and in Transjordanian and Egyptian occupied areas (the West Bank and Gaza Strip respectively), UNRWA officials provided education, health, relief and social services.

Simultaneously with UNRWA setting up facilities – at first minimal – to take over the refugee camps beyond Israel's borders, Israeli legislation made any realistic prospect of the return of more than half a million people increasingly unlikely. The first such legislation had been the Abandoned Areas Ordinance (1948). As the number of refugees grew, through the natural increase of births over deaths, so the legislation designed to make the return of the refugees more difficult, if not impossible, also increased. This legislation included the Emergency Regulations Concerning the Cultivation of Waste Land (1949), the Absentees Property Law (1950) and the Land Acquisition Law (1953).

Within a decade of its passage, this last law was to transfer to Jewish ownership and cultivation between 3,200 square kilometres (the Israeli figure) and 4,600 square kilometres (the United Nations figure) of Arab-owned land. Among those who protested was the German-born Jewish philosopher, Martin Buber, who, in answer to the Israeli government's correct insistence that the Custodian of Absentee Property was granting leases rather than titles to land and houses, wrote to the Speaker of the Knesset (on 7 March 1953), 'We know well, however, that in numerous cases land is expropriated not on grounds of security, but for other reasons, such as expansion of existing settlements etc. These grounds do not justify a Jewish legislative body in placing the seizure of land under the protection of the law. In some densely populated villages, two thirds and even more of the land have been seized.'

Within three years of independence, and two years before Buber wrote his letter of protest, 1,400,000 people, a quarter of the Jews of Israel, were housed on 'absentee' Arab property. Golda Meir, who shared Ben-Gurion's outlook in this regard, commented without prevarication, 'We used the houses of those Arabs who ran away from the country whenever we could for new immigrant housing,' and she added, with a reference to the policy of Jordan, Syria and Egypt towards the dispossessed Palestinians, 'We did not keep our refugees in camps.'

It was the Arab refugees who were kept in camps by their Arab hosts, and deliberately so, in order to keep the Palestinian plight in the mind of the international community. The Palestinian refugees therefore remained under UNRWA supervision. In 1996, forty-eight years after the end of the War of Independence, UNRWA was responsible for 1,832,000 Palestinian refugees and their descendants in Jordan; 1,200,000 in the West Bank (including refugees from the 1967 war); 880,000 in the Gaza Strip; 372,700 in Lebanon; and 352,100 in Syria. As negotiations with the Palestinians began in the 1990s, the return of Palestinian refugees was an important – and intractable – item on the negotiating agenda.

Within weeks of the establishment of the State of Israel, Jewish immigrants were arriving in enormous numbers, day by day, in a steady, increasing and powerful stream. One of the first acts of the provisional government was formally to abrogate Britain's Palestine White Paper of 1939, the most recent legislation with regard to immigration. The abrogation of these restrictions automatically enabled any Jew to enter, as had the original terms of the British Mandate, and, in effect, the Israeli Declaration of Independence.

Between 15 May 1947 and 31 December 1951 a total of 686,739 Jews had arrived in Israel, mostly by sea. They came from seventy different countries, and constituted, in terms of the size of the population of the State to which they came, the largest single migration of the twentieth century. Within four and a half years the population of the State had been doubled.

To persuade countries to let Jews leave could be a difficult and long drawn out process. In Iraq, where the Jewish community had long been harassed despite its considerable contributions to Iraqi society, commerce and government, clandestine methods were employed to win government support. An Iraqi-born Jew, Shlomo Hillel, travelled from Israel to Baghdad in the guise of a British businessman. His 'taxi driver' was a leading Iraqi-born Zionist also living in Israel, Mordechai Ben-Porat. After secret negotiations in Baghdad with the Iraqi Prime Minister, Tawfiq al-Suwaidi, Hillel and Ben-Porat succeeded, on the basis of a substantial cash payment, in securing a law allowing Iraqi Jews to emigrate to Israel. The Bill to this effect was introduced to the Iraqi parliament on 2 March 1950. It became operational a week later. Jews could leave provided they left behind all gold, jewellery and valuables, and provided also that they gave up Iraqi citizenship (the Soviet Union was to make the same stipulation for Russian Jews two decades later).

On the first day of the law allowing Iraqi Jews to leave for Israel, 3,400 of Iraq's 130,000 Jews registered for emigration. The Iraqi government had expected that no more than 8,000 would want to leave. But on the following day, a further 5,700 registered, and within five weeks, 50,000 Jews had declared their desire to leave. By the end of three months, the figure had risen to 90,000. The emigrants had to give up their homes and their livelihoods. So large were the numbers, and so important did Israel consider the airlift, that the Israeli chief rabbinate allowed the flights to take place on the Sabbath. The Iraqi Jews had been allowed to go eastward, across the border into Iran, from where a less hostile administration allowed planes to take them westward to Israel.

As the emigration from Iraq gathered momentum there were five bombing incidents, in one of which a Jewish teenager was killed and twenty-five people injured outside a synagogue in Baghdad. The American Information Office in Baghdad and three Jewish businesses were also bombed, though without any fatal casualties. Some said the Iraqi government was responsible, hoping thereby to encourage the emigration of wealthy Iraqi Jews in order to have possession of their property. Others said that the bombings were the work of the extreme nationalist Istiklal Party, angry that the Jews were being allowed to leave rather than being continually chastised and gradually impoverished. In fact, the bombings were the work of Israeli agents hoping to stimulate the emigration of all Iraqi Jews, rich and poor, leading civil servants and merchants alike.

Two Iraqi Jews, Shalom Salakh Shalom and Yosef Abraham Basri, were arrested, convicted and publicly hanged in Baghdad on 19 January 1952. Two emissaries from Israel were also arrested, Yehuda Tajir – who spent the next ten years in Iraqi prisons – and Mordechai Ben-Porat, who managed to escape and get back to Israel (where he later entered the Knesset).

The largest single influx of Jews to Israel came from Iraq, 123,371 in all. They were Arab speaking and, in the main, had felt alienated and hostile to the Arab culture in which, while some of them had flourished, most had been very conscious of being second-class citizens. The second largest group of new immigrants were the 120,000 and more survivors of the Holocaust (108,404 of them from Poland), many of whom had been held in British detention camps in Cyprus, or in the DP camps throughout Germany and Austria, or who made their way from their former home towns in Central Europe, in which Jewish life had been destroyed.

The pre-war Zionist hope of bringing trained agriculturalists and pioneers zealous to establish new settlements and to till the soil and raise cattle and cultivate their fields and orchards had to be set aside in the realization that most of the survivors of the Holocaust were people who had no such ideals, had never intended to go to Palestine, and would require a world of their own in which to recover as best they could from the physical and mental devastation of their wartime torment. Many of the tens of thousands of survivors who had wanted to get to Palestine were quite untrained, or broken by their terrible experiences in the war. The teenagers among them had lost up to six years' schooling and education in the ghettos and concentration camps. But they could not be excluded, and the Jewish Agency had to adjust its expectations and prepare for the arrival of those who would need considerable support.

Under the auspices of Youth Aliyah, which had been established in 1932 to help youngsters from prewar Germany, tens of thousands of children were introduced after 1948 to a new and settled way of life, something many of them had never known. More than 15,000 young survivors of the Holocaust had been brought from Europe in the three years preceding the State by Youth Aliyah. After the establishment of the State, Youth Aliyah undertook to be responsible for child immigration and care from every land of the Diaspora. Its head from 1948 to 1964 was Moshe Kol, a member of the Provisional Council of State, who, having been a leader of a Zionist youth movement in Poland as a young man, had come to Palestine in 1932 at the age of twenty-one. Up to the end of 1970 Youth Aliyah had taken in, educated and prepared for adult life more than 125,000 children.

Typical of those who made their main concern the needs of immigrants from difficult or deprived backgrounds was Salomon Adler-Rudel, who had worked in Berlin between the wars on behalf of the many Eastern Jews in the city, and had worked in London during the Second World War as the administrator of the Central British Fund, responsible for tens of thousands of German and Austrian immigrants (he himself was from the Austro-Hungarian city of Czernowitz). Settling in Israel, Adler-Rudel was made director of the Jewish Agency department of international relations, and prepared a series of agreements with the International Refugee Organization for the transfer to Israel of what were known as the 'hardcore cases' in the DP camps in Europe.

Roumania, under Communist rule since the end of the war, had at first declined to let its Jews go. But in December 1949 it relented, encouraged to do so by generous payments, and suddenly 117,950 more Jews were on the move to Israel. Many of these Roumanian Jews had also spent the war years in ghettos and slave labour camps, and on death marches of the utmost savagery. The American Jewish Joint Distribution Committee – The Joint – which had been founded in 1914 to help Jews displaced in the First World War, played a crucial role in enabling these Jews to be brought to Israel. Joint funds, as well as its diplomatic skills and tightly knit organization, ensured that the work of rescue went ahead on a massive scale.

From Morocco, Tunisia and Algeria – lands where the forces of militant Islam had always constituted a threat even in times of relative calm and prosperity, and where The Joint was also active – 45,336 Jews made their way to Israel. Like the Jews from Iraq, they formed part of the Sephardi community, feeling themselves often relegated to menial and unappreciated tasks, and suspicious of the political ascendancy of the predominantly European Ashkenazi Jews who had formed the bulk of the prewar Zionist structure.

Almost all Bulgaria's Jews made their way to Israel without let or hindrance: 37,260 in total. Theirs was a community that had actually increased in numbers during the Second World War, the Bulgarian King and his parliament having refused to agree to German demands for the deportation of the Jews to German-occupied Poland, and the death camps.

From Turkey, which had been neutral in the war, 34,547 Jews came to Israel in its first thirty months of statehood, more than a third of Turkish Jewry, drawn by proximity and identity to the country which had once been a remote provincial backwater of the Ottoman Empire. From Yugoslavia a further 7,661 Jews set off through the Adriatic, the Aegean and the Eastern Mediterranean for Israel.

Catching the imagination even of a people that was becoming used to stories of extraordinary flight and rescue was the fate of the Jews of Yemen. Their community was an ancient one, and also one that had sent members to Palestine in the Turkish and British time. But with Israel's independence there had been anti-Jewish riots in the Yemen, and deaths, and a large community, numbering more than 45,000, was suddenly on the move, making its way to collection points organized by The Joint. At the collecting points, doctors and nurses from Israel did what they could to prepare the Yemen Jews for the journey. The operation was given the name 'Magic Carpet': because Egypt had closed the Suez canal to all Israeli shipping, the normal way by sea was barred; instead, Israel mounted an airlift of remarkable proportions, flying 45,640 in transport planes fitted out with seats, row on row as in a rural bus, but with 500 to 600 in every plane. A further 3,275 Jews were flown in from Aden.

'Sometimes I used to go to Lydda,' Golda Meir has recalled, 'and watch the planes from Aden touch down, marvelling at the endurance and faith of their exhausted passengers. "Had you ever seen a plane before?" I asked one bearded old man. "No," he answered. "But weren't you very frightened of flying?" I persisted. "No," he said again, very firmly. "It is all written in the Bible, in Isaiah, They shall mount up with wings of eagles." And standing there on the airfield, he recited the entire passage to me, his face lit with the joy of a fulfilled prophecy – and of the journey's end.'

The passage which the old man had recited, from Isaiah chapter 40, reads:

> Hast though not known? Hast thou not heard, that the everlasting God, the Lord, the Creator of the ends of the earth, fainteth not, neither is weary? There is no searching of his understanding.
>
> He giveth power to the faint; and to them that have no might he increaseth strength.
>
> Even the youths shall faint and be weary, and the young men shall utterly fall.
>
> But they that wait upon the Lord shall renew their strength; they shall mount up with wings as eagles; they shall run, and not be weary; and they shall walk, and not faint.

Golda Meir's visit to the airport to greet the Jews of Yemen and Aden was not mere curiosity. She was entrusted by Ben-Gurion with the task of going to the United States and finding the money needed to create minimally acceptable housing for the mass of new immigrants. Her appeal was a forerunner of many appeals, hundreds by her, thousands by others of her ministerial colleagues, that over the years acquired for Israel the financial resources with which to build up the country. Her theme on this first peacetime fund-raising visit was that the money was needed 'not to win a war but to maintain life'. As she told her American Jewish listeners at her first official public appearance in the United States:

> I went to our Parliament two weeks ago last Tuesday, and presented a project for 30,000 housing units by the end of this year. Parliament approved it, and there was great joy in the country. But actually I did a strange thing: I presented a project for which I didn't have the money.

What we want to do is to give each family a luxurious apartment of one room; one room which we will build out of concrete blocks. We won't even plaster the walls. We will make roofs, but no ceilings. What we hope is that since these people will be learning a trade as they build their houses, they will finish them, and eventually, one day, add on another room. In the meantime, we will be happy, and they will be happy, even though it means putting a family of two, three, four or five into one room. But this is better than putting two or three families in a single tent . . .

It is an awful thing to do – to forge a signature to a cheque, but I have done it. I have promised the people at home and the people in the camps that the government is going to put up these 30,000 units, and we have already started to do so with the little money we have. But there isn't enough for these 30,000 units. It is up to you either to keep these people in camps and send them food packages, or to put them to work and restore their dignity and self-respect.

The money was forthcoming. The generosity of American Jewry, and of Western Jewry in general (British Jewry particularly prides itself in this respect) has never failed the Israeli public. But even so it was not enough, and by October 1950 only one-third of the hoped-for housing units had been built. Precious funds had been spent to cover emergency building of metal huts, that succeeded in keeping out the winter rains better than tents could do, but were furnaces in the heat of the summer. As the numbers coming in far exceeded the ability to house them, the expedient of nailing canvas sheets to wooden frames provided temporary walls and a roof for tens of thousands of newcomers. Even the $100 million bank loan from the United States for the absorption of new immigrants, granted by Congress at President Truman's urging, was insufficient for the massive task in hand.

It was a time of great hardship. A nation of 600,000 people had fought for its life. It had then to absorb an even larger number of immigrants. Most of those who arrived had been through terrible hardships, some as Holocaust survivors, others as refugees. Most of them arrived with almost no possessions, and no livelihood. 'A lot of the misunderstanding, the social hatred, began at that time,' Atara Mintz, who had been born in Palestine and served as a combat nurse in the War of Independence, later recalled.

Yet the immigrants could not be kept away, and every effort was made to find them food and shelter. All over Israel, enormous tented camps were set up for them. Known as ma'abarot (singular, ma'abara) they were the first stage whereby the massive influx was housed, albeit in extremely cramped and primitive conditions. One of the largest of the ma'abarot was set up in the former Arab village of Ras al-Ain, renamed Rosh Ha-Ayin, which had been captured from the Iraqis in the War of Independence. A large British army camp, with many wooden and corrugated-iron army structures, served

as the main part of the ma'abara. The newcomers were almost all Jews from the Yemen, more than 5,000 of them. Largely by natural increase, the population grew to more than 11,000 by 1970. It was the only large Jewish urban centre in which almost all the inhabitants were from a single country of origin.

The burdens on Israel of the immigration process were enormous. 'Absorbing these immigrants would have been beyond the ability of a well-established, prosperous country, let alone one newly born and struggling to defend itself,' Chaim Herzog has written. 'Hundreds of thousands of men, women and children were concentrated in tents in transit camps known as ma'abarot. There was no privacy and conditions were appalling. An embryonic educational system for children was established, but they suffered from the intense heat and dust in the summer, and wallowed in the mud in winter rains. At one stage there was no grain whatsoever left and no money to pay for a shipment due to arrive from America.'

It was David Ben-Gurion, Herzog recalled, 'who not only held the struggling nation together but also found solutions to these overwhelming dilemmas. At our Prime Minister's urging, Yadin moved the army toward performing functions over and above purely military duties. It became involved in instructing farmers. Agricultural programmes were instituted in which the army engaged in the actual production of fruits and vegetables to help relieve the food shortage. Staff officers were allocated to deal with the absorption of the immigrants, and the army even developed its own radio service.'

Yigael Yadin even tried to persuade Ben-Gurion to make use of the new medium of television to serve as a unifying and educational focus for the new immigrants, and as a means of communicating to them en masse. But Ben-Gurion resisted the introduction of television not only for new immigrants, but for the country as a whole, believing it to be uncultured and uncivilized; it was only after his political ascendancy was ended, in the mid-1960s, that television came, belatedly, to Israel.

The ma'abara was in every sense a refugee camp, inhabited mainly by Jewish refugees from Arab countries. Golda Meir once remarked that Israel's propaganda mistake was to do away with its own refugee problem, by treating the refugees as human beings and making every effort to give them a decent home and jobs. The world looked with far greater sympathy on refugees who were kept in their camps (as happened to the Palestinian Arabs in Lebanon, Syria, Egypt and Jordan) and given the minimum of help, or even no help, to integrate into society. The ma'abarot had disappeared within a decade; many of the Palestinian Arab refugee camps remain to this day.

Throughout the Diaspora, appeals were made to help the immigrants. Wealthy philanthropists contributed towards schools and clinics and baby

homes. The smaller donations of those who could afford only a little – for in Europe it was still a time of austerity – combined to make a further significant contribution. In the United States and Britain the annual appeal for Israel became a challenge for hundreds of devoted Zionists. Jews in every trade and profession vied with each other to be the most generous. Israeli personalities began to travel and speak at fund-raising events, not merely at times of emergency but on a regular basis, focusing on the needs of the immigrants.

Inside Israel the sight of the mass immigration inspired the poet Nathan Alterman, after a visit to a factory making parachute flares for the military industries, to write:

> And when I saw the African Jew bending over the furnace,
> To draw out the ingot of red-hot steel,
> And passing it with his tongs to the immigrant from the Balkans,
> I saw a people standing firm on its foundations.

Alterman recorded with care the countries and towns of origin of those whom he saw at work:

> They are Jews from Tripoli, Turkey, San'a and Lvov,
> From Sofia and Yassi, clean-shaven, heavy-bearded.

A vast programme of public works was initiated by the Israeli government to provide employment for the refugee immigrants, who were mainly unskilled. The Jewish National Fund also embarked upon a nationwide afforestation project, employing the immigrants to prepare the stony soil and to plant forests throughout Israel on land that had been ravaged by the elements over many centuries, and which had not hitherto been regarded as capable of sustained planting and growth. Through these forests the landscape was transformed. Only forest fires, as in the Jerusalem Corridor in the summer of 1995, reveal the dull, rounded hills as they were during the Ottoman time when trees were only regarded as something to be cut down.

From the first months of independence, the Israeli government was concerned that, despite the massive influx of Jewish immigrants, in Galilee the Arab population far outnumbered the Jewish. Many new settlements were therefore established, from the very first months, in what had hitherto been a largely Arab area. In February 1949 kibbutz Lavi was founded in eastern Lower Galilee, eight miles west of Tiberias. The kibbutz based its economy on field crops, vineyards and dairy cattle. Its first settlers were members of the British Bnei Akiva religious Zionist movement. Later they established a wood factory specializing in synagogue furniture. The newcomers took their name Lavi (Lioness) from the Hebraized version of the Arab 'Lubiya', the name of the village that was on the site until the flight of its inhabitants during the War of Independence.

Another of the new Galilee settlements that year was Kfar Rosh Ha-Nikrah (Village of the Headland of the Cleft) founded in 1949 on the Lebanese border, overlooking the Mediterranean. Its founders were entirely Israeli-born youth, known colloquially as Sabras.

Also founded in Galilee in 1949 was Almah, less than 4 miles south of the Lebanese border. Its first inhabitants were Jews from Libya, who had been part of the massive influx of Jews from Arab lands that year. Their livelihood in that hilly countryside came from vineyards, fruit orchards and vegetables. The name of their moshav came from a town mentioned in the twelfth-century travels of Benjamin of Tudela, as a place where Jews resided; its actual historical location is unknown. But Jews had lived in Upper Galilee in those ancient days, forming a slender but continuous link of Jewish settlement from biblical to modern times.

Less than two miles from the Lebanese border, in rugged hill terrain, kibbutz Sasa was another of the settlements founded in 1949. It lay 2,550 feet above sea level on the site of a Jewish village that had existed in Roman times, and of an Arab village – Sa'sa' – which had been abandoned a few months earlier, during the War of Independence. Many of those who founded the kibbutz had come from North America. The United States and Canada were never to provide the numbers that came from other countries, but the Jews from North America always prided themselves on their modernity and practicality.

Another of the Galilee settlements of 1949, just north of Acre, was kibbutz Lohamei Ha-Getta'ot (The Ghetto Fighters). It was unusual in that its founders were Jews who had come from Poland and Lithuania after the Second World War, having been active during the war in the armed Jewish resistance against the Nazis inside the ghettos. Hence the name of their kibbutz. Among the first settlers were two of the leaders of the Warsaw ghetto revolt of 1943, Yitzhak (Antek) Cukierman and Zivia Lubetkin.

The security of Israel's borders with Lebanon, Syria, Jordan (formerly Transjordan) and Egypt – none of whom had been willing to sign a peace treaty with her in 1949 – was in the hands of the Israel Defence Forces. It was Yigael Yadin who, on a visit to Switzerland shortly after the War of Independence, had seen how a small country could create, through the combined and skilful use of a small professional army and compulsory nationwide military service, a single, large and permanent reserve of military manpower. The professional army would be in place at all times, training and guarding. The reserves – every able-bodied former national serviceman – would be ready at any time to be mobilized.

Reserve duty and emergency call-up could between them ensure that the frontiers were always manned and sudden danger met. Reflecting on the system thus created, the American general and observer of the Israeli scene, S. L. A. Marshall, wrote in 1961:

Here is the smallest standing professional force serving any modern state. Israel's sword and shield is a citizen army which may be summoned from the streets, shops and olive fields within a few hours.

The population is approximately one million less than in the Detroit metropolitan area. But this trained and ready civilian reserve, male and female, approximately equals in numbers the total reserve of the United States Army, exclusive of the National Guard. The West German contribution to NATO field strength after five years of effort is also roughly equivalent.

'Here then,' Marshall added, 'we have the only full measure of what a free people may do to bring forth a sufficient military manpower from a civilian mass when freedom's survival may well depend on other communities seeing the light.'

The history of Israel, like that of pre-State Palestine – and of the often divided capital, Jerusalem – is one scarred by violence and bloodshed. But any account that does not give some impression of the normality of life even under the shadow of conflict will be a false one. 'The human spirit is amazing,' Chaim Herzog, the Chief of Israeli Military Intelligence in 1949, has written. 'Even amid extraordinary pressure and exhaustion, it strives for normalcy. Which I suppose is why our first son, Joel, was born on 3 October 1949. While it may seem a bit strange to bring a new life into a world where the struggle to survive is overwhelming, it made perfect sense and gave me an even greater reason to continue. I had seen my dream come true – the Jewish State existed. Now my dream for my children and for their children was to live in it free of the strife and hatred and suffering and fear that my generation had known.'

Three weeks before Joel Herzog's birth, the Knesset had passed a law for free and compulsory education.

Under the Israel–Jordan armistice agreement, the Arab villages of the 'Little Triangle' were included inside Israel, in order to give Israel road access from the coast to the Jezreel Valley. This added to the Arab population of Israel. The main village in the triangle was (and is) Al-Tayyiba; twenty years after its annexation, its population had reached 10,000. Today it is a town.

The drawing of the Israel–Jordan border created several anomalies. In the Jerusalem region, those Arab villagers of Beit Safafa who lived along the railway line were incorporated inside Israel. Those to the south of the line came within Jordanian jurisdiction. At Tulkarm, the line was drawn in such a way that while the town was part of Jordanian territory, the railway and many fields were on the Israeli side of the line. Tulkarm was also typical of towns on the West Bank, in that half its inhabitants were refugees from Arab towns and villages in the coastal plain (in 1967 the figure for Tulkarm was 5,137 locals and 5,020 refugees). Jordanian policy was to keep the refugees separate, thus exacerbating the sense of continual grievance against Israel.

As more and more immigrants flocked to Israel, unimpeded as hitherto by any immigration restrictions, the need for more settlements intensified. On a hillside just west of Jerusalem, the new settlement of Beit Zayit (House of the Olive Tree) was founded in 1949 as a moshav, by immigrants from Yugoslavia, Roumania and Hungary. Fruit orchards, vegetables and poultry were its livelihood. Nearby, a dam was built to store winter flood water. The moshav took its name from the extensive olive groves all around it – owned before 1948 by the nearby Arab villages.

Those abandoned or deserted Arab villages that had not been destroyed, and the former Arab towns such as Ramle and Lydda (given the Hebrew name Lod), were quickly settled by Jewish immigrants. In Beisan (given the Hebrew name Beit Shean), where 3,000 Arabs had lived until 1948, 1,200 Jews, mostly from Morocco, were settled. Within twenty years, the Jewish population of Beit Shean multiplied tenfold, to more than 12,000, with Jews from other Muslim countries, including Turkey, Iran and Iraq, joining the Moroccans. Initially, the town's livelihood derived from hired farmwork in the nearby Jewish settlements, and from small trading. Later a textile mill, clothing factory and plastics plant provided more work, though the social and living standards of the town were cause for concern right through to the 1970s. Its favourite son, David Levy, himself Moroccan-born, was twice Foreign Minister of Israel.

The coastal town of Acre was another former Arab town that grew into a large Jewish metropolis. About 8,000 Arabs had remained there after the war, their main mosque, the al-Jazzar, being the largest within Israel's borders. Jewish immigrants were at first a minority of the population, but grew within seven years of the end of the War of Independence to a majority, more than 20,000 settling in the town during that period. Today the Jews and Arabs of Acre live peaceably together, as they do in nearby Haifa, across the bay.

The Arab town of Yibna, south of Tel Aviv, all of whose 4,000 inhabitants had fled during the May 1948 battle, was near the Jewish settlement of Yavne, and was repopulated, as Beisan had been, with new immigrants (today many army and air force officers live there). In 1949 three new farming settlements were founded near Yavne: Ben Zakkai, Beit Gamliel and Benayah. In its early years, Yavne was poor, living standards were low, and the housing was below even the hard standards of the time. Many of the new immigrants worked as hired labourers in farming and industry.

Among the Arab villages abandoned during the War of Independence was Caesarea, which had been established seventy years earlier, on the ruins of one of the most prosperous Roman cities, by Bosnian Muslims who had left

Bosnia after the Austrian occupation in 1878. The area became an immigrant camp.

The Israeli political argument at the time of the resettlement of so many Arab towns and villages by Jews was that a de facto exchange of populations had taken place, brought about by a war that the Arabs had launched. In Abba Eban's words at the United Nations in 1949, 'Every war produces its crop of refugees.' Hundreds of thousands of Arabs had fled from Palestine, and hundreds of thousands of Jews had come from Arab lands where they had been, in the main, unwelcome. On this basis, it seemed that what had taken place, despite the harshness of the circumstances on both sides, was a legitimate exchange of population (of the sort that had taken place between Greece and Turkey in 1922).

To enable housing to be built for immigrants on a scale hitherto unimaginable in Zionist planning, the Israeli government set up in 1949 a national immigrant housing company, Amidar. It was a wholly national – and nationalized – institution, the government controlling 75 per cent and the Jewish Agency 25 per cent of its shares. Amidar constructed the apartment blocks, administered them, and was responsible for assigning tenants and maintaining and improving the properties. It also organized communal activities around them.

With the scale of immigration so great, and the background of the immigrants so mixed, Amidar undertook basic civic and municipal tasks, trying to instil in its tenants the need to take care of their apartments and communal property, and to take responsibility for its cleanliness and repair. For immigrants renting an Amidar apartment the rent was initially fixed at between 7 and 10 per cent of their average monthly earnings. When the immigrants became more settled – this scheme was instituted in 1955 – and their earning power was more steady, they would be encouraged to buy the apartment from Amidar with a down payment of between 10 and 20 per cent, and the rest on a twenty-five-year mortgage, at an interest no greater than 4.5 per cent.

There was one remote part of the area within Israel's borders that had no road: the western shore of the Dead Sea. In 1949, Nahal – that branch of the army that trained young soldiers for settlement on the land, particularly in remote border areas, once their military service was over – also undertook the building of a road along the Dead Sea shore from the potash works in the south to the freshwater springs of Ein Gedi, in the Judaean desert. This road passed below the massive rock of Masada, where, 2,000 years earlier, the Zealots had held out against Rome until the imminent success of the Roman siege led to their mass suicide.

As well as settlements, schools and educational institutes were being established throughout the country. On 2 November 1949, the thirty-second anniversary of the Balfour Declaration – to which Chaim Weizmann had contributed so much – the prewar Daniel Sieff Institute at Rehovot was re-inaugurated, much enlarged, as the Weizmann Institute of Science. It was to become a magnet for scientists from all over the world, and to bring to Israel from all over the Diaspora Jews who would make their contribution to every branch of scientific inquiry, most recently the harnessing of the sun's rays for domestic energy. In November 1949 it had a scientific staff of fifty, including teachers and research students. Twenty years later there were nineteen research departments, more than 400 scientists, and 500 students.

Among the settlements founded in 1950 was Kfar Mordechai, three miles south of Yavne, named after Israel's first Minister-Plenipotentiary to Britain, Mordechai Eliash. Its first settlers were from Britain. Later they were joined by others from English-speaking countries, in particular Australia, and also by a number of native-born 'Sabras'. Citrus and dairy cattle farming were their main activities.

Also founded in 1950, on the eastern slopes of the Meron mountains in Galilee, was Amirim (Summits). Six years after it was established, it was taken over by a group of vegetarians and naturalists. Eschewing livestock, it grew – and still grows – apples, apricots, peaches, nuts and vegetables.

In the coastal plain near Petah Tikvah immigrants from South Africa founded Gannei Yehudah (Gardens of Judaea) in 1950. Their main liveli-hood was to come from the cultivation of citrus fruits. That same year, on the very border with Jordan – facing the Arab town of Tulkarm – kibbutz Yad Hannah was founded by immigrants from Hungary and Israeli-born youth. It was named after the Hungarian-born Jewess, Hannah Szenes, one of the Jewish parachutists from Palestine who had been caught and killed by the Germans in 1944.

In an isolated spot in the Arava Valley, a desert experimental farm, Ein Yahav (the Yahav Spring), was founded in 1950. Thanks to the spring from which it took its name, the farm was well supplied with water, albeit partly brackish. The combination of water and the hot climate enabled fruit, vegeta-bles and flowers to be grown in midwinter, leading to the development of a flourishing export trade.

As part of the Israel government's plan to secure the Jerusalem Corridor, which before the War of Independence had been dominated by more than twenty Arab villages, an immigrant transit camp (ma'abara) was set up at Beit Shemesh, where the Tel Aviv to Jerusalem railway enters the Sorek gorge. A year later a permanent urban settlement was begun, and light industry brought to the town, including textiles, steel-chain manufacture, bicycles, stoves, refrigerators, diamond polishing and printing. Later facto-ries for the manufacture of Murano glass and jet plane engines were built.

A new settlement established in the Jerusalem Corridor in 1950 was Aminadav (named after a prince of the tribe of Judah). Many of the first settlers were new immigrants from Turkey and Morocco, employed as labourers by the Jewish National Fund on land reclamation projects in the Judaean Hills.

New immigrants from Morocco were also among the first Jews to settle in the development town of Kiryat Shmonah. Its name means 'City of Eight' and commemorates the eight defenders of nearby Tel Hai who were killed during an Arab attack in 1920. The idea of development towns was as important to the building up of Israel after the War of Independence as the kibbutzim had been in the years leading up to it. They became the spine or backbone of modern Israel. Within four years there were 3,300 Jews in Kiryat Shmonah; within twenty years the population had risen to more than 15,000.

During the early years of statehood, the purchase of arms abroad was a high priority. One area in which Israel felt vulnerable was at sea. The Egyptian naval advance towards Tel Aviv during the War of Independence had nearly led to disaster. To rectify Israel's weakness in this area, Shimon Peres, then aged twenty-six, was given the responsibility of purchasing torpedo boats, corvettes and submarines. These were acquired in Canada. Ceremonial aspects of a nation's armed forces had also to be catered for. 'We even managed,' Peres has written, 'to furnish the officers with silver-plated insignia of rank, no mean feat in those days'.

When an earthquake struck Greece, Israel's fledgling navy was able to join in the assistance sent by the other navies of the Mediterranean. This help led to an episode which, ever since, has raised a smile on the lips of those Israelis who remembered the British Mandate. As Peres recalled:

> When the King of Greece held a reception for all the visiting officers, our men were able to turn out proudly in their dress uniforms and shining metal epaulettes. At one point, a senior British officer approached them. 'What country are you from?' he asked, looking uncertainly at their resplendent uniforms.
>
> 'Israel.'
>
> 'Where the hell's that?' asked the distinguished Englishman.
>
> 'Israel is what used to be Palestine,' our officer replied.
>
> 'Ah yes, Palestine,' the Royal Navy man replied. 'Are the Jews still making trouble there?'

A new country had many institutions to create, and also many diplomats to find for the overseas postings crucial to the representation of the needs of statehood. A British Jew, Walter Eytan, who had been educated at Oxford University and who served during the Second World War in one of the most sensitive areas of British Intelligence (Signals Intelligence, dealing with the Ultra secret code) had established a school for diplomats in Jerusalem in the last year of the Mandate. When statehood was proclaimed, a Foreign Office had been set up and ambassadors appointed. So tiny was the budget of the new service that it is said it did not have the money to telegraph to all the capitals in the world that Israel had been born.

Those who became Israel's first diplomats had often been active in some form of representation in the pre-State years. Thus in 1950 Russian-born Avraham Nissan was appointed Israel's Minister to the Scandinavian countries. Thirty years earlier (as Avraham Katznelson) he had been head of the Palestine Office in Constantinople. Russian-born David Hacohen, who had been taken to Palestine by his parents in 1907, at the age of nine, and had been the Haganah liaison officer with the British army in the Second World War – and one of those interned by the British at Latrun on 'Black Saturday' in 1946 – was Israel's first ambassador to Burma. The Polish-born Chaim Pazner, a former head of the Palestine Office in Geneva during the Second World War, served as Israel's representative in Argentina (four decades later his son Avi served as Israel's ambassador first in Rome and then in Paris). Yitzhak Chizik, a brother of Sarah and Ephraim who were killed in the riots of 1920 and 1929, respectively, and whose family had been among the agricultural pioneers, served as Israel's chargé d'affaires in Liberia and then as consul in Chicago.

On the forty-sixth anniversary of the death of Theodor Herzl, 5 July 1950, the Government of Israel chose to enact what is arguably its single most important piece of legislation, the Law of Return. The law guaranteed the right of every Jew, wherever he might live in the world, to enter Israel as an immigrant and to become a citizen immediately on arrival. Three previous documents were referred to in the preamble as providing a precedent: the Basle Programme of 1897, Article Six of the Mandate for Palestine of 1922, and Israel's own Declaration of Independence of 1948. There was also the precedent of the first legislative act of the Government of Israel in coming into existence, which had abolished all restrictions on Jewish immigration. The new law, Ben-Gurion explained, 'lays down not that the State accords the right of settlement to Jews abroad, but that this right is inherent in every Jew by virtue of his being a Jew, if it but be his will to take part in settling the land'. The right of settlement, Ben-Gurion added, 'preceded the State of Israel; it is that which built the State.'

Under the Law of Return, those judged to be a danger to public health or security were excluded. This exclusion was extended four years later to include 'a person with a criminal past, likely to endanger the public welfare', which led to the denial of entry to Meyer Lansky, one of the notorious American gangsters of the inter-war years. A later judgment of the Supreme Court (1962) stated that the Law of Return did not apply to a person who although born a Jew had subsequently converted to Christianity. This decision followed the attempt of a Polish-born Jew, Samuel Rufeisen, who had survived the Holocaust in White Russia, and who subsequently converted to Roman Catholicism and became a Carmelite monk, to be allowed to become a citizen on his arrival. He was refused citizenship, but was allowed to live in Haifa, where he lives, as Father Daniel, to this day.

Under the Law of Return, not only individual Jews and families, but whole communities, made their way to Israel, including virtually all the Jews of Bulgaria and the Jews of Yemen; and, after the fall of Communism in 1991, the Jews of Albania.

In contrast to the Law of Return, the refusal of Israel to take back the Arab refugees displaced in 1947 and 1948 created a source of friction in international relations, with Israel under pressure to take the refugees back. But more than 300 of their villages – including several dozen in the Jerusalem Corridor – had been deliberately and systematically bulldozed to the ground, and many of their homes elsewhere – including the former Arab suburbs of Jerusalem – were being lived in by immigrants, many of them Jews from Arab lands.

In 1951 John B. Blandford, the director of the United Nations Refugee and Works Administration, UNRWA, announced a plan to spend $100 million in twelve months to resettle between 150,000 and 250,000 of the Arab refugees in various Arab countries. The Arab States rejected Blandford's plan, and he resigned. While Israel struggled to absorb the Jewish immigrants into Israeli society, of the Arab States with Palestinian Arab refugees only Jordan (as Transjordan had been generally known since June 1949) granted them full citizenship; they gradually and successfully participated in Jordanian life, holding some of the most senior offices of State.

Since the early 1980s, it has been assumed that in some future peace negotiations both sides to the Arab–Israel conflict will present their compensation demands with regard to the refugees of 1948. To this end, the Israeli-based Association of Jews from Arab Countries has drawn up its own evaluation of abandoned Jewish properties in Arab countries, and has presented it officially to the Government of Israel. According to its calculations, the value of the properties and possessions lost by the Palestinian Arab refugees is in no way more than the value of all that was lost by Jews fleeing from Arab lands.

In 1951 the Government of Israel focused its attention on the vast area of land running south of Beersheba to the tiny port of Eilat. It constituted a formidable desert, the Negev, replete with desolate stretches of wilderness, deep canyons, almost barren scrubland, hidden streams and rock pools. Shimon Peres later reflected, 'The Negev is the future of Israel, and Ben-Gurion had to decide: either we shall overcome the desert, or the desert will overcome us.'

On its eastern edge, the Negev ran along the border with Jordan. It was there, in the Arava Valley – on a stretch of road more than 100 miles long with only three freshwater wells – that the Ein Yahav experimental farm had been established in 1950. A year later the government established a second settlement along the border. Twenty-six miles north of Eilat, it was called Yotvata after a biblical location referred to in both Numbers and Deuteronomy. The first settlers at Yotvata lived in simple huts. Most of them were soldiers – young men and women – serving in Nahal. Many of them were Israeli-born graduates of the youth movements.

Well-shafts sunk at Yotvata had revealed ample supplies of fresh water. This was essential in order to wash the salt out of the soil, a painfully slow process. Once the soil had been desalinated, and thus effectively sterilized, special crops had to be planted to provide it with nutrients. Only then could commercial crops be planted.

Like Ein Yahav, Yotvata had no frost. Crops could be planted and harvested in the Arava Valley while elsewhere in Israel they were still being tended. Seasonably early vegetables and flowers could command good prices in the markets of Tel Aviv and Jerusalem. By exporting to northern Europe, even higher prices could be obtained. Not only out-of-season vegetables and flowers, but dates and tropical fruits, grew in the irrigated strips within a few yards of the sandy, saline soil of the desert that lay around. An experimental agricultural station was founded; in due course this would be of benefit to other settlements in the area. Israeli archaeologists also had a part to play in the desert region; modern agricultural techniques developed in the Negev included the restoration of ancient Nabatean farming methods, in particular for the cultivation of fruit trees. Today, the milk products and chocolate of Yotvata make it an almost compulsory stopping place for Israeli families driving to Eilat.

Another border area in which a Nahal group established a kibbutz in 1951 was on the eastern edge of the Gaza Strip. Named Nahal Oz (from the same root – 'oz' (strength) – as the Hebrew word for Gaza) the soldier-settlers were later joined by Jewish youth from South America. From the kibbutz gate can be seen the minarets of Gaza City.

Immigrants continued to be settled in villages from which the Arabs had fled in 1948. One of these, just over three miles east of Lod, was al-Haditha. In 1951 sixty Yemenite immigrant families were located here. They called their settlement Hadid, the name of one of the towns mentioned in the Bible to which the Babylonian exiles returned. It was also a city in Hasmonean times (and appears on the ancient mosaic map in the Jordanian town of Madeba). Most Yemenite immigrants were brought in 1951, as a first step, to a large, abandoned British army barracks at Rosh Ha-Ayin, which had guarded the prewar water pumping station for the pipeline up to Jerusalem. While most of the new immigrants then went elsewhere, some remained in Rosh Ha-Ayin, where they became farmers.

That same year, immigrants from Kurdistan in northern Iraq who had found work in the stone quarry just to the west of Jerusalem – the quarry belonged to the building cooperative Solel Boneh – were housed near the deserted Arab village of Kastel, the scene of fierce fighting in 1948. Their settlement was called Ma'oz Zion (Stronghold of Zion).

Another town that was being built in 1951 was Or Akiva (The Light of Rabbi Akiva), named after one of the great Jewish sages of ancient times. The purpose of the new town was to house the immigrants who were then living in a vast immigrant camp on the sand dunes at Caesarea – a Roman port city with a magnificent ancient theatre. Carpet and silk weaving were the main productive enterprises in the town.

New immigrants, mostly from Roumania, Iraq and North Africa, were settled in 1951 at a specially constructed rural settlement, Yeruham, in the northern Negev hills. The site had been used during the British Mandate period for the drilling of oil, but without success. When the farming failed, through lack of water and poor soil, the immigrants took to road building, and the mining of the glass sand and ceramic clays which were to be found in the nearby Great and Ramon craters. Instead of a farming community, an urban community came into being. There was considerable unemployment, and for several years the people of Yeruham lived in temporary wooden huts.

Since 1927, the largest group of kibbutzim had been the left-wing *Hakibbutz Hameuhad* (United Kibbutz), most of whose members belonged to Mapai. In 1951 there was a split in the movement. The Mapai members had grown disillusioned with Soviet Communism, and were distressed by the growing anti-Jewish policies being pursued by Stalin. Mapam still retained a faith in the Soviet system. The two factions decided to go their respective ways. This decision led to the actual physical splitting of many settlements between Mapai and Mapam members.

In 1951 Ben-Gurion visited the United States. His visit was private. He was never, in fact, to be invited officially to Washington at any time in his premiership, a reflection of the distance that the United States establishment, and in particular the State Department, kept from Israel in the early days of statehood when American interests seemed to be most closely aligned with Saudi Arabia and the Muslim oil-producing nations of the Middle East.

While in the United States, Ben-Gurion went to see the Israeli air force students who were training in California. He also visited a company manufacturing aircraft parts, which had been established by Al Schwimmer, a former American volunteer in the Israeli War of Independence. For two years Schwimmer had been converting whatever spare parts and metal 'junk' that he could acquire into operational aircraft, and also working to improve aircraft design.

To rebuild old Mustang and Mosquito aircraft that had been bought as war surplus, Israel made use of the facilities at Fairbanks, California. These aircraft were then flown to Israel by American Jewish volunteers, via Europe. Shimon Peres later recalled, 'One of these Mosquitoes got lost and I headed a group that went on to search for the plane and the pilot. We spent seven days and seven nights watching with blinding eyes, without any success.' It was then that it was decided to build an aircraft industry in Israel itself.

Chaim Herzog, who had visited Al Schwimmer's factory a year before Ben-Gurion's visit, and who had accompanied Ben-Gurion to it, later wrote, 'Shimon Peres, always a man of vision, was with him and set out to convince Ben-Gurion to move the entire group to Israel. He had visions of Schwimmer's group becoming the nucleus of an aircraft industry that would be the most important between the Far East and Western Europe. Peres was attacked, ridiculed and criticized, but Ben-Gurion believed him. Israel Aircraft Industries is now a $1.5 billion industry. We build our own fighter planes and space satellites and are self-sufficient in many aspects of the aircraft industry.'

In addition to building aircraft, the Israel Aircraft Industries undertook military research and development which trained generations of scientists, engineers and technologists, and served as a basis for the electronically based high technology of the 1980s and 1990s.

Israel was always dependent for its security on the stability and caution of the Arab and Muslim States around it. In the 1950s this stability was under constant threat. On 7 March 1951 the Prime Minister of Iran, General Ali Razmara, was shot and killed by an Islamic fundamentalist. On 15 July 1951 the Lebanese Prime Minister, Riad Bey e-Solh, was assassinated in Beirut by a member of a Syrian extremist nationalist group that wanted Lebanon to be an integral part of Syria. While attending a memorial service for e-Solh at the al-Aksa mosque in Jerusalem on July 20, King Abdullah of Jordan was himself assassinated. His assassin was a Palestinian. The King was accused by Palestinian extremists of wanting to come to an amicable arrangement with Israel. In his memoirs, published in 1946, the King had written, 'I was astonished at what I saw of the Jewish settlements. They have colonized the sand dunes, drawn water from them, and transformed them into a paradise.' His most recent statement, that angered many Palestinians, was that the Jews had rights to their Holy Places in the Old City. These rights had been an integral part of the Armistice Agreements of 1949. They were never honoured, beyond the fortnightly convoy to the Mount Scopus enclave.

King Abdullah's grandson, the seventeen-year-old future King Hussein, personally witnessed the assassination of his grandfather, and was to refer to it in moving tones more than forty years later, when he attended, also in Jerusalem, the funeral of the assassinated Israeli leader Yitzhak Rabin.

In Egypt, King Farouk, who had ruled since 1936, was deposed on 23 July 1952 by a group of army officers headed by General Muhammad Naguib and Colonel Gamal Abdel Nasser. Another Egyptian officer who took part in the coup was Lieutenant-Colonel Anwar el-Sadat. The anti-Western and pro-Soviet leanings of the new rulers boded ill for Israel. Egyptian control

of the Gaza Strip constituted a danger to Israel should Egypt decide to embark
on hostilities.

In Israel, the second General Election in the history of the State was held on
30 July 1951. It was to determine whether the predominantly secular, and
moderate Socialist, orientation of the State was to continue. In a nationwide
campaign, Ben-Gurion was the main speaker for Mapai, and the bluntest.
'You can't stop speaking Hungarian overnight,' he told a meeting of the
Hungarian Immigrants Association in Tel Aviv on July 21, 'but you should
think of yourselves in the first place as Jews and citizens of Israel.' Mass
immigration, Ben-Gurion told them, 'would not cause unemployment,
because immigrants brought with them their creative abilities. The lag
between present needs and rising production would be overcome with the
help of Jews in the Diaspora.' Levi Eshkol, Ben-Gurion's Minister of Finance
from 1952 to 1963 (and himself an immigrant from Russia in 1914), also saw
the positive side to the economics of mass immigration, commenting that
'every immigrant brought in his suitcase jobs for two more'.
   The extraordinary nature of mass immigration was understood by all who
witnessed it and guided its steps. The Minister of Agriculture, Pinhas Lavon,
described it as 'a bloodless revolution, proceeding at a much lower price
than any revolution in history. There could be no slackening in the
Ingathering process.' Moshe Sharett, addressing a mass rally of new immigrants at a housing project near Rishon le-Zion, gave five separate addresses
in five languages – Yiddish, Turkish, Arabic, French and Hebrew.
   The result of the election was the retention of power by Mapai, the main
political Party of the Labour movement, with forty-five seats, a loss of only
one. Further to the left, Mapam secured fifteen seats, a drop of four, and the
start of what was to be a steady decline. The General Zionists, fighting on a
non-Socialist platform, obtained twenty seats, almost trebling their number
in the first Knesset (seven) and raising hopes, among themselves at least,
that they might one day be able to challenge Mapai as coalition builders.
   The right-wing Herut, whose representation dropped from fourteen seats
to eight, remained excluded by Ben-Gurion's insistence from any place
inside the coalition, as did the Communists, whose seats rose from four to
five. The United Religious Front, which had gained sixteen seats in the first
Knesset, broke up into its various component parts. Agudat Yisrael (established in Germany in 1912) gained three seats. The other religious parties
between them could only muster ten seats This blow to the religious bloc
led to an eventual reorganization of the religious parties under a single
banner, the National Religious Party (NRP) in time to fight the next election (under a briefly different name, the National Religious Front), and to
secure the political influence on the basis of which to obtain the
sectional patronage for their religious institutions. It was Labour's boast that
there was a 'historic alliance' between itself and the religious Zionists; it

was an alliance that was to last for almost three decades.

The philosophy of Agudat Yisrael, the principal religious Party, was expressed many years later to Shimon Peres (who frequently had to negotiate with it during coalition building) by its spiritual leader, Rabbi Shach: 'What is a Jewish State? Every nation has a State. What is a Jewish Agriculture? Every nation has agriculture. What is a Jewish Army? Every nation has an army. What is the Jewish Industry? Every nation has an industry. There is just one thing that the Jews have that the others don't, and that is the Torah. The Torah keeps the Jews alive.'

In later years Ben-Gurion reflected on the proliferation of Parties, which had survived and flourished during the first three years of statehood, and whose changing patterns (though he could not know it) were to be a feature of the next fifty years. 'The Parties', he wrote of the situation in 1951, 'had not yet overcome their unfortunate heritage of exile and a life of dependence; morbid and excessive fragmentation derived from a life devoid of independence and national responsibility. They had not yet surmounted traditions and habits detrimental of the life of the State. Not all of them understood that the State, far from making our lives easier, would impose upon its inhabitants heavy burdens and tremendous responsibilities such as they had not previously known.'

In forming his new government (which was presented to the Knesset on 7 October 1951), Ben-Gurion held fifty-five meetings with the different Parties, seeking to balance their needs and demands. This experience confirmed him in his advocacy of the British two-Party system, which he had seen at first hand before the Second World War, but which he was unable to introduce to Israel. The many smaller Parties understood then, as they do today, that their factional power could only be retained under the existing proportional representation system.

All efforts since Ben-Gurion's time to introduce electoral reform along two-Party or constituency lines have failed. The small Parties, who under any constituency system would be unable to secure a single seat, are able, under the Israeli usage of proportional representation, to get two or three seats (even one seat) which can give them a stranglehold on the process of forming a government, increasingly so as the main political Parties decline in popularity, and the smaller Parties proliferate. Their coalition-making (and coalition-breaking) powers make the smaller Parties all the more determined to extract political concessions. In the case of the religious Parties, this often takes the form of a diversion of State funds to their own institutions, out of proportion to their size or needs.

Among the political Parties represented at the establishment of the State was Agudat Yisrael. Founded before the First World War as a movement to preserve Jewish Orthodoxy by the very strictest adherence to religious laws, the Party had for many years been hostile to Zionism, stressing that it was only by divine agency that Jewish statehood could be achieved. Before the

Second World War its followers in Palestine had been deeply opposed to the Zionists. In 1935, however, as a result of pressure from its headquarters in Poland, it had formed its own agency to deal with immigration and absorption. The Polish and Central European communities that had formed the heartland of Agudat Yisrael were destroyed in the Holocaust. Shortly after the outbreak of war, the leader of Agudat Yisrael in Poland, the Hasidic rabbi of Ger, arrived in Palestine, and cooperation began with the national Jewish institutions.

The Agudah never swerved from its ideological non-Zionism, and even anti-Zionism. With the exception of Ben-Gurion's first Cabinet, it has never agreed to the appointment of a Cabinet Minister of its own, even though being a member of a coalition. To be a Minister would mean taking the oath of office to a secular State (in the Netanyahu government of 1996 an Agudah leader, Rabbi Menachem Porush, became Deputy Housing Minister, thus avoiding the need to take the oath, but was granted 'full Ministerial powers' in order to satisfy his, and his Party's, political demands).

From the moment of the establishment of the State, Agudat Yisrael was represented in all national and local organizations, and in the Knesset. In Israel's first government the Party leader, Rabbi Y. M. Levin, was made Minister of Social Welfare, a position he held for the first three years of statehood. From the outset, Agudat Yisrael made every effort to obtain the observance of Jewish religious principles and practice in Israeli public life. When the government introduced (in 1953) a unified school system and free national education, Agudah Yisrael established an independent school system of its own, which today accommodates about a third of the school population. Because of Agudah pressure, Yeshiva students, as well as other ultra-Orthodox youngsters, were excused, and remain excused, from national military service.

The architect of the unified school system was Benzion Dinur. Lithuanian-born, he had settled in Palestine in 1921 and later served as head of the Jewish Teachers' Training College in Jerusalem. A historian of distinction, he had been elected to the first Knesset on the Labour (Mapai) list, and in 1951 was appointed Minister of Education and Culture. In bringing to an end the hitherto Party-based educational system, he sought to ensure that no one Party, and no one grouping, whether religious or secular, could distort the evolution of national unity. It was to take forty-six years before the religious school system, set up to satisfy a religious minority, was to find itself part of a national education system controlled – as a result of electoral considerations – by the religious Parties.

Among the settlements founded in 1952 was Orot (Lights), in southern Israel. Its members came from the United States. The cultivation of field and garden crops, and the raising of dairy cattle, provided their livelihood. Further south in the central Negev hills, eleven miles south of the new settlement of

Yeruham founded a year earlier, the pioneer outpost of Sde Boker (Cattle Rancher's Field) was founded in 1952. Surrounded by desert, and on the edge of the spectacular canyon of the Wilderness of Zin, its first settlers were mostly *vatikim* – those who had lived in the country for many years, working the land elsewhere.

Sde Boker began as a horse-breeding community. Later, sheep were bred. Slowly, the desert soil was reclaimed and orchards planted. Ben-Gurion, who had always looked to the Negev as pivotal to the growth of Israel, was much attracted to Sde Boker and was to make it his home within a year of its establishment.

Throughout the first four years of the State, the President, Chaim Weizmann, had lived at Rehovot, in the grounds of the scientific institute that bore his name. The Prime Minister had not had the time or inclination to take Weizmann into his confidence, or even to keep him in touch with the inner activities of the government. One of the few people at the centre of affairs who did take the trouble to go down to Rehovot and to give Weizmann an account of State policy was Yigael Yadin, who, after a successful military career culminating as Chief of Staff, was about to turn to archaeology.

Weizmann felt isolated at Rehovot, and his thoughts were at times tinged with bitterness. 'He chafed', wrote Abba Eban, 'at his own inability to impress the new society with his own message of social progress, intellectual integrity, aesthetic refinement, and manifest dedication to peace. Israel had been born in violence and conflict; it continued to live an embattled existence.' There were times, Eban continued, 'when Weizmann was seized by a poignant concern for Israel's inner quality; but whenever he fell into doubts and regrets he looked through his window at Rehovot upon the verdant rolling plains and rich orange groves surrounding the scientific laboratories established under his inspiration. On a clear day his gaze would go as far as the Judaean Hills. The landscape in between was dotted with villages and townships indicative of the new impetus given to Jewish national vitality. And then a deep contentment would come upon him, and his mind would become serene, as befitted a man who to a degree unshared by any figure in contemporary history had seen an improbable vision translated, largely through his own effort into vibrant and solid reality.'

After a long and painful illness, Weizmann died on 9 November 1952, a week after the Balfour Declaration anniversary – the thirty-fifth – when the thoughts of those for whom Jewish history lived had always turned to him.

CHAPTER SIXTEEN

Conflicts and achievements
1952–1955

Throughout 1951 the governments of Israel and Germany had discussed an agreement whereby the German Federal Republic (West Germany) would pay reparations to Israel, for the persecution of the Jews by Germany during the Second World War. The agreement had been principally negotiated on the Jewish side by Nahum Goldman, the New York based Chairman of the Praesidium of the Conference on Jewish Material Claims against Germany, known as the Claims Conference. It was opposed by a small but vociferous group in Israel, led by Menachem Begin and his Herut Party, who wanted to have no dealings at all with Germany. Ben-Gurion persevered; he felt strongly that the Germany with which he was dealing was very much a different Germany from that which had embarked on war and persecution only thirteen years earlier.

On 7 January 1952 Ben-Gurion presented the agreement for ratification by the Knesset. To a crowded chamber – in the building in King George V Street (Froumin House) that was being used until a permanent Knesset building could be constructed – he reviewed the history of Israel's approach to the four occupying powers in Germany – the United States, Britain, France and the Soviet Union. The Soviet Union, he said, had not replied, nor had any indication had been received about the view of the East German government. The three Western Powers, he went on, had expressed support for the principle of reparations, but said that they were bound by existing treaties not to make additional claims upon Germany either for themselves or on behalf of others.

Ben-Gurion went on to explain that 'under the pressure of public opinion and after friendly intercession by official British circles and others', the West German Chancellor, Konrad Adenauer, had written to him a few weeks earlier, on behalf of his Government, to say that they were ready to discuss 'with the State of Israel and Jewish representatives the question of reparations on the basis of the claims set forward by Israel in her note of 12 March 1951'. The Government of Israel, Ben-Gurion continued, regarded itself 'as bound, together with representatives of world Jewry and without undue delay to make every effort to restore as quickly as possible the maximum indemnification to individual Jews, and to the Jewish people'. Ben-Gurion added, in a final defence of his desire to accept reparations, 'Let not the murderers of our people be also their inheritors' – a reference to the biblical rebuke (1 Kings 21:19), 'Hast thou killed and also taken possession?'

The decision to accept reparations from Germany was attacked by three Opposition speakers, two of whom had lost their parents in the Holocaust. Elimeleh Rimalt, of the General Zionists, argued that 'as Hitler's extermination of the Jews had not been logical, so was the current issue not one of logic but of emotions', and he went on to declare that the argument that murderers should not inherit 'did not apply here because the Nazis had not murdered for the legacy'. He then quoted his small son as asking him, 'What price will we get for grandpa and grandma?'

A Mapam speaker, Yaakov Hazan, said that reparations could be justified 'if it were possible to restore the property', or if the Germans themselves had destroyed the Nazis. The whole issue, he said – speaking from the perspective of his small but vigorous left-wing Party – was 'a trick by the Western powers designed to facilitate the grooming of Western Germany as the spearhead of new hordes to attack Russia'.

As the Knesset debate continued, Menachem Begin – who until that day had defiantly refused even to take his seat in the chamber – and the former Herut candidate for the Presidency of the State of Israel, Joseph Klausner, addressed a large gathering in Zion Square, from the balcony of the Tel Aviv Hotel. The *Jerusalem Post* reported:

> Mr Begin spoke with emotion, frequently shouting, interspersing his words with many biblical quotations. He referred to the Government statement in support of German reparations discussions as the culmination of the policies of 'that maniac who is now Prime Minister'.
>
> Midway through his harangue, Mr Begin pulled a note from his pocket, held it aloft dramatically and said: 'I have not come here to inflame you; but this note which has just been handed to me states that the police have grenades which contain gas made in Germany – the same gas which was used to kill your fathers and mothers. We are prepared to suffer anything, torture chambers, concentration camps and subterranean prisons – so that any decision to deal with Germany will not come to pass.'

'This will be a war of life and death,' Begin told his supporters. 'Today I shall give the order: "Blood!"'

The gathering in Zion Square ended with the singing of the National Anthem. Then, the *Jerusalem Post* reported:

> . . . groups of youths led the march up Ben-Yehuda Road in the direction of the Knesset building. A number carried haversacks loaded with stones. Many openly bragged that they had come from Tel Aviv and Haifa and had brought 'our arms with us'.
>
> Earlier in the day, police had cordoned off a large section of the city's centre, running from Jaffa Road to Terra Sancta College. Barbed wire concertinas blocked the roads, and bus routes were temporarily changed. Pedestrians with business in the area were permitted to pass the barriers, although they were kept away from the immediate Knesset environs. Heavy detachments of police, estimated to number over 600, patrolled the cordoned area. Most were armed with shields, batons, steel helmets and gas-mask kits.
>
> The lower barrier on Ben-Yehuda Road at the corner of Rehov Hapo'alim was broken through in short order, with little police resistance apparent. As the crowds of demonstrators swelled, however, and violence became evident, groups of police went to the roofs of the nearby buildings and lobbed down tear-gas bombs in an attempt to disperse the mob.
>
> District Police Superintendent Levi Avahami reported, however, that the tear-gas was first employed by the demonstrators who, by this means, were able to break past the first barrier. The light wind wafted the gas into the faces of the police, away from Ben-Yehuda Road, and in the direction of the Knesset.
>
> The shrieking sirens of Magen David Adom ambulances, the billowing clouds of tear-gas and the ring of pistol shots fired by police above the heads of the mob soon gave the area the semblance of a street battle.

It was a traumatic moment, with the passions of Begin's supporters high, and with the violence in the street again raising the grim spectre of Jew against Jew in the Jewish State. The *Jerusalem Post* account continued:

> As road blocks were removed forcibly by the marchers, the police, who had been ordered to observe extreme self-restraint in dealing with the demonstrators, fell back to positions around the Knesset. The crowd showered the police with stones; and even Magen David Adom ambulances rushing first aid to the injured were stoned and halted. Attempts were also even made to drag out the injured.
>
> During the first hour of the demonstrations, thirteen policemen were injured among them Deputy Superintendent Israel Maiber and Assistant Superintendent Moshe Ayalon, who sustained a serious eye injury from a flying rock.

A car parked outside the Knesset was overturned by the demonstrators. As the gasoline poured out of the tank, a tear-gas bomb apparently ignited it. The blaze was put out by the Jerusalem Fire Brigade. A number of cars parked along King George Avenue, between Jaffa Road and the Jewish Agency compound, were also damaged.

At this point no one was permitted to enter or leave the Knesset building, which was in fact in a state of siege.

Among those who treated the wounded was Ben-Gurion's wife Paula, who had been a nurse before the First World War. As violence continued in the street outside, the scene inside the debating chamber became chaotic, the *Jerusalem Post* reporting:

> The shouting of a mob not far off, the intermittent wail of police cars and ambulance sirens, sporadic explosions of gas grenades and the glow of flames from a burning car came through the windows of the Knesset building, and later the window panes were splintered by rocks, and fumes of tear-gas bombs from the battle-scarred street outside permeated the chamber. One member was hit in the head by a stone.
>
> Through all this disturbance, the meeting went on. The section of the hall where stones and glass splinters fell . . . was vacated and members stood around elsewhere.

Begin had made his way from Zion Square to the Knesset which he then addressed. As he was speaking, stones continued to come through the windows. In answer to Ben-Gurion's accusation that the 'wrath of the people' had been stage-managed, Begin read out a list of rabbis, scholars and poets who had signed a petition opposing negotiations with Germany. Ben-Gurion, who, the *Jerusalem Post* reported, 'had been remarkably quiet throughout, rose from his place and pointed to the windows. "They are not identified with your hooligans in the street", he said.'

Begin then called out to Ben-Gurion, 'You are a hooligan.' This, the *Jerusalem Post* noted, 'had not been the worst epithet hurled across the floor of the house'. Not only did Begin refuse to withdraw his words, but he then 'declined to leave the platform when ordered to do so by the Deputy Speaker amidst an uproar'. The newspaper account continued:

> This marked Mr Begin's return to the political scene from his semi-retirement which followed the setback to his party in the last Knesset elections. It was his first appearance in the Second Knesset, and in fact, he took the oath of office only some two hours before he began his speech.
>
> After the recess, Mr Begin returned to the platform and apologized.

He added that he was waiving his Knesset immunity and that this would be his last appearance in the Knesset, and made what most listeners thought was a threat to go underground if an attempt was made to negotiate with Germany.

'Some things are dearer than life,' Begin declared. 'Some things are worse than death. We are willing to leave our families, and die. People went to the barricades for lesser things. I know we will be dragged to concentration camps. We will die together.'

This extremist rhetoric could neither deter Ben-Gurion in his course nor affect the balance of political and voting power in the Knesset. After a ten-hour debate, the principle of reparations was accepted, and negotiations with Germany continued. They reached their conclusion on 11 September 1952 in Luxembourg when the Bonn government agreed to pay, and Israel agreed to accept, 3,450 million marks ($865 million) 'as reparation for the material damage suffered by the Jews at the hands of the Nazis'. The preamble to the agreement read:

> Whereas unspeakable criminal acts were perpetrated against the Jewish people during the National Socialist regime of terror
>
> and whereas by a declaration in the Bundestag on 27 September 1951, the Government of the Federal Republic of Germany made known their determination, within the limits of their capacity, to make good the material damage caused by these acts
>
> and whereas the State of Israel has assumed the heavy burden of resettling so great a number of uprooted and destitute Jewish refugees from Germany and from territories formerly under German rule and has on this basis advanced a claim against the Federal Republic of Germany for global recompense for the cost of the integration of these refugees,
>
> now therefore the State of Israel and the Federal Republic of Germany have agreed . . .

The Luxembourg signing ceremony was a remarkable one. The hostility of extremists in Herut caused it to be held in secret. The *Jerusalem Post* reported: 'The two delegations sat opposite each other across a long table, Foreign Minister Moshe Sharett and Chancellor Konrad Adenauer facing each other in the centre. Mr Sharett had seven advisers and Dr Adenauer five. Before entering the room, the two Ministers met briefly and shook hands. The ceremony, which lasted thirteen minutes, was conducted in complete silence. Not a single word was exchanged across the table until, the signing over, Dr Adenauer leaned forward and suggested to Mr Sharett that they should rise.'

Following the signing of the agreement, Adenauer and Sharett were joined by Nahum Goldman, and two further protocols were signed. The first dealt with the extension of compensation and restitution legislation for individuals then in force in Germany. Germany undertook 'as soon as possible' to take all steps within her power to ensure the implementation of principles applying to compensation and restitution and to carry out the entire programme in not more than ten years. The second protocol provided for the payment of 450 million marks to Israel for the benefit of the Claims Conference.

The *Jerusalem Post* noted: 'A small group of reporters representing world news agencies, who had been conveyed to the building in official cars without being told of their destination, stood at the end of the room to witness the signing. Stringent security measures were imposed throughout the ceremony to meet a possible threat to the lives of the two statesmen. After signing the treaty, Mr Sharett, Dr Adenauer and Dr Goldman talked in a small room for twenty minutes. No one else was allowed in the room which was guarded by armed detectives.'

That same day, in New York, an agreement was signed by Abba Eban and representatives of the Claims Conference whereby Israel would pay to the Claims Conference a proportion of the German reparations payments – about 15 per cent of the total sum – for the 'relief of victims outside Israel'. Speaking after the signing, Eban described the reparations payments as 'a moral victory, a victory of conscience over the dictates of brute force', and he added, 'The Germans, who wanted to wipe the Jews off the face of the earth, are now signing a contract of compensation with the Jewish State.'

In a broadcast over Israel Radio, the Minister without Portfolio, Pinhas Lavon, noted that the benefits the nation could draw from the $715 million in reparations to be paid directly, and the further $110 million given to world Jewry, were as great as the $800 million that Israel had received in the previous four years from all sources – appeals, private and national loans, grants, private investments, the property of new immigrants, and free imports. 'With the latter sum,' Lavon pointed out, 'we have fought and were victorious in the War of Liberation; we re-equipped our Defence Forces; we brought 650,000 Jews from different countries and assured accommodation to hundreds of thousands; we established hundreds of agricultural settlements and equipped older settlers with all needed equipment; we built new cities in various parts of the country; we laid pipelines and built irrigation works on hundreds of thousands of dunams; and we enlarged Haifa harbour and built Kishon port.'

Lavon stressed that the goods to be received from West Germany under the reparations agreement would further the process of development. 'They would permit us to exploit the Negev mines, broaden agriculture, develop transportation, shipping and fishing, increase electrical output, develop basic industries and build homes for ma'barot residents. The people who wanted to exterminate us,' Lavon said, 'is forced to bear some of the burden involved in creating a new Jewish centre of strength and a place of rebirth.'

In 1952, as in future years, the West German policy-makers did not hesitate to acknowledge a 'special responsibility' in their dealings with the Jewish State. The first reparations payments took the form of fuel oil, bought with West Germany's Sterling balances, and sent to Israel within weeks of the signing ceremony.

Most Israelis realized how important the reparations agreement was to the economy and safety of Israel; but still survivors and non-survivors alike felt the pain of the Holocaust even more during the reparations debate because of the very notion that money could in any way compensate for the suffering. In its own gesture of monetary justice, on October 9 the Government of Israel agreed to release the blocked accounts of Arab refugees in Israeli banks.

Only six weeks after the signing of the reparations agreement in Luxembourg, Israel was plunged into another crisis relating to the Jewish past and the position of Jews in the Diaspora. Once more, the small State felt linked, by race, religion, history and national sentiment, to the plight of Jews elsewhere in the world. The country concerned was Czechoslovakia which since 1948 (at the time of helping the new Israel with arms supplies) had been under a strict Communist regime. A number of political leaders had been brought to trial and condemned to death. As Moshe Sharett explained to the Knesset on 24 November 1952, the majority of the accused were Jews, 'and the prosecution has spared no pains in highlighting their racial origin and in tracing their alleged crimes to this prime cause'. The indictment, the court proceedings themselves, as well as the publicity given to the trial in the official Czechoslovak press, 'are all permeated with a spirit of rabid anti-Semitism'. Sharett went on to explain:

> The prosecution has unfolded a dark screed of criminal plotting, of acts of conspiracy and subversion that these Jewish enemies of the Czechoslovak people have either perpetrated or sought to perpetrate.
>
> Not satisfied with impugning their Jewish origin, it has denounced them as Zionists. Men who had never had any connection with the Zionist movement – some of whom, indeed, have persecuted Zionism with a vengeance – have themselves been branded with that stigma.
>
> The Zionist movement as a whole, that movement of liberation and of return to the Homeland, has been smeared and slandered as a band of intriguers and spies, bent on undermining the very foundations of the Czechoslovak regime, seeking the ruin of that country, and dishonestly exploiting the property of its citizens for its own nefarious purposes. The attempts of the Jewish survivors of the Nazi inferno to recover a tiny fraction of the immense quantity of Jewish property that had been plundered, have been denounced as acts of deceit and robbery.

The principal accused was Rudolf Slansky, former Secretary of the Communist Party of Czechoslovakia, a Jew, and a committed anti-Zionist who had been opposed to the sale of arms to Israel at the time of the War of Independence. At his trial in Prague, Slansky was accused of 'Trotskyite-Titoist-Zionist' activities. When, in the Knesset, a Communist member, Meir Wilner, echoed the Communist line by calling Slansky one of a 'gang of spies and traitors', he was loudly booed.

The Czechoslovak government had also accused Israel itself of having conspired with the 'enemies of Czechoslovakia' in order to destroy the Czech economy and undermine her security. Two Israelis, Mordechai Oren, a Mapam leader, and Shimon Ohrenstein, were arrested in Prague, and 'admitted' Zionist espionage activities. 'The Government of Israel,' Sharett declared, 'holds it utterly superfluous to attempt any detailed and factual denial of the tissue of libels and fabrications regarding activities of its members and emissaries, produced by the fertile imagination of the Czechoslovak Secret Police and Public Prosecution,' and he went on:

These slanders are self-contradictory in the light of simple reason. Their falsity is obvious to the naked eye, refuted by patent facts. Israel has always entertained a sincere sympathy for the Czechoslovak people.

Israel has sought to establish and foster friendly relations with the present Czechoslovak state. Israel obtained valuable aid from Czechoslovakia during her War of Independence for which she paid in full. This was arranged with the full knowledge and authority of the heads of the Czechoslovak state, some of whom still occupy high seats of power.

At one stage Israel concluded a commercial agreement with Czechoslovakia, upon terms that proved suitable to both parties. Israel admitted thousands of Jews officially authorized to leave Czechoslovakia after they left most of their property behind them.

Israeli representatives had 'never served as agents or spies of any foreign power', Sharett added. 'Only those to whom espionage and sabotage come naturally as a matter of daily practice can invent such stories of sinister plotting about the ministers of Israel.'

In the mid-1950s, Oren and Ohrenstein were released and returned to Israel. A decade later the Czech government officially exonerated those who had been indicted. A heavy burden had been inflicted throughout the crisis on the Israeli Ambassador to Czechoslovakia, Ehud Avriel.

The plight of Jews outside Israel was to remain a central feature of the concerns of its citizens, and of the activities of the State. One area which was of particular concern to Ben-Gurion was that of the Jews of the Soviet Union, numbering, it was believed, at least two million. Western experts on the Soviet system were convinced that it would never allow these Jews to express their Jewish identity, or to espouse Zionism, or to emigrate to Israel. But Ben-Gurion, who had been born in Tsarist Russia, and knew how deeply rooted Jewish traditions had been before 1917 (and in the Baltic States as late as 1940) was certain that links could be built up, and signals of support sent, and that one day the gates would open.

Immigration continued to be the great challenge facing the State throughout its first decade. Many of the immigrants came from countries where the way of life had been poor in the extreme. Many came from remote rural communities which knew nothing of twentieth-century urban life. There was considerable resentment among many immigrants at the extent to which life in the ma'abarot in the 1950s was dominated by Ashkenazi attempts to impose Socialist doctrines: the doctrines that were at the forefront of the ideology and practice of the government of the day. Yet places were found throughout Israel in which the new immigrants, leaving their ma'abarot, could settle down, and where they could preserve as much as possible of their original communal way of life, albeit often with enhanced agricultural methods. An example of this was the settlement in 1953 of a group of immigrants from the Yemen at Kfar Jawitz, about an hour's drive (in those days) north-east of Tel Aviv.

Named after the historian Ze'ev Jawitz – a member of the pre-First World War committee responsible for developing Hebrew as a modern language – Kfar Jawitz was originally founded as a moshav in 1932. It had suffered at the start from insufficient arable land and, in 1936, from Arab attacks. During the War of Independence it was in the battle zone. After the war, its members dispersed. The arrival of Jews from the Yemen gave the moshav a new lease of life. It flourished, with citrus groves, cattle and poultry the three main areas of activity.

The Nahal soldier-settlers also continued to create new settlements. In 1953 they established a kibbutz on the Dead Sea shore, at the remote and stunningly beautiful Ein Gedi, to which, four years earlier, another Nahal group had constructed a road from the south. The combination of the hot desert climate, and an abundance of fresh water from the perennial springs, made for successful farming. Bananas, tropical fruit and date-palms all flourished. But the kibbutz was on the Jordanian border, with all the anxiety which that entailed, and the need for vigilance.

Also founded in 1953 was the artists' colony of Ein Hod, on the slopes of Mount Carmel. The village was the brainchild of the Roumanian-born painter Marcel Janco, who had fled to Palestine from Roumania in 1941. Janco obtained the government's permission to set up his home, and to invite other artists to do likewise, in the abandoned Arab village of Ein Hod. The studios and galleries of the village soon became a tourist attraction.

The immigrant transit camps were slowly being closed down, and the immigrants sent to villages set up specially for them. One of these, established in 1953, was Hazor Ha-Gelilit, in Upper Galilee, near the ancient site of King Solomon's city of Hazor. The immigrants came from the nearby camp at Rosh Pinah. But, as also happened elsewhere, conditions in the new village were poor, and many of the immigrants left for the towns – in particular Tel Aviv and Haifa. It was to take twenty years before the fortunes of Hazor Ha-Gelilit improved, and a degree of prosperity arrived.

In January 1953 the Asian Socialist Conference met in Rangoon. The head of the Israeli delegation was the Foreign Minister, Moshe Sharett. Other than Burma, Israel was the only country in Asia where a Socialist Party was in power. Ben-Gurion was especially keen to develop good relations with the Asian countries. He had studied their national movements – India and Pakistan had achieved independence from Britain less than a year before Israel, Burma only five months before – and he had also studied the oriental religions.

Israel lies on the western rim of Asia. It constitutes a minuscule 0.005 per cent of the Asian land mass. When Israel became independent, Asia lay quite 'beyond our ken', as Walter Eytan expressed it. But Ben-Gurion saw beyond the Arab States that formed Israel's immediate neighbours, and looked to Asia as a natural hinterland. As a result of his efforts, Burma sent farmers to Israel to be trained in the modern agricultural techniques that Israel was pioneering. For its part, Israel sent officers and material to modernize the Burmese army. When the Burmese Prime Minister, U Nu, visited Israel in 1955 he received a rousing welcome.

Israeli diplomatic missions were opened in Burma, Japan, the Philippines and Ceylon. A goodwill mission was even sent to Communist China; the head of the mission, David Hacohen – Israel's Minister in Burma – was well received. But the growing closeness of the Soviet bloc to Egypt, and Egypt's own growing hostility to Israel, brought that eastern avenue of opportunity to an end.

On 19 August 1953 a law was passed in the Knesset that established a new institution, the aim of which was 'the commemoration in the Homeland of all those members of the Jewish people who gave their lives, or rose and fought against the Nazi enemy and its collaborators', and to set up 'a memorial to them, and to the communities, organizations and institutions that were destroyed because they belonged to the Jewish people'.

The new institution was given the formal title The Holocaust Martyrs' and Heroes' Remembrance Authority, but became known as Yad Vashem (a monument and a memorial), from the verse in Isaiah, 'Even unto them will I give within My house and within My walls a monument and a memorial . . . I will give them an everlasting memorial, which will never be cut off'.

The survivors were to find many havens in Israel. The remnants of once-vibrant communities formed associations in which they could meet. Some published memorial books about their home towns. Monuments were erected in cemeteries. Following the establishment of Yad Vashem, many thousands of survivors gave testimony. Others submitted the names of non-Jews who had saved them, and whose deeds were recognized by the State of Israel in a special ceremony and the planting of a tree in the Avenue of the Righteous at Yad Vashem.

The remembrance of prewar Jewish life in Europe was to take many forms. On the campus of Tel Aviv University a museum, Beit Hatefusot (House of the Diaspora) set out the story of Jewish life in the Diaspora, complete with replicas of many fine synagogues, and artefacts brought from Europe. There were also in Israel a few survivors from the great Hasidic dynasties, whose followers had been all but wiped out in the war. Thus Judah Moses Tiehberg, the grandson of the Aleksandrow Rebbe – who was murdered at the Treblinka death camp in 1942 – re-established the dynasty in the ultra-Orthodox town of Bnei Brak, just outside Tel Aviv.

Most Hasidic dynasties had reassembled in the United States after the war (almost exclusively in New York). In the first post-Holocaust decade it was assumed that the ultra-Orthodox way of life had all but died out with the virtual destruction of their Eastern European communities, and of so many of their leaders. Slowly, however, the numbers and religious strength of Hasidism returned. Within fifty years of the establishment of the State of Israel, the ultra-Orthodox community there had become as powerful, both numerically and politically, as it had ever been. Adherents of the ultra-Orthodox – *haredi* – lifestyle and beliefs, which in the early days of the State were not regarded as of any relevance to the daily conduct of life, and whose revitalization was not foreseen, have restored and revitalized themselves to the point where they have become a dominant force influencing the path of Jewish religious orthodoxy. This is despite their self-ghettoized character, as seen most strongly in Mea Shearim in Jerusalem and Bnei Brak near Tel Aviv.

On the day on which Yad Vashem was established in Jerusalem as a Holocaust memorial, an Arab infiltrator from the Gaza Strip reached Ashkelon, where he killed a restaurant owner, Yeshayahu Frankman, and severely injured his twenty-five-year-old daughter Zipporah. Ten days later, on the night of August 28–29, a recently established army reprisal force carried out its first action. The force was known as Unit 101. It consisted entirely of volunteers, and was led by Ariel Sharon, who had fought at Latrun during the War of Independence, and been wounded.

In its first combat operation, Unit 101 crossed the Gaza border and attacked the house in the Bureij refugee camp that was suspected of being an infiltrators' base. Three Unit 101 squads, about fifteen men in all, took part. Sharon's squad was identified prematurely by Arabs and failed to reach its objective. One of the other squads rescued Sharon's men, who made their escape. The third squad blew open the gate to the house of Major Mustafa Hafez, the head of Egyptian Military Intelligence in the Gaza Strip, injuring several civilians. But Hafez was not home. During this attack, twenty Palestinian refugees were killed, including seven women and five children, and twenty-two were wounded. Two Unit 101 soldiers were wounded.

The Bureij raid was followed by soul-searching debates by the soldiers of Unit 101. One of the participants wrote, 'Is this screaming, whimpering multitude the enemy? How did these *fellahin* sin against us? War is indeed cruel. The depression is general. No one tells stories. All are silent and self-absorbed.' In the Cabinet, the raid was criticized by the Social Affairs Minister, Moshe Shapira.

The acting director of UNRWA, Leslie Carver, wrote, 'Incident has caused intense alarm and unrest in the whole Strip.' Carver went on to urge that the United Nations protest strongly to Israel against the 'unprovoked attack upon harmless and defenceless refugees'. Israel denied responsibility, in such an emphatic way that leading diplomats and officials came to the conclusion that 'Israeli settlers' or 'a local kibbutz' had carried out the raid on their own initiative.

Pinhas Lavon, who had recently become the Acting Minister of Defence, had apparently been deceived about the Bureij raid by the army High Command, both before and after the event. Moshe Dayan, then head of the Operations Branch of General Headquarters, had apparently asked Lavon for permission to mount a small raid to ambush a car. After the raid, Lavon was told that the raiders had accidentally encountered an Egyptian patrol and been forced to withdraw into the Bureij refugee camp, where they had been forced to kill civilians.

As mass immigration continued, and the demand was for towns to replace the ma'abarot, a joint venture by the South African Jewish community (many of its members immigrants from Lithuania, and strongly Zionist) and the Israeli Ministry of Labour was begun in 1953. This was the creation, next to the ruins of the ancient Philistine city of Ashkelon, of a model township. Known as Afridar, it was built on the sand dunes at the edge of the sea. Not only small houses with gardens but six tourist hotels were built there, including a Holiday and Vacation Village sponsored by a French travel agency. One of the hotels was named after Dagon, the Philistine sea god whose temple in nearby Gaza was pulled down by the blinded Samson.

The River Jordan is one of Israel's few water resources. Its major headwaters originate in Lebanon and Syria. In September 1953 Israel began work on a canal just south of the B'not Yaakov (Daughters of Jacob) bridge on the river Jordan, the area that had been returned to Israel by Syria in the cease-fire agreement. As Syria still controlled the east bank of the river – which at that point on its course was little more than a stream, though at times very fast-flowing – Israel wanted to have a waterway that would be entirely within her territorial and security control. More than that, Israel was even then setting up a National Water Carrier, hoping to take water from the River Jordan to the coastal plain and northern Negev.

As soon as digging the course of the canal began, Syria objected. Through its Chief of Staff, General Vagn Bennike, the United Nations Truce Supervision Organization (UNTSO) called for an immediate suspension of the work until an agreement could be reached with the Syrians. Israel complied. President Eisenhower then sent a special emissary to the Middle East, part of a persistent American effort at facilitating Arab–Israeli agreement that was to span the first half century of Israel's existence as an independent State. The envoy, Eric Johnston, was authorized to secure agreement between Israel, Jordan, Lebanon and Syria on the utilization of the water resources of the River Jordan and its tributaries. It was intended that Johnston would address himself, not only to the question of the canal, but to the wider issues involved in the Arab–Israel conflict that had halted innumerable potential water projects requiring cross-border cooperation.

After more than a year of talks, Johnston produced a Unified Water Plan, which would have resulted in a major improvement in the availability of water. Israel accepted his proposals. The Arab League refused to accept them, and both Lebanon and Syria turned their back on any further talks. Israel and Jordan, however, proceeded with their own respective water development works in accordance with the plan, opening the way to the prospect of closer Israel–Jordan contacts.

Work on the B'not Yaakov bridge canal was never restarted, though Ben-Gurion was much criticized in the Israeli press and Parliament for his alleged pusillanimity in the matter. Israel had then to find another starting point for the National Water Carrier, and did so by installing a pumping station on the northern shore of the Sea of Galilee (which lies below sea level), and pumping the water uphill to a point 500 feet above sea level from which it could enter a carefully graded canal, and flow steadily westward towards the coast, and then southward as far as the Negev, enriched by sweet water springs along its path. The water remained that of the River Jordan, but was extracted beyond the range of Syrian guns.

The United States agreed to provide financial help to both Israel and Jordan in pursuance of their respective water plans, the National Water Carrier in Israel and the Ghor Canal in Jordan. This enabled both countries to establish round-the-year irrigation crops, improving their respective export potential.

The water dispute in the north was not the only dilemma facing Israel in the autumn of 1953. In central Israel, where the border with Jordan was in places hardly ten miles from the sea, there had been an escalation of border infiltrations, and a rise in the number of Israeli civilians killed, often while tilling their fields. Over the five-year period from 1951 to 1956, there were more than 6,000 recorded border crossings (an average of almost three a day) and 400 Israelis had been killed.

On 13 October 1953 an Israeli mother and her two children – the youngest child was only eighteen months old – were killed by infiltrators from across the Jordan border. On the following night Unit 101, together with a unit of regular paratroopers, carried out a reprisal raid on the Arab village of Kibya, only a few hundred yards from the border. Sharon commanded the raid. In his memoirs he recalled, 'The orders were clear. Kibya was to be a lesson. I was to inflict as many casualties as I could on the Arab home guard and on whatever Jordanian reinforcements showed up. I was also to blow up every major building in the town.' The written instructions to the group from army headquarters were 'to carry out destruction and maximum killing in order to drive the villagers from their homes'. During the raid, 69 Arabs were killed, most of them women and children, and 45 Arab houses blown up.

The world outcry was enormous, and inside Israel there was much heart-searching. A public cover-up attempt was made by Ben-Gurion to present the raid as a spontaneous reprisal by outraged Israeli civilian farmers. Moshe Sharett called the raid 'a serious error'. Henceforth, the nature of reprisals changed. Israeli forces were ordered to strike at Arab military targets rather than villages or towns.

In 1953 the Hebrew University and all institutes of higher learning opened their doors to Israeli Arab students. Many Israelis commented with pride that the position of the Arabs inside Israel was not only better than it had been since 1948, but better than that of the Arab refugees living in Arab lands. This view did not go unchallenged. 'What is the use of opening the university to Arab youth,' asked the journalist Ze'ev Schiff in *Ha'aretz*, 'if the military governor can delay for months the permit of a youth studying in Jerusalem to prevent him from leaving his village because his father quarrelled with the government?'

On 6 December 1953, at Ben-Gurion's insistence, Moshe Dayan was appointed Chief of Staff. He was to make his mark in one unexpected way. 'He decided all officers should be trained to parachute, and he was going to do it, too,' his daughter Yaël later wrote. 'Jump he did, and fractured his knee in the night jump, but if some were surprised or saw in it some typical Dayan stunt, we did not. He would never send a soldier where he wouldn't go himself, and if the ability, the courage, or whatever it took, was to be the standard of an Israeli officer, it was natural for him to undergo it himself. The day he got his paratrooper's wings, pinned to his shirt by Sharon, then the paratroopers' commander, he felt proud and joyous like a little boy, and we felt as proud. For myself, I felt invigorated and truly privileged at this sensation of belonging, allowed to share, not only witness, this world of purposefulness.'

The appointment of Dayan as Chief of Staff was the last official act of Ben-Gurion's premiership. On the following day, December 7, to the amazement of the Israeli public, Ben-Gurion resigned as Prime Minister and retired to the small farming community of Sde Boker, in the Negev. He had been worn down by successive political crises of the previous two years, in which one small Party or another had left the coalition requiring him to find new political partners, each with their own specific demands in terms of patronage and policies. A year earlier he had been forced to dissolve the government after the resignation of three members of the religious Parties. 'Each of the government crises has its special causes,' he said at that time, 'but they all stemmed from one source: the crisis that had marked the State from the first moment to this day – the one inherent in the inflated and morbid fragmentation we have inherited from our unhappy past, from a life of dependence devoid for thousands of years of statehood and national responsibility. In this Knesset we have no fewer than eight small parties totalling only thirty-four members, without taking into account five factions, that have only one or two members each.'

Before announcing his retirement from the premiership, and from politics, Ben-Gurion explained his decision to President Ben-Zvi (his friend and colleague since Ottoman Turkish times). It was not only since the establishment of the State six years earlier, he wrote, but since he had become Chairman of the Jewish Agency in 1936, that he had been working 'under the greatest tension'. His letter continued:

> You know that in my world there is nothing dearer than the State of Israel, nor any privilege greater than to serve it faithfully. It is not only a privilege but a duty, and I am not afraid to say a sacred duty, a duty that must be done until one's last breath.
>
> However, for a year now I have felt unable to bear any longer the psychological strain under which I work in the Government – the tension without which I cannot and am not entitled to work.
>
> This has not been just ordinary fatigue; on the contrary, when I leave my work in the Government for a few days I feel practically no tiredness and am capable of working, from both the physical and psychological standpoints, as I did twenty or thirty years ago. But there seem to be limits, at least in my case, to the psychological effort one can make. I have come to the unfortunate conclusion that I have no choice but to leave this work for a year or two or more.

Ben-Gurion will always be remembered in Israel as the man who transformed a factional, disparate people into a nation, through his insistence on *mamlachtiyut* – the primacy of sovereign institutions. These included (and still include) the independence of the courts of law, the unity of the Israel Defence Forces, and the primacy of the Knesset and other institutions of Israeli democracy. For having done this, he was rightly called the Father of the Nation. Those who, in most recent years, have belittled the courts of law and mocked at the institutions of statehood when they fail to serve a sectional or political purpose, endanger the structure Ben-Gurion so tenaciously created and sustained.

Ben-Gurion was succeeded as Prime Minister by Moshe Sharett, who retained the Foreign Affairs portfolio he had held since 1948. In Sharett's government, which was formed on 6 January 1954, Pinhas Lavon became Minister of Defence. Golda Meir remained Minister of Labour, the post she had held since 1949.

While he lived at Sde Boker, Ben-Gurion's security was the responsibility of Southern Command. A special unit was assigned to protect him. 'Every day, he insisted on a vigorous walk,' Chaim Herzog, who was Chief of Staff of the command for several years, later recalled, 'and on a number of occasions I accompanied him, wheezing, gasping, and trying to keep up with his pace. What impressed me most were his willpower, stubbornness and refusal to back down. But he also had a mischievous side. In the late 1960s, on a farewell international tour of Jewish communities, he went to Johannesburg, where the pro-apartheid mayor gave a lunch in his honour to which the Calvinist clergy were invited. Turning to the clergymen, Ben-Gurion asked, "How do you explain the fact that Moses married a black woman?" To say there was total consternation around that table would be a serious understatement.'

Ben-Gurion's reference was to the Kushite woman whom Moses married after the Jews left Egypt. Both Miriam and Aaron had objected to this. God, in punishing Miriam for her racism, gave her an ailment that turned her skin white.

On 17 March 1954 an Israeli bus was travelling up the precipitous hairpin bends of the Scorpion Ascent in the Negev when it was attacked by a group of Arab gunmen, who had infiltrated from Jordan. They first killed the driver, Ephraim Fistenberg, and then ten of his passengers, including his wife.'The inside of the bus', wrote a reporter for the *Jerusalem Post*, 'was littered with coral and other souvenirs which tourists usually bring from Eilat.'

An official statement issued that night by the Government of Israel expressed its 'revulsion and horror' at the attack. The statement continued:

> The Government of Israel regards this attack as a clear warlike act, responsibility for which falls squarely upon the Government from whose territory this unit of murderers was sent forth across the border into Israel territory to carry out this dastardly deed.
>
> The series of wanton acts of hostility by neighbouring Arab states against Israel has reached a new climax with this latest outrage.
>
> The Government of Israel will take all necessary measures within its power to ensure that the blood of Israeli citizens will not be shed with impunity. All steps will be taken to safeguard the security of travellers

along the roads of the Negev so that movement will continue without hindrance.

Other than more army patrols on remote roads, there was nothing the government could do. The criticisms of the reprisal taken against the Arab village of Kibya a year and a half earlier undoubtedly helped to inhibit action. At the same time, the Jordanian government insisted that the killings were 'an incident in which Jordan could not possibly have been involved'.

Life in Israel has constantly been punctuated by incidents of terror. Each one casts a pall, and creates a sense of unease. In a small country, many people know the victims, or family and friends of the victims. The newspapers give accounts of the lives (often all too short) of those who have been killed. Daily life, with its own problems of earning a livelihood, bringing up children, caring for elderly parents, and enjoying the holidays which follow both the changing seasons and past historical events – all these constitute the pattern of life in many nations. In only a few, however, are these human pursuits, challenges, hardships and joys paralleled year after year by internal and external dangers, and by internal and external forces hostile to the impulses of statehood and growth.

The new State's search for its roots made every major archaeological discovery a matter of national importance. In 1952 Yigael Yadin had resigned as Chief of Staff to return to his first love, archaeology, and to follow in the footsteps of his distinguished father, Eliezer Sukenik. Since before the establishment of the State, Sukenik had worked on two scrolls which had been found by a Bedouin shepherd in a cave at Qumran, overlooking the Dead Sea. Sukenik died in 1953. Within a year, his son was involved in an extraordinary mission, following the appearance of four more scrolls in the United States. Yadin coordinated the purchase of these scrolls in a dramatic series of meetings and messages with the Israeli Ambassador in the United States, Avraham Harman, and the head of the Prime Minister's office in Jerusalem, Teddy Kollek.

In one telegram, sent on 11 June 1954, Yadin told Kollek that four Dead Sea Scrolls, including the Book of Isaiah, had been brought to the United States and were offered for sale. 'They can be bought at once for $250,000,' Yadin reported. 'No need to stress the importance of the scrolls and the unrepeatable opportunity. Any delay may ruin our chance. Have already probed several important donors, and consider it certain that the sum may be collected within a year. A guarantee from the Treasury for the whole sum is imperative. Request your immediate intervention with the PM and Minister of Finance. Harman who is near me, doing his best to help. I rely on you and expect a positive answer. Secrecy imperative.'

Moshe Sharett gave the go ahead, and Avraham Harman was able to arrange a loan in the United States. Today the Dead Sea Scrolls are a central feature of the Israel Museum in Jerusalem: their content is still a matter of intense academic study.

The year 1955 was an important one in terms of settlement, the already eighty-year-old imperative of Zionist endeavour. It saw the establishment in the southern coastal plain of Kiryat Gat, a development town that was to be at the centre of a new concept of regional planning, based on a comprehensive technique whereby whole regions were created around a central urban hub, in this case Kiryat Gat itself. Around the hub, clusters of villages were established, all served by the hub, and serving it.

The new town was named after the Philistine city of Gath, believed to have been nearby. Another ancient site, Lachish, gave its name to the Development Region, in which cotton became the main crop. Within three years, Kiryat Gat had a population of 4,400, almost half of them immigrants from North Africa. This was to multiply fourfold within the following decade.

On 31 January 1955 the Egyptian authorities hanged two Jews in Cairo. They were Dr Moshe Marzouk and Samuel (Shmuel) Azar. Both had been found guilty of being spies for Israel. Marzouk, a surgeon at the Jewish Hospital in Cairo, was the first to be led to the gallows, his legs in shackles. He was followed half an hour later by Azar, a teacher. Both men were in their twenties. Eight other Jews had been charged with spying and given long prison sentences. Two more had committed suicide.

In Ramle, the Israeli town with the largest concentration of immigrants from Egypt, flags were flown draped in black. In several towns, streets were named after the two executed men, and hundreds of boys born soon afterwards were called 'Moshe' and 'Shmuel'. In Beersheba, immigrants from Egypt paid their respects to Azar's sister. In Tel Aviv, the Secretary of the Municipal Labour Council declared, 'Those who spilled the innocent blood of others should not think that the blood-spilling finished there.'

The executions were cruel, but the two executed men were in fact not innocent of spying. Together with those who had been imprisoned, and the two who had committed suicide, they were part of an Israeli sabotage team that had been set up in the summer of 1954. One of those who committed suicide was an Israeli Intelligence officer, Max Bennett. The aim of the operation – which was evolved in the Ministry of Defence, where Pinhas Lavon was Minister, and with the active participation of the chief of Israeli Intelligence, Benjamin Gibli – had been to give the impression that anti-British saboteurs were in action in Egypt and, by this means, to persuade Britain to delay its proposed withdrawal from Egypt, a withdrawal that Israel regarded as detrimental to its interests.

The sabotage actions were carried out in July 1954. The historian Shabtai Teveth, who later examined this episode in detail, has written, 'These efforts bore no political fruit, achieving nothing beyond: two hangings; two suicides; the extended incarceration of three young men and a young woman; and ignominy heaped upon Israel and its intelligence service.' Teveth went on to ask, 'Who advocated this course? When and where was it debated? Who authorized it? Who ordered its implementation? These questions were to rock Israel with tempestuous discord that sapped the young state's foundations, exposed Ben-Gurion and Lavon to private and public travail, sundered Mapai and reduced the political arena to utter chaos. Israel would plunge into agonized strife over the relentless question: "Who gave the order?"'

The debate that raged in subsequent years was to tear at the heart of Israeli politics. Gibli later claimed that Lavon (who had been Minister of Defence since January 1954) had given him the order directly, and in private. Lavon always denied this. The actual orders to act were transmitted over the radio by Israeli Military Intelligence in the form of cooking recipes for housewives. The acts of sabotage themselves were small and of no political (and hardly any physical) consequences: a parcel office mailbox set on fire at the central post office in Alexandria, and two small explosive charges set off in American libraries in Cairo and Alexandria on July 14, and four small explosive charges set off in cinemas in Cairo and Alexandria nine days later.

The first political repercussions in Israel came six months before the executions, when the news of the arrests of the saboteurs had been announced by Cairo Radio. It was then that Moshe Sharett, who had been Prime Minister at the time of the fiasco, told Golda Meir, the Minister of Labour, that it was Lavon who had given the instructions to 'liven things up' in the Middle East in order to create tension between the Arabs and the British and Americans.

Detailed enquiries began at the highest level, and in strictest secrecy, even before the executions in Cairo. What was later to become very public indeed, and known to all Israelis as the Lavon Affair, was rumbling on ominously. On 2 February 1955, the day after the news of the Cairo executions was published in the Israeli newspapers, Lavon offered to resign. That day, Sharett and Golda Meir travelled to Sde Boker to see Ben-Gurion. Lavon's resignation was accepted on February 18. That day, Ben-Gurion agreed to come back from his retirement as Minister of Defence: four months later in June, he again assumed the post of Prime Minister, replacing Sharett who remained Foreign Minister.

Throughout the winter of 1954–55 small groups of armed Palestinians – the fedayeen – crossing the Egyptian border from the Gaza Strip, had caused concern and anger inside Israel. At regular intervals, often of only a few days, individual Jews were attacked and killed while walking along the roads or tilling their fields. On 28 February 1955, in retaliation for these raids, Israeli troops launched an attack on an Egyptian army base near Gaza. Thirty-eight Egyptians and two local Arabs were killed. The response of the fedayeen was to shell the Jewish settlements across the border.

The effect of the reprisal raid on the last day of February was monitored by the adjutant to the Ministry of Defence, Nehemia Argov. As Moshe Sharett, who as Foreign Minister as well as Prime Minister would have to defend the action to the outside world, awaited Argov's call, he was – he wrote in his diary – 'gnawed by anxiety'. His account continued:

> At seven precisely the telephone rang. From the tone of the first word Nehemiah uttered I could tell that he had bad news. He started by saying that the action had been carried out, but that we had lost eight dead, and about the same number of wounded. Egyptian resistance had been extremely stiff – they rained a deadly fire; our units had never experienced anything like it.
>
> However, our troops too fought in an unprecedented way: commanders who had seen action in the War of Independence testified that there had been no assault like this.
>
> Despite the enemy fire, the mission was carried out to the last letter – all structures were blown up and the waterworks and camp completely destroyed. The Egyptians are estimated to have lost about fifteen dead. Despite the heavy losses, the men returned to base singing and in high spirits.
>
> I began to digest the shocking news. I telephoned and asked to be put on to Nehemiah again. I asked him if all the dead and wounded had been brought out. Yes, this had been done, although their extrication had been fiendishly difficult. Their bodies had been dragged on foot for seven kilometres. The dead were taken to Tel-Hashomer and the wounded were brought at dawn to Kaplan Hospital, and surgery began at once.

Going to the Foreign Ministry Sharett learnt that thirty-five Egyptians had been killed in the raid. 'I was aghast,' he wrote. 'The figure changes not only the dimensions of the incident, but its nature. It turned the incident into an event which could bring complications and grave dangers in days to come, political dangers and threats to our security.' Without consulting with the Foreign Ministry, the army spokesman had put out a press release with 'a false version of what took place – apparently instructed by the Minister of Defence: our unit, attacked (it seemed) by an Egyptian force in our own territory, returned the fire and went into a battle which flared up and reached the scale it did.' Sharett added, 'Who will believe that we have told the truth? There might have been some short-term logic in publishing a story of this sort, but it is evident that in the long run we shall not be able to maintain it.'

Sharett noted in his diary, 'The operation had been approved on the basis of an estimate of no more than ten dead. The number of killed was actually four times that figure. I said that we had to prepare our missions to accuse Egypt, and not simply to defend ourselves.'

Attempts at amelioration of the Arab–Israel conflict were often made, and sometimes bore fruit. In April 1955 an agreement came into force in Jerusalem, whereby the District Commanders on both sides of the barbed wire and concrete divide could make contact with each other through a direct telephone line. The agreement, reached between the governments of Jordan and Israel, authorized the respective commanders to 'discuss and settle questions regarding the maintenance of peace in Jerusalem'. The commanders also met face to face. Indeed, long before the telephone link they had exchanged daily newspapers on an informal basis. One of them, Chaim Herzog, has recalled:

> My meetings with the Jordanian commanders usually took place at a table in Mandelbaum Square, the main crossing point between the Israeli and Jordanian sectors. We dealt with such practical day-to-day matters as the occasional shots fired from Jordanian positions toward the Israeli positions and the throwing of stones by the Israeli troops. There was also the problem of guerrilla infiltrators.
>
> The Jordanians, maintaining that the Egyptians were activating the guerrillas, claimed they were doing their best to prevent them. This was a tricky problem for them since public opinion in Jordan heavily favoured the Egyptians. I agreed to keep them up-to-date on the infiltrators, and we arranged to eliminate the firing on individuals who had entered no man's land and to return people who had lost their way.

Herzog also had cause to realize how strongly King Hussein's regime was opposed by many Palestinians. On one occasion his Jordanian opposite number, Lieutenant-Colonel Sa'adi, told him, 'As fate would have it, the country most interested in Jordan's independence is the State of Israel.' It was not so much Hussein's rule over the Palestinians living in Jordan, as over those in the West Bank, that was resented by the Palestinians. The West Bank town of Nablus was a particular hotbed of Palestinian nationalism. Just as the Egyptians, led by Colonel Nasser (Egypt's Prime Minister from 1954, and later President), were continually trying to destabilize Jordan by stirring up the Palestinians, so Israel made every effort to maintain the King on his throne, fearing that Palestinian nationalism would gain a base in Jordan from which to use the West Bank in order to attack Israel. A senior Jordanian Cabinet Minister once remarked, 'What saved Jordan from Egypt was the few miles of Israeli territory at Eilat standing between itself and Egypt.'

The establishment of the Academy of the Hebrew Language in 1955 was the climax of an almost century-long effort to create a modern spoken Hebrew that, while making the fullest use of biblical forms, would also incorporate new words, chosen to replace foreign words that had not been in use at the time of the Bible. Among those who were present at the opening of the academy was Joseph Klausner, whose devotion to the Hebrew language had long preceded his immigration to Palestine in 1919; he had published his first Hebrew article in Odessa in 1893. Among the words which Klausner himself introduced to modern Hebrew were *iparon* (pencil), *hultzah* (shirt), *hagdarah* (a definition) and *petel* (raspberry).

In his memoirs, Klausner reflected on what he had seen achieved in Palestine and Israel in four decades, not only in the linguistic field. 'I am happy that I have seen my people awaken to a new life,' he wrote, 'even if everything is not as I had dreamed, its language alive and spoken even if not to the hoped-for degree, and my land being rebuilt as a Jewish State even though more slowly and not, as I see it, in the proper direction.' As a life-long Revisionist, the Revisionist candidate for the presidency in 1948, Klausner could never accept the Labour predominance or ideology.

In April 1955 twenty-nine newly independent Afro-Asian countries met at Bandung, in Indonesia. Israel was not invited to attend. The aim of the meeting was to create a bloc which would be independent of both the Soviet Union and the United States. But among the participants anti-Americanism was stronger than anti-Communism. At the conference Chou en-Lai, the Chinese Prime Minister, arranged a meeting between Egypt's Prime Minister, Colonel Nasser, and the future Soviet Foreign Minister, Dmitri Shepilov. It was a meeting that led to the shipment of the most up-to-date Russian-made arms to Egypt. These included jet aircraft and heavy artillery, and immediately altered the balance of power against Israel.

Israel reacted by turning to both Britain and France for help. On May 2, Ben-Gurion wrote in his diary, 'Shimon Peres is going to England and France tomorrow on an arms mission. He believes it is a good time to buy from France because of "pique over Bandung".'

Peres made immediate contact with the French Minister of Defence, Pierre Koenig (after whom a street in Jerusalem is named). Within a few days the two men had signed an agreement whereby France would sell Israel light tanks, field artillery and, most importantly for Israel, twenty-four Mystère-2 jet warplanes. Subsequently, Peres had to use all his skills as a diplomat to negotiate through the labyrinthine complexities of successive and rapidly changing French administriations; something he did with consummate success.

On 26 July 1955 elections were held for the third Knesset. Mapai once more secured the largest single number of seats, forty, and Ben-Gurion was once again able to form a coalition from which Herut, with fifteen seats (almost double its 1951 number), was excluded. Once more, Ben-Gurion brought in the religious Parties, then fighting under the single banner of the National Religious Front, and winning eleven seats. In bringing in the religious Parties, Ben-Gurion was acting as much from principle as on political expediency, reflecting his belief in the 'historic alliance' between secular and religious Zionism; one of his closest associates was Rabbi Yehuda Leib Maimon, of Mizrachi, the religious Zionist movement established in Eastern Europe in the first years of the century.

The General Zionists, who continued to campaign on a non-Socialist platform, obtained thirteen seats, ending their earlier hopes of being able to challenge Mapai as coalition builders. The ascendancy of Mapai was thus maintained, and Ben-Gurion, and those who had worked with him since the formation of his first government in 1949, continued to direct the affairs of State.

On July 27, the day after the elections, at the height of the East–West Cold War, an El Al airliner on its way from Vienna to Tel Aviv was intercepted near the Bulgarian–Greek border by Bulgarian warplanes and ordered to change course and land at Sofia. As it was descending it was shot down with the loss of all fifty-one passengers and seven crew members. The episode sent shock waves through Israel, and aroused fears of a campaign against its civilian airlines; but there was no repetition of the incident. Despite obstacles raised by Bulgaria, the bodies of those killed were taken back to Israel and the crew were buried at a memorial site in Kiryat Shaul cemetery, Tel Aviv. The shooting down of the plane led critics to renew criticism of a national airline as 'an unnecessary luxury'.

In the Cabinet which he formed on 3 November 1955, Ben-Gurion retained the Defence Ministry. His Director-General at the Defence Ministry was Shimon Peres, entrusted with the task of arms procurement, the foundation of the aviation and electronics industries, and the building up of a nuclear energy programme. In 1956 Israel's first nuclear reactor was established in Nahal Sorek, near the abandoned Arab town of Yibna. Going 'critical' on 16 June 1960, the Nahal Sorek reactor was to become a major centre of research and teaching.

Following the Israeli raid on Gaza in February, Colonel Nasser had assured the Egyptian people that no future Israeli army attack could go unanswered. On August 22 the Israelis carried out a reprisal raid into the Gaza Strip. A dozen fedayeen squads, some sixty men in all, were made ready to go across the Gaza border with orders to kill and to carry out sabotage.

Four squads crossed the border on the night of August 25–26: one ambushed an Israeli jeep, killing a watchmen at kibbutz Yad Mordechai, and then planting mines near the Erez crossing point; the others blew up water installations. Two nights later, several squads penetrated as far as the Ashkelon area, where they ambushed and killed a soldier and wounded two people. On August 28 four Israeli soldiers died and two were wounded when their vehicles hit mines that had been laid near the Gaza border. On the night of August 29–30, an Israeli family was attacked outside Rehovot, 25 miles north of the Gaza Strip. One man was killed and four women and children were wounded.

That same night, another squad killed four farmers near moshav Beit Oved; a radio transmitter was blown up near Qubeiba; a military vehicle was ambushed near Bilu junction, injuring one soldier; kibbutz Erez was briefly mortared; and an Egyptian fighter aircraft penetrated Israeli airspace north of the Gaza Strip, but was driven off. On the following day another Israeli army vehicle hit a land-mine, and two soldiers were injured. In an ambush north of Beit Guvrin, a fedayeen squad killed four civilians in a car. That night, August 30–31, two residents of Ramle were killed when their civilian truck was ambushed. The fedayeen squad responsible for this then threw hand-grenades into a house in moshav Nahala, wounding a woman. In another ambush, an army officer was injured.

In September 1955 a new town, Dimona, was founded in the central Negev. The reason for Dimona was to give the workers at the potash works at the Dead Sea, and at the Oron phosphate field, somewhere to live that was healthier than where they worked, but not too far away. In a bare desert landscape, an urban infrastructure was created, and parks and gardens planted. Within five years, 3,500 people were living in Dimona. But life was hard there, with few civic amenities and as far away as any town in Israel from the centres of cultural and social activity.

Ten years after the founding of Dimona, after Israel's second nuclear reactor was established nearby (it became operational late in 1964), the town's population had grown to 20,000. Industry was also created, including two textile mills, and immigrants from North Africa were encouraged (or directed) to settle there, making up 65 per cent of the population by 1968. But life remained a hard one; Dimona was not a place to which many Israeli went unless they had specific work to do there.

The most southerly kibbutz in Israel had its origins in 1955, when a group of pioneers established a small fishing camp on the Red Sea shore near the tiny town of Eilat. Members of the kibbutz took their part in building the town itself. Eight years later they moved to the nearest high ground north of the town, where they established kibbutz Eilot, planted date-palm orchards, and grew vegetables, melons and flowers in deepest winter. For many years they faced all the uncertainties of being located only a few hundred yards from a hostile frontier. Today their fields face the Arava border crossing between Israel and Jordan opened after the peace treaty of 1995.

In the Negev Desert, the town of Ofakim (Horizons) was founded in 1955. It was settled mostly by new immigrants, of whom almost three-quarters were from Morocco and Tunisia. Others came from India, Iran and Egypt. The town did not flourish at first, and there was considerable social as well as economic hardship. It was to be a decade before, with the start of diamond polishing and textile industries, the fortunes of the town improved.

During 1955, twelve Jewish children were brought to Israel from Ethiopia. They were followed a year later by fifteen more. These children were black – members of the Falasha community tens of thousands strong whose history had been traced back to the fourteenth century, and some believed much earlier than that. Certainly, they had preserved Jewish practices dating back at least a thousand years. Various emissaries had made contact with them and were confident that they were Jewish by belief, tradition and descent. The most recent emissary had been sent in 1954 by the Jewish Agency; he had set up a school for Ethiopian Jewish children in Asmara. The first children to be brought to Israel in 1955 were from that school. They were taken to a youth village, Kfar Batya, to learn Hebrew and to be taught a trade. Most of them returned to Ethiopia, though a number stayed in Israel. It was to be twenty years before a significant effort was made to bring a substantial number, initially – due to the hostility of the prevailing Ethiopian regime – by mounting secret operations to bring them out.

The election of Isaac Nissim as the Sephardi Chief Rabbi of Israel in 1955 marked a step towards toleration, and was a relief to those who feared an increasing polarization between Orthodoxy and the rest of the community. It was a polarization that was to be felt bitterly even forty years later. Nissim, an Iraqi Jew, had been born in Baghdad in 1896 and had settled in Palestine in 1925. As Chief Rabbi he took the unprecedented step of visiting the predominantly Ashkenazi, left-wing kibbutzim as a gesture of goodwill. He was also emphatic that the Bene Israel community of India should be recognized as Jews.

The Second World War had ended ten years earlier, but the traumatic legacy of the Holocaust was ever present. In June 1955 a survivor of the Holocaust, Malkiel Gruenwald, accused a fellow Hungarian Jew, Israel (Rezsö, Rudolph) Kastner, the former head of the Zionist Rescue Committee in Hungary, of having done a deal with the Nazis in 1944 in order to save his own family and 1,700 other Jews, while three-quarters of a million Hungarian Jews were deported to Auschwitz and their death.

Kastner was a senior official in the Israeli Ministry of Industry and Trade, and a former candidate for a Mapai seat in the Knesset. The story of the train that he had managed to persuade the Nazis to send to Switzerland had always been controversial, but had been accepted on balance as having been a desperate initiative that did at least save 1,700 people who would otherwise have been murdered. To protect Kastner, the State brought an action for slander against Gruenwald. At the trial Kastner argued that he had negotiated with the Nazis in order to save all the Jews of Hungary, in return for material to be provided by the Allies to the SS; it was the Allies, he pointed out, who had refused to follow up this deal, or to ascertain if it was genuine.

In an extraordinary court drama, Gruenwald's defence counsel, Shmuel Tamir, used the trial to put Kastner on trial for having abandoned the Jews of Hungary to their fate. Tamir was so successful in stirring up feelings against Kastner that Gruenwald was acquitted of slander and the judge, Benjamin Halevi, found Gruenwald's accusations of collaboration proven, and told the court, 'Kastner has sold his soul to the devil'.

What gave the trial an added drama was that while Kastner was an active member of the ruling Mapai Party, Tamir – a former member of the Irgun – had been an activist in the Herut Party: here were the establishment and the opposition locked in battle over events that had taken place eleven years earlier, on another continent. Judge Halevi later resigned from the bench and became a Herut Member of Knesset.

The Kastner debate did not end with Gruenwald's acquittal. When Herut brought a motion of no-confidence against the government, the General Zionists abstained, and the government resigned. It was then re-formed without them. Nor did Gruenwald's acquittal go unchallenged; the Attorney General appealed against it, his appeal was upheld by the Supreme Court and Gruenwald was convicted of slander. By the time of Gruenwald's acquittal, Kastner was no longer alive: he had been assassinated by a Jewish nationalist extremist for his alleged collaboration with the Nazis. It was the first assassination of a Jew by a Jew in the history of the State.

The security of Israel seemed to enter a new and dangerous phase in October 1955, when details became public of a substantial Czechoslovak arms deal with Egypt. This was Moscow's attempt to leap over the barrier created earlier that year by President Truman's containment of the Soviet Union through the Baghdad Pact, whereby Turkey and Iraq, and later Pakistan and Iran, aligned themselves in the American defence sphere.

The Soviet bloc, which in the aftermath of the creation of Israel had looked with favour on the Jewish State, had decided to put its full weight behind the Arab side in the Arab–Israeli dispute. Influence in the Arab world was a prize for which the Soviet Union fought with all its arsenal of economic and ideological inducements. In response, Israel turned towards the United States, seeking both arms and a security agreement. On 30 October 1955, Moshe Sharett met the American Secretary of State, John Foster Dulles, in Paris, and told him that if a security agreement with the United States, which Sharett had hoped to negotiate, was impossible, it would be better to 'stop talking about it'. But he wanted Dulles to know that Israel had 'lost everything' with regard to its former good relations with the Soviet Union, 'without gaining a thing' from the United States.

The only reason that Israel maintained an embassy in Moscow, Sharett told Dulles, was 'in order to encourage the Jews of Russia to hold out – so that they can see before them a mark and token that the day will come when their link with Israel and the Jewish people will be renewed'. But, Sharett added, the Russians had told Israel that there was 'no point' in making concessions on the Soviet Jewry issue 'as long as we declare our willingness to enter a military alliance with the USA, which to them means subservience, bases, and every other abomination'.

The United States decided not to proceed with the security agreement, not wishing to alienate those countries in the Arab world with whom she still hoped to have good relations, but Dulles did promise Sharett that 'Israel should trust that the USA will not abandon her'.

On the day of Sharett's discussion with Dulles in Paris, 30 October 1955, an Egyptian army unit crossed the Israel border near Nitzana and penetrated several miles into the Negev. As a reprisal, Israeli troops crossed into Sinai, attacking the Egyptian army base at Kuntilla. Ten Egyptians were killed.

In the centre of Israel, only 8 miles south-east of Tel Aviv, terrorist raiders killed five school children and their teacher at Kfar Habad, a village set up six years earlier on the initiative of the American-based Lubavich Hasidic Rabbi, Joseph Isaac Schneersohn. His original aim was to make it a centre for Hasidic Jews emigrating from Russia (he himself had left Russia after the revolution). The number of Russian immigrants being minimal at that time, Kfar Habad became a centre of the Lubavich Hasidic movement in Israel, with a religious academy, a teachers' seminary, and a printing school.

The killings at Kfar Habad renewed the sense of vulnerability that many Israelis felt, and heightened the desire to find some way of bringing cross-border terror attacks to an end. It was not reprisals, however, that were looming on the horizon at the end of 1955, but war.

CHAPTER SEVENTEEN

Paths to war
December 1955–October 1956

On 12 December 1955 – during the Jewish festival of Hanukkah – the Syrians on the north-eastern shore of the Sea of Galilee opened fire on an Israeli patrol boat on the lake. Such episodes were not unusual. Patrol boats escorting fishing vessels were frequently fired on from the Syrian shore. What made this episode different was that, instead of using rifles or light machine-guns, the Syrians had opened fire with their artillery.

A year earlier, a similar attack had been answered by an Israeli artillery bombardment on the Syrian artillery post at Kursi. This time, it was decided to launch a land attack. Three hundred Israeli soldiers were sent into action. Starting as darkness fell, the troops in the first echelon made their way down to the banks of the Jordan River, some distance north of where it flows into the lake. Crossing the river was difficult. An eyewitness reported in the *Jerusalem Post*: 'Several soldiers forded the swift, ice-cold current, to stretch ropes across. The rest of the unit crossed the river holding fast to the rope, the shorter men struggling, submerged to their necks. Guns and ammunition were also transported by rope.'

The second echelon followed further south, fanning out to cover an 11-kilometre front, and walking 8 kilometres to their objective, the Syrian fortifications near the village of Nukeib, along the shore of the lake. Explosive charges were detonated against the fortifications on the shore, no part of which actually belonged to Syria (whose border lay slightly inland). There were about 200 Syrian defenders, who opened fire on the Israeli force, and battle was joined. Six Israeli soldiers were killed. The Syrian dead were stated by the United Nations Truce Supervision Organization (UNTSO) to be forty-three. The French government issued a statement saying that it 'deplored' Israel's action.

The operation had ended with the dawn. That morning the Israeli inhabitants of kibbutz Ein Gev, just to the south of the area of the battle, which had been shelled by Syrian artillery during the night, carried on with its regular daily life. The eyewitness reported:

> Ein Gev settlers this morning observed Syrian forces returning to the positions which were blown up by an Israeli force last night. They began rebuilding their destroyed posts and dug communication trenches. They also moved up cannon and mortars and tested the range with machine-gun fire.
>
> The settlement was calm today after a wakeful night. The children and infants were taken to shelters yesterday. The older children were in high spirits this morning when the full news of the successful operation became known. Because of the sleepless night, only the most essential work, including the cowshed chores, was done this morning.
>
> Reinforcements arrived from Jordan Valley settlements to help the tired men. The settlers wistfully recalled today that they were to have held their traditional Hanukkah celebration last night.

*   *   *

In January 1956 indirect talks were initiated in strictest secrecy between Egypt and Israel. The intermediary in the talks, chosen by President Eisenhower, was Robert B. Anderson, a former Secretary of the Navy (and later Secretary of the Treasury). At their first meeting, President Nasser told Anderson of his inability to restrain the Palestinian Arab refugees in the Gaza Strip. He was in favour of a settlement with Israel, Nasser said, but went on to warn: 'The refugees were extreme and would raise a lot of noise against any settlement.' Even if it came to the signature 'of final peace', Nasser said, 'there was a danger of incidents by the refugees – until they were dispersed and settled.' But Egypt had refused to allow the refugees to disperse or settle inside Egypt, nor had it any intention of doing so. There had been several incidents inside the Gaza Strip when refugees had attacked Egyptian soldiers, claiming that the Egyptians were not doing enough for them.

In his talks with the American emissary, Nasser appeared conciliatory on a number of issues. Israeli shipping could use the Suez Canal, he told Anderson, once the 'state of belligerency' between Egypt and Israel was ended. An agreement might be reached on the territorial partition of Jerusalem between Jordan and Israel, rather than the 'internationalization' of the city, for which the United Nations had pressed. Nasser accepted that Ben-Gurion had not returned to government 'to create incidents' between Israel and Egypt.

Anderson flew from Egypt to Israel, landing at a small airfield near Petah Tikvah (Gate of Hope) to ensure that secrecy was preserved. He was not to know that there had been more shooting from across the Gaza border even during his mission. When he reported Nasser's remarks, Ben-Gurion was surprised by their conciliatory nature, telling Anderson, and his own closest advisers, Moshe Sharett, Teddy Kollek and Yaakov Herzog (Chaim Herzog's brother), who alone knew of the talks:

Now I want to think aloud. We carry a terrible responsibility. Nasser is a dictator. Our colleagues do not know and will not know about these discussions of ours, but the questions are fateful. You have undertaken a mission of hope and tremendous danger. I will not speak about secondary matters. Is there hope, and is it near, and what must we do to realize it?

Everything can be explained in two ways. It is possible that Nasser is really afraid of his people and the Arab countries – Iraq and its Syrian ally and the Saudi Arabians. If there is hope for peace – that would be the greatest thing in our lives.

Several years before the establishment of the State I tried to formulate our principal aims: 1) the security of our people; 2) the founding of a Jewish State; 3) a Jewish–Arab alliance. There is a connection between the three. I always see a Jewish–Arab alliance as the ultimate goal. And therefore, if there is the slightest possibility of achieving peace with Egypt, we must do everything towards that end. But there are things that make it difficult to see hope. I have tried to dissipate all doubts – but if you cannot stop the shooting.

The Emissary: Has there been shooting during the last few days?

Ben-Gurion: Yes, every day.

The Emissary: Have people been killed?

Ben-Gurion: No, but rifles and machine-guns are fired every day. I have suspicions, but I am prepared to understand. We have to understand. We have to understand Nasser's difficulties and do all in our power to make it easier for him to meet us and try to reach a settlement.

As a democratic leader I find it very difficult not to tell these things to my colleagues. In the United States it is easier, for there is a president there. But if this is only a matter of secrecy, we can wait a month, a year, five years.

The ideal is so great that we must do nothing to jeopardise peace. But your mission also involves the danger of war. It is possible that there will be no peace – and yet no war either.

Ben-Gurion told Anderson that he wanted to believe Nasser, and why. 'I want to believe him,' he said, 'and one should, in order to bring peace nearer. It is possible that he is sincere, but there are forces that are brought to bear on him. If I told our people about your mission, they would be glad. Nasser cannot do that, just as he is unable to bring to trial men who commit acts of violence.' With reference to the recent Soviet-sponsored arms deal with

Egypt, which had come to dominate Israeli military thinking, fearful that the arms balance was tipping dramatically in Egypt's favour, Ben-Gurion told Anderson that in another six months, Nasser 'will be very strong. Three days ago Egypt, Saudi Arabia and Syria declared that they would fight if we continued work at B'not Yaakov. We can understand that there is no contradiction between this and Nasser's attitude. But in another six months, when he and his army feel that they are strong enough to defeat Israel, the pressure on him is liable to be great.'

Ben-Gurion had correctly assessed the situation. Israel, he explained to Anderson, needed arms so that she would not appear vulnerable to the Arab States. But she had no intention of using those arms in an aggressive war. Indeed, he saw the possession of adequate arms, and of an Israel that was less vulnerable, as a means whereby the peace process could begin. As he told Anderson:

> Let us suppose that the danger of war has been eliminated. As the first step on the road to peace, the shooting must be stopped. There is shooting every day, and it is a miracle that people are not being killed. If anything is required of us, it will be done. I give you a categorical undertaking. In our country only the army bears arms. There will be no shooting from our territory, and none of us will cross the border.
>
> If Nasser issues an order to cease fire, you can give a categorical undertaking in our name. If that is done, the atmosphere will be changed for the better. Next, communications should be established to deal with minor matters – if anything happens, he could transfer information to me and I to him. Although this is difficult for him, we must arrange direct communications for the transfer of messages from both sides.
>
> Then the third stage should come: a meeting between representatives. No mediation, not even the most ideal could bring about what the two parties can achieve in a direct meeting.

If the time came for a meeting between Egyptian and Israeli representatives, Ben-Gurion told Anderson, then 'questions will be asked and it will be necessary to make decisions'. He went on to examine how the discussions might develop:

> Nasser is interested in the refugee question and in territory. I have explained our attitude, but I may possibly have some ideas for him – for the attainment of the things he wants without harming us. Then I shall have to tell my colleagues.
>
> There will be a need for contact between people on both sides who are capable of bearing responsibility. It will take perhaps a year

to achieve: 1) elimination of the danger of war; 2) a cease-fire; 3) the establishment of communications; 4) a meeting of representatives; 5) a meeting of persons with the authority to decide.

If you succeed in ensuring peace, you will have done one of the greatest things in the history of our time. We are grateful to the President, and we must do all in our power to help you.

Our greatest difficulty is the danger of war. Time is the main factor, and if within four or six weeks we do not get a minimum of planes and tanks we shall not be able to do what is needed for a settlement.

Anderson continued his mediation. Ben-Gurion offered to go to Cairo. Nasser replied (as Anderson reported to Ben-Gurion) that he believed he would 'make peace with Ben-Gurion in one meeting, but the next day he would be assassinated'. Ben-Gurion offered to meet Nasser anywhere. Nasser drew a map of Egypt's basic demands: they included the transfer of almost all the Negev to Egypt, to within a few miles of Beersheba. Ben-Gurion concluded from this that Nasser had no intention of making peace.

Although war broke out before the end of the year, the concept of talks, and the content – and above all the spirit – of the talks as outlined by Ben-Gurion, was to be the precursor of the talks between Israel and Egypt in 1977, which led to the Egyptian–Israel Peace Treaty, the first between Israel and any of its Arab neighbours.

Throughout her short history as an independent State – and in February 1956 she was not yet eight years old – Israel was in a greater need of allies than most nations. Each one of her four neighbours – Lebanon, Syria, Jordan and Egypt – still considered itself in a state of war, and looked for active support both from the United Nations and, increasingly, from the Soviet Union. Israel's period of good relations with the Soviet Union was long past. Gromyko's support at the time of the Declaration of Independence had become something of a historic curiosity, raising a smile on the lips of Israeli diplomats whenever they recalled it at receptions where the Soviet representative was frosty. Israel's sense of vulnerability was accentuated by the Soviet arms deal with Egypt, as the flow of arms was seen to be both substantial and continuous.

The fedayeen raids across the border from Gaza were also continuous, bringing pain almost every week to some family and town. The killers struck at the heart of Israel, which appealed to the United Nations to use its influence to secure a halt to the raids and the murders. The United Nations Secretary General, Dag Hammerskjöld, who made a number of visits to Israel and Egypt, did manage to arrange for a cease-fire, but it lasted for only a few days. Golda Meir, who had just become Israel's Foreign Minister, later wrote of how, when the raids resumed, Hammerskjöld 'let it go at that and didn't return to the Middle East'. She was later to write:

I know that today a small cult has grown up around the personality and fine perceptions of Mr Hammerskjöld, but I am not a party to it. I used to meet with him often, after he had seen Ben-Gurion and talked to him for an hour or two about Buddhism and other philosophical topics, in which they had a common interest. Then he and I would discuss such commonplace subjects as a clause in the armistice agreement with Jordan that was being contravened or some complaint we had against the United Nations.

No wonder Hammerskjöld thought that Ben-Gurion was an angel and that I was impossible to get along with. I never considered him to be a friend of Israel and although I tried hard not to show it, I expect he sensed my feeling that he was less than neutral as far as the situation in the Middle East was concerned. If the Arabs said no to something – which they did all the time – Hammerskjöld never went any further.

*   *   *

On the morning of 5 April 1956, Egyptian artillery, opening sustained and systematic fire from the Gaza Strip, bombarded settlements in the Negev. Defensive fortifications, recently built in the settlements by volunteers, protected the settlers, none of whom were killed, though four soldiers and two civilians were wounded. Later in the morning, Egyptian mortar fire was directed on the regular Israeli army patrols that monitored the border. At the same time, Egyptian infiltrators crossed the border near Be'eri, but ran back when they were spotted by an Israeli patrol. Egyptian artillery then opened fire again. Three Israeli soldiers were killed.

Shortly after midday, Israeli artillery returned fire, trying to hit the Egyptian gun positions inside the Gaza Strip. In the course of the Israeli barrage, sixty-six Arab civilians were killed. An Israeli Foreign Ministry spokesman, while speaking of the 'regrettable loss' of civilian life, added that it was 'the inevitable boomerang effect of Egyptian reckless folly'.

Later that afternoon, Egyptian mortar fire was directed against five Israeli settlements: Kisufim, Ein Ha-Shlosha, Nahal Oz, Yogev and Nirim. There were no casualties. The settlers had become used to descending into their shelters. But the constant infiltrations were a burden both physically and mentally. In April, Nahal Oz – from whose fields one could see, and can still see, the minarets of Gaza City – was again the target of an attack. During a clash with the infiltrators, twenty-one-year-old Ro'i Roitberg was killed. At his funeral, on May 1, Moshe Dayan spoke the eulogy. During the course of his remarks, he expressed an understanding of the Arab perspective which was both unusual and moving. 'Let us not condemn the murders today,' Dayan said. 'What do we know of their fierce hatred for us? For eight years they have been living in refugee camps in Gaza while right before their eyes we have been turning the land and the villages in which they and their forefathers lived into our own land. We should demand Ro'i's blood not from the Arabs in Gaza but from ourselves, for closing our eyes to our cruel fate and the role of our generation.'

Dayan continued, 'We are a generation of settlers, and without the combat helmet and the barrel of a gun, we will not be able to plant a tree or build a house. This is the fate of our generation, and the choice before us is to be ready and armed, strong and hard, or to have the sword snatched from our hands and be cut down.'

Life in the border settlements could be hard, especially those like Nahal Oz, in the south near the Gaza Strip, and in the north, below the Golan Heights. But every year saw the founding of new settlements throughout Israel, and the extension of the cultivated area of the State.

On 26 July 1956, President Nasser nationalized the Suez Canal. Hitherto, the defence of the canal had been Britain's responsibility. The British Prime Minister, Anthony Eden, was outraged by Nasser's move, and, after several months of fruitless negotiation with Egypt to restore international control for the canal, began to make plans with France to seize the canal from Egypt. The French were particularly interested in some form of Israeli participation in the military side of the operation.

For several years, Shimon Peres had taken a lead in developing close relations with France, and in securing French arms and military equipment. On a visit to Paris not long after the Suez Canal had been nationalized, he was at a meeting with the French Minister of Defence, Maurice Bourgès-Manoury. 'He suddenly turned to me,' Peres later wrote, 'and, out of the blue put this question: "How much time do you reckon it would take your army to cross the Sinai Peninsula and reach Suez?" I replied that our army people estimated it would probably take from five to seven days.' Bourgès-Manoury was 'somewhat amazed by this answer', Peres recalled, ' and asked whether we were not being too optimistic, for the accepted estimate by his own advisers was that it would require at least three weeks. He went on to ask whether Israel was thinking of taking action at some time along her southern borders, and if so where. I replied that our "Suez" was Eilat, our southern-most port, now under blockade through the closure of the Straits of Tiran. We had never reconciled ourselves to this situation, and any Israeli operation would be aimed at freeing the Straits and liberating Eilat.'

Peres returned to Paris in August. Nasser's speeches were becoming more and more extreme in their hatred of Israel, threatening its annihilation. The Egyptian blockade of the Straits of Tiran was preventing merchant ships from reaching Eilat; since Nasser's nationalization of the canal in July, an Egyptian economic boycott of Israel had prevented ships bound for Tel Aviv or Haifa from using the canal. From Bourgès-Manoury, Peres learned that Britain and France were seriously considering military action against Egypt. This news was far from unwelcome.

Visiting London later that month, Peres found a similar determination to act in British government circles. There was no attempt in either capital to enlist Israeli support, but he was asked many questions 'about the strength of the Israeli Army and Israel's general attitude'. The Israeli perspective was different, but also tending to a confrontation with Egypt. The fedayeen raids into Israel from the Gaza Strip, which Egypt had occupied in 1948, had never ceased. Acts of sabotage remained a regular occurrence, as did Israeli reprisal raids. 'This escalation of sabotage and reprisal,' Peres later wrote, 'brought the Middle East – so it seemed to the Israelis – to the edge of full-scale war, and the question seemed to be not if but when. Nasser strengthened this impression by a succession of inflammatory speeches, his tone of confidence much buoyed up by heavy supplies of modern Soviet weapons which had already reached Egypt.'

Irrespective of whatever plans Britain and France might be making to regain control of the Suez Canal, Israel faced the prospect of a full war with Egypt. It was to acquire the weaponry needed for such a war that Peres returned to Paris in mid-September, and, he recalled, 'proposed to Bourgès-Manoury that an informal meeting be held at the highest level to exchange views on the situation, and to see what could be done to inject immediate additional strength into Israel. It was reasonable to presume that in such grave circumstances, our French friends would be interested in increasing the power of the Israeli Army, even without any reference to their own designs. The French Defence Minister promised to consider my proposal, and I flew on to New York.'

Here at last was a means to redress the arms imbalance created by the Soviet arms deal with Egypt. French arms soon became predominant in the Israeli Defence Forces, especially in the air force. On reaching New York, Peres found a telegram summoning him back to Paris. Bourgès-Manoury wished 'to continue our talks as soon as possible'. Peres flew back through the night, reaching the French capital on the morning of September 23. It was a Sunday. France and Britain were in the final stages of preparing Operation Musketeer, a joint Franco-British plan to land in Egypt and seize the Suez Canal. The French were anxious to draw Israel into the discussions. Peres promised to convey the request to Ben-Gurion at once, and flew back to Israel.

On September 23, the day on which Peres was in France talking to Bourgès-Manoury, a Jordanian soldier at a border post north of Bethlehem opened fire with a Bren gun on a group of a hundred Israeli archaeologists who were examining (as part of an international archaeological conference) the ancient ruins excavated during the previous two years at Ramat Rahel, the southernmost point of Jewish Jerusalem. Four of the archaeologists were killed. Two days later, also near Jerusalem, a Jordanian patrol crossed the border into Israel near the Israeli village of Aminadav, and opened fire on a group of women picking olives. One of them, Zohara Umri, a Jewish immigrant from Yemen, was shot dead. In the Beisan Valley, an infiltrator from Jordan murdered an Israeli tractor driver.

On September 25 the Israeli Cabinet discussed a reprisal action to take place at night. Ben-Gurion wanted 'vigorous' retaliation. It was agreed that any reprisal action should avoid civilian casualities. Only one Minister, Moshe Shapira, the Minister of Religion, opposed retaliation altogether. Dayan originally hoped to choose a target deep inside Jordanian territory, but was prevailed upon to attack one of two police forts not far across the border. He chose the police fort at the village of Husan, the inhabitants of which had taken part in the murder of the thirty-five Palmach soldiers sent as relief to the Etzion bloc settlements in January 1948, and had also taken part in the final attack on the bloc which had led to the massacre and mutilation of dozens of its inhabitants.

When Lieutenant-General E. L. M. Burns, the Chief of Staff of UNTSO, learnt of the impending attack he tried to contact Ben-Gurion or Golda Meir in order to argue against it, but neither was available to see him. On the morning of September 26 Moshe Dayan drove with Shimon Peres to the headquarters of Colonel Ariel Sharon, the officer commanding the paratroops, who had been instructed to carry out the reprisal. 'We left Dayan's car at the border farm-settlement of Mevuot Betar,' Peres later wrote, 'and continued another few hundred yards on foot. The fighting had already started when we reached the HQ and there was a good deal of mortar and machine-gun fire. (Dayan's driver was among the first to be wounded).'

The objective was reached and the police fortress destroyed. Thirty-seven Jordanian Legionnaires and National Guardsmen, and two civilians, were killed, and a school building in the village of Wadi Fukin blown up. The Israeli force then withdrew. But ten Israeli soldiers had been killed, more than the number of Israelis killed in the recent spate of cross-border incidents. Peres noted: 'As we trudged back on this cold September morning, Dayan observed: "We're reaching the end of the beginning".'

The American ambassador to Israel, Edward B. Lawson, told Ben-Gurion that he doubted the Husan reprisal would be effective. His doubts were justified. On October 4, eight days later, a group of Bedouin from the West Bank ambushed five Israelis near Sdom and murdered them. Dayan was anxious to mount yet another reprisal raid for the Sdom killings, but on October 6 Ben-Gurion wrote to him urging restraint. Preparations were underway for a major military operation against Egypt.

On October 7 the Cabinet gave its approval to Ben-Gurion's pragmatic policy of restraint. Two days later, on October 9, a group of infiltrators reached the very centre of Israel, nine miles from the border, and murdered two Israeli workers at Even Yehuda, cutting off their victims' ears, it would seem in order to prove to those who had sent them that they had succeeded in their mission. At an emergency Cabinet meeting on the following day, Dayan urged an immediate and substantial reprisal, the destruction of the Kalkilya police fort ten miles from Even Yehuda. The troops were instructed to avoid harming civilians.

The Kalkilya reprisal was led by Major Mordechai Gur. The police fort was blown up, but eight Israeli troops were killed during the battle for it. A second Israeli force, which had been sent east of Kalkilya as a 'blocking force', was itself attacked by Jordanian Legionnaires. Another five Israelis were killed. By the end of the day eighteen Israelis had been killed: the heaviest loss of life in any reprisal action since 1951. More than seventy Jordanians were also killed: policemen, National Guardsmen and Legionnaires. 'The reprisal raids', wrote Yitzhak Rabin (then head of Northern Command) in his memoirs, 'had proved ineffective in dealing with the problem of terrorism.' The Israeli army, Dayan reflected a year later, had 'reached the end of the chapter of night reprisal actions'.

One phase of the Arab–Israel conflict was indeed ending, and another was about to start. Arms were arriving from France in French ships, reaching the port at Haifa in the middle of the night and amid strict secrecy. Shimon Peres, who had been responsible for negotiating the arrival of these arms, was present one night to watch them being unloaded. He took with him the poet Nathan Alterman, who was inspired to write about his experience in verse:

> I dreamed last night of steel, much steel, new steel.
> The bearer of laden canisters, ringing on iron chains,
> Arrives from afar, sets foot on the shore and, as imagination turns into
> reality,
> With the first touch of the land he becomes the expression of the
> power of the Jews.

*   *   *

On September 28 an Israeli delegation left Lod airport in strictest secrecy for Paris. It was headed by Golda Meir, whose son's wife was the daughter of one of the archaeologists killed at Ramat Rahel five days earlier. Also on the delegation were Moshe Dayan, Shimon Peres and the French-speaking Moshe Carmel, the Minister of Transport. To maintain secrecy, the Israelis flew, not on an Israeli commercial airliner, but on a French bomber 'of World War Two vintage', Peres recalled, 'which had been refitted to accommodate passengers'. They flew first to Bizerta, in North Africa, where they stayed overnight with the French commander of the town, before flying on to Paris.

The French were much influenced at this time by their own confrontation with Arab nationalism in Algeria. In the French perspective, Nasser was the embodiment of Pan-Arabism, which they regarded as also fermenting the troubles in Algeria. A pro-Israeli stance was particularly noticeable in French military circles.

Discussions between the Israeli delegation and the French Foreign and Defence Ministers took place in Paris throughout September 30. It was clear that the French wanted to take action against Egypt, and were worried about United States efforts to find a solution without war; a solution that would effectively leave Nasser in control of the canal, albeit with guarantees. Golda Meir warned the French, however, that Israel could not afford to quarrel with the United States. As the talks continued, 'it became clear', Peres wrote, 'that France would support Israel if Israel acted alone. This would not mean support by the French Army, but only help in strengthening the weaponry of the Israeli Army.'

On October 1 the Israeli delegation returned from France (still in strictest secrecy). Following the Kalkilya attack, King Hussein of Jordan asked the commander of the British forces in the Middle East to send British aircraft to Jordan under the Anglo-Jordanian Defence Treaty. Activating her treaty with Iraq, Jordan also summoned Iraqi troops to come to her aid. On October 12 the British government informed Israel that, in accordance with its treaties with both Jordan and Iraq, Britain would go to the aid of both these countries if they were attacked by Israel.

As Iraqi troops prepared to enter Jordan, Golda Meir issued an official statement that any such move was 'part of the plan to advance the territorial ambitions of Iraq and to bring about far-reaching changes in the status of the region. It constitutes a direct threat to the territorial integrity of Israel by an Arab State which invaded her borders in 1948 and subsequently refused to sign an armistice agreement.' The Government of Israel would 'frustrate this hostile design'. Iraq had been the only Arab belligerent not to sign an armistice after the War of Independence.

Ben-Gurion, speaking in the Knesset on October 15 – the day on which the Iraqi troops were to enter Jordan, and the day on which a partial mobilization was ordered in Israel – stated with solemnity: 'Israel reserves to herself freedom of action.' The situation was not only grave, but also complex, and almost absurd. Only a few days before a decision was to be taken on whether Britain and France, and possibly Israel, would be at war with Egypt, 'it was not at all clear', Shimon Peres later wrote, 'who would be fighting whom, or when'. No one knew which option might be the actual one: 'Israel against Jordan alone? Israel against Jordan and Iraq and, indirectly, also against Britain? Israel against Egypt alone? Or Israel together, indirectly, with Britain against Egypt – perhaps at the same time as she would be fighting Jordan, Iraq and, indirectly, Britain on the eastern front! There were enough arguments to make feasible any one of these possibilities. What was clear was that war was inevitable.'

In the third week of October, Nasser moved part of his army into the Gaza Strip, where so many fedayeen had their bases. Fire was opened on Israeli settlements near the border, making farming a hazardous occupation. Crossing the Sinai border, Egyptian troops harassed Israeli army patrols. Nasser also began to move troops, and weapons, into the Sinai. These weapons were recently supplied by the Soviet Union. They were up to date and likely to be effective. 'There was no doubt in Israel's mind,' Shimon Peres recalled, 'that Nasser was about to invade at any moment.'

Another secret Israeli mission to France was undertaken, this time headed by Ben-Gurion. The objective was to coordinate with Britain and France a triple military attack on Egypt. The meetings were held outside Paris, in a villa at Sèvres, starting on the afternoon of October 22. The British government was represented by the Foreign Secretary, Selwyn Lloyd, the French by their Prime Minister (Guy Mollet), Foreign Minister (Christian Pineau) and Minister of Defence (Maurice Bourgès-Manoury).

As well as Ben-Gurion, both Moshe Dayan and Shimon Peres were present. Discussions continued throughout the following day. Ben-Gurion was nervous about starting any military action before the American Presidential election on November 6; he wanted to secure American support, and knew that it would not be forthcoming during an election campaign. He was also deeply concerned about a possible Egyptian air raid over civilian targets in Israel, and was anxious not to provoke it. The British were worried lest a prolonged war would unite the Arab world against Britain, endangering her oil and other commercial interests. The plan that emerged was that Israel would attack Egypt and advance into the Sinai, towards the canal. Britain and France would then intervene militarily in order to protect the canal, would occupy the canal, and would then demand a cease-fire. Israel, having achieved its own objectives in Sinai, would comply. Britain and France would be in occupation of the canal. Nasser, his regime humiliated, would fall.

On the morning of October 24, while still at Sèvres, Ben-Gurion asked to see Dayan and Peres. Peres later recalled:

> When we arrived, we found Ben-Gurion sitting with his military aide
> under the thick branches of a fine old tree. It was a sunny autumn day,

soft and clear. 'What, here so quickly?' exclaimed Ben-Gurion. A second later he was asking Dayan to sketch once again the operational plan he had suggested.

There was no map of the Middle East in the garden, so Dayan asked us to imagine it. We did not think that was good enough, so I took out a packet of cigarettes, carefully tore off the printed parts to leave as much blank space as possible, and handed what was left to Dayan. Upon it he proceeded to draw an outline of the Sinai Peninsula and marked it with three arrows. The centre one with dotted lines ran through the middle of Sinai, showing the direction of the planes which would land paratroops at Mitla, near the Suez Canal. The arrow above it marked the westward movement of the armour, along the coastal road towards the Canal. The southern arrow, which Dayan traced smoothly over the flat surface of the cigarette box – the terrain is in fact mountainous and almost impassable – ran along the Gulf of Akaba, the point of the arrow impaling Sharm el-Sheikh.

Ben-Gurion wrote down ten questions. From these it was clear that he had in mind no territorial additions, except perhaps for a narrow strip that would link Eilat with Sharm el-Sheikh. It was also apparent to me that he had made his decision.

A British participant in the meeting on October 24, Donald Logan, later recalled, 'We pressed Dayan for assurance that the Israelis understood that unless their military action posed a threat to the Canal, British forces would not act. It did not come easily. The Israelis did not conceal that their main objective would be Sharm el-Sheikh on the Straits of Tiran to enable them to maintain passage for their ships to the port of Akaba. We emphasized that a move in that direction would not pose a threat to the Canal. Eventually sketch maps were made and we were assured that there would be military activity in the region of the Mitla Pass. More than that we could not get, but the Mitla Pass being reasonably close to the Canal we concluded that the Israelis sufficiently understood the British position though they remained suspicious of our intentions.'

The Sèvres meeting was over. For many years, all the participants were to deny that it took place at all. The Israelis returned home on October 25. That day, the government ordered a partial call-up of the reserves. It also alerted a battalion of paratroops to be ready for action in four days' time. On the morning of October 28 the Cabinet met and agreed that Israel would cross the Egyptian border into Sinai.

Shortly after midday on the following afternoon, October 29, four Israeli piston-engined P-51 Mustang fighters, veteran aircraft of the Second World War, flew across the Egyptian border into Sinai and, descending to the incredibly low altitude of 12 feet above the ground, cut with their propellers and wings all overhead telephone lines in the Sinai connecting the Egyptian army and air force headquarters with the Egyptian units in Sinai. Two hours later, at 3.20, sixteen Dakotas, carrying the paratroopers who had been alerted four days earlier, took off from their bases in Israel and flew towards Egypt. The Sinai campaign had begun.

CHAPTER EIGHTEEN

The Sinai campaign

The start of the Sinai campaign came, in many ways, as a relief to the Israeli public, despite the inevitable uncertainties and casualties of war. The Czech arms deal with Egypt, the Suez Crisis, and Nasser's fulminations, had left Israel feeling isolated and threatened. The continuing terror raids across the border, and the speeches and statements emanating from Egypt threatening Israel's existence, amounted, in Israeli eyes, to an abrogation of the 1949 armistice agreement.

The objective of the Israeli paratroop attack on 29 October 1956 – the opening move of the Sinai campaign – was the eastern entrance to the Mitla Pass, 156 miles into the Sinai, and 45 miles from the Suez Canal. The drop was made at five in the afternoon, by 395 paratroopers, commanded by Lieutenant-Colonel Rafael (Raful) Eitan. 'The move,' wrote Chaim Herzog in his book *The Arab–Israeli Wars*, 'was a classic application of some of the basic principles of war. The element of surprise was complete – indeed so great was the surprise that for twenty-four hours the Egyptians were kept guessing: what was the real purpose of the operation? Was this merely a reprisal raid or all-out war? If so, where would the main Israeli attack fall? The Israelis were in a position to retain the initiative and maintain momentum, while the enemy forces were still not in a position to realize what was in fact happening.'

To establish a supply line across the expanse of desert between the Israeli border and the entrance to the Mitla Pass – which was still under Egyptian control – a group of paratroopers commanded by Colonel Ariel Sharon crossed the border into Sinai and advanced overland. There were three Egyptian military positions between the border and the pass; each one had to be fought for, and each one was overrun.

On the afternoon of October 30, in London, the British government handed an ultimatum to the representatives of both Israel and Egypt. They were to withdraw their forces back from the canal to a depth of at least 16 kilometres (10 miles). As Egypt was then in full occupation of the canal, and Israeli forces were at least 50 kilometres (30 miles) from it, the ultimatum was, as intended, an order to Egypt to leave the canal. Egypt rejected the ultimatum at once. Israel made no reply.

During October 30 President Eisenhower informed Ben-Gurion that the United States would refrain from censuring the Israeli action if Israel declared that the purpose of the attack into Sinai was not territorial expansion. Ben-Gurion replied on the following day that the aim was 'to convert the existing situation into one of peace', which, he explained, must include the disbandment of the fedayeen, the annulment of the Egyptian economic boycott against Israel, the guarantee of freedom of navigation through the Straits of Tiran, and the end of Arab military alliances against Israel.

The residents of Haifa were woken up at 3.30 on the morning of October 31 by gunfire from an Egyptian frigate, the *Ibrahim Al Awwal*, which fired more than 200 shells into the town. It was then attacked and driven off by the French destroyer *Crescent*. Some two hours later the Israeli destroyers *Eilat* and *Jaffa* opened fire. At dawn two Israeli fighters – French Ouragons – attacked and damaged the *Ibrahim Al Awwal* with their rockets. With no hope of escaping, her captain ran up the white flag of surrender at 7.10 a.m. His ship was then boarded, captured, and towed into Haifa harbour.

On October 31 British and French warplanes bombarded the Egyptian airfields. But it was to be another six days before any Anglo-French military landings took place. Meanwhile, Israel alone was engaged in ground fighting against Egypt. Sharon had been ordered not to attack the heavily defended Mitla pass; such an attack was superfluous to the plan Dayan had devised. He did obtain permission to send a limited 'patrol' into the pass, provided it avoided any involvement in serious combat. Sharon then sent a large detachment, commanded by Major Mordechai Gur. It was caught in an Egyptian ambush, and for seven hours held out against intense firepower. Sharon then had to commit the rest of his brigade to extricate the trapped force. After 38 men were killed and 120 wounded, the pass was captured. More than 200 Egyptian soldiers had been killed.

Chaim Herzog has commented, 'This tragic operation, which had been completely unnecessary from a tactical or strategic point of view, brought in its wake some serious recriminations between Dayan and Sharon. Dayan maintained that he had been misled by Sharon, who had requested and received permission to send a patrol into the Pass. Taking advantage of this approval, the paratroop commander had engaged in what Dayan termed "a subterfuge" by calling the operation a patrol in order to get the approval of the General Staff. Dayan was heavily criticized at the time for not disciplining Sharon.'

Reflecting on the character and career and Israel's most controversial general, Herzog added:

A very independently minded and assertive character, Sharon was later in his political career to be accused of dictatorial tendencies by his opponents. He was to be accused, both in this and later campaigns, of insubordination and dishonesty. He can best be described as a Patton-like, swashbuckling general, who rose in the ranks of the Israel Defence Forces, proved himself to have an uncanny feel for battle, but at the same time to be a most difficult person to command.

Few, if any of his superior officers over the years had a good word to say for him as far as human relations and integrity were concerned, although none would deny his innate ability as a field soldier. Probably because of this, he never achieved his great ambition, to be Chief of Staff of the armed forces.

Also on October 31, Israeli troops attacked the Egyptian fortified positions at Abu Ageila, only 15 miles from the Israeli border. This position had to be overrun if Israeli troops were to be able to maintain their hold on central Sinai without the danger of an attack from the rear, and to control the coastal area around El Arish. The task was entrusted to Colonel Yehuda Wallach, a veteran of the War of Independence (and later to be Professor of Military History at Tel Aviv University).

The Egyptians fought at Abu Ageila with great tenacity. In one of the tank battles, every Israeli tank was hit. For two days the battle raged. One brigade commander, whom the Chief of Staff, General Dayan, felt had not made the necessary effort to enter the combat, was replaced. Another commander, Colonel Shmuel Galinka, was killed leading a frontal attack against a fortified position. It was not until nightfall on November 1 that the Egyptians, recognizing that they could not maintain their positions for much longer, retreated eastward. Israel could now supply its troops in the whole of central Sinai without danger of attack.

That same day, November 1, the coastal town of Rafa, at the entrance to the Gaza Strip, and little more than 2 miles from the Israeli border, was attacked with the aim of cutting off the Gaza Strip from Egypt, and controlling the coastal highway. Two Israeli brigades took part in the assault, one commanded by Colonel Benjamin Gibli, a former Director of Military Intelligence (who had been forced to resign from his intelligence work at the time of the Lavon affair), the other by Colonel Chaim Bar-Lev, who had played a leading part in the conquest of the Negev in the War of Independence. As at the Mitla Pass and Abu Ageila, the Egyptians fought a fierce defensive action.

At one point in the battle, because the explosive equipment given to them did not work properly, the Israeli soldiers had to cut their way through a series of barbed wire emplacements under heavy machine-gun fire. As at the Mitla Pass, there was much hand-to-hand fighting. For the Israeli attackers, the turn of the tide came when Bar-Lev's brigade linked up with

Gibli's. Dayan, who accompanied it, described in his diary, 'We fell into each other's arms in the classic tradition of a Russian movie.' Half an hour later, Bar-Lev's brigade was on the move again, pushing eastward to the town of El Arish, the principal town of northern Sinai.

Even as the Sinai war was being fought, the fedayeen raids inside Israel had continued, several gangs roaming almost at will in the northern Negev. On November 1 a car in which members of kibbutz Erez were travelling hit a mine laid by fedayeen: three passengers were killed.

The commander of the Jerusalem District, Chaim Herzog, was alerted to the task of defending the Mount Scopus enclave should the Jordanians join the war and try to seize it. Should such an attack take place, he had the High Command's authority to counter-attack. 'But the Jordanians did not intervene or ever give us cause for concern,' he later wrote. 'Indeed on one occasion, a Jordanian soldier fired his gun, and within minutes the Jordanian commander called on our direct phone line to apologize and emphasize that it was a stray shot.'

The imperative of securing the border by settlement continued even during the war. Just north of Kastel – the Crusader ruin and former Arab village overlooking the road from Jerusalem to the coast, which had been captured by Israel in the War of Independence – was the bare hillside of Sheikh Abdul Aziz, also under Israeli control. As part of the 'Jordanian diversion', Herzog recalled, an orchard was planted there, right up to the Jordanian border. 'We awaited the Arab Legion's response. Knowing that we were likely to be attacked, we sent soldiers dressed as farmers to plant the trees.'

A new settlement was in being. Later it became the village of Mevasseret Yerushalayim (Herald of Jerusalem), and served as a garden suburb for labourers in the factories in the western edge of Jerusalem. Seven years later its inhabitants merged with the workers' suburb of Ma'oz Zion, across the Tel Aviv–Jerusalem highway to the south, and formed a single municipality, Mevasseret Zion (Herald of Zion).

During the Suez campaign, Jordanian artillery did open fire on the planters at Mevasseret. They also opened fire on United Nations Truce Supervision soldiers who came to examine Israeli complaints that fighting had broken out. But the planting went on, and the firing ceased.

On November 2 the spectre of a wider war opened up when the Syrian embassy in Washington informed the United States government that Syria had 'decided to implement immediately' the joint Egyptian–Syrian defence pact. The note went on to say that all Syria's armed forces had been placed under the command of the Egyptian Chief of Staff, General Abdul Hakim Amer, and that all Syria's resources were, from that moment on, 'devoted to the common cause'.

One Israeli fear was that Egyptian bombers would bomb the main cities. There were a number of bombing raids during the first two days of the war, but they were not effective. 'On several occasions,' Shimon Peres has written, 'their Ilyushin bombers penetrated Israel air space, but their pilots for the most part jettisoned their bombs on open fields where they ran no danger of counter-fire from harmless blades of grass. In each case, Cairo Radio announced that the targets – giving the names of the cities – had been attacked, just as they kept claiming "successes" and "victories" throughout the campaign.'

During November 2 the Israeli forces in the northern Sinai entered El Arish – the capital of the region – ensuring their control of the coast and completing the isolation of Gaza. In his diary, Dayan described the scene in the military hospital at El Arish, with the bodies of Egyptian soldiers 'who had been abandoned in the midst of operation and treatment' littering the building. Sniping was still taking place as he went through the wards and, as he stood for a moment at an open window, looking out at the street below, an Egyptian soldier opened fire with a machine-gun. Dayan's signalman, who was at his side, was killed.

That evening, an Israeli task force reached a point only 15 kilometres (10 miles) east of the Suez Canal. There, it halted. Among the booty of war which it acquired on its way were forty Soviet-built T-34 tanks and sixty armoured cars, a substantial acquisition to Israel's armoury.

November 2 also saw the Israeli attack on the Gaza Strip, where 10,000 Egyptian troops were stationed. The assault was made from the south, across the Ali Montar ridge. It was this ridge which, forty years earlier, had caused the British such high casualties in their two unsuccessful attempts to drive the Turks from Gaza during the First World War. The ridge was strongly defended, but was breached in its south-west corner by an Israeli tank squadron, which then moved rapidly northward, pushing into the Gaza Strip at Beit Hanun.

At midday on November 2, as a result of the mediation of the United Nations Mixed Armistice Commission, which had remained in Gaza City (and which was mindful of the fate of more than 200,000 Palestinian Arab refugees in the various camps around Gaza City), the Egyptian Governor of Gaza City surrendered, and then persuaded the substantial garrison to lay down its arms. A few hours later, the Egyptian Governor of the Gaza Strip also surrendered. Towards the southern end of the Strip, Khan Yunis was occupied by Israeli forces that same evening.

During November 2 the Secretary-General of the United Nations, Dag Hammerskjöld, informed Israel that the General Assembly had passed a resolution calling for a cease-fire. On the following day, November 3, Israel informed Hammerskjöld that she accepted. Israeli troops had already halted

15 kilometres east of the canal, in accordance with the original Anglo-French ultimatum (which in fact specified 16 kilometres). The General Assembly went on to create a United Nations Emergency Force (UNEF) to replace the withdrawing troops.

As news of the mass of weaponry and stores captured by the Israeli army were announced on November 3, there was shock inside Israel at the announcement that among the 'standard equipment' of the Egyptian officers captured was a two-volume paperback edition in Arabic of Hitler's *Mein Kampf.*

For six days, Israel had fought its war and overrun Sinai. The final military assault was a dramatic 155-mile dash along the western shore of the Gulf of Akaba, with the aim of capturing Sharm el-Sheikh and the Egyptian guns dominating the Straits of Tiran. The attack was begun from just north of Eilat by an infantry brigade commanded by Colonel Avraham Yoffe. This was a reserve brigade, raised from the farmers of the Jezreel Valley. Yoffe, himself a farmer and a noted authority on wildlife, was subsequently the head of Israel's Nature Reserve Authority.

Advancing with 1,800 men and 200 vehicles, the brigade negotiated some of the roughest terrain encountered by Israeli forces. Chaim Herzog has described it as 'an area of saw-tooth ridges dropping straight into the sea; an area strewn with large boulders, deep sand, and ravines; excruciatingly hot, devoid of water, and above all, a camel route that had not been designed for passage by a fully motorized infantry brigade.' Shortly after passing the oasis of Ein Al-Furtaga, the brigade came to what Herzog has described as a 'major physical obstruction – a steep uphill climb through Wadi Zaala, with deep, boulder-strewn, powdery sand in which most of the vehicles sank and had to be pushed, pulled or shifted by hand.'

At the Dahab oasis the Israeli troops fought their first battle, having been ambushed, as a result of lack of caution, by the camel-riding section of the Egyptian Frontier Force – the desert police force that had been the main instrument of control over the Bedouin of Sinai. A struggle ensued, and there were several casualties, but victory was secured. While at Dahab, the battalion received the fuel needed to continue its southward march from landing craft sent by sea from Eilat.

At Sharm el-Sheikh, an Egyptian garrison of 1,500 men awaited the attackers behind its stone and concrete fortifications, well equipped with artillery. But as Colonel Yoffe's men drew near on the afternoon of November 4, men from Sharon's parachute brigade were also making their way along the coast of the Gulf of Suez towards Sharm el-Sheikh, effectively doubling the size of the attacking force. The main attack by Colonel Yoffe's brigade began while it was still dark, in the early hours of November 5. Egyptian artillery caused heavy casualties, and at one point the assaulting battalion had to withdraw, taking its wounded with it. After four hours of continuous fighting the Egyptian command post was overrun. It was nine o'clock in the morning. Half an hour later, the Egyptian guns overlooking the Straits of Tiran – which had been the main instruments closing the straits to Israeli shipping – had been disarmed. Israel had achieved its principal war aim.

The Anglo-French attack on the Suez Canal took place on the morning of November 5, when French paratroops seized Port Said. British and French troops then proceeded to move southward along the canal. The Egyptians resisted their advance. Then, under American pressure, the Anglo-French troops were forced to halt, and to promise to withdraw.

On the day of the Anglo-French landings, the Soviet Prime Minister, Nikolai Bulganin, sent a message to Israel which had ominous implications. The Soviet Union, he said, was in possession of ballistic missiles 'capable of reaching any point on the globe'. Fearful of a possible direct Soviet intervention, Ben-Gurion sent Golda Meir and Shimon Peres to France to find out whether Israel could count upon French support in the event of Soviet intervention. Talks took place in Paris on the morning of November 6, when the French Foreign Minister, Christian Pineau, told the two Israelis that although France would support Israel 'with everything we've got', and would even share with Israel her military resources, 'one could not escape the fact that Russia was much more powerful than France, and commanded missiles and non-conventional weapons'. Peres later noted laconically, 'One was constrained to draw the inevitable conclusions.'

Golda Meir and Shimon Peres flew back to Israel on the afternoon of November 6. As they were en route, a message from Eisenhower reached Ben-Gurion, demanding that Israeli forces stop fighting 'immediately' and withdraw from Sinai. An emergency Cabinet meeting was called at three in the morning of November 7 (it was still the afternoon of November 6 in Washington). It was agreed that Ben-Gurion should tell Eisenhower that Israel was not prepared to withdraw from Sinai without safeguards. But she would stop fighting, advance no further, and abide by the United Nations call for a cease-fire.

Israel's first war since the War of Independence eight years earlier was over. It had lasted for a week. Her losses were 172 killed and 817 wounded.

Israel was much condemned in the international community, particularly by the United States, for having taken the initiative against Egypt. The American Secretary of State, John Foster Dulles, threatened economic sanctions. Two senior Israeli emissaries were despatched to America to try to redress the balance. One was the commander of the Jerusalem District, Chaim Herzog, who received his summons to go to the United States while he was touring the newly captured Sinai Desert. He was, indeed, at Mount Sinai itself when the message reached him. At the State Department in Washington, Herzog faced the American government's clear disapproval of the Israeli attack in Sinai, which was condemned as an offensive action by Israel. Earlier in his visit he had put Israel's position to the leading American-Jewish fund-raisers. 'Pointing out that I was the second Jew in history to receive a message on Mount Sinai,' he later wrote – Moses being the first – 'I explained the background of the Sinai Campaign, our alliances with France and Britain, and our reasoning for going forward into Egypt. Away from the slanted scrutiny of the media and the politically motivated response of government officials, people seemed to accept and understand our position. I made the same case to State Department officials. I spoke to both officials and civilians from the heart and pulled no punches.'

The second Israeli emissary to be sent, and the more senior, was Golda Meir, who told the United Nations General Assembly, with regard to the 'rights of war' and 'belligerent status' that had been accorded to the fedayeen in the public mind outside Israel: 'A comfortable division has been made. The Arab states unilaterally enjoy the "rights of war": Israel has the unilateral responsibility of keeping the peace. But belligerency is not a one-way street. Is it then surprising if a people labouring under this monstrous distinction should finally become restive and at last seek a way of rescuing its life from the perils of the regulated war that is conducted against it from all sides?'

On November 14 the Knesset agreed to an Israeli withdrawal from all the territory captured in the Sinai campaign, provided that a satisfactory arrangement could be worked out with the United Nations Emergency Force (UNEF) to prevent Gaza being used by Egypt either as a jumping off point for terrorist raids, or to re-close the Straits of Tiran.

During a final debate on the withdrawal, held in the Knesset on 5 March 1957, Ben-Gurion was heckled and taunted by members of Herut from the moment he began to speak. The withdrawal took place on 16 March. As proof that the straits would be kept open, on April 24 an American cargo ship, *Kernhills*, sailed through the Straits with a cargo of crude oil and docked at Eilat. An Egyptian administration returned to the Gaza Strip, but not the Egyptian army.

The initial Israeli understanding had been that Egypt would not return to the Gaza Strip, since this was not considered Egyptian territory: it had been conquered and occupied by Egypt in 1948, having previously been part of Mandate Palestine. There was uproar in Israel when the Strip was immediately handed back to Egyptian administration and military control.

By the Sinai campaign, Israel shattered its earlier image of a helpless, besieged mini-State living on borrowed time, and dependent for its survival upon the will and whim of its largest Arab neighbour. After eight years of statehood, a very brief span of time indeed in the history of nations, it no

longer appeared vulnerable and temporary. Its international standing improved dramatically as a result of the campaign, as did the prowess and
reputation of its citizen army, the Israel Defence Forces. Israel's perceived
and actual military strength, became its most important asset in its dealings
with the Arab world for the next ten years.

CHAPTER NINETEEN

A State in being
1956–1963

For Israel, the war of 1956 had lasted for a hundred hours. Immediately after it, a strenuous effort was made by Israel to persuade the United States and the United Nations that an Israeli withdrawal from Sinai and the Gaza Strip would be dangerous in the extreme, opening Israel to the possibility of a direct Egyptian attack. In December 1956 the Foreign Minister, Golda Meir, was sent to New York to argue the case in the United Nations. Before leaving, she decided to visit the Sinai and Gaza Strip and see for herself the situation there. Later she recalled:

> I shall never forget my first sight of the elaborate Egyptian military installations – built in defiance of the United Nations itself – at Sharm el-Sheikh for the sole purpose of maintaining an illegal blockade against our shipping.
>
> The area of Sharm el-Sheikh is incredibly lovely; the waters of the Red Sea must be the bluest and clearest in the world, and they are framed by mountains that range in colour from deep red to violet and purple. There, in that beautiful tranquil setting, on an empty shore, stood the grotesque battery of huge naval guns that had paralysed Eilat for so long. For me, it was a picture that symbolized everything.
>
> Then I toured the Gaza Strip, from which the fedayeen had gone out on their murderous assignments for so many months and in which the Egyptians had kept a quarter of a million men, women and children (of whom nearly 60 per cent were Arab refugees) in the most shameful poverty and destitution.
>
> I was appalled by what I saw there and by the fact that those miserable people had been maintained in such a degrading condition for over eight years only so that the Arab leaders could show the refugee camps to visitors and make political capital out of them.

Those refugees could and should have been resettled at once in any of the Arab countries of the Middle East – countries, incidentally, whose language, traditions and religion they share. The Arabs would still have been able to continue their quarrel with us, but at least the refugees would not have been kept in a state of semi-starvation or lived in such abject terror of their Egyptian masters.

I couldn't help comparing what I saw in the Gaza Strip to what we had done – even with all the mistakes we had made – for the Jews who had come to Israel in those same eight years.

With these images and thoughts in the forefront of her mind, Golda Meir flew to New York. There, on 5 December 1956, she expressed her feelings before the General Assembly. 'Israel's people', she said, 'went into the desert or struck roots in stony hillsides to establish new villages, to build roads, houses, schools and hospitals; while Arab terrorists, entering from Egypt and Jordan, were sent in to kill and destroy. Israel dug wells, brought water in pipes from great distances; Egypt sent in fedayeen to blow up the wells and the pipes. Jews from Yemen brought in sick, undernourished children believing that two out of every five would die; we cut that number down to one out of twenty-five.' While Israel 'fed those babies and cured their diseases, the fedayeen were sent in to throw bombs at children in synagogues and grenades into baby homes.'

These were angry words. But they reflected a perspective that was felt in every Israeli home.

While in the United States, Golda Meir also held discussions with the American Secretary of State, recalling 'the hours of difficult fruitless conversation with Mr Dulles, the equally difficult and fruitless behind-the-scenes-negotiations we had held with the delegates of other powerful states, the incessant efforts we had made – without the slightest success – to explain that there was only one way to ensure peace in the Middle East: not by continuing to appease the Arabs at our expense but by insisting on a non-aggression pact between the Arab states and Israel, on regional disarmament and on direct negotiations.'

On 1 March 1957 the Government of Israel, having bowed to the international pressure for a withdrawal from Sinai and from the Straits of Tiran, and having accepted with reluctance a United Nations military presence there as a form of protection, announced that it would regard 'any interference with shipping' through the Straits 'as an aggressive act against which Israel is entitled to exercise self-defence.' Whether this announcement would serve as a deterrent to a future Egyptian government, no one could tell. But it was a statement behind which stood the full determination of Israel not to allow herself to be subjected to territorial or maritime threats by any other Power. It was also supported that day by a strong statement from the French delegate to the United Nations, Georges Picot, that if Israel were to suffer Egyptian 'encroachments' either at Gaza or the Straits of Tiran, 'it would be fully entitled to use its right of self-defence'.

In terms of borders, Egypt was still sovereign in the Sinai, and was again in occupation of the Gaza Strip. Nothing was changed, the American general S. L. A. Marshall wrote four years later, 'except that settlers in the Negev at last knew blessed sleep, free of terror that came by night'. In the decade following 1956 there were practically no fedayeen raids into Israel. 'These were the best ten years from the point of view of daily security,' Shimon Peres recalled. There was also the assurance of the guarantee to keep open the Straits of Tiran.

The American hostility during the Suez War, which had forced the British and French to halt their operation before it could achieve its objectives, had created bitterness in Israel. This was accentuated by the fact that after the war Israel still had to continue paying off a large loan that had earlier been received from the United States Import-Export Bank. Golda Meir later recalled how, when this was discussed in the Cabinet, the government was 'greatly tempted' to ask for a postponement of the large repayment that had become due at the end of 1956. 'A recession had been going on in Israel then,' she wrote, 'and it was really very hard for us to scrape up the money we owed. But we weighed the pros and cons and finally decided that, no matter how difficult it was, we were not going to be in default, even by a day. And I shall never forget Eban's description of the surprise on the usually guarded faces of the people at the Import-Export Bank in Washington when he walked in – at exactly the specified time and date – and presented our cheque.'

In 1957 the difficulties of agriculture in the Adullam region (west of the Hebron Hills) led to its being designated a special area. It was clear that land reclamation had to precede settlement activity. The Jewish National Fund undertook the task, using money raised largely in Britain and the United States, and with it reclaiming potential farming land by terracing and stone clearing and by draining the stony soil. Deep wells were drilled to improve the water supply. Three rural centres were created to advance the scheme, from which eleven villages were the beneficiaries. Five of these villages were new, established by the Jewish Agency's Agricultural Settlement Department.

Orchards and vineyards were introduced on the higher ground to the east, and poultry farming provided the main livelihood. In the lower, more westerly region, wheat, cotton, sunflowers, tobacco, vegetables and sorghum were grown, and sheep and cattle raised.

Northern Galilee, in which the Israeli Arabs formed a majority of the population (until the mid-1990s) also saw a growth of settlement activity after the Suez War. In 1958 Adamit was founded on the Lebanese border, accessible only by a specially built, 3-mile-long serpentine road that led up to its isolated mountain plateau. Its first settlers were mostly Israeli-born.

They took the name of their kibbutz from the biblical town of Adami, one of the towns of the tribe of Naphtali, which is thought to have been located in the region.

The years following the Sinai campaign were years of growth and relative tranquillity. It was also a period when immigration rose again, largely as a result of the hostility generated by the Sinai campaign in Egypt. In 1956 more than 55,000 immigrants arrived, and in 1957 a further 70,000, most of them from Egypt and North Africa.

Under Ben-Gurion's influence from the first years of statehood, Israel had sought good relations with the new States of Asia and Africa. On 6 March 1957, Ghana became independent, and the British flag was run down, as it had been in Palestine, to signal the end of the colonial era. Three weeks later, Israel established its embassy in Accra. This was the beginning of a sustained Israeli effort not only to open up diplomatic relations but to reach agreements on agricultural cooperation with twenty-one African States during the course of the next decade. The first of these agreements was signed with Mali, followed by Upper Volta, Madagascar and Dahomey. More than 3,000 Israeli experts made their way to Africa during the next decade to advise on the introduction of new technology and crops, the establishment of agricultural farms and training centres, the organization of rural institutions, and the planning of comprehensive regional rural development projects. Israelis were also active in helping local medical, public health, educational and construction activity.

In Cameroon, Israelis helped to create a vegetable growing and marketing programme; in Senegal a bee-raising programme; and in Uganda experimental citrus farms. Israeli military advisers were also active throughout the continent; by 1966 they were training the armed forces of Ethiopia, Ghana, Sierra Leone, Tanzania, Uganda and Zaire. Israelis also played a leading part in building hotels; and in three countries – Dahomey, the Central African Republic; and Togo – it was Israelis who helped to establish national lotteries.

Israel's growth in the second decade following the establishment of the State, and her slow but steady economic growth, was always conducted under the shadow of the declared hostility of its Arab neighbours. The fact that Israel had no non-Arab neighbours made this hostility doubly harsh. Israeli farmers tilled the soil right up to the armistice lines of 1949. But Arab recognition of these borders was never obtained. An example of this uncompromising stance came during the Law of the Sea international conference in 1958, when Saudi Arabia was represented by a Palestinian Arab, Ahmed Shukeiry, who told the conference with regard to Israel's narrow outlet on the Red Sea, 'We have made no mention of Israel as a bordering state on the Gulf of Akaba, not out of political reasoning, neither was it a mere forgetful omission. It is with full purpose, and the reason is one of law and not of politics.' Shukeiry went on to explain, 'The States in the region do not recognize Israel, be it the existence, territory or the boundary, if any. Second, Israel's foothold on the Akaba Gulf, apart from its illegal origin, is based on Armistice Agreements which by their character and express provisions vest no sovereignty whatsoever.'

May 1958 marked the tenth anniversary of Israel's statehood. 'The first decade hands on unfinished tasks,' wrote Abba Eban at the time. 'But it will never be eclipsed in Israel's memory; for in the eyes of an aged weary people it rekindled the splendour of a youthful dawn.'

On 1 April 1959 there took place an episode which the then head of Southern Command, Chaim Herzog, has called 'an embarrassing foul-up'. A number of officers on the General Staff, without consulting their superiors, decided to institute a mock mobilization of Israel's reserves. The idea was to see what the reaction would be among the Arab States. The method used for mobilization was, as hitherto, to broadcast over the radio the code-words indicating the units being mobilized. It was a move, Herzog recalled, 'creating apprehension and reminding people that they might be facing full-scale war. To do it as an unannounced test is not just foolish, it is extremely dangerous and leaves the nation vulnerable, frightened and confused.'

When the code-words were being broadcast, Ben-Gurion was entertaining Queen Elizabeth, the Queen Mother of Belgium. He was greatly angered. The commander-in-chief of Northern Command, Yitzhak Rabin, was at a concert in Tel Aviv, together with the Chief of Staff, Chaim Laskov. As the hall 'filled with music', Rabin later wrote, 'frantic messengers called the Chief of Staff to step outside. He was followed by a train of senior officers. Were we at war?' Not only inside Israel, but around the world, the news spread that Israel was preparing for war. 'Dramatic radio announcements in dozens of languages galvanized the IDF reserves as though we faced a grave state of emergency,' Rabin has written. 'Israel held its breath, and the whole world with it.'

The two senior officers responsible for the mock mobilization, General Meir Zorea, Director of Military Operations, and General Yehoshafat Harkabi, Director of Military Intelligence, were dismissed from their posts. As the new Director of Military Operations, Ben-Gurion appointed Yitzhak Rabin. Chaim Herzog was brought back to his former job as Director of Military Intelligence. Within a year, another crisis confronted the Israeli military establishment. After Syrian artillery had opened fire on Israeli settlements just below the Golan Heights – an all-too-frequent occurrence – Israel increased the number of its troops in the north. In a cynical move, on 15 January 1960, the Soviet Union informed Syria that this movement of troops was part of a plot between 'the Imperialists and the Zionists' to initiate an Israeli attack on Syria.

Egypt and Syria – which then constituted a single unit, the United Arab Republic – put their armies on full alert two days later. At the same time,

President Nasser notified the United Nations forces stationed in the Gaza Strip and Sharm el-Sheikh that there was a possibility of Egyptian action through Sinai. On January 19 he began to send troops and tanks eastward across the Suez Canal. Yitzhak Rabin, as Director of Military Operations, had to know exactly what was happening in order to gauge Israel's response. 'On January 23', he later wrote, 'we sent an aerial reconnaissance mission over the Canal – a complex mission and the first of its kind in a long time – but it failed to locate the Egyptian units. We therefore assumed that the divisions had crossed the Canal and advanced eastward into Sinai. Indeed before long we discovered that the main body of the Egyptian army had crossed the whole of the Sinai peninsula and was massed near the Israeli border! We were taken completely by surprise and had no more than twenty or thirty tanks in the area under the jurisdiction of Southern Command.'

An emergency meeting of senior army officers was held at General Headquarters on the evening of January 24. As Chaim Herzog was giving his intelligence assessment on the reconnaissance flight and its findings, Rabin passed a note to the Air Force Commander, Ezer Weizman: 'We've been caught with our pants down. For the next twenty-four hours, everything depends on the air force.' It would take at least twenty-four hours before Israel could get enough troops to its Sinai border to halt an Egyptian attack. In a state of near panic, more than a hundred Israeli tanks were ordered south, to the border.

On January 25, Rabin and Weizman flew by helicopter to Southern Command headquarters. 'Our helicopter was carrying no less than four generals, including me,' Rabin has written. 'At one point two of our fighter planes came flying toward us, and their pilots had obviously not spotted us. Ezer Weizman was the first to sight them and shouted at our pilot, "Dive!" Having avoided a mid-air collision we could afford to joke about the rapid promotions that would ensue if all four of us had been eliminated.'

The two armies faced each other, but neither was prepared for war, or intended to go to war. There was no shooting, and gradually the respective forces withdrew from the border. But for the Israeli military leaders, there was distress that the Egyptian troop movements had been discovered so late. 'Alarm bells jangled at GHQ,' Rabin later wrote, 'for it was clearly an urgent priority to improve our early-warning system.' Thirteen years later that system was to fail again.

In 1960 Ben-Gurion met the German Chancellor, Konrad Adenauer, in New York. It was a historic moment, the meeting of the Prime Minister of the Jewish State with the leader of the country that, less than twenty years earlier, was endeavouring amid extreme barbarity to destroy the Jews altogether.

At their meeting, Adenauer promised Ben-Gurion that he would continue financial aid to Israel after the end of Germany's reparations commitment. He also agreed, some time later, to supply arms to Israel. But the opening of diplomatic relations was delayed; Adenauer feared that if West Germany were to recognize Israel, the Arab States might recognize East Germany.

Internally, Israel in 1960 was involved in a debate as to what the future nature of the State would be. Thirty-three years after the death of Ahad Ha-Am, those who followed his philosophy of a national spiritual centre for Jewish society, embodying in the pattern of its life and thought 'the true spirit of the Jewish people', still held to his vision. 'The young State came into existence, and has struggled through the difficulties of its early years, in conditions not conducive to the development of those elements in the national consciousness which alone had value in Ahad Ha-Am's eyes as specifically Jewish,' wrote Leon Simon, a British Zionist and one of Ahad Ha-Am's biographers, and he added:

> It is difficult to conceive conditions in the Middle East, or indeed in the world at large, less favourable than those of our epoch to the universalistic nationalism of the Hebrew prophets, which in Ahad Ha-Am's view is the kernel of the Jewish tradition handed down through the centuries of dispersion; and in any case it is not easy for the free-born *sabra* to recognize the essentials of his tradition through the wrappings in which the generations denied freedom have been compelled to swathe them.
>
> Yet there is no reason to assume that Israel is destined to become 'like all the nations', distinguished from the rest by nothing more than its territory and its language.

Simon believed that what he called 'the isolated position of Israel in the world of international politics' had, along with the obvious drawbacks of such isolation, 'the advantage of reinforcing the State's sense of belonging to the whole scattered family of the descendants of Jacob. The land itself, and the Hebrew language, are charged with universalist associations.' In the educational system of Israel, Simon argued, the Bible, 'as the nation's most precious literary possession, occupies a place such as could not be assigned to it in the state-controlled schools of any other people'.

There was nothing, Simon believed, to suggest 'that the Government of Israel, for all its inevitable preoccupation with military and economic needs, has any bias against the spiritual conception of nationalism for which Ahad Ha-Am stands: rather the reverse. Some of his essays are still, as they were in pre-State days, compulsory reading in the schools, and many of the teachers are Ahad Ha-Amist in outlook.' A recent decision to introduce instruction in 'the Jewish consciousness' into Israeli schools, 'as a means of counteracting the tendency of the younger generation of Israelis to turn its back on the Diaspora and to lose the sense of belonging to the historic Jewish people, is rightly regarded as a triumph for the ideas of spiritual

Zionism; and whatever may be its effects in the long run, it is at least significant as a pointer.'

The State was only twelve years old. What its spiritual orientation would be three or four decades hence was unclear. There seemed to be 'no reason', Simon felt, 'why Israel should not become capable of exercising the influence of a spiritual centre – "a power-house of the national spirit," to use one of Ahad Ha-Am's metaphors – in the more strictly spiritual sense as well as in the psychological. If this does come about – and clearly we are in a region of speculation in which the last word must be with faith rather than with reasoned argument – Ahad Ha-Am's campaign for the ideas of spiritual Zionism will be entitled to rank among the decisive episodes in modern Jewish history.'

The Jewish propensity for optimism was strong; the frictions within a society almost overwhelmed by the problems of a diverse and largely impoverished immigration, and by the Ashkenazi–Sephardi divide, were seen as capable of being overcome by faith, if not by 'reasoned argument'.

At four o'clock on the afternoon of 23 May 1960 the seats in the Knesset were filled to capacity, as was the public gallery. Word had spread throughout the building that Ben-Gurion was about to make an extraordinary statement. No one knew what it was about. All that was known was that half an hour earlier Ben-Gurion had called a special meeting of his Cabinet, to tell them what he was about to tell the parliamentarians and the nation. The Knesset session opened at four. Called to the rostrum by the Speaker, Ben-Gurion announced that Adolf Eichmann, one of those 'responsible for what they call the final solution of the Jewish question, that is, the extermination of six million Jews of Europe', had been found, brought to Israel, was under arrest, and would shortly be brought to trial.

'The House was electrified,' wrote Moshe Pearlman, an Israeli writer and journalist who was present. 'And for several seconds there was a stunned silence. Suddenly, from all parts of the Chamber, came a roar of applause. Rarely had the Knesset been so unanimous. Rarely had its members been so moved. The murderer of their people had been caught. He would be brought to justice.' The emotion of that moment, Pearlman reflected, 'transcended mere desire for revenge or retribution. It was a reinforcement of faith in ultimate justice. The man who personified the forces of darkness responsible for the extermination of millions of Jews would now be tried by due process of law by the Courts of the Jewish State.'

Eichmann, who had been living in the Argentine, had been tracked down by Israeli agents. For a year he was questioned about the war years, and gave copious testimony of his activities. His trial, which began on 11 April 1961, was held in Jerusalem. Evidence was given, not only about Eichmann himself, but about every aspect of the Holocaust. More than a hundred survivors gave evidence. Eichmann himself called sixteen witnesses in his defence, six of whom were in prison in Germany, Austria and Italy for war crimes, and who were examined abroad (Israel had offered them safe conduct to Jerusalem, which they declined). The published transcript of the trial ran to 6,000 pages. Although courts of law do not allow foreign advocates to appear before them to plead on behalf of a defendant, Israel allowed Eichmann to be defended by Dr Robert Servatius, a German counsel who had appeared for the defence in the Nuremberg Trials fourteen years earlier.

The trial lasted fourteen weeks. All Israel was riveted by the proceedings, none more so than the survivors who gave testimony, attended the trial, or listened to it every day on the radio, or read of its progress in the newspapers. A generation of young Israelis for whom the Holocaust was remote and unclear learned for the first time in detail, and from eyewitnessses, of what had occurred. In conversation with a young Israeli diplomat, Yehuda Avner, Ben-Gurion explained that he had wanted Eichmann brought to trial in Jerusalem so that Israeli youth should learn the truth of what had happened to the Jews of Europe between 1933 and 1945. 'Israeli youth had been sceptical and critical about the Holocaust,' Avner noted, 'because they charged that the Jews of Europe allowed themselves to be led "like sheep to the slaughter". Why did they not fight back? We did fight back, they argued, when we had no planes or tanks, against an enemy that did. It was a fact of history, and the Eichmann trial finally stilled this criticism in all but a few.'

On December 11 Eichmann was found guilty. His appeal to the Supreme Court, which was heard three months later, was rejected. On 29 May 1962 he appealed for clemency to the President. His appeal was turned down, and he was hanged three days later. Neither before then, or since, was the death penalty carried out in Israel; even in the case of the worst terrorist actions, the death sentence has always been commuted.

Scandal struck Israel in 1960, with the extraordinary revival of the Lavon affair. After Lavon's resignation as Minister of Defence (following the explosions carried out by Israeli agents in Egypt in 1954), he had re-entered mainstream Labour politics, and became Secretary of the Histadrut, the General Federation of Labour. Benjamin Gibli, the former head of Intelligence, who, with Lavon, had been held responsible for giving the order for the sabotage actions, was serving as the Israeli Military Attaché in London.

When Lavon learned that in the original inquiry the documents which made it seem that he had initiated the sabotage scheme were forged, he asked Ben-Gurion to hold a new inquiry. The documents seemed to make it clear that when Gibli had learned of the failure of the operation, for which he had not received any prior orders from Lavon, he sent Lavon a request to carry out the operation (which had already failed, unknown to Lavon), and Lavon then gave permission, thereby making himself responsible, on paper, for something that had already taken place.

Having ascertained through his military adjutant, Colonel Chaim Ben-David, that Lavon's charges were correct, Ben-Gurion agreed to an inquiry. It was headed by one of the most distinguished Israeli lawyers, Supreme Court Justice Haim Cohn. But Lavon was unwilling to see his reputation wait upon the outcome of a judicial inquiry: he wanted Ben-Gurion to clear his name at once. This Ben-Gurion refused to do. In the event, Haim Cohn and the two army colonels who were serving with him on the judicial inquiry concluded that the Intelligence documents implicating Lavon had been forged. Lavon wanted Ben-Gurion to announce that he had been vindicated. But Ben-Gurion insisted that the findings of the Cohn Committee were not sufficient to clear Lavon's name, and that a fuller investigation was needed. Lavon felt betrayed.

There followed many months of acrimony in which Ben-Gurion seemed to many of his colleagues to have become obsessed with insisting that a full inquiry alone could clear Lavon's name. They, and the newspapers, felt that Ben-Gurion was showing himself insensitive and indifferent to Lavon's just demands for rehabilitation. Lavon was, after all, Secretary-General of the Histadrut, a senior position in the economic well-being of the State. But Ben-Gurion continued to argue that it was not he, but only a full inquiry, that could clear Lavon's name. In this way, Teddy Kollek – then Director-General of the Prime Minister's Office – has written, the Lavon affair became 'the instrument that would eventually deflect and debilitate Ben-Gurion's career'.

In his impatience, and desperation, Lavon revealed secret details of the affair to the Knesset Foreign Affairs and Security Committee. He had not given the order, he told the committee, but had fallen 'victim to a trap' as a result of the forgery of documents by the Intelligence service. This was the first time these details had been made public. Ben-Gurion was incensed, not only by what he saw as a breach of the necessary secrecy regarding Intelligence operations, but because the opposition Parties were represented on the committee.

The leaders of Mapai feared that a judicial inquiry of the sort insisted upon by Ben-Gurion would not only create a prolonged crisis, but possibly reveal far too many controversial and potentially harmful details. To try to prevent Mapai from exposing itself and even tearing itself apart in public, Levi Eshkol set up a ministerial committee headed by the Minister of Justice, Pinhas Rosen. This Committee of Seven, as it was known, decided that Lavon had not issued the order, and that the acts of provocation in Egypt had been carried out 'without his knowledge'. This did not satisfy Ben-Gurion, who was convinced that the committee's decision was a whitewash, and that Eshkol's aim in setting it up had been to put the affair to rest, and to protect the Mapai leadership. When the Cabinet endorsed the findings of the Committee of Seven, Ben-Gurion resigned.

Eshkol and the other Mapai leaders begged Ben-Gurion to retract his resignation, but he was adamant; there were two points which he reiterated again and again: that Ministers cannot become Judges, and that the truth must overshadow all other considerations.

Despite the conclusion of the Committee of Seven, the Mapai Central Committee, meeting on Eshkol's initiative, and without Ben-Gurion being present, discussed the dismissal of Lavon, in order to persuade Ben-Gurion to return. The vote was secret, with 159 members in favour of Lavon's dismissal, and 96 against. Lavon was sacrificed to retain Ben-Gurion. When the coalition partners refused to accept this outcome, Ben-Gurion – back in the Prime Minister's Office – called new elections, two years before they were due.

The elections, held on 2 November 1961, confirmed the Labour predominance. Although Mapai lost five seats it remained by far the largest Party, with forty-two seats, The Herut Movement won seventeen seats, the National Religious Party (with which Ben-Gurion formed the basis of his coalition) twelve, and Mapam nine.

The Lavon affair did not end with Ben-Gurion's victory. The point he continued to make was that justice had not been done. For two years he conducted a sniping war of words and vitriol against Lavon and all who supported him. Pinhas Rosen, one of Ben-Gurion's oldest colleagues, became an implacable enemy. Several attempts by intermediaries to persuade Ben-Gurion to be reconciled with Rosen were in vain. Reflecting on the Lavon affair, Teddy Kollek has written:

> One of the reasons why the Lavon Affair has been misunderstood is that so much emphasis was placed on Ben-Gurion's behaviour that the real issue was lost in the shuffle. Ben-Gurion's point was that nobody should be above the law. Nobody. He insisted that only the courts could rule.
>
> This was not some freak, isolated issue for him. It was part and parcel of his general belief that Israel would continue to exist only if we preserved our moral qualities as a nation of virtue. His greatest fear was of levantinization, and again he saw many years ahead.
>
> Ben-Gurion often believed that once he had formulated and analysed a problem, it was up to others to deal with solving it. So the fact that he took the Lavon Affair completely on himself – whether or not he handled it properly – proves how seriously he took the problem.
>
> If we laughed then at the concept of a nation of virtue, we lived to see how profound his vision was and how important it was to follow his lead. Ben-Gurion knew we would always be only 3 or 5 per cent of the surrounding Arab population, and if we failed to maintain our moral and intellectual integrity, we would not be able to stand up to our adversaries. But in the case of the Lavon Affair, he failed to convey this issue to the public. He often gave the impression that he was trying to settle a personal score with Pinhas Lavon, and his enemies within the party saw to it that this impression was magnified.

As Ben-Gurion lost ground, he became increasingly frustrated and less articulate. Those of us around him were unable to change that pattern.

That the man who had come to be seen as the founding father of the State should become so obsessed made a painful impression on his colleagues, and on the public. Those closest to him were appalled. 'He lost a lot of his self-confidence, and his contact with people, as though he had lost the ability to talk to them,' Kollek later wrote. 'He knew his cause was just, but when he saw that he was unable to convince others, he became exasperated. In his fury of frustration, he began to attack his colleagues personally.' No one, Kollek added, liked seeing Ben-Gurion 'descend to such depths. It was embarrassing. People couldn't accept his attacks – especially the lashing out at the men he himself had raised, like Eshkol. And every time Ben-Gurion attacked Lavon personally, he lost support.'

For two years the bickerings over the Lavon affair continued, injecting a poisonous note into the daily political scene. 'So Ben-Gurion's political opponents within the party and without latched on to the opportunity to bring him down,' Kollek has written. Golda Meir was also a witness to the effect of the Lavon affair on Ben-Gurion and those who had been his closest colleagues and supporters, herself included. In her memoirs, she explained her own perspective:

> I couldn't forgive Ben-Gurion either for the ruthless way in which he was pursuing Eshkol or for the way in which he treated and spoke about the rest of us, myself included. It was as though all the years that we had worked together counted for nothing. In Ben-Gurion's eyes we had turned into personal enemies, and that was how he behaved towards us.
>
> We didn't see each other for years after that. I even thought that, feeling as I did, it wouldn't be right for me to attend Ben-Gurion's eightieth birthday party in 1969 (from which Eshkol had been excluded), though Ben-Gurion had sent a special emissary to invite me. I knew that it would hurt him very much if I refused, but I just couldn't get over it.
>
> If we were really as stupid as he had said we were, well, when people are born stupid, not much can be done about it, and it isn't anyone's fault. But no one is born corrupt, and that is a terrible accusation! If other party leaders were willing to overlook the fact that Ben-Gurion thought, or at any rate said, that they were corrupt, well and good. Eshkol wasn't, and neither was I. I couldn't pretend that it never happened. I couldn't rewrite history, and I wouldn't lie to myself.
>
> I didn't go to that birthday party.

*   *   *

In 1961 an extraordinary archaeological discovery was made in the Judaean desert near Ein Gedi, only a mile from the Jordanian border. During an expedition led by Yigael Yadin, an almost inaccessible series of caves were found to have in them artifacts and documents from the time of Bar Kochba's revolt against Rome. Among the finds was a letter, clearly legible after two thousand years, written in Bar Kochba's name, describing him as 'President over Israel'. There were many other letters, a marriage contract, bronze jugs, a virtually intact willow basket, and a carefully wrapped package filled with glass plates and a large glass bowl.

'We opened the package later in the camp in the presence of the Prime Minister, Mr David Ben-Gurion, who had come to visit us that day,' Yigael Yadin recalled. 'When we took out the plates we could scarcely believe our eyes. The glass was as translucent as if it had just been manufactured. I must confess that a terrible thought flashed through my mind that if I should find "Made in Japan" stamped on the bowls I would collapse on the spot. Subconsciously I turned the bowl over, but there was no such inscription.' It was common, Yadin added, 'to see ancient glass with patina created by dust and humidity throughout the ages; in fact we like ancient glass for that very reason. But here in the cave, because of the absolute lack of humidity, no patina formed, and the glass was preserved exactly as it was two thousand years ago'.

The announcement of the discovery of the papyrus letter with Bar Kochba's name on it was made by Yadin at a meeting called specially in the house of President Ben-Zvi. Many distinguished guests, Ben-Gurion and many of his Cabinet Ministers among them, had been summoned for an unexplained revelation. When Yadin produced a slide showing Bar Kochba's name 'for a moment', he recalled, 'the audience appeared to be struck dumb. Then the silence was shattered with spontaneous cries of astonishment and joy. That evening the national radio interrupted its scheduled programme to broadcast news of the discovery. Next day the newspapers came out with banner headlines over the announcement.'

'Why was a whole nation elated over the discovery of a name on a fragment of papyrus?' Yadin asked, and he went on to explain: 'The answer lies in the magic of the name, a name treasured in folklore but almost lost to authenticated history, and the realization at this meeting that after nearly two thousand years the desert had given up factual links with the man who led the last attempt of his people to overthrow their Roman masters.'

The part which archaeology, and the Bible, played in the mind and activities even of secular Israel was profound. Some of Israel's most respected secular leaders were the descendants – even the grandsons – of distinguished rabbis. The fascination of the Bible was enhanced by proximity to the towns and villages, valleys and battlefields, which are described in it. Shimon Peres has recalled an occasion in 1961 when (at a time when the

Lavon affair was still rumbling) Ben-Gurion summoned a meeting of journalists in Sokolow House, the journalists' headquarters in Tel Aviv. 'This must have been the most extraordinary speech ever made by a Prime Minister to an audience of journalists' anywhere,' Peres wrote, 'The central issue under discussion was most topical – the Exodus from Egypt.'

Ben-Gurion began his address with the following words: 'The establishment of the State of Israel and the War of Independence have shed a new light on my understanding of the Bible, and after examining it afresh in the light of the reality of the War of Independence and the resettlement of Israel in our time, I have been moved to pose questions to which biblical commentators throughout the ages have paid insufficient attention.' He then put forward what he called 'thirty intriguing questions', beginning with: 'Why did Terah and his family leave Ur of the Chaldees, a developed and fertile country, to go to the land of Canaan?' and ending with: 'How could the grandchildren of the seventy couples who went down into Egypt have taken possession of the lands of the Amorite King Sihon and Og King of Bashan to the East of the Jordan, and defeated the majority of the peoples of Canaan in battle? Such a thing would have been impossible. They were obliged not only to conquer territory but also to settle in it. How could they settle? How many were they at the time?'

As the discussion continued, Ben-Gurion was 'not satisfied', Peres recalled, until he had proved that the 'six hundred thousand men, excluding children' who left Egypt and went to the land of Canaan 'were not six hundred thousand but six hundred families, since the word meaning "thousand" is also used in the Bible for the concept of "family".' Ben-Gurion then cited an ancient rabbinical source, Rabbi Kimhi, who said: '*Alfey* (thousands) is related to *aluf*, which means lord and master.' From which, Ben-Gurion said, it followed, according to Rabbi Kimchi, that 'the heads of thousands of Israelites were the heads of families of Israelites'.

The pursuit of truth, Peres commented, made Ben-Gurion 'a great student, perhaps the greatest student produced by our people among the leaders of the new era'.

In 1961 Yigal Allon, the former commander of the Palmach, was appointed Minister of Labour. In his seven years in office he did much to improve the use, training and productivity of manpower in State enterprises. He also extended the road network, which had become inadequate for the growing population and increasing amount of road transport. Labour courts were established to help resolve disputes that led to strikes and lockouts. At the same time, he set up liaison bureaus to make it easier for Israelis who had gone abroad to study for lengthy periods to return and be reintegrated into society. Social insurance was extended, and the first steps taken to introduce automation into industry. A new town, Arad, was also founded in 1961 in the desert overlooking the Dead Sea. It was built by engineers and architects who lived on the site. One innovation was interlocking apartments with private courtyards and the maximum of shade. The town's economy was based on the Dead Sea potash works and several nearby industrial plants.

In order to create an Israeli presence along the borders, Nahal (Pioneer Fighting Youth) outposts continued to be set up by army units which then became settlements in their own right. One of these was established in 1961 at Almagor, in Upper Galilee, across the River Jordan from Syria. A number of its soldier-settlers were killed in Syrian ambushes in the vicinity; today there is a monument to their memory.

Also founded in 1961 was the kibbutz of Jotapata, near the ruins of an ancient town of that name where Jews had settled after the Roman destruction of the Temple. The new settlers were Israeli-born youngsters, mostly from Haifa. It soon became a predominantly vegetarian kibbutz, with fruit orchards as the main produce of its fifty inhabitants.

These were also the years of the beginning of the establishment of industrial enterprises in kibbutzim throughout Israel. Many of them later grew into major industrial plants. Another imaginative commercial enterprise, launched in 1961, was an eighteen-hole golf course, club house, villas and hotel, aimed at western European visitors and immigrants. It was created on the sand dunes next to the remarkable ruins of Roman and Crusader Caesarea.

Living in the new development at Caesarea, but working in the nearby immigrant town of Or Akiva, was Hannah Billig, whose brother Levi had been killed in the Arab riots in 1936, and who had herself won the George Cross – the highest British civilian decoration for bravery – for her work as a doctor in the London Blitz. Reaching Palestine after the Second World War, Hannah Billig devoted her life to the poor.

A new political Party had been formed in 1961 among those who were always hoping to have a greater influence on policy, and a place in the government coalition despite being outside the establishment ring of Mapai and the Labour movement. Known as the Liberal Party of Israel, it was formed by the merging of the Progressive Party – established in 1949 and commanding between four and six seats in successive elections – and the General Zionists – established before the Second World War, winning twenty seats in 1951, and thirteen in 1955, declining to eight in 1959. At its formation, the new Liberal Party could command seventeen seats in the Knesset. It favoured a private economy with the minimum of government participation, and a reduction in the power of the Histadrut by separating its production and trade union functions.

The closeness of the Liberal Party values to those of Herut was to lead within four years to the merger of the two, as the Gahal bloc. In the 1965 election, the Liberals were to hold eleven out of Gahal's twenty-six seats. A year after the merger the Progressive Party wing of the Liberals broke away from Gahal, to form the Independent Liberal Party. The breakaway group had five seats. With these, it entered the government (on 16 January 1966) as part of the predominantly Labour coalition. Within a decade, the seats of this breakaway Independent Liberal Party had dwindled to one, still tenaciously inside the Labour bloc. The Liberal Party from which it had broken away continued to be allied with the right opposition, and was eventually (in 1977) to enter government as an integral part of the Likud bloc, with fifteen out of Likud's forty-three seats.

The splitting and regrouping of political Parties was a feature of Israeli politics that arose in part from the system of proportional representation, which gave encouragement to small Parties and necessitated the formation of coalitions made up of many Parties. Even a very small Party could attain a controlling position when the numerical battle of coalition forming narrowed down to the search for a majority of two or three. The fragmentation of Parties also arose from a national tendency to be disputatious and contentious, which was itself the result of so many different political and regional traditions being thrust together in a single system. In the 1960s it was the Ashkenazi and Labour-oriented parties that held the dominant role in Israeli life, but the Sephardi voters also looked for the means of attaining greater political influence, as did the former Revisionists, against whom Ben-Gurion and the bulk of the pre-war Jewish Agency politicians still looked with contempt, scorn or indifference.

The debate over the electoral system in Israel is one that was to continue to the present day. Among those who have favoured a British-style two-Party system were two former senior military men, Yigael Yadin and Chaim Herzog.

The first Israeli census held under normal conditions took place on 22 May 1961. Of the previous two, the first had been held during the final stages of the War of Independence, when it was impossible to make a proper count of the Arab population, and the second in 1950, when Jewish immigration was so rapid that the figures became obsolete even as the counting was in progress. The 1948 census had given a Jewish population of 716,678, with more than 65,000 Arabs. The 1950 census gave 1,029,000 Jews and more than 150,000 Arabs.

The census of 1961 gave a total of 1,932,400 Jews (Just over 88 per cent) and almost 250,000 non-Jews. The non-Jews were divided between Muslim Arabs, Bedouin and Circassians (170,800), Christian Arabs (24,000) and Druse (24,000). The Israeli Arab village population lived mostly in Galilee, and in the Triangle area. There were also two Arab cities in Israel, Nazareth and Shfaram, and other cities with Arab as well as Jewish populations, principally Jaffa, Haifa, Acre, Ramle and Lod. The Israeli Arabs did not serve in the army – though both the Druse and the Circassians were recruited, and many Bedouin volunteered.

The Government of Israel tried not to neglect the rights and well-being of its Arab minority. But its relationship with the Israeli Arabs was always subject to strain because of the ongoing conflict with the Arab States across Israel's borders. A Military Administration had been established over the Arab-inhabited areas of Israel in the first days of statehood. Its aim was to prevent hostile activities against the State from within. It was opposed by the Left in Israeli politics as being anti-democratic and unnecessarily oppressive. Ben-Gurion had been its strongest advocate; when Levi Eshkol was Prime Minister he relaxed its implementation.

There were five regulations on which the Military Administration pivoted: Regulation 109 which permitted the arrest of a person for being in a prohibited area; Regulation 110 which allowed police supervision over a person for up to one year; Regulation 111 which provided the legal basis for administrative detention by military commanders; Regulation 124 which provided for house arrest; and Regulation 125 which permitted military commanders to declare certain areas closed areas, persons entering or leaving which had to possess a special permit. Gradually the areas to which these regulations applied were reduced. In 1962 the Druse were exempted from the regulations. Even when the Military Administration was lifted altogether in 1966, the Emergency Regulations which they embodied could still be implemented by the heads of the various military commands.

Despite the Military Administration, the Arab population not only had the right to vote in Israeli elections, but also had several Arab parties to vote for. At the government's initiative, a year after the 1961 census, a regional centre for Arab vocational training was opened at Tamara, in Western Galilee, as well as an Arab vocational high school in Nazareth and a Muslim orphans' home in Acre.

At the end of 1962 Ben-Gurion approached President Tito of Yugoslavia, through a Yugoslav-born Israeli emissary, in the hope of enlisting Tito's support as a mediator between Israel and its Arab neighbours. The emissary was Shaike Dan, who had been one of the wartime parachutists behind German lines. Ben-Gurion proposed either 'open or secret' negotiations between himself and Tito, 'to discuss ways of advancing peace between ourselves and our neighbours in the Middle East'. In a letter to Tito on 28 December 1962, in which he set out his hopes that it might be possible for Egypt and Israel to be brought together in negotiations, Ben-Gurion explained to his Communist interlocutor his view of Israel's position on the political spectrum:

> I will not discuss here the special difficulties and grave problems that
> we have confronted since the renewal of our independence, but despite
> them all we have succeeded in establishing in Israel a progressive and
> democratic country, which has many achievements to its credit in

agriculture, industry, education, scientific research and social progress.

I cannot yet call Israel a Socialist country, for we also encourage private capital, but in agriculture, industry and transport we can claim Socialist achievements, and even the beginnings of a truly communist society in the labour settlements known as 'kibbutzim', to which – so far as I know – there is no parallel anywhere else in the world.

One of the factors that hinder economic development in our area (including, to no small extent, our own country) is the burden of defence and the arms race between our neighbours and ourselves – especially between ourselves and Egypt. And I am convinced that nothing would more effectively further the development and progress of the Arab countries in our area – of Egypt, especially – than peace and co-operation between them and Israel.

Ben-Gurion was proud of the efforts on which Israel had embarked to open up relations and trade with the new States of Africa and Asia, and to provide them with the benefits of Israeli technical expertise, whether in the building of ports or roads, or hotels or hospitals. He saw these efforts as much more than a commercially driven enterprise, or even a means to widen Israel's support in the United Nations. It was above all, for him, a sense of solidarity with those States which, like Israel, had recently broken away from foreign rule, and, despite their poverty and lack of vast natural resources, were inspired by the same sense of national identity and purpose.

Something of Ben-Gurion's pride in this regard came through when he told Tito: 'Our country is both small and poor, but through the pioneering and Socialist initiative of our Labour movement, which has held the leading place in the State since its establishment, we have succeeded in overcoming our internal difficulties and created a progressive society, unique in the whole of the Asian continent, as well as making a modest contribution to the advancement of the new States in Africa and Asia – and I have not the slightest doubt that under conditions of peace and co-operation with our neighbours, we could offer no little assistance to their progress.'

Returning to Israel's relations with Egypt, Ben-Gurion told Tito that, given Nasser's desire for social progress in Egypt, 'I assume that he understands the importance of peace with his neighbours as a force making for progress.'

Tito did not reply to Ben-Gurion's letter for more than three months. When he did so, on 14 April 1963, his answer was a disappointment to the Israeli leader. Yugoslavia's contribution could best be made, Tito averred, through the United Nations. Like Winston Churchill when he approached the Soviet Union in the opening months of 1955, Ben-Gurion had hoped to use his authority and experience, gained over many years, to create a breakthrough in the relations between old enemies. Tito's negative response put an end to that hope, just as, in 1955, the Soviet leaders had rebuffed Churchill's request for talks to break the impasse.

On 17 April 1963, four days after Tito sent his letter, Egypt, Syria and Iraq signed a treaty of alliance in Cairo, in which they put their three armies under a unified command, and proclaimed that the aim of the new alliance and its armed forces was the 'liberation of Palestine'.

Only Jordan remained outside the ring of the fanatically anti-Israeli Arab States. Inside Israel, senior officials not only recognized that King Hussein needed peace with Israel in order to preserve the stability of his regime, but actively sought to engage the King in dialogue. A series of meetings between the King and Israeli leaders was organized by the Cabinet Secretary, Yaakov Herzog. On one occasion, the King came in the utmost secrecy to Tel Aviv, where he met Golda Meir. On another occasion, the King and Golda Meir met in London, in the consulting rooms of the doctor who administered to Marks and Spencer's head office.

Israel's attempt to create a working relationship with Jordan had some success. Chaim Herzog, Commander of the Jerusalem District in 1955 and 1956, has recalled how, after a number of fedayeen attacks in the Jerusalem area, his then opposite number, Colonel Ahmed Bublaan, 'begged us not to retaliate against Jordanian police buildings and headquarters; that would only help break down their structure of law and order. We held them responsible for everything that happened over the border, but it was obvious they did not control the guerrilla extremists.'

Amid the hostility of the Soviet Union, its Communist-bloc allies, and the Arab and Muslim nations, Israel strove for the friendship of the United States. Given the importance of successive, and very different, Presidents, in shaping policy, efforts were made to put Israel's case to them at the highest and most personal level possible. In 1963 the Israeli Foreign Minister, Golda Meir, was in Florida when President John F. Kennedy, who was on holiday there, asked to see her. The first-ever shipment of American arms – Hawk missiles – had been approved by the Kennedy administration.

'We talked for a very long time, and very informally,' Golda Meir later wrote. 'We sat on the porch of the big house in which he was staying, and I can still see him, in his rocking chair, without a tie, with his sleeves rolled up, listening very attentively as I tried to explain to him why we so desperately needed arms from the United States. He looked so handsome and still so boyish that it was hard for me to remember that I was talking to the President of the United States – though I suppose he didn't think I looked much like a Foreign Minister either!'

Golda Meir told Kennedy that the Government of Israel was in many ways 'no different from any other decent government. It cares for the welfare of the people, for the development of the state, and so on.' But, she continued, 'in addition, there is one other great responsibility, and that is for the future. If we should lose our sovereignty again, those of us who would remain alive – and there wouldn't be very many – would be dispersed once more. But we no longer have the great reservoir we once had of our religion, our culture, and our faith. We lost much of that when six million Jews perished in the Holocaust.'

Golda Meir then told Kennedy that what was 'written on the wall' for Israel was: 'Beware of losing your sovereignty again, for this time you may lose it for ever'. If that should happen, she added, 'then my generation would go down in history as the generation that made Israel sovereign again, but didn't know how to hold on to that independence.'

'When I finished,' Golda Meir recalled, 'Kennedy leaned over to me. He took my hand, looked into my eyes and said very solemnly, "I understand, Mrs Meir. Don't worry. *Nothing* will happen to Israel." And I think that he did truly understand.'

A few months later, Kennedy was assassinated. He had opened the way for what was to be Israel's overriding foreign policy and defence orientation for at least thirty-five years: the closest possible links with the United States.

CHAPTER TWENTY

Years of growth
1963–1966

Israel is often rocked by disputes and scandals in a way that involves the whole nation, and seems to absorb the whole energies of the nation. The Kastner affair had been one such scandal, ending in Kastner's assassination. The Lavon affair had led to recriminations and resignations. In 1962 another such all-pervasive scandal came to dominate the political life of the country. It was revealed that year that German scientists had been working in factories in Egypt, developing missiles that had a range of up to 350 miles. Ben-Gurion asked his deputy at the Ministry of Defence, Shimon Peres, to raise the matter with the German government. Peres wrote to the German Defence Minister, Franz-Josef Strauss, asking him to take 'immediate action' against the scientists. Strauss asked for details, but with the exception of a single scientist, there was no serious information.

Unwilling to accept this situation, the Mossad, the Israeli secret service, took independent action, seeking out the daughter of one of the scientists, and trying to persuade her to convince her father to leave Egypt. The Mossad agents had spoken to the daughter in Switzerland. She told the Swiss authorities of the encounter, and the Israeli agents were arrested. Suddenly the Israeli newspapers were filled with lurid details of the German scientists' work and its threat to Israel. The Mossad hoped, in this way, to secure the release of its agents.

In March 1963, during a debate in the Knesset, the Germans were accused of continuing to threaten the Jews with 'extermination'. The government's critics, led by Menachem Begin, accused Ben-Gurion of being soft on the Germans. They also denounced his decision to sell Israeli-made 'Uzi' sub-machine-guns to Germany, and even abused him for maintaining diplomatic relations with Germany. When the arms sales to Germany were opposed by the National Religious Party members of his Cabinet, Ben-Gurion asked angrily, 'In which religious code of laws is it written that it is forbidden to sell arms to Germany?'

Ben-Gurion, not wanting to harm German–Israeli relations which, under his careful nurturing, were steadily improving, and knowing the facts, insisted that the public had no reason for alarm, and that the missiles around which the storm had arisen were unusable. The German scientists in Egypt were not top class, nor was their work of any danger to Israel. The head of the Mossad, Iser Harel, insisted that this was not so, and that the threat was real. When he told Ben-Gurion the source of his information, Ben-Gurion was astounded. He had met the same man, the head of the German Parliament, who had told him an entirely different story. The head of the Mossad was duly fired.

The founding of new towns, and the settlement of Arab towns and quarters, continued apace. In 1962 the South-African-born Joan Comay, who was writing a guidebook – about what Ben-Gurion called in his Foreword 'the dynamic quality of a new State turning deserts into gardens and welding heterogeneous immigrant groups into a sturdy nation' – wrote of the former Arab town of Jaffa. Of the 65,000 Arabs living in Jaffa in 1948, only 5,000 had stayed behind to rebuild their lives inside the Jewish State. Within fifteen years, 100,000 Jews had come to live in the town. None of these newcomers, Joan Comay wrote, had been in Israel before 1948:

> In Jaffa the most ardent tourist need not worry about remains of the past, but can simply relax and enjoy the cosmopolitan human scenery of the present. An extraordinary medley of languages bubbles up from the pavements or is scrawled on the stores, and just as extraordinary a variety of national dishes can be sampled in the little neon-lighted cafés and eating places.
>
> In a single swift leap, the young children have become Israelis. To their parents, they talk the tongue of the country from which the family came, whether it is Yiddish, French, Bulgarian, Arabic, or what you will; but in their street games, they scream at each other in Hebrew.'

*   *   *

On 16 June 1963 Ben-Gurion resigned as Prime Minister. He had been a powerful figure for more than a decade in the pre-State era, and the dominant figure in the first fifteen years of the State: dominating not only because of his political skills and determination, but because of his character and ideals. The news of his resignation came as a shock to the senior army officers, who sent a delegation, headed by the Deputy Chief of Staff, Yitzhak Rabin, to see him. 'We told him of our sense of dismay over his decision,' Rabin has written, 'and tried to persuade him to remain at his post for the sake of the army. I spoke at length of the military situation. France was edging away from us and the Arab countries were growing stronger. Was this a time for Ben-Gurion, who was unique and unrivalled in the eyes of the army, to leave the helm and abandon the IDF? Ben-Gurion did not confide in us the reasons for his resignation, but he spoke with great feeling of his affection for the IDF and the men and women who comprised it. His words were emotional, and I had the impression that he was deeply touched by our appeal.'

The appeal was, however, unsuccessful, and, for the second and last time, Ben-Gurion retired from public life. Levi Eshkol became Prime Minister and Minister of Defence.

On 14 November 1963 Yigael Yadin was digging on the summit of Masada when, his biographer Neil Silberman writes, he was 'summoned urgently from another excavation area, made his way cautiously down a makeshift wooden stairway to the lowest level of the palace, built on a small, square terrace overlooking the arid gullies and salt flats of the western shore of the Dead Sea. It was a moment of high drama and unforgettable emotion. Yadin watched in silence as the volunteer diggers cleared the debris from a small bathing pool, whose smoothly plastered surface was darkly stained with what seemed to be blood. Nearby were the jumbled bones of two young men and a young woman, whose long hair, still neatly braided, had been miraculously preserved in the dry atmosphere of the region around the Dead Sea for almost two thousand years.'

This was one of a series of dramatic discoveries that made the excavations remarkable, and electrified the Israeli public. The events described by the Roman-Jewish historian Josephus Flavius nearly two thousand years earlier were being confirmed almost every day. Here was evidence that the Jewish Zealots, in the final year of the eight-year-long Jewish revolt against Rome, besieged and trapped, had committed mass suicide rather than fall victim to a Roman triumph.

In July 1962 a twenty-one-year-old Mexican tenor, Plácido Domingo, was given a contract to sing with the Hebrew National Opera in Tel Aviv. He was to remain with the Tel Aviv company for three years. It was, he recalled, a 'dream contract' of £1,000 a month: but it emerged that these were not British pounds, but Israeli pounds, quite another matter: he was to give ten performances a month at £16.50 a performance.

While still in Mexico, Domingo married the singer Marta Ornelas; that December they arrived in Israel together. 'The human and artistic encounters there,' he later wrote, 'probably could not have been equalled anywhere else in the world. One could get as much out of Israel as one's own mental and spiritual resources allowed.' His first role was that of Rodolfo in *La Bohème*; Marta's début was as Micaela in *Carmen*.

Conditions were not ideal. The summers were 'terribly hot', Domingo recalled, 'especially in the dressing room – a single large area with a curtain down the middle to separate the ladies from the gentlemen. The room would become so stuffy that we could hardly breathe; then we would walk onto the stage and get hit in the face by the air-conditioned atmosphere of the auditorium. That dried our throats out. Having sung under those conditions, I knew that I could sing anywhere.'

Domingo remembered the variety of singers in those early days of opera in Tel Aviv: *La Traviata* in which the singer playing Germont sang in Hungarian, Violetta in German, Alfredo (himself) in the original Italian, and the chorus in Hebrew. 'The public was fantastic,' he wrote. 'Such a mixture of cultures! There were always people from Poland, Russia, Roumania, Yugoslavia, Bulgaria, Germany, Czechoslovakia, Austria, Hungary and elsewhere in the audience. One Roumanian Jew, Lazer, used to tell us that he had traded the outsides of animals for the insides: in Europe he had been a furrier, but in Tel Aviv he had become a butcher. He was a tremendous opera lover and used to come to many performances, always dressed in a tuxedo. In return for our invitations to the opera, he would save us the very best meat – including pork – at the Carmel Market, the Shook Carmel.'

In his two and a half years in Israel, Domingo sang ten different operas. He also learned English, French, Italian and Hebrew. A high point of his operatic career in Israel came with the performance of Bizet's *Pearl Fishers* in Jerusalem. Six months later he left Israel for an international career that was to see him celebrate – in 1992, at a concert in Toledo – the five hundredth anniversary of the expulsion of the Jews from Spain. Many of those had made their way to the Ottoman Empire, including Palestine.

In January 1964, at a summit meeting of Arab countries, the Government of Syria proposed making use of the 75,000 Palestinian refugees who had been in Syria since 1948 and the far larger number living in other Arab lands, and to do so in a way which could destabilize Israel. The Syrian proposal was accepted: the Arab summit formally authorized the Palestinians, wherever they might be living, to 'carry out their role in liberating their homeland and determining their destiny'. Later that year the Jordanian government allowed an 'Assembly' of Palestine Arabs to meet in East Jerusalem. As a result of its discussions a new organization was created, the Palestine Liberation Organization. The aim of the PLO, according to its founding manifesto, was to 'attain the objective of liquidating Israel'. To this end, a Palestine Liberation Army was established. The PLO would also receive financial backing from the Arab governments. President Nasser, hoping to win patronage of the Palestinians from Syria, placed both Sinai and the Gaza Strip at the PLO's 'disposal'.

A direct confrontation between Israel and Syria flared up suddenly at the end of 1964. The issue was again water. Syria was determined to prevent Israel from using the waters of the River Jordan for its National Water Carrier.

Two of the sources of the Jordan, the rivers Hazbani and Banias, had their sources in Syria, and flowed across the Syrian–Israeli border into the river Jordan. Plans were made to direct the waters of the Hazbani before it reached Israel, and to build a canal that would take the waters of the river Banias from Syria to Jordan, bypassing Israel.

A former Chief of Staff, Moshe Dayan, was consulted. His solution was to seize the Syrian territory through which the proposed canal was to run. But the new Chief of Staff, Yitzhak Rabin, opposed this. 'I hoped to find an answer to the problem that would not entail war,' he later wrote.

On November 3 the Syrians took the initiative. Two Syrian tanks, from the security of their bunkers on the Golan Heights on the high ground of Tel Nahila north of kibbutz Dan, fired with their machine-guns on an Israeli tractor working just below them. Ten Israeli tanks, which undetected by the Syrians had made their way to the border twenty-four hours earlier, moved out of their protected positions and opened fire on the Syrian tanks, who responded in kind. Firing lasted for several hours. None of the tanks was hit on either side, but the confrontation marked a serious escalation of the border tensions.

Firing was renewed ten days later, on November 13, when Syrian tanks again opened fire with their machine-guns, this time on an Israeli troop carrier that was patrolling the border road. Israeli tanks then fired at Syrian tanks. When the firing ended, three Israeli soldiers were dead. Two of the Syrian tanks were destroyed. In an attempt to deter the Syrians from further action, Rabin asked Eshkol to authorize an air strike. Eshkol gave his approval and that same day Israeli pilots attacked Syrian artillery positions on the Golan Heights. It was their first such mission. 'The results were excellent,' Rabin later wrote, 'and thereafter the Syrians were more prudent about using their artillery.'

Syrian efforts to divert the waters of the Hazbani and Banias rivers did not stop, however. When bulldozers appeared, clearly visible less than a mile away from the Syrian side of the border, to construct the diversionary canal, Rabin summoned Chaim Bar-Lev, the Director of Military Operations, and David Elazar the head of Northern Command (both of which positions he had earlier held himself), as well as the head of the Armoured Corps, General Yisrael Tal, and asked them, 'Can our tank guns hit the Syrian earth-moving equipment at a range exceeding 1,300 yards?' Rabin's aim was to show the Syrians that their equipment could be destroyed without Israeli forces having to cross the actual border.

General Tal did the necessary research. 'He instituted up-to-date ballistic training,' Rabin has written, 'undertook a thorough study to find the shells most suitable for targets of this nature; worked out new firing techniques for tank guns; and improved the marksmanship of the gunners to a standard that was to yield splendid results in the course of the Six Day War.' The plan was then taken to Eshkol. 'After being briefed on every detail of our plan,'

Rabin wrote, 'Eshkol gave us the go-ahead, and on the appointed day I set out to observe the combat zone. The tanks assigned to destroy the Syrian equipment were manned by our finest officers and gunners. Within a short time the Syrian earth-movers were wiped out by direct hits with incendiary shells. A renewed attempt to send in earth-moving equipment had the same results. When the Syrians took the hint and shifted their operations further back, we extended our range. The ability of our tanks to score hits at a range of up to two and a half miles finally convinced the Syrians that their diversion scheme was doomed.'

The world expected war to break out, but Egypt pressed Syria not to embark on general hostilities, arguing that the time was not yet ripe. But it was not by all-out war alone that Syria sought to get the better of Israel, and to prepare for its destruction. Terror was also a weapon to be tried. On 1 January 1965, while the Christian world celebrated New Year's day, a new organization was born, with Syrian encouragement: the Movement for the National Liberation of Palestine, quickly known by its Arabic acronym Fatah – Victory. Its leader was a Palestinian Arab who used the name Abu Ammar. His real name was Yasser Arafat. From that moment, for the next thirty years and more, his fate and Israel's were to be intertwined.

Fatah commandos, usually working in small teams of three or four men, crossed into Israel and planted bombs. Syria, as a precaution against retaliation, would not allow them to cross into Israel from Syria. Their first action, which was boasted to the world by a 'victory' communiqué that New Year, was not in fact successful: the commandos were stopped by a Lebanese policeman on the Lebanese–Israeli border. On 28 February 1965 another commando team crossed the border from Jordan, penetrated three miles into Israel, and blew a hole in a grain storage silo.

In May there were repeated Fatah crossings of the border, in the south and the north. These included shots fired at a farm truck on the Arava road, a grenade thrown at a chemical tanker in the Lachish region, the damaging of two houses in Ramat Ha-Kovesh, when four kibbutz members were injured. In retaliation for the attack on Ramat Ha-Kovesh, an Israeli commando unit crossed the border with Jordan opposite Kalkilya and blew up a fuel distribution centre.

Founded in 1932 by pioneers from Eastern Europe, Ramat Ha-Kovesh (Height of the Conqueror) was no stranger to attacks. In the Arab riots in 1936 fifteen members had been killed. It had also been in the front line during the War of Independence. After 1965 the kibbutz, with more than 500 inhabitants, had flourishing citrus groves and a large bakery.

In May 1965 an event took place in Jerusalem which marked an important cultural and national step forward for the still-young State. Hitherto, no national museum had existed. That month saw the culmination of five years' intensive planning and fund-raising, coordinated by the director of the Prime

Minister's Office, Teddy Kollek. Some years earlier, while serving in the Israeli Embassy in Washington, Kollek learned that a well-known Jewish banker had just promised his important art collection to Harvard. When they met, Kollek asked him why he had not considered Israel. The banker replied that there was no museum in Israel 'fit to receive the paintings'. Kollek was determined to rectify that situation. He was supported, among other influential figures, by the then Mayor of Jerusalem, Gershon Agron, and the former Chief of Staff turned archaeologist, Yigael Yadin.

Almost exactly five years before the opening of the museum, Ben-Gurion had asked the Knesset for a budget allocation for the 'National Museum of Israel', telling the legislators (on 30 May 1960): 'As befits an ancient people, dedicated to the values of the spirit throughout its tortured history and now reviving its independence in its ancient land, Israel, in its twelfth year of statehood, is about to establish a National Museum.' It would be built in Jerusalem, 'city of King David, amidst the timeless Judaean hills'. Despite Israel's 'daily preoccupations with defence and security, economic and social development, and housing the newcomers,' Ben-Gurion added, 'it has been resolved to spend part of our resources, energy and talent in what is destined to become the most impressive cultural centre in the country'.

The museum was located at Neve Sha'annan – the Hill of Tranquillity – on the (then) western outskirts of Jerusalem, within sight, to the south, of the Jordanian border. 'A few days before the opening,' Teddy Kollek has written, 'things looked so chaotic that people were betting the museum would not be ready in time. I was sure it would all come together, and took every bet. And I won them. Everything clicked into place, including the May sun and breezes.' An art student, Martin Weyl, who was himself later to become the museum's director, was present at the opening ceremony. 'It was with great awe,' he later wrote, 'that I witnessed, on 11 May 1965, the opening of the Israel Museum. As a young art student supporting my studies by working on the building of the Museum, I stood at the side with several other construction workers and maintenance people watching the many VIPs I had known only from the newspapers – famous artists, government dignitaries, and elegant-looking contributors from abroad. I shall never forget that beautiful day filled with flags, ceremonies and speeches, a day pervaded with a feeling of happy celebration, the glory of culture, the pride of achievement and confidence in a better future. More than anything else, it probably had to do with the privilege of witnessing a historic moment – the establishment of a national museum in our new homeland.'

On 19 May 1965, Damascus, the Syrian capital, became the main focus of Israeli attention. That day the Syrian authorities hanged Eli Cohen in front of a vast crowd gathered to witness the scene in Martyrs' Square. Cohen was a Jew who had been born in Egypt. In 1948 he had emigrated to Israel. Recruited by Israeli military intelligence, he pretended to be a Chilean citizen of Arab descent. In this disguise, he lived in Syria and became a popular figure in the Syrian military establishment, whose decisions he was able to send back to Israel. He also sent back photographs which he himself had taken of the Syrian defences on the Golan Heights. In a final letter to his family, Cohen asked his wife to give their two young children a good education, and he added, 'The day will come when they will be proud of me.'

Eli Cohen's public hanging was made an occasion for an upsurge in Syrian denunciations and vitriol against Israel. It took place at night, the square illuminated by searchlights. Syrian television covered the execution live. Afterwards Cohen's body was left hanging for six hours as crowds filed past. The next day the West German Ambassador in Damascus left for Bonn: Syria had broken off diplomatic relations with West Germany because it had established relations with Israel.

Fatah continued to try to gain attention, and do harm. On May 25 – six days after Cohen's hanging – a small commando unit penetrated across the River Jordan to the town of Afula. There it blew up a house, badly injuring a woman and her two children. The Israeli response was swift: two days later a paratroop attack across the River Jordan sought out and destroyed two Fatah bases on the West Bank. The next day, May 28, the Israeli Chief of Staff, Yitzhak Rabin, announced that Israel would take further, similar action, unless the Jordanian government acted to curb the Fatah attacks. 'This was no punitive expedition,' Rabin told military journalists who were summoned to his 'war room' at general headquarters. 'It was a warning strike intended to express the seriousness with which we regard Jordanian indifference towards the activities of El Fatah gangs operating from their territory. We intended to stress that our patience is limited, and that borders can be crossed in two directions.'

Fatah waited scarcely forty-eight hours to challenge and taunt Rabin. On June 1 a small squad crossed the Lebanese border and planted bombs. No one was killed, but Fatah was able to congratulate itself on the audacity of the raid, and to threaten many more.

To the alarm and annoyance of the Labour establishment, Ben-Gurion, from his retirement in Sde Boker, on the edge of the Wilderness of Zin, raised once more the spectre of the Lavon affair. Ten years had passed since Lavon had resigned as Minister of Defence following the Israeli Intelligence operation in Egypt which had been a costly failure. Asking an Israeli journalist, Hagai Eshed, to prepare the material for him, Ben-Gurion then used this material – which included information that Lavon had in fact given the order for the sabotage operation in Egypt – to demand a full judicial committee of inquiry. Eshkol, who had succeeded Ben-Gurion as Prime Minister, and was trying to bring Lavon back (for the second time) into public life, told Ben-Gurion that he had no desire to reopen the affair.

Ben-Gurion decided to pursue his own inquiries even further, and to apply to the Attorney General, Moshe Ben-Ze'ev, for an investigation. The Attorney General found the material convincing enough to recommend opening an investigation. Eshkol, who feared he would not be able to persuade a majority of Mapai to oppose Ben-Gurion's new tactic, resigned as Prime Minister. Mapai prevailed upon Eshkol to return to office, then voted against any further inquiry. Eshkol's new government ignored Ben-Gurion's continuing and frenetic appeals. But at the Mapai Party convention, held in February 1965, Ben-Gurion managed to put the Lavon affair back on the agenda, again demanding a full judicial inquiry into who had given the order a decade earlier. Moshe Sharett, who had been Prime Minister at the time of the affair, was then dying; but he appeared at the convention to plead with the Party members not to proceed with the dispute any further. Ben-Gurion was forcing the Party to deal with a matter 'of which the public wished to hear no more,' Sharett declared.

A secret ballot was held to decide whether to ask the State judicial bodies to examine the affair. Sharett's plea was successful: 841 members (40 per cent) voted for Ben-Gurion, but 1,226 (60 per cent) against him. So incensed was Ben-Gurion that he mounted a nationwide campaign against Eshkol, asserting that he was not fit to lead the government. He even challenged Eshkol's leadership of the Party, but at the Mapai Central Committee where the leadership question was put to the vote, Ben-Gurion received only 36.5 per cent as against 63 per cent for Eshkol.

In his fury, Ben-Gurion left Mapai and formed his own Party, Rafi – *Reshimat Po'alei Yisrael U'bilti Miflagtiyim*, the List of the Israel Workers and Non-Partisans. Those Mapai activists who followed Ben-Gurion into Rafi included a future Prime Minister, Defence Minister and Foreign Minister – Shimon Peres – and a future Defence Minister and Foreign Minister – Moshe Dayan. There were also two future Presidents, Yitzhak Navon and Chaim Herzog. They were dubbed 'youngsters' because, relative to the veterans who dominated Israeli politics at that time, they were indeed young – in their forties and fifties. Another Rafi activist, who was just thirty when Rafi was founded, was Gad Yaacobi who, as Minister of Economics and Planning after 1984, was to be responsible for Israel's first five-year economic plan.

Those who joined Ben-Gurion in Rafi realized the enormous challenge they were making to the Labour establishment. Peres, who was present at the meeting in Ben-Gurion's house when the decision to break away from Mapai was taken, recalled how 'to part company with colleagues and close friends who had decided not to join us; to take the field with such an impromptu force, without organization, without resources, without prior reconnaissance, to go to war against the establishment and demand its reform – it all seemed too hasty, too rash, a gesture of protest with no prospect of success.'

Mapai took what action it could to express its anger at the breakaway. Ben-Gurion and those who had broken away with him insisted that they were still a part of Mapai. He and ten others were put before a Party trial and formally expelled. Their prosecutor, the jurist Yaakov Shimshon Shapira, called Ben-Gurion a 'coward' and those who had gone with him 'a neo-Fascist group'. He subsequently became Minister of Justice.

The Lavon affair was the catalyst for the founding of the new Party, but it was not merely a Party of protest. The founders of Rafi had high ideals. They were motivated by a desire to see in Israel a more forward looking, less politically driven system, and emphasized the need of 'modernization and scientification'. Directly challenging the traditional Labour high ground of Mapai, they sought to combine economic conservatism and social liberalism with a pledge, as Chaim Herzog wrote, 'to avoid the infighting and self-interest that were already downgrading Israeli politics and the Knesset'. Among Rafi's promises were unemployment insurance, national health insurance, and free compulsory education from the age of three to sixteen.

Peres has described the first days of the new Party's activities, after Ben-Gurion had returned to the Negev, and Moshe Dayan was away. 'A friend put at our disposal two or three rooms in a building in the centre of the city,' he recalled, 'but the rooms were unfurnished and empty – as were our pockets. Gradually things began to arrive: one lady brought two stools from her house, another provided curtains; somebody brought a Chagall print – two green doves on a light blue background; a husband and wife, both lawyers, offered their services as drivers – they had a car!'

At this point, help came from Moshe Haviv, a lawyer and retired army officer, who had fought in the War of Independence – at one point leading a contingent of new immigrants in the conquest of Beersheba, and the opening of the road to Sdom. Known as 'Mosh', he quickly captivated those whom he had joined. 'It turned out that Mosh could do anything,' Peres later wrote. 'A book needed publishing – he was a publisher; there was a law-suit to be contested – he was the advocate; funds had to be raised – he organized it; colleagues became estranged – he conciliated them; branches needed opening – he founded them; a speech had to be made – Mosh dazzled his audience; somebody needed a heart-to-heart talk – Mosh could spare the time.'

Two years later, on the eve of the Six Day War, Peres, and their mutual friend Gad Yaacobi, tried to dissuade Moshe Haviv from returning to the army. 'Look,' they said to him, 'we went to war in 1947–8, and again in 1956; now once more we are going out for a meeting with fate. Don't you think that it is a bit too much?' Haviv returned to the army, and was in action on the Golan Heights. He was fighting there on the last day of the war. 'When I heard a rumour that Mosh had fallen,' Peres later wrote, 'I checked all the relevant data – the place, the vehicle, the number, and I knew that Mosh was indeed no more. I was overcome by a sense of loneliness such as I had not known for many years.'

In order to strengthen its depleted ranks after the formation of Rafi, Mapai joined with another left-wing Party, *Ahdut Ha'avodah-Po'alei Zion* (which had been in existence in various forms since its earlier split with Mapai in 1944) to form a new, left-of-centre Party. It took the name *Hama'arach Le'achdut Po'alei Yisrael* – the Alignment for the Unity of Israeli Workers, known more commonly as the Ma'arach (or the Alignment).

Rafi's social policies had a strong appeal, with which, together with the magnetism of Ben-Gurion's personality, it was hoped to secure many seats in the next Knesset election. In the elections held on 2 November 1965 the Alignment won forty-five seats, making it by far the largest single party in the Knesset. Its nearest rival, the right-wing coalition Gahal (the Herut–Liberal bloc established just before the election), won twenty-six, the National Religious Party eleven, Mapam eight.

Rafi won only ten seats. That the name 'Ben-Gurion' could pull no more than that was a disappointment to those who had such high hopes. Rafi remained with the opposition Parties, depriving men like Peres, Dayan and Herzog of the political positions they might have occupied in government had they remained with Mapai. One of Ben-Gurion's biographers, Michael Bar-Zohar, has written, 'A group comprising several of Israel's most gifted leaders found itself in barren opposition. Ben-Gurion was an ageing lion whose strength was on the wane and whose roar was progressively fading. Beginning as the just fight of a courageous and honest leader, his battle had ended in a shameful defeat which heralded his final decline.'

The Israeli military administration, which had ended in most of the Arab inhabited areas of Israel in 1963, continued in the Negev until 1965. The situation of the 26,000 Negev Bedouin was precarious. Since the ending of the War of Independence a sustained policy had been adapted by the government of Israel to move them out of most of the areas in which they lived and to concentrate them in the north-eastern part of the Negev and to the north of Beersheba.

Those Bedouin who lived south of the Wadi Paran had left during the fighting and were not permitted to return. Others had fled to Jordan and into the Egyptian Sinai and the Gaza Strip, becoming refugees. During the period when Moshe Dayan was commander of the Southern Front he worked closely with Yosef Weitz, who, as in 1948, saw the Israeli imperative as acquiring land for Jewish settlement. One area in which the Bedouin might have been settled – the Lachish region – was coveted by Weitz and the Jewish Agency for Jewish settlement. Considerable land was also expropriated from the Bedouin, using the argument that it was needed for defence purposes. General nervousness inside Israel with regard to Egypt's intentions along the Sinai border rendered this argument sufficient for the

Israeli public. Compensation was offered for land expropriated, though it was less than that considered fair by the Bedouin themselves.

Unlike the general Arab population, but like the Druse, many Bedouin served (and still serve), in their case as volunteers, in the Israeli army, their role as trackers being particularly prized. One of the most celebrated Bedouin soldiers, Abd el-Majid Khidr, known by his Israeli name Amos Yarkoni, founded an élite commando unit in the late 1950s. He was to lose an arm in one military action and a leg in another; in 1967 he was appointed the first Israeli governor of Central Sinai.

For many years the Bedouin were not involved in Arab nationalist activity. In the 1990s, however, a Bedouin member of the Knesset, Talib El-Sane, manifested some of the bitterness that had accumulated as a result of the long delay in accommodating Bedouin aspirations. Under Labour, a paramilitary Green Patrol had been established, which participated in a series of actions and pressures designed to push the Bedouin off the land in the Negev and into towns. This was done by outlawing goat grazing which was the main source of Bedouin livelihood.

As the Green Patrol acted with increasing severity the Bedouin found a champion in Clinton Bailey, an Arab-speaking Israeli academic whose life's work was the recording and rescue of Bedouin culture as preserved through a vigorous oral tradition. Bailey was instrumental in obtaining legal aid for many Bedouin, and in getting injunctions which prevented them from being evicted from the land at least until their cases had been heard in court. The climax of Bailey's activities was a 150-day sit-down strike by the Bedouin in front of the Prime Minister's office in Jerusalem (from 1 September 1993 to 30 January 1994). This succeeded in increasing the number of proposed compulsory Bedouin townships from seven to at least fifteen, and in maintaining the possibility of the Bedouin pursuing their traditional livestock-raising, and agricultural activities. 'I was simply a researcher of the Bedouin culture,' Bailey later commented. 'I had known these people for a good number of years and was aware that they had traditional claims to the land – and I wanted them to have a fair deal.'

Israeli and Hebrew literature received worldwide recognition in 1966, when Shai Agnon was awarded the Nobel Prize. A Jew from Galicia, born Samuel Josef Czaczkes, he had first gone to Palestine in 1907, at the age of nineteen, and had finally made his home there in 1924. In his novel *A Guest for the Night*, which had first been published in 1938, he had written of a return visit to his home town (Buczacz) and of the spiritual and moral desolation of the prewar Jewish world, both in Europe and in Palestine. His Nobel Prize was the first to be awarded to a Hebrew writer.

In presenting the prize to Agnon, the King of Sweden asked him where he had been born. Agnon answered, 'Through a historical catastrophe – the destruction of Jerusalem by the emperor of Rome – I was born in one of the cities of the Diaspora. But I have always deemed myself a child of Jerusalem – one who was in reality a native of Jerusalem.'

Israel's artistic initiatives flourished. Also in 1966, the America Palestine Fund (renamed the America-Israel Cultural Foundation in 1948) opened the America-Israel Culture House in New York, exhibiting Israeli arts and crafts, and holding cultural programmes to show the range and achievements of Israeli artistic activity. The fund made considerable donations to cultural institutions in Israel, and provided scholarships to enable Israeli artists to bring orchestras, theatre groups, dance ensembles and art exhibits to the United States. It also encouraged the transfer of art collections to Israel, including the works for the Billy Rose Sculpture Garden at the newly established Israel Museum, and gave financial support to theatres, music academies, concert halls and cultural centres in rural areas throughout Israel.

Britain under a Labour Prime Minister, Harold Wilson (from 1964) was showing a willingness to put aside the bad memories of Israel's clash with the postwar Labour government. The United States under Johnson was expressing sympathy. But it was with France that the main links had been forged, and from France that the main arms purchases, so crucial for Israel's security, were being made. Yet the new Israeli Foreign Minister, Abba Eban, was uneasy at the nature of France's commitment. 'When I reached Paris in February 1966,' he later recalled, 'our military and supply missions had shown me impressive lists of helpful French actions in fields vital to our security. Economic, technological and cultural cooperation were also in full spate. What did protocol matter in comparison with such things?' Several 'eminent Frenchmen' concerned with Israel's relationship with France, Eban noted, 'were even more emphatic to me in their reassurance. They thought that any expressions of nervousness from us would create the very situation that we wanted to avoid. To put de Gaulle's assurances in doubt would invite resentment – and not from him alone. Verbal gestures were admittedly scarce, but aircraft and other equipment were flowing copiously.'

A new agreement for fifty Mirage V aircraft had been signed in the summer of 1966. This, Eban wrote, 'seemed to show that France valued the substance of things above their form. True, de Gaulle had reacted with irritation in May 1963 to Ben-Gurion's plea for an official alliance, but a year later he had reiterated to Eshkol that France was still Israel's "ally and friend". If the French relationship was rich in content but sparse in outward expression, was this not better than if the opposite were true? I had scarcely had time to pursue my doubts or to put the reassuring counter-arguments to test by the time the 1967 crisis came.'

On 8 April 1966 a Fatah squad crossed into Israel from Syria and planted a mine. It killed an Israeli farmer. Another farmer was killed when a mine blew up on May 16. Then, on July 12 and 13, more Fatah units crossed the border from Syria and planted four more mines. As a result, two Israelis were killed. This led Israel to retaliate by air strikes.

A pattern of violence had begun to scar the Israel–Syrian border. On August 15, after another Fatah raid across the border – in which no one was killed – Israel attacked Syrian positions from the air. The Syrians returned fire. Planes, artillery, tanks and Israeli gunboats on the Sea of Galilee all participated. There was another Fatah crossing of the border on September 6. Israel complained to the United Nations, but there was no response. Two days later an Israeli patrol caught a Fatah unit which had crossed from Syria into Upper Galilee. In the gun battle that followed, two of the Fatah were killed. A day later, on September 9, three Israeli soldiers were killed when a Fatah-laid mine detonated in their path. The Fatah was gaining its 'martyrs', and Israel its victims.

Two more Fatah cross-border attacks were launched on October 8. In one of them, three Israeli civilians were hurt by bomb explosions in West Jerusalem, and four more by a mine explosion south of the Sea of Galilee. When the Israeli government protested to Syria, the Syrian Prime Minister replied, 'We are not sentinels over Israel's security.' The temperature in the Arab world was rising. King Hussein of Jordan declared that if Israeli troops attacked Syria, Jordan would open a separate (eastern) front against Israel. On October 12 the Soviet Union said that Syrians could count on its support if Israel began hostilities. On the following day, three Israeli soldiers were killed in a clash with a Fatah unit that had crossed into Israel from Jordan. Anticipating an Israeli reprisal raid on Syria, eleven Arab States declared their 'solidarity' with Syria. But the Arab alarms and threats and plans for joint action had no basis in the Israeli reality; Israel had no intention of attacking Syria.

Syria, her Golan Heights strongly fortified against any Israeli attack, signed a mutual defence pact with Egypt on 4 November 1966. These were the years when Nasser's Pan-Arabism – his attempt to create a single Arab national outlook led by Egypt – was at its height. The Egyptian leader also hoped to rally the Arab States to a final assault on the hated Jewish entity in their midst. For Israel, the Syrian–Egyptian pact created a two-front challenge which it did not have the military capacity to meet. On paper, and presumably on the ground as well, the combined forces of Syria and Egypt were overwhelming. That was without whatever contribution the Soviet Union might decide to make. Soviet propaganda against Israel, portraying her as the imminent aggressor, was intensifying.

As well as Fatah, other Palestinian Arab groups were beginning to cross the border from Jordan, Lebanon, Syria and Gaza, to attack Israeli civilians. Most of those crossing the border were recruited in the refugee camps. These camps existed outside most West Bank towns, and had been kept carefully segregated by the Jordanian authorities from the local Palestinian Arab population.

A debate was taking place inside the upper echelons of the Israeli armed forces, as to what Israel's response to the continuing cross-border infiltration and acts of terror ought to be. Military action against Syria or Egypt was judged too risky, especially with the two countries linked in a military alliance. Jordan was the weakest neighbour militarily. Yet Fatah and other groups were coming across Jordan's border as well. The border between Israel and Jordan was Israel's longest with any of its neighbours, stretching from Eilat in the south to the Yarmuk River in the north.

An influential voice for tough action was that of the former head of the Israeli air force, Ezer Weizman, who had just become head of the General Staff Branch of the armed forces. His argument, as he recalled in his memoirs, was that 'in 1966 we can't carry out a 1955-style reprisal raid, going in at night, laying a few pounds of explosives, blowing up a house or a police station, and then clearing off. When a sovereign state decides to strike at its foes, it ought to act differently. We have armour, and we have an air force. Let's go in by day, operating openly and in force.'

Many senior officers shared Weizman's argument in favour of a substantial retaliation. A plan was drawn up by the General Staff, and taken by the Chief of Staff, Major General Yitzhak Rabin, to the inner Cabinet – known as the War Cabinet – presided over by Levi Eshkol. Rabin argued strongly in favour of retaliation. The Cabinet members, aware that something on a large scale had the backing of a considerable segment of public opinion, gave the plan its approval. No date was set. Then, on November 12 a truck containing an Israeli border police patrol struck a mine that had been laid on the Israeli side of the Israel–Jordan border, at the southernmost point in the West Bank. Three border policemen were killed. The reprisal action was ordered at once. It had a clear purpose: to cross the Jordanian border and reach the village of Samua – from which several Fatah raids had been launched – and blow up forty of the buildings there as a punishment. The inhabitants would be warned by loudspeaker to leave the area where the demolitions would take place.

On the morning of November 13 an Israeli parachute battalion and ten light tanks crossed the border into Jordan north-east of Beersheba. Destroying a Jordanian police post on its way, it advanced to Samua, 3 miles inside the West Bank. Entering the village, it gave its warnings and began to demolish the empty houses. Then disaster struck. A Jordanian Army Infantry Brigade based on Hebron, which had been told that the Israeli force was attacking the nearby town of Yata, drove through Samua on its way to the supposed scene of the attack. Driving straight into the Israeli battalion, the Jordanians were caught completely unawares. Fifteen of their twenty trucks were destroyed, and they withdrew to take up positions elsewhere in the village. A full-scale battle then ensued, with Jordanian aircraft summoned to help, and Israeli aircraft being sent up to intercept them. After three hours, the Israeli battalion withdrew back across the

Israel–Jordan border. Fifteen Jordanian soldiers and three civilians had been killed. One Israeli officer was killed, and one of the attacking Israeli pilots was injured (he was half-blinded).

On November 25, less than two weeks after the Samua raid, the United Nations Security Council, which had said nothing after the repeated Fatah raids on Israel – seventeen in less than a year – condemned Israel for the raid on Samua. Within the Arab world, the Egyptians were made to feel bad that they had not activated their agreement with Jordan, whereby they should have come to the assistance of the Jordanians immediately Jordan was attacked by Israel. Syria saw Israel's decision to attack into Jordan and not into Syria as a sign that Israel was unwilling to take on the Syrian army. This view was confirmed, to the satisfaction of Damascus, a month later, when a clash between Israeli and Syrian border forces in Upper Galilee, after an Israeli patrol had found a recently laid land mine, escalated from machine-gun fire to mortars to tank guns to artillery, without Israel crossing the border into Syria.

CHAPTER TWENTY-ONE

Nasser's challenge

The first three months of 1967 were marked by repeated Syrian artillery bombardments and cross-border raids on the Israeli settlements in the north. Israeli air raids against Syrian positions on the Golan Heights would result in a few weeks' quiet, but then the attacks would begin again. On 7 April 1967 Syrian mortars on the Golan Heights began a barrage of fire on kibbutz Gadot, on the Israeli side of the B'not Yaakov bridge. More than 200 shells were fired before Israeli tanks moved into positions from which they could reach the Syrian mortars.

As the Israeli tanks opened fire, the Syrian artillery did likewise. Firing quickly spread along the border to the north and south of Gadot. Then Israeli warplanes – Mirage fighter-bombers purchased from France – flew over the Syrian border and over the Golan Heights, strafing several Syrian strongholds and artillery batteries. Fifteen minutes later Syrian warplanes – Soviet MiG-21s – took on the Israeli planes in aerial combat. Within a few minutes, six MiGs had been shot down and the rest chased eastwards to Damascus. The citizens of Damascus could see the Israeli planes between the capital and the snow-capped peak of Mount Hermon, Syria's highest mountain. One Israeli plane was shot down.

Syria protested that Israel was preparing for war. In public, the Soviet Union supported this claim. In private, the Soviet Ambassador in Damascus warned the Syrians to restrain the Fatah raids into Israel. The Egyptian Prime Minister, visiting Damascus to boast of the creation of a common front against 'Israeli aggression', likewise warned the Syrian government not to provoke Israel into going to war. For her part, Syria complained bitterly to Egypt that she had not rushed to help her ally, particularly in view of the Syrian–Egyptian defence pact signed the previous year.

It was the Soviet Union that sought, most publicly, to condemn Israel's action (or, more properly, her reaction). On April 26 the Soviet Ambassador in Tel Aviv, Dmitri Chuvakhin, protested to Levi Eshkol that Israel was indeed planning a war, telling the Israeli Prime Minister, 'We understand that in spite of all your official statements, there are, in fact, heavy concentrations of Israeli troops all along the Syrian borders.' Not only did Eshkol deny the allegation; he offered to accompany Chuvakhin on a fact-finding trip along the whole Israel–Syrian border. The Ambassador declined. The Syrians took the Soviet claim seriously.

Following the Gadot clash, Fatah renewed its campaign inside Israel, using the Syrian border as a conduit. On April 29 a water pipeline was blown up, and a few days later mines were laid on the main road leading north from Tiberias, damaging an Israeli army truck.

Gradually, during May, President Nasser emerged as a champion of the Syrians – or rather of the Arab world generally, the leadership of which he so wished to assert. Beginning on May 13, Egyptian troops moved in large numbers into the Sinai, from which Israel had withdrawn nine years earlier and which had been demilitarized as security for Israel after her withdrawal. As the Egyptian troops moved forward, Cairo Radio set the tone of a propaganda war that became Egypt's daily barrage: 'Egypt, with all its resources, is ready to plunge into a total war that will be the end of Israel,' the radio declared.

Israel then made what has since been judged a psychological mistake. Hoping to assert her peaceful intentions, and to calm the jittery atmosphere created by Arab – and Soviet – accusations of an imminent Israeli attack on Syria, she held her May 15 Independence Day parade without the usual large numbers of tanks and heavy artillery. The parade took the form of a night-time military tattoo held in the stadium of the Hebrew University on Givat Ram. The full parade was not held because of an agreed limitation of tanks in Jerusalem, as laid down in the armistice agreement with Jordan, and because Israel did not wish to exacerbate tension on the Jordanian front.

Noting the lack of heavy armour, the Egyptians at once accused Israel of having sent the 'missing' tanks and other weaponry to the north. Egypt also named May 17 – a mere two days away – as the day on which Israel would invade Syria.

On the following day, May 16, Egypt acted to raise the temperature in the region still further and to threaten Israel. That day, at ten in the evening, Nasser ordered the United Nations to remove its forces from Sinai. Since 1956 a United Nations Emergency Force (UNEF) of 3,400 men had been stationed in the Gaza Strip and at Sharm el-Sheikh, at the southern tip of the Sinai peninsula, with the internationally approved task of monitoring the Egyptian–Israeli cease-fire. It had been able to take up its monitoring positions on Egyptian soil only with Egyptian consent. That consent was suddenly withdrawn.

Israel expected the United Nations Secretary General, U Thant, to ask at least for a period of time in which to delay. The Egyptians themselves expected that the demand for the withdrawal of the Emergency Force would be challenged by the Security Council. But U Thant did not even call the Council. Instead, he accepted Nasser's demand at its face value and ordered the troops to pull out at once. The troops began moving within twenty-four hours of Nasser's demand, and on May 19, only three days after the demand had been made, the last of the troops sailed away.

Egypt was in total military control of Sinai. Nasser had seen an international organization, hitherto committed to maintaining the cease-fire, turn tail and run. On May 20 Israeli reserves – the basis of the citizens' army – began partial mobilization, leaving their homes and their workplaces and hastening to their camps and assembly points. As the bulk of Israel's armed forces are drawn from the civilian reserve, full mobilization means a virtual stop to the Israeli economy.

That day, May 20, the Egyptian Minister of War, Field Marshal Abdul Hakim Amer, travelled to Gaza to inspect the Egyptian troops that had replaced the United Nations contingent. Alongside the Egyptians were soldiers of the Egyptian-sponsored Palestinian Liberation Army.

Inside Israel, at the highest level, a fierce debate was taking place. Senior army officers and leading politicians argued for a pre-emptive military strike against both Egypt and Syria. But the Chief of Staff, Yitzhak Rabin, and the Prime Minister, Levi Eshkol (who was also Minister of Defence), felt that any such action would be unwise. It was not certain, they argued, that either Damascus or Cairo was determined on war.

As the crisis intensified, the Secretary-General of the French Foreign Ministry, Hervé Alphand, visited several Arab capitals. While in Beirut he stated publicly that there was no contradiction between France's recognition of 'Israel's existence' and France's friendship with the Arab States. Considering that, eleven years earlier, France had been Israel's active ally and co-belligerent, and was still its major arms supplier, the relegation of France's commitment to the mere 'existence' of Israel struck a sour note in Tel Aviv and Jerusalem.

With each day that passed, the Israeli leaders discussed what they regarded as the terrible prospect of war. On May 21 the Foreign Minister, Abba Eban, asked the Chief of Staff, Yitzhak Rabin, what likelihood there was of Egypt trying to close the Straits of Tiran, and thereby denying Israeli ships access to the port of Eilat through this international waterway. 'Rabin was very tense,' Eban later recalled, 'chain-smoking all the time. He pointed out that Israel's military preparedness had always been related to the northern and eastern fronts, with little attention to the south. When I asked him what the diplomatic establishment could do to help, he said to me, "Time. We need time to reinforce the south".'

At a meeting of senior Israeli Cabinet Ministers later that day it was agreed that one indication of Nasser's true intent would be if he closed the Straits of Tiran to Israeli shipping. But that evening, only an hour after the inner

Cabinet agreement to treat the closure of the Straits as a cause of war, Levi Eshkol spoke over the radio to the Israeli public, and in a conciliatory speech, stressing Israel's desire not to go to war, made no mention of the Straits of Tiran or the blockade or freedom of passage for Israeli ships to Eilat. Nasser read Eshkol's speech and drew what seemed to him to be the obvious conclusion – the same conclusion that Syria had drawn after the artillery duel at Gadot six weeks earlier – that Israel was not prepared to go to war.

During his speech, Eshkol had difficulty deciphering a word in his script. His stumbled sentence left a terrible impression on his Israeli listeners, who feared that he was breaking under the strain. As a result, there was immediate talk of replacing him as Defence Minister. As Israelis' public confidence in their government fell, Nasser's confidence rose. On the following day, May 22, he announced that Egypt was reimposing her blockade of the Straits. He made his announcement – which suddenly raised the spectre of war with Israel – at an Egyptian air force base at Bir Gafgafa, in the Sinai, a hundred miles from Israel's Negev border.

Abba Eban, who listened to a recording of Nasser's speech an hour later, described it as offering Israel a choice, 'slow strangulation or rapid, solitary death'. Nasser told his pilots, and the waiting world:

> We are in confrontation with Israel. In contrast to what happened in 1956 when France and Britain were at her side, Israel is not supported today by any European power. It is possible, however, that America may come to her aid.
>
> The United States supports Israel politically and provides her with arms and military material. But the world will not accept a repetition of 1956. We are face to face with Israel. Henceforward the situation is in your hands. Our armed forces have occupied Sharm el-Sheikh . . . We shall on no account allow the Israeli flag to pass through the Gulf of Akaba. The Jews threaten to make war. I reply: 'Ahlan wa sahlan' – 'Welcome!' We are ready for war . . . This water is ours.

Nasser had committed his nation to war. 'Turning his back on a whole decade of prudence,' Abba Eban has commented, 'he now uttered a courtly and exultant welcome to the approaching war: "*Ahlan wa sahlan*". It was as if he were greeting the unexpected appearance of a beloved and long-absent guest.'

Israeli ships would no longer be able to sail from Eilat into the Red Sea and Indian Ocean. Yet this narrow waterway, Israel's commercial lifeline to the east, had been guaranteed by, among others, the United States, Canada, Britain and France. The reaction of the guarantors was immediate, but far from unanimous or clear-cut. French Foreign Ministry officials went so far as to intimate that there were 'judicial obscurities' about Israel's position.

They spoke, Abba Eban has written, as if Ambassador Georges Picot's speech of 1 March 1957 – supporting free Israeli maritime passage through the Straits of Tiran – 'had never been made', and in official talks 'we had been asked if the economic value of our Red Sea outlet was really enough to justify war.'

On the morning of May 23 the Israeli Cabinet and military leaders were summoned to Tel Aviv for an emergency meeting in the Ministry of Defence. As he drove down from Jerusalem, Eban noted that those who greeted him from their cars, or from the roadside, 'managed to give their gestures an implication of anxiety'. Only a month earlier, 'the national mood had been as close to normalcy as could be expected by a people born in war and nurtured in siege. Now the crisis was upon us. As countryside and townships sped past the window, I was gripped by a sharp awareness of the fragility of all cherished things. For the whole of that day in Tel Aviv, and far into the night in Jerusalem, our minds revolved around the question of survival; so it must have been in ancient days, with Babylon or Assyria at the gates.'

The emergency meeting convened at nine that morning. There were, Eban recalled, 'no cheerful faces' around the table. Levi Eshkol had invited the leader of the Herut opposition, Menachem Begin, to be present, as well as the senior military trio, the Chief of Staff, Yitzhak Rabin, the Chief of Operations, Ezer Weizman, and the head of Military Intelligence, General Aharon Yariv. Eshkol opened the meeting with a blunt and unemotional statement: 'We have heard the news on the political front. I don't know if you have all heard it. It requires consultation and, probably, action as well.' Eban later commented, 'The peril was taken for granted; it stood in no need of rhetorical adornment. The accent was placed on clarity of decision. A great doom was in the making and it seemed to be coming on relentlessly.'

General Yariv made his report. The Egyptian battle order in Sinai was not yet complete. Although the airfields in Sinai were being made ready for combat, their technical preparedness was 'still deficient'. On the Jordan front, no movement of troops had taken place towards the border. The Syrians were not making moves of any particular military urgency. But in many of the cities of the Arab world, vast crowds were demonstrating against Israel, calling for her destruction. Eban recalled the mood of the meeting:

> There was no doubt that the howling mobs in Cairo, Damascus and Baghdad were seeing savage visions of murder and booty. Israel, for its part, had learned from Jewish history that no outrage against its men, women and children, was inconceivable. Many things in Jewish history are too terrible to be believed, but nothing in that history is too terrible to have happened. Memories of the European slaughter were taking form and substance in countless Israeli hearts. They flowed into our room like turgid air and sat heavy on all our minds.

Before my turn came to speak, I noticed that our military colleagues had made no proposals for immediate action.

Rabin was then asked for his opinion. In 1956, he pointed out, Egypt was Israel's only adversary. And Israel had been allied to two major powers. This time Israel would be alone, while Egypt might have Syria, Jordan and contingents from other Arab States fighting with her, as well as the full support of the Soviet Union. Rabin was confident of military victory, but warned that it would be 'no walkover'. Ezer Weizman also spoke. 'Our military advisers', recalled Eban, 'could make no comforting predictions about the scale of Israeli losses. The candour of their words left a chilling aftermath.'

Eban read out to the meeting a telegram from Washington, reporting a request from the United States that Israel make no decision for war for forty-eight hours, to allow diplomacy to seek a way out of the impasse. In addition, the Cabinet was told, President Johnson 'would take no responsibility for actions on which he was not consulted'.

Rabin and Weizman agreed that Israel would lose no military advantage by agreeing to a forty-eight-hour delay before any decision for military action. No one at the meeting made any proposal for an immediate military response. But it was agreed that the occupation by Egypt of the Straits of Tiran should be considered by Israel an act of aggression, and that this should be conveyed to all foreign governments. A formal statement, prepared by Eban, represented the official Israeli stance. It read: 'The Government of Israel decides to give effect to the policy which it announced on 1 March 1957, namely, to regard any interference with shipping as an aggressive act against which Israel is entitled to exercise self defence.'

This left it open for Israel to take action as and when it chose, not as an initiating act of war, but as the response to the act of war by another.

There then began a flurry of international diplomatic activity aimed at averting war. On May 24 – within a few hours of the arrival at Cairo International Airport of an armoured brigade from Kuwait, as a gesture of support for Egypt – the Secretary-General of the United Nations, U Thant, flew from New York to the same airport to persuade Nasser not to commit his forces to war. Nasser would make no such commitment, and on the following day U Thant flew back to New York, his mission a failure.

On May 24 Abba Eban flew to Paris where President de Gaulle warned him that it would be 'catastrophic' if Israel were to attack first. The dispute must be resolved by the Four Powers, Britain, the Soviet Union, the United States – and France. 'Don't shoot first.' When Eban pointed out that in Israel's view the first act of war had already taken place in Egypt's closing of the Straits of Tiran, de Gaulle refused to accept this. He was upset, feeling that if Israel sought his advice, it ought to take it. Opening hostilities meant, in his view, firing the first shot. As for the pledge by Georges Picot that the

Straits of Tiran should be kept open, that, said de Gaulle, 'was correct juridically, but 1967 was not 1957'. Picot's statement had reflected the 'particular heat' of 1957.

From Paris, Eban flew to London. The Labour Prime Minister, Harold Wilson, took a far more pro-Israeli stance; indeed, he seemed to delight in turning his back on the traditional British Foreign Office reserve towards Israel. The Cabinet had just met, and had decided, Wilson told Eban, 'that the policy of blockade must not be allowed to triumph'. Britain would work with other like-minded nations to open the Straits of Tiran, and was already in negotiations with the United States to find a common policy, and an effective one.

In the United States, the Israeli Ambassador, British-born Avraham Harman, went on May 24 to see former President Eisenhower at his home in Gettysburg, Pennsylvania. After the fighting in Sinai in 1956 Eisenhower had made a United States commitment to keep open the Straits of Tiran. Harman was anxious that Eisenhower would continue to support that pledge, and do so publicly, should President Johnson renew it. Eisenhower told Harman (as Eban recounted in his memoirs), 'that he was not accustomed to making statements, but if asked by newspapermen, he would say that the Straits of Tiran was an international waterway. This had been determined in 1957. He would repeat the attitude which he and Secretary Dulles had then taken. He would add that a violation of the rights of free passage would be illegal.' Eisenhower went on to tell Harman, 'His friends in the Republican Party had already been in touch with him and he was going to tell them exactly what he thought.' Eisenhower then strongly criticized the United Nations role in recent weeks. Referring to U Thant's current conversations in Cairo, he said that Nasser had created 'an illegal position and there should be no compromise with illegality'.

Eisenhower then asked Harman about the positions of France and Britain. Reflecting on the past, he said that he 'still regretted that they had not taken steps of a concrete nature in the Suez Canal similar to those which had been adopted with regard to the Straits of Tiran'.

On hearing Harman's report of a speech by President Johnson on May 23 in which he had condemned Egypt's blockade as 'illegal and fraught with danger', Eisenhower said that he hoped the President's position 'would be strongly maintained'. He said that when he was President, 'the Russians tended to believe his strong statements because he had been a military man'. Eisenhower then told Harman, as his concluding words: 'I do not believe that Israel will be left alone.'

From the Israeli perspective, all would depend on the attitude and actions of the United States. As Eban flew from London to New York during the morning of May 25, President Johnson publicly denounced the blockade as 'illegal' and 'potentially disastrous to the cause of peace'. That day, an even more alarming development took place, as seen from Jerusalem: Egyptian armoured units crossed the Suez Canal and took up positions inside the Sinai. In Israel airfields had been put on high alert in case of a surprise Egyptian attack.

From all that Eban could gather, particularly after a late-night talk with the American Secretary of State, Dean Rusk – who had spoken at length with President Johnson – the United States was prepared to take a strong stance in Israel's favour. As Eban reported to Levi Eshkol in Jerusalem:

> In my view, the President was likely to discuss a programme for opening the Straits by the maritime powers led by the United States, Britain and perhaps others. The plan in its present form was based on the idea of a joint declaration of maritime States, including Israel, concerning their resolve to exercise freedom of passage. The second stage, according to what had been said to us, would be the dispatch of a naval task force which would appear in the Straits.
>
> Some officials had predicted that the President would make a pledge that the Straits would be opened, even if there was resistance. Some press reports were appearing in the same sense. I told them that after my talk with the President, I would fly home at once and bring the thoughts of the United States government to the knowledge of my colleagues; in the meantime, I had no authority to define any attitude during the present short visit. My efforts were limited to inducing them to make their proposals in the fullest detail, including a timetable and a method of carrying out any plan, so that our government should be able to determine its attitude one way or the other.
>
> I had emphasized that in the absence of an immediate plan for opening the Straits, there would, in my opinion, be no escape from an explosion. Since their plan included a certain reliance on the United Nations, I expressed a deeply sceptical appraisal of its effectiveness.

On the evening of May 26, Eban saw President Johnson, who started the conversation with the words, 'I am not a mouse from Washington, I am a lion from Texas.' Once it became apparent, Johnson said, that the United Nations could not keep the Straits of Tiran open 'then it is going to be up to Israel and all of its friends, and all those who feel that an injustice has been done, and all those who give some indication of what they are prepared to do, and the United States would do likewise. The United States has had some experience in seeking support of friendly states, but Israel should put its embassies to work to get support from all those concerned with keeping the waterway open. The British are willing and the United States is trying to formulate a plan with them.'

Meanwhile, the President added, it would be 'unwise' of Israel to 'jump the gun'. He stressed this point by repeating three times during the course of his conversation with Eban: 'Israel will not be alone unless it decides to go alone.' It was also the point stressed by the note which Johnson then handed to Eban, setting out the position of the United States. The note was short, and blunt. Regarding the Straits of Tiran, it read, 'we plan to pursue vigorously the measures which can be taken by maritime nations to assure that the Straits and the Gulf remain open to free and innocent passage of all nations. I must emphasize the necessity for Israel not to make itself responsible for the initiation of hostilities. Israel will not be alone unless it decides to do it alone. We cannot imagine that Israel will make this decision.'

Eban prepared to leave the White House, with this mixed message. The United States would work to reopen the Straits of Tiran, but it did not want Israel to take unilateral action, despite Israel's argument (which Eban had made to Johnson, as to de Gaulle) that it regarded the closure of the Straits in itself as an act of war. As Eban and Johnson walked out of the Oval Office, Eban asked him: 'Again, Mr President, can I tell my Cabinet that you will use every measure in your power to ensure that the Gulf and Straits are open for Israeli shipping?' 'Yes', replied Johnson. Eban recounted, 'He shook my hand with such a paralysing grip that I doubted that I would ever regain the use of it.'

While the diplomats, Presidents and Prime Ministers took up their respective positions, Nasser was raising the temperature of the crisis. 'The battle will be a general one,' he declared on May 26, 'and our basic objective will be to destroy Israel. I probably could not have said such things five or even three years go. Today I say such things because I am confident.' In the Security Council the Soviet Union made it clear that it would veto any proposal that might not be in accord with the wishes of Egypt or Syria. Flying from Moscow to Cairo, the newly appointed Soviet Minister of Defence, Marshal Grechko, brought Nasser a personal message of encouragement from the Soviet Head of State, Alexei Kosygin.

Nasser was confident of a military victory that would end the existence of Israel. Golda Meir, reflecting on her efforts after 1956 to convince the Americans that Israel would be in danger if it withdrew from Sinai, later asked in anguish, 'Why had it seemed so simple and so obvious to us but so impossible of attainment to everyone else? Hadn't we explained the realities of life in our part of the world properly? Had I made some dreadful mistake or left something crucial unsaid? The more I thought about those months of 1956 and 1957, the more apparent it became to me now that nothing at all had changed since then and that the Arabs were once again being permitted to delude themselves that they could wipe us off the face of the earth.'

This was not merely an Israeli perspective. Following Nasser's speech of May 26, one of his close allies, Mohammed Heykal, wrote in the Cairo newspaper *Al-Ahram* that an armed clash between Israel and Egypt was 'inevitable'. It would come because of the inexorable logic of the situation:

Egypt has exercised its power and achieved the objectives at this stage without resorting to arms so far. But Israel has no alternative but to use arms if it wants to exercise power. This means that the logic of the fearful confrontation now taking place between Egypt, which is fortified by the might of the masses of the Arab nation, and Israel, which is fortified by the illusion of American might, dictates that Egypt, after all it has now succeeded in achieving, must wait, even though it has to wait for a blow.

Let Israel begin; let our second blow then be ready. Let it be a knockout.

The Israeli War Cabinet met on May 27 to decide whether or not to take military action against Egypt, using the continued closure of the Straits of Tiran as the reason. The delay requested by the United States to enable diplomacy to work had gone by. In Israel's own inner counsels, a short but intense period of doubt by Yitzhak Rabin had passed. Rabin had regained his confidence that Israel could be successful on the battlefield without a long period of bloodletting.

As the discussion around the Cabinet table continued, it became clear that the Israeli Ministers were evenly divided. No head count was taken, but it seemed that about nine Ministers favoured immediate military action against Egypt, and about nine were keen that the process of international diplomacy should be allowed to continue. Eban, who had just arrived from the United States – he was driven straight from the airport to the Cabinet Room – proposed a forty-eight-hour 'disengagement', after which Ministers should meet again and decide on military action or further diplomacy. Levi Eshkol suggested a much shorter pause – enough for Ministers to sleep for the rest of the night, and to reassemble the following afternoon.

The new situation which confronted the Israeli Ministers when they reassembled on May 28 was an apparent strengthening of the resolve of the United States to reopen the Straits of Tiran. The American Ambassador to Israel, Walworth Barbour, passed on the information that the United States and Britain were even then looking into the military and naval aspects of an international naval task force (which later came to be known as the 'international flotilla'), in which British and American warships would play a major part. Harold Wilson had been particularly emphatic on British participation, having just made a visit to Canada to enlist wider support. The Canadian and the Dutch governments had both agreed to join in such a force.

The sense of foreign support at this level of seriousness influenced the Israeli Cabinet not to decide on immediate military action. Levi Eshkol proposed a two-week pause, to see what the multinational naval task force could do, assuming that it came into being. The thought of such a long period before Israel initiated action was appealing to most – indeed to almost all – the Ministers. Abba Eban has recalled how 'the expectation of victory was overshadowed by fear of terrible casualties'. Zalman Aranne, the Minister of Education, 'had spoken eloquently' – Eban wrote – 'of the fearful toll of war and of the moral need to do everything possible to avoid it. He was a man of refined consciousness and strong individualism. He was always more likely than anyone to give utterance to feelings which other Ministers held discreetly in their hearts. The Minister of the Interior, Moshe Shapira, was in consultation with Ben-Gurion, who also thought that a military challenge by Israel without allies by her side would be exorbitant on blood.'

Only one Minister, Moshe Carmel, the Minister of Transport, argued in favour of an immediate Israeli attack at the Cabinet meeting of May 28. His worry was that with every day that passed the Egyptians would be in a stronger position to launch a surprise attack.

The decision to wait beyond May 28 was not understood by some of the Israeli public, estimated at 24 per cent, who felt too endangered to wait any longer, and had no faith in international – or United States – support. The memories of America's hostility in 1956 were still vivid. Moshe Shapira later defended the decision to delay in the following words:

> If the war came it was essential that Lyndon Johnson should not be against us. If we had not waited we would still have conquered in the field of battle; but we would have lost in the political arena. The United States would not have stood by our side in the way that she did. We must remember that the general mood in those days was that we could not reasonably expect the emphatic victory which ensued.
>
> I said then and it is clear to me today that if we had begun war too early, we would have shown a lack of responsibility for our future. This has now been proved. The United States is giving us support such as we have never known before. I believe that a Superior Force directs our history. There is a destiny that shapes our ends.

Eban, who was later criticized for having pushed the argument for delay, wrote in his defence, and that of Eshkol, that both men were 'using time as currency to secure ultimate political support'. It was the American hostility in 1956 that weighed on both men, and on many of their colleagues. Johnson's support for a naval initiative could not be allowed to be nipped in the bud. As Eban later wrote:

> Either the multilateral naval action would collapse, in which case the United States would have little right or cause to restrain Israel's independent action, or if it succeeded, Nasser would, for the first time, believe that Israel had political backing as well as military strength. We must remember that our only aims in the Egyptian context were to break the blockade and disperse the troop concentrations.

The idea of a new boundary for Israel was not in the air at the end of May; it was only later that the Jordanian and Syrian interventions brought the whole Arab–Israel territorial structure under question. To defeat Nasser's blockade and troop concentrations in May by a combination of military preparedness and political pressures would be no less honourable, and in the long run, no less significant than to bring him low by an actual trial of strength.

The seventeen Ministers who supported the delay were 'dominated', Eban added, 'not by "confusion" or panic but by a mature political calculation'.

Immediately after the Cabinet meeting Eshkol saw the military leaders. He was accompanied, and supported, by Yigal Allon, then Minister of Labour. Some of the army men expressed their fears that unless Israel took military action at once, Egypt would secure the advantage. They were confident of success only if the Egyptians were attacked at once. Eshkol told them, bluntly, 'You are exaggerating quite a lot.' This was also Eban's view. 'Our military advisers were now fervent in the promise of victory,' he later wrote. 'True, their buoyancy was somewhat deflated by the contrary thesis that a brief delay would convert certain triumph to certain ruin. To me it seemed unlikely that we could be assured of utter victory if we acted on May 28 – and of complete rout if we waited a few days.'

That evening, immediately after the seven o'clock news, the Israel Broadcasting Service transmitted the first of what were to be sixteen daily commentaries by General Chaim Herzog. It was hoped that these commentaries would help to calm the public mind, and they did so. In his first broadcast, Herzog stressed the problems facing Nasser in trying to hold the vastness of the Sinai, should Israel be 'called upon to react to an Egyptian attack'. Herzog told his listeners, 'If one takes into consideration the strength and preparedness of the Israel Defence Forces, the Egyptian commander has his problems.' These were morale-boosting words. They were badly needed.

Starting on May 29, there came a period known in Israel as the *hamtana* (waiting). Without that waiting period, Yitzhak Rabin later wrote, 'it is doubtful if Israel would have been able to hold firm at the cease-fire lines and in the political arena two years after the war.' At the time, however, it was not the eventual political advantage, but the daily fears and anxieties, that dominated the politicians and the people. On May 29 Nasser spoke to the members of the Egyptian National Assembly. 'The issue is not the question of Akaba, the Straits of Tiran or the United Nations Emergency Force,' he said. 'The issue is the aggression against Palestine that took place in 1948.'

In 1948, Nasser insisted, Israeli 'aggression' had been carried out with the 'collaboration' of Britain and the United States. It was Britain and the United States who now wanted to 'confine' the issue to the Straits of Tiran, the United Nations Emergency Force, and the right of passage through the Gulf of Akaba. But, Nasser declared, 'We are not afraid of the United States and its threats, of Britain and its threats, or of the entire Western world and its partiality to Israel.'

In fact, the multinational naval force which the United States and Britain had hoped to assemble was not coming into being. Even Britain, which was so keen on it, discovered – to Harold Wilson's mortification – that there were no British warships near enough to reach the area for several days. Instead of concerted naval action to end Egypt's blockade, a coalition of a different sort was being created. On the day after Nasser's speech of May 29, King Hussein of Jordan flew to Cairo. In explanation of his journey, the King later wrote, 'The desire to meet Nasser may seem strange when one remembers the insulting, defamatory words which for a whole year the Cairo radio had launched against the Hashemite monarchy; but from every point of view we had no right nor could we decently justify a decision to stand aside in a cause in which the entire Arab world was determined unanimously to engage itself.'

At the meeting, Nasser produced a file containing the Syrian–Egyptian defence pact which had been signed a month earlier. 'I was so anxious to reach an agreement,' the King later wrote, 'that I contented myself with a rapid perusal of the text and said to Nasser, "Give me another copy; let us replace the word Syria by the word Jordan and the matter will be arranged.'

Israel was suddenly confronted by the possibility of a war on two fronts – or, if Syria were to honour its pact with Egypt and join in, by a war on three fronts. On May 31, the day after the Egyptian–Jordanian pact was signed, troops from Iraq reached Egypt, eager to join the battle. In Israel, it became clear that the two-week respite on which the Cabinet had agreed three days earlier was probably drawing to a premature close. There was an added dimension of danger, and fear, in Israel when Ahmed Shukeiry, the commander of the Palestine Liberation Organization – whose headquarters was in Cairo – flew back to Amman with King Hussein, at Nasser's insistence, and then appeared in Jordanian East Jerusalem to breathe fire and death against Israel. After the Arab victory, he said, those Israelis who had been born elsewhere would be 'repatriated'. When he was reminded that more than half of the Israeli population had been born in Palestine and (after 1948) Israel, he replied, 'Those who survive will remain in Palestine, but I estimate that none of them will survive.'

From Cairo came a further ominous threat: 'The occupation of the Israelis of the harbour of Eilat was illegal.' This was a clear intimation that Egyptian forces, already stationed within a few miles of Eilat, at the Taba border post, intended to occupy Eilat itself, Israel's southernmost town.

There remained the Anglo-American search for an international naval task force to reopen the Straits of Tiran. Eighty countries were asked to support this action, but only two – Canada and Denmark – were willing to do so without equivocation. From Washington, Avraham Harman reported that President Johnson 'could see no way out of the crisis'. Five days earlier,

Johnson had urged Israel, in no uncertain terms, not to fire the first shot. But on June 1 the Secretary of State, Dean Rusk, when asked if the United States would take measures to restrain Israel from precipitate action, replied, 'I don't think it is our business to restrain anybody.'

During June 1 the mood in Israeli government circles was changing. If the United States was no longer seeking to restrain Israel, then perhaps the time had come to take the military initiative. This was the view of many of those whose temperament was for conciliation and diplomacy, men like Arthur Lourie, one of Eban's advisers at the Foreign Office. Eban, who had initially proposed the two-week pause, felt strongly that there ought not to be any further delay. In his memoirs he wrote, of the turning point for him on June 1:

> I went to the Dan Hotel for a conversation with the most intimate of my advisers, Arthur Lourie, who urged me strongly along the course which I was contemplating. Both of us thought that the hour was now ripe to pick up the fruits of our patient efforts of the past ten days.
>
> I returned to our Tel Aviv office and asked the Director-General, Aryeh Levavi, to accompany me across the lawn to a meeting with the Chief of Staff, General Rabin, and the chief of military intelligence, General Yariv. I told them that I no longer had any political inhibitions to such military resistance as was deemed feasible, necessary and effective, and that if we were successful, I believed that our political prospects were good. We would not be set upon by a united and angry world as in 1956.

Eban's account continued, 'It took but a few sentences for me to say to the two generals what was on my mind. I told them, without specific details, that I believed the waiting period had achieved its political purpose; that its advantage would unfold in the coming days and weeks; that there was nothing now for which to wait; that the need to withstand the throttling grip of Arab aggression was paramount; and that any decision on methods and timing should now be reached on military grounds alone.'

Eban, Levavi, Rabin and Yariv then discussed 'possible times and occasions' when Israel might strike, 'all of them close at hand, at which Egyptian pressure would invite total response'. The meeting over, Eban left the generals' room and returned across the lawn towards the Foreign Office building. 'My step was lighter than when I had entered,' he later wrote.

Levi Eshkol was waiting on the lawn. When Eban told him that he had told the generals that the next step must be taken and timed on military grounds alone, Eshkol's 'relief' – Eban has recorded – 'was unconcealed'.

The mood in Israel was one of grim determination. Golda Meir has recalled how, at the same time that the soldiers were mobilized, 'the over-age men and women and children of Israel buckled down to clean out basements and cellars for use as makeshift air-raid shelters, to fill thousands of sandbags with which to line the pathetic home-made trenches that fathers and grandfathers dug in every garden and schoolyard throughout the country and to take over the essential chores of civilian life, while the troops waited, under camouflage nets in the sands of the Negev – waited, trained and went on waiting. It was as though some gigantic clock were clicking away for all of us, though no one except Nasser knew when the zero hour would be.' Golda Meir's recollection continued:

> By the end of May, ordinary life – as we had known it in the previous months – came to an end. Each day seemed to contain double the normal number of hours, and each hour seemed endless.
>
> In the heat of the early summer, I did what everyone else was doing: I packed a little overnight bag with a few essential belongings that might be needed in the shelter and put it where it could most easily be grabbed as soon as the sirens started to wail. I helped Aya make identification discs out of oilcloth for the children to wear and blacked out one room in each house so that we could put on the light somewhere in the evenings. I went to Revivim one day to see Sarah and the children. I watched the kibbutz that I had known from its first day calmly prepare itself for the Arab onslaught that might turn it into rubble, and I met with some of Sarah's friends – at their request – to talk about what might happen. But what they really wanted to know was when the waiting would end, and that was a question I couldn't answer. So the clock ticked on, and we waited and waited.
>
> There were also the grim preparations that had to be kept secret: the parks in each city that had been consecrated for possible use as mass cemeteries; the hotels cleared of guests so that they could be turned into huge emergency first-aid stations; the iron rations stockpiled against the time when the population might have to be fed from some central source; the bandages, drugs, and stretchers obtained and distributed.
>
> And, of course, above all, there were the military preparations, because even though we had by now absorbed that fact that we were entirely on our own, there wasn't a single person in Israel, as far as I know, who had any illusions about the fact that there was no alternative whatsoever to winning the war that was being thrust upon us.
>
> When I think back to those days, what stands out in my mind is the miraculous sense of unity and purpose that transformed us within only a week or two from a small, rather claustrophobic community, coping – and not always well – with all sorts of economic, political and social discontents into two and a half million Jews, each and every one of whom felt personally responsible for the survival of the State of Israel

and each and every one of whom knew that the enemy we faced was committed to our annihilation.

Inside Israel distress at the apparent inability of the government to take a decision had spread throughout the society. Public trust in the government was at its lowest ebb. The tens of thousands of reservists who had been called up felt that there was no leader. There was a growing public call for Moshe Dayan, a former and highly regarded Chief of Staff, to be brought back to serve at the highest level. Eshkol was ready to appoint him Commander of the Southern Front, facing Egypt.

This was not enough for those demanding a clear lead. There were large public demonstrations demanding that Eshkol give Dayan his Ministry of Defence portfolio. Among the demonstrations calling for change was a women's march in Tel Aviv. Most of the marchers were the wives of reserve officers (they quickly became known, in an unusual moment of whimsy during such worrying times, as the Merry Wives of Windsor). Eshkol bowed to the storm and the Cabinet that he then created was one that reflected the public mood and the coming emergency.

The national unity government, the first in Israel's history, was a historic turning point, a consequence of the public trauma of the impending war. Menachem Begin – head of the Herut opposition – was brought into the government as Minister without Portfolio. The ten-member Rafi Party, which had broken away from Mapai two years earlier as a result of the Lavon affair, and been in opposition since then, returned to the government, enabling Moshe Dayan to be made Minister of Defence.

On the first evening of his appointment, Moshe Dayan asked the Chief of Staff, Yitzhak Rabin, and the General Officer Commanding Southern Command, Brigadier-General Yeshayahu Gavish (known in the army as Shaike) to present their war plans. The two men were the authors of two very different plans. Rabin had for more than two years been the advocate of a limited war, with Israel restricting its aim to the conquest of the Gaza Strip and using the Strip as a bargaining counter to force Egypt to re-open the Straits of Tiran. Gavish had devised a more ambitious plan, to strike deep into Sinai and to attack and defeat the Egyptian forces there.

As Dayan faced the two generals, Rabin asked Gavish to explain 'the plan' to the new Minister. 'Which plan?' asked Gavish, assuming that Rabin would press his own more limited scheme. 'The second', was Rabin's reply. He had opted for Gavish's more ambitious project. Dayan gave it his approval.

After the seven o'clock news on June 1, the Israel Broadcasting Service transmitted the fifth daily commentary by General Herzog. 'Obviously every precaution must be taken,' Herzog told his listeners – who were fearful of an Egyptian air bombardment – 'but I must say in all sincerity that if I had to choose today between flying an Egyptian bomber bound for Tel Aviv, or being in Tel Aviv, I would out of a purely selfish desire for self-preservation, opt to be in Tel Aviv.' Rabin later wrote of the importance of these words, their content and their tone, spoken as they were 'when hundreds of thousands of mothers in Israel were engaged in pasting protective strips of paper or material on windows, and when their children were busy digging slit trenches in the backyards.'

From Cairo, Nasser continued to exacerbate the situation by his public declarations. On June 2, as a threat to the Anglo-American efforts to create a naval task force, he warned, 'If any Power dares to make declarations on freedom of navigation in the Straits of Tiran, we shall deny that Power oil and free navigation in the Suez Canal.'

Israel's military position was, on paper, precarious. On the Egyptian front at least 100,000 troops and 900 tanks were deployed in Sinai. On the Golan Heights Syria had more than 75,000 men and 400 tanks ready for action. The Jordanians had 32,000 men under arms, and almost 300 tanks. This made a total force of 207,000 soldiers and at least 1,600 tanks. A further 150 tanks were moving into Jordan from Iraq, which was determined to join what was being called in the Arab world 'the final battle'. Should it become necessary Egypt was able to send from the west of Sinai a further 140,000 troops and 300 tanks into that battle.

Against this substantial Arab force, Israel had, with the full mobilization of the civilian reserves, 264,000 soldiers and 800 tanks. An estimated 700 Arab combat aircraft were also ready for action. Israel had only 300.

During a meeting at the Foreign Ministry on June 2, Eban told his officials that it was his reading of American policy that 'if we were successful, the United States would feel relieved at being liberated from its dilemma, and would not support international pressures against us'.

Moshe Dayan spent most of Saturday June 3 preparing Israel's war plan. There was no possibility of a traditional Sabbath day of rest for him, his planners or his commanders. And yet, as is the nature of the Sabbath in Israel, the day of rest imposed its own characteristics. 'The beaches and the picnic grounds', Eban has written, 'were crowded with officers back on short leave from the front.' This was a deliberate ruse to mislead the Egyptians with regard to the imminent Israeli attack.

During the Sabbath the Israeli Ambassador to Washington, Avraham Harman, flew back to Israel. Driving straight from the airport to Eban's official residence in Jerusalem, Harman reported on his most recent conversation – the previous day – with Secretary of State Dean Rusk. From what Rusk had told him, it was clear that there was 'even less international disposition' to act against Nasser than there had been a few days earlier. The most that could be expected was a Vice-Presidential visit to Cairo. Rusk had told Harman that measures to be taken against Egypt by the maritime powers were still under consideration, but that 'nothing has been firmly decided'.

This, Eban noted, was 'a far cry' from Rusk's own statement five days earlier through the American Ambassador to Israel, that the military preparations of the maritime powers had 'reached an advanced stage'.

That evening Eban took Harman with him to see Eshkol. An impressive trio of former generals was also there: Dayan, Yadin and Allon, as well as several other senior officers. Unanimity prevailed as to the position of the United States. Harman's 'realistic report', as Eban described it, 'strengthened our certainty that there was nothing for us to expect from outside' – unlike the Anglo-French cooperation in 1956. 'It was now clear,' Eban recalled, 'that the United States was not going to be able to involve itself unilaterally or multilaterally in any enforcement action within a period relevant to our plight. But we all felt that if Israel found a means of breaking out of the siege and blockade, the United States would not now take a hostile position.'

Those meeting at Eshkol's house were also clear, Eban wrote, that Israel's military plan was 'concerned with Egypt alone; we would not fight against Jordan unless Jordan attacked us'. The meeting then dispersed. 'As I walked the short distance to my residence in the still night,' Eban wrote, 'I came across groups of workers building shelters near the schools. In conformity with the general mood, my wife, son and daughter had put sticking tape inside the windows of our home, as protection against explosions. Everyone in Jerusalem was doing this, but I had to ask my long-suffering family to spend some hours peeling the tape away since television teams were going to arrive to record interviews with me: I thought that visible evidence of defence preparations in the Foreign Minister's own house would give too sharp a hint of impending war.'

On the following morning, Sunday June 4, the national unity Cabinet met, presided over by Eshkol. For seven hours Dayan set out his military proposals. 'The atmosphere was now strangely tranquil,' Eban has written. 'All the alternatives had been weighed and tested in recent days; there was little remaining to do except plunge into the responsibility and hazard of choice.' The most frightening factor was the information reaching Israel of the mood in Egypt and throughout the Arab world. There were reports, Eban recalled, which made clear there was 'a higher morale than the Arab world had known in all our experience', and he went on to explain, 'The frenzy in the Arab streets belonged to the tradition of hot fanaticism which, in earlier periods of history, had sent the Moslem armies flowing murderously across three continents. Reports were reaching us of Egyptian generals and other leaders straining hard against the tactical leash which Nasser had imposed upon them. His idea of absorbing the first blow and inflicting a "knockout" in the second round was receding before a simpler impulse which told Egyptian troops that a first-blow victory was possible and that there was no need to "absorb" anything.'

That the Arab 'street' was clamouring, and eager, for war was clear. Dayan then presented his war plan. Israel could win a war, he told his ministerial colleagues, if it were to embark on it sooner rather than later. Every day saw the Arab forces gaining in strength and readiness. For Israel, the 'optimum moment' had arrived. He had one, overriding request: that he be allowed to send the army into battle at a time to be chosen secretly by himself and Eshkol. When Eshkol asked the Cabinet for a show of hands, there was no dissent.

After the Cabinet broke up, Dayan saw Eshkol alone. The time he proposed to launch an Israeli attack was, Dayan said, 7.45 the next morning, Monday June 5. Eshkol agreed. Israel would take the military initiative against those who were threatening her annihilation.

CHAPTER TWENTY-TWO

The Six Day War

It was not until after midnight on the first day of the war that the Israeli public were told, in a radio broadcast by the Chief of Staff, Yitzhak Rabin, and the commander-in-chief of the air force, Mordechai Hod, the astounding news that more than 400 Egyptian, Syrian and Jordanian aircraft had been put out of action, many while still parked on their runways, and that Israel had mastery of the air from the Sinai border to the Golan Heights. 'The Six Day War was won in the first two hours,' Shimon Peres has commented. Golda Meir later recalled:

> I had been kept informed all day of the general situation, but even I had not quite grasped the implications of what had happened until after the broadcast. I stood alone for a few minutes at the door of my house, looked up at the cloudless and undisturbed sky and realized that we had been rescued from the terrible fear of air raids that had haunted us all for so many days.
>
> True, the war had only started: there would still be death and mourning and misery. But the planes that had been readied to bomb us were all mortally crippled, and the airfields from which they had been about to take off were now in ruins. I stood there and breathed in the night air as though I had not drawn a really deep breath for weeks.

The Six Day War was, when it ended, a spectacular victory for Israel. But like all wars, however short in duration, its outcome was neither inevitable nor free from cost. Even the lightning strike at Egyptian airfields with which Israel launched its military operations was not free from risk. Only twelve fighter planes had been left in Israeli air space: all the rest set out on the mission to destroy Egyptian air power at its source.

The Israeli Chief of Staff, General Rabin, and the High Command of the

Israeli forces, spent much of the morning of June 5 at the air force command post. Tension was high, and anxiety almost tangible, until the moment when the items of news – always fragmentary in war, and never entirely clear – made it finally certain that the Egyptian air force had indeed been disabled, according to plan. One-third of all Egypt's war planes had been destroyed on the ground. Most of the runways at the main Egyptian air bases had been rendered unusable. It was eleven o'clock that morning when Mordechai Hod told his incredulous colleagues that at least 180 Egyptian war planes had been destroyed, and that all the communications installations of the Egyptian air force were out of operation for at least a few hours – crucial hours in the unfolding battle.

It was with the war against Egypt moving into the phase of ground attack that Israel approached the Jordanians to urge them not to enter the conflict. Three separate channels of communication were used. Israel's representative on the Israel–Jordan Mixed Armistice Commission (which had existed since 1949 and met on many occasions) passed the message to his Jordanian counterpart, for transmission to Amman. General Odd Bull, the Norwegian head of the United Nations Truce Observation teams, which had also been along the cease-fire lines since 1949, was contacted. And the American Embassy in Tel Aviv was asked to be a conduit.

The message passed to each of these three was the same: even though Jordanian artillery had opened a sporadic fire on Jewish Jerusalem, and along other parts of the armistice line, if firing stopped, and Jordan 'refrained from any other warlike acts', Israel would commit herself 'to honour the armistice agreement with Jordan in its entirety'.

The choice was that of King Hussein. Pressed by Nasser to take advantage of the imminent defeat of Israel, he made no answer to the Israeli offer. Instead, he ordered his troops in Jerusalem to attack across the armistice line. His air force was also alerted for action. Over the telephone, Nasser told Hussein – the conversation was intercepted by Israeli Intelligence – that most of the Israeli air force had been destroyed. 'He was bluffing, and being bluffed,' Shimon Peres later commented.

The next initiative was taken simultaneously by Syria, Iraq and Jordan, whose war planes launched a series of attacks on targets in Israel. It was 11.50 in the morning. Within two hours, Israeli aircraft had shot down and driven back the attacking forces, and then proceeded to destroy the Syrian and Jordanian air forces both in the air and on the ground. The main Iraqi air base, at H3, was also destroyed. Israel was no longer vulnerable from the air. 'In all', Rabin later wrote, '400 enemy planes were destroyed on the first day of the fighting; and that in essence decided the fate of the war.' The Arab air forces still had some 280 planes, 'but they were no longer a factor to be reckoned with during the remaining five days of battle. Moreover, the elimination of Arab air power was of decisive importance for morale. It undermined the fighting spirit of the Arab military leadership, as well as that of officers and men in the field, while precisely the opposite happened within Israel's political and military leadership and its combat units. Still, however, we were not over-confident and were certainly not itching for a fight merely to demonstrate our prowess.'

The most intense ground fighting of the first day of the war took place in Jerusalem. The Israeli government had decided, both in deference to the religious feelings of Christians and Muslims, and to avoid the losses that might be incurred in house to house fighting, that the Old City would not be attacked. But Jordanian artillery continued to hit buildings in Jewish Jerusalem from the Jordanian areas of the city. The Knesset was among the buildings hit: its members continued their deliberations in the shelter.

Brigadier General Uzi Narkiss, commanding Central Command, told Colonel Uri Ben-Ari, the commander of the armed forces in the Jerusalem area: 'This was to be a revenge for '48. We had both fought here: that time we had been defeated.' Rabin had also fought for Jeruslaem in 1948; he too had been born in Jerusalem. That afternoon Israeli troops captured Government House, and the fortified zone behind it, from the Jordanians. Eight Israeli soldiers were killed during the assault. Narkiss ordered the Israeli flag to be flown from the building – from which before 1948 Britain had governed Palestine. That night the Chief Rabbi of the Israeli forces, General Shlomo Goren, told Narkiss, 'Your men are making history. What is going on in Sinai is nothing compared to this.' Narkiss told Goren to prepare his trumpet.

Seeking to avoid a clash of arms in the narrow alley-ways of the Old City the Israeli army then launched two pincer attacks, one to the north and the other to the south. The objective of the northern attack was the Jordanian fortified position near French Hill (named after a British army officer who had been stationed there in 1918). The southern objective was Government House, hitherto serving as the United Nations headquarters (since 1949), and in one of the no man's lands established under the armistice, but occupied earlier that day by Jordanian troops. It was hoped that if these two Jordanian positions could be overrun, thereby cutting the Old City off from reinforcements from the north and south, the Old City would then surrender.

To prevent Jordanian tanks entering the Jerusalem battle from the Jordan Valley, Israeli air attacks concentrated that afternoon on the Jericho–Jerusalem road, which had only recently been improved by King Hussein, having previously followed the more winding and in places precarious bed of the British (and before that the Turkish) road. The Jordanian tanks and armoured vehicles were held up sufficiently to enable Israel to win over French Hill, and the nearby Ammunition Hill, before Jordanian reinforcements could join the battle. Nevertheless, the fighting was heavy, and the Jordanian defence impressive.

In Sinai the capture of Abu Ageila enabled the Israeli army to break through into the entire peninsula. An Israeli armoured column commanded by General Yisrael Tal reached the sea at El Arish, cutting off the Gaza Strip from all contact with Egypt. That night, Nasser telephoned Hussein, and, in an attempt to explain why the Egyptian air force (and the Jordanian) had been knocked out of the war, he told the King that American planes from the United States Sixth Fleet, and also British warplanes, had taken part in the defence of Israeli air space and in the destruction of the Arab air forces. This was quite untrue. To this day it is not known whether Nasser knew it to be a lie, or whether it was the only way he could explain how such a severe defeat had been inflicted so quickly. At six o'clock on the morning of June 6, the Supreme Command of the Arab armed forces broadcast the story to the world.

The Israeli Minister of Defence, Moshe Dayan, was a voice urging restraint in the north as battle raged in the south. He was determined if possible never to fight on more than one front at any given time. To repeated requests to open an attack against the Syrians on the Golan Heights, he refused to authorize action. He also repeated on June 6 the Israeli Cabinet's decision of the previous day not to try to capture the Old City of Jerusalem. What did concern Dayan was the possibility that the war might end – particularly as the Soviet Union was already calling for a cease-fire – before Israel had secured Sharm el-Sheikh, the Egyptian position dominating the Straits of Tiran, the closure of which had precipitated the war. On the morning of June 6 he summoned Rabin to the Ministry and asked him, 'What about Sharm el-Sheikh? We'll find the war coming to an end before we get our hands on its cause. Get to Sharm and establish our presence there, irrespective of the fighting in Sinai!'

Rabin acted on Dayan's instructions. A plan was made to seize Sharm el-Sheikh on the night of June 7 by a combined airborne and naval assault. The Israeli ships would sail down the Gulf of Akaba from Eilat, the port whose freedom of seaborne trade would be secured by the opening of the Straits. But before the assault could be launched, the Egyptians gave a general order to their troops in Sinai to fall back to the Suez Canal, and Sharm el-Sheikh was evacuated on the night of June 6, together with all but a holding Division along a defensive line on the eastern approaches to the Gidi and Mitla Passes.

On the morning of June 7 the United Nations Security Council called for a cease-fire. 'We saw in the sand the political hour-glass beginning to run out,' Rabin later wrote, 'and it was vital to speed up our operations. I therefore issued orders to move up our assault on Sharm el-Sheikh, but when the navy got there (before the other units) the flotilla's commander reported, "There's no one to fight! Sharm el-Sheikh had fallen without a single shot."'

On the previous evening, both Yigal Allon and Menachem Begin had pressed Eshkol to order Israeli troops into the Old City and to regain the Jewish Quarter lost in 1948. At seven o'clock on the morning of June 7, with an imminent United Nations cease-fire a possibility, Eshkol deferred to their urgings, and Dayan gave orders for Israeli troops to occupy the Old City of

Jerusalem, and to do so as quickly as possible. A paratrooper for whom this was his first battle recalled, as his unit fought from house to house:

All of a sudden I saw this man coming out of a doorway, this gigantic Negro.

We looked at each other for half a second and I knew that it was up to me, personally, to kill him, there was no one else there. The whole thing must have lasted less than a second, but it's printed in my mind like a slow-motion movie.

I fired from the hip and I can still see how the bullets splashed against the wall about a metre to his left. I moved my Uzi, slowly, slowly, it seemed, until I hit him in the body. He slipped to his knees, then he raised his head, with his face terrible, twisted in pain and hate, yes, such hate. I fired again and somehow got him in the head. There was so much blood.

I vomited, until the rest of the boys came up. A lot of them had been in the Sinai Campaign and it wasn't new to them. They gave me some water and said it's always like that the first time, not to worry. I found I had fired my whole magazine at him. It's true what they said: you grow more and more callous as you go along, and at the same time, you get used to the gun and miss less.

But I'll never forget that moment. It just goes slowly through my mind all the time.

*   *   *

On June 7, the day on which the paratroopers were conquering the Old City of Jerusalem, there was a period at Israeli headquarters of what Rabin recalled as 'sheer terror'. In his memoirs, written six years later, he wrote of how, 'I was seated in my office at the GHQ command post when I received a message that sounded odd: explosions had been reported in the El Arish area. By that time El Arish was in our hands and our forces had advanced eighty or a hundred kilometres beyond it along the northern route.' His account continued:

An initial guess was that the Egyptians might be coming in from the sea to attack our units in the town, so I ordered the navy and air force to look into the matter.

A second report, which arrived an hour later, led to a change in our assessment: a ship had been sighted opposite El Arish. Following standing orders to attack any unidentified vessel near the shore (after appropriate attempts had been made to ascertain its identity), our air force and navy zeroed in on the vessel and damaged it.

But they still could not tell us whose ship it was. Then a third message removed all doubts, but it sent our anxieties rocketing sky-high. Our forces had attacked a Soviet spy vessel!

Rabin reported to Eshkol and Dayan, and called in the senior headquarters' commanders for urgent consultation. 'It was vital to make preparations,' he wrote, 'but no one was prepared to articulate exactly for what. We did not dare put our fears into words, but the question that hung over the room like a giant sabre was obvious: are we facing massive Soviet intervention in the fighting?'

A Soviet spy vessel was indeed one of seventy warships which the Russians had introduced into the Eastern Mediterranean in the three weeks preceding the war. But the vessel that had been attacked was not a Soviet one. As Rabin wrote:

> While we were discussing the matter a fourth report came in and finally clarified the situation. The vessel was American – amazing but true. Four of our planes flew over it at a low altitude in an attempt to identify the ship, but they were unable to make out any markings and therefore concluded that it must be Egyptian. They notified the navy of their attack, and one of our ships finished the task by firing off torpedoes at the *Liberty*, leaving the vessel heavily damaged.
>
> I must admit I had mixed feelings abut the news – profound regret at having attacked our friends and a tremendous sense of relief stemming from the assumption that one can talk with friends and render explanations and apologies.
>
> The frightful prospect of a violent Soviet reprisal had disappeared. After consultation with the premier and the defence minister, we reported the mishap to the American embassy, offered the Americans a helicopter to fly out to the ship, and promised all the necessary help in evacuating casualties and salvaging the vessel. The Americans immediately accepted our offer, and one of our helicopters took their naval attaché to the ship.
>
> The scene aboard the *Liberty* was dismal: there were many wounded and some thirty-two dead, including a number of American Jews serving in the crew because of their command of Hebrew.
>
> The vessel's task had been to monitor the IDF's signals networks for a rapid follow-up of events on the battlefield by tracking messages transmitted between the various headquarters. The Sixth Fleet declined our services, evacuated their own wounded and towed the vessel to Naples (one of its home ports) for repairs.

During Rabin's term as Ambassador to the United States (1968–73) he learned that the United States government had instructed the Sixth Fleet to move its vessels away from the Israeli coastline once war broke out, 'but due to a bureaucratic blunder the order failed to reach the *Liberty*'.

Rabin also learned, from President Johnson's memoirs, that the United States had believed that the planes attacking the *Liberty* were Soviet. The incident, Johnson wrote, was one of the 'most critical moments' in his life. He faced, or believed that he faced, the decision of ordering United States warplanes to attack the Soviet Fleet in the Mediterranean. 'I encountered a fascinating parallel,' Rabin later wrote. 'Just as we were relieved to learn that the ship was American rather than Soviet, Johnson and the heads of the American armed forces were reassured upon learning that the attackers were Israelis.' Israel eventually agreed to pay $13 million as compensation to the families of the Americans killed or wounded in the attack. She refused the American request to pay for the repair of the ship (a far larger sum) on the grounds that it was not Israel that bore the responsibility for the Americans' own error in not getting the order to the *Liberty* in time.

Although the confusion over the identity of the attacked ship provided several worrying hours for the Israeli leaders, the news from Jerusalem gave them cause for relief. The Mount of Olives was taken, and the order given by General Motta Gur to enter the Old City. Late on the morning of June 7 Gur reported that his troops had penetrated the Lions' Gate and were approaching the Dome of the Rock. Dayan and Rabin flew from Tel Aviv to Jerusalem and hurried to the Wailing Wall (usually referred to by the Israelis as the Western Wall – of the ancient Temple enclosure). 'When we reached the Western Wall I was breathless,' Rabin later wrote. 'It seemed as though all the tears of centuries were striving to break out of the men crowded into that narrow alley, while all the hopes of generations proclaimed, "This is no time for weeping! It is a moment of redemption, of hope".'

In the Sinai, and on the West Bank, Israel was all but victorious. The Egyptian and Jordanian armies were in retreat. The Security Council had ordered a cease-fire to come into effect at ten o'clock that evening, June 7. On Israel's northern front, Syria, Egypt's ally, had made no move to cross the border. Moshe Dayan had refused to allow the Israeli commander in the north, David (Dado) Elazar, to try to capture the Golan Heights. There was fear at headquarters in Tel Aviv that any Israeli attack on to the Heights would bring in the Soviet Union as an adversary. The shock of a Soviet missile ship having possibly been attacked was too near, and too vivid, to allow complacency.

It was Nasser who decided not to accept the cease-fire resolution. He had wanted any such resolution to include a demand for the simultaneous withdrawal of Israeli forces from the territory they had occupied in Sinai, but the resolution did not demand this (it had been one of the conditions laid down in the cease-fire resolution in 1956).

For twenty-four hours Nasser continued the war. But he was unable to regain any part of his lost territory, and at midnight on June 8 he accepted the cease-fire. It was too late to save his army: Dayan had already given the order that Israeli troops were to advance as far as the Suez Canal and they were within striking distance of the waterway. The Egyptians fought hard to hold the Israelis back, but they were overwhelmed by the Israeli armour, and the sense of victory that accompanied the Israeli infantrymen as they drew closer and closer to the canal. The scale of the Egyptian losses was high: 15,000 dead were left in Sinai. Tens of thousands of Egyptian troops tried to flee westward across the desert to the canal. 'It was a sight that even the victors did not savour,' Rabin wrote twelve years later. 'Ragged and barefoot and terrorized, the troops left their shattered illusions behind and fled back to their homes at the mercy of a triumphant enemy. In order to forestall any errors, I issued explicit orders against opening fire on Egyptian soldiers who surrendered themselves. Of those who fell into our hands only the officers were to be kept in detention; the rest would be allowed to cross the Canal and return home. (This order was issued at a time when we already held between 5,000 and 6,000 Egyptian prisoners.) Our far-flung forces were already facing supply difficulties, and there was no point in burdening them further with thousands of prisoners.'

On the evening of June 8 General Elazar travelled from Galilee to Tel Aviv to see Rabin. He was determined to be given the opportunity to drive the Syrians from the Golan Heights. His appeal was supported by the representatives of the northern settlements, who had lived for so many years in the shadow, and often within the fire, of Syrian guns. They pressed Prime Minister Eshkol to authorize an attack on the Golan, and the capture of the fortified positions from which they had been shelled and fired on. The Israeli army had defeated 'our enemies to the south and east', they said. 'Are we going to remain at the mercy of the Syrian guns?'

An emergency meeting of the Ministerial Committee on Defence was summoned to hear the settlers' case. Rabin put forward the operational plan which would enable the Heights to be conquered. But Dayan reiterated his opposition to any such attack. The Soviet Union, he said, would come to Syria's aid and Israel would be in grave danger. Dayan's argument made its impact; the Ministers agreed that there should be no attack. Shortly before midnight Rabin telephoned Elazar – who had returned to the north – to tell him of this decision. Elazar was distraught. 'What has happened to this country?' he expostulated. 'How will we ever be able to face ourselves, the people, the settlements?' Reflecting on Syria's policy to Israel, and actions against Israel, since 1949, Elazar asked rhetorically, 'After all the trouble they've caused, after the shellings and the harassments, are those arrogant bastards going to be left on the top of the hills riding on our backs? If the State of Israel is incapable of defending us, we're entitled to know! We should be told outright that we are not part of the State, not entitled to the protection of the army. We should be told to leave our homes and flee from this nightmare!'

Syria had joined Egypt and Jordan in expressing its willingness to accept a cease-fire. The war had lasted four days. Egypt and Jordan had been defeated on the battlefield, and both the west bank of the Jordan (including Arab East Jerusalem) and the entire Sinai, were under Israeli military control. At two o'clock that night, Rabin, the victorious Chief of Staff, went home to bed.

While Rabin slept, Dayan had a change of mind. At six o'clock on the morning of June 9 he reached the Command Centre – known as 'the pit' – and was given details of the total disintegration of the Egyptian army. He was also told that Syrian units on the Golan Heights 'were crumbling and their soldiers had begun fleeing' – even though there had been no Israeli attack, as Dayan had insisted. At a quarter to seven Dayan telephoned General Elazar and told him to attack the Golan Heights 'immediately'.

At seven o'clock, fifteen minutes after Dayan's order, Rabin was woken up with news of Dayan's decision. He hurried to the pit in time to give orders for reinforcements to be sent to the north from the victorious Central Command. He also telephoned Elazar with a warning, 'The Syrian army is nowhere near collapse. You must assume that it will fight obstinately and with all its strength!'

Rabin then flew by helicopter to the north. By the time he reached Elazar's headquarters, the northern prong of the Israeli attack had begun to break through the Syrian defences. But the fighting was severe. Only by hand-to-hand fighting were the Syrian fortified positions overrun. Together with Ezer Weizman, Rabin watched the Israeli planes attacking the Syrian positions. 'I have never seen Ezer in such a state of inner turmoil,' Rabin later wrote. 'He murmured the pilots' names as though he were directing the air battle from our vantage point, and he begged them to protect themselves from harm. When one of our planes was hit and went up in flames and Ezer learned that the pilot was one of his many favourites, his features twisted into a grimace of agony.'

As the Israeli troops drove deeper and deeper across the Golan Heights, Rabin, who had returned to Tel Aviv, telephoned Elazar and ordered him to send a military force to capture the town of Kuneitra, the one large town in the region, which was located less than 40 miles west of the capital, Damascus. Hardly had Rabin put down the telephone than he was told that Dayan had ordered a halt to all military operations by the following morning (June 10) at the latest. Rabin at once telephoned back to Elazar, but it seemed it was too late to countermand the order. 'Sorry,' Elazar told Rabin. 'Following your previous order, they began to move off, and I can't stop them.' In fact, the troops involved had not yet received their orders to move; they did so soon after.

During the night of June 9 the Syrian forces, which had fought tenaciously throughout the day, began to fall back. That night a discussion took place inside the Israeli Foreign Ministry in Jerusalem to alert the Israel Defence Forces, in the event of an occupation of Damascus, to the need to effect the rescue of the city's Jewish community, some 15,000 Jews in all, whose right to leave Syria had long been refused by the Syrian government.

Worried lest Israel might indeed reach Damascus (for which there were in fact no Israeli military plans), the Syrian High Command ordered its troops on the Golan to pull back to the east of Kuneitra, abandoning the town in order to try to save the capital.

During the night the Syrians appealed to Egypt to send its commando units that were then stationed in Jordan to the Syrian front. Not a single Egyptian commando moved. In Nasser's mind, the war was over. All he could offer Syria was a pious hope. 'If Israel does not honour the cease-fire,' he said, 'Egypt will renew the war.' While Egypt certainly had the ability to send its commandos from Jordan to Syria, to join the battle there, it certainly did not have the ability to 'renew the war'. It was an empty pledge.

Fighting on the Golan continued into the early hours of June 10. 'Our forces were exhausted,' Rabin has written, 'and their heroic fighting in the course of the breakthrough had taken its toll.' Pressure from the United States led Dayan to order an end to the fighting at eight that morning. Kuneitra was still in Syrian hands. Air operations had to end by two in the afternoon. Rabin ordered Elazar to 'forgo the occupation of Kuneitra'. But at 8.30 that morning – within an hour of Rabin's order – Syrian radio announced that Kuneitra had fallen to the Israelis. The news was false. It may have been intended as a means of pressure on the Soviet Union to enter the war, or pressure on the Security Council to enforce the cease-fire. Its effect was to cause panic among the Syrian troops near the town, who struggled to get back towards Damascus. As soon as Dayan learned that the Syrian soldiers were in precipitate retreat he extended the order for an end to the fighting until two that afternoon (when air operations would also cease). Elazar's soldiers continued their advance. Kuneitra was occupied by Israel. Its defenders were gone and its inhabitants had fled.

The cease-fire on the Golan Heights came into effect at 6.30 on the evening of June 10. The Six Day War was over. Two days later, Israeli troops were flown by helicopter to take control of the deserted Syrian stronghold on the summit of Mount Hermon. This was the only territorial gain made after the cease-fire. 'We could have extended the area under our control,' Rabin later wrote. 'There was no Egyptian force capable of halting the IDF had we intended to occupy Cairo. The same held for Amman, and on June 11 it would not have required much effort to take Damascus. But we had not gone to war to acquire territories, and those we already occupied presented enough of a burden.'

It was a burden that was to weigh heavily on Israel for the next thirty years. Rabin expressed it succinctly twelve years later. 'Israel now faced three major problems,' he wrote, 'two of which have troubled us continuously from the Six Day War right up to the present day. The first was that overnight we found ourselves in control of an enormous expanse of territory. The area occupied by the IDF was three times the size of the prewar State of Israel, and we had difficulties in stabilizing new defence lines on all three fronts (particularly on the Suez Canal). We had never before thought of distance in terms of hundreds of miles. Moreover, we had to overcome the resultant logistic and transport difficulties with the help of limited manpower, since tens of thousands of reservists had returned to their fields and factories, schools and offices.'

Israel had also obtained control of a million Palestinian Arabs, including hundreds of thousands who were refugees from the fighting in 1948.

The number of Israeli soldiers killed in the Six Day War was 777, far fewer than had been expected when tensions mounted so sharply on the eve of war, but a heavy blow to a small, tight-knit community of less than three million people. The paratrooper who had killed his first enemy in the house-to-house fighting in the Old City of Jerusalem recalled:

> I came back without any joy. The victory didn't mean anything to me. None of us could even smile, though the people were cheering us when we came through the Mandelbaum Gate. But we had lost 50 per cent of our company. Another company – fifty men – came back with four alive. I never want to go back. I've had enough of the place.
>
> I'll tell you in two words what the battle was: murder and fear, murder and fear. I've had enough, enough.
>
> We had to do it, though. That's all I know. But it must never, never happen again. If it doesn't then perhaps it will have been worthwhile. But only if it never happens again.

*   *   *

In order to express its gratitude at the army's victory in restoring access to the Mount Scopus enclave, the Hebrew University of Jerusalem offered Yitzhak Rabin an honorary degree. The ceremony was to be held in the open air theatre, overlooking the Judaean desert and the Dead Sea, where the university's opening ceremony had been held forty-two years earlier.

At the ceremony, Rabin was also asked to speak on behalf of the other recipients of honorary degrees that day. In his memoirs he recalled the long and difficult process of preparing that speech, in the presence of those for whom the victory over Egypt, Syria and Jordan had been such a spectacular one:

> Again and again my thoughts were drawn to the phenomenal swiftness of it all, which had both positive and negative ramifications. Obviously, there was never any question – or desire – that the war would drag on for weeks, with all the accompanying tension and losses. Yet we had been so busy deciding whether or not to fight that few had taken the time to think out the consequences of victory.

We were left in a somewhat unexpected, if not downright vexing situation; and it took years for merely one of our adversaries – Egypt – to agree to start unravelling that tangle of circumstances.

The lightning pace of the war inevitably gave rise to myths – both in Israel and abroad – about the character of the fighting during those six days. The IDF was portrayed as steam-rolling its three adversaries in a breath-taking dash across desert sands and up rocky cliffs. But there was a double edge to that kind of myth-making, for it simultaneously portrayed the Israeli army as 'invincible' and the Arab armies as a 'pushover'. Nothing could have been further from the truth.

Before 5 June I repeatedly warned that if we opted for war we must know that it would be a difficult undertaking. The fact that our successes exceeded all expectations did not in any way disprove that prediction. A number of infantry and armoured assaults – in the Sinai as well as on the Golan Heights – were particularly difficult, and our men fought with determination and great valour.

Motta Gur's paratroopers carried on a stubborn and bloody battle for East Jerusalem at a terrible toll in lives on both sides. In those heady days of euphoria after the war, while the mass media around the world were active creating legends, I wanted to keep things in their proper perspective.

Of course we were proud, and we had every right to be – not because we were 'invincible' and not because our adversaries were tin soldiers, but because the IDF had earned the praise being showered upon it by its professionalism, creativity and sheer obstinacy. We had earned the right to feel confident in our military prowess without denigrating the virtues of our adversaries or falling into the trap of arrogance.

This sense of the fragility and loss of war was reflected in Rabin's remarks. One passage in his speech struck a particularly strong chord with many of his listeners, so many of whom had served in the army, and were still serving – and so recently fighting – in the reserves. It was a passage which marked, in retrospect, a turning point in his own career, which had hitherto focused on the battlefield. 'We find more and more,' Rabin told his audience, 'a strange phenomenon among our fighters. Their joy is incomplete, and more than a small portion of sorrow and shock prevails in their festivities, and there are those who abstain from celebration. The warriors in the front lines saw with their own eyes not only the glory of victory but the price of victory: their comrades who fell beside them bleeding, and I know that even the terrible price which our enemies paid touched the hearts of many of our men. It may be that the Jewish people has never learned or accustomed itself to feel the triumph of conquest and victory, and therefore we receive it with mixed feelings.'

CHAPTER TWENTY-THREE

The dilemmas of victory

In the immediate aftermath of the Six Day War, a debate began inside Israel that was to continue for the next three decades: how to rule, and for how long to rule, the Palestinian Arabs. The Cabinet position was that for 'full peace' Israel would be ready to withdraw to the borders of before the war. With regard to Syria, the Cabinet stressed the added importance of ensuring that security and water problems were first resolved. This was not a partisan decision: the National Unity Government formed by Eshkol on the eve of the war was still in place, with Begin as the senior Likud representative.

A few people argued with urgency that the West Bank and Gaza Strip ought to be given back as quickly as possible, that even a temporary occupation would hold grave disadvantages to the occupying power. But these were very much minority voices. The State was not yet twenty years old, yet its conquest of the West Bank and Gaza Strip seemed something that would last for a long time; and it was certainly to overshadow, and at times to dominate, all earlier purposes and ideals, and still to be a contentious and painful issue thirty years later.

Even as the war had been fought, the question of how to rule over so many Arabs was being asked, and answered. On June 7, Moshe Dayan had been asked by Chaim Herzog – newly appointed Military Governor of Israel's military forces on the West Bank – 'What is our policy? What am I supposed to do?' Dayan's reply (which Herzog later recalled in conversation with Dayan's biographer, Robert Slater) might have served as a blueprint for the years ahead: 'You are supposed to see that everything returns to normal,' Dayan told Herzog. 'But don't try to rule the Arabs. Let them rule themselves. Of course, they have to know you're the boss. It's enough that we suffer from Israeli bureaucracy. They don't deserve it. In a nutshell, I want a policy whereby an Arab can be born, live, and die in the West Bank without ever seeing an Israeli official.'

At first such a policy was put into effect, with Israeli military command posts being located away from the main Arab thoroughfares in the West Bank and Gaza. When Dayan learned that a third of the Palestinian homes in Kalkilya had been blown up by Israeli forces as a reprisal for Arab sniping at Israeli soldiers, he ordered the destroyed buildings (800 in all) to be rebuilt, and their inhabitants, who had fled to Nablus and to various villages or were living in the fields, to be allowed back. This was the exact opposite to the policy pursued, including by Dayan, in the War of Independence. The Six Day War was not intended to see a second round of flight and expulsion, or the creation of empty Arab towns which would then be filled by Jews. The West Bank and Gaza Strip were to remain Arab, and to be administered, not annexed or settled; but until when they were to be occupied no one knew.

In the immediate aftermath of the Six Day War, the Palestinian population behaved as if in a state of shock. It seemed amenable and cooperative to its new rulers, and there was a general feeling among Israelis that theirs was indeed an enlightened and benign occupation. Israelis were shocked to find how restrictive Jordanian rule had been, and how the Jordanians had very much relegated the West Bank to subordinate status within the Hashemite kingdom, neglecting its economy and failing to encourage its local institutions and aspirations.

If Jordan had not been a particularly forward-looking occupying power, despite its Arab, Muslim and even Palestinian affinities, Israel had confidence that it could be a benign occupier, capable of enhancing the prosperity of the whole region. In addition, to many Israelis, Dayan among them, for whom biblical locations and episodes were second nature, the West Bank was not just occupied land but a part of the historic Land of Israel. Even if they had no desire to rule it indefinitely, or to annex it, they felt an affinity with it. Direct links abounded: Abraham, Isaac and Jacob were buried in it (at Hebron), David had ruled over it (at first from Bethlehem), two Jewish kingdoms had been established over it, in Judaea and Samaria.

Dayan, with his memories of the British Mandate, did not want a repetition of the bitterness of ruled against their rulers that he had seen intensified in the years before 1948 in the relationship between the Jews and the British. In interviewing the officers being appointed as governors and deputy governors of the West Bank towns, he had a clear picture of the type of man he did not want. Referring to the first Governor of the Gaza Strip, Dayan said, 'I don't need someone like Motta Gur, who is a good officer, but when he enters the room all the notables of Gaza have to get up as if he were a British governor-general. We are not British here. This isn't what I need. I need someone who is admired but still respected by the Arabs.'

Whether occupiers could ever be respected, whether occupation could even be benign, was a question seldom asked in Israel during the early days of occupation. But even in those early days there were those who asked it, and who were worried at the effect of occupation, both on the anger of the ruled, and the arrogance of the rulers. Interviewed by the American journalist Joseph Alsop, Dayan himself made the point: 'We must not interfere, become involved, issue permits, make regulations, name administrators, become rulers. For if we do, it will be bad for us.'

There were few who thought that the occupation would be permanent. When Alsop asked Dayan how long the system of non-interfering occupation would work, Dayan replied that it would be for no more than two to four years. What would replace it, he did not say, and did not know. But when the question arose at the end of the year of allowing a local Palestinian, Hamdi Ka'anan, to administer the West Bank city of Nablus, Dayan saw virtue in the proposal, telling a Cabinet Committee on November 17, 'We need a manoeuvring option, so that we do not have only King Hussein to negotiate with. Such a Palestinian representation is needed for our manoeuvrability, but, who knows, maybe something will really come out of it in the form of a Palestinian entity.'

Dayan went on to explain to his colleagues, 'It is the government's policy, unanimously accepted, that Israel should not behave either as a "Russian commissar" or as a "British Mandatory official" in the West Bank and in the Gaza Strip. A "Russian commissar" is a system where the central government has spies and representatives supervising every decision and every act at every level. A "British Mandatory official" is a system where there is no foreign presence or interference at the lower echelons of the administration, but from a certain level to the top, all is in British hands.' With this in mind, Dayan continued, 'it is in our interest to have Mr Hamdi Ka'anan running things in Nablus. Even if he is not "Zionist" and he openly says so, he is always preferable to an Israeli officer in charge.'

Beyond such Arab administrative responsibilities, Dayan was not prepared to go. When the former Jordanian military governor of Jerusalem, Anwar Khatib, suggested an independent Palestinian 'entity' on the West Bank – under Israeli sovereignty – both Chaim Herzog, the Israeli military governor of the West Bank, and Yigal Allon, the Deputy Prime Minister, favoured the idea. But Dayan refused to consider it, fearing that any such Palestinian entity would fall under the control of the head of the Palestine Liberation Organization, Ahmed Shukeiry. Palestinian autonomy was no more acceptable to Dayan than a Palestinian entity. When General Uzi Narkiss proposed Palestinian autonomy on the West Bank, with Israel retaining control of defence, foreign affairs and finance, 'Dayan almost threw me out of the room,' Narkiss later recalled.

On 22 November 1967, the United Nations Security Council passed Resolution 242, which was to have an important impact on Arab–Israel relations in the years ahead, and on Israel's relations with the wider international community. Twenty-five years later it was to govern the whole reconciliation process, being the origin of the formula 'land for peace'.

Sponsored by Britain, and adopted unanimously, Resolution 242 called for the withdrawal by Israel 'from territories occupied' as a result of the Six Day War (described in the resolution as 'the recent conflict'). This call was interpreted differently by Israel and the Arab States. For the Arab States, it meant that 'all' occupied territories had to be evacuated. For Israel, the word 'territories' meant some, but not necessarily all, of them (the word 'the' had been deliberately excluded by Israel from the wording of the resolution, though it did appear in the official French translation, at French insistence). Israel stressed that no territories could be evacuated except in the context of a general peace agreement, which the resolution implied was its purpose.

There was more to Resolution 242 than the question of the occupied territories, which formed the first part of a two-part clause requiring 'the establishment of a just and lasting peace in the Middle East'. That peace was also to be based, according to the resolution, on 'Termination of all claims or states of belligerency and respect for and acknowledgement of the sovereignty, territorial integrity and political independence of every state in the area and their right to live in peace within secure and recognized boundaries free from threats or acts of force.' This applied, of course, to the Arab States as well as to Israel, something they had hitherto refused.

One aspect of Resolution 242 was distressing to many Israelis, who shared Ben-Gurion's long-held suspicion towards the language of any agreement which did not mention 'Israel' by name. The resolution spoke only of the recognized and secure boundaries of all the 'States' in the region, not mentioning Israel as such. When would the Arab States agree to mention Israel by name? This was a question many Israelis asked. Most were doubtful that they would ever see that moment come to pass.

The resolution went on to affirm 'the necessity (i) For guaranteeing freedom of navigation through international waterways in the area; (ii) For achieving a just settlement of the refugee problem; (iii) For guaranteeing the territorial inviolability and political independence of every State in the area, through measures including the establishment of demilitarized zones.'

There was no mention in Resolution 242 of the Palestinians. The phrase 'refugee problem' was intended to cover them. Many Israeli leaders, most notably Golda Meir, challenged the very notion of a Palestinian people. But the Israeli occupation of the West Bank served as a powerful catalyst to Palestinian nationalism, awakening many dormant aspects of national feeling.

The territories occupied by Israel were vast. They also contained an enormous Arab population, a million Arabs in all. The Rakah Party (*Reshimah Kommonistit Hadashah* – New Communist List), which had won three seats in the Knesset election of 1965, and whose voters were predominantly Arab, demanded complete Israeli withdrawal from the West Bank, and advocated the right of the Palestinian Arabs to establish a State.

Rakah also argued that the terrorist activities of the Palestine Liberation Organization were a legitimate means in what it described as a national guerrilla war. Many Israeli Arabs supported Rakah, not because they were Communists, but because of the party's espousal of Arab national identity.

The debate inside Israel about the future of the occupied territories was continuous. Among those who took a particularly strong view against giving them back to Jordan, or giving them up in any way, was Yitzhak Tabenkin, the eighty-four-year-old veteran Labour Zionist leader, who re-entered political life in order to help establish the Land of Israel Movement. The central credo of the movement, which was founded in August 1967, was that Israel retain all the territories conquered during the war. Thirty years earlier Tabenkin had been a vociferous opponent of the Peel partition plan because it meant what he regarded as the unnatural division of the Land of Israel.

Tabenkin was not the only Labour leader to support the Land of Israel Movement from its outset. Another was the novelist Moshe Shamir, once a figure very much on the left of the Israeli political spectrum. Among Revisionists, the movement was encouraged by Menachem Begin and his Herut Party. It was also supported by Israel Eldad, one of the leaders of the Stern Gang during the Second World War. Another prominent supporter was a former general, Avraham Yoffe, whose brigade had made the dramatic dash to Sharm el-Sheikh in 1956. Among the movement's first public demonstrations was a protest against the return to the United Nations of its former headquarters in no man's land in Jerusalem, which had been occupied by Israel during the 1967 war, following its seizure by Jordanian forces. The protest failed.

Shortly after the end of the Six Day War the Israel Defence Forces commissioned a song for one of its recreational music units. It was called 'The Song of Peace' and ended with the appeal:

> Do not whisper a prayer
> Better sing a song for peace
> With a great shout.

The song, which was promoted by the chief of army education, caused controversy among the upper echelons of the army for its pacific tone. One general refused to allow its performance by any of the soldiers under his command, claiming that it might subvert the morale and 'soften the hearts' of his soldiers. When the question was brought before Yitzhak Rabin, he supported his education chief Mordechai Bar-On, who took the view that 'Israel will never reach peace unless it has a strong army, but the army will not be strong unless its combatants are convinced that the ultimate goal of all their endeavours and sacrifices is to reach peace'.

In the aftermath of the Israeli victory of 1967 many difficult questions were being asked. One of them was whether Jews could, or should, remain an occupying power, ruling over Arab people and land. Another was a question about the nature of a society which had fought three wars in less than twenty years. One of the few Palestinian Jewish parachutists to survive the missions behind German lines in 1944 had stood up at during a public discussion on the war at her kibbutz in Upper Galilee to make a short interjection. 'There is one question that gives me nightmares and I would like to ask it,' she said. 'How many wars will our boys fight before they will become animals?'

At the end of the Six Day War, President Nasser had announced his resignation, but soon withdrew it in the face of popular demand. The Soviet Union immediately began to rearm Egypt and Syria even more massively than before, sending arms, and also 'observers' to train the Egyptians in the use of them. On October 21, scarcely four months after Israel's victory over Egypt, an Egyptian missile opened fire on the Israeli warship *Eilat*, which was then more than 13 miles from Port Said, and outside Egyptian territorial waters. The *Eilat* was sunk, and forty-seven Israelis killed. In reply, Israeli artillery opened fire along the whole Suez front, the oil refineries of Suez City were set on fire, and tens of thousands of Egyptian civilians had to be evacuated from Suez City and Ismailia.

In autumn 1968, artillery duels across the Suez Canal started up again. In one of them, ten Israeli soldiers were killed. 'We must reply with a fighting refusal to any effort to push us off the cease-fire line,' was Dayan's reaction. Israeli aircraft then bombed bridges over the Nile, and Israeli paratroops, making an attack deep inside Egypt, blew up a large power station. Israel began to build fortifications along the whole front line, with a series of strongholds designed to repulse any Egyptian attack across the Canal. The line, completed in March 1969, was named after the Chief of Staff, Chaim Bar-Lev, who in 1956 had advanced deep into Sinai to cut off the Egyptian forces in Gaza.

Following the Six Day War, all the abandoned settlements of 1948 were again under Israeli control. Beit Ha-Arava, on the northern shore of the Dead Sea, was found to have been completely razed to the ground by the Arab Legion, so that its soil, brought into productivity by such hard toil, had become saline again. Indeed, almost no vestige of the village could be made out. In 1968 a Nahal group set up a new settlement, Nahal Kaliya, on the approximate site of the old. Working on land with a high salt content, in summer conditions of intense heat, they slowly washed and irrigated the soil, and began to create a flourishing fruit and vegetable farm.

The Independence Day parade in May 1968 was the first to be celebrated by Israel with its new borders. 'It was widely assumed,' Walter Eytan – the first Director-General of the Israeli Foreign Office – wrote five years later, 'even by members of the Israel government, that this third round of fighting (after 1948 and 1956) must be the last, and that peace could be around the corner. It was taken for granted that, in this event, at least the bulk of the territories which had been Egypt's, Syria's and Jordan's would be evacuated and restored to these countries.' Hardly had the war ended when, on 16 June 1967, the Israeli government conveyed far-reaching peace proposals to Egypt and Syria, through the good offices of the United States. These proposals included a readiness to withdraw from most of the West Bank, with only minor adjustments in the border. But when the Arab answer came, from the Khartoum Arab summit conference on September 1, it was a resounding negative, three negatives in fact: 'No peace. No negotiation. No recognition.'

Israel's hopes of entering negotiations with the defeated Arabs were dashed by the Khartoum resolutions, which encapsulated the refusal of Egypt or Syria to contemplate any discussions whatsoever with Israel. Jordan did not have the authority in the Arab world to open negotiations on its own. Western commentators well-disposed to the Arab point of view stressed how hard – indeed impossible – it was for the Arab character to accept such a 'humiliation' as talks following defeat. The Soviet Union took a lead among the non-Arab States in demanding immediate, unilateral Israeli withdrawal from the occupied territories. 'When it became clear that peace, if it came at all, would be a long time coming,' Walter Eytan later commented, 'the administration of these territories at once began to be thought of not in terms of weeks or months, but of years. Steps were taken to establish it imaginatively and dynamically in the interests of the inhabitants no less than those of Israel herself.'

Israel was determined to be a very different sort of occupying power from those under whom the Jews themselves had suffered so much in the past. 'This was, almost from the start, no mere caretaker administration whose object was confined to safeguarding law and order and ensuring the operation of essential public services,' wrote Eytan. 'Although it was more than likely that the greater part of these territories would not remain permanently under Israeli rule, large sums were invested in their development. Roads were built, the educational system extended, modern methods of agriculture introduced, elections held for city and village councils, the export and import of goods encouraged, homes built for formerly nomadic Bedouin tribes.'

As the occupation of the West Bank and the Gaza Strip – known in Israel as the 'administered territories' – continued, there was also an unexpected economic effect. For Walter Eytan, who remembered the original United Nations partition plan's emphasis on the economic union between the Jewish and Arab States to be created in Palestine, this was far from unwelcome. 'The under-employment which had sapped the morale of the West Bank under the Jordanian occupation ended abruptly,' he wrote six years later. 'Scores of cars from Gaza can be seen daily in Tel Aviv, and hundreds from all over "the territories" on every Israeli highway. The Israeli market buys up furniture (including the latest "Louis XV") as fast as Samarian carpenters can turn it out. West Bank shops are stocked, in the main, with Israeli-made goods.'

When it had resolved in 1947 on the establishment in Palestine of separate Jewish and Arab States, the United Nations General Assembly, had envisaged an economic union between them, believing that neither could be viable on its own. 'Since the Arabs opposed with all their power the establishment of the Jewish State, and the Arab State itself never came into existence,' Eytan reflected, 'there could naturally at the time be no question of economic union. By a curious quirk of fortune, such a union has rapidly been taking shape *de facto* since the Six Day War, dictated by the interest of Jew and Arab alike. It is naturally not called by this name, and it does not exist in any formal sense, but in practical terms it is there for all to enjoy and is expanding fast. The economies of Israel and the formerly Arab-administered territories are in fact quite largely complementary, and trade, which normally moves along the channels of least resistance now flows strongly between them.'

These words were written early in 1973. The six years following the Six Day War seemed to hold many avenues of hope of a brighter, more peaceful future. Dayan instituted an 'open bridges' policy whereby the Arabs of the West Bank could cross into Jordan and sell their produce throughout the Arab world. Many Israelis who had left the country earlier, finding greater opportunities overseas, returned to Israel. Among them was Yedidyah Admon, who had just spent thirteen years in the United States, and before that had lived for many years in Paris.

A composer and songwriter, born in Ukraine just before the turn of the century (his family name was then Gorochov), Admon helped to create a specifically Israeli form of song, weaving in four very different traditions: the music of the Oriental Jewish communities from Persia and Yemen, together with the Arab music so popular among Jewish immigrants from North Africa, and Hasidic music, much of it from the region of his own birthplace (he had been brought to Palestine by his parents at the age of nine), and biblical melodies. All this, Admon wove together in accordance with the rhythm of the Hebrew language, thereby achieving for song what Eliezer Ben-Yehuda and his language committee had achieved for speech several decades earlier. Admon's music included scores for theatrical performances on biblical themes, and a symphonic poem 'The Song of Deborah'.

The Palestinian Arabs were articulating their own national culture, much of it expressing hostility to Zionism, Israel and the occupation. When Dayan invited the Palestinian poetess Fadua Toukan to his home, many Israelis were shocked that the Minister of Defence could show hospitality to someone whose verse was so strident in its hostility to Israel. 'It was not I who made Fadua Toukan a poet,' Dayan replied, 'nor did I dictate her nationalistic poems to her. But since there is a Palestinian public, and since this public has poets, I suggest that the Israeli public listen to those poets who are popular among the Arabs – in order to understand them; even if they are themselves not willing to try to understand who we are and what the Zionist movement is.'

To a considerable extent, that Zionist movement was in crisis. 'It has to be said', the Israeli diplomat Yehuda Avner has written, 'that the post Six Day War period witnessed the retreat of the pioneer ethos (except for the religious Gush Emunim) and the rise of the entrepreneurial culture, resulting in the gradual erosion of power of the Histadrut, and ultimately of the Labour Party itself.'

On 18 March 1968 a bus carrying Israeli schoolchildren on a holiday trip in the Negev went over a mine that had been laid by Palestinian terrorists. Two of the schoolchildren were killed and twenty-seven wounded. At General Staff Headquarters in Tel Aviv, Dayan and his advisers planned a reprisal attack. It was made against the Jordanian town of Karameh, the town closest to the Israel–Jordan border area from which the Palestinians had crossed.

Even as the attack was in its final stages of preparation, Dayan went on an archaeological dig (his hobby) at Azur, near Holon. While he was digging inside a trench, it collapsed on top of him, burying him completely. After being dug out of the trench, he was taken to hospital. The doctors at first despaired of his life. The day after the accident, as Dayan lay in pain and danger, the raid on Karameh took place. The Fatah were incredulous at the story of his accident. 'Who would believe that the war criminal who was then preparing his abortive military operation, was practising his archaeological hobby at that time in particular?' asked the Middle East News Agency. According to Fatah, Dayan had been seriously injured during a Fatah machine-gun and grenade attack near Holon, while he was supervising the preparations for the Karameh raid.

There was another Israeli reprisal action nine months later, following an attack by the PLO on an El Al plane at Athens airport when an Israeli was killed. Israeli commandos (among them the nineteen-year-old Benjamin Netanyahu, later Prime Minister) landed at Beirut airport and succeeded in blowing up thirteen planes on the tarmac.

The Land of Israel Movement, founded in August 1967, whose first demonstration – to protest against the return to the United Nations of its former headquarters in no man's land in Jerusalem – had failed, was active after the war in supporting the renewal of Jewish settlement in Hebron. Not only did the movement provide the leader of the settlers, Rabbi Moshe Levinger, with essential financial and moral support, but it used its political influence to intercede on Levinger's behalf with the authorities.

The return to Hebron took place on 4 April 1968, on the eve of Passover. Jews had lived in the town – where the graves of the Patriarchs and Matriarchs are located, in a site holy to both Jews and Muslims – since time immemorial. Ten Israeli families, pretending to be Swedish tourists, registered as guests in the Park Hotel in Hebron. Later that day, Levinger announced that the group were reviving Jewish settlement in Hebron, abandoned after the riots and killings of 1936. The two men who might have moved the settlers out of Hebron without delay, Moshe Dayan and Shlomo Gazit (the coordinator of activities in the occupied territories), were both unable to come to the city: Dayan was still in hospital after his accident, and Gazit was at home observing the traditional eight-day mourning period for his father. The settlers remained in the hotel.

The government, led by Eshkol, might still have taken a strong stance, but there seemed to be considerable public support for a return to Hebron. Within the Cabinet, Yigal Allon was particularly active in persuading the government to let the settlers stay. It was he who reached the agreement with them that, while they must leave the centre of the city, they could live in an Israeli army base on the outskirts. Dayan was furious, and urged his colleagues to send the settlers away altogether, but his advice was not taken. When the settlers were eventually given permission to build a Jewish suburb outside the town – known as Kiryat Arba (an alternative biblical name for Hebron) – Dayan begged the settlers not to seek conflict with their Arab neighbours: 'Don't raise your children to hate them,' he said.

As the Israeli occupation of the West Bank began, Dayan came to favour some sort of Jewish settlement throughout the occupied areas. In 1969 he proposed the building of four Jewish cities along the mountain ridge that stretches the whole length of the West Bank, dominating the coastal plain. His idea was that these cities would break up the existing contiguity of Arab settlement between Jenin, Nablus and Ramallah, and between Bethlehem and Hebron. But it was never part of the Dayan doctrine that Jewish settlements would be built in the heart of the Arab population centres, as was to happen a decade later. This would have been contrary to his overriding belief in 'live and let live'.

In explaining his doctrine to a group of students in Haifa on 2 April 1969, Dayan told them that in order to 'initiate changes in the basic situation, changes in structure', Israel should establish 'Jewish and Israeli possessions in the administered areas throughout, not just in the Golan, and not just with the intention of withdrawing from there. These should not be tent camps which are set up and taken down. With this in mind, we should establish possession in areas from which we will not withdraw in accord with our view of the map.'

In Dayan's 'view of the map', new administrative areas would be drawn, in which the Jewish town of Afula and the Arab town of Jenin would be part of a single unit; the same would be true of the Jewish town of Beersheba and the Arab town of Hebron. This would abolish the 'green line' border of the West Bank, and create quite different internal borders that would eliminate altogether the West Bank as an entity.

The Deputy Prime Minister, Yigal Allon, who was also Minister of Immigrant Absorption, had a different view from that of Dayan. Over a number of years he developed the plan that was to bear his name. Under the Allon Plan, Israel was to remain in control of the Jordan Valley and of the first mountain ridge to the west of it, where it would maintain settlements and early-warning systems against a possible attack from the east. East Jerusalem, the Etzion bloc area (known in Hebrew as Gush Etzion) and Kiryat Arba, the new Jewish settlement on the outskirts of Hebron which Allon had agreed to establish, were to remain under Israeli sovereignty. A few minor border changes were to be introduced, one of them around Latrun. The area in the West Bank to be returned to Arab sovereignty was to be connected to Jordan by means of a wide corridor around Jericho. The Gaza Strip was to be connected to the West Bank by means of a highway.

In the Sinai, Israel would retain control over the Rafa Salient, as well as over the west bank of the Red Sea from Eilat to Sharm el-Sheikh. Israel would also keep two military airfields constructed close to the former international border. The rest of the Sinai peninsula would be returned to Egypt. Most of the Golan Heights would also remain in Israeli hands.

The Allon Plan was based on the premise that for Israel to remain a Jewish and a democratic State in which the principle of Jewish labour was considered a basic value, it could not continue to rule over the 1.2 million Palestinian Arabs who lived in the territories occupied by Israel in 1967. Allon also stressed Israel's security concerns, which called for the demilitarization of all Arab territories west of the Jordan River and Israel's maintaining a military and civilian presence in the Golan Heights, along the Jordan Valley and along the Red Sea.

Although the Allon Plan was never formally adopted as the official policy of the Government of Israel, during the Labour-led governments until 1977 all the Jewish settlement activities in the West Bank, the Gaza Strip, the Golan Heights and the Sinai were limited to those areas to be retained by Israel. In 1973 the principles of the Allon Plan were included in the Labour Party platform as what was called the 'territorial compromise.' But the plan was never accepted by any of the Arab States.

As well as the rapid growth of the Land of Israel Movement after the Israeli occupation of the West Bank in 1967, religious Zionism substantially transformed itself into a movement of settlement in the West Bank, impelled by a messianic vision that the Jewish occupation of the whole of the Land of Israel was a decisive step on the road to Redemption. The Jewish historian Ehud Luz, in his book on religion and nationalism in the early Zionist movement (1882–1904) wrote, twenty years after the Six Day War, of how since 1967 'religious Zionism has begun to demonstrate self-confidence and feelings of superiority, whereas secular Zionism has retreated to apologetics.'

After 1967, Luz added, the religious public 'displayed tendencies to self-segregation and extremism, in both the religious and nationalistic-messianic directions, and shunted aside values like spiritual openness, tolerance and cooperation with the secular community in all matters relating to the State'. These were strong words; but they reflected a growing feeling among secular Israelis that the religious establishment, the religious leaders, and the religious parties, were moving slowly but steadily to the right, and that the Sephardi–Ashkenazi divide that had troubled the first two decades of statehood would, in due course, be replaced by as deep, or an even deeper divide between the religious and secular. In the aftermath of the Six Day War, and under the impact of divided attitudes towards the West Bank, the historic alliance between religious Zionism and the Labour movement broke down. Thirty years after the Six Day War, the divide had become a harsh fact of Israeli life.

In 1968, Avraham Harman, the former Israeli Ambassador to the United States, became President of the Hebrew University. He was to hold the position for fifteen years. Under his leadership the Mount Scopus campus, which had been cut off and abandoned in 1948, was rebuilt; a massive task and one which revitalized the university. Considerable funds were obtained from Jews in the Diaspora through the Friends of the Hebrew University (the American Friends had been established in 1931).

Despite a period of economic difficulty, four separate campuses were maintained: the rebuilt Mount Scopus, the Givat Ram campus which had been built after 1948 on a hill in western Jerusalem, the Ein Karem medical campus at the Hadassah Hospital, which had also been forced to move after 1948 to western Jerusalem, and the Rehovot campus in the coastal plain, where the Faculty of Agriculture was located.

In February 1968 Levi Eshkol met Lyndon Johnson at the American President's ranch in Texas. During their talks, Eshkol persuaded Johnson to sell Israel American phantom jets, in the light of the massive Soviet build-up of the Arab arsenals, and a French arms embargo which had been imposed on Israel after the Six Day War. The Eshkol–Johnson agreement marked the beginning of an arms race which transformed the United States into Israel's major arms supplier, and pitted the United States against the Soviet Union through their respective client States in the Middle East.

For much of that year Levi Eshkol was ill with heart trouble. The question of the succession was much discussed in inner government circles. Golda Meir was also unwell, so much so that she resigned as Foreign Minister. It was at that time, Abba Eban later recalled, that Pinhas Sapir, the Minister of

Finance, engaged him in an 'intimate conversation', telling him 'that it was unlikely that Eshkol would survive another heart attack. As responsible Ministers we should think of the succession.'

Sapir told Eban that he believed that 'only the appointment of Golda could prevent a suicidal clash in the Labour Party between Dayan and Allon'. The two men had very different visions of the future of the West Bank, with Allon's Plan looking forward to the return of most of it, and of all the Arab-inhabited areas, to Arab rule. There were also many points of personal rivalry. On the eve of the Six Day War, Allon had been in Leningrad: he had been unable to get back to Israel in time to press his candidature for the Ministry of Defence, which had gone to Dayan.

Sapir had sounded Golda Meir out about the succession, Eban added, 'during one of her visits to Switzerland where she was taking an uncharacteristic rest. Sapir had told me that a condition of her acceptance would be that all the other Labour Ministers remain exactly in their places, thus avoiding the tensions of a reshuffle. He assured me that Golda, whose health was also variable, would take the office only for the year remaining until an election would be necessary. I had given my consent.'

On 13 February 1969, Levi Eshkol died in Jerusalem. Golda Meir's succession was attractive to her rivals on several grounds. She was not well, and would not remain at the helm very long. Her becoming Prime Minister would avert – or at least postpone – a damaging struggle for succession between Moshe Dayan and Yigal Allon. The transition period was, however, 'less smooth', Eban recalled, than that which Sapir had suggested to him. 'There were some weeks,' Eban later wrote, 'during which Yigal Allon, as Deputy Prime Minister, took over the leadership of the Cabinet. He did this with such modesty and skill that leading Party members asked themselves why he should not continue in the Prime Minister's office.'

Dosh, a leading cartoonist in the widely circulated Israeli newspaper *Ma'ariv*, published a cartoon showing Yigal Allon sitting in the Prime Minister's chair. The caption was: 'Why not leave him there?' Eban recalled:

> In one of our intimate party gatherings, two veteran members, Shaul Avigur, the brother-in-law of Moshe Sharett, and Yitzhak Ben-Aharon, the Histadrut leader, had openly supported Allon for the premiership.
>
> I considered that this was good advice. Allon's leadership would balance what I already felt to be the inordinately strong influence of Moshe Dayan, whom Golda both resented and loyally followed. Allon, unlike Dayan, respected Arabic culture and spoke Arabic fluently. His personality was expansive, cheerful, and optimistic. These qualities had inspired his leadership of our armed forces in the War of Independence.
>
> I asked my trusted political secretary, David Rivlin, to explore directly with Allon whether he would allow his candidacy to be presented.

Rivlin visited Allon in the Education Ministry and added on my behalf that I would support him, both in his campaign for leadership of the party and for his premiership if he were to attain that peak.

It later emerged that Allon, while grateful for the support of his friends, was so intimidated by the party veterans supporting Golda, especially Sapir and Galili, that he recoiled from the challenge. Harold Wilson, the British Prime Minister who knew Yigal and me intimately, said that we both lacked 'the killer instinct'.

Golda was not slow in discovering that I had been willing to support Allon's candidacy. This meant that both he and I would thereafter incur her suspicious gaze whenever we entered the Cabinet room.

Golda Meir became Prime Minister of Israel, the first – and to date the only – woman to hold this position. It was soon said that she was 'the only man in her Cabinet'. The weight of the responsibilities on her, Eban narrates, 'seemed to liberate new energies within her. The idea disseminated by Sapir that she would be content to hold office for a year and then pass the burden to younger hands was sheer illusion. Her defiant personality certainly gave the nation's military and political struggle a strong dimension. My own feeling was then, and remains to this day, that the Labour movement and the country would have benefited from the fresh wind that Allon's appointment would have brought to the nation's helm.'

In her memoirs, written six years later, Golda Meir recalled her feelings when, on 7 March 1969, the Central Committee of the Labour Party voted to nominate her as Prime Minister:

I have often been asked how I felt at that moment, and I wish that I had a poetic answer to the question. I know that tears rolled down my cheeks and that I held my head in my hands when the voting was over, but all that I recall about my feelings is that I was dazed. I had never planned to be Prime Minister; I had never planned any position, in fact.

I had planned to come to Palestine, to go to Merhavya, to be active in the Labour movement. But the position I was now to occupy? That never. I only knew that now I would have to make decisions every day that would affect the lives of millions of people, and I think perhaps that is why I cried. But there wasn't much time for reflection, and any thoughts I had about the path that had begun in Kiev and led me to the Prime Minister's office in Jerusalem, had to wait.

Today, when I can take time for those reflections, I find I have no appetite for them. I became Prime Minister because that was how it was, in the same way that my milkman became an officer in command of an outpost on Mount Hermon.

Neither of us had any particular relish for the job, but we both did it as well as we could.

Golda Meir's first announcement as Prime Minister was to say that 'we are prepared to discuss peace with our neighbours, all day and on all matters'. Within three days, President Nasser replied that 'there is no voice transcending the sounds of war' and 'no call holier than the call to war'. In March 1969 – following Nasser's dictum 'what was lost in war must be restored by war' – Egyptian artillery opened fire on the Israeli forces stationed on the east bank of the Suez Canal. Israel returned the fire with alacrity. What became known as the War of Attrition had begun.

On April 6 Golda Meir spoke to 3,000 teenagers in Jerusalem, expressing her 'absolute faith' that peace would come. Until then, she said, the country must continue to be prepared to defend its existence. 'Many dreams, once considered impossible, have been made to come true by the nation,' she said, 'and I have no doubt that the dream of peace will also be realized.'

During a five-hour attack on July 17, Israeli aircraft bombed military targets between Kantara and Port Said. In the air battle that developed five Egyptian planes were shot down, and two Israeli planes. Two weeks later, another twelve Egyptian planes were shot down. A further five Egyptian planes shot down at the end of July were being flown by Soviet pilots.

The Israeli–Egyptian border remained on the Suez Canal; but on either side of it men served in battle stations, died or were wounded. The main objective of Israeli counter-bombardment was the Canal city of Ismailia – no Israeli cities lay within range of Egyptian artillery. Recalling the armed conflict of those days and months, Ezer Weizman wrote, 'We had showed no mercy for Ismailia. We'd bombarded the city incessantly, devastating it from the air as well as with land-based artillery. The aerial photographs of Ismailia that reached my desk then showed its western portions resembling the cities of Germany at the end of World War II. Ismailia had been pulverized and destroyed; not a single house was left standing in its eastern quarters. During one of our bombardments of the city, an Israeli shell had killed the Egyptian commander in chief – a severe blow to the Egyptian army's morale, which was already at a low point.'

In June 1969 Golda Meir offered to go personally to Egypt to seek a compromise agreement with Nasser. This offer, which Nasser rejected, led to derisory comments in the Arab world. One Jordanian newspaper wrote, 'Mrs Meir is prepared to go to Cairo to hold discussions with President Nasser but, to her sorrow, has not been invited. She believes that one fine day a world without guns will emerge in the Middle East. Golda Meir is behaving like a grandmother telling bedtime stories to her grandchildren.'

The vision of 'a world without guns' was one which Moshe Dayan addressed that August when he told the graduates of the Israel Defence Forces Staff and Command School that 'the question "What will be the end?" has been with our people for four thousand years,' and he added:

It can be said that the concern over 'what will be' is an organic part of our people. You undoubtedly noticed that in my last sentence I omitted 'the end'. I said 'What will be?' not 'What will be the end?' I did so because I feel that the emphasis should be placed on the road we take, and not on the final destination.

Rest and peace for our nation have always been only a longing, never a reality. And if from time to time we did achieve these goals, they were only oases – a breath that gave us the strength and the courage to take up the struggle again.

I therefore think that the only answer we can give to the question 'What will be?' is 'We shall continue to struggle.' I feel that now, as in the past, our answer must centre upon the assurance of our ability to face difficulties, our ability to fight, more than on expectations of final concrete solutions to our problems.

We must prepare ourselves morally and physically to endure a protracted struggle, not to draw up a timetable for the achievement of 'rest in peace'.

During the winter of 1969 the Israeli Cabinet discussed the question of seeking to deter the Egyptian bombardments by launching a series of in-depth bombings against Egyptian military targets on and even beyond the River Nile. On December 21 the Israeli Ambassador in Washington, the former Chief of Staff, Yitzhak Rabin, was summoned back to Israel to discuss this proposed change of policy – an escalation of the war. 'His strong views in favour of this course', Abba Eban recalled, 'were already well known from his telegrams, some of which were formulated in terms of sharp rebuke to his government for failing to understand what he termed "an irrevocable opportunity". We in the Cabinet had a deep interest in hearing his views on the potential response of the United States to an escalation of our pressure against Egypt. The fact that Rabin had held the highest authority for the nation's physical security would surely enable him to take a balanced view. I found nothing in Rabin's report that would justify the conclusion that the United States had any interest in the escalation of the fighting in or near the Canal area.'

On 4 January 1970, following an Arab summit in the Moroccan city of Rabat, Abba Eban published in the London *Sunday Times* an appeal for peace between Israel and its Arab neighbours, in which he wrote:

The Arab leaders who met at Rabat last week represent fifteen sovereign States with a population of 100 million, an area of four million square miles, vast mineral wealth and opportunities of creative growth which only they can fulfil or squander. In a world in which few nations achieve their full ambition the Arab nation has come off better than most.

On the other side of the conflict is the small State of Israel. Arab nationhood and sovereignty are thus assured beyond doubt or challenge; and Israel is the only nation which stands or falls in history by the way in which the conflict is resolved. Thus Israel's security is the overriding moral imperative in this dispute.

'Sooner or later,' Eban wrote, 'it will be evident to the Arab States and Israel that they must look more and more towards each other for what they want, and less and less towards the external forces which have traditionally governed Middle Eastern history; 1970 can be a turning point if free negotiation by the States within the area supersedes the mad hope of another war and the illusion of settlements imposed from outside.'

Eban then contrasted what he called the 'distortion and prejudice' in Western newspaper reports of the situation on the West Bank and in Gaza by noting that, 'side by side with tensions and intermittent disorder, there is a great area of peaceful encounter on the human level. Tens of thousands of Arabs pass freely across the Jordan in both directions. Thirty thousand permits of return have been granted. Schools are open and commerce is in full development. Trade between the eastern and western banks of the Jordan is expanding; and thousands of West Bank Arabs find employment in the Israeli labour market. The total picture is so remote from the inferno described by some writers that it is hard to deny that predilection and emotion are often hard at work.'

For the Arab States, Eban argued, the 'central issue' was not Palestinian self-determination but 'the duty of peace with Israel which would, of itself, enable both nations in the area to face each other in freedom and equality'. The area of the original Palestine Mandate, he pointed out, included both sides of the River Jordan. In that area 'there are two, not three nations. There is Israel and an Arab people which for twenty years expressed its political life through a Jordanian State of which the Palestine Arab formed the great majority.' The establishment of peace, Eban asserted, would leave the Arabs 'to the east of Israel's boundary' (that is, in Jordan) 'free to determine the structure, name, regime and institutions which they desire'. This would not be the case for those who lived to the west of the boundary, in the West Bank and the Gaza Strip. What their future would be he did not say.

The inhabitants of the occupied territories were even then adopting a new identity, not that of Arabs, or of Palestinian Arabs, but of 'Palestinians'. The days when the Jews themselves were 'Palestinians' – under British rule – with Palestinian passports, citizenship and institutions, were over. The new Palestinians had entered the national and international arena as a force to be reckoned with, even if some of the earliest Israeli reactions were to deny their very existence as a recognizable entity. Golda Meir had gone so far to say that there was 'no such thing as a Palestinian'.

* * *

During his visit to Israel for consultations, Yitzhak Rabin repeated a phrase that he had often used as Chief of Staff, that the Israeli army refused 'to have the rules of the game dictated to it'. On January 5 a nine-man Egyptian unit crossed the Suez Canal and, supported by artillery fire from the western bank, attacked an Israeli army post. All nine Egyptians were killed. Two days later the Israeli in-depth bombing of Egypt began. Military camps north and south of Cairo were hit. Some planes flew more than 80 miles into Egypt. At the same time, there were Israeli air strikes on terrorist bases in Lebanon and Jordan. 'In the Metulla area,' reported the *Jerusalem Post*, 'people watched the Israeli planes go into action against the Mount Hermon terrorist bases, swooping repeatedly on their targets and seeing plumes of smoke rising.' Twenty miles east of the Israeli–Syrian border, three Syrian MiG fighter planes were shot down: this made seventy-nine Egyptian and Syrian warplanes destroyed in aerial combat since the end of the Six Day War two and a half years earlier.

Ezer Weizman's son Sha'ul was among those serving in one of the forts on the Bar-Lev line along the canal that had been completed ten months earlier, fortified positions that were intended to prevent any Egyptian recapture of the eastern bank of the canal, or advance into Sinai. The father, who had been so brave in war, was full of anguish that his son had to carry on the burden. 'Where did we go wrong?' Ezer Weizman asked his wife Reuma when their son enlisted. 'Why do our sons have to go to war?'

In June 1970 Sha'ul Weizman was wounded – shot in the head by a single shell from an Egyptian sniper. The Chief of Staff, Chaim Bar-Lev, telephoned the news to Ezer Weizman while he was lunching at a Jerusalem restaurant. Weizman later wrote:

> During the few seconds it took me to race to the phone, I reassured myself that Sha'ul had not been killed. If he were dead, the news would not have been conveyed by telephone; as I knew from sad experience, IDF tradition would have required me to be notified personally.
>
> Several times during Sha'ul's period of service, I found myself starting up from a nightmare in which officer friends were standing motionless at the gateway to my home, not daring to enter because they lacked the courage to tell me the news. Indeed, there had been occasions when I, too, found myself incapable of entering a home – of a superior or of one of my subordinates – when it was my duty to tell the family of the calamity that had befallen them.
>
> Sha'ul was wounded. Ever since that day, my family has lived in the shadow of his injury and the physical and emotional infirmities it brought upon him.

Two decades later Sha'ul Weizman was killed in a car crash together with his wife.

*   *   *

Israel's victory in the Six Day War had one quite unforeseen impact. Inside the Soviet Union, Jews who had no contact at all with Israel, or with their Jewish roots and traditions, or with the outside world, felt a sudden surge of Jewish and Zionist belonging. The victory stimulated the wish of tens of thousands of Soviet Jews to emigrate. But emigration was not Soviet policy for any of its citizens, and very few of those who sought exit visas were granted them.

On 15 June 1970 eleven Soviet citizens, nine of them Jews, tried to force a Soviet aeroplane to fly them out of the country. They were arrested before the plane could leave the ground, and brought to trial. Two of them, Eduard Kuznetsov and Mark Dimshits, both Jews, were sentenced to death. There was an outcry throughout the Jewish world at this, and the death sentences were commuted to fifteen years in prison. As a result of the trial and the sentences, the focus of many Jewish communities in the Diaspora turned to the plight of those Jews in the Soviet Union who wanted to leave, and to live in Israel.

To the surprise of many the Soviet Union, which in 1970 had allowed 1,027 Jews to leave, let 12,819 emigrate in 1971. Of these, all but fifty-eight went to Israel. A new, and unexpected, source of Jewish immigration had begun. In 1972 the figure rose dramatically, to 31,652. Of these, only 251 went anywhere other than to Israel. Suddenly, immigration from behind the Iron Curtain exceeded, by far, that from any other single country.

In June 1970 as the Israeli bombing of the area west of the Suez Canal continued, the American Secretary of State, William P. Rogers, tried to broker a cease-fire. His idea was that Israel and Egypt would talk under the auspices of a Swedish diplomat, Dr Gunnar Jarring. Under Rogers's proposal there would first be a ninety-day cease-fire between Egypt and Israel. Then talks would begin, based on Israel's 'withdrawal from territories occupied in the 1967 conflict' (the formula in Security Council Resolution 242) and on 'mutual acknowledgment of each other's sovereignty, territorial integrity and political independence' – as also laid down in Resolution 242.

The right-wing members of the National Unity Government, the Gahal bloc, headed by Menachem Begin, rejected any proposal that involved even contemplating a withdrawal from the 1967 cease-fire lines. Not one inch of the Sinai, they insisted, must be given up. Golda Meir has recalled her conversation with Begin on this issue, and her reflections on Gahal's point of view:

> 'But we won't have any cease-fire unless we also accept some of the less favourable conditions,' I tried to explain repeatedly to Mr Begin. 'And what's more, we won't get any arms from America.'

'What do you mean, we won't get arms?' he used to say. 'We'll *demand* them from the Americans.'

I couldn't get it through to him that although the American commitment to Israel's survival was certainly great, we needed Mr Nixon and Mr Rogers much more than they needed us, and Israel's policies couldn't be based entirely on the assumption that American Jewry either would or could force Mr Nixon to adopt a position against his will or better judgement.

But *Gahal*, intoxicated by its own rhetoric, had convinced itself that all we had to do was to go on telling the United States that we wouldn't give in to any pressure whatsoever, and if we did this long enough and loud enough, one day that pressure would just vanish.

I can only describe this belief as mystical, because it certainly wasn't based on reality as I knew it, and today I shudder to think what would have happened in October 1973 if we had behaved in 1969 and 1970 as defiantly and self-destructively as *Gahal* wanted us to. There might well have been no US military aid at all from 1970 on, and the Yom Kippur War would then have ended very differently.

Golda Meir persevered with negotiations for a cease-fire, using the good offices of Dr Jarring. When the agreement on a cease-fire was signed in August 1970, the Gahal Ministers, led by Begin, resigned from the government. In just over a year, 593 Israeli soldiers and 127 Israeli civilians had been killed in the Egyptian bombardments: almost as many as had been killed during the Six Day War. A month later, on 24 September 1970, Golda Meir flew to the United States. With the obstacle of American hostility to the War of Attrition overcome, she hoped to persuade the Americans to take a more positive attitude towards Israel's needs. She had never met President Nixon before, and had no clear indication of what his attitude to Israel would be. 'But I was sure of one thing,' she later wrote. 'Whatever impression I might make on him, I had to lay before the President all of our problems and difficulties, quite candidly, and try to convince him, beyond a shadow of doubt, that there was a great deal that could be asked of us by way of compromise and concessions, but that we could not be expected to give up our dream of peace or to withdraw a single soldier from one inch of land until an agreement could be reached between the Arabs and ourselves. And that was not all. We desperately needed arms, and I felt that I should ask him for them myself. On the face of it, they weren't very complicated messages, but I think that I would have been less than human and a fool if I hadn't been extremely nervous about delivering them.'

In Israel it was said that Golda Meir was going to America 'with her shopping bag'. Her first stop was Philadelphia, where she was to be met by members of the local Jewish community. She was worried that the enthusiasm of American Jews for Israel might have waned in the two years since the ending of the Six Day War. Her worries were unfounded. As she later recalled:

> At the Philadelphia airport, a crowd of thousands was waiting for me; hundreds and hundreds of schoolchildren singing *Hevenu Shalom Aleichem* (We Bring Peace Unto You) were waving and carrying banners. I remember one of those banners read, 'We dig you Golda', and I thought it was the most charming expression of support for Israel – and perhaps for me – that I had ever seen.
>
> But I had no idea, other than smiling and waving at them, how to make clear to those youngsters that I certainly 'dug' them too. So I just smiled and waved and was delighted when I spotted my own family.
>
> At Independence Square, an even larger crowd greeted me – 30,000 American Jews, many of whom had been standing there for several hours to see me. I couldn't get over the sight of all those people pressing against the police barricades and applauding. I spoke to them only very briefly, but as someone said to me, 'You could have read a page from the telephone book, and that crowd would still have cheered'.

Reaching the White House, Golda Meir found that her second worry, a coldness on the part of the President, was likewise unfounded. 'The President and I stood on a raised platform covered by a red carpet while a Marine band played the two national anthems,' she later recalled. 'I listened to "Hatikvah" and although I made an effort to look perfectly calm, my eyes filled with tears. There I was, the Prime Minister of the Jewish state, which had come into existence and survived against such odds, standing to attention with the President of the United States while my country was accorded full military honours. I remember thinking: "If only the boys at the Canal could see this", but I knew, at least, that in Israel thousands of people would be watching the ceremony on television that evening and be as profoundly moved and heartened by it as I was.'

Golda Meir added, 'Perhaps other nations take these ceremonies for granted, but we don't yet. In fact, it was all a little like a dream, the kind of dream I used to dream with my friends years ago when we sometimes talked about what it would be like when we actually had not only a State of our own but also the trappings of statehood.'

During the negotiations with Nixon, Golda Meir put forward the argument that there were already two States between the Mediterranean Sea and the borders of Iraq, in what was once Palestine. These, she said, were Israel and Jordan (the original area of Britain's Palestine Mandate had included the territory of the Hashemite kingdom). 'There is no room for a third. The Palestinians must find the solution to their problem together with that Arab country, Jordan, because a "Palestinian State" between us and Jordan can only become a base from which it will be even more convenient to attack and destroy Israel.'

The year 1970 was dominated by the attempt by Palestinian radical elements to bring about a revolution in Jordan. They competed in spectacular acts of international terrorism, culminating in early September in the hijacking of four international civil airlines, three of which were brought to a desert airstrip in northern Jordan, where they were blown up. That month, known as Black September, the Palestine Liberation Organization attempted to overthrow King Hussein. The King's ruthless response, in which 2,000 PLO fighters and several thousand more Palestinian civilians were killed, resulted in a Syrian threat to invade Jordan. This threat was halted when Israel concentrated its tanks on the Golan Heights, threatening Damascus. Israel felt able to act in this forceful way in support of Hussein because of an American guarantee to come to Israel's aid should the Soviet Union intervene.

The second impact of King Hussein's reaction was the flight of the PLO to Lebanon, the southern region of which it soon took over.

Golda Meir was adamant that there could not be a Palestinian State on the West Bank. But she was equally adamant that Israel sought a way forward to peace, and had a vision of what that peace could be. As she told the biennial convention of American trades unionists (the AFL-CIO): 'It will be a great day when Arab farmers will cross the Jordan not with planes or tanks, but with tractors and with their hands outstretched in friendship, as between farmer and farmer, as between human beings. A dream it may be, but I am sure that one day it will come true.'

The element of dreaming reached a high point in the early 1970s. The State was only twenty-two years old, but many of its citizens had already fought in three wars. The poet Chaim Guri, himself a veteran of the War of Independence, wrote, in an anguished prose passage, 'When, my friends, have we last seen peace? The soil is insatiable. How many more graves, how many more coffins are needed until it will cry out – enough, enough!' The Israeli writer and journalist Amos Elon (a biographer of Herzl) has commented that the non-Israeli, the 'outsider', cannot fully realize the 'depth and extent 'of the sadness generated by war throughout Israeli society without first considering 'the time sequence and the toll', and he went on to write:

> Many thousands must go through life haunted by the harrowing sights, sounds and smells, which have been the recurrent features of their youth and manhood; the groans of wounded and those dying in their arms; the screams of fear more piercing than the thunder of nearby explosions; the sight of uncountable corpses littering vast flat expanses of sand; the wretched refugees walking off into an unknown distance; the machines of war ablaze like huge torches against the darkening

desert skies, the stench of burning fuel and incandescent rubber mixing with the reek of roasting human skin and flesh.

Between June 1967 and March 1971, 120 Israeli civilians and 180 soldiers were killed by Palestinian terrorists inside Israel. During that same period, Israeli troops killed 1,873 terrorists within Israel. 'Our basic aim is to liberate the land from the Mediterranean Sea to the Jordan River', Yasser Arafat declared at the beginning of August 1970. 'We are not concerned with what took place in June 1967 or in eliminating the consequences of the June war. The Palestinian revolution's basic concern is the uprooting of the Zionist entity from our land and liberating it.'

Not only inside Israel, but outside it, there were repeated acts of terror, including nine successful hijackings. In Zurich, forty-seven passengers and crew were killed when a Swissair plane was sabotaged: seventeeen of the passengers were Israeli. That same day (13 February 1970) seven elderly Jews were killed during an attack on an old people's home in Munich. After the hijacking of a Sabena plane in 1972, Benjamin Netanyahu was among the Israeli commandos who carried out a rescue mission, and was wounded.

As Israel approached the fifth anniversary of the occupation of the West Bank there were repeated calls from the United Nations, and also from President Nixon in Washington, for a complete Israeli withdrawal from the occupied areas. Inside Israel there was a realization, among those who had been active in serving the State before 1967, that some form of compromise was needed. On 22 March 1972 in an article in the *Jerusalem Post* Walter Eytan, Israel's former Ambassador to Paris (and subsequently head of the Israel Broadcasting Authority), wrote, 'If a referendum were held in Israel on the simple question: "Are you ready to give up the West Bank?", I have no doubt that the answer would be overwhelmingly No. But if the Israeli voter were asked if he agreed to withdrawal from the West Bank in return for a true and lasting peace, the answer would be at least as overwhelmingly Yes.' Eytan continued:

> The borders which were laid down in the armistice agreements (and which remained unchanged till 1967) were regarded at the time as an historic achievement – and let us not forget that these agreements predicated 'the return of permanent peace'.
>
> The government has refused consistently to 'draw maps' (or anyway to publish such maps as presumably have been drawn), and in this it has undoubtedly been right. There are a great many modifications, some of them quite unexpected, that can be agreed upon in the course of negotiation; and there is no point in making either minimal or maximal claims known in advance.
>
> But it should be stated quite bluntly – and explained to the electorate

– that Israel deliberately never annexed the West Bank (that is, has never claimed it as her own), and is not thinking of annexing it now, and it should be stated equally bluntly that peace is more important than the West Bank.

On 10 May 1972, three Japanese gunmen, members of the radical Red Brigade, armed with a machine-gun and working for Palestinian terrorists, opened fire in the arrival hall at Lod airport. Twenty-seven passengers were killed, of whom twenty-one were Christian pilgrims from Puerto Rico. Among the other dead was one of Israel's leading scientists, the polymer chemist Aharon Katzir, who in 1925 had been brought to Palestine from Poland as a boy of eleven. In the first years of the State he was at the Weizmann Institute of Science at Rehovot, and for six years he had been President of the Israel Academy of Arts and Sciences.

There was yet another act of terror in 1972. At the Olympic Games, held that year in Munich in September, eleven Israeli athletes were seized, held hostage, and then killed. German police killed five of the terrorists. Inside Israel there was considerable distress that, after the massacre, the Games had gone on. The Mossad, then headed by Zvi Zamir, undertook to track down and kill the remaining perpetrators of the massacre. They fulfilled their objective.

In London, two weeks after the Munich massacre, Ami Shachori, the Agricultural Attaché at the Israeli embassy was killed by a letter bomb.

The Israeli capture of the Golan Heights in 1967 had liberated the many settlements that had been within its shadow from the constant danger of Syrian rifle fire, shelling and ambush. After 1967 these settlements began to flourish, and to find new activities. Almagor, north of the Sea of Galilee, established a rest house for tourists and sailing facilities.

Following the general guidelines of the Allon Plan, new Israeli settlements were established on the Golan Heights. The Labour government also encouraged settlements along the length of the Jordan Valley. The River Jordan was declared to be Israel's 'security frontier', irrespective of what arrangements might in due course be made for the West Bank. A pattern was emerging, whereby Israel would expand both on the Golan Heights, from which all the Syrian villagers had fled in 1967, and in the largely unpopulated and desert Jordan Valley. There would be no drive to set up Israeli settlements near the Arab-populated areas and towns on the West Bank or in the Gaza Strip. The Palestinians would conduct their own municipal affairs, have their own mayors, teach according to Arab educational curricula, maintain their links with Jordan in the cultural and economic spheres, and have their security controlled by Israel.

While there was no terror and no violence against Israeli army patrols, or against Jewish travellers, there would be little visible in the towns and villages of Israel's presence; and what there was would be benign. At the same time, Israelis and Palestinians would mix in market places and shops, visit each other's parts of the county; Palestinian Arabs would swim on Israeli beaches, Israeli Jews would explore the ancient Jewish sites on the West Bank, or visit the Cave of the Patriarchs at Hebron, eating in the local restaurants and pulling in for petrol at Palestinian garages.

This live-and-let-live attitude did not lead to close ties or large scale mingling between the two peoples, but there were those on both sides of the divide who hoped it might. Given time, there was no reason why a peaceful occupation should not evolve. Those who warned that all occupations would in the end be resented and fought, were isolated voices, unheard – or unwelcome – Jeremiahs.

Within Israeli society the principal struggle remained the integration of the 'oriental Jews' from North Africa, the Sephardi poor who had been dispossessed before leaving their former homes, and felt discriminated against in their new ones. In 1971 Fodor's guide book *Israel 1971* presented the issue with considerable understanding. The greatest difficulty, it wrote, lay 'in providing income-producing work, especially among the Jews from North Africa, who despise manual labour – unlike the early European settlers, who idealized it and made it the cult of the *kibbutz*. Whereas they came to Palestine drawn by an ideal, the present Orientals have come as more or less passive victims of circumstance. To adapt at all they must learn a new manner of living, a new language, how to read, and new agricultural or manual skills they never knew before, a task beyond the capacity of most of them. For teenage immigrants, however, the period of military service, which provides as much classwork as drill, is an effective forcing house. Mixing with the native-born *sabras*, they learn to speak Hebrew and feel Israeli very soon.'

Fodor admitted that 'antagonism between Orientals and Europeans certainly exists', and went on to explain:

> The latter, who led the return and reclaimed the country have made Israel, despite geography, predominantly Western in ideas and habits. They are not particularly happy about the flood of darker-skinned people, whom they yearn to see balanced by a portion of their three million compatriots still locked up in Russia. (The Soviet government refuses to allow a general exit, because it would annoy their Arab friends and because voluntary departure would reflect poorly on the Soviet paradise.)
>
> The Orientals resent the fact that the earlier comers hold the better houses and jobs and, on the whole, the direction of the country (although there are two Cabinet ministers of Oriental origin). They are burdened with all the frustrations and troubles of a group which feels

itself inferior. Israel has an integration problem, but it does not have a deep or hardened segregation pattern to overcome. But with both will and need working for a rapid solution, Israelis talk of absorbing their Oriental citizens into the society within two generations.

Fodor then addressed itself to what it described as 'The Making of an Israeli':

Efforts are concentrated on the children, whose problems are many but whose inner transformation into Israelis can be quick and visible. The process can be seen at work in the schools of Beersheba, where both difficulties and progress are as sharply exhibited as anywhere in Israel.

As explained by the woman principal of one elementary school, the absolutism of the Oriental father, particularly the Moroccan, collapses in Israel. The parents lose prestige, and the children, quickly feeling ashamed of them, look for revenge and become discipline problems. Those more adaptable or more ambitious, who conform well in school, return home each night to the constant struggle of a crowded, ill-lit room without the privacy, often without table or lamp, above all without the parental encouragement to do their homework.

Yet the pervasive influence of the drive toward nationhood reaches through to them and somehow they battle through.

*   *   *

In May 1973 Israel celebrated its twenty-fifth anniversary. Walter Eytan wrote in a special anniversary article: 'Israel is 25 years old, but also 4,000. Here, in particular, all is mysterious. We are as conscious of our youth as of our age: each is equally real to us, and there is no contradiction between them. We take as much pride in being the descendants of the Prophets as we do in what Israel has achieved in this crowded quarter-century of her new-found independence. At the same time, we labour under a heritage which in some ways we should be happy to shake off, or at least amend, just as we are struggling and shall have to struggle for a long time to come with the difficulties, some of them of our own creation, which have accompanied the rebirth of our State. The old and the new, almost inextricably, combine to form the reality in which the people of Israel live in their own country, and in which their country lives in the world.'

In this same article, Eytan reflected on the situation of the Arabs who had come under Israeli rule six years earlier, in 1967. In the West Bank and the Gaza Strip, the 'administered territories' as they were known officially in Israel, the Arab population, he wrote, 'is more prosperous, and probably freer, than at any time before, bound by increasingly close economic and personal ties with Israelis. Something like 40,000 of these Palestinians work each day in Israel.' The situation of Gaza had been transformed. 'Where formerly unemployment was endemic and terrorism rife, today every able-bodied person can find work either in Israel or in the Gaza Strip itself (where in fact a labour shortage prevails at the present time), while terrorist action for the most part belongs to a nightmare of the past. Every Palestinian Arab has been able to see for himself the benefits of normal, constructive co-existence with Israel, and it is daily experience of good neighbourly relations in practice that is more likely than all the resolutions of the United Nations or the best-intentioned mediation of outsiders to bring about peace between Israel and the Arab world.'

Conflict, terrorism, economic hardship: to have relegated these torments to the 'nightmare of the past' was a hope that offered the prospect of continuing growth and prosperity for Jews and Arabs in Israel. But beneath the surface, the Arabs resented occupation, and fostered their own national ambitions.

Within Israel the future of the West Bank had become a contentious issue. The National Religious Party, which since the foundation of the State had always put its Knesset seats behind the Labour-led coalitions of the day, voted at a Party convention to resign from any government that agreed to give up any part of the 'inheritance of the Patriarchs' as part of an agreement with Jordan. On May 10, in a letter to a kibbutznik in the Jordan Valley, Moshe Dayan, the Minister of Defence, advocated Jewish settlements in 'specific places' in the West Bank. 'We have to extend settlement,' he wrote, 'extend Jewish presence in agriculture, industry, urban population', as well as 'public and private sector and cooperative undertakings with Arabs' on the West Bank. 'I think we should be doing more than just striving for this,' Dayan added. 'In my opinion, the thing can be done, too.'

Questioned about the permanence of the West Bank settlements, Dayan told a BBC television interviewer on May 14, 'I do think that Israel should stay for ever and ever and ever and ever in the West Bank, because this is Judaea and Samaria. This is our homeland.'

Dayan wanted the Labour Party platform to incorporate his ideas, and was prepared to press them forward in both public and political forums. He advocated a substantial Jewish urban and industrial expansion around Jerusalem, and he wanted individual Jews to be allowed to buy land and to build homes and settlements anywhere on the West Bank. Only the Israeli government had asserted the right to buy land since the 1967 conquests, and had acquired substantial building land, especially in eastern Jerusalem, whose municipal boundaries it extended to include formerly Arab areas, and on which it had begun to build several large suburbs.

Dayan wanted Israeli citizens to be free to buy land anywhere on the West Bank. In an attempt to win support for his view, he spoke throughout Israel in favour of land purchase. Nor were Dayan's plans for Israeli settlements and buildings restricted to the West Bank. In the Rafa Salient, on the northern coast of Sinai, which had been captured from Egypt in 1967, he wanted the construction of a port city, to be called Yamit, which would serve as a major Israeli urban and commercial centre in northern Sinai, cutting Gaza off from Egypt, eclipsing Gaza economically, and establishing Israel permanently in Sinai.

The widespread scale of settlement that Dayan was advocating was opposed by the influential Finance Minister, Pinhas Sapir, who took the view, also shared by Yigael Yadin and Yigal Allon, that if Israel incorporated the occupied territories, adding more than a million West Bank and Gaza Arabs to the existing 400,000 Israeli Arabs, Israel would cease to be a predominantly Jewish State. Given any form of democratic system of elections and representation, or the extension of Israeli–Arab voting rights to the as yet unfranchised West Bank and Gaza Strip, the Arabs would then have at least a third of the seats in the Knesset, and a proportionately high influence on policy. Yet the essential aim of the founders of Israel was that it should be a State for Jews, a homeland for the dispersed and the dispossessed among the Jewish people, and a place where Jewish values, and Jewish aspirations, would be the State's object and theme.

In March 1973 Yitzhak Rabin returned to Israel from Washington, his period as Ambassador over. 'The Israel I came home to', he later wrote, 'had a self-confident, almost smug aura to it, as befits a country far removed from the possibility of war.' The possibility of war seemed remote. That April, speaking to army graduates on Masada, Dayan spoke of 'the superiority of our forces over our enemies, which holds promise of peace for us and our neighbours'. During a visit to the Suez Canal by members of the Knesset Foreign Affairs and Defence Committee, in answer to a question of what would happen if the Egyptians tried to cross the canal, Dayan replied, 'We'll step on them, I will crush them, let them come.' Dayan envisaged any Egyptian troops who crossed the canal being encircled by the second day, then destroyed on the third day by Israeli tanks. On that assumption, the army asked for a 'war budget' based on just five or six days' fighting.

That May the Egyptians put their army on high alert. When the Chief of Staff, David Elazar, responded by a partial mobilization of Israel's reserves, he was widely thought to have reacted with undue panic, wasting public money. When, after the Egyptian alert had been called off, the High Command in Tel Aviv was told by one of its Intelligence sources that a planned canal crossing had been put off until October, no alarm bells rang; the information was discounted and filed away. That summer, the Israel Defence Forces announced that they were planning to reduce the length of compulsory military service from thirty-six months to thirty-three months. Reserve duty would also be cut, from sixty days a year to thirty. Defence spending, which had been 40 per cent of the national budget in 1970, and had dropped to 32 per cent in 1973, was planned to fall to 14.6 per cent in the budget of 1977.

On June 13 the new head of the Israeli air force, General Benjamin (Benny) Peled, expressed his concerns to Dayan that without a pre-emptive strike in the event of an Arab war plan, Israel would be at a serious disadvantage – 'this whole "Opera" will be conducted according to plan only if we are the initiators' – Dayan reassured him that 'if we believe that the Arabs are planning to attack us' the air force would get approval for a pre-emptive strike, as had happened in 1967.

Throughout the summer of 1973, any sense of danger was muted, almost non-existent. On July 15, General Ariel Sharon, who had decided to go into politics – in the ranks of the opposition – handed over command of the Southern Front to General Shmuel Gonen. When, a few hours before the formal handover, Sharon warned Dayan that 'if we have a war here, and we might have one, Gonen does not have the experience to handle it', Dayan replied: 'Arik, we aren't going to have any war this year. Maybe Gonen is too inexperienced. But he'll have plenty of time to learn.' Three weeks later, on August 9, at a lecture to the Staff College, Dayan told the officers: 'The overall balance of forces is in our favour, and this is what decides the question, and rules out the immediate renewal of the war.'

The unlikeliness of war did not weaken Dayan's popularity; indeed, it was the superiority of strength that he was credited with having created for Israel that enhanced his popular appeal. Sensing this, he decided to challenge the Labour Party leadership on a central issue, that of the expansion of Jewish settlements on the West Bank, and into Sinai. Two of Labour's so-called 'doves', Pinhas Sapir and Abba Eban, opposed settlement expansion. But Sapir told Eban that if Dayan were to leave the Labour Party, he might take with him as many as fifteen Knesset seats, making the Party vulnerable, for the first time since 1948, to the challenge of the right-wing opposition parties (Menachem Begin had never given up waiting in the wings for the call to power).

The Minister without Portfolio, Israel Galili, one of Golda Meir's most trusted colleagues, was asked to try to produce a document that would be acceptable to both sides of the divide, and maintain Party unity. It was published on August 23. Two of Dayan's wishes were set aside: the building of a major port at Yamit (postponed for three years) and the free sale of Arab land on the West Bank to Jewish settlers (the land could be sold, but under strict supervision). The Galili Document – as it became known – did however authorize the building up of Yamit town, as opposed to the port, and the construction of 800 housing units (for 3,000 residents) within the next five years. The existing forty-six West Bank settlements were to be expanded to seventy-six, but none was to be built in densely populated Arab areas. In any future redrawing of Israel's borders, all of these settlements would, at Dayan's insistence, be 'inside Israel's final borders'.

Israel's annexation of much of the West Bank seemed suddenly to be much nearer. Even Sapir had been prevailed upon – in the interest of heading off a Dayan revolt – to provide $300 million over the coming four years to finance the West Bank building.

On 13 September 1973 Israeli and Syrian jets clashed over the Mediterranean, off the Syrian port of Latakia. Thirteen Syrian aircraft were shot down, for the loss of a single Israeli plane. When, shortly after that, the Syrians mobilized their armed forces and began to increase the number of troops on the Golan Heights, this was interpreted by Israeli Intelligence as a reaction to the air battle, not as anything presaging war. Further troop exercises by the Egyptians on the western bank of the Suez Canal seemed just that: exercises, with no more sinister intent than to keep the Egyptian army in a state of readiness – for defence.

CHAPTER TWENTY-FOUR

The October War
Yom Kippur 1973

On 4 October 1973 the Israeli newspapers reported that Colonel Gadaffi of Libya was sending 'terrorist squads' to stage acts of terrorism in both Israel and Jordan. In Israel, the Cabinet met that day to discuss another crisis. The Austrian government – following an attack by two Palestinian gunmen on a train carrying Soviet Jewish emigrants from Moscow to Vienna – had decided to close down the refugee camp at Schoenau, where many Soviet Jews were waiting for their onward journey to Israel. The United Nations had just refused to take over responsibility for the camp, on the grounds that, as the refugees held valid documents to enter Israel, they were not refugees in the accepted sense of the word, but people in transit with somewhere to go.

The Palestinian gunmen had taken five Jews and an Austrian customs official hostage. When the Austrian government (headed by Bruno Kreisky, himself a Jew – with a brother living in Israel) announced that it was closing down the transit camp, the hostages were released and the gunmen disappeared. Many Israelis were outraged at what they saw as a craven submission to terrorism. The Israeli Prime Minister, Golda Meir, flew first to Strasbourg to address the Council of Europe, and then on to Vienna, to try to persuade Kreisky to reverse his decision. Kreisky would not do so, and Golda Meir returned to Israel. 'Kreisky did not even offer me a glass of water,' she reported back in Jerusalem.

Such items of news – terror, and the threat of terror, and the pusillanimity of a European government – items which might have created a considerable stir in Denmark or Luxembourg, were a regular part of life in Israel. When, that week, an Israeli police officer was killed in Gaza by a grenade thrown into his vehicle, that was not a major headline. Of seemingly greater importance was a decision of all eighteen member States of the Arab League to launch a worldwide political campaign aimed at preventing the Soviet Union from allowing more Jews to emigrate to Israel. The Arab League was also urging the countries of Europe not to open alternative camps for the Soviet Jews who were being forced to leave Austria.

For Israel, these activities were pinpricks. So, too, were two items of information, reported that morning by Reuters news agency: that Syria had decided to resume diplomatic relations with Jordan; and that President Mobutu of Zaire had broken off relations with Israel 'until Egypt and other Arab countries recover the territories they lost in the Six Day War'.

At a meeting with Golda Meir and several of her senior advisers on the morning of October 3 – just after her return from Vienna – Dayan said that recent Egyptian and Syrian military concentrations on the Suez Canal and the Golan Heights were 'unusual'. But there was no sense at the meeting that war was imminent, or in prospect. Indeed, it was agreed, at Yigal Allon's suggestion, that there was no need to call the full Cabinet together in advance of its next scheduled meeting, on October 7, in four days' time.

On October 4, Dayan lunched with General Rehavam (Gandhi) Ze'evi, who had just retired as commander of the Central Front. Their conversation, as recounted by Ze'evi to Dayan's biographer Robert Slater, contained the following revealing exchange:

> Dayan: 'What's the matter?'
> Ze'evi: 'I suspect we are moving towards war. And I will not be part of it.'
> Dayan: 'What are you talking about? There's not going to be a war. Not this summer and not this fall. You're not talking to the point.'

That morning, as reports reached him that Soviet advisers and their families were being evacuated from Syria (Nassar's successor, Anwar Sadat, had earlier rid himself of their dominating influence), Dayan ordered the air force on full alert. The army was placed on a 'C' alert, the highest alert short of calling up the reserves. But without calling up the reserves, and bringing the productive enterprise of the State to a virtual halt, there was no way that Israel could repel a sustained attack on her borders.

On the Northern Front, on the Golan Heights, the Israeli commander, Major-General Yitzhak Hofi, who had earlier expressed his unease at Syrian troop concentration to Dayan, asked Dayan for help. The Syrians had 1,200 tanks available for action; hundreds of them had been brought up towards the 1967 cease-fire line, along a forty-mile front – in June 1941 the German army had deployed 1,400 tanks against the Soviet Union, along a 1,000-mile front.

Dayan agreed to let Hofi move up more than 100 tanks from the reserves that were then in southern Israel. This was to increase the number of Israeli tanks on the Golan on the morning of October 7 from 60 to 170. It was still a relatively small number, less than one-tenth of the number of tanks that Syria was preparing to throw into battle, but it was to prove decisive.

*   *   *

At 9.40 a.m. on the morning of Friday, October 5, the few Cabinet Ministers who had not already returned to their homes throughout Israel in preparation for Yom Kippur met at the Prime Minister's Office in Tel Aviv. The Chief of Military Intelligence, General Eli Zeira, told the gathering that, in his view, it would be possible for Egypt and Syria to open hostilities against Israel without any further warning. The Chief of Staff, General Elazar, disagreed. Only when further evidence of an impending Arab attack had been brought forward, he said, should Israel mobilize its reserves.

Yet the unease and uncertainty at the meeting were palpable. When Dayan suggested that, in the event of fighting breaking out on the following day – Yom Kippur – Golda Meir should be authorized to mobilize the reserves on her own, the Cabinet Ministers present agreed. To Golda Meir's suggestion that she spend Yom Kippur at her daughter's kibbutz, Revivim, in the south, and that if needed in Tel Aviv she could always come back by helicopter, Dayan said, 'If there is a war, we might not be able to get you back by helicopter.'

The evening of Friday, October 5, was the beginning of Yom Kippur, the Jewish Day of Atonement, a day of fasting and prayer when most Jews, and most Israelis, spend the evening and much of the next day in synagogue. In Israel, as darkness fell, there came, as every year, the strange calm and deep quiet of the holiest day in the Jewish calendar. But something was happening that began to disturb that calm. There were several ominous signs. For example, Yitzhak Rabin – who had recently entered politics as a potential Labour Party Member of Knesset – recalled how, during that Friday, his son Yuval, who had arrived home for his very first weekend leave from the Navy Training Centre where he was serving, had no sooner reached home than he received orders to return to his unit immediately. Rabin added, 'My son-in-law, who was a captain in a tank battalion in Sinai, was supposed to enter hospital the next day. But when he learned that Yuval had been recalled from leave he likewise returned to his unit immediately. One didn't have to be a former Chief of Staff to realize that the IDF was on alert footing, but even so I never expected war.'

Nor did the Israeli military leaders. Although such Intelligence information as they had indicated that something serious might be afoot, the consensus of expert opinion was that war was not imminent.

At 8.30 on the Saturday morning, October 6, Rabin received a telephone call asking him to join the other former Chiefs of Staff at a meeting with the Minister of Defence at three that afternoon. Yigael Yadin was also asked to attend. At noon, as Yadin was preparing to leave from his Jerusalem apartment, Moshe Pearlman, who had been the army spokesman in previous wars, came in from a synagogue nearby to report – with some alarm – that the man sitting next to him in synagogue had received his mobilization papers, and that throughout the morning prayers similar notices were being brought to people while they prayed. 'This is impossible, unprecedented, something must be up,' was Yadin's response. Equally unprecedented for Yom Kippur was the sudden appearance of cars on the street.

At 2 p.m., Yadin left his apartment to drive down to Tel Aviv for the meeting at the Ministry of Defence. As he started up his car, the noise of his car engine was overlaid by another, more strident noise, that of an air raid siren. No such siren had been heard wailing in Jerusalem since the Six Day War more than six years earlier. Throughout the country, as the sirens were turned on in every city and town, people turned on their radios to hear the news. But there is no broadcasting on the Day of Atonement. The radios were silent. The sirens had fallen silent too.

At 2.30 p.m. those who had kept their radios open in the hope that broadcasting would be resumed heard a brief Hebrew announcement: 'The sirens are not a false alarm. When the siren sounds again, everyone must go to their shelters.' That was all: there was no explanation of what was happening, or of why the sirens would sound on the Day of Atonement. After the announcement, there was music: Beethoven's *Moonlight Sonata*.

An hour later, at 3.30 p.m., after what had seemed an eternity of waiting for those glued to their radios, there was a second announcement. It was also a brief one. 'Egypt and Syria have attacked. Partial mobilization has been ordered.' Then the sirens wailed again, rising and falling, and hundreds of thousands of Israeli citizens made their way to the shelters which are an integral part of every house and apartment building. Ten minutes later, the sirens wailed again, this time a loud, continuous hum – the all-clear.

In the streets of every town, vehicles hurried hither and thither, many with their lights on and with pieces of hastily cut out card placed in their windscreens indicating that they are being used as army vehicles. They were taking soldiers to their bases. It was 4 p.m. Another news flash gave two staccato orders to the citizens of Israel: all inessential traffic was to keep off the main roads. Petrol stations were opening at once. Then the classical music returned, still Beethoven. Nothing was said of what was happening on the borders.

At 4.20 p.m. the radio announcer spoke again. All public road transport was to start at once. An emergency hospital had been opened for military casualties. All non-emergency patients in hospitals must return home. With that reference to 'military casualties', it became clear that something very serious was taking place along the frontiers. Twenty minutes later came confirmation from the announcer that a serious situation had developed: 'There is fierce fighting in Sinai,' he said. 'The Egyptians have crossed the Suez Canal and are on the east bank.'

The unthinkable had happened. Six months earlier, when the Minister of Defence, Moshe Dayan, had taken a group of Diaspora Jewish leaders on a tour of Israel's Suez Canal defences along the east bank of the canal, one of them had asked whether the Egyptians could launch a surprise attack across it. 'The moment we see the glint of war in their eyes,' Dayan had replied, 'we will blow them away.'

The news flashes were coming every few minutes. At 4.45 p.m. the announcer came on again to say: 'The Egyptians have crossed the canal at several points. There is fighting on land and in the air.'

It was at 2 p.m. on October 6 that the Egyptian and Syrian forces had attacked Israel simultaneously, according to a joint plan that had been kept secret with remarkable success. It later emerged that 95 per cent of the Egyptian officers taken prisoner by Israel in the weeks ahead had only known on the morning of October 6 that the military exercises in which they were taking part were in fact a prelude to war.

Such warning signs as there had been, and the way in which they had been interpreted – or misinterpreted – were to be studied in detail after the war by a special commission of inquiry headed by the President of the Supreme Court, Dr Simon Agranat. It concluded that the Israeli army chiefs had held the 'conception' that Egypt would not attack Israel without sufficient air power to attack in depth and to dislocate the Israeli air force (something Egypt could not do on October 6), and that Syria would not attack without Egypt. Such detailed Intelligence as had arrived in the forty-eight hours before the attack ought to have served as a warning, the Agranat Commission reported, but due to the 'conception' was evaluated incorrectly.

According to the Agranat report (which was partially made public within the year), Lieutenant Benjamin Siman Tov, the order-of-battle officer in Southern Command Intelligence, had, on October 1, submitted a document to Lieutenant-Colonel David Gedaliah, Intelligence Officer of Southern Command, in which he analysed the Egyptian deployment on the western side of the canal. That deployment was, he wrote, an indication of preparations to go to war. This, Siman Tov added, was despite the 'military exercise' that was taking place. Two days later, on October 3, Siman Tov had submitted a second report, pointing out a number of factors that indicated the military exercise might be a cover-up for preparations for war.

Colonel Gedaliah did not distribute the lieutenant's evaluation, and it was omitted from the Southern Command Intelligence report to General Headquarters. As a result, the Director of Military Intelligence, General Zeira, did not learn about Lieutenant Siman Tov's evaluation until five months after the war, during the Agranat Commission hearings (whereupon he asked Siman Tov, who had been removed from Southern Command Intelligence, to come to his office; he listened to his story and promoted him to the rank of captain).

On October 5 the division manning the Israeli defences along the Suez Canal requested reinforcements. Its commander wanted additional troops to man the strongholds along the canal itself, as well as further forces for deployment in the Sinai passes twenty miles to the east – the passes that had been the scene of fierce fighting seven years earlier. In reply, he received a signal from Southern Command Headquarters repeating a signal from General Headquarters in Tel Aviv, that the Egyptian military exercise on the canal was nearing its conclusion.

Despite not having been sent Lieutenant Siman Tov's Intelligence evaluation of October 3 that Egypt might be preparing for war – and despite General Headquarters' rejection of a similar assessment by the senior Intelligence Officer of the Israeli navy, who was worried by the sudden withdrawal of Soviet warships from Alexandria and Port Said on October 4 – the Chief of Staff, David Elazar, had been uneasy at the sanguine interpretations of the Syrian and Egyptian troop movements. He had therefore given instructions, on Friday October 5, that the standing army was to be put on the highest degree of preparedness. At the same time, he had ordered a limited mobilization of reserves in certain units, and in the air force. This, however, was a long way from the degree of mobilization that was needed in the event of a full-scale attack.

At four o'clock on the morning of Saturday October 6 an Intelligence message reached General Zeira that war would break out later that day – 'towards sundown'. He at once telephoned Dayan, Elazar, and the Vice Chief of Staff, General Tal. There could be no doubt that the assessment was authentic, as it came in the form of a coded telegram from Zvi Zamir, the head of the Mossad, who had recently gone in the utmost secrecy to Paris, to obtain first hand, incontrovertible knowledge.

Elazar's response to Zamir's telegram was to ask Dayan for permission to mount a pre-emptive strike against Syria. Dayan said no: the United States had warned Israel not to attack first. Elazar then asked Dayan if he could order full mobilization, in order to be able to mount an immediate counter-attack once hostilities had begun. Again, Dayan refused. It would be impossible to mobilize the whole army at once; it would take two days before a complete mobilization could take place – a far-reaching decision that could give the impression that Israel intended to attack first. There could be partial mobilization, Dayan said, but only for defensive purposes. He did not want to do anything before the attack that might be interpreted as Israeli provocation. If Israel started with a large-scale mobilization, and carried out a pre-emptive strike, the Arabs would accuse Israel of being the aggressor. He wanted to make it crystal clear that Israel was attacked, and not attacking.

Unwilling to accept Dayan's decision as final, Elazar went with Dayan to see Golda Meir in her office in Tel Aviv. He took with him his request for full mobilization – and also the Mossad chief's telegram reporting that war would break out that same evening. At the meeting, Dayan reiterated his point of view, and proposed an upper mobilization limit of 50,000 men. Elazar pushed for 150,000. At Golda Meir's suggestion, a compromise was reached: 100,000 men would be mobilized. The order was issued at 10 a.m.

Elazar went away to give the necessary orders, and, on his own responsibility – convinced that full mobilization could in no way be 'provocative', as Dayan believed – raised the number authorized by the Prime Minister to the number he had originally requested. That was the mobilization which was taking place while most of the soldiers were at home or in synagogue later that Saturday morning. The fact that so many were in synagogue or at home saved the State.

The Egyptian onslaught was swift and successful. At 2 p.m. – three hours at least before the time given to General Zeira by Israeli Intelligence that morning – 240 Egyptian warplanes flew over the canal and attacked Israeli positions in Sinai. At the same moment, 2,000 Egyptian guns opened fire, in an intensive artillery barrage, during the first minute of which more than 10,000 shells fell on the Israeli positions. The barrage continued for fifty-three minutes.

The Egyptian strike aircraft were back over Egyptian soil at 2.15 p.m. At that moment 8,000 Egyptian assault troops crossed the canal. Mostly avoiding the Israeli fortifications, they pushed eastward. The Egyptian planners had estimated up to 10,000 Israeli dead by the end of the day. The actual death toll was 208. In places the Egyptians were amazed at the tenacity of the Israeli defenders. One Egyptian general, the commander of an infantry division in the first assault, later told the story, which Chaim Herzog repeated in his book *The Arab–Israeli Wars*, of how 'a lone Israeli tank fought off the attacking forces for over half an hour, causing very heavy casualties to his men when they tried to storm it. When they finally overcame it, the Egyptian general recounted how, to his utter amazement, he found that all the crew had been killed with the exception of one wounded soldier, who had continued to fight. He described how impressed he and his men were by this man, who, as he was being carried away on a stretcher to a waiting ambulance, saluted the Egyptian general.'

There were sixteen Israeli fortifications on the eastern side of the canal, built seven to eight miles apart. They were manned by a total of 436 soldiers, reservists of the Jerusalem Brigade who were doing their annual reserve duty in Sinai. Many of these soldiers were new immigrants. Almost none of them had combat experience. In some of the fortifications the officer in charge was killed in the early stages of the Egyptian attack and a corporal, even a private, took command. The names of the defenders of these forts have become legendary in Israel: among them Corporal Or-Lev, (the 'Ketuba' fort), Lieutenant Shlomo Ardinest (the 'Quay' fort), and Captain Motti Ashkenazi ('Budapest'). Ashkenazi's fort, which was set back some five miles from the canal, held out until the end of the war, the only one of the sixteen to do so. Or-Lev and his surviving defenders managed to break out and reach the Israeli lines. Ardinest and his men surrendered. Chaim Herzog has written:

After holding out for a week against forces of the Egyptian Third Army, when the garrison was left with only twenty hand-grenades and a few belts of light-machine-gun ammunition, the fortification was authorized to surrender via the Red Cross at 11.00 hours on the Saturday morning.

Ordering his troops to wash themselves in the few drops of water left in the jerry cans and to change their battle-soiled clothes, Lieutenant Ardinest paraded his men and marched into captivity, led by a soldier carrying a Torah scroll from the fortification.

The thousands of Egyptians surrounding the position watched the proceedings in awe.

After the evacuation, the Egyptian officers searched high and low for non-existent heavy machine-guns in the position, unwilling to believe that the garrison had held out for a week with only four light machine-guns.

*   *   *

At the time of the Egyptian crossing of the Canal, there were only three Israeli tanks on the waterfront. Many of those hurried forward were knocked out of action. As Dan Shomron, the commander of one armoured brigade, drove from the Mitla and Gidi Passes towards the canal, with orders to try to reach the forts, he lost seventy-seven of his 100 tanks.

As the Egyptians consolidated their hold on the eastern bank of the Suez Canal, in the north of Israel the Syrians attacked on the Golan Heights. One of their first objectives, the Israeli radar and Intelligence gathering centre at the Israeli summit of Mount Hermon, was defended by one officer and thirteen men: its defences had been so neglected that the main gate, which had been damaged some time before, swung open on its hinges unrepaired. Four Syrian troop-carrying helicopters landed on the peak. One exploded and its passengers were killed. The soldiers from the other three seized the peak and its sensitive equipment. 'For the Soviet advisers of the Syrians,' writes Chaim Herzog, 'the electronic equipment captured there was of singular value.'

That afternoon, a total of 1,400 Syrian tanks advanced, in the centre, south and north, driving the Israeli defenders back across the Golan Heights. Breaking through the Israeli border position at Rafid, in the centre, the main force of Syrian tanks, 600 in all – together with a Moroccan brigade – swept forward along the Tapline road, the pre-1948 Iraq–Haifa oil pipeline access road. To hold back the 600 was a force of fifty-seven Israeli tanks. Rapidly the Syrians advanced almost halfway across the Golan, which, even from Rafid to the River Jordan, its widest point, is only fifteen miles wide.

An Israeli lieutenant, Zwicka Gringold, who had been among those who had hurried back to his unit from leave when the sirens first sounded, arrived to find many of the tanks knocked out and their crews dead inside them.

Removing the bodies of his former companions-in-arms, he improvised a small force of four tanks – known as Force Zwicka – and advanced against the Syrian armour. Fighting at night, he destroyed many Syrian tanks, and created the impression that his was a much larger force. Only when his tank alone was left of the fighting four did he return to base at Nafekh. He arrived at the very moment when the base was being attacked and the first Syrian tank had broken through the camp defences. He destroyed that tank.

An attempt by the remaining tanks near the base to advance along the Tapline road and drive the Syrians back was a failure. All the remaining Israeli tanks were destroyed. But the twenty-four-hour battle had broken the initial momentum of the Syrian attack and Nafekh, although entered by the Syrians, was never completely overrun. South of Rafid, the Syrians reached as far as Ramat Magshimim and then pushed on another five miles to El Al, less than six miles from the Gamla Rise overlooking the Sea of Galilee. In the centre, they reached Snobar, a mere ten minutes' tank driving time from the River Jordan at the B'not Yaakov bridge.

In the northern sector, Syrian tanks had also pushed forward, outnumbering the Israeli defenders. Here too, as along the Tapline road, a determined individual helped to blunt the impetus of their forward thrust. He was Lieutenant-Colonel Naty Yossi, who had been on honeymoon in the Himalayas, and had managed by considerable effort to get back to Israel in time to reach the Golan Heights just as the Israeli tank forces were about to make a major withdrawal. Taking thirteen damaged tanks that had been taken back several miles for repairs, and organizing crews for them, many of whom were wounded men who discharged themselves from hospital, he took his force forward and, linking up with the seven still-operating tanks at the front, brought the Syrian attack to a halt. In all, 500 Syrian tanks were destroyed in the north sector.

Israeli reinforcements had been rushed north and, first at El Al and then in the more northerly sectors, began to push the Syrians back. The Syrian forces were never again during the course of the next two weeks of battle to reach the positions they had reached in the first twenty-four hours. But for the Israeli public, their success in pushing so far westward had been a terrifying spectacle.

The events of October 6 were traumatic in the extreme for those in Israel who had to remain at home, their only contact with the unfolding events being the radio and television. At 5 p.m., fifteen minutes after the radio announcer had reported the Egyptians had 'crossed the canal at several points' and that there was 'fighting on land and in the air', a news flash stated that 'Syrian planes are in action in Upper Galilee. A fierce air battle is in progress.' Syrian shells had fallen in the Huleh Valley. Israeli outposts on the Golan had been attacked.

At 5.10 p.m. a further radio announcement told all citizens to black out their windows. Three-quarters of an hour later another message advised people to stick tape over all windows, mirrors and pictures, advice that raised the spectre of Syrian bombing raids on the main cities. Five minutes after this announcement, Golda Meir spoke over the radio. At two o'clock that afternoon, she said, while the Israeli Cabinet was in session to discuss 'reports of a possible invasion', both Egypt and Syria crossed the cease-fire lines and opened hostilities on land and in the air.

As Golda Meir spoke, her words were interrupted by a high-pitched wail, followed by a mobilization message in code (such coded messages were being broadcast repeatedly). The members of some units were being ordered to their base. After the message she continued with her speech. 'I have no doubt that no one will give in to panic,' she said, and went on, in grim but firm tones, 'We must be prepared for any burden and sacrifice demanded for the defence of our very existence, our freedom and independence.'

At 6.05 p.m. the radio announced that there were three Egyptian bridgeheads over the Suez Canal. It added that eleven Egyptian helicopters carrying troops had been shot down. It was the first direct intimation of loss of life since the radio had come on three hours earlier. At 8 p.m. a news flash reported that all women and children had been evacuated from the Golan Heights. It was also announced that Dayan would speak on television in an hour's time. At 8.50 p.m. there was another news flash: everyone was asked not to use the telephone 'unless it is absolutely necessary'. The sense of isolation of hundreds of thousands of families was suddenly increased.

At 9 p.m. Dayan spoke on television. He spoke of a 'dangerous adventure' undertaken by President Sadat. Israel, he said, had 'suspected' such an attack in advance, but had decided not to open fire first because, although this meant losing the military advantage, it also made sure, for the outside world, 'that the picture will be known, the true one, that the Egyptians and Syrians started the war'. This attack, Dayan continued, was 'the start of all-out war again'.

Dayan then spoke of Israel's lack of readiness. If the Suez Canal line was always fully manned, he said, 'then we cannot have normal life in the country' – too many young people have to be under arms. When the Egyptians had attacked that afternoon, Israel had 'only small forces on the spot'. The reserves were still being mobilized. 'We do not have enough forces today', he said, 'i.e. enough forces to meet the enemy and the control of the enemy there. That is one problem we have to face.'

As Dayan was speaking, despite the problem of lack of adequate forces, he did not seem too pessimistic. 'Some of the Egyptians did manage to cross the canal', he said, but although there were Israeli casualties, they were 'not significant ones'. As to Sinai, it was, he stressed, 'far away, and it is a big desert'. On the Golan, 'the Syrians lost much more heavily than we did'. Their objective, 'to kick us out of the Golan Heights – will fail'.

Dayan then turned to the future course of the struggle. 'I suppose', he said, 'that during the night they will have more forces crossing, and tomorrow morning we shall see more Egyptians on our side of the Suez

Canal. Then we can start the real war the way we see it and not only try to stop them.'

Questioned by journalists, Dayan said that the Israeli dead could be counted 'by the tens; it wouldn't be hundreds'. As to aeroplane losses during the day, these, he said, were 'very few'. As for the Golan: 'I don't think it was a bad day for us.' With regard to Jordan, Dayan issued a warning. 'I hope that Jordan will be clever enough, wise enough, not to interfere, not to join the Egyptians and Syrians,' he said. 'They had their Black September. I don't think they want a Black October'. As to the future, 'I believe the war will end up with no Egyptians staying on this side of the Canal.' It would not be too long, he said – 'it won't take months or weeks, before we manage to wipe them out.'

After Dayan had spoken, an Englishman caught in Israel on the outbreak of war noted in his diary, 'Confident words. But he looks more worried than he sounds.' That night, details emerged of the extent of the Syrian shelling in the north. Some shells had fallen on Migdal Ha-Emek, in the centre of Israel, less than fifteen miles from the Mediterranean Sea. Shells had also hit the northern town of Kiryat Shmonah, and one person had been killed. Syrian artillery fire had destroyed the dining halls of two Jewish settlements in Galilee, Kfar Giladi and Mahanayim. But there were no casualties: because it was a day of fasting, the dining halls were empty.

Over the radio, the mobilization messages had continued all evening. By midnight, more than 200,000 reservists had been called to their units. Many had been sent straight into battle. The rest were expected to be at their battle stations by the following evening. At many of the mobilization points and equipment stores there was chaos. Much of the equipment proved not to be operational.

Confronting the 200,000 Israeli reservists were as many as 300,000 Syrian and more than 850,000 Egyptian soldiers. Even if no other Arab country joined in the attack, Israel could be outnumbered by six to one. But it was clear that not all these troops would be committed to battle. Most of them would be required to defend the approaches to Cairo and Damascus, once Israel felt able – as she already did on the Golan Heights – to launch a counter-attack. 'Once the front is stabilized,' Yigael Yadin commented that night (in private) after his return from General Headquarters, 'we will surely try to counter-attack. After all, once the Egyptians attacked across the Canal they initiated a situation in which subsequent decisions would no longer be made by them alone. They will not be able to dictate the nature of the counter-attack.'

In the Sinai, as on the Golan, the counter-attack had begun. It too began from a situation of peril for the Israeli forces. By the morning of October 7, the 290 tanks which had been operational the previous morning had been reduced by two-thirds. But two extra divisions were quickly sent forward, one commanded by Major-General Ariel Sharon, the other by Major-General Avraham ('Bren') Adan, and committed to preventing an Egyptian advance any deeper into Sinai.

After Dayan had visited the Sinai front on October 7, he recommended a withdrawal as far back as the Mitla and Gidi Passes, a line, he believed, that would be more easily defended. But General Elazar did not accept this. He was convinced that the Egyptians could be held much closer to the canal, enabling a counter-attack to be mounted more effectively. It was not clear when that counter-attack could come. General Sharon wanted to try to reach the east bank of the canal that day, but Elazar would not allow this. He did, however, authorize the counter-attack to begin on the following day, October 8.

The strategic debate took place against a background of continuous and bloody fighting. Tremendous efforts, superhuman courage, heavy loss of life, and many moments of uncertainty, marked the continuing course of the battle. At noon on October 8, an Israeli armoured brigade commanded by Colonel Gaby Amir launched a tank attack towards the canal. The attack was driven back by several hundred Egyptian infantrymen firing anti-tank weapons at short range. Twelve Israeli tanks were destroyed. That afternoon, another Israeli tank attack by Colonel Natke Nir's armoured brigade came to within 800 yards of the canal. Once again hundreds of Egyptian infantrymen appeared, firing anti-tank weapons, and eighteen tanks were destroyed. Chaim Herzog has written, of the brigade's commander:

> Nir is an unusual example of perseverance and courage overcoming disability. While serving as battalion commander in the Six Day War, he had been seriously wounded in his legs, and had undergone more than twenty operations. He refused to be retired from combat duty, and by sheer perseverance was awarded a fighting command in a reserve tank brigade, despite his disability.
>
> Like the knights of old, he had to be assisted or hoisted in order to mount or dismount from a tank, but his bulldog nature brought him to a combat command.
>
> When he withdrew his forces from the inferno in which they found themselves in attacking in the direction of the Canal, only four tanks were capable of withdrawing with him.

During October 8 four-fifths of Israel's armoured forces that were still operational were committed to the battle. That day, all the Israeli reservists from fighting units reached the front. One of the morning news broadcasts – they were being transmitted at half-hourly intervals – told of Australian Jews arriving as volunteers to help with civilian work in the cities, as so many able-bodied men were mobilized. American Jews had raised $100 million for the Israeli war effort. Algerian air force units had arrived in Egypt in order,

Cairo radio reported, 'to support the fight against Israel'. King Hassan of Morocco had appealed to Moroccan reservists to register as volunteers for the 'battle of destiny'. President Idi Amin of Uganda had ordered all Ugandan officers who were then in Egypt to join in 'the Arab war against Israel'. The Prime Minister of Bangladesh, Sheikh Mujib, had telegraphed to Egypt and Syria: 'Seventy-five million people in Bangladesh support you in your just cause.'

In the 11 a.m. news broadcast, the Israeli Soldiers' Welfare Committee appealed over the radio for the public to send items needed by 'our soldiers at the front'. These included transistor radios, light reading in Hebrew – detective stories and suspense stories in particular – and games. The games mentioned were chess, draughts, dominoes and sheshbesh. During that same broadcast, an Israeli civil defence spokesman asked every family that did not have a shelter at home, or that had 'too small a shelter', to dig a slit trench in the yard or garden. These trenches should be deep enough to enable a person to stand up in them 'without being seen'.

Serious hoarding of food supplies was reported from Tel Aviv. Rumours also abounded: on the afternoon of October 8 it was rumoured in Jerusalem that Israeli forces had entered Damascus. They were in fact making small, and costly advances, thirty miles from the Syrian capital. But the sense that the danger to Israel was passing – even that it had already passed – was gaining ground that evening. It was enhanced at 8.30 p.m. when General Elazar gave a televised press conference. The Egyptians, he said, had in some places been 'driven back to the Suez Canal' (this was not so; perhaps, like Yadin's broadcast about the encirclement of Egyptian forces near Isdud during the War of Independence, it was intended that the Egyptians should believe it to be so).

Elazar added that the Syrian penetration of the Golan 'is over and done with'. Almost everywhere, the Syrians had been driven back to the cease-fire lines of 1967 (this was true). 'This war is serious,' Elazar added. 'The fighting is serious. But I am hoping to tell you that we are already at the beginning of the turning point, we are already moving forward.' Asked by a journalist when the fighting would end, Elazar replied: 'I will predict one thing: we will continue to attack and beat them, give them a real clobbering.'

Heavy casualties were suffered by Israeli forces, seeking, in vain, to reach the east bank of the canal on October 8. At a press conference on the following day, Dayan astounded journalists by his pessimism. He seemed to be thinking along the lines of an Israeli withdrawal from the whole of western and central Sinai, with a defensive line to hold the coast of the Gulf of Akaba from Eilat to Sharm el-Sheikh. When Dayan said he would address the Israeli public on television that night, Golda Meir was so alarmed that she instructed another officer, General Aharon Yariv, to appear instead.

During the morning of October 9 the Israeli news broadcast made it clear, for the first time, that Egypt was in full control of the eastern bank of the

Suez Canal, and that the Israeli front line was some three to five miles east of the canal. It also reported continued Syrian long-distance shelling of the Jezreel Valley: at Nahalal, the kindergarten and the school had been hit, but no one had been killed. That evening the six o'clock news reported that Israeli warplanes had hit Syrian military concentrations in and around Damascus, and that civilian buildings 'may also have suffered'. Half an hour later, it was announced that the Soviet Cultural Centre in Damascus had been hit. There had been some loss of life.

At 9.30 p.m. General Yariv gave the televised press conference that Golda Meir had forbidden Dayan to give, fearing that it would be too negative. Speaking in English, Yariv began by pointing out that for the previous six years there had been 'a gigantic flow of armaments' from the Soviet Union to Egypt and Syria; that Israel had done her 'level best' to counter this influx both by her own production and 'with getting what we could from our friends'; but that when war came the discrepancy in material was a 'pronounced' one.

The Syrian army, Yariv warned, was not yet broken. Israeli casualties had been high. 'But we have dealt them a very, very severe blow.' Israel would make it clear 'that the game of breaking the cease-fire is a two-sided one, and we are going to press and we are going to push and we are going to bomb and we are going to punish them as much and as long as we can until the other side does understand the rules of the game'. The situation on the Syrian front, was, he added, 'stable', even though 'many things can still happen'.

With regard to the battle in the south, Yariv confirmed the morning's news of an Israeli line several miles east of the canal. He called it 'a firm base for operations'. But he went on to warn, 'There is still a way ahead of us which will not be easy in which we will have to do a lot of fighting, in which our nerves will be probed and tested.'

Yariv then invited the journalists to question him. 'And now ladies and gentlemen', he said, 'the Syrians have fired, the Egyptians have fired, we have fired, and now you fire.' The journalists began their questions. During his answers, Yariv told them, 'Israel is insecure as long as we have not licked the Egyptians.'

On October 9 a former Chief of Staff, General Bar-Lev, who was Golda Meir's Minister of Trade and Industry, was asked by General Elazar to go to the Southern Front and to take charge as the personal representative of the Chief of Staff. Finding that General Sharon wanted to take his own military initiatives in advancing against the Egyptians Bar-Lev went so far as to recommend that Sharon be relieved of his post. Dayan vetoed the proposal, arguing that it would create 'unnecessary internal political problems'; Sharon was the most prominent field commander identified with the opposition Parties in the Knesset.

During the afternoon of October 9 there was concern in Israel that King Hussein might join in the attack. But the bridges over the Jordan remained open, and, as in peacetime, Arab residents of the West Bank moved freely across in both directions: a thousand were said to have done so on the previous day.

That afternoon, the Egyptian attempt to reach Ras Sudar, on the Gulf of Suez, was finally repulsed. But in a private briefing at 5.30 p.m., Yigael Yadin spoke of how difficult it was proving to bring down the Russian-made anti-aircraft missiles that were providing a 'protective shield' for the Egyptian troops on the Sinai bank of the canal. These missiles, he explained, were easy to operate. Manned by a small crew, they used a radar beam to lock on approaching aircraft, and they were very mobile. 'Of course,' Yadin adds, 'any Egyptian tanks that try to advance into Sinai beyond the protection of the radar screen are vulnerable to Israeli planes. But under the screen they are very hard to dislodge.'

On the Golan Heights, during the morning of October 10, the Israeli counter-attack reached the lines from which the Syrians had launched their offensive four days earlier. By midday there was not a single Syrian tank in fighting condition west of that line. 'The Hushniya pocket,' writes Herzog, 'in which two Syrian brigades had been destroyed, was one large graveyard of Syrian vehicles and equipment.'

General Elazar was prepared to push beyond the line, deep into Syria, but, at a General Staff conference that evening, Dayan hesitated. Damascus was only thirty miles to the north-east. Were the Israeli units to get too close to it, he argued, it might draw the Soviet Union into the war. Elazar countered that for security reasons Israel should advance about 12 miles, to a point where it could establish a new defensive line, while having the ability to bombard Damascus with artillery fire, should the Syrians try to renew the offensive.

To resolve the dispute, Dayan took Elazar to see Golda Meir. It was she who, after listening to the arguments on both sides, decided in favour of attacking across the 1967 cease-fire line into Syria.

At 8.30 p.m. on October 10, Golda Meir spoke on television to the Israeli people. Her face lined with worry, her head bowed, she spoke bitterly about Soviet arms supplies, telling her listeners that, even as she spoke, 'Russian missiles of the latest type are being flown into Syria continuously', and that both the Syrians and the Egyptians were using the most modern weapons. 'Everything that is in the hands of the Syrian and Egyptian soldier', she said, 'all this comes from the Soviet Union.'

Golda Meir then appealed to King Hussein of Jordan not to join the attack. Surely, she asked, he had learnt his lesson in 1967, when he was 'goaded' to attack by false tales from Egypt that the defeat of Israel was imminent, and when he lost, as a result, the West Bank of the Jordan and East Jerusalem. Israel, said Mrs Meir, would not allow itself to be wiped out. The Jews must not allow themselves the 'luxury' of despair. 'I have but one prayer in my heart, that this will be the last war.'

An hour after Golda Meir had spoken, and having talked to a number of people who were equally dependent on news broadcasts and television for their knowledge of what was happening, the English visitor wrote in his diary:

> There is great gloom everywhere. The confident expectations of Monday have simply faded away.
>
> Everyone realized that Egypt and Syria have been equipped by Russia to such an extent that Israel was outnumbered in armour by at least four to one. Some people realized that the quality of the Russian equipment was far superior to what Russia had provided before, and superior to the Israeli armament.
>
> But what has shaken people now is that Russia is still, while the battle rages, sending in massive supplies – the most modern tanks, the most effective anti-tank missiles, and aeroplanes.
>
> It looks as if any Israeli success on the battlefield will be offset by material flown in from Russia during the morning.

*   *   *

The morning of October 11 brought news of an Egyptian bombing raid on Abu Rudeis, their former oil installations on the Gulf of Suez which Israel had captured in 1967. Casualties had been inflicted, the 8 a.m. news bulletin reported, but the oil installations had not been hit. In the Mediterranean, Israeli warships had sunk two Syrian missile boats off the Syrian port of Latakia.

The Israeli attack from the Golan Heights into Syria – to which Golda Meir had agreed on the previous evening – was launched on the morning of October 11. In the northern sector of their front, as elsewhere, the Syrians defended their positions fiercely, but were driven back. The commander of the Syrian forces there, Colonel Rafiq Hilawi, was a Druse. A few days later, wrote Herzog, he was 'paraded in a camp near Damascus: his badges of rank were torn off him as, with eyes blindfolded, he faced a firing squad. He had been court-martialled and sentenced to death for withdrawing, his guilt having been compounded by the intense suspicion with which the Syrian regime regarded the Druse people.'

A Syrian counter-attack in the northern sector was, briefly, successful, but then the Israelis advanced again. The attack on the Druse village of Tel Shams was led by Lieutenant-Colonel Naty Yossi – the officer who had hurried back from his Himalayan honeymoon on the first day of the war to be with his men. 'Covered by a heavy artillery bombardment,' Herzog wrote,

'Yossi led his small force and stormed Tel Shams, with two tanks covering the attack and six attacking.' As they neared the top of the hill, 'a hidden anti-tank battery opened fire, destroying four of the attacking tanks. Yossi himself was thrown out of his tank and lay wounded among the rocks. The covering force on the main Damascus road endeavoured to extricate Yossi but failed. Ultimately under cover of darkness a special paratroop unit, led by a young officer called Yoni, made its way through Syrian-occupied territory and, in a dramatic rescue operation, evacuated Yossi from under the noses of the Syrian forces on Tel Shams.'

The 'Yoni' of the rescue operation at Tel Shams was Yoni Netanyahu, who, three years later, was to lead the Entebbe rescue raid. During the October War, Yoni Netanyahu also led an assault with eight men against twelve Syrian commandos, all of whom were killed. 'This is a scene I shall never forget,' his deputy at Entebbe later recalled, 'Yoni attacking, shooting and leading his men into battle, leading them, not giving orders from behind.'

At eight in the evening of October 11, an Israeli news broadcast revealed that Israeli forces were ten kilometres inside Syria, 'on the road to Damascus'. Two hours later, Dayan spoke on Israeli television. He was in a more pugnacious mood than hitherto, telling his listeners: 'The Syrians must learn that the road from Damascus to Israel is also the road from Israel to Damascus.'

In the twenty-four hours ending on the morning of October 12, eighty Soviet Antonov transport aircraft landed in Syria, flying in arms and missiles. That morning, President Bourghiba of Tunisia urged his fellow Head of State, King Hussein of Jordan, to join the battle against Israel. But Hussein, who had lost the whole of the West Bank and all of East Jerusalem to Israel in 1967, made no move. It was the Iraqi government that, outside the direct combatants, was making the largest contribution to the Arab war effort. As many as 15,000 Iraqi troops had reached the Syrian front.

Towards sunset on October 12, as the second Sabbath of the war was about to begin, a doctrinal issue came unexpectedly to the fore. The Sephardi Chief Rabbi, Ovadia Yosef, announced that it was wrong to bake bread on the Sabbath, even in wartime. The Ashkenazi Chief Rabbi, Shlomo Goren – who in 1967 had been Chief Rabbi to the Forces – disagreed. It was absolutely in order, he said, to break the Sabbath rules in wartime by baking bread.

It was on October 12 that David Elazar and Chaim Bar-Lev explained to Dayan their plan for crossing the Suez Canal. Dayan, so often the voice of caution in the previous week, opposed the plan. Driving the Egyptians back across the canal, he said, 'would not decide anything, nor would it bring the Egyptians to ask for a cease-fire'. Elazar persisted that the attack should be made. The dispute was brought before a meeting of various Cabinet Ministers, chaired by Golda Meir. While the meeting was in progress, Intelligence information was received that Egyptian tanks and armoured vehicles were crossing the canal from west to east, constituting an added danger to the Israeli positions. It was obvious that this new Egyptian move had to be countered before any Israeli attempt to cross the canal could be made.

The Egyptian reinforcements began to reach the eastern bank of the Canal on October 13, and to move forward to attack the Israeli positions. At six o'clock on the following morning they launched their main attack. Their objectives were the Mitla and Gidi Passes, and the eastern side of the Gulf of Suez (controlled by Israel since 1967) as far south as Ras Sudar. The Israelis strained all their resources to prevent this Egyptian breakthrough. In the tank battle which followed, more tanks were employed – 2,000 in all – than in any of the great tank battles of the Second World War with the sole exception of the Russo-German battle at Kursk in 1943.

On General Sharon's front, where in places tank fought tank at the incredibly close range of a hundred yards, 110 Egyptian tanks were destroyed, and the Israeli line was unbroken. Further south, on that part of the front held by General Kalman Magen – who had replaced General Avraham (Albert) Mandler, killed by an Egyptian shell – 60 Egyptian tanks were destroyed. The Mitla and Gidi Passes remained under Israeli control. In announcing General Mandler's death, Israel radio stated that no other names of those killed would be released.

In all, 264 Egyptian tanks were knocked out that day. Israeli losses totalled only ten tanks. Recognizing that the Egyptian losses would have been a blow, not only to their fighting ability but to their morale, Elazar issued orders for an Israeli crossing of the canal the following night.

In the north, Israeli forces continued to advance deeper into Syria throughout October 13. An Iraqi brigade, which had been hurried forward into the battle, and for some hours threatened to break the flank of the Israeli advance, was annihilated, and every one of its eighty tanks destroyed. A second Iraqi brigade, which had been ordered to come to the assistance of the first, was unable to do so: an Israeli commando unit which had infiltrated 100 miles east of Damascus, on the Damascus–Baghdad road the previous night, destroyed the first tank of the second brigade, causing momentary havoc, and blew up the road bridge along which the rest of the force would have to pass.

Shortly after midday, Radio Amman announced that Jordan had sent some of its 'élite forces' to the Syrian front. Two minutes after the Jordanian radio announcement, Israeli air raid sirens sounded over Jerusalem. 'Can it be the start of a Jordanian attack?' The British civilian observer asked in his diary, 'Or is it Iraqi planes flying south from Syria?' It was in fact a false alarm: a Jordanian plane had been seen on Israeli radar flying near the border, but did not, in fact, fly across it. Jordan would help Syria to defend Damascus, but it would not declare war on Israel, or risk a direct confrontation along the common border.

As Soviet supplies continued to fly into Damascus – 'as if it were Stalingrad and not Damascus', commented Yigael Yadin – the Israeli government decided that it would be too great a risk of war with Russia if it were to shoot down these planes in the air. But two of the Antonov transport planes were destroyed on the ground. The Israeli argument was that, as Syria was at war with Israel, she could not expect her airfields to be free from attack. That day, the Israeli advance in the north was so rapid that Israeli artillery was able to shell Syrian army barracks in the suburbs of Damascus.

The Soviet Union viewed Israel's advance towards Damascus with alarm. On October 11 the Soviet Ambassador to the United States, Anatoly Dobrynin, indicated to the American Secretary of State, Henry Kissinger, that Soviet airborne forces were already on the alert to move to the defence of Damascus. Soviet warships were ordered to the Syrian coastal towns of Latakia and Tartus. The Chairman of the Soviet Communist Party, Leonid Brezhnev, sent a message to Houari Boumedienne of Algeria, reminding him of his 'Arab duty'. Soviet tanks were taken by sea from Yugoslavia, across the Mediterranean to Algiers, for Algerian units that were to be sent to the Egyptian front.

On the Golan Heights, the Israelis continued to advance, the troops under Major-General Dan Laner reaching to within eighteen miles of Damascus, just south of the village of Knaker. But on the following morning, October 12, they were confronted by the first fifty of more than 300 Iraqi tanks which had been brought across the desert on transporters and were being sent into action against the southern flank of the Israeli thrust. The Israelis waited until the lead tanks of the first assault were a mere 300 yards from them, and then opened fire. Seventeen tanks were knocked out, and that assault came to an end. General Laner had only a few hours before the bulk of the Iraqi tanks would attack.

The Iraqi tanks advanced after darkness on the evening of October 12. Then they halted at the very edge of an ambush that General Laner had prepared for them between Nasej and Tel Maschara. But throughout the night the Iraqis made no further move. It was as if they knew of the trap that awaited them. Then, at three in the morning of October 13, they advanced again right into the trap. When the Iraqi tanks were only 200 yards from the front line of Israeli tanks, the Israelis opened fire. Eighty Iraqi tanks were destroyed, and the Iraqis pulled back. Not one Israeli tank had been hit.

During the fighting on the Golan that day, the Israelis captured several Soviet-made Sagger anti-tank missiles, intact. Each one was about two and a half feet long, gleaming like a child's toy. It was being used by infantry units, and could be operated by a three-man team. To be fired, it did not need either a vehicle or a special launching site. Each system consisted of four missiles, with a separate control box and control stick. The movement of the stick altered the aim of the missile. It had a range of 3,000 metres. During its flight, the missile was directed by a tracer, which emitted an infra-red light guiding the missile to its target. It had a speed of 110 metres per second. The warhead had the power to break through the armoured plate of a tank, and then to shoot out splinters and shrapnel inside the tank, caus-ing devastation. It had never been used, or seen, in war before October 6. The Israelis had not yet found a way to neutralize it.

On October 13 President Nixon, after much hesitation, finally ordered an airlift of United States military supplies that would enable Israel to sustain its subsequent military initiatives in the face of overwhelming odds.

At 8.30 on the evening of October 13 Golda Meir spoke on television for the second time in three days. The occasion was a press conference for foreign journalists, and she spoke in English. In her answers to questions, she stressed Israel's 'hatred of war and hatred of violent death'. After emphasizing her government's desire for direct negotiations with the Arabs, she said, 'And here we are again, in a war which we did not want, which we did not initiate.' She went on, speaking in a sombre but not frightened voice, 'We know that giving up means death, means destruction of our sovereignty and physical destruction of our entire people. Against that we will fight with everything that we have within us.'

On Sunday October 14 the Israeli newspapers reported that, according to a source in Washington, during the past week the Soviet Union had flown at least 1,000 tons of weapons and ammunition to Egypt and Syria. Among the war materials flown in were surface-to-air missiles and anti-tank weapons. On their flight path to the Middle East, the Soviet planes had landed at airports in Yugoslavia. Some Soviet aid was also arriving by sea; a Soviet freighter had been sunk in the Syrian port of Tartus.

At 5 p.m. that Sunday, Israel radio announced the first casualty figures since the war had begun eight days earlier. The number of Israelis killed in action stood, thus far, at 656. 'Everywhere I go there is deep gloom,' the British observer noted in his diary. 'In eight days, Israel has lost on the battlefield more than 10 per cent of the total Israeli war dead in all eight months of fighting in the War of Independence; and almost as many dead as in the Six Day War. And the fighting goes on, on both war fronts.'

As almost every able-bodied postman was among the reservists called to action, schoolgirls were asked to go to the post office and collect telegrams for delivery. On the evening of October 14, the observer noted: 'Met a schoolgirl volunteer. Today she found herself delivering a telegram announcing "Your son is dead". She watched in horror as the bereaved mother beat her head on the floor in grief.' Within a few days the delivery of such telegrams was entrusted, wherever possible, to women who had lost either a husband or a son in earlier wars.

That night it was announced over Israel Radio that the Egyptian offensive on the Suez Canal on the previous day had been repulsed, that more than 200 Egyptian tanks had been destroyed, and that as many as 1,000 Egyptian soldiers had been killed during the day's fighting. No mention was made, of course, of the imminent Israeli plans to cross the canal from east to west, a military operation filled with tension and danger.

In preparation for the crossing, a 190-yard long pre-constructed bridge had been assembled in Sinai and was being brought across the sand on rollers, pushed forward by a dozen tanks. The initial crossing of the canal was entrusted to General Sharon. The first troops to go forward towards the canal were those commanded by Colonel Amnon Reshef. As he moved forward at 4 p.m. on October 15, he was at first successful in finding a way towards the canal that avoided any Egyptian troop concentrations. But, Herzog has written:

> Unknown to Reshef, his force had moved into the administrative centre of the Egyptian 16th Infantry Division, to which the 21st Armoured Division had also withdrawn after being so badly mauled on October 14.
>
> His force found itself suddenly in the midst of a vast army with concentrations of hundreds of trucks, guns, tanks, missiles, radar units and thousands of troops milling around as far as the eye could see. The Israeli force had come up through the unprotected southern flank of the Egyptian Second Army at the junction of the Egyptian Second and Third Armies – had entered by the 'back door', as it were – and had suddenly found itself plumb in the centre of the administrative areas of two Egyptian divisions and literally at the entrance to the 16th Infantry Division headquarters.
>
> Pandemonium broke out in the Egyptian forces. Thousands of weapons of all types opened fire in all directions, and the whole area seemed to go up in flames.

By 9 p.m., with the Israeli forces still not at the canal, there were many Israeli casualties, and the Egyptians were planning a counter-attack. Some of the Israeli battalions were down to a third of their strength. Parallel to the canal lay the Lexicon road. Between the Israeli forces and the canal lay the Missouri fortified position. By 10 p.m. the battle was raging for both of them. Some Israeli units were pulling back. Herzog wrote:

> The scene in the area was one of utter confusion: along the Lexicon road raced Egyptian ambulances; units of Egyptian infantry were rushing around in all directions, as were Egyptian tanks.
>
> The impression was that nobody knew what was happening or what to do. On all sides, lorries, ammunition, tanks, surface-to-air missiles on lorries and radar stations were in flames in one huge conflagration which covered the desert. It was like Hades.

Days later, the entire area between the Canal and 'Missouri' was to appear from the west bank as one vast, eerie unbelievable graveyard. As a background to this scene, the concentrated forces of artillery on both sides fired with everything they had.

Again and again, the reserve paratroop battalion under Reshef attacked the crossroads, the paratroopers not realizing that they were attacking a major, concentrated Egyptian force of at least a division in strength.

During the attack, part of the paratroop force was trapped and, despite several attempts to extricate them, was wiped out.

That night, the Israelis evacuated their wounded. Another attack by Colonel Reshef at four in the morning was a failure. At dawn on October 16 he moved to high ground to survey the scene of the night's battles. 'In all directions,' Herzog writes, 'the desert was covered with a vast fleet of burning and smoking tanks, vehicles, guns, transporters. Dead infantry lay everywhere. It seemed as if there was not a single item of military equipment that had escaped destruction: there were SAM-2 missiles, mobile kitchens.' The remnants of the Israeli forces were there too, 'and frequently the distances between them and the Egyptians were no more than a few yards'.

More than 150 Egyptian tanks had been destroyed during the night's battle. But Sharon's division had lost more than 300 men killed, and more than eighty tanks. The Israeli commanders were determined, however, to cross the canal, and called forth the greatest exertions from their men. The actual task of crossing the canal was given to a parachute brigade that was then at the Mitla Pass, commanded by Colonel Danny Matt, who moved forward with his men on the evening of October 15.

It was at 1.35 a.m. in the early hours of October 16 that the first Israeli troops crossed the Suez Canal. Within seven hours a three-mile bridgehead had been established.

The news of the crossing of the canal was kept a close secret from the Israeli public, as the battle to hold and enlarge the bridgehead on the western bank began. Other concerns crowded in on the cities on October 15, as the war entered its ninth day. Food stocks had begun to run low, there being few spare trucks to take food from the farms and supply centres to the cities and local shops. A radio announcement at nine in the morning of October 15 reported that there would be no more cottage cheese available 'until the end of the war', and that, the blackout having disrupted chicken batteries, egg-laying had been brought almost to a halt.

At the Israeli Foreign Ministry, the concerns were of a less domestic nature. 'Will Hussein feel that he has to attack before the Syrians are defeated?' asked one senior official. France's role in supplying Israel's enemies was also the subject of angry Foreign Ministry comment that day. In Sinai, the Israeli air force had shot down two Mirage fighters which, it emerged, had been sold by France to Libya and then sent by Libya to the Egyptians. But the greatest anger among the diplomats was reserved for Britain, which had announced that it would withhold military supplies 'equally' between the belligerents. This embargo meant that crucial supplies for the Centurion tanks which Israel had bought from Britain would not be sold to Israel. But Jordan, the only other country to suffer from the Centurion ban, was not a belligerent, so that the British restrictions on Jordan did not affect in any way the military potential of Israeli's attackers.

The continuing training of Egyptian helicopter pilots in Britain was also held up as a violation of the British embargo. When the British Foreign Secretary, Sir Alec Douglas-Home, was asked about this by the Israeli Ambassador in London, he replied that it was better for Israel that these Egyptian pilots should be in England than at the front.

At General Headquarters it was calculated that Israeli armoured forces would be able to cope with a preponderance of four to one; that forty Egyptian tanks could be knocked out by ten Israeli tanks, and that this ratio 'is being maintained'. The arrival of military supplies to Israel from the United States was also judged 'very good'; this was despite the British government's refusal – Edward Heath was then Prime Minister, at the head of a Conservative administration – to allow the American cargo planes to land and refuel at the British sovereign air base at Akrotiri in Cyprus. The Portuguese government had stepped into the breach.

More alarming to General Headquarters, the amount and quality of Soviet war supplies to Egypt and Syria 'does not seem to diminish'. As of October 15, more than a hundred Soviet troop transporters had brought tanks and missiles to Syria; Soviet ships had also been used for the transport of arms. In Sinai, the most recent Egyptian thrust had been backed by Soviet weapons and tanks of the most advanced type. Russia's SAM-6 surface-to-air missiles along the whole western bank of the canal had provided Egypt with a protective 'umbrella' stretching 20 miles eastward into Sinai. Throughout this missile band, Israeli war planes were at risk.

It was a priority of the Israeli troops that had just crossed the canal to try to destroy as many of these missile bases as possible. To this end, a force of twenty tanks, commanded by Colonel Chaim Erez, set off in the early hours of October 16, driving west of the canal to search out and destroy the missile bases which made it almost impossible for Israel to secure mastery of the air above the canal. East of the canal, however, the Egyptians were still fighting for the fortification known as Chinese Farm (Japanese military instructors had taught there before 1967; their wall signs had appeared Chinese to the Israelis who were stationed there after 1967).

If the Egyptians could not be dislodged from Chinese Farm, it was difficult to see how the Israelis could remain on the far bank of the canal. The pre-constructed bridge had not yet arrived. Dayan, visiting the southern front, proposed pulling back the paratroopers from the west bank. 'We tried,' he said. 'It has been no go.' The crossing should be given up, he said, otherwise, in the morning, the Egyptians would 'slaughter' those Israelis on the other side. Sharon, however, argued that even before the bridge could be thrown across the canal, men should be sent across on rafts, and the troops on the western side ordered to push on. Bar-Lev turned down Sharon's proposal.

Meanwhile, as fierce fighting continued on the eastern side of the canal, and the bridgehead was maintained on the western side – but under continuous and intense Egyptian artillery fire – the pre-constructed bridge was being brought slowly but steadily forward on its rollers.

By midday on October 16, the unit on the western bank of the canal commanded by Colonel Erez – 20 tanks and 7 armoured personnel carriers – had destroyed 2 Egyptian surface-to-air missile bases, as well as 20 Egyptian tanks and 12 armoured personnel carriers, for the cost of one man wounded, and had reached a point fifteen miles west of the canal. On the eastern side of the canal, it took two days of intense fighting before the Egyptians were driven from the Chinese Farm. That day, the Egyptian Commander-in-Chief, General Shazli, was relieved of his duties by President Sadat, who at the same time asked the Soviet Prime Minister, Alexei Kosygin, who was in Cairo on October 16, to convene the United Nations Security Council and to seek a cease-fire.

The Soviet Union had watched with growing alarm the failure, and the destruction, of the massive amount of armaments that it had sent – and was continuing to send – to Egypt and Syria. Inside Israel, anti-Soviet feelings were high. A Foreign Ministry official, Aluf Har Even, put it strongly and succinctly on October 16, in a message to be passed back to the British government. 'Since 1955,' he said, 'the Soviet Union has effectively closed the option for peace in the Middle East. Each of the four wars was caused by this fact: the war of 1956, after which we bowed to American pressure to return to the insecure frontiers; the war of 1967, when we were blockaded, and only averted a massive attack by our pre-emptive strike; the War of Attrition, in which through continual frontier bombardments more people were killed than in the Six Day War; and now this war.' Har Even continued:

The Russians gave the Arabs so many arms, arms of such a massive scale, that the Arabs said: 'Why should we negotiate with Israel when we can always make war against her?' Thanks to Russia, and only to Russia, Egypt had 4,900 tanks at the beginning of this war: Soviet tanks with Soviet ammunition brought on Soviet ships.

How could Russia expect Egypt to stay quiet after giving them armaments on this scale? In June 1941, the Germans deployed 1,400 tanks against Russia – on a one thousand mile front. Yet even the Syrians on their small front of fifty miles in the Golan, had over 1,200 tanks at the beginning of this war – again, given to them by Russia.

The question I am asking is, what role does the Soviet Union wish to play here in the Middle East? At least, in 1967, you could argue that the Cold War was still going on.

But now, in a world of detente, what possible reason can you give for sending, as if on a conveyor belt, these terrible weapons of war to nations who could gain at the negotiating table far, far more than at the battlefield? How can Russia say that she is for peace, and at the same time send over 5,000 tanks to an area in conflict? I can also tell you that since the outbreak of the war, over 100 Soviet planes have been flown into Syria carrying tanks and missiles, and now they are sending ships as well.

In Israel, Har Even continued, 'we are convinced that the Soviet Union knew the war was coming; that they did nothing to stop it; and that from the moment it started they began to fuel it, and continue to fuel it even now, while we are talking here'. What were the concrete advantages that the Soviet Union had gained here in the Middle East since 1955? 'I think it is very little. What have they got in exchange for their vast military expenditure – over ten milliard dollars. Both in 1967, and in 1970 their action got them into a direct confrontation with the United States. In 1967 they had a terrible loss of face. We captured great amounts of their equipment. Their involvement was exposed. Now too, the war may not end in a very favourable way for them. So what is it that they stand to gain by stoking once more the cruel fires of war?'

Har Even went on to point out, in tones of deeply felt sorrow:

> It is not after all, the life object of young Israeli or Egyptian students to die in the desert, or be blinded, or mutilated for life by Soviet missile. Instead of sending 5,000 tanks to Egypt why not send 50,000 tractors – or leave Egypt to be looked after by the wealthy Arab states, and provide the tractors instead for their own agriculture? In fact, the cost of 5,000 tanks is far more than that of 500,000 tractors, and I am not talking at all about the added cost of ammunition, and fuel, and human life.
>
> The Soviet action in sending arms on a scale hitherto unknown in war to Egypt and Syria is totally irresponsible, on a purely human level, quite apart from the international implications. Egypt and Syria are essentially poor countries. There is a shocking contradiction between the extremely expensive arms provided by the Soviet Union, and the desperately poor villages through which these arms are sent.
>
> Then there is the terrible wastage on the battlefield. For what? Is this the way a responsible world power should operate, turning an area that desperately needs tractors, irrigation, seeds, and saplings, a share of technical skills, a pool of agricultural knowledge into a cockpit of bloody war?

The Soviet Union must have known, Har Even pointed out, that all the arms and weaponry it was sending to Egypt and Syria 'would seem an invitation to the Arabs to turn their propaganda and their hate against Israel, and to prepare for war'. Before the war began, Egypt had a standing army of over 800,000 men, 'men who could have been cultivating the fields, harvesting the waters of the Nile, studying peaceful techniques – all of them under arms twenty-four hours of the day, and taught to look forward to war with Israel'. Har Even's plea for British involvement continued:

> We have always refused to have a large standing army. We do not want to be a military nation. America supplied us with arms – albeit on a far smaller scale than the Soviet supplies to the Arabs – knowing fully that Israel would not use those arms to open war. We knew that we would lose hundreds of lives by allowing the Arabs to attack first on the Day of Atonement. But we decided against it.
>
> We lived in the hope that they would not really attack, that they would see the futility of seeking a military solution, that they would shun the waste and slaughter which war always brings with it. We were wrong. They did choose to attack. But I am still certain we were right not to try to anticipate or provoke the attack in any way.
>
> If only the Soviet Union could feel that there was strong pressure upon them to behave in a responsible way in the Middle East. They have encouraged two poor nations to go to war, and if their policy is not changed, there will be protracted war, fed by the Soviet Union. I ask myself, if the Soviet Union operates against the spirit of detente in the Middle East, in a year's time they could do it elsewhere. If they exploit poor countries here and feed them with weapons of war, they could do the same anywhere in the world.

In a conversation that afternoon, Shimon Peres, the Minister of Transport and Communications, reiterated this theme. 'We must try to persuade the Russians to stop pushing the Arabs forward. If the Arabs would regain their prestige and honour, then I say, thank Heaven. But if the Russians continue to provide arms – and a ship with a hundred Soviet tanks has just arrived in Latakia – then we will have to think now of yet another war, the fifth war. With so many planes and tanks being sent in every day, a fifth war might not be so far away, even though we are still in the middle of the fourth one.'

Peres was also worried on October 16 that if the Jordanians 'see us hesitate on the battlefield, they will surely come in against us'. The 'whole Arab world', he pointed out, had mobilized against Israel. 'King Feisal has put on the table a cheque for a billion dollars.' The 'unholy alliance' of the Arabs and Russia had not really been challenged by either the United Nations or Britain. Peres added, laconically, 'As someone said to me this morning – a left-winger, no less – "Thank God for the reactionary United States".'

That evening Golda Meir spoke in the Knesset. She too praised American magnanimity in providing supplies when they were most needed, and damned British hypocrisy in maintaining the arms embargo. The death of every individual soldier, she said, 'affects everyone in the State'. But Syria and Egypt had not yet been 'hit hard enough' to accept a cease-fire. They still felt that they could defeat Israel. 'Their true objective is to destroy the State of Israel altogether.' Their demand for the pre-1967 borders was 'ridiculous'. In an interview with the Israeli Ambassador to Britain, however, the Foreign Secretary, Sir Alec Douglas-Home, said with vigour, 'We told you so. You asked for it. I forecast that with the present frontiers the Arabs would have to get back their lost lands.' Britain was proposing that the European Common Market unite behind the call for a cease-fire, on the basis of the pre-1967 borders, and with a British peace-keeping force.

On the Golan Heights, two tank battles were fought on October 16. One was at Tel El-Mal, in the very region where the Iraqis had been beaten three days earlier. This time it was Jordanian tanks – British made – that, having come to the aid of Syria, advanced against the southern flank of the Israeli bulge towards Damascus. In the first moments of the battle, twenty-eight Jordanian tanks were hit, and the Jordanian brigade withdrew. Then, four miles to the east, in a move that was apparently not coordinated with the Jordanians, the Iraqis moved forward again, from Kfar Shams to Tel Anter, hoping to break into the Israeli bulge at Tel El-Alaliekh. Another tank battle ensued. After sixty Iraqi tanks had been destroyed, the Iraqis withdrew. The Israeli bulge, which had held for two days only eighteen miles from Damascus, was undented.

The Israelis in the bulge suffered a setback on the following night, however – October 17 – when eight Syrian tanks, equipped with optical night-finding equipment – the latest Soviet technology – attacked Israeli troops who had captured the village of Um Butne, which formed a small Syrian indentation on the southern flank of the Israeli bulge. The Syrians were driven off, but there had been many Israeli casualties.

Throughout October 17 Soviet war supplies continued to reach Syria and Egypt by air: 200 plane loads in the previous seventy-two hours. American supplies to Israel were likewise continuous. Neither superpower had sent troops, but it was made known that day that Henry Kissinger had warned the Soviet Union that if it were to send troops to the region, the United States would do likewise.

Not for the first time, the talk in Israel was of the lack of preparedness on the first day, and of bitterness against the government for not having been prepared. This bitterness had become a daily feature of conversation, swelling in volume and intensity. On October 17 a British writer, David Pryce-Jones, told a group of Israelis in Jerusalem about the Israeli soldiers on the first day of the war in the north of which he had been a witness. 'They came streaming off the Golan Heights,' he said, 'without even their weapons, angry because the Israeli government allowed their country to be caught unawares and unprepared.'

That day, in Sinai, sixty Egyptian commandos, members of a number of units that had tried to operate behind Israeli lines, gave themselves up. They had run out of water. They came out of the desert and surrendered.

On the evening of October 17, Uzi Narkiss gave a briefing for the foreign press corps in Tel Aviv. Twelve days earlier he had been Director-General of Immigration. Six years before that he was one of the commanders in the Six Day War. He spoke of how the Israeli Task Force on the western bank of the Suez Canal had managed to bring its tank fire to bear on concentrations of Egyptian artillery, anti-aircraft batteries, ground-to-air missiles, command posts, installations, and infantry units. During the day, there had been Egyptian counter-attacks 'on both sides of the canal', all of which had been broken, and 110 Egyptian tanks destroyed. Twenty Egyptian aeroplanes had also been shot down.

On the evening of October 18 the pre-fabricated bridge reached the Suez Canal. Shortly after midnight it was operational, though subjected to repeated Egyptian air and artillery attacks. Of the task force under Colonel Jackie Even that had brought the bridge forward, 100 men were killed – forty-one in one night alone.

Even after the bridgehead had been established on the western side of the Canal, substantial Egyptian forces remained on the eastern bank. For the Israeli troops there was as no swift advance and no clear cut sweep to victory as there had been in 1956 and 1967. Unlike those two wars, the October battle was one in which, at many moments, the tide could have turned against the Israeli forces. It was only a sustained and unremitting effort that enabled them to push the Egyptian and Syrian tanks and infantry back to a point where they could no longer threaten to break through the Israeli lines, and where the Egyptian forces still on the eastern bank were powerless to take offensive action. The talk inside Israel had become increasingly critical of the government, both among civilians still dependent on newspapers, radio and television for their news, and among the soldiers on the battlefield, some of whom were now beginning to have brief periods of leave at home. On the morning of October 19 an Israeli accountant, Dan Bawli – whose father, Lazar, had been one of the driving forces of Jewish settlement in the Jezreel Valley in the 1920s – and who, as a reservist, had been mobilized by the Army Press Office to brief journalists during the war, spoke with passion at a briefing in the Kibbutz Ginossar guest-house on the Sea of Galilee. He said that combat soldiers were saying to him, 'First we shall settle with the Syrians, and then we shall settle with our leaders'. He then spoke of the 'great bitterness' among the soldiers because the government had failed to mobilize well in advance, or at least to reinforce the frontiers. He had heard one soldier call the war 'the war of the government's cocksureness'. Major

Bawli added, 'I tell you when the war is over, then the struggle will begin inside Israel. We were really eroding before the war. The fat around us, it was awful. Thank God the Arabs did not wait for another two or three years, our alertness might have been completely eroded away.' When the war is over, Bavli declared with vigour, 'we shall have to build a new Israel'.

That day, on the Golan Heights, Israeli forces took Mazraat Beit Jan, which had considerable water resources, and helped lessen the logistical burden of holding the 'fist' (as it was called) that had been punched into Syria towards Damascus. The Israeli front line reached as far as the outskirts of the Syrian town of Sassa, less than twenty miles from Damascus. There were no plans to take the town, which was heavily fortified; nor did it seem wise to provoke the possibility of direct Soviet intervention by trying to push any closer to Damascus. Israeli troops were also only five miles from the road running south from Damascus into Jordan. Here too the advance had halted.

During October 19 there was sustained Syrian artillery fire, especially on the Israeli-held section of the road in front of Sassa, as Syrian, Iraqi and Jordanian troops attempted to break into the 'fist' from the north and south. To support the attack, Syrian jets continued to fly over the Israeli lines – almost all of them were shot down. By two in the afternoon the Israelis had beaten off the attack. Both during the battle, and after it, Israeli jets bombarded the former Israeli observation post on Mount Hermon which the Syrians had seized on the first day of the war and from which Syrian observers still maintained a commanding view of the battlefield.

News of the Syrian execution of the Druse commander, Colonel Rafiq Hilawi, had reached the Druse villages of the Golan on October 18. It was the belief of the Druse inside Israel that his execution was not because of any failure on his part on the battlefield, where he had commanded a Syrian brigade, but as revenge for the support given by the Druse inside Israel to the Israeli war effort. That day the Syrians had bombarded the Druse village of Majdal Shams, on the slope of Mount Hermon, and ten villagers were killed.

During October 19 an Israeli officer who visited the nearby Druse village of Masada, a few miles west of the 'fist' – to the background booming of heavy artillery, the villagers were harvesting the grapes of their late harvest – recalled that in 1967 Yigal Allon had wanted the Israeli forces that had captured the Golan to advance eastward as far as the Jebel Druse mountains, where the main Syrian Druse population lived. 'Perhaps we should go in now,' the officer commented, 'and set up a Druse State, as an independent buffer State between us and Syria. The Druse would welcome independence. We could protect them.' It was scarcely thirty miles across the desert from the 'fist' to the Druse mountains.

From across the Lebanese border, firing had continued each night since the start of the war on October 6. This was the attempt by Fatah to show that it had participated in what its spokesmen called 'the Holy War against the Jews'. No serious damage was done, but the Israeli farmers and villagers in the northern settlements, and in the town of Kiryat Shmona, were forced to spend their nights in underground shelters. The shelling from 'Fatahland' added to the Israeli sense of having the whole Arab world against them. The Lebanese government, however, maintained a strict neutrality.

Even as the war was still being fought, its effect on Israeli life and morale was being much discussed. A reservist captain, called back from civilian life to serve as a liaison officer with foreign journalists, commented on October 19 – as the day's battle on the Golan Heights died down – that he hoped a cease-fire would come into effect soon: 'I would rather spend six years talking, at least six years, than go to war again,' he said. 'Let the talking be a protracted business, so what?' and he went on to explain:

> Israel has a standing army of about 80,000 men. Yet to fight a war she must mobilize over 300,000, and even then she is outnumbered numerically, perhaps by as much as two to one. Once mobilized, as now, the civilian life grinds almost to a halt, and Israel's workers, her students, her shopkeepers, put on their uniforms and go to the front.
>
> Israel cannot be mobilized twelve months of the year. A state of war eats into our daily life. Mobilization throws a terrible burden on our economy. We are a civilian nation. We want peace with our neighbours, because we want to be able to go back to our civilian life, to go on trying to build up a good society.
>
> It is good that the line has been pushed back. But when we hear of the boys who have been killed, how can we rejoice at any successes in the field?
>
> No names of the dead have been announced, but among the old families who know each other, who lost brothers in the war of independence twenty-five years ago – as I did, as Yigael Yadin did – the word soon goes round. It sounds such a small figure, 650 dead, but it is not so small for us. We are a small nation.

*   *   *

On the evening of October 19, the 5 p.m. news reported that the counter-attack that day on the Golan had been halted. 'The Syrians made no territorial advances and exhausted themselves.' Thirty Jordanian and Iraqi tanks were destroyed. Two Syrian planes were shot down. On the Suez Canal, seventy Egyptian tanks were destroyed, and fourteen Egyptian planes shot down. It was the thirteenth day of the war.

President Sadat's appeal to the Soviet Union to use its good offices to bring about a cease-fire was energetically followed up. On October 19 Israel radio announced that Alexei Kosygin had returned to Moscow after three days in

Cairo, trying to persuade the Egyptians to accept a cease-fire. Russia did not want to see its Egyptian or Syrian allies totally defeated. After the intervention of the Russian Ambassador in Washington, Henry Kissinger – a Jew who had gone to the United States before the Second World War as a teenage refugee from Nazi Germany – flew to Moscow. There, with the Soviet leader, Leonid Brezhnev, he agreed that the United States would join the Soviet Union in urging both sides to end hostilities. In a message to Sadat, Brezhnev promised that Soviet troops would be sent to Egypt to help maintain the cease-fire.

Throughout October 20 the fighting continued. A Syrian plane, trying to reach and bomb the oil refineries at Haifa, was shot down before it reached its target and crashed on the coastal town of Nahariya. Damascus radio, anticipating a successful raid, reported that the refinery was in flames. On the southern front, Israeli forces continued to advance and consolidate their positions on the western side of the Suez Canal. That afternoon, Yigael Yadin, returning for a few hours from General Headquarters in Tel Aviv to his home in Jerusalem, set out his assessment of the situation in the south:

> Our advance inside Egypt is more important, deeper than the Egyptians admit – perhaps deeper than they know. The Egyptian high command are now beginning to lie to their commanders in the field. We control several of the main arteries from the Canal to Cairo. We have made a very serious breach in their lines. They had most of their forces on *our* side of the Canal when we broke through across it, and they cannot easily bring them back.
>
> The Egyptian high command do not entirely realize the seriousness of what we have done. They are not giving the right orders to meet such a serious situation. Some of their commanders are now trying to retreat. But they are telling them to advance. They cannot advance. All they can do is to dig in. They are paralysed. Even the Egyptians in the northern Suez sector are under fire. Militarily, it does them no good to hold the Sinai bank.
>
> In the next forty hours things should be clearer. We are well, well inside Egypt, on a wide front. They are trying to hide this from their own troops, their very own people. They are trying to hide from the Russians that they are weak, that they do not have a good bargaining position.

The Israeli leaders doubted if a cease-fire for which both Russia and America were pressing would come into effect all that soon. At a press conference on October 20, Moshe Dayan told the journalists that he saw no prospect of it. When the Deputy Prime Minister, Yigal Allon, visited Sharon's division on the west bank of the canal on October 21, he assured them that they had 'ample time' and that there was 'no hurry'.

On the Golan Heights, fighting had continued throughout the previous two days, as Syrian, Jordanian and Iraqi troops tried yet again to break into the Israeli bulge from the south. Heavy losses were incurred, with the battle being particularly fierce on Tel Anter and Tel El-Alaliekh, where in some instances there was only five yards between the Iraqi and Israeli tanks as they fought for control of the hills. By the end of the battle, sixty Iraqi tanks had been destroyed, and hundreds of Iraqi infantrymen killed. The Israeli bulge had held.

For several days the Israeli fortification and signals station on Mount Hermon which the Syrians had captured in the opening hours of the war had been pounded by Israeli aircraft. Then, at two in the afternoon of October 21, Israeli paratroop forces were lifted by helicopter to retake it. For twelve hours the soldiers advanced along the ridge, 8,200 feet above sea level, and above the snow line. One Israeli, an officer leading one of the assault parties, was killed. One by one the Syrian positions were overrun. When the attackers reached the Syrian command post they found twelve Syrian soldiers dead inside. They had been killed by Israeli artillery. When the main Syrian position was reached, it was found to be empty. But in the struggle for the lower slopes of Mount Hermon, fifty-one Israeli soldiers were killed. One of those who survived the battle told Israeli television viewers a few days after the battle, 'We were told that Mount Hermon is the eyes of the State of Israel, and we knew we had to take it, whatever the cost.'

On October 21, Henry Kissinger flew from Moscow to Tel Aviv, where he obtained Golda Meir's agreement to a cease-fire. It was this Kissinger intervention that saved Egypt's Third Army, which was surrounded on the eastern bank of the canal, from annihilation. This angered many Israelis, but it was Kissinger's belief that unless the Third Army was saved, Sadat would feel humiliated and his honour demeaned and that he would therefore be unlikely to agree to a political deal on the road to peace: the only way forward, in Kissinger's view, if some form of lasting agreement in addition to a cease-fire was to be reached.

When the Security Council met at dawn on October 22 in New York, it passed Security Council Resolution 338, calling for a cease-fire, 'on the basis of the line of the front', within twelve hours and not later than 6.52 p.m. that evening. It also called for 'the beginning of negotiations' between the parties and a 'just and durable peace'.

The call for a cease-fire gave the Israeli forces on the western bank of the canal less than a day to complete whatever advance they were capable of towards the canal city of Ismailia. Most of the Egyptian ground-to-air missile sites had been knocked out, so that Israel at last had virtual mastery of the air above the canal, on both sides.

At 6.50 p.m. on October 22, Israel radio announced that the Israeli government had agreed to implement a cease-fire 'in two minutes' time'.

This was the last possible moment indicated in the Security Council resolution. The announcer added that Egypt had also agreed to honour the Security Council resolution. Colonel Gadaffi of Libya was said to be 'furious' that Egypt had accepted. The Syrian government was 'still considering the question'. At 6.52 p.m. the radio was silent. At 7 p.m. the announcer stated: 'We have no news as to whether the cease-fire has started or not.' From southern Lebanon, the Palestinian Fatah declared that they would 'continue to fight'. In their cross-border bombardments since the war began – more than 150 attacks in all – they had killed twenty people in northern Galilee.

It was not until 9 p.m. that Israel radio was able to report that fighting had not ceased, but was continuing on both war fronts. On the Golan, Syrian artillery continued to bombard Israeli positions inside the 'fist'. West of the Suez Canal, Israeli troops, commanded by General Sharon, pressed north of their bridgehead, reaching the Ismailia–Cairo road only a few miles south of Ismailia. Also west of the canal, Israeli troops under Dan Shomron (who was later to command the Entebbe rescue raid) drove southward and cut the Suez–Cairo road at kilometre 101. They then continued their advance until they reached the western shore of the Gulf of Suez near the port of Abadiah. Two Egyptian torpedo boats that tried to escape from the port by speeding out to sea were sunk by Israeli tank fire from the shore.

Israel was in control of a substantial area on the western – Egyptian – bank of the Suez Canal, from just south of Ismailia to the Gulf of Suez, an area of 1,000 square miles. But on the eastern bank – in Sinai – two Egyptian armies still held out. Although cut off by Israeli troops from any retreat or escape, they maintained a precarious Egyptian hold on Sinai in a narrow band both north and south of the Israeli crossing point.

During the night of October 22, rockets continued to be fired into northern Israel by Fatah from its positions in southern Lebanon. Eleven Israeli towns and settlements were hit. On the following morning, President Idi Amin of Uganda – who had done military training with the Israeli army – appealed to all Arab nations not to accept the cease-fire. That morning Egyptian artillery bombarded Israeli positions on the western bank of the canal. Chaim Herzog broadcast over Israel radio at 1 p.m. to say that the Israeli breaches of the cease-fire along the canal are 'the usual process of clearing up.' Half an hour later, Israel radio announced that Israeli forces had 'resumed their operations' on the canal, and that on the Golan, where the Syrians had not accepted the cease-fire, twenty Syrian planes had been shot down during the morning.

In Washington, Henry Kissinger was pressing the Israelis to carry out the cease-fire. Inside Israel, after Golda Meir spoke in the Knesset in support of a cease-fire, Menachem Begin, speaking on behalf of the coalition of right-wing parties led by his own Herut, denounced the government for accepting a cease-fire. It should not be agreed to, he said, 'while the battle is not yet over, while the enemy is still on our territory east of the canal'.

At 2 a.m. on the following morning, October 24, the Knesset endorsed the cease-fire. As it did so, the Security Council, meeting in New York, passed a second cease-fire resolution. At 4 a.m. Middle East time, making use of the good offices of the United Nations headquarters in Jerusalem, Dayan suggested to the Egyptians that the cease-fire should come into force in three hours' time, at seven in the morning. The Egyptians agreed.

At 7 a.m., Israel radio announced that as Egypt and Syria had accepted the new cease-fire resolution, 'the cease-fire comes into force at this very time, seven o'clock'. Shortly afterwards, Egypt mobilized its militia, in order to repel what Cairo radio called Israel's 'barbaric invasion west of the Suez Canal'. That morning, an Israeli drive into the centre of the town of Suez was repulsed, and eighty Israeli soldiers killed. It was the last serious military operation of the war. At General Headquarters in Tel Aviv, there were those who argued that it was still not too late to continue the fighting for another day, and to defeat the Egyptians more decisively. They argued that the Egyptian troops trapped on the eastern bank of the canal had less than two days' water supply. If the fighting could go on for another twenty-four hours, they would have to surrender.

At 1 p.m. Israel radio announced that the cease-fire had come into effect 'on all fronts'. Two hours later, Chaim Herzog told radio listeners that the Egyptian forces on the eastern bank of the canal 'are simply disintegrating'. At 5.54 p.m. the blackout that had been imposed throughout Israel since the war began was ended. That night, the lights came on in the streets for the first time in eighteen days. On the Suez front, United Nations observers arrived at sunset to plant their cease-fire flags between the armies.

The war was over. The casualties had been heavy on all sides. On the Golan Heights, the Syrian war dead numbered 3,500, and the Israeli war dead, 722. Clusters of bodies still lay on the terrain.

Syrian tank losses on the Golan amounted to 1,150. The Iraqis lost more than 100 tanks and the Jordanians fifty. Of the Arab tank losses, more than a third were burned-out wrecks, but 867 were still capable of being put back into combat order. Israel lost 100 tanks on the Golan Heights. A further 250 Israeli tanks that had been damaged on the Golan were later repaired (some, indeed, were repaired during the course of the fighting, and went back into action).

The Israeli military death toll on both fronts in the October War, on land and in the air, was 2,522. This was just over three times that of the Six Day War. The figure was one which was to be much spoken about in Israel in the weeks and months ahead. At first, before the official announcement, 1,500 was spoken off as the death toll, and that figure had seemed far too high. When the actual number was announced, a pall fell over Israeli society. So many families were bereaved, so many settlements had lost young men whom they had known and loved, schoolchildren had lost teachers, fathers who had fought in the wars of 1948, 1956 and 1967 had lost sons: often, fathers and sons were both in action in 1973. 'This is the first war in which fathers and sons have been in action together,' Yigael Yadin commented in the war's final days. 'We never thought that would happen. We – the fathers – fought in order that our sons would not have to go to war.' That was a sentiment widely expressed as the frightening uncertainties of the battle gave way to the no less frightening certainties of counting the cost.

In the aftermath of the October War, Israeli women soldiers called on those who had lost fathers and sons, and handed over the dead soldiers' possessions, in most cases their kitbags and the contents. One of those who carried out this task was a nineteen-year-old officer, Sarah Oren. 'I had the kitbags in my room,' she later recalled. 'I couldn't sleep at night. I was haunted by it.' One of those to whom she took a kitbag was a soldier's wife who had given birth to her child after her husband had been killed in action.

The consequences of the October War were of crucial importance to Israel. First, it revealed her growing dependence on the United States. This dependence manifested itself on the diplomatic front in terms of military supplies and economic aid, both of which had greatly increased during the war and were to continue at a high level after it. Since the war cost Israel the equivalent of a whole year of its Gross National Product, her external indebtedness reached such high dimensions that she had no choice but to depend on a vast amount of American economic aid. Indeed, Israel became by far the largest recipient of foreign aid from the United States, currently in excess of $3,000 million every year.

The second consequence was that the Third World countries broke off diplomatic relations with Israel, thereby intensifying her international isolation (the Communist Bloc had broken off relations after the Six Day War). The hostility of the Third World was particularly distressing for Golda Meir, who had invested enormous energy over many years in a programme of Israeli assistance to newly emerging countries, especially in Africa and Asia, based, not on the prospect of commercial gain, but on the ethos of 'Israel sharing its nation-building experiences with other new nations'. All that contribution, partnership and expectation was destroyed overnight.

A third consequence of the war was that, although from a strictly military point of view Israel had undoubtedly won an impressive victory, Egypt's initial success in crossing the canal and breaking through the Bar-Lev line created a psychological breakthrough that eventually led to an Egyptian–Israeli political settlement. Egypt had regained its honour: this was the object of the attack in the first place, rather than any sweeping forward through Sinai to the borders of Israel itself. For Egypt the October War was a war not of conquest, but of self-assertion.

Beyond Egypt the oil-producing Arab States had found a new strength. An oil embargo on those States that supported Israel was followed by the rise of the Organization of Petroleum Exporting Countries (OPEC), dominated by Saudi Arabia, which used its ability to raise oil prices at will, and to raise them to unprecedented heights, to cast the Western economies into chaos. This spawned what came to be known as 'petro-dollar diplomacy', which led to pressure being put on Israel by those States which felt threatened (and were threatened) by a rise in oil prices if they did not act against Israel in the diplomatic arena.

Many of the soldiers who fought in the October War were veterans of the Six Day War. Others were the sons of those who fought in 1948 and 1956. Although all of Israel's wars had been relatively brief, the horrific nature of war affected all those who fought in it, as well as the wives to whom the soldiers returned in the immediate aftermath of the fighting. The impact of war remained long after the guns had fallen silent on the battlefield, and the deafness brought on by the guns had disappeared. Naomi Shepherd had emigrated to Israel in 1957, where she married Yehuda Layish. In her memoirs she recalled the aftermath of the war:

> The only evidence of what Yehuda had experienced in Sinai was his sudden deafness, which was gone by the next time I saw him, and something else which was to recur at intervals during the years that followed: it was the sound of an unfamiliar high-pitched voice in the bed next to me.
>
> Yehuda's normal voice was deep and measured, but when he woke one night in terror, it was to scream in a whimper, kicking convulsively, trying to escape from some horror that pursued him. When I had woken him from this nightmare for the third or fourth time, he told me what it was. There was a young reservist with whom he had shared a tent in Sinai, a pleasant young man who had been married only a few days before the war. One day he had gone out and been blown to pieces by a shell. Yehuda, together with the unit's chaplain, had collected the remnants of the bridegroom's body and once more they shared a tent: the plastic body-bag lay under Yehuda's camp-bed for days, because the shelling was too heavy for it to be sent, with the unit's other corpses, to the rear.

Domestically, the consequences of the war for Israel were profound. The war shook the confidence of Israeli society, so much so that its reverberations were felt for the next two and a half decades. Its immediate impact in this regard was to sound the knell of the decline of the Labour Alignment as the historical and natural governing party of the State.

CHAPTER TWENTY-FIVE

The fourth postwar era

Negotiations for the disengagement of forces on the Suez Canal began on 11 November 1973, at one of the closest points to Cairo that Israeli soldiers had reached, kilometre 101 on the Suez–Cairo road. But the two countries had very different ideas of what the disengagement should be. Israel was prepared to withdraw all its forces back across the canal, to a line 10 kilometres to the east. Egypt wanted all Israeli forces to be withdrawn to a line drawn through Sinai from El Arish to Sharm el-Sheikh, putting the Gidi and Mitla Passes under Egyptian control for the first time since 1967. At first it seemed that Egypt would be prepared to accept the Israeli proposal, at least as a first stage, but on November 28, following an Arab summit meeting in Algiers, the negotiations broke down.

For six weeks Israeli forces remained on the western side of the canal, the bulge on the map representing a permanent threat to Cairo and a humiliation to Egypt. Nor was there any Israeli withdrawal from the Syrian front. A war of words followed the war of guns, as each side put forward its objectives. 'We stick to our stand that Israel should withdraw from all Arab territories she occupied since June 1967,' King Hussein of Jordan declared on December 2, 'and say that there can be no peace in this area without complete withdrawal. We also need not say that Arab Jerusalem, that precious jewel on the forehead of this homeland, will never and under no circumstances come under any sovereignty other than absolute Arab sovereignty.'

Golda Meir was not averse to some form of territorial compromise, but she had strong views as to what could not be given up. 'We will not descend from the Golan,' she said on December 29, 'we will not partition Jerusalem, we will not return Sharm el-Sheikh, and we will not agree that the distance between Netanya and the border shall be eighteen kilometres.' But she went on to say that, 'if we want a Jewish State, we have to be prepared to compromise on territory.'

On December 21 a conference was convened in Geneva to try to effect agreement in the Middle East. Security Council Resolution 338, which had called for the cease-fire, had also called for 'negotiations between the parties concerned, under appropriate auspices, aimed at establishing a just and durable peace in the Middle East'. The conference was convened jointly by the United States and the Soviet Union, and met under the auspices of the United Nations. It was called the Geneva Peace Conference.

Egypt and Jordan agreed to go to Geneva on condition that they would not have to negotiate directly with Israel. Israel, which had never been given by the Syrians a list of Israeli soldiers captured in the October War, said that it would not sit down with the Syrians until they handed over the list: hundreds of Israeli parents and families were in deepest anguish at not knowing whether their sons were prisoners of war, or dead. Syria refused to participate at all. The Palestinian Liberation Organization was not invited to Geneva. When the first session opened on December 21 the delegations that had agreed to come sat at separate tables arranged as a hexagon to accommodate the Arabs' refusal to sit with Israel. One table was left empty for the Syrians. After each delegation made its opening speeches the conference was adjourned until after the Israeli elections, which, postponed from October, had been called for December 31.

There had not been an election in Israel since October 1969, four years earlier. The Israeli public, Yitzhak Rabin has written, 'was still stricken when it went to the polls'. The Labour Alignment, which had been in power when the war broke out, won fifty-one seats, a loss of five on the election of October 1969, but still constituting the largest single block of seats in the Knesset. The Likud, standing for the first time as a unified bloc of right-wing Parties (and brought together under the impetus of General Sharon, who had resigned from the army immediately after the war in order to enter politics), won thirty-nine seats. It was thus the second largest Party – or, more properly, like the Labour Alignment, group of Parties.

Labour was once again able to form a government with ease, the National Religious Party having won ten seats, and hitherto always having cast its lot with (and received its political patronage from) the Labour-led coalitions that had governed Israel since 1949. Golda Meir remained Prime Minister, and prepared to form a government. Rabin (who had been elected to the Knesset for the first time, having returned from his ambassadorial post in Washington), commented in his memoirs:

> Considering that they were exhausted, mourning their dead, and having difficulty in digesting recent events or comprehending their significance, the voters were merciful toward the Labour Party. Not that there was any inclination to forgive the party's leaders or absolve them of their sins of omission. But many people were entranced by the prospect of the Geneva Conference and fervently clung to the hope that it would bring the peace they were yearning for.

Labour supported the conference; the right-wing Likud vigorously opposed it, and in doing so I don't think any Israeli political movement ever erred so seriously. The Geneva Conference was inaugurated on December 21 – nine days before the elections – and it failed to bring peace to the Middle East. But it did rob the Likud of its chance to gain power. The Labour Alignment lost a few Knesset seats, but it had clearly won the elections.

At the same time, Labour's relative success failed to salvage the authority of the party's leading figures. Deep personal differences between the party's leaders – hitherto swept under the carpet – now came out into the open as the short-lived, grass-roots protest movements raised an angry outcry against Golda Meir, Moshe Dayan and others.

Confusion and depression reigned at the frequently convened sessions of the Labour Party's Central Committee. It was against this background that Golda Meir embarked upon the task of setting up a coalition and building a Cabinet.

The Geneva Conference never reconvened. For just over two months, Golda Meir struggled to form a government. At first the Rafi faction of the Alignment, led by Moshe Dayan and Shimon Peres, refused to accept the idea of a Labour-led, narrow-based administration, arguing that the times were so dangerous that an all-Party government was needed, and that Likud, the second largest Party, should not be excluded.

On March 5, at a meeting of the Central Committee of the Labour Party, Yitzhak Rabin, who disagreed with the concept of an all-Party coalition, urged the formation of a government without any more delay. 'Elections are no game,' he said. 'If we want to halt the process of internal disintegration, we must form a Cabinet!' His words were met with prolonged cheers.

The Party meeting was followed almost immediately by a meeting of the Knesset Foreign Affairs and Defence Committee, at which reports were produced that the Syrians intended to renew the fighting on the Golan. Once again, there was an acute sense of danger. Once again, reservists, some of whom had only recently been discharged from the army, were called back to their units. Once again, fears spread through the populace that war would start up once more. Five days later, Golda Meir formed her new Cabinet. In view of the emergency, Dayan and Peres agreed to become part of a narrow-based Labour coalition.

The government that was formed on 10 March 1974 was the shortest lived in the history of Israel. The Agranat Commission, set up to examine the lack of readiness on the eve of the October War, published its interim report on April 1. Only forty pages long, it was political dynamite. It recommended the dismissal of the head of Army Intelligence, the removal from active duty of the Commander of the Southern Front, General Shmuel Gonen, and, most serious of all, the removal of the Chief of Staff, David Elazar. The case made against Elazar was that he had not made his own Intelligence evaluations, had failed to prepare a detailed defence plan, and had been 'overconfident' of the ability of the Israeli army to push back the enemy with regular forces only.

Because political responsibility was not in its terms of reference, the commission found no fault with either Golda Meir or Moshe Dayan. But its criticism of the military leadership was so severe that, on April 11, only ten days after the interim report was published, Golda Meir resigned. The public criticism of her alleged failures on the eve of war to read the Intelligence correctly or to take effective action to meet the threat of war had bowed her down and driven her from office. She was seventy-five.

At the instigation of the Israel Labour Party, Golda Meir was succeeded as Prime Minister by Yitzhak Rabin. He was the first native-born Israeli to lead his country. Moshe Dayan also resigned: he had been as much reviled in the aftermath of the war as he had been lauded before it. He was replaced as Minister of Defence by Shimon Peres.

Throughout the tense winter of 1973–4 Henry Kissinger acted as the main mediator in bringing about a disengagement agreement between Egypt and Israel. On 18 January 1974 the Egyptians, having abandoned their claim for an immediate Israeli withdrawal from more than half of Sinai, agreed to what had essentially been Israel's original proposal. The Israelis then withdrew from across the canal and took up positions five kilometres inside Sinai. The eastern bank of the canal, which Israel had held since 1967, was transferred to Egyptian control – most of it having been under Egyptian army control since the first stage of the war.

No disengagement agreement seemed possible on the Golan. On February 2, there was Syrian shelling of Israeli military positions and civilian settlements on the Heights. Following Israeli counter-barrages the Syrian shelling ceased. But on the following day the Syrian Foreign Minister, A. H. Khaddam, announced that Syria was carrying out a 'continued and real war of attrition'. The aim of this war, he explained, was 'keeping Israel's reserves on active duty and paralysing its economy'. The Syrians knew full well that the reservists were men drawn from across the whole spectrum of Israeli society and productive enterprise.

Any prolonged emergency would have the effect of harming the economy. There was also the question of morale: the Israeli public had suffered the shock of the coming of the war, of its traumatic moments, and of the continuing high level of mobilization after it. That February, the inquest into the war was still continuous in many homes. So too was the distress of further call-ups, as men who had served throughout the war were taken away once more from their work and their families, to meet the renewed threat.

For two months the Syrians refused even to enter into negotiations with

Israel unless she withdrew from the Golan completely. On March 31 the Israelis were shocked by a statement from Washington, that there was a 'foreign legion' serving inside Syria that consisted of troops from Kuwait, Morocco and Saudi Arabia, as well as a number of North Korean pilots, and an armoured brigade from Cuba with more than 100 tanks.

There were two more events in 1974 that deepened the Israeli sense of bitterness and siege, while the PLO won international recognition. The first of these events was on the night of May 14, when three Palestinian terrorists seized the school in Ma'alot, in northern Israel. Twenty-two pupils from a religious school in Safed, who were sleeping there during an excursion, were murdered. It was discovered that the PLO faction behind the operation were the so-called 'moderates' who had been liaising with Israeli 'doves' – a small but determined group who were desperately seeking a way forward through dialogue.

On their way to the school at Ma'alot the terrorists came to a private apartment belonging to Fortuna (Hanna) and Yosef Cohen, and their three children, Beiah, Eli and Itzhik. The terrorists knocked on the door. The father, Yosef, answered. The terrorists dragged him to the stairwell and shot him with a Kalashnikov. He died immediately. The mother was awakened by the sounds of the shots. Two of her children, Eli, her three-year-old son, and Beiah, her six-year-old daughter, ran to their mother's room. Itzhik, who was one and a half years old, a deaf-mute infant, slept in a crib next to his parents' bed. His mother, with great presence of mind, took him out of his crib and hid him under her bed. Since he could not hear a thing, he continued to sleep.

Subsequently the terrorists entered the bedroom, and murdered both the mother and Eli. Beiah was badly wounded. Itzhik, who continued to sleep under the bed, was unhurt.

On October 25 an Arab summit at Rabat in Morocco decided to recognize the PLO as the 'sole representative' of the Palestinian people. These manifestations of Palestinian extremism and support in the Arab world fed the mood in Israel that gave increasing support for its settlement activity on the West Bank, and encouraged the opposition Parties of the right, Labour's opponents, in their determination to come to power after thirty-six years in the political wilderness.

A year had passed since the October War. David Harman, an educator in civilian life, who had seen the horrors of war at first hand on the Golan Heights, wrote, shortly after the Jewish New Year as the anniversary of the war approached, 'The holidays are drawing to a close this year. They passed peacefully but sombrely. Memories of last Yom Kippur are still too ripe and, when coupled with the rather murky prospects for the immediate future did not combine to make a very happy New Year season. Israel is today plagued by many of the same maladies that afflict other countries – a deteriorating economic situation, poor and lack-lustre leadership – and, in addition, a disturbing internal social situation readying itself for eruption. Superimposing upon this the unstable defence and foreign conditions certainly does not make the best combination for a happy year.'

On 22 November 1974 the United Nations General Assembly, meeting in New York, voted to accept the Palestine Liberation Organization as an observer at all United Nations meetings. It would be present as the 'representative' of the Palestinian Arabs, whose rights to 'national independence and sovereignty', and to return 'to their homes and property', the Assembly asserted by a substantial majority. 'We have entered the world through its widest gate,' Yasser Arafat told the United Nations that day, with a gun in his belt. 'Now Zionism will get out of this world – and from Palestine in particular – under the blow of the people's struggle.'

Arafat received a standing ovation at the United Nations. Inside Israel, his emergence as an accepted – and by many a welcomed – figure in the international forum was a deeply distressing development. His commitment to Israel's destruction had seemed one of the few certainties of the Fatah and PLO cause. Article 22 of the Palestine National Covenant, drawn up seven years earlier, had described Zionism as 'a racist and fanatical movement in its formation; aggressive, expansionist and colonialist in its aims; and Fascist and Nazi in its means.' That covenant assumed an added importance, and constituted an intensified danger for Israel, with Arafat's new-found international status. Only eleven days before his appearance at the United Nations, Arafat had declared, 'We shall never stop until we can go back home and Israel is destroyed. The goal of our struggle is the end of Israel, and there can be no compromise or mediations. We don't want peace, we want victory. Peace for us means Israel's destruction, and nothing else.'

For a whole year, the PLO lobbied to have Zionism officially condemned. Its first successes were at the International Women's Year Conference, held in 1975 in Mexico, and at two United Nations bodies, the International Labour Organization (ILO) and the United Nations Educational, Scientific and Cultural Organization (UNESCO). As a climax of Arafat's campaign, on 10 November 1975, the United Nations General Assembly was asked to condemn Zionism as 'racism'. Thirty-five countries protested against this equation, among them Britain, France, West Germany, the United States and Canada. Daniel Moynihan, the United States representative at the United Nations, told his fellow delegates during the debate, 'The United States will not abide by, it will not acquiesce in, this infamous act. A great evil has been loosed upon the world. The abomination of anti-Semitism has been given the appearance of international sanction.' Moynihan added, 'It is an attack not on Zionism but on Israel, as such, it is a general assault by the majority of nations on the principles of liberal democracy, which now are found only in a dwindling number of nations.'

Chaim Herzog, then Israel's Ambassador to the United Nations, concluded his defence of Israel by tearing up the proposed resolution on the podium of the General Assembly.

When the vote was taken the thirty-five opponents of the resolution found that seventy-five States went into the lobby against them, including seventeen Arab States, thirteen Communist States, twenty-two African States and twenty others – among them Mexico, Brazil, Portugal and India. The President of the United Nations Assembly, Gaston Thorn, from Luxembourg – which also opposed the resolution – told the Assembly, 'The spirit of the United Nations has been jeopardized by the adoption of a resolution that was stupidly and needlessly pressed to the vote by extremists who did not know when they had gone too far. I fear the evil consequences of this vote will appear only too quickly.'

The London *Observer* saw another flaw in the resolution. Israel's constitution, it wrote on the day before the resolution was passed, 'guarantees equality of citizenship to its Arab minority – a vital difference. Its insistence on being a distinctively Jewish State does not make it any more racist than countries like Pakistan, Saudi Arabia or Mauritania, which constitutionally call themselves Islamic States.'

During 1975 Yitzhak Rabin took the initiative, with Henry Kissinger's support, for the negotiation of an interim agreement over Sinai with President Sadat. It was to this end that Kissinger embarked on his famous shuttle diplomacy. The first attempt, in March 1975, failed. This led President Ford to declare a 'reassessment' of American policy which was very much hostile to Israel, and meant, in practice, a suspension of United States aid. In August, however, Kissinger's second attempt led to the signing of an agreement between Israel and Egypt at Geneva on September 4. This agreement was accompanied by a far-reaching American 'memorandum of understanding', in which Washington committed itself, *inter alia*, not to treat with the PLO, it being a terrorist organization that did not recognize Israel's right to exist.

For Rabin, this agreement was fashioned on his 'peace doctrine' of five phases: Disengagement, Defusion, Trust, Negotiations, Peace (the Oslo Agreement was later to follow the same model). The agreement provided for Israeli withdrawal on the Suez front to the eastern ends of the strategic Gidi and Mitla Passes, with most of the vacated area becoming a United Nations buffer zone. It also provided for the Israeli evacuation of the Abu Rudeis oil fields, and of a narrow strip of territory along the Gulf of Suez connecting the oilfields with Egypt. No Egyptian troops were to be stationed in this evacuated area. Egypt was to allow Israel to purchase the oil at a fixed and modest price.

Rabin also secured an Egyptian undertaking to refrain from the use of threat of force or military blockade (this undertaking was reciprocal).

Passage was secured for non-military Israeli ships through the Suez Canal. Finally, American-manned electronic early-warning stations were set up in the area of the Gidi and Mitla Passes.

In December 1975 the nature of Israel's position in the West Bank was changed, on a small scale, but irrevocably. For the previous two years a fanatical extremist religious group calling itself Gush Emunim (Bloc of the Faithful), headed by Rabbi Moshe Levinger – who had re-established an Israeli presence just outside Hebron in 1968 – pressed for the widest possible settlement of the whole West Bank. The aim of Gush Emunim was not merely to rebuild settlements where earlier ones had been destroyed – in 1929, 1936 and 1948. Gush Emunim was inspired by Rabbi Zvi Yehuda Kook (son of Rabbi Abraham Isaac Kook) who had taught – in the words of Israel Medad, one of his followers – 'that the main purpose of the Jewish people is to attain both physical and spiritual redemption by living in and building up an integral Eretz Yisrael. The territory of Eretz Yisrael is assigned a sanctity which obligates its retention once liberated from foreign rule, as well as its settlement, even in defiance of government authority.'

While Kook's philosophy was totally religious, that of Medad and those who were with him in Gush Emunim was nationalistic. Kook, like his father, taught that Zionism was possessed by a dynamic, based on an unfolding process of Divine Redemption. Inherent in this theory was the conviction that Zionism is undergoing a transformation, and that ultimately religious legitimacy will replace the secular political leadership. In Kook's vision the relationship between the Jewish people and the Land of Israel was endowed, as were the people of Israel themselves, with an innate holiness. Hence the imperative to create settlements. In this way, the religious beliefs of Kook (father and son) and the nationalism of Medad came to the same conclusion.

For Gush Emunim, the Palestinian Arab presence in the West Bank, pre-dating modern Zionism by several centuries, was 'foreign rule'. The nature of 'government authority' was also well understood. Under the Allon plan the government encouraged Jewish settlement in the essentially unpopulated Jordan Valley, to form a defensive line of settlements between Israel and Jordan. Gush Emunim rejected the restrictions imposed by the geography of the Allon Plan. It saw every square mile of the unpopulated areas of the West Bank and Gaza Strip as areas of predestined and preordained settlement, however close they might be to Arab villages or centres of Arab population.

Gush Emunim organized a series of marches and protest demonstrations in favour of unrestricted settlement throughout the areas conquered in 1967. At several sites which it had chosen, its members organized sit-ins until they were pushed away by the Israeli army. The government was determined not to give in to irresponsible tactics but its hands were somewhat tied as the demonstrations occurred on the same day that had been declared 'Jewish solidarity day', to protest against the United Nations 'Zionism is Racism' resolution. Because of this, the government was reluctant to turn the image of solidarity into an open confrontation.

One group, determined not to give in to government or army pressure, set up its tents at Eilon Moreh near the ancient city of Sebastia, a few miles north of the Arab city of Nablus. Seven times they were forcibly removed by Israeli troops. Then, in December 1975, after negotiations with the Minister of Defence, Shimon Peres, and Prime Minister Rabin's Intelligence Adviser and a former Likud member of the Knesset, Ariel Sharon (a believer in substantial Israeli settlement on the West Bank and in the annexation of most of the West Bank to Israel), Gush Emunim was given permission to move 'temporarily' into a former Israeli army camp at Kaddum. It had won the first round. Soon afterwards it moved to another site, Mount Kabir, east of Nablus. It was never forced to withdraw. 'To his last breath Rabin regretted having surrendered to the Gush over Eilon Moreh,' one who knew him well, and worked with him closely, later wrote. 'He saw the Gush as a threat to Israel's democracy.'

The impetus given to Gush Emunim by the government's compromise was enormous. Within ten years there were 40,000 Jewish settlers in several dozen such settlements throughout the West Bank. Twenty years later the number had risen to 140,000.

The temporary agreement with regard to Eilon Moreh not only became permanent, but settlements proliferated in a way which those who devised the Eilon Moreh compromise for the government had not imagined possible. Some settlements were still strategic, but none of these were inside Arab-populated areas. Ma'ale Adumim (Red Heights), established in the Judaean desert just over four miles east of Jerusalem on December 5 as a satellite suburb of Jerusalem, was an example of this. It was not intended as a messianic thrust into the West Bank, but, because the site lay above the Jerusalem–Dead Sea road, it was considered a strategic necessity.

Yet even Ma'ale Adumim began before the government was ready when its plans were hijacked by Gush Emunim and, to Rabin's great anger, a small group of men and women – in this instance twenty-three families – obtained permission, after considerable pressure, for a 'residential camp' for workers in the area. They took the name of their settlement from a place in the Book of Joshua described as the border area between the tribes of Judah and Benjamin. The rocks in the region of the new settlement were an unusual red hue (it lay only a mile west of the Inn of the Good Samaritan of Christian fame).

Ma'ale Adumim grew fast. Within two decades it had become a major urban centre, with a population exceeding 20,000, an industrial zone with 100 factories, and linked by a fast road direct to Jerusalem. Unusually for an Israeli town, 17 per cent of the population is of English-speaking (mostly American) origin. Building remains continuous, with a further 10,000 people expected to purchase homes there by the year 2,000. Because it is in the West Bank, it was defined in 1996 (by the Netanyahu government) as within Development Zone A, receiving special financial incentives; hence its continuuing high rate of expansion.

On 30 March 1976, six Israeli Arabs were killed when Israeli troops fired on a demonstration during an Arab strike protesting the confiscation of Arab land in Galilee. Henceforth, March 30 was to be celebrated by the Palestinian Arabs both in Israel and in the occupied territories as Land Day. The confiscation of land was justified by Rabin on the grounds that it was needed for Jewish regional development, and that the Arabs of the region as well as the Jews would in the long run benefit from that development.

Rabin promised that as little agricultural land as possible would be confiscated, and that compensation would be paid to the owners of the land that was taken from them. But he was angry that the predominantly Arab Rakah Party, the New Communist List, which had four seats in the Knesset, encouraged the protests and the strike. Many of the supporters of Rakah saw the party not so much as a Communist one, but as one which supported their rights as Arabs within Israel, and also the rights of the Palestinians to establish a State on the West Bank and Gaza.

When municipal elections were held throughout the West Bank and Gaza Strip, they took place without Israeli interference or pressure. Many PLO supporters were elected mayors. Peres, who was the Defence Minister, was much criticized by the Likud for allowing this.

An unusual experiment was embarked upon in 1976, and succeeded. At the northern Israeli town of Metulla, a crossing point was made, known as the Good Fence. Created by Shimon Peres, it was a physical and psychological opening in the highly fortified border. Several tens of thousands of Christian and Muslim Lebanese were to make use of this crossing point to enter Israel. Some made the journey in order to receive medical treatment at the Good Fence itself from Israeli doctors, or to go into Israel for treatment in hospitals there. Others came to trade. Cars with Lebanese number plates could be seen even in the streets of Jerusalem.

Israeli government ministers do not meet on Saturdays – the Sabbath – any more than British, French or American Cabinets meet on Sundays. But on Saturday 3 July 1976 it was not only the Israeli Cabinet that was meeting, but the military leaders and intelligence experts. A week earlier, on June 27, an Air France passenger airliner on its way from Tel Aviv to Paris had been hijacked by Arab terrorists shortly after takeoff from Athens and flown to Benghazi in Libya. There, the Israeli and Jewish passengers had been separated out by the hijackers, and the non-Jews had been released. The Jews – ninety-eight of them – were flown on to Entebbe in Uganda, 4,000 kilometres from Israel, where they were held hostage at the airport.

For a week there was consternation in Israel, and a sense of terrible impotence. The balance of public opinion was that Israel must secure the freedom of the hostages by complying with the hijackers' demands, including the release of Fatah terrorists being held in Israeli prisons. That Saturday morning the Cabinet had been called in order to make a final decision. 'The decision which was taken was in clear opposition to what had recently been the prevailing mood within the Government,' Peres later wrote. 'For a week it had sat and debated the question of what should be done to free the passengers kidnapped in the Air France plane and flown to Entebbe in Uganda. The balance of opinion seemed to be in favour of exchanging the hostages for Fatah terrorists. As Minister of Defence I had worked with my colleagues throughout that week to find another solution – a means of freeing the kidnap victims by an armed intervention of our own.'

That armed intervention was among the most remarkable episodes in the history of Israel, an airborne commando raid, carried out by a unit composed entirely of volunteers. Rabin, as Prime Minister, insisted on the highest degree of planning. He also brought Menachem Begin, as Leader of the Likud opposition, into the secret. Enormous efforts were made, night and day, by the Chief of Staff, General 'Motta' Gur – who in 1956 had commanded the battle for the Mitla Pass – and the heads of every branch of military, strategic and logistic planning. When the plan was finally presented to Rabin he was impressed by how the different elements involved in preparing it – among them Military Intelligence and the Mossad – had coalesced so successfully. 'Every detail, every phase of the operation, was etched in my brain,' he later wrote. Peres was present as the raid began:

> The young men, armed to the teeth, began climbing the gangways of the giant Hercules transports, engines already running, bound for a faraway destination, four thousand kilometres from home, there to tackle the unknown.
>
> The officers and men were in high spirits, although it was clear that they were aware of the importance of the mission and the difficulty of the test that awaited them.
>
> We shook hands with the officers. They told us not to worry and strode confidently towards the aircraft. At the steps they turned back, waved their hands and disappeared into the belly of the gigantic bird.

Four large carrier aircraft took part in the mission, flying from Sharm el-Sheikh directly to Entebbe. Radio silence was imperative throughout the long flight. Only when the flight was over did a brief radio message come through to the command headquarters in Israel from the commander of the operation, General Dan Shomron: 'We've landed. Don't worry. If anything goes wrong, I'll let you know.'

Almost nothing did go wrong. Only one of the hostages was killed: an elderly woman, Dora Bloch, who choking on some meat caught in her throat had been taken to the local hospital before the rescue raid. She was murdered after the raid, while still in the hospital. The terrorists – Arabs and Germans – were killed during the raid, as were three of the civilian hostages caught in the crossfire. There was one more death. Peres has recorded how he and his colleagues learned of it:

> When the mission had been completed – one of the most daring operations in military history – we received a further report: 'We have a casualty.'
>
> There was silence in the room. Nobody dared to ask the identity of the casualty, or what his condition was. Security considerations and premonition combined to prevent the asking of this question. There were a number of generals in the room. Two of them had sons taking part in this perilous mission. The expression on their faces did not change – at such a time all men are fathers to all sons.
>
> At three o'clock in the morning Motta Gur, the Chief of Staff, came into my room and said: 'It's Yoni!' We had not speculated on the identity of the casualty when we heard the news over the airwaves, but even then we knew in our hearts that the slight tremor in the voice reporting 'We have a casualty' concealed some particularly precious name.
>
> Both Motta and I felt lumps in our throats. There was nothing to add.

'Yoni' was Yonatan Netanyahu. One of his friends wrote of him, 'Yoni – is a perpetual fight against sleep, fatigue, idleness, forgetfulness, inefficiency, helplessness, deceit. Yoni – is the turning of the impossible into the possible.' He was killed by a bullet through the heart.

Twenty years after the Entebbe raid, Yoni's brother Benjamin – who also served as a commando – was to become Prime Minister.

The Entebbe raid not only rescued the hostages, and was a blow to terrorism, it was, in the words of Shimon Peres, 'An operation unique in military history. It proved that Israel is capable not only of maintaining defensible frontiers, but also of taking decisive action in defence of her interests. Against a background of international terrorism of the most extreme kind, terrorism aided and abetted by the army and the President of Uganda, at a distance of more than four thousand kilometres from home, in one short hour, the stature of the Jewish people was enhanced – as was the stature of free and responsible people throughout the world.'

Yitzhak Rabin, reflecting on the rescue, wrote, 'We do not bask in the glory of such victories. We do remember, in sorrow and pride, the men who gave their lives – at Entebbe and elsewhere in the defence of Israel – that this nation should live in peace and dignity in its land.'

*　　*　　*

In December 1976 the first delivery of a new generation of warplanes reached Israel from the United States: three F-15 fighter jets. They had been refuelled in the air while on their journey and arrived on a Friday afternoon, shortly before the Sabbath. Because of the importance of the occasion, a ceremony was arranged at the airport, which several senior officials attended. The ceremony ran over into the Sabbath, rousing the anger of the religious Parties, one of which, the Po'alei Agudat Yisrael, with two seats in the Knesset, brought a no-confidence motion against the government, which was dependent for its majority (as all Labour-led governments had been since 1948) on the National Religious Party seats.

Replying to the motion of no-confidence Rabin said that the entire motion was based on an erroneous assumption. The ceremony had finished at 4 p.m. – seventeen minutes before the start of the Sabbath. He personally, as well as the military correspondent of an orthodox newspaper (*Hatzofeh*) arrived home before the Sabbath. At the same time, Rabin pointed out, the government had officially expressed its regrets, should any 'desecration' have been entailed.

There were other facts which Rabin then recounted. Only 290 guests were invited, of whom 200 were air force officers, 'because the ceremony had been downgraded from a state event to an intra-military event'. The rest of the participants lived on the base.

Then, to the amusement of his supporters in the Knesset and the galleries, Rabin recalled that when the first Phantom F-4 jets had arrived – on Friday, 5 September 1969 – at a nearby air base, they touched down only twenty minutes before the Sabbath. On that occasion the intra-military ceremony was attended by the Prime Minister, the Defence Minister and the Foreign Minister. 'The Likud men who were in the Cabinet then made no complaints about the twenty-minute time span before the Sabbath. Yet today they are supporting no-confidence although the planes landed forty-seven minutes before the Sabbath.'

In the vote of no-confidence all ten National Religious Party members of Knesset abstained. Rabin, livid, took the opportunity to remove the NRP Ministers from the government, and then to present the resignation of his whole government to the President (on December 22), and call early elections. Shimon Peres was angered that the government had allowed the crisis over the planes to drive the National Religious Party out. 'The ceremony could have easily been postponed until Sunday,' he wrote twenty years later. 'Clearly, the planes could have landed on Sunday without violating the Sabbath.' The decision 'to get rid of the NRP was a tragic mistake,' he added. 'Ben-Gurion never governed without the NRP and we since then could not assemble a real majority without them.'

Rabin, however, reflecting on the dismissal of the National Religious Party members of his Cabinet, told his wife, 'I'm tired of being squeezed and blackmailed by them on every occasion.' The power of the religious political pressure had been ever-present throughout his premiership, and it had been growing, characterized by the blocking of projects such as a building in which Reform Jews could pray, and a public swimming pool that would be open on the Sabbath.

Rabin remained Prime Minister at the head of a caretaker government. With the elections only three months away, he was challenged for the Labour Party leadership by Shimon Peres. The contest between the two men was fierce, and the margin close. At the Labour Party convention, where the matter was put to the vote on February 22, Rabin beat Peres by the narrowest of margins, 1,445 to 1,404, with sixteen abstentions. The Rabin–Peres vendetta came to plague the Labour Party until the last years of Rabin's life, when the two men finally made their peace with each other at a crucial moment for the peace process in which they both invested great energy, and for which Rabin gave his life.

The aftermath of the October War continued to exercise its influence in the widespread feeling that Labour was no longer fit to rule, that in the three and a half years since the war had ended it had failed to revitalize its policies or its institutions. The leader of the opposition, Menachem Begin, after nearly thirty years in opposition, derided by Labour for the extremism of his opinions, lost no opportunity in the Knesset to attack Labour policies and failures. Following a financial scandal, the Minister of Housing, Avraham Ofer, committed suicide (a leading member of the Labour establishment, Yaakov Levinson, had also committed suicide as a result of rumours of financial impropriety).

The National Religious Party, angered at its dismissal from government and at the loss of the political patronage it had enjoyed as Labour's partner since the foundation of the State, turned both politically and emotionally towards the opposition. Religious leaders spoke openly of the 'lack of Jewishness' of those in Labour who had maintained a secular lifestyle. The fact that men like Shimon Peres were deeply versed in the Bible and steeped in Jewish tradition was ignored. A rift was opened between the secular and the religious, and between the left-wing Parties and the religious Parties, which has grown to terrible proportions today.

On the West Bank, across the Green Line in the coastal plain, the new settlement of Elkana (the husband of the childless Hannah of the Bible) was founded on 5 January 1977. On February 10 a new religious settlement, Netzer-Hazany, set up independently of the government, was founded in the Gaza Strip; its climate was later described in its prospectus as 'hot and humid summer, very humid winter'. The establishment of these settlements was accepted by the Labour government, though it was criticized by the United States.

On February 14, three weeks after the inauguration of Jimmy Carter as

President of the United States, the Secretary of State, Cyrus R. Vance, flew to the Middle East. In Jerusalem, Rabin told him that for 'real peace' Israel would 'make territorial compromises in all sectors'. Any agreement would have to be with Jordan, as the PLO would not be 'an acceptable partner' for Israel. Travelling to Egypt, Vance found that President Sadat was, as recalled by one of the Americans present, William B. Quandt, 'anxious to see a negotiating process resumed with active American involvement'. Sadat's response revealed the maturing of trust that had followed the Rabin–Sadat interim agreement of a year and a half earlier.

Back in Washington, both Vance and Carter decided to proceed on their course as mediators, but to do so with caution; neither man wanted to create the impression among the American Jewish leadership that it was intended to put pressure on Israel. The merest hint of such pressure in the past had led to criticism of successive administrations for not supporting Israel fully (Israel was then in receipt of almost $2 billion a year from the United States).

In March 1977 Rabin made his first visit as Prime Minister to Washington, for talks with Jimmy Carter. Leah Rabin later recalled how, after a long day of negotiations on March 7, Rabin was shocked when Carter dismissed his other dinner guests and said to Rabin, 'Now Mr Prime Minister, tell me what you really think.' Rabin replied, 'Mr President, we had long talks today. I don't have two agendas or opinions. I have only one and you heard all about it today.' This was typical of Rabin's bluntness, and his intellectual integrity.

Among the points that Rabin had made during the day was one that held out the prospect of an agreement between Israel and Egypt. The 'bulk of Sinai' could be returned to Egypt, he said, in return for peace. Despite earlier pronouncements by Moshe Dayan that Israel must never give up Sharm el-Sheikh, Rabin told Carter that although Israel would need to control the tip of the Sinai peninsula overlooking the Straits of Tiran, she had 'no need for sovereignty' there. Rabin then told Carter that Israel did not want to 'come down' from the Golan Heights. As to the West Bank, Rabin spoke in terms of an eventual Israeli withdrawal, though not, he said, 'total withdrawal' to the 1949 cease-fire lines.

Rabin was adamant that what he called 'a Jordanian-Palestinian State' was acceptable to Israel, but that a third State in the area (that is, a purely Palestinian State) was not. To pave the way to a Jordanian-Palestinian State, Rabin told Carter, there should be direct negotiations between Israel and Jordan, rather than a summit between Israel and all the Arab States; in support of his view, Rabin recalled his own positive experience of direct negotiations with the Egyptians at Rhodes in 1949, when the armistices were concluded and signed.

The talks between Rabin and Carter continued on March 8. To Rabin's surprise, Carter described the Israeli settlements on the West Bank as illegal, and he went on to tell Rabin, 'Your control over territory in the occupied regions will have to be modified substantially in my view. The amount of territory to be kept ultimately by you would only, in my judgement, involve minor modifications in the 1967 borders.' This was a repetition of the June 1970 Rogers Plan which called essentially for Israel's withdrawal to the 1949 armistice lines 'with minor modifications'. This proposal had failed to be acceptable to Israel then, and was no more acceptable seven years later.

Rabin returned to Israel. A week later, on March 16, Carter spoke at a meeting in Massachusetts of the need for an Arab commitment to peace, and of the need for negotiations over borders. Both these 'needs' were acceptable to Rabin. But Carter then went on to say that there was 'a third ultimate requirement for peace', the need 'to deal with the Palestinian problem', and he expressed this in words that went further than Rabin was then prepared to go. 'There has to be a homeland provided,' Carter said, 'for the Palestinian refugees who have suffered for many, many years.'

Yasser Arafat and the PLO were encouraged by Carter's remarks. Rabin feared that they would harden opinion inside Israel and have an adverse effect on Labour in the coming election.

While she had been in Washington with her husband that March, Leah Rabin intended to close the small bank account that they had maintained jointly while he had been Ambassador there, more than four years earlier. It was not a large account, never exceeding $20,000, and standing at $2,000 when she decided to close it. Although her husband was a co-signatory, she alone had used it. It was entirely legal for Israelis living abroad (as the Rabins were, when he was Ambassador) to hold such accounts, but it was illegal for them to do so while living in Israel, as they had been since 1973. The account had therefore been illegal for the previous four years.

As Leah Rabin had been about to set off to go to the bank to close the account, she was urged by the security people to proceed to the next official engagement with her husband, time was pressing; so she decided to close the account on her next visit. On March 15, within a week of the Rabins having returned to Israel, an Israeli journalist, Dan Margalit, the Washington correspondent of *Ha-Aretz*, broke the news of the Rabin bank account. Throughout the following month, the air of scandal was maintained in the Israeli newspapers.

On April 7, ten days before his wife was to be tried for the illegal bank account, Rabin resigned as Prime Minister. Shimon Peres became acting Prime Minister in his place, at the head of a caretaker government that was on the eve of elections. On April 17, Leah Rabin was brought to trial for the illegal bank account, and fined a sum ten times the amount of the account at its highest. Three weeks before the elections, Peres was elected leader of the Labour movement.

Such hopes as Rabin had harboured for progress on both the Jordanian and Egyptian diplomatic fronts, with or without the goodwill of the United States, had to be set aside, as electioneering, under another Labour leader, gathered momentum, and as the Likud adopted as part of its election platform a hard line with regard to any border modifications. In its fifteen-point programme, the Likud adopted an outspoken stance, calling for an increase in the 'setting up of defensive and permanent settlements, rural and urban, on the soil of the homeland': that is to say, the whole of the West Bank: what the Likud called 'Western Palestine' – from the Mediterranean Sea to the River Jordan.

The Israeli elections in May 1977 were to prove a watershed in the history of the State. During the election campaign, the rule of the Left, and of the established Parties of government – as represented during the election in their most recent manifestation, the Alignment – was being seriously challenged for the first time since 1948. From both the centre and the right came opposition to the Labour establishment more formidable than any faced since Independence.

In the centre, the archaeologist Yigael Yadin stood at the head of a new political Party, the Democratic Movement for Change (DMC, or Dash). Among those who were active in the leadership of Dash were Professor Israel Katz, a leading social reformer, General Meir Zorea (Zaro), a veteran of the War of Independence, Stef Wertheimer, a leading industrialist, Shmuel Tamir, who had been expelled from Herut in 1966 for criticizing Begin's leadership, and several former Labour Party activists, including Meir Amit, a former head of the Mossad (1963–68).

Modernization was one of the main platforms of the new Party, which favoured electoral reform (a single-member constituency system to replace proportional representation and block party electoral lists), decentralizing government power and strengthening the regional administrations, and the establishment of a Ministry of Welfare in order to focus all the authority needed for social reform in such a way as to ensure far greater government effort on behalf of the dispossessed and the outsiders in Israeli society. A cardinal principle in Dash was social integration and the closing of social gaps.

With regard to the Palestinians, Dash wished to seek an accommodation with them that would preserve 'the Jewish character of the Jewish State': that is, to have a Jewish State which would not contain within it a substantial Palestinian population (on the West Bank and in the Gaza Strip) that, sooner or later, would have to be given voting rights and equalities that would – if Jews and Palestinians were still living within the same jurisdiction – eventually dilute the Jewishness of the administration and society.

Among those who saw Dash as offering a way forward for the underprivileged sections of Jewish–Israeli society was the young leadership from the development towns, who felt that the needs of their essentially poor and neglected communities had not been addressed by Labour. Many of the Moroccan youth of the Oded movement, which had been founded in 1962 to redress the balance of discrimination against the Moroccan community in Israel, were also drawn to Dash, and joined its ranks.

Dash was also joined by Shinui (Change), a centrist, liberal, political protest movement. Led by Amnon Rubinstein, and formed three years earlier, Shinui had been dedicated to healing the rifts in Israeli society between the haves and the have nots, the Ashkenazi and the Sephardi, the religious and the secular. Yet for all that, Dash remained a largely Ashkenazi-dominated Party.

The election for the ninth Knesset was held on 17 May 1977. The Democratic Movement for Change did exceptionally well, for a new Party, winning fifteen seats. Its members hoped, in the main, to become an essential part of a Labour-led coalition, and to influence that coalition towards social and electoral reform. In previous elections fifteen seats would have given it the power to negotiate with Labour over coalition building, and would have secured several senior Cabinet posts. But Labour itself – still suffering from the stigma of the October War – did less well than at any time in its history, winning only thirty-two seats.

Overwhelmingly, the Sephardi voters, still feeling the effect of what they saw as Ashkenazi élitism, decided to support Menachem Begin. Although Begin was himself an Ashkenazi (born in Russian Poland), he was able by his own past as an outsider, and by his skilful rhetoric, to appeal to the Sephardi voters, many of whom, surrounding him at election meetings with great enthusiasm, would call out in unison, 'Begin, King of Israel!' The children of the ma'abarot, who had reached voting age, flocked to Likud, and it was Likud that emerged as the electoral victor, with its highest ever number of seats, forty-three.

Shimon Peres resigned as Prime Minister, and was succeeded by Menachem Begin. Begin had come a long way since Ben-Gurion's emphatic and effective declaration that he would rule 'without the Communists and without Begin'. The former head of the air force, Ezer Weizman – a leading Herut activist who was credited with being the architect of the Likud victory as head of Begin's electoral campaign – became Minister of Defence. In a political coup for Begin, Moshe Dayan, once a stalwart and bastion of the Labour Alignment but relegated to the political wilderness following the October War, agreed to join the Likud-led coalition as Foreign Minister.

Begin, himself an unknown on the international stage, wanted Dayan in his Cabinet because of the international eminence that still clung to him, despite the setbacks and disasters of the October War. Many in the Labour movement were distraught that Dayan agreed to join Begin. But their time of power had gone. After almost thirty years of unbroken dominance of the Israeli political scene, they were in opposition.

One new Party that found itself on the right side of the new political divide was Shlomzion, founded by Ariel Sharon, willing to give the Palestinians of the West Bank self-government provided Israel remained sovereign and had military control, but arguing that the most effective solution would be the fall of the Hashemite Kingdom of Jordan and its replacement by a Palestinian State east of the Jordan (the Party promoted the slogan 'Jordan is Palestine'). Shlomzion gained two seats in the election. When it became clear that Likud was going to form a government, Sharon returned to his former Party.

By bringing in the National Religious Party, with its twelve seats, and the Agudat Yisrael with four, Menachem Begin was able – with the numerical addition of Dayan and Sharon – to form a Likud-led coalition on June 20 without calling on Dash at all. Sharon was brought back to the Cabinet as Minister of Agriculture.

In the revolutionary changeover of political mastery in Israel, it was not clear what would happen in the civil service. Most senior civil servants were Labour men, used to serving Labour masters, and those allied to Labour, with ease and over many years – since the founding of the State almost thirty years earlier. In the immediate aftermath of the election, the senior official in the Israeli Foreign Office asked a British visitor what would happen to him under the British system. Would he be sacked? Would he have to resign? Would someone with Likud sympathies be brought in to take his place?

Begin had in fact already decided, and indeed insisted, that the integrity of the professional civil service must be maintained, on the British model, despite the political upheaval. He therefore took it as axiomatic that the senior civil servants in the Prime Minister's Office would remain, as they did. So too did the anxious Foreign Office official; he was later Israeli Ambassador to the United States.

One of the first things Begin did on assuming office was to introduce the ancient biblical names for the West Bank: Judaea and Samaria. He refused to use any other term. With regard to the Palestinian Arabs, the hard-line nature of the new government was revealed within a few days when Begin went with his close associate, Shmuel Katz, to Washington. Katz, an old-time Irgunist, was originally from South Africa. Begin had appointed him to a newly created post, adviser on information. Not only did Katz speak, on July 19 in the Cabinet Room at the White House, of the right of the Jewish people to the whole of 'Western Palestine' – of which the West Bank was a part – but he stated (to President Carter's amazement) that were Israel to refrain from building new settlements it would 'prejudge the outcome of negotiations'. Not to build would be tantamount to accepting in advance one of the demands that might be put forward; and one, incidentally, that Israel, under its new leadership, might well reject.

To a point made by Katz, that an Arab entity of any kind west of the River Jordan would be a threat to Israel, and that whatever the Arabs might say, what they actually believed was that Israel must be eliminated, Carter noted: 'I see no moderation here.'

Four months had passed since Rabin's visit to Washington, with its intimation of Israeli territorial concessions. But after Katz had spoken, Begin told Carter of the 'danger' that would face Israel if she were to return to the pre-1967 borders. 'Men would not be able to defend women and children,' he said. Those present thought that tears were about to come into Begin's eyes as he spoke. Shmuel Katz then spoke, at Begin's request, of how the Palestinians already had their homeland, on the east bank of the Jordan, in the Hashemite Kingdom.

When, in talks with Cyrus Vance, Begin was asked if he would accept an Israeli withdrawal to 'mutually agreed and recognized borders on all fronts, phased over years in synchronized steps, and with security arrangements and guarantees,' Begin replied, 'In the whole world, there is no guarantee that can guarantee a guarantee.' Elaborating on what he had earlier told Carter, Begin then told Vance that a Palestinian State on the West Bank would be a 'mortal danger' to Israel. Such a State, Begin warned, would become a Soviet base with Soviet military aircraft arriving from Odessa in two and a half hours, and Soviet generals stationed in the West Bank. Jerusalem would be under crossfire 'from three directions' (presumably, Jordanian, Palestinian and Soviet).

In a private talk with Carter that evening, Begin revealed a flexibility that he had not shown before. He was making 'tentative plans', he said, to meet Sadat. He would try to 'accommodate' the United States on settlements, though it would be 'difficult' for him, politically, to restrict new settlers to the existing settlements. Israel would stay on the Golan (Rabin had also said this). As to Sinai, Begin told Carter (this was also what Rabin had said) that he was prepared to contemplate 'substantial withdrawals'.

Begin totally accepted two facets of the Rabin doctrine: the importance of keeping in step as much as possible with the Americans, and recognizing that Egypt, the largest and strongest of the Arab States, was the key to a comprehensive Middle Eastern peace.

It seemed in Washington that, with regard to Egypt, progress towards peace might be made. On the Palestinian track, the gulf between Begin and Carter (and between Israel and the Palestinians) seemed unbridgeable. On August 8, in discussion with Cyrus Vance in Jerusalem, Begin tried to bypass the American desire for a territorial arrangement for the Palestinians – whether in the pre-1967 borders or in an adjustment of them – by suggesting that Israel might consider offering 'our neighbours in Judaea, Samaria and Gaza full cultural autonomy'. Not only would this be within Israel's borders but, Begin explained, it would be in conjunction with the possibility of Israeli citizenship and full voting rights.

The Americans were still trying to prevail upon Israel to halt settlement building on the West Bank. On August 17 the Israeli Cabinet approved the establishment of three new settlements. In a message to Begin on the following day, in a personal note sent through the American Ambassador,

Carter reiterated his belief that settlement building was illegal. 'These illegal, unilateral acts in territory presently under Israeli occupation create obstacles to constructive negotiations,' the message read. 'The repetition of these acts will make it difficult for the President not to reaffirm publicly the US position regarding 1967 borders with minor modifications.'

For many Israelis, Begin's offer of Arab voting rights raised the spectre of more than a million Arabs having a predominant say in the policies of the State, and the possibility of the third largest bloc of seats in the Knesset, something that would not have been acceptable to his Herut followers, or to the Labour opposition. Yet Begin had no intention of ceding the West Bank or the Gaza Strip to the Palestinians. The way to avoid continued American pressure with regard to the West Bank, he felt, was to take an active part in the search for a peace treaty with Egypt. If peace could be made with Egypt, the United States would be pleased, and no harm would be done to the 'integrity' of the West Bank. When President Sadat saw Cyrus Vance that August, he said he wanted to meet Begin; Vance at once passed this news back to Begin, who expressed interest.

Israel still hoped to reduce Palestinian pressure for a West Bank negotiation by persuading the Jordanians to be Israel's partner in all matters relating to what had been, from 1949 to 1967, a Jordanian-ruled region. On August 22, Dayan met secretly with King Hussein in London: it was the first time that any member of the Begin government had held direct talks with any Arab leader. But Hussein made it clear that he was not prepared to speak on behalf of the West Bank Palestinians, for whom the Arab summit at Rabat three years earlier had stated that the PLO was their 'sole representative'; nor was he prepared for any division of the West Bank between Israel and Jordan as the basis for an agreement with Israel. It was essential, the King said, for Israel to withdraw fully from all occupied Arab territory, including East Jerusalem.

A meeting with Sadat was at the top of Begin's foreign policy priorities. Hardly had Dayan returned from London and reported on Hussein's refusal to contemplate a deal behind the backs of the Palestinians and the PLO, than Begin flew to Roumania (the only Communist State that maintained full diplomatic relations with Israel). As Sadat was due to visit Roumania shortly afterwards, Begin asked the President, Nicolae Ceausescu, to help arrange a meeting between them. On September 4, five days after Begin's return from Roumania, Dayan flew to Morocco, where, in a secret meeting with King Hassan, he asked the King to help expedite a Begin–Sadat meeting. The Americans were not informed of this Israeli initiative.

On September 16 Dayan returned to Morocco, where, as a result of King Hassan's efforts, he met the Egyptian Deputy Prime Minister, Hassan Tuhami. This was the decisive meeting. Tuhami made it clear that Sadat was prepared to negotiate directly with Israel, that he did not insist on a conference with the other Arab States (America's position), and that he would most probably accept an Israeli withdrawal from Sinai in return for an Egyptian–Israeli peace treaty, without insisting on Israel withdrawing at the same time from the Golan Heights, the West Bank, or even the Gaza Strip (which Egypt had ruled from 1949 to 1967). The way was set for a Begin–Sadat encounter, and for a separate deal between Israel and Egypt.

There was another catalyst affecting both Begin and Sadat. Their secret initiatives took place against a background of President Carter's attempts to revive the Geneva Peace Conference. These included a State Department draft joint statement, to be issued with the Soviet Union, whereby the two superpowers would commit themselves to a comprehensive settlement incorporating 'all parties' and 'all questions'. The words 'all parties' were interpreted by both Begin and Sadat to mean the PLO, and 'all questions' to refer to the West Bank and Jerusalem. Sadat liked this as little as Begin. In Sadat's eyes, the initiative would benefit Syria, Russia's special protégé, who would bring back the Soviet Union into the politics of the region. This was not a congenial prospect for Sadat. Syria had also become the PLO's special patron in Lebanon.

Sadat extricated himself from the possibility of American and Soviet pressure by 'taking the bull by the horns' (as one Israeli diplomat expressed it) and deciding to take a direct and separate initiative of his own – a visit to Jerusalem.

That September the Christian militia in southern Lebanon were engaged in a series of military confrontations with PLO units under Arafat's command. Israeli forces crossed the border and took part in the fighting alongside the Christian soldiers. Begin had not informed Carter of this in advance, a violation – the Americans pointed out – of Begin's promise not to take military action inside Lebanon without 'prior consultation' with the Americans. In addition, the Israelis were using in Lebanon armoured personnel carriers that had been sold to Israel by the United States on condition that they would only be used for 'legitimate self-defence'.

The Israeli government denied that any of these carriers was being used in Lebanon. By resorting to what William B. Quandt called 'new and exotic technology', the Americans were able to give the lie to the Israeli assertion. On September 24 Carter expressed his displeasure in a tough letter that was delivered to Begin by hand. 'I must point out', Carter wrote, 'that current Israeli military actions in Lebanon are a violation of our agreement covering the provision of American military equipment and that, as a consequence, if these actions are not immediately halted, Congress will have to be informed of this fact, and that further deliveries will have to be terminated.'

Begin gave way. Israel, he said, would withdraw its forces from Lebanon within twenty-four hours. Meanwhile, new Jewish settlements continued to be set up on the West Bank, despite Carter having made clear his continued opposition to such settlements when he met Dayan in Washington on September 19, and Dayan having promised Carter that for one year there would be 'no new civilian settlements'. Begin insisted that this was an unauthorized undertaking. Ten days after this pledge, on September 29, the new civilian settlement of Tekoa was established just east of Bethlehem. It was named after a biblical town believed to have been located nearby.

A new source of immigration was opened up in 1977. Up to that year, no more than 150 Jews had made their way from Ethiopia to Israel. The youth village experiment of twenty years earlier had not been repeated. Ethiopia having been invaded by the pro-Soviet government of the Sudan, which had occupied the Ogaden desert, the Ethiopian government did not look with favour on emigration. But in 1977 an arrangement was made, in the utmost secrecy, whereby Israel would airlift military supplies to Ethiopia and, in return, the Ethiopian government would allow the planes to return to Israel with immigrants. The scheme began well, with 121 Ethiopian Jews being flown in, but then, when Moshe Dayan revealed the arms component of the deal, there were protests from the Israeli public about arming the Ethiopian regime – for whatever reason – and the deal was brought to an immediate halt. Three years were to pass before the Ethiopian government, itself in disarray, agreed to relax its emigration procedures.

Although, in June, Begin had been able to form a government without Dash, on October 24 the new Party agreed to join him. Yigael Yadin believed that both the Party and he personally could have an influential part to play, both in social reform policy and in any future negotiations with the Arabs: it was Yadin who had helped the crucial negotiations with King Abdullah during the armistice negotiations with Jordan in 1949.

Yadin had worked hard while establishing his Party to prepare a blueprint for action in Israel's poorest neighbourhoods. It was called Project Renewal. The idea was to encourage local leaders and local youth to establish community centres, and much improved municipal, health and leisure facilities, working with volunteers from the Diaspora who would adopt communities and help develop them. Money would be raised from Diaspora Jewry for specific towns. The Israeli government would provide the money for infrastructure, to help rebuild sewers, pavements and water systems. Yadin's vision was that with physical would come spiritual renewal, raising the neighbourhoods from despised and hated slums to suburbs and townships whose local councils would take a lead, and take pride, in the task of self-improvement.

At first Project Renewal worked well: at Ashkelon, on the Mediterranean, mainly British volunteers brought medical, dental and town-planning expertise as well as great enthusiasm for working with the local people. Some of the volunteers even decided to stay, becoming an integral part of the community they had come to serve. Two extremely poor and run down inner city slums, one in Jerusalem and the other in Tel Aviv, were also renewed. Unfortunately for Yadin, within two years of Project Renewal getting under way, and beginning to show impressive results, a new Finance Minister, Yigal Hurvitz, whose personal and political priority was to allocate funds to West Bank settlements, declared that Project Renewal was a 'dangerously inflationary' undertaking, and froze all the government funds that had been allocated to it.

The increase in West Bank settlement building after Begin's electoral victory was one of the obstacles to Yadin and his new Party entering the Begin government. As a condition for joining, Yadin had been assured by Begin that he could personally appeal in Cabinet against any decision to build a West Bank settlement in densely populated Arab areas, something to which Yadin and his Party were implacably opposed. But those who had the main ministerial authority with regard to settlement building, the Agriculture Minister Ariel Sharon and the Defence Minister Ezer Weizman, had more political weight than Yadin, who had for many years been an outspoken opponent of the right-wing politicians, and indeed of Begin himself.

When Yadin protested at the new settlements, Sharon replied that these were not new settlements at all, but 'expansions' of existing settlements. Yadin called this device a 'shameless deception', and clashed openly with Sharon in Cabinet over it. But his threats to resign were brushed aside, and he remained in the Cabinet – unable to prevent the creation of several dozen West Bank settlements – in the hope of continuing Project Renewal, and improving the social structure and morale of Israel's many poor neighbourhoods.

As leader of the Labour movement, Shimon Peres faced the task – unique in the history of the Labour movement in Israel – of leading an opposition rather than heading a government. Peres inherited a disillusioned Party, which he rebuilt with skill and patience. One of the burdens that he was to face while leader was the publication (in August 1979) of Yitzhak Rabin's memoirs, in which Peres was described as 'an inveterate schemer', and his judgement repeatedly belittled. Rabin wrote of Peres that he had not been above working 'against me' through leaks, when Rabin was Prime Minister and Peres was Minister of Defence. 'He not only tried to undermine me,' Rabin wrote, 'but the entire government, trusting in the old Bolshevik maxim that "the worse the situation, the better for Peres". He spread lies and untruths and wrecked the Labour Party, thereby crowning himself as leader of the opposition.'

The publication of the memoirs stirred further controversy when the *New York Times* revealed (on October 23) that Rabin gave details, in a passage which a board of censors of Cabinet Ministers had deleted, of the expulsion of 50,000 Arabs from Ramle and Lydda in 1948. According to Rabin's account, when Yigal Allon asked Ben-Gurion, in Rabin's presence, what was to be done with these Arabs, 'Ben-Gurion waved his hand in a gesture which said, "Drive them out."' Rabin's account continued:

'Driving out' is a term with a harsh ring. Psychologically, this was one of the most difficult actions we undertook. The population of Lod did not leave willingly. There was no way of avoiding the use of force and warning shots in order to make the inhabitants march the ten or fifteen miles to the point where they met up with the Legion.

The inhabitants of Ramle watched and learned the lesson. Their leaders agreed to be evacuated voluntarily

Yigal Allon denied Rabin's account of the use of force, but the damage was done. Subsequently the historical record was to bear Rabin out.

On October 30 the settlement of Mevo Dotan was founded on the West Bank. The settlers were secular. Their aim was to strengthen Jewish settlement in northern Samaria. Two days later, another new settlement, Beit El (The House of God) was founded on the West Bank, in the hills just north of the Arab town of Ramallah. The basis whereby the founders came together was religious: 'To settle the land of Israel and to base it on the power of Torah and Zionism.' Among the religious articles that they were to make in the settlement, and sell commercially, were phylacteries and audio cassettes of the holy books. They also established an aluminium factory and a factory producing cartons. Twenty years later, there were 230 families (1,500 people) living there, three synagogues and a Yeshiva high school.

On November 6, five days after the founding of Beit El, Arab gunners, members of the PLO, fired Katyusha rockets from across the Lebanese border at the seaside town of Nahariya. One Israeli woman was killed. She was a survivor of the Holocaust. The Minister of Defence, Ezer Weizman, had to decide whether, and how, to respond to the attack. 'Like other native Israelis, or Sabras, I did not personally experience the Holocaust, nor was I closely involved,' Weizman later wrote. 'But – again like most of my generation – I live in its shadow. There is no avoiding it in Israel. The more we try to repress it, the more it springs up out of the depths of our unconscious.' Weizman continued:

I found myself contemplating the fate of that woman. She had been rescued from the very jaws of purgatory – only to find her death in Nahariya. On hearing the first volley of Katyushas, she had run outside to call her two children to the shelter – and was killed on the spot, before their eyes. That scene would probably haunt them for the rest of their lives.

A Defence Minister is supposed to make his assessments and reach his decisions by coldly rational thinking. As a rule, there is no room for sentiment. It is his duty to take into account a whole range of considerations: Is a military strike necessary? What will happen in consequence? What will the UN say?

But a Defence Minister is only flesh and blood. His judgement cannot be totally divorced from his personality, his feelings – perhaps even his philosophy.

During my period as commander of the air force, I had frequently advised the political echelons to employ our airplanes against enemy targets; and then, when my counsel had been adopted, I'd proceeded to send our pilots on their way.

As in all democracies, the elected government is in full control of the military forces, and the air force commander reports to the defence minister. But now the final responsibility was mine, and I found myself torn. There were considerations that made me reluctant to use the air force. The Likud government had a hawkish image, which I had no desire to stress further.

There was also a substantive consideration: any action on our part was likely to provoke the terrorists into exacting vengeance upon our northern settlements. Such a response from the terrorists was bound to entail a further strike on our part. This exchange would lead to a deterioration of security in a densely populated area.

I had to bear in mind that our pilots are no supermen. It needed nothing more than one bomb falling off target to deliver yet another blow to Israel's tattered image in world opinion. It's hard to cope with television clips of Lebanese children scrabbling in the rubble, searching for their parents – the terrible and sometimes unavoidable result of war.

'However,' Weizman concluded, 'in my mind, I saw that woman from Nahariya. There had been no bloodshed in Sinai for more than three years. But in the north a citizen of Israel could still be mowed down in daily life. There could be no mercy for the terrorists. With the approval of Prime Minister Begin, I gave instructions for the planes to take off on their mission.'

The Israeli retaliatory raid took place on the following morning. Later that day, President Sadat, in a speech at the opening of the Egyptian parliament, said that he wanted to address members of the Knesset in Jerusalem. 'I am willing to go to the ends of the earth for peace,' he said. 'Israel will be astonished to hear me say now, before you, that I am prepared to go to their own house, to the Knesset itself, to talk to them.'

'I didn't believe a word of it,' Ezer Weizman recalled. 'I thought his statement about coming to Jerusalem was no more than meaningless lip service, and I utterly mistrusted it.' That evening, Begin replied publicly that if Sadat came, he would be welcome. The official invitation was delivered to him through American diplomatic channels. Sadat accepted.

Israel was in a turmoil of anticipation and disbelief. In the Cabinet, Ariel

Sharon proposed that, on his arrival, Sadat should be offered, as a unilateral gesture by Israel, the civilian administration of El Arish (Sinai's administrative capital, captured by Israel in 1967) and the right of passage by road across the Sinai to the city. Begin was attracted to this plan, but Dayan opposed it, arguing that Sadat was surely not coming to Israel to secure a 'morsel' of land, and that, in addition, he would be accused by the Arab world of separate peacemaking with Israel.

On November 14, as preparations for Sadat's visit were under way – he was due five days later – the Chief of Staff, General Motta Gur, gave an interview to the newspaper *Yediot Aharanot*, without first seeking the customary approval of his immediate superior, the Minister of Defence. In the interview, after hinting at Sadat's deceptive intentions in coming to Israel, he said, 'We know that the Egyptian army is in the midst of preparations for launching a war against Israel towards 1978.'

Furious, Weizman summoned Gur to the Defence Ministry in Tel Aviv and told him that he intended to cut short Gur's term as Chief of Staff by four months: he must leave that December. Gur said that if that was to be his fate, he would rather resign straight away, or be sacked. Weizman set off by car to Jerusalem to put the issue before Begin. On the journey, Weizman's driver swerved to avoid a person on the road and Weizman was badly injured.

By the time Weizman regained consciousness in hospital, the preparations for Sadat's visit were much further advanced. 'In these epoch-making days, of such great importance in the chronicles of our times,' Gur wrote to Weizman, 'it would be unforgivable if two men like us were to engage in trivialities instead of rising to the occasion.'

Gur remained Chief of Staff. Sadat finalized his plans for the historic journey. Ezer Weizman, forced to remain in his hospital bed, watched the event on television:

> There were my Cabinet colleagues strutting about the runway, rigged out like bridegrooms, sporting their finest plumes and feathers and blown up with their own self-importance. When the camera zoomed in on Motta Gur, clad in festive attire, I almost went out of my mind: pressed and polished, he was standing in line along with the rest, waiting to shake Sadat's hand.
>
> I alone was condemned to be trussed up in bed.
>
> The plane door opened. When no one appeared, I held my breath. But a fraction of a second later, I could make out the familiar figure of the Egyptian President. Our enemy set foot on Israeli soil. The unbelievable was happening.

Rabin had been in New York on a lecture tour when news of Sadat's visit was announced. As a former Chief of Staff and former Prime Minister, he was invited to be present at the welcoming ceremony. He was just able to catch a plane that would bring him back to Israel in time. 'When I greeted President Sadat on the receiving line,' he later wrote, 'there was no time for anything more than a brief exchange of pleasantries. Yet it was the first time I had ever seen him in the flesh, and I was enormously impressed by the poise with which he handled himself in such a unique situation. Here he was meeting all his former arch-enemies, one after another, in the space of seconds, and he nonetheless found a way to start off his visit by saying exactly the right thing to each and every one of them.'

Sadat greeted Begin with the words: 'No more war. Let us make peace.' When Golda Meir shook Sadat's hand she said, 'We've been waiting for you a long time', to which Sadat replied, 'The time has come.'

Sadat's welcome by the people of Israel, who lined the streets of Jerusalem waving the Egyptian flag, was symptomatic of the historic, psychological breakthrough that was occurring. While Sadat was in Jerusalem he stayed at the King David Hotel. Before going to pray at the al-Aksa mosque he performed his private devotions in his hotel room. On November 20 he addressed the Knesset. He had not come to make a separate peace between Israel and Egypt, he said, but to end the state of belligerency between the two countries. He then urged the complete withdrawal of Israeli troops from all occupied Arab lands, and the establishment of a Palestinian State. As he said these words, Ezer Weizman wrote a brief note which he handed to Dayan: 'We've got to prepare for war.' But Sadat went on to pledge an end to fighting. 'Tell your sons that the past war was the last of wars and the end of sorrows,' he said. Dayan was impressed.

The American government's immediate reaction to the Sadat visit was surprise, bewilderment, and even apprehension that Israel and Egypt might cut a deal and leave American interests stranded. Secretary of State Cyrus Vance made a rapid visit to both Jerusalem and Cairo, where he was given assurances that close American participation in the process would be welcomed.

In the immediate aftermath of his visit, Sadat suggested a second secret meeting, in Morocco, between Dayan and Hassan Tuhami, the Egyptian Deputy Prime Minister. Begin agreed and, on December 2, when Dayan met Tuhami he gave the Egyptian a proposal for the restoration of Egyptian sovereignty over the whole of Sinai. The area between the Mitla and Gidi Passes and the Israeli border would be demilitarized. Israeli settlements in Sinai would remain, under the flag of the United Nations.

To Dayan's chagrin, Tuhami was not pleased. Any arrangement between Egypt and Israel, he said, would have to include 'a resolution of the conflict' with all the other Arab States. Three weeks later, on December 25, Begin, Dayan and Weizman went to Egypt, invited by Sadat to visit him at his holiday residence. 'Look,' Dayan said to Begin as they drove there from

Ismailia, 'not a single Israeli flag. Not even a banner to welcome us.'

During the talks, Sadat rejected the territorial proposals that Dayan had put to Tuhami at the beginning of the month. Israel must withdraw from all occupied lands, he said, and must agree to Palestinian self-determination. There could be no separate Egyptian–Israeli peace agreement. Begin and Sadat agreed to set up several Egyptian–Israeli 'working committees'; but five days after a difficult meeting of the political committee at the Hilton Hotel in Jerusalem, Dayan told Israel television, 'If Sadat really wants an Israeli commitment to withdraw from all territories occupied in '67 before negotiations begin, then the situation is at a dead end.'

For six months, there was deadlock. Early in 1978 a group of 350 army reserve officers wrote a letter to Begin, urging him to pursue 'the road to peace'. Their letter was the first manifestation of a new group, Peace Now, that for the next decade consistently pressed for negotiations with the Palestinians, holding many public rallies to protest whenever it felt that government policy was neglecting the peace option.

'Our enemy set foot on Israeli soil,' Ezer Weizman had reflected on Sadat's visit. 'The unbelievable was happening.' But other enemies were still active. On 11 March 1978 a PLO unit in Lebanon landed on the Israeli coast. Their first victim was Gail Rubin, an American-Jewish woman photographer who was working at the point where they landed. The unit then hijacked a bus on the Haifa to Tel Aviv highway and forced the driver to continue driving south. Shortly after the bus went through Herzliya it was stopped, and in the ensuing struggle thirty-nine of those on the bus were killed.

Israel retaliated four days later by invading southern Lebanon – known in Israel as Fatahland – and sweeping northward as far as the Litani river. Several dozen PLO soldiers were killed or captured, and a 10-kilometre strip of territory along the border was occupied by Israeli troops. All PLO installations were systematically destroyed. Three months later, on June 13, the Israeli forces withdrew and United Nations troops, the United Nations Interim Force in Lebanon (UNIFIL), took over the policing of the region.

Five days after the Israeli withdrawal from Lebanon, negotiations between Israel and Egypt, which had languished since the Ismailia summit six months earlier, seemed to take on a new momentum when high-level talks were resumed at Leeds Castle in the south of England. Moshe Dayan led the Israeli delegation, Egypt was represented by its Foreign Minister, Mohamed Ibrahim Kamel, and the United States by the Secretary of State, Cyrus Vance. But the Egyptians still wanted the Palestinian issue to be an integral part of any agreement.

Following Israel's withdrawal from southern Lebanon, the Lebanese border returned to a more peaceful state. This was not surprising: until the PLO had moved into the border areas seven years earlier, the Lebanese–Israeli border had been the most tranquil of them all.

In the Lebanese conflict itself, Israel supported the coalition of Christian leaders – formed in 1976 – known as the Lebanese Front, which was struggling to prevent Syrian domination of Lebanon. Israel not only gave the Lebanese Front arms and ammunition, but provided training for its soldiers, and flew warning and reconnaissance missions over Beirut.

Amid the drama created by Sadat's visit and its aftermath, and the constant worries about PLO activity and the Lebanese imbroglio in the north, the artistic and cultural life of Israel continued on its usual imaginative course. On 15 May 1978 the Diaspora Museum – Beit Hatefutsot – was opened on the campus of Tel Aviv University. As well as a host of Israeli dignitaries, headed by President Katzir (whose brother had been killed in the Lod airport massacre in 1972), the overseas guest list was headed by Jacqueline Kennedy.

It was the United States that continued to take the initiative in trying to get Israel and Egypt to reach agreement before the impetus of Sadat's visit to Jerusalem was entirely lost. On 4 September 1978 a meeting began in Maryland, USA that was to transform the map, and in many ways the atmosphere, of the Middle East. The place in which the negotiations were to take place, Camp David, was the retreat used by American presidents to escape from the crush and pressure of Washington. In President Roosevelt's day it had been known as Shangri La.

To everyone, not least to the prestige of President Carter, the meeting at Camp David was an enormous gamble. The Israeli participants who gathered on September 4, headed by Menachem Begin, included the Foreign Minister, Moshe Dayan, and the Minister of Defence, Ezer Weizman. Throughout the discussions, President Carter served as a mediator. President Sadat was there, one of several Arab leaders who, it was hoped, would participate in the search for a 'comprehensive peace'; indeed, it was only on the basis that the purpose of the meeting was a comprehensive peace encompassing all Israel's Arab neighbours that Sadat agreed to attend. The leaders of the other Arab countries – Jordan, Syria, Saudi Arabia as well as the Palestinians – refused to join him, however. But Sadat persevered. Whatever agreement he could reach, he would present as one that had been reached on behalf of the absentees as well. It would, in his words, be 'a first step to comprehensive peace'.

After thirteen days of intense negotiation, agreement was reached. This would not have happened but for the patient, persistent and personal efforts of President Carter, who engaged directly in every aspect of the negotiations, and wrote the final draft himself. This was not a peace treaty, but 'A Framework for Peace in the Middle East', to be elaborated into full treaties through future negotiation.

The Camp David Accords were signed by President Carter, President Sadat and Prime Minister Begin on September 17. Two accords were signed. The first accord called for the implementation of an autonomy plan for the West Bank and Gaza Strip, to be followed after five years by a 'permanent settlement'. Negotiations regarding the permanent status of the West Bank and Gaza Strip 'must also recognize', the accord stated, 'the legitimate rights of the Palestinian people and their just requirements'.

This was the first time that Israel had conceded what were essentially the national aspirations of the Palestinians, a people hitherto regarded as either former Jordanians, or as Arabs who happened to live in and around the cities of Nablus, Hebron, Ramallah, Jenin, Bethlehem and Tulkarm whose loyalties were primarily to their families and regions. The acceptance of a Palestinian identity, and of the 'legitimate rights' of the 'Palestinian people', was a major step forward for Israel, and one which the Palestinians were in due course to use to the fullest advantage.

The first Camp David Accord also called for the 'normalization' of relations between Israel and Egypt, to be followed by similar agreements with Israel by Jordan, Syria and Lebanon. At the time of signing, none of these three countries showed the slightest interest in pursuing such a call.

The second Camp David Accord was a framework for the conclusion of a peace treaty between Israel and Egypt. This was to be based upon the complete withdrawal of Israel from Sinai, 'up to the international recognized border between Egypt and Mandatory Palestine'. To replace the security Israel had sought in its air bases in Sinai – at Refidim, Eytam, Etzion and Sharm el-Sheikh – the United States had agreed to establish two early-warning stations just to the east of the Gidi and Mitla Passes, and to use these to alert Israel in the event of any build-up of Egyptian forces over and above the levels set in the accords.

Israel also agreed to give up all its settlements in Sinai. In a conversation with Dayan when the talks had reached an impasse, Sadat had implied that withdrawal from these settlements was the condition for opening full diplomatic relations with Israel and signing a peace treaty. During the negotiations, one of the Americans had asked Dayan why he had urged the building of the largest of the Sinai settlements, Yamit. Dayan replied, 'We never thought the Egyptians would make peace. Now that we are making peace, we have to give it up.'

In giving up Sinai, Israel also gave up the airfields it had built throughout the peninsula. In return, the United States agreed to build two military airfields for Israel in the Negev, and to do so before the final Israeli withdrawal from Sinai was completed. Although Israel's demand for the demilitarization of Sinai was rejected, Sadat did agree to a large buffer zone between the two armies.

The second accord laid down that once Israel had withdrawn from Sinai, normal relations were to be established between Egypt and Israel – which had been in a technical state of war with each other for exactly thirty years.

Such relations were to include the establishment of diplomatic, cultural and economic relations. Egypt was also to agree to 'the termination of economic boycotts' on Israel. These were a feature of the policy of every Arab State, and had forced many Western companies to choose between trade within Israel and trade with the far larger and more economically tempting Arab world (the companies almost invariably chose to forgo trade with Israel).

The Camp David Accords also touched on the question of Jerusalem. When, towards the end of the negotiations, Begin had insisted that he would sign nothing that meant 'signing away Jerusalem', it was Carter who came up with the idea of dealing with Jerusalem, not in the body of the text, but in a separate exchange of letters between Begin and Sadat. Both men agreed to this. Begin's letter declared that Jerusalem was both 'indivisible' and 'the capital of Israel'. Sadat's letter declared (with equal solemnity) that Arab East Jerusalem was an 'indivisible' part of the West Bank, and that it should be returned to 'Arab sovereignty'.

Constitutionally, Begin was not obliged to get Knesset approval for the Camp David Accords. He sought it, however, because of the historic importance of the issue. The Egyptian–Israeli peace was a turning point in the Arab–Israeli conflict. The cycle of violence of the multiple Arab military alliance was over. No war in the past had ever been launched against Israel without Egypt being the leader, and then the other Arab States following. No war against Israel had ever ended without Egypt being the first to pull out, and then the others following.

There was a debate inside the Labour Party on the terms of Camp David. Yigal Allon argued that they could not support that part of the agreement which called for the dismantling of the settlements in Sinai. In response to this, Peres argued that the agreement could not be divided: 'We have to work either for it or against it.'

The Camp David Accords were voted on by the Knesset on September 27. Eighty-four members gave their approval, including the majority of the Labour members, who thereby secured Begin his majority. Nineteen members voted against. There were seventeen abstentions, among them a member of the National Religious Party to whom Dayan sent a laconic note: 'By abstention shall the righteous live.' Among those who voted against were Yigal Allon and Moshe Arens, a leading Likud figure and recent chairman of the Herut Central Committee; he was opposed to the withdrawal from all of Sinai. A leading Labour politician, Shlomo Hillel – Minister of Interior and the Police in the previous Labour government – also voted against for the same reason. Among those who abstained was Begin's eventual successor as leader of the Likud, Yitzhak Shamir.

The signing of the Camp David Accords created serious divisions inside Likud. Those who had supported Israeli settlements beyond the Green Line were outraged that Begin had agreed to give up Sharm el-Sheikh, at the entrance to the Gulf of Akaba, whose closure by Egypt had twice in the past led to war. There was equal, and for some even greater outrage, that the Rafa Salient on the northern coast of Sinai, in which, since 1967, eleven Israeli settlements had been established, the home of 2,000 Jews, was to be given back to Egypt. The largest of these settlements, Yamit, was on the verge of a considerable expansion. Begin had also agreed to give up three settlements on the Gulf of Akaba – Neviot, Di Zahav and Ophira – home to 800 Jews and, with their unspoilt beaches and coral reefs, magnets of Israeli tourism.

Those who had been uneasy with Begin's commitments at Camp David began to speak up. One of them was Shmuel Katz, who had been Begin's adviser on information. Another was the former left-wing novelist, Moshe Shamir, whose brother Elik had been killed in the War of Independence. Moshe Shamir had become a Likud member of the Knesset in 1977. Another Likud member of the Knesset to break away in protest against Camp David was Geula Cohen who as a young woman had been the Irgun's radio broadcaster; captured by the British she was sentenced to nine years in prison but escaped and returned to broadcasting. Cohen and Moshe Shamir, while retaining their seats, formed (as was common in Israeli political life) a separate Party. It was called Banai: *Brit Ne'emanei Eretz Yisrael* – Alliance of the Land of Israel Faithful.

Those who rejected the Camp David Accords were on the far right of the Israeli political spectrum. The fact that the accords had been negotiated and signed by the most dominant figure on the right, Menachem Begin, made the accords all the more galling. In an attempt to form an effective political opposition, three right-wing groups – the Loyalist Circle in Herut, the Land of Israel Movement, and Gush Emunim – decided to establish a nationalist Party, in opposition to the Begin government. In October 1979, joining with Banai, it took the name Tehiyah (Revival).

The principal platform of Tehiyah was the abrogation of the Camp David Accords. Its ideological base was the 'redemption' of the Jewish people in the whole of the Land of Israel – never exactly defined, but always including the whole of the West Bank. To the assertion in the Camp David Accords that the Palestinians had legitimate rights, Tehiyah responded that such rights as the Palestinians possessed (and all human beings were possessed of rights) were as individuals, not as a people. One of Tehiyah's parliamentary successes was a law proposed by the Party and passed by the Knesset a year later, authorizing the application of Israeli law to the Golan Heights.

How far the agreement with Egypt could survive was constantly debated in Israel. On November 10, Hannah Ruppin, the widow of Arthur Ruppin, who had lived in Jerusalem since before the First World War, wrote in a private letter, 'As I am a Pessimist, I don't believe we will have peace.' On the practical question of Egypt's insistence on having the future autonomy of the West Bank as an integral component of the agreement, a member of the legal department of the Israel Foreign Ministry confided, 'At this point in time I don't see how we are going to get out of the basic linkage issue; do the Egyptians make their agreement conditional on progress on the West Bank or not? My personal feeling was that a "fig leaf" would be sufficient, but I am not so sure now.'

In the event, the Egyptians accepted Begin's 'fig leaf' and went ahead with their Sinai agreement (as did Israel) without any substantive progress towards any form of Palestinian self-government on the West Bank. The Jewish settlers were also active in the wake of the Israel-Egyptian agreement. During 1979 the ideology and activities of the Gush Emunim settlers' movement received a further boost – the settlement at Eilon Moreh in 1975 had been its first – when Rabbi Moshe Levinger, his wife, and twenty other women slipped into a building in the centre of the Arab city of Hebron and refused to move. The building they chose was the Hadassah clinic which had been one of the centres of Jewish life in Hebron before the Arab riots of 1929, the massacre of several dozen Jews, and the evacuation of the Jewish community.

Rabbi Levinger and his followers set up living accommodation in the building and remained stubbornly in the midst of a hostile Palestinian population until they were granted, by Begin's government, legal status, and the protection of the Israeli army.

The final negotiations for the peace treaty between Israel and Egypt took place during a visit by President Carter to Jerusalem in March 1979. The United States gave Israel a guarantee that for the next fifteen years it could purchase oil from the Egyptian oilfields in Sinai, which Israel had developed during its six-year occupation. After those fifteen years, if Israel could not get oil from other sources, the United States would provide it.

In the final tense hours of the Jerusalem negotiations, Israel refused Sadat's demand for Egyptian liaison officers to be stationed in Gaza. It was Carter's direct intervention that persuaded Sadat to drop this demand (Carter telephoned Begin from Cairo airport to say that Sadat had given way). On March 20 the Knesset agreed to the peace treaty by 95 votes to 18. The treaty was signed in Washington on March 26: Sadat, Begin and Carter were the three signatories.

On 10 December 1979, Menachem Begin travelled to Oslo. There, he received the Nobel Prize for Peace together with Anwar Sadat.

CHAPTER TWENTY-SIX

To the Lebanese war and beyond

The year 1980 saw the largest increase in settlements in Israel for thirty years. The hundred settlements of 1950 had been set up as a result of the mass immigration movement, and were of course all within the pre-1967 borders. The thirty-eight settlements of 1980 were almost all across the Green Line, in the West Bank and Gaza Strip, some of them in heavily Arab-populated areas. This was the largest number of new settlements in any one year in Menachem Begin's government: there had been forty-nine over the previous three years. The aim was to establish a Jewish presence throughout the occupied territories, and it succeeded.

On 30 July 1980, Begin secured the passage through the Knesset of a Basic Law on Jerusalem, originally proposed by the right-wing Tehiyah Party, which had opposed the Camp David Accords with Egypt. Although, more than thirty years earlier, Ben-Gurion had stated that 'the State of Israel has had and will have only one capital – the eternal Jerusalem', this was nowhere enshrined in law, despite an attempt by Begin himself to do so at that time. Political power now enabled Begin to return to what he felt was an essential need in the legality of Israel's status in the city; a lawyer in pre-war Poland, Begin always regarded legal forms as important.

The Basic Law put the status of Jerusalem in the same category as a constitutional amendment in the United States. Although Israel had no constitution (and has none to this day), a series of Basic Laws had been given the status of constitutional amendments. The Jerusalem Law was in this category. It laid down that the city would never be divided, that it would be the seat of all the national institutions (some of which still remained in Tel Aviv), that access to the Holy Places was guaranteed to members of all religions (something Israel had been practising since 1967) and that the city would take priority in the national development budget.

As proof of his own determination to secure Jerusalem's indivisibility,

Begin announced that a new government centre would be built just over the former divide, in East Jerusalem, and that he himself would move his office there. The centre was built, but, largely due to United States pressure behind the scenes, Begin (and his successors) kept the Prime Minister's Office in West Jerusalem. Begin's timing had been poor. He made the announcement of the move at the height of talks with Egypt on autonomy for the Palestinians: a follow-up of Camp David.

The Herut movement, of which Begin had been the head since the War of Independence, had its own rural settlements, thirty-two by 1980. Most of them were villages established near the Jordanian border before 1967, including Mevo Betar and Bar Giyora in the Jerusalem Corridor, and Nordiya in the coastal plain, as well as several youth farms. Fourteen of the Herut villages were located in the occupied territories. Since 1977, the imperative of settling in the occupied territories had attracted several thousand Herut supporters, determined to see come to fruition the dream of their founder and mentor, Vladimir Jabotinsky, of an Israel stretching 'from sea to sea' – from the Mediterranean to the Dead Sea, including the West Bank. Nevertheless, the overriding drive for new settlements came from the National Religious Party.

During 1980 there had been a renewal of the immigration of black Jews from Ethiopia: 209 arrived that year. But this was only the beginning of what was to become one of the main immigrations of Israel's fourth decade. In the following year the number rose to 956; the year after it was 891. At the very moment when immigration from the Soviet Union had fallen to its lowest in a decade (1,314 in 1983, 896 in 1984), that from Ethiopia took its place. Most of these new immigrants came from the Tigre province: within five years, almost all of the 4,000 Jews of Tigre had reached Israel. From other provinces, the rate of immigration was steady at about 200 a month.

Because of the political chaos in Ethiopia, most of the Ethiopian immigrants made their way across the border into the Sudan. Many of them left illegally, recalled Yehuda Avner, the Chairman of the Inter-Ministerial Committee that oversaw the operation. By 1984 as many as 10,000 were in refugee camps in the Sudan. On their journey to these camps, across rough country and constantly beset by brigands as well as hunger, over 2,000 died. The Israeli government, confronted by this death toll, and the possibility of it rising still further, decided to abandon the policy of gradual, phased immigration, and to set up a massive airlift. Within a two-month period, starting in mid-November 1984, more than 6,500 Ethiopian Jews were flown to Israel. The planes were chartered in Belgium.

In order to prevent action by the Ethiopian and Sudanese authorities to stop the airlift, which was clearly a violation of local air space, the Israeli government kept the operation – code-named Operation Moses – a secret. But in order to fund it, money was sought from wealthy Jews in the Diaspora;

the appeal for funds, while kept as confidential as possible, inevitably leaked out. Newspaper reports began to appear in the United States about the rescue activities.

The Israeli government denied all knowledge of an airlift. But as the publicity grew, details were published in the Israeli newspapers, and on 5 January 1985 the airlift had to be suspended. As a result, several hundred Ethiopian Jews were stranded in the Sudan. Enlisting the help of the American Central Intelligence Agency (the CIA), the Israeli government was able to arrange for one more airlift, Operation Sheba. The total number of Ethiopian Jews who had reached Israel in the thirteen years from 1972 to 1985 was 14,305. Many of them had never seen running water or electricity before they reached Israel.

On 18 January 1981 the Begin government announced that elections would be held in six months' time. One certain casualty in the political alignment was going to be the Democratic Movement for Change (Dash). Having joined Begin's government amid high expectations of being a catalyst for social change, and also a curb on West Bank settlements, it had been virtually ignored by the Prime Minister and his Likud colleagues. One by one the Dash Cabinet Ministers drifted away from the mainstream of their earlier idealism, and on April 2 the members of the Party voted to disband.

A brave political experiment, the creation of a third force of political moderation, was ended. Labour supporters bitterly resented the fact that the fifteen Dash seats that had been taken away from them in 1977 had ended Labour's political ascendancy. Yadin returned to archaeology. Project Renewal was transferred to the Ministry of Housing. In vain, Yadin tried to convince the Minister of Housing, David Levy, that it was not primarily a building and construction project, but a social project intended to heal rifts in society and give the weakest members in the urban hierarchy a chance to rebuild their lives and surroundings by their own efforts. Of his failure to change political attitudes, Yadin later told an interviewer, 'I couldn't play the game – by the rules of the game of the politicians. I suddenly found myself in the jungle, and in the jungle there are the rules of the jungle. And if you can't play by the rules of the jungle, then you are prey for other animals.'

In June 1981, a few weeks before Israel went to the polls to elect its tenth Knesset, Begin ordered an Israeli air force unit to bomb the nuclear reactor that was being built in Iraq. The reactor, the uranium for which was supplied by France, would, in the not too distant future, have had the capacity to manufacture material needed for nuclear weapons. The pinpoint raid was totally successful. Inside Israel it was criticized by Shimon Peres, who described it as an election ploy. Rabin supported him. When confronted by public hostility to his criticism, Peres announced that, if Labour were to form a government after the election, Rabin would be his Minister of Defence.

The election was held on 30 June 1981. The daring raid deep into Iraq certainly helped Begin at the polls. Once more Likud won the largest single number of seats, forty-eight – a gain of five on the previous election – and was able to call on the two main religious Parties to make up the necessary majority. The Labour Alignment, no longer facing the challenge of the Democratic Movement for Change, which had disintegrated, won forty-seven seats, a gain of fifteen – the exact number of seats won by Dash in 1977 – but it remained in opposition. Begin remained Prime Minister. Shimon Peres remained leader of the opposition.

On 6 October 1981 the Egyptians commemorated the eighth anniversary of their crossing of the Suez Canal, at the start of the October War. At a military parade to mark the occasion, President Sadat took the salute as units of his army passed by. As artillery pieces reached the reviewing stand, a small group of Muslim fundamentalists, posing as soldiers, jumped from one of the vehicles towing the artillery, and ran towards the stand, firing with automatic weapons. Sadat was assassinated. Among his alleged 'crimes' was having made peace with Israel.

Sadat's successor as President, the former Vice President, Hosni Mubarak, took severe action against the fundamentalists. Israel's peace with Egypt survived. But it had already become what was called a 'cold peace', lacking the hoped for expansion of trade and tourism, and seeing nothing of the opening up of each country to the cultural contributions or economic enterprise of the other. Israeli tourists continued to flock to Cairo and to visit the pyramids, including those said to have been built with the slave labour of their distant ancestors – Begin believed that these were indeed the pyramids built by the ancient Hebrew slaves; he was not to be dissuaded when Yadin said this was not so. But despite the Israeli tourist influx into Egypt, tourism in the other direction was sparse. Some university and research institute links were established, but little was built on them.

Israel had continued to fulfil its obligations under the peace treaty with Egypt, withdrawing not only from the whole Sinai, but from the Rafa Salient on the north coast of Sinai, a region which Dayan, among others, had been so determined to keep. It was in the Rafa Salient that the town of Yamit was being built at the time of the October War. Its abandonment was an essential feature of the Israeli–Egyptian accords. The residents of the Salient formed their own group, the Stop the Withdrawal Movement, and appealed for support from their fellow-Israelis. But the Begin government insisted that the settlers leave.

When the time came for the withdrawal from the Salient, many of the settlers refused to go, and had to be taken away by force. The settlers at Hatzar Adar hid in the sand dunes, and returned five times to their homes, only to be evicted each time by the Israeli army. Millions of television viewers around the world saw the Israeli army hosing the Yamit settlers down as they clung to their rooftops. The buildings of Yamit, its greenhouses and its orchards, were then systematically destroyed by the Israeli authorities, 'in order' (said a government spokesman) 'to deprive members of the Stop the Withdrawal Movement of hiding places'. Bitterly, one of the settlers, Motti Ben-Yannai, asked, 'Why was it necessary to uproot orchards? Why was a green and fertile area turned back into desert?' The Israeli government's reason for the destruction of crops and farming facilities was to prevent the Arab population from establishing itself so close to the border.

A group of settlers locked themselves in a bunker in Yamit, and threatened to blow themselves up if any attempt was made to evict them. Ariel Sharon, who was responsible for the operation, asked Begin what to do. Begin ordered the eviction to go ahead. Eventually the demonstrators left their bunker.

One tiny corner of Sinai (a few hundred square metres) was not returned by Israel to Egypt. This was Taba, a small bay on the Gulf of Akaba, just across the Egyptian border from Eilat. Israel claimed that it was within its borders. A long drawn out, and at times acrimonious dispute developed. Israel was accused of moving the border stone – set up in 1949 when the border was delineated as part of the Israeli–Egyptian armistice agreement – which made it clear that the bay had been included in Sinai at that time. The original border had been indicated by the British on a map of 1906. Frantic searches were made in the British archives by both sides for the original.

Since 1967 the bay had become a tourist paradise, with a beach village and a newly built holiday hotel. There was nothing more to it: no hinterland and no economic value beyond tourism. As the negotiations continued, and the Egyptians refused to accept the alleged location of the border stone, Israel claimed that the 1949 border had only been a 'temporary dividing line' between the Egyptian and Israeli armies. The Egyptian–Israeli peace treaty called for 'a process of negotiation and conciliation' to be followed, if that process failed, by arbitration. Reluctantly, after more than two years of futile negotiations, Israel agreed to arbitration. Then, just as it looked as if a settlement might be reached, five Israelis were killed and several more injured by an Egyptian soldier who opened fire from his border post at Ras Burka, just south of Taba. There was deep anger inside Israel that the Egyptian authorities appeared to hold up efforts to help the injured.

Arbitration finally took place, but not before it had been opposed by a number of Ministers, led by Shamir and by Ariel Sharon. A majority in favour was only secured when a leading Likud activist, David Levy, who always adopted a moderate stance in foreign affairs, joined the Cabinet. As a result of the arbitration, Taba was transferred to Egypt. For its part, Egypt agreed that Israeli tourists could cross the border to visit the beach and hotel with the minimum of formalities.

In March 1982 the co-chairman of the World Zionist Organization, Matityahu Drobles, announced that in the coming twelve months there would be a substantial increase in settlement building on the West Bank and Golan Heights, and in the Gaza Strip: £33 million had been put aside by the Israeli government to build sixteen new civilian and fourteen Nahal settlements.

A public challenge to the harsh nature of Israeli rule over the occupied territories during Sharon's tenure of the Defence Ministry came on 10 May 1982. Six senior reserve officers, who had recently been among those responsible for carrying out his policies on the West Bank, decided to speak out against the policy that was leading to repeated episodes of ill-treatment. Tsali Reshef, one of the founders of Peace Now, later recalled, 'These kind of stories, repeated with variations by other reserve soldiers, caused us to take an unprecedented step. After Sharon had refused our request to meet with me and receive the information we had gathered, Abu Vilan and Yarum Adiav of Peace Now initiated a press conference in which the six officers (including Benny Barabash, Gaby Bonoit and Yuval Neriah) each testified to the events which occurred in their presence during their tour of duty in the Territories.' Yuval Neriah had won the Medal for Bravery on the Sinai Front during the October War.

The six officers spoke forcibly about what they called the 'atrocities' committed by Israeli soldiers as a consequence of the 'repressive measures' instituted by Sharon on the West Bank, and blamed the Defence Minister for 'corrupting' young Israeli soldiers by ordering them to clash with unarmed Palestinian civilians. According to the six officers, the government's policies were resulting in army 'brutality, triggering copy reactions and indiscriminate collective punishment'. The officers stressed that while public attention was aroused from time to time by specific sensational events, 'the daily reality in the territories is one of violence and brutality'.

One of the officers, Benny Barabash, told the press conference, 'we are gradually losing our humanity. The local population are gradually becoming objects in our eyes – at best mere objects, at worst something to be degraded and humiliated.' One episode which the officers related was that of a soldier who wrote the identity numbers of detainees on their forearms so he could identify them. By coincidence this happened on Holocaust Remembrance Day. 'And the soldier was not even aware of the implications of what he was doing,' one of the officers said.

Yuval Neriah described how small patrols of between five and eight soldiers find themselves in almost constant confrontation with the local population – 'generally women, youths and children, because the men are away working. The soldiers are pelted with rocks and bottles and the army does not provide us with adequate means to react. In the end all that we have are our rifles. The result is frustration, despair and fear.' Neriah added that while what often happened was contrary to the instructions of senior commanders, control broke down at the local level. Soldiers often witnessed and sometimes followed the example set by those who served constantly in the area – officers of the military government and border policemen.

The six officers stressed that none of them was considering refusing to serve in the occupied territories, but that their concern was with the morality of actions undertaken by the army and with preserving the Israel Defence Forces ideal of 'purity of arms'. In Tel Aviv, more than 80,000 demonstrators marched in protest to denounce Sharon's 'iron fist' policy.

Tsali Reshef added, 'The initiative paid off: investigations were opened against the parties involved, and some of them were put on trial and convicted. But when Yuval, Benny, Gaby and their friends had decided on this move, they did not know that this would be the army's reaction, and were apprehensive about the conflict that could arise between their military roles and their involvement in a protest movement with decisive political and moral attitudes. Today their deliberation seems easy compared with the conflicts we had to face only a few weeks after that press conference, with the outbreak of the Lebanon War.'

In April 1982, bowing to pressure from the six members of the National Religious Party in his coalition (the Party had never had the political leverage to flex its muscles to this end in the years when it was joined with Labour), Begin agreed to discontinue all El Al flights on the Sabbath. Hitherto the national airline of Israel had not only flown on the Sabbath but had derived almost a quarter of its revenue from those flights.

When the Knesset began its summer session on May 3, there was heckling and derision from the Alignment benches at this decision, which effectively halted most Friday flights as well as those on Saturday. As Begin tried to explain the decision, Alignment members called out: 'Why didn't you stop Sabbath football games?', 'What about television?' and 'Why were you silent for the last five years?' Aryeh Rubinstein, the *Jerusalem Post* Knesset reporter, wrote that 'even a number of Likud members were obviously uneasy over Begin's pathos-laden defence of the sanctity of the Sabbath, including a quotation of the Fourth Commandment. During this part of his speech there were smiles on the faces of several Liberal MKs.'

As the catcalls continued, Begin tried to put his decision on the highest possible level. 'Is it conceivable,' he asked, 'that in the reborn Jewish State monetary profit and loss will outweigh the religious, social, national, historical and moral value of the Sabbath in Israel?' Addressing the El Al workers whose wages would suffer as a result of the loss of work, Begin told them that a profit-and-loss reckoning had 'no place' where 'an eternal value of the Jewish people' was concerned.

The decision not to allow El Al to fly on the Sabbath marked a turning point in the balance between the secular and religious sections of Israeli society. Secular Israel, which had been so much a part of the vision of its founders, was being eroded. The determination of the founders not to be beholden to religious law in the day-to-day working of a modern State was being challenged. The earlier acceptance by the religious section of the population, and by its political representatives, that Israel was essentially a secular State that respected their religious convictions, was giving way to a search for the greater imposition of religious law and custom on the daily conduct of life. The closure of the El Al Saturday flights was the first change of substance that showed the religious Parties which way they might go if they were to continue to assert, and to increase, their political power within the parliamentary system.

On Independence Day, which in 1982 fell on 28 May, Sharon gave his approval to the establishment of yet more new settlements on the West Bank. Hitherto the growing peace movement, led by Peace Now, had not demonstrated in the occupied territories against settlements. It planned to do so against Sharon's new move, but he took preventative action, temporarily closing the territories. Some demonstrators were able to avoid the road blocks, reach the sites, and disrupt the inauguration ceremonies.

At the entrance to a settlement near Hebron, Israeli troops blocked the demonstrators' way, then used tear gas to force them to disperse. Mordechai Bar-On, a leading figure in Peace Now, later wrote of how 'Palestinians from the outskirts of Hebron, who had experienced this suffocating feeling many times before, gave the demonstrators onion peels and drinks to help alleviate the effects of the gas.'

In 1982 a young Israeli Arab, Jawdat Ibrahim, was among the students who enrolled at Neve Shalom, a school not far from his home town of Abu Ghosh. The purpose of the school was (and still is) to foster good relations between Jews and Arabs of high school age. Jawdat stayed at Neve Shalom. 'I learned there that we can live together – Jews, Muslims and Christians.'

On 3 June 1982, in London, a Palestinian gunman, Hassan Said, opened fire on the Israeli Ambassador, Shlomo Argov, as he was leaving a printing and publishing dinner at the Dorchester Hotel in Park Lane. Surgeons were able to save Argov's life, though he was left totally paralysed. The attack on Argov coincided with renewed PLO shelling, from bases in southern Lebanon, of Israeli settlements along the border in northern Galilee. No Israelis were killed, but the shelling renewed calls in Israel for some tough action to be taken against the PLO bases. The attempted assassination of Argov heightened the anger felt among the Israeli public at their government's apparent impotence with regard to PLO actions, even though Israeli intelligence told Begin that the gunman was not a PLO member but belonged to the small Iraqi-supported Palestine National Liberation Movement, headed by Abu Nidal, a fierce opponent of Arafat's leadership. Begin withheld this information from his Cabinet.

On June 6, three days after the attack on Shlomo Argov, Israeli troops crossed the Lebanese border and drove northward. This was the beginning of the Lebanon War, the fifth war in Israel's thirty-four year history. It was the first war in Israel's history for which there was no national consensus. Many Israelis regarded it as a war of aggression.

The name given by the Begin government to the attack was 'Operation Peace for Galilee'. For three years it was to draw in more and more Israeli soldiers to Lebanon for the task of conquest and control. Of all Israel's wars in the first half century of its existence, it was the one that generated the most internal controversy as 'the unnecessary war'.

Reflecting on the name Operation Peace for Galilee, the Israeli writer Amos Oz commented, 'Wherever war is called peace, where oppression and persecution are referred to as security, and assassination is called liberation, the defilement of the language precedes and prepares for the defilement of life and dignity. In the end, the state, the regime, the class or the idea remain intact where human life is shattered. Integrity prevails over the fields of scattered bodies.'

The initial cause of the war – the growth of PLO bases in southern Lebanon – was thought by many to have called for some form of response. Using Russian-made Katyusha rockets, the PLO was systematically lobbing shells into the settlements in upper Galilee, forcing the residents to spend many nights in deep shelters. Groups of PLO fighters also repeatedly crossed the border into Israel to carry out attacks on property and civilians.

The Minister of Defence, Ariel Sharon, assured the Cabinet, at an emergency meeting on June 5, that Operation Peace for Galilee would be localized, that the troops would not advance more than 25 miles into Lebanon, and that the operation would be completed within two to three days. On this basis the Cabinet approved it, as did the opposition leaders, Rabin and Peres, whom Begin consulted. Sharon was convinced, however, that a military operation into southern Lebanon should do more than drive the PLO out of its bases. If Israeli troops could continue to Beirut, they could destroy the PLO headquarters there – and drive out Yasser Arafat – ending the PLO's ability to strike at Israel across the Lebanese border.

The plan devised by Sharon, and carried out under the Chief of Staff, General Rafael (Raful) Eitan – who in 1956 had led the parachute drop over the Mitla Pass – went in the event beyond the limited objectives that Sharon had described to the emergency Cabinet. The wider plan known as Operation Big Pines encompassed the military defeat of the PLO and the installation of Bashir Jemayel, the Lebanese Christian Phalangist leader, as President of Lebanon, and through him, the conclusion of a peace treaty between Lebanon and Israel. The northern border would then be secure, and Israel would have acquired a second Arab leader willing to make peace.

By the end of the first day of fighting, Israeli forces had reached the Litani River, 17 miles north of the border. This gave them control of the area from which the PLO was firing its rockets into the northern towns and settlements.

That evening Israeli forces captured the PLO artillery stronghold located in the Crusader fortress of Beaufort, while on the Mediterranean coast the Israeli navy landed tank and infantry forces just north of the mouth of the Awali River. On June 6 the United Nations Security Council passed a resolution demanding Israel's withdrawal from the Lebanon. Begin ignored it.

The Israeli advance continued during June 7, reaching to the outskirts of the coastal town of Damour, only fourteen miles south of Beirut. There were also several clashes on the second day of the war between Israeli forces and the Syrian army stationed mostly in the Beka Valley in the east of Lebanon, along the Syrian border. Israeli and Syrian jets also clashed over Beirut, and in the central sector west of Jezzine there was an exchange of tank fire between Israeli and Syrian tanks.

Throughout June 8 – the third day of the war – groups of PLO soldiers behind the advancing Israeli troops maintained a strong resistance in the two coastal towns of Tyre and Sidon, and in the refugee camps along the coastal road. That night, Israeli troops advancing in the central sector reached to within 4 miles of the Beirut–Damascus highway.

The danger of Israel being drawn into war with Syria was intensified during the fourth day of the war when on June 9 the Israeli air force attacked and destroyed seventeen of the nineteen SAM (Russian) anti-aircraft batteries which Syria had installed, and was manning, in the Beka Valley. There was an air battle between Israeli and Syrian warplanes, in which twenty-nine Syrian MiG-21 and MiG-23 Soviet-built jet fighters were shot down. Israeli and Syrian tanks and troops were also in direct conflict that day. But neither side had the desire to declare war on the other. Lebanon would serve as a proxy battlefield.

Along the coast, Israeli forces continued their advance, overrunning Damour and reaching the village of Khalde, only two miles south of Beirut international airport, on June 9. The Israeli public, which had been told that a limited operation was in progress to drive the PLO from southern Lebanon, was astounded that its troops were within striking distance of Beirut. It was also alarmed by the escalation of the conflict with Syria. On June 10 there was a sustained tank battle between Israeli and Syrian forces north and north east of Lake Karun. The Syrians were beaten back. In the air that day, another twenty-five Syrian MiG fighters were shot down.

The United Nations Security Council was calling for a cease-fire. Israel agreed to this on June 11, and it came into effect at midday that same day. Sharon had achieved a remarkable territorial conquest. On the coast, Israeli troops were dug in two miles south of Beirut airport. Further east, they were only a few miles from the Beirut–Damascus highway. In the Beka, they were less than four miles from the Syrian border.

The PLO remained in Beirut. On the afternoon of June 13, Israeli troops reached the eastern entrance to the city. Sharon then instituted what became known as the 'crawling stage' of the war. By the evening of June 14 Israeli troops had moved forward around Beirut to encircle the city by linking up with the Phalangist forces in the east of the city. Israeli troops also captured and crossed the Beirut end of the Beirut–Damascus highway. In the vicinity of the Lebanese presidential palace at Baabde, on the eastern outskirts of Beirut, Israeli and Syrian tanks were in battle on June 14. Beginning that day, and continuing for two months, 400 Israeli tanks and 1,000 guns bombarded West Beirut, while Israeli warplanes systematically struck at all known PLO strongholds in the western part of the capital. By the end of June more than 500 buildings were in ruins. 'Never in their worst nightmares,' wrote the historian Howard M. Sachar, 'had the PLO leaders imagined the Israelis capable of this kind of warfare.'

The crawling stage of the war continued on June 15 when Israeli forces extended their control along the Beirut–Damascus highway, cutting off any chance of Syrian reinforcements using the road to enter or reinforce the PLO in Beirut. The Syrians strengthened their hold on two of the villages along the road, Aley and Shtoura, and held them against repeated Israeli attack. Meanwhile, on June 19, Israeli artillery bombarded PLO military positions inside Beirut. 'When I heard about the raid on Beirut,' Peres later wrote, 'I called up Begin and to my surprise I discovered that he was unaware of it.'

On June 20 the PLO and Syrian forces withdrew from Beirut airport. Israeli forces entered the terminal and advanced along the runways without fighting. On the following day, fighting broke out again between Israeli and Syrian forces along the Beirut–Damascus highway. It was a bitter battle that continued for almost a week. When it ended, Israel's control of the road was significantly extended, but Aley and Shtoura remained in Syrian hands.

Inside Israel, on July 2, at the ancient site of Herodion on the West Bank (east of Bethlehem), the supervisor of the 2,000-year-old site, a twenty-seven-year-old new immigrant from the United States, David Rosenfeld, was found stabbed to death in his office. The father of two children, he was a member of the nearby settlement of Tekoa.

Two Arabs from a village within sight of Herodion confessed to the murder. In Rosenfeld's memory another new settlement, El David (To David), was established on the following day at the base of Herodion. Its purpose was to be a mixed community of religious and secular Jews 'living together with mutual respect'.

Outside Beirut a number of Israeli officers were shocked by the instructions they had received to prepare for a possible offensive against the city's western, Muslim sector. One of these officers was Colonel Eli Geva. Aged thirty-two, the son of a distinguished retired general, Geva was the youngest brigade commander in the Israeli army. It was his tank column that would lead any assault on West Beirut. Geva asked General Eitan to relieve him of his brigade command and allow him to serve as a mere tank officer.

Should orders be issued to lead his unit into the city, Geva told Eitan, he could not as 'a matter of conscience' expose his troops, and Beirut's civilian population, to the heavy casualties that were sure to ensue.

Shocked by Colonel Geva's request, first Eitan, then Sharon, and finally Begin himself sought to dissuade him. They failed to do so, and Begin had to accept Geva's effective resignation. The episode deepened public reservations about the war. 'Slowly the reality strikes home,' one non-commissioned officer recalled, of the capture of the Lebanese coastal city of Tyre. 'It's not possible we did such devastation! No shop not blasted, buildings crumpled under their roofs, boats sunk in the bay . . . and above all the overwhelming stench . . . of rotting bodies in the village.'

When a visiting Cabinet Minister, Yaakov Meridor – a founding member of the Irgun – arrived in Israeli-occupied Sidon, his recommendation for dealing with the Palestinian refugees was, 'Push them out and don't let them come back.' To many reserve officers this approach reflected the apathy, even callousness, that seemed to have infected the military command and government alike. While on leave, a number of those officers set about forming a movement, 'Soldiers Against Silence', to demand an end to the war altogether.

On July 3 Peace Now organized a demonstration in Tel Aviv: an estimated 100,000 citizens participated. No such protest had been mounted before in wartime. Writing in *Ma'ariv*, Abba Eban declared that 'these six weeks have been a dark age in the moral history of the Jewish people'. Following the Peace Now rally, Likud organized a counter-demonstration in Tel Aviv. As many as 200,000 attended it, many brought in by bus from throughout the country. Addressing the rally, Begin put the burden of dissent entirely on the Labour opposition. Sharon went further; denouncing Labour for the dissent within the army, he sought (unsuccessfully) to censor all newspapers that criticized the war.

Inside Beirut, throughout the first days of July, the PLO defended themselves tenaciously. Although Israeli forces did not enter the city, they were in the hills above it and besieged it. Day after day Israeli artillery pounded PLO centres inside the city. On July 4 at a meeting with Menachem Begin, Yitzhak Rabin gave his support to a plan – put forward by Sharon – to tighten the siege of Beirut by cutting off the city's water supplies, which Israel then controlled. 'The advice to close the water taps was to haunt Yitzhak,' his widow later wrote. 'People were astonished by his position and wondered how Yitzhak Rabin could support this horrible war.' Rabin was not a supporter of the war, she added, 'but he was aware that the IDF had already lost hundreds of lives, and he didn't want this ill-advised campaign to be a total disaster. He believed that closing the taps might help end it faster.'

On July 23, after a month of siege, the Israeli navy and army intensified its bombardment of PLO positions. On August 1 Sharon ordered fourteen hours of non-stop air, naval and artillery bombardment. Two days later Israeli troops entered south-east Beirut and sought to engage the PLO forces in direct combat. On August 4 the Burj al-Barajneh Palestinian refugee camp, in which the PLO were strongest, was completely encircled. For twenty hours the Israeli infantrymen fought without respite. But they were able to advance only a quarter of a mile into the camp. That day Israeli troops took control of Beirut Airport.

The Lebanese war was the most televised war in history up to that time. The daily television transmission of Israeli artillery bombarding Beirut, the columns of smoke, dust and fire rising in the air, and the close-up pictures of the destruction, including on one occasion serious damage to a hospital, caused immense harm to Israel's international image, and much anguished discussion within Israel itself.

The frustration of the Israeli High Command at being unable to flush out the PLO was considerable. An American envoy, Philip Habib, was using his good offices to try to effect a cease-fire and settlement, even while the Israeli bombardment continued. On August 11 the Israeli air force began a two-day aerial bombardment of the PLO-held areas of Beirut. By the criteria of its objective, the bombardment was a success; on August 12, following a further round of negotiations with Habib, the PLO agreed to leave the city.

A multi-national force, including 1,800 American Marines, was assembled to protect the PLO as it prepared to leave Beirut for Tunis. Its departure was to be by sea. Begin, in an act of chivalry, agreed to let the PLO leave carrying its side arms. On August 30, as Yasser Arafat prepared to leave the city an Israeli sniper had him in his sights. But the Israeli government had specifically ordered that he was not to be killed, and the moment passed. Arafat left Beirut that day, and by September 4 the PLO was gone.

On September 1, as the PLO evacuation of Beirut was in full spate, Bashir Jemayel, the Christian Maronite President-elect of Lebanon, flew to Nahariya on the Israeli coast, where Begin urged him to sign a peace treaty with Israel. He refused to do so, proposing a more limited non-aggression pact instead. When Sharon, who was also at the meeting, told Jemayel that Israel had Lebanon in its grasp, and that he would be wise to agree to Begin's request, Jemayal was incensed, holding out his arms and crying out, 'Put the handcuffs on. I am not your vassal.'

Later that day the 'Reagan Plan' was announced by Washington. It included the principle of 'self-government for the Palestinians of the West Bank and Gaza in association with Jordan'. This new plan was presented by the Americans as the 'next step' in the Camp David peace process. But the Begin government was resolved not to relinquish Israeli control of the West Bank (or, in its nomenclature, Judaea and Samaria) and dismissed the American initiative.

Following the departure of the last PLO troops from Beirut, the Syrian troops in the city also left. All was set for Israel's Lebanese ally, the Phalange, to take control. But on September 14 the President-elect, Bashir Jemayel, was assassinated. His Christian Maronite followers were incensed against the Palestinian Muslims, who claimed responsibility for the assassination.

The Israeli troops were in a situation, far from their own borders, which they could not control. On September 15 they occupied West Beirut 'in order', Menachem Begin explained, 'to protect the Muslims from the vengeance of the Phalangists'. The main concentration of Palestinian Muslims was in the Sabra and Chatila refugee camps. Israeli forces sealed off the two camps from the outside world: 'hermetically sealed' was how General Rafael Eitan later described it. From the Israeli perspective the Phalangist soldiers would seek out the Palestinian fighters, not the civilian population: to this end, Israeli military searchlights illuminated the camps at night. On September 17, Phalangist forces entered these two camps and massacred the inhabitants, fighters and civilians alike. According to an Israeli army report, 474 Muslims were killed. The Palestinian Muslim dead included 313 men, 8 children and 7 women. The Lebanese Muslim dead included 98 men, 8 children and 7 women. Also killed in the two camps were 21 Iranian Muslims, 7 Syrians, 3 Pakistanis and 2 Algerians who had thrown in their lot with the PLO. When Sabra and Chatila were opened and newspaper reporters from around the world entered the two camps, the bodies (many of them mutilated) of 2,300 Palestinian men, women and children were found.

The massacre at Sabra and Chatila shocked the world, and shocked many in Israel, although no Israeli troops were involved in the killings. As far as the Israeli soldiers were concerned, the Christian Phalangists were to undertake the mission of clearing out suspected nests of Palestinian fighters. What the Christian soldiers did followed the earlier pattern of the Lebanese civil war, the wreaking of vengeance on the civilian population – vengeance, in this case, in the immediate aftermath of the assassination of President-elect Bashir Jemayel.

Those Israelis who had opposed Operation Peace for Galilee from its first day, or had been against its extension beyond the first 25 miles (the originally declared objective), were outraged that the massacre had not been foreseen and forestalled by the senior miliatary authorities in the area. On September 25 there was a mass demonstration in Tel Aviv, in which 400,000 people – more than 10 per cent of the total population – participated. It was the largest ever demonstration held in Israel. 'No such explosion of public outrage had ever occurred in Israel's history,' wrote Howard M. Sachar in 1996. Politically, the demonstration was supported by the Alignment (which had forty-seven seats in the Knesset, as against Likud's forty-eight), Shinui (with three seats in the Knesset), the Civil Rights Movement (led by Shulamit Aloni, its one Member of Knesset), and Peace Now.

The Tel Aviv demonstration called for a judicial commission of inquiry into what had happened at Sabra and Chatila, and in particular any Israeli involvement. For several days the government strenuously resisted this demand, but on September 29 it gave way. The inquiry was headed by Supreme Court Justice Yitzhak Kahan. Its report, made public on 7 February 1983, concluded that 'no intention existed on the part of any Israeli element to harm the non-combatant population in the camps.' But the report went on to criticize the Director of Israeli Military Intelligence, Yehoshua Saguy, for 'closing his eyes and blocking his ears' to what was clearly happening inside Sabra and Chatila once the Christian Phalangists entered.

There was also a strong criticism of Ariel Sharon in the Kahan report. It was felt that he should not have allowed the Phalangists to enter the Palestinian refugee camps without giving 'suitable instructions' to prevent or reduce the danger of a massacre. Humanitarian obligations, the report concluded, 'did not concern him in the least'. As Defence Minister, Sharon was found responsible 'for disregarding the danger of acts of revenge and bloodshed by the Phalange against the population in the refugee camps' following the assassination of President-elect Bashir Jemayel; and also 'for not taking this danger into consideration when he decided to let the Phalange into the camps', and 'for not taking appropriate measures to prevent or limit this danger'.

Sharon strenuously denied this verdict and stressed that he had not foreseen that there would be slaughter. Two years later he was to bring a libel action against *Time* magazine which claimed that he had encouraged the Jemayel family to take revenge for Bashir Jemayel's assassination. During the libel suit the defence argued on behalf of *Time* that the un-published annexes of the Kahan Commission Report contained information which supported the magazine's claim that Sharon had encouraged Bashir Jemayel's family to take revenge for his killing. However, those who were allowed to look at the secret annexes did not find any proof for this allegation. Although the jury found the accusation unfounded and the article defamatory, it decided it had been published without malicious intent, and Sharon was denied his $50 million compensation claim against *Time*.

Menachem Begin was also criticized by the Kahan Report for what was called unjustifiable 'indifference', and 'for not having evinced during or after the Cabinet session any interest in the Phalangists' actions in the camps'.

Public opinion in Israel was never more intensely expressed than during and after the Lebanese War. There were repeated demonstrations and counter demonstrations.

On the ninth day of Israel's invasion of Lebanon, Peace Now placed an advertisement in the newspapers:

> In this war, the Israeli Army is proving once again that Israel is powerful and self-confident. In this war, we are losing brothers, sons and friends.

In this war, thousands are being uprooted from their homes, and towns are being destroyed. Thousands of civilians are getting killed.

What are we getting killed for? What are we killing for? Has there been a national consensus for going into this war? Has there been an immediate threat to Israel's existence? Will it get us out of the cycle of violence, suffering and hatred?

We call upon the Government of Israel: Stop! Now is the time to invite the Palestinian people to join in negotiations for peace. Now is the time for a comprehensive peace based on mutual recognition.

During a Peace Now demonstration which was held outside the Prime Minister's office in Jerusalem three days after the publication of the Kahan Report, calls were made for its implementation. During the meeting, someone who was never identified threw a grenade into the crowd. One of the Peace Now demonstrators, Emil Grunzweig, was killed.

Grunzweig was a thirty-three-year-old kibbutz member and paratroop officer who had fought in Lebanon. His funeral the next day in Haifa, attended by over 10,000 sympathizers, became a scene of even more furious anti-government recrimination. 'The nation was polarized more acutely than at any time since the German reparations crisis of 1952,' wrote Howard M. Sachar. 'Never before had a war been debated in purely Jewish – in contrast to Israeli – terms. Begin continued to defend it, and the casualities inflicted on a civilian Arab population, by repeatedly invoking the images and memories of the Second World War and the Holocaust, by equating the PLO with the Nazis. But the analogies had worn too thin. Characteristic now was an open and widely publicized letter to Begin from a bereaved father who had lost his only son in Lebanon.' The letter read:

> I, remnant of a rabbinical family, only son of my father, a Zionist and Socialist who died a hero's death in the Warsaw ghetto, survived the Holocaust and settled in our country and served in the army and married and had a son. Now my beloved son is dead because of your war.
>
> Thus you have discontinued a Jewish chain of age-old suffering generations which no persecutor had succeeded in severing. The history of our ancient, wise and racked people will judge you and punish you with whips and scorpions, and let my sorrow haunt you when you sleep and when you awaken, and let my grief be the Mark of Cain upon your forehead for ever!

In its conclusion the Kahan Report had stated that Sharon had not fulfilled his duty and called for him to 'draw personal conclusions'. But he did not resign from the government; instead, after giving up the Ministry of Defence he remained in the Cabinet as Minister without Portfolio. Yehoshua Saguy resigned from Military Intelligence.

*   *   *

In the immediate aftermath of the Sabra and Chatila massacre, the United States had urged Israel to withdraw from Beirut. The withdrawal began on September 19. The multinational force, including the American Marines, which had earlier supervised the PLO departure, took up defensive positions in the capital.

The cost of the Lebanese War had been high for all the combatants. An estimated 6,000 PLO troops had been killed, as had 460 Lebanese civilians outside Beirut, 600 Syrian troops, and 368 Israeli soldiers. The Israeli journalist Ze'ev Schiff, who had broken the story of the Sabra and Chatila massacre in *Ha-Aretz*, reflected (in 1985) on the impact of the Lebanon War on Israel:

> Israel must now bear the cost of its venture in Lebanon: hundreds of dead, thousands of wounded, and the shattering of a long-standing consensus on security that has produced an alarming rift within Israeli society.
>
> From the sobering consequences of Operation Peace for Galilee one may be forced to conclude that a country can be victorious on the battlefield but lose a war strategically; that a small nation whose leaders fail to appreciate the limits of military power is doomed to pay dearly for their arrogance; and that a democracy like Israel, whose defence is based on a militia army, cannot possibly win a war that lacks not only broad public support but even the slimmest national consensus regarding its very necessity.

The loss of Israeli life in Lebanon did not stop with the end of hostilities. On November 4 a suicide operation carried out by Shi'ite Muslims (with the support of Syria) destroyed the Israeli military headquarters in the southern Lebanese port of Sidon, killing thirty-six Israelis.

Direct talks between Israel and Lebanon began on December 28. Within five months, on 17 May 1983, an agreement was signed. All foreign forces – Israeli and Syrian – would withdraw from Lebanon. But Syria refused to accept this, and Israeli troops remained in place throughout the summer. It was not until September 3 that Israel agreed to withdraw south of the Awali River; but she insisted on retaining control of Jebel Barouk, the strategic height only 15 miles from Beirut.

The massive task of the occupation of southern Lebanon involved the presence of more than 30,000 Israeli troops and administrators. Most of these troops were reservists, called up from their civilian jobs to undertake several months' military service. Army duty in Lebanon was not popular. There were instances of soldiers who thought that Israel should not be an occupying power in Lebanon declining to serve there. Over a nine-month period, sixty Israeli soldiers were imprisoned in Israel for refusing to do their reserve duty in Lebanon.

Elsewhere in Lebanon, other forces controlled different parts of the country. There were more than 40,000 Syrian troops stationed in the Beka Valley. Their original armament of 350 tanks before the war was increased to 1,200. Some 26,000 Lebanese militiamen from various pro-government factions maintained what order they could in and around Beirut. An estimated 10,000 PLO supporters were still under arms in southern Lebanon, despite the evacuation of their leadership from Beirut to Tunis, where Arafat set up his headquarters.

The policy of the Begin government to allow a proliferation of settlements on the West Bank and in the Gaza Strip was coming rapidly to fruition. Sharon was a leading supporter of such a policy. Between 1977 and 1983 a total of 186 new settlements had been established, forty-one of them in 1983. This brought the number of settlers to 20,000, living near, and sometimes very close indeed, to an Arab population of more than a million: 721,700 on the West Bank and 464,300 in Gaza.

During 1982 the government had announced plans to raise the settler population to 100,000 by the end of the decade. The Labour opposition was against a policy that would, after a while, make it difficult if not impossible to separate Jews and Arabs in such a way as to enable some form of autonomous Palestinian entity to come into being. Some of the settlements were very small, with fewer than thirty families, living in caravans and capable, should the policy change, of being moved. The main fear voiced by Labour was that as the settlements grew they would become more permanent.

The increase in West Bank settlements in the 1980s was stimulated by financial inducements given by the government to anyone who moved to the West Bank. They were offered larger houses, lower house prices, lower mortgages, and greater tax benefits than they could get elsewhere in Israel. This made the new settlements particularly attractive to a number of diverse groups: Russian immigrants, many of whom had little money and a hostile attitude towards Palestinians; right-wing fanatics from the United States, who saw themselves as blazing a new frontier; people with no ideological reason for living on the West Bank but for whom the subsidized housing was a major inducement, especially young Israeli couples for whom the cost of apartments in Tel Aviv or Jerusalem was prohibitive.

Labour promised to remove the financial inducements to settle on the West Bank, and were to do so when they came to power a decade later, in 1992, as part of the attempt to win Palestinian confidence for the peace process; but in 1996 these inducements were restored by the Likud-led Netanyahu government, making it once more economically beneficial to move to the West Bank, especially for families struggling to earn a living. Ideology and economic advantage made a powerful combination.

Some of the settlers displayed an attitude to the Arabs which was racist.

In the settlements established in heavily populated Arab areas the settlers lived almost besieged, armed, encroaching on their Arab neighbours, and sometimes attacking them. The Palestinians also used violence. Each time a settler was wounded by stone throwers, or killed while travelling through Arab areas or walking in a remote spot, the indignation of the settlers intensified, and the clashes became more frequent.

Quite apart from the settlements, the continued Israeli occupation of the West Bank and Gaza Strip had created, after sixteen years, deep resentment inside the occupied areas, and a gradual acceptance of occupation by many Israelis; an eighteen-year-old Israeli soldier – man or woman – had only been two years old when the areas came under Israeli rule; a twenty-one-year-old Israeli university student had only been five years old: neither knew any situation except that of seeing more than a million Arabs as subject people. In the higher echelons of the Israeli government there was talk of annexation (though to this day the West Bank has not been annexed by Israel). In June 1983 Abba Eban, then a Labour Member of Knesset and an opponent of annexation (and of perpetual occupation), wrote:

> Not a single country in the world community, including those most in favour of Israel, was prepared to support the idea that Israel's security required the imposition of permanent Israeli jurisdiction over a foreign nation. At least half the Israeli nation opposed the idea of the incorporation of the West Bank and Gaza into Israel.
>
> There does not exist on the surface of the inhabited globe a single State that resembles what Israel would look like if it were to incorporate the West Bank and Gaza coercively into Israel. A democratic country ruling a foreign nation against its will and against the will of the world would be a unique reality.

Eban noted both the demographic and democratic dangers that annexation, and even continuing occupation, would create. 'If the 1.3 million Palestinians were to be integrated into the central political system of Israel,' he wrote, 'they would take the balance of power in all Israeli decisions, and Israeli politics would become a constant, restless pursuit of their votes. If, on the other hand, the 1.3 million Palestinians are denied the equal possibility of electing and being elected to the Israeli parliament, Israel will face the implications of becoming a State whose inhabitants have two different levels and categories of rights and obligations: Jews and Israeli Arabs who can elect and be elected and, a few metres away, Palestinian Arabs who would be held down by military force in an allegiance to which they give no real devotion.'

Eban went on to warn, 'The choice between maintaining the present territorial breadth at the expense of Israel's democratic vocation and accepting a more compact structure for the sake of national and social harmony will be Israel's most fateful decision in the 1980s.' And, indeed, in the 1990s.

In a book which he published in the following year, *Heritage: Civilization and the Jews*, Abba Eban stressed what he saw as the dangers of continuing occupation by Israel of the 1,300,000 Arabs in the West Bank and Gaza Strip. 'Control of these territories would add to Israel's strategic depth. But control of the inhabitants, who have all the attributes of a separate, national entity totally alienated from Israel's flag and legacy, would make Israel different from what it had always aspired to be. It would be a society held together by physical power rather than by voluntary and consensual allegiance.' Eban added that, 'for this very reason' the story he was telling ended 'still in suspense', and he went on to forecast:

> While there are those who aspire to perpetuate Israeli control of the heavily populated Arab areas, at least as many probably welcome compromise under which Israel's boundaries would be improved to some extent while the heavily populated Arab areas would be free to pursue an Arab destiny in association with Jordan, thus solving the Palestinian question in a non-coercive atmosphere.
>
> Yet the fact remains that Israel is one of the few countries in the world with a mark of interrogation posed over its dimensions, configuration, and human composition.

*　　*　　*

On December 6, six Israeli civilians, including two children, were killed when a bomb went off in a bus in Jerusalem. The PLO claimed responsibility. The Lebanese War had failed to end PLO terrorism or to curb their activities. For Menachem Begin, public criticism of the Lebanese War had been compounded by a sense that he had been deceived by Sharon as to the scale and intention of the war. With continuing criticism of that war and following his wife's death, Begin fell into a sharp decline. On leaving office, he became a virtual recluse, living in an apartment on the edge of Jerusalem until his death in 1992. Begin's son, Benjamin, when himself a leading Likud Member of Knesset, was to accuse Sharon of not having revealed to his father the plan to thrust as far as Beirut. Sharon denied this. The acrimony between the two men, erupting in 1997, brought the dispute over the Lebanese War into a new era.

On 10 October 1983 Yitzhak Shamir became Prime Minister. During his period of office, Islamic terror intensified against the American and French peacekeeping forces in Lebanon, culminating in the bombing of the American Marine headquarters in Beirut and the deaths of 241 Americans and Lebanese. This gave Shamir the impetus to secure an agreement for strategic cooperation with the United States.

Shamir was determined to retain Israeli military control over southern Lebanon until such time as 'adequate security arrangements' could be put in place. Shimon Peres, the leader of the opposition Labour Party, believed that the search for such arrangements would fail, and that Israel's security interests along its northern border would be best served by a speedy withdrawal from southern Lebanon. As this debate intensified, Shamir's government was defeated on a vote of no confidence on April 4. As preparations were put in place for a general election, Shamir remained Prime Minister at the head of a caretaker government.

Elections were held on 23 July 1984. The Labour Alignment won 44 seats and Likud 41. The right-wing Tehiyah–Tsomet alliance won five seats. All other Parties (twelve, including the once-powerful coalition builder, the National Religious Party) could muster no more than a maximum of four seats each. It was a hung Knesset. In the search for a working majority neither of the main Parties was able to form a coalition.

To resolve the political impasse, which threatened to paralyse all government, in September 1984 a National Unity Government was formed. It was to be based on a rotating premiership and the sharing of power between the two great rival political blocs. Shimon Peres became Prime Minister, with the outgoing Prime Minister, Yitzhak Shamir, as Deputy Prime Minister and Foreign Minister. Under the rotation agreement, which was an integral part of the formation of the National Unity Government, Shamir would succeed Peres as Prime Minister in two years' time.

One anomaly of the votes cast in 1984 was the election of Meir Kahane. His Kach party was anti-Arab in both word and aspiration. The two Hebrew letters of the Party's name had been chosen, apparently, without particular thought, thought most people believed that he had taken them from the Irgun slogan *Rak Kach* – 'only thus!'

Kahane had spent most of his political career in the United States, where he was an ordained rabbi, and was an outspoken advocate of Jewish self defence. In 1971 he had been imprisoned for illegal possesion of weapons. He had campaigned in Israel in both the 1973 and 1981 elections, on a platform of the removal of all Arabs from Israel, the West Bank and the Gaza Strip (by financial inducements if they would accept them and by expulsion if they would not), but had failed to acheive the one per cent of the total votes cast needed to secure a seat. As a result of his election in 1984, the Knesset passed a law preventing lists with racist platforms from running for the Knesset. It also passed an amendment to the penal code making racial violence a punishable offence.

In 1988 Kahane was disqualified from running for the Knesset again. After addressing a gathering of his supporters in New York City two years later, he was assassinated by an Arab of Egyptian descent. Kahane's funeral, in Jerusalem, was the occasion for an outburst of anti-Arab and anti-leftist rioting. Kach members remained active in the decade following their leader's murder; a film made in the Jewish section of Hebron in the summer of 1997 showed how strong their hatred of Arabs remained.

Under the leadership of Shimon Peres, Israel's priority was to tackle the burden of a massive and unprecedented rise in inflation, which had reached 35 per cent a month. In his memoirs Peres recalled how 'some of my friends had urged me to decline the Prime Ministership because of the state of inflation. They said that there was no way I could tackle it effectively without losing every shred of popularity. I resisted this advice – perhaps because I did not fully understand all the insidious and destructive properties of inflation.' Peres added, 'Inflation is like drugs: it destroys the internal organs while giving its victim a wonderful, floating feeling in the head. It gives rise to two new classes of society: those who wax suddenly fat on inflationary speculation, and ordinary people and pensioners who find themselves suddenly poor as their real income is eroded. Inflation is paradise for the foxes and purgatory for the ants. The sight of shop assistants marking new, higher prices for basic foodstuffs every day filled me with trepidation.'

Peres's economic advisers recommended, as he recalled, immediate and drastic action: a devaluation of 30 per cent, 'which would have resulted in widespread unemployment'; a reduction of up to 50 per cent in defence spending, 'which would have seriously weakened Israel and exposed it to grave military dangers'; sweeping budgetary cuts in all government departments; a price freeze; and across-the-board wage cuts. 'Even if these ideas made economic sense,' he wrote, 'I knew that politically, and indeed socially, they were non-starters. If I tried to introduce them, they would lack the minimal support needed to make them work.'

Determined to defeat inflation, Peres resolved that, 'alongside our broad-based political coalition, we needed a coalition of all the major economic interests in the country: the Government, the unions, the industrialists and the academic community'. He created an Economic Council, comprising the Secretary General of the Histadrut Trades Union Confederation and his deputy, the leaders of the Employers' Association, the Minister of Finance, and himself, as well as senior professors of economics. After prolonged and intense discussions, Peres's first 'remedial action', as he described it, took the form of a social contract – or a 'package deal', as it was called – between the government, the Histadrut and the Employers' Association, whereby prices and wages were pegged for three months. This was subsequently extended for a further three months. And yet, Peres recalled, 'by mid-May 1985 it was clear that this temporary remedy had run its course. Inflation for the month of April alone was more than 15 per cent; we were back on the road to economic ruin.'

To tackle the new crisis, Peres convened an emergency late-night meeting in his home in Jerusalem of the country's leading academic economists, together with the Finance Minister, Yitzhak Modai, his top aides, and his own economic adviser, Amnon Neubach. From their discussions emerged the Economic Stabilization Plan, or ESP. Its main elements were a major devaluation of the shekel, coupled with an announcement by the government that the new rate would remain in force indefinitely. The system of almost daily devaluations was to be ended. Also part of the package were steep rises in prices, coupled with a government announcement that the new prices would be frozen until further notice. Existing price control mechanisms were to be toughened to ensure that the freeze was rigorously enforced.

The government also announced that subsidies for basic commodities were to be cut dramatically by $750 million. The government deficit was to be further reduced through spending cuts in individual ministries. In addition, the mechanism for calculating incremental rises in the monthly cost of living, which was responsible for linking wages with prices, was temporarily suspended, so that wage-earners would not be compensated for the effect of the price rises and the subsidy cuts. A severe erosion in the real value of wages was anticipated, and a restrictive monetary policy introduced.

After two weeks of intensive negotiations, Peres was able to convince the leaders of the Histadrut that this was the only possible solution. They agreed with his proposal that wages would be allowed to rise again by the end of the year, and that during 1986 living standards would gradually return to what they had been before the Economic Stabilization Plan went into effect.

'People hardly believed that I had taken all these draconian steps,' Peres later wrote. 'I had not previously been thought of as wicked or unfeeling. But I was convinced that these harsh measures were the only way to win back the nation's confidence. And frequent visits to shops and factories in the months after the ESP went into effect strengthened my conviction and gave me, too, the confidence to continue.'

In August, the month after the Economic Stabilization Plan was implemented, inflation, which had been at 28 per cent in July and 20 per cent in June, plummeted to 2.5 per cent. By the end of 1984 the monthly figure was averaging 1.5 per cent.

Inflation had been successfully challenged, at the very moment when it was undermining the economic viability of the State from within, as much as the Arab armies had once threatened it from outside. The problems of defence in the National Unity Government became the concern of the new Minister of Defence, Yitzhak Rabin – the former Prime Minister and Chief of Staff, who had never before held the Defence portfolio, but who brought to it immense experience.

Under Rabin's guidance, the military withdrawal from Lebanon, a matter of urgency, was begun. The first phase was to strengthen the Christian Lebanese militia in southern Lebanon, the South Lebanese Army (SLA). This was to be armed and supported militarily by Israel. The second phase was to pull back all Israeli troops to a Security Zone along Israel's northern border, a strip of Lebanese territory 3 to 4 miles deep. The Israeli withdrawal began in January 1985. It was completed within six months.

Rabin also took an initiative in the early months of 1985 to try to find some area of possible agreement with Jordan. At a secret meeting in London with King Hussein, he presented the King with an Israeli-made Galil assault rifle in a specially crafted olive wood case. The meeting was courteous, but no more. 'It was delightful,' Rabin told his wife, 'but there were no results.'

As Minister of Defence, Rabin had to take a decision that caused considerable controversy – and anguish – inside Israel. In Lebanon, three Israeli soldiers were being held captive by the extremist Palestinian organization led by Ahmad Jibril (who as a twelve-year-old boy had been among the 50,000 Arab refugees from the Lydda–Ramle region in 1948). Negotiations were begun, and, in exchange for the three Israeli soldiers, 1,150 Palestinians were released from prisons inside Israel. Rabin was convinced that, even at the risk of some of these Palestinians joining terrorist ranks, the three soldiers had to be freed. He was determined to maintain the record of the Israeli army, that it had never abandoned even a single soldier to a fate from which he could be saved.

In September 1985, after three Israelis had been killed in Cyprus by the PLO, Israel responded with a bombing raid on PLO headquarters in Tunis. During the raid, seventy-three Palestinians and Tunisians were killed.

The nature of Israeli society was an unusual, perhaps a unique mixture of peoples and customs. The unifying factor was their Jewishness, but the many strands of Judaism and Jewishness that had evolved in the Diaspora since the destruction of the Second Temple by the Romans 2,000 years earlier made for complex and often conflicting lifestyles and attitudes. Unlike any other State, a majority of Israel's inhabitants were immigrants, and they came from very different linguistic, social and cultural backgrounds.

In 1985 Israel's Jewish population consisted of only 18.5 per cent native-born Israelis. Immigrants from Asia, mostly Russians, constituted 21.3 per cent, immigrants from Africa, 22 per cent, and immigrants from Europe and the Americas, 38.2 per cent. One divide was between the predominantly north European and North American Ashkenazi Jews, and the Sephardi Jews who came in most instances from North Africa, Iraq, Persia and the Balkans. Pnina Morag-Talmon, a lecturer in sociology at the Hebrew University in Jerusalem, has written, 'Generally unifying characteristics among the Ashkenazim are: their relation towards elements of the Christian culture among which they lived, desire for technological progress, secular education, trends of emancipation, and experience in composite political frameworks. The eastern communities, on the other hand, lived in an Islamic culture; their inclination towards technical progress and secular education developed later and was limited in comparison to the Ashkenazim, and they were accustomed to traditional conservative political and social frameworks.' This division was seen most clearly in higher education, where the percentages in 1985 were, for Israelis from the north European and American immigration, 20.1 per cent, native-born Israelis, 14.4 per cent, and Israelis from Asian and North African countries, 4.7 per cent.

The unifying factor for the great diversity of immigrants and backgrounds is a specifically Israeli culture. This culture, Pnina Morag-Talmon has written, 'draws on the various sources which developed in the Diaspora, but its principal strength comes from the new reality, from the society which is gradually consolidating itself in Israel'. Kindergarten school and compulsory military service each play a part in creating the unifying elements, as does the Hebrew language, and the yearly cycle that focuses around the Jewish festivals and Holy Days 'uniting the whole nation in festivity, commemoration, and ceremony'.

At Passover all Israelis celebrate the exodus from Egypt under the leadership of Moses. The Jewish New Year is a time to take stock of the past, and make plans for the future. The Day of Atonement (on which Egypt and Syria attacked Israel in 1973) is a time of reflection and memorial. Hanukkah is a time to recall the heroic deeds of the distant past. On Holocaust Memorial Day the tragedies and heroisms of the Holocaust are recalled. On Independence Day the deeds of those who fell in Israel's wars are recounted. Every Saturday – the Sabbath that stands at the centre of Jewish family life and religious observance – the Sabbath candles are lit, the blessings spoken over wine and bread, and twenty-four hours of rest and reflection are inaugurated. Even secular Jews share in this time of national calm. Buses do not run, fewer cars are on the streets, and those Israelis who do not go to synagogue find pleasing, restful pursuits, of which, for all but the winter months, picnicking and going to the beaches are the most common.

As an 'Israeli' style of life develops, other divisions survive, or emerge more strongly, often exacerbated by extremist political Parties, and by those – often by origin from Arab lands – who are ill at ease with the search for peace and reconciliation with the Arabs.

For more than eight years Anatoly Shcharansky, a young Soviet Jew born in Russia four months before the establishment of the State of Israel, had been in prison in the Soviet Union. One of the leaders of the Refusenik movement which demanded the right of Jews to emigrate from the Soviet Union should they wish to do so, he had been sentenced in 1978 to thirteen years' deprivation of liberty: three in prison and ten in strict regime labour camp. Every year on his birthday, January 20, demonstrations were held throughout the Jewish world by those who were campaigning for the free emigration of

Soviet Jews and for the release of more than twenty 'Prisoners of Zion' of whom Shcharansky was serving the longest term. Many hundreds of Soviet Jewry activists, among them Irwin Cotler, Barbara Stern, Genya Intrator and Wendy Eisen in Canada, Michael Sherbourne, Nan Greifer and Rita Eker in Britain, Jacob Birnbaum, Irene Manekovsky, Glenn Richter and Lynn Singer in the United States, and Enid Wurtman in Israel, worked tirelessly within the Soviet Jewry organizations of which they were a part, to organize demonstrations, educational programmes, public appeals and international support.

The Israeli government, operating through a bureau known as 'the office with no name', helped to maintain contact with several thousand Refuseniks, organizing an almost daily stream of visitors to the Soviet Union, many of them Israelis travelling on non-Israeli passports. Although virtually unknown to the Israeli public, Nehemia Levanon served as the inspiration and stimulus to a remarkable organization of contact and support. Through Levanon and his emissaries throughout the Diaspora, and through the public Soviet Jewry organizations with their wide grass roots support, pressure was put on the Soviet government at every level, in every capital city, and at every international gathering.

On 11 February 1986 a vast crowd gathered at Ben-Gurion airport to welcome Anatoly Shcharansky who had been released that same morning from the Soviet Union. Shcharansky's release did not end the Soviet policy of imprisoning Jews who were active in demanding the right to live in Israel, or to study Hebrew and Jewish history. Nor did it mark an opening of the gates of Soviet Jewish emigration, which that year stood at 914, almost the lowest for two decades. But it did provide an upsurge of hope and determination that the thirteen other Jews still in prison might be eventually released, sooner rather than later, and that the two million and more Jews of the Soviet Union might, in due course, be allowed to leave. Shcharansky himself, taking the Hebrew first name Natan, was the driving force a year and a half later in organizing the largest ever Soviet Jewry solidarity rally – 400,000 people in all – held in Washington on the eve of the second Reagan–Gorbachev summit. That year the number of Jews allowed to leave the Soviet Union for Israel rose to almost 6,000; a year later, it reached 17,000, and the year after that (1989) an unprecedented 58,961. Israel was about to witness a demographic revolution.

Among the Jewish immigrants of the earlier mass immigration from Morocco was Mordechai Vanunu, born in Marrakech in 1954, who had come to Israel with his parents in 1963. Soon after reaching Israel he had joined the Likud. 'I went with the flow,' he told an interviewer (Ilana Sugbaker) in the summer of 1985, and 'when I started to think for myself I changed my views'. For ten years he studied at Beersheba University. Asked what he would do if he were Prime Minister, Vanunu told his interviewer, 'First I would create real equality between Jews and Arabs in all areas, in education, construction, housing. Secondly, I would go for peace with the Arabs and of course it would have to include a solution to the fundamental question of the Palestinian refugees. Even if Israel returns all the territories and makes peace with all the Arab countries, the problem of the refugees will remain; 200,000 people have to be rehabilitated and settled where possible, whether on the left bank of the Jordan, in Gaza or in Israel.'

In 1976 Vanunu took a job as a nuclear technician at the Israeli Atomic Research Reactor at Dimona in the Negev. During his nine years working there he studied philosophy and geography. After his job at Dimona ended, and before arriving in London in September 1986, Vanunu converted to Christianity. Once in London he had a number of meetings with reporters from the *Sunday Times* and gave them details of the work being done at Dimona. He told the newspaper that he was doing this for reasons of conscience, in order to promote a public debate on Israel's nuclear programme. On September 30, before any of his revelations were published, he disappeared from view; he had in fact been lured to Italy by the Mossad and kidnapped. On the following Sunday, while his whereabouts were still unknown, the *Sunday Times* published a full account of what he had told them, the centrepiece of which was that Israel not only had the atom bomb, but was the sixth most powerful nuclear power (after the United States, the Soviet Union, Britain, France and China). According to the information that Vanunu reported, Israel had possessed the bomb for more than two decades; nuclear scientists consulted by the *Sunday Times* calculated on the basis of Vanunu's evidence that at least a hundred, and possibly as many as two hundred nuclear weapons had already been assembled.

The *Sunday Times* article was published on October 5. Vanunu had already been taken to Israel, although his presence there was not acknowledged by the Israeli government until November 9. His trial began the following August in the Jerusalem District Court; he was charged with 'intent to assist an enemy in war against Israel' and 'intent to impair the security of the State'. The trial itself was held without any journalists or reporters being allowed into the courtroom. An Amnesty International delegate was refused permission to attend the trial.

Vanunu was sentenced to eighteen years in prison, and held in solitary confinement. More than a decade into his sentence, he is allowed two hours' exercise a day, but no contact with other prisoners. His subsequent petitions to change the nature of his confinement were rejected. The fact that Vanunu had been a 'whistle blower' and not a spy did not help him. Efforts to reduce his sentence, or to end his solitary confinement, have been led by the British-based Campaign to Free Vanunu and for a Nuclear-Free Middle East, in which the British playwright Harold Pinter and actress Susannah York were prominent. A Nobel Prize winner, Professor Joseph Rotblat, appealed in his

Nobel lecture in December 1996, 'What, in fact, can be the purpose of continuing to keep him in prison? One explanation is that he is being used as an example; if so, spending ten years in prison should be enough to serve as a deterrent. A more widely perceived explanation is that he must be kept in isolation in order to prevent him from revealing further information harmful to Israel. If this refers to new nuclear secrets, then it is nonsense. Having followed developments in nuclear weapons throughout the years, I am firmly of the opinion that there is nothing of significance that he could tell.' As late as February 1997, a decade after Vanunu's incarceration, the President of Israel, Ezer Weizman, ruled out the possibility of a pardon.

On 4 March 1987, while Mordechai Vanunu was awaiting trial in Israel, Jonathan Pollard, an American Jew working as a United States intelligence analyst, was found guilty of spying for Israel and sentenced to life in prison. In Israel, there was public pressure for Pollard's release, but in 1993 President Bush rejected appeals for clemency. In an attempt to show Israeli government support for the convicted spy, on 23 November 1995 the newly appointed Minister of the Interior, Ehud Barak – a former Chief of Staff – granted Pollard Israeli citizenship. But in 1996 and 1997 further appeals for clemency were rejected by President Clinton. Many Israelis call for Pollard's release; few have any sympathy with Vanunu's plight.

In October 1986, in accordance with the rotation agreement Shamir succeeded Peres as Prime Minister. Within the National Unity Government, Shimon Peres was determined to move the Arab–Israel conflict forward from its parlous condition, typified by the continuing state of war with Syria, Lebanon and Jordan, and exacerbated by growing Palestinian unease at the Israeli occupation of the West Bank and Gaza, to the level of negotiation, compromise and agreement. Underlying Peres's approach was the conviction that Israel's Arab neighbours, and the Palestinians, were equal participants in the problems and opportunities of the region.

On 11 April 1987, following secret talks with King Hussein in London, Peres and Hussein reached an agreement outlining the method whereby a peace treaty could be negotiated between Israel and Jordan. The agreement went even further: the first part laid down that the Secretary-General of the United Nations would invite the five Permanent Members of the Security Council (the United States, the Soviet Union, Britain, France and China), and all the parties involved in the Arab–Israel dispute, to negotiate a settlement based on Security Council Resolutions 242 (Israeli withdrawal from territories occupied in 1967) and 338 (the implementation of Resolution 242).

As envisaged by Peres, the London Agreement offered a chance to break a forty-year deadlock. But Shamir, who opposed it, sent Moshe Arens, whom he had appointed to be in charge of Arab affairs, to Washington. Arens informed the Secretary of State, George Schultz, that the adoption of the agreement by the Americans would be seen by Shamir as an 'intervention in the internal affairs of Israel'. The agreement was also leaked to the Israeli press, with the result that King Hussein – who had insisted that it be treated with 'great confidentiality' – denied its existence. When Peres brought it to the Israeli Cabinet for approval, the Likud members, including Shamir, opposed it, and the vote was a tie. The plan was therefore rejected.

Shamir had been furious over Peres's initiative, and with Peres personally. 'The whole episode', recalled Yehuda Avner, the Israeli Ambassador to London at that time, 'was ultimate proof to Prime Minister Shamir that the rotation experiment had now spent itself and was a flop.'

The West Bank settlement movement was given a boost on 27 May 1987 at a dinner held in New York by the American Friends of Ateret Cohanim (Crown of the Priests), a group, founded in 1979, that considered it imperative for the Jews to settle throughout the occupied territories. The Parisian-born spiritual leader of the movement, Rabbi Shlomo Aviner, expressed its attitude succinctly when he said, 'We must settle the whole land of Israel, and over all of it, establish our rule. The Arabs are squatters. I don't know who gave them authorization to live on Jewish land. All mankind knows that this is our land. Most Arabs came here recently. And even if some Arabs had been here for 2,000 years, is there a statute of limitations, that gives a thief a right to its plunder?'

At the New York dinner, attended by 500 American Jewish enthusiasts, money was raised for the Ateret Cohanim's Jerusalem Reclamation Project – a scheme for the purchase of property throughout Arab East Jerusalem. The main speaker that night was the Israeli Ambassador to the United Nations, Benjamin Netanyahu, who expressed his support for the idea that Jews could live anywhere within the post-1967 boundaries of the State. Within three years of the dinner, Ateret Cohanim had raised enough money to take over the St John's Hospice in the very centre of the Old City. 'These activities,' Israel Medad on the right wing of the political spectrum has written, 'enjoyed the moral support of Ariel Sharon, who as Minister of Construction and Housing was able to give them also material support.' Sharon also acquired a house for himself in the Muslim Quarter.

Terrorist activity against Israel continued. On 25 November 1987 a member of the Syrian-backed Popular Front for the Liberation of Palestine – General Command flew silently across the Lebanese border in a hang glider, landing just outside the town of Kiryat Shmonah. Entering a nearby Israeli army camp, in a burst of gunfire he killed six Israeli soldiers before being killed. Less than two weeks later a Jewish civilian was stabbed to death in the centre of Gaza City. Those in Gaza who rejoiced at these two episodes, and those in Israel who mourned, were not to know that a wave of sustained violence, unprecedented since 1967, and transforming the nature of life in the occupied territories, was imminent.

40. David Ben-Gurion, Shimon Peres and Moshe Dayan speaking for their political party, Rafi (founded two years earlier), December 1967.

41. Abba Eban and Yitzhak Rabin on the Suez Canal, December 1967.

42. New immigrants from Russia reach Israel, 1970.

43. Ariel Sharon and Moshe Dayan in the Gaza Strip, 1971.

44. A hasid dancing on a roof, overlooking the rebuilding of the Jewish Quarter, Jerusalem.

45. Children on a kibbutz, looked after by their guardians while their parents work in the fields.

46. The October War, 1973: Israeli troops in the Sinai Desert.

47. The October War: Israeli artillery on the Golan Heights.

48. The October War: a wounded man taken from the Sinai battlefield.

49. The October War: Golda Meir visits the wounded in hospital.

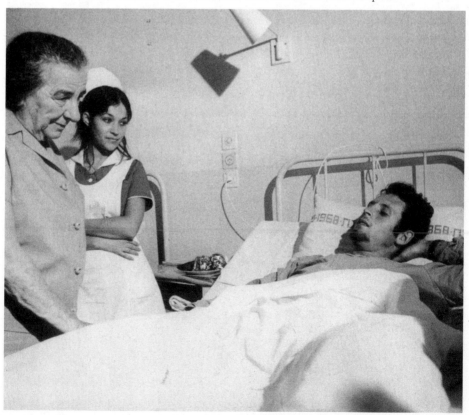

50. The October War: Egyptian prisoners captured in Sinai.

51. After the October War: Golda Meir visits the troops. On her right, Moshe Dayan.

52. Chaim Herzog (top right) at the United Nations in New York, November 1975, listening to the condemnation of 'Zionism as racism'.

53. The Entebbe hostages return to Israel, 4 July 1976.

54. President Anwar el-Sadat of Egypt at Ben-Gurion airport, 19 November 1977. Next to him, Golda Meir. Behind her Yehuda Avner. On Avner's right, General Motta Gur (with beret). Next to General Gur, Yitzhak Rabin.

55. Sadat in Jerusalem, 22 November 1977, with Moshe Dayan and Menachem Begin.

56. Agreement is reached at Camp David, 17 September 1978: Sadat applauds as Carter and Begin embrace. Behind them, the American, Egyptian and Israeli flags.

57. The Cairo talks, March 1979: Begin, Sadat and Mubarak. Between Begin and Sadat, Yehuda Avner. Just behind Mubarak (in glasses), Boutros Boutros Ghali.

58. Yamit: a Jewish settlement under construction on the Mediterranean shore of Sinai.

59. Yamit being dismantled, 5 April 1979.

60. Ariel Sharon explains the advance into Lebanon, 11 June 1982.

61. Beirut under Israeli bombardment, 14 June 1982.

62. A demonstration against the war in Lebanon, 25 August 1982. Some of the Hebrew banners read: 'Peace Now', 'Stop Sharon', 'Sharon, resign', 'Make way for peace', 'We are not the policemen of the Middle East', and 'The Government's policy in Beirut has brought about the perversion of our image as chosen people.'

63. Shimon Peres and Yitzhak Shamir check their watches, 14 August 1984, for the precise moment at which Shamir must hand over the premiership to Peres.

64. The intifada, Ramallah, 11 March 1988: the Palestinian flag flies on the pole.

65. The East Room of the White House, Washington DC, 28 September 1995: King Hussein, Prime Minister Yitzhak Rabin, President Bill Clinton, Chairman Yasser Arafat, Foreign Minister Shimon Peres (standing) and President Hosni Mubarak during the signing ceremony of the Oslo Accords.

66. Tel Aviv, 4 November 1995: Shimon Peres and Yitzhak Rabin wave to the crowds at the end of a rally for peace: a few moments later, as he made his way to his car, Rabin was assassinated. Also on the podium, the mayor of Tel Aviv, Shlomo Lahat.

67. Prime Minister Benjamin Netanyahu and Palestinian Authority Chairman Yasser Arafat in Washington, 2 October 1996, after a difficult meeting.

68. Ehud Barak, former Chief of Staff and Foreign Minister, newly elected leader of the Labour Party, visits the Wailing Wall, 4 June 1997.

69. Ariel Sharon and Ehud Olmert, 4 January 2006.

70. Israelis capture 50 tons of munitions, including several hundred 107mm Katyusha rockets, being smuggled by ship to Lebanon, 4 January 2002.

71. A Katyusha rocket hits a house in Safed, northern Israel, 13 July 2006.

72. Shimon Peres with the Egyptian and Jordanian Foreign Ministers, President's residence, Jerusalem, 25 July 2007.

73. Mahmoud Abbas and Ehud Olmert, Jericho, 6 August 2007.

CHAPTER TWENTY-SEVEN

Intifada

Israel's occupation of the West Bank and Gaza had been met since 1983 by many episodes of stone throwing and tyre burning, of flag-waving demonstrations and verbal abuse by Palestinian Arabs against the Israeli soldiers patrolling the streets of every Arab town. None of these episodes involved firearms. But on 9 December 1987, after four Arab workers from the Gaza Strip were accidentally run over and killed by an Israeli truck, violence broke out in which Israeli soldiers were attacked throughout the occupied areas first with stones, then with Molotov cocktails and finally, in some instances, with guns. It was a popular uprising, and it took even the PLO in Tunis by surprise.

Many Arab villages sought to block the entry of Israeli soldiers. Arab schoolchildren were encouraged by their elders to hurl stones and abuse. Strikes were declared, trade between Israel and the West Bank was brought to a halt, and pamphlets calling for a continuous struggle were circulated. Arab demonstrators raised the Palestinian flag on pylons and rooftops. This was declared illegal under Israeli military law and the flags were pulled down. Sometimes those trying to raise them were shot and killed.

This was the 'intifada'. It drew much of its numerical strength from the twenty-seven Palestinian refugee camps inside Israel. Not allowed by the Jordanians to integrate into Jordanian society before 1967 and remaining for the most part in their camps after 1967, or in the poorest of housing elsewhere in the West Bank, the refugees of 1948 contained a bitter hard core of extremists who were prepared to face Israeli bullets in order to defy the occupiers and assert their national identity. When the intifada began, in addition to half a million native Gazans, there were 445,397 Palestinian refugees (many of them the children and grandchildren of the refugees of 1948) in the Gaza Strip, just over half of them still in refugee camps. In the

West Bank, of the 372,586 refugees nearly 100,000 were still in camps.*

In an attempt to suppress the intifada, the Israeli army used tear gas, rubber bullets, plastic bullets and – increasingly, when under extreme provocation – live ammunition. There were many examples, one of them filmed, of prolonged and savage beatings. Several Israeli soldiers were found guilty by Israeli courts of abuse of their powers. Rabin expressed the dilemma confronting the Israeli soldiers – most of them eighteen- and nineteen-year-old conscripts – when he told an Israeli television interviewer: 'It is far easier to resolve classic military problems. It is far more difficult to contend with 1.3 million Palestinians living in the Territories, who do not want our rule, and who are employing systematic violence without weapons.'

In addressing army officers about the use of force against the Palestinian stone and rock throwers, Rabin, who was Minister of Defence, told them that it was appropriate to use force to bring a demonstration under control, but that force should not be used recklessly. If, when confronted with a demonstrator throwing a Molotov cocktail, the choice was between shooting at him with a rifle and striking him with a club, then it was better to use the club. This advice was essentially one of moderation: it was better to break a demonstrator's bones than to shoot him, risking serious injury or death. But the advice, 'striking with a club', appeared in newspapers around the world as 'break their bones', which appeared both cruel and callous, and Israel was held up to considerable obloquy.

During a visit to Britain while the intifada was at its height, Rabin met Margaret Thatcher for breakfast. The British Prime Minister told Rabin, in her characteristically blunt manner, that Israel 'mistreated' the Palestinians, and she went on to express her concern about their misery and poverty. Rabin replied as bluntly as he had been spoken to. 'Mrs Thatcher,' he said, 'we would never object to any kind of support that Britain might like to offer to the people of Gaza. Would you like to build a school, a hospital? We would welcome it very much.' Mrs Thatcher's reply is not recorded, but Leah Rabin later wrote, 'Yitzhak said it was the shortest breakfast meeting he ever had.'

The intifada was to last for almost five years. It was to lead many Israelis to question whether the occupation could possibly go on. At first, the seriousness of the uprising was not appreciated. Within two months, however, at least fifty-one Arabs were killed by Israeli soldiers responding to being attacked. Hundreds more Arabs were wounded. In a protest against what was called Rabin's 'iron fist' policy, one of Labour's Arab members of the Knesset, Abdel Wahab Darawshe, left the Labour Party and founded an independent Arab parliamentary list, the Arab Democratic Party.

* Outside Israel, there were 278,609 Palestinian refugees in Lebanon (143,809 in camps); 257,989 in Syria (75,208 in camps); and 845,542 in Jordan (208,716 in camps).

The Arab Democratic Party was the first purely Arab Party to enter the Knesset (Darawshe retained his seat after resigning from Labour, and stood and won a single seat in the next Knesset, and two seats in the following Knesset). The Party recognized the Palestinian Liberation Organization as the 'sole representative of the Palestinian people' and supported the establishment of a Palestinian State in the West Bank, the Gaza Strip, and East Jerusalem. It also demanded 'complete equality' for the Arabs of Israel, and the return of all land confiscated from its Arab owners.

The continuation of the intifada led to the deepest of heart-searchings inside Israel, as more and more young men were sent, as part of their three-year compulsory military service, to the West Bank and Gaza Strip. Instead of defending the borders of the State against external attack, they were policing the cities, villages and roads of the country itself, acting at times – as occupying powers inevitably do – with harshness. On 22 December 1987 the Security Council debated a resolution denouncing Israeli violence in the occupied territories. It was assumed by many in Israel, and by many leading Jewish figures in the United States, that, as with so many resolutions critical of Israel in the past, the United States would cast its veto to prevent it being passed. But the United States did not, and Israel stood condemned by the United Nations. Yet the Israeli judiciary continued to uphold the rule of law. On Independence Day 1987 four border policemen had seized and tortured four Arab hotel workers in Tel Aviv. In finding the four policemen guilty, and sentencing one of them to a year and the others to eight months in prison, Judge Moshe Talgam told the court, 'These acts cause me to shudder by the associations they raise precisely because I am a Jew. The punishment that is required, given the defendants' jobs as policemen, sworn to uphold law and order, but who acted in accordance with deviant norms, is a normative punishment that will emphasize to them and their fellow man how much opposed their acts are to the basic norms of law, order and fairness.'

Among the Arab States, the Security Council resolution critical of Israel served as a spur to their own collective action. When the Arab Foreign Ministers met in Tunis in January 1988, 'Palestine' was at the top of their agenda. A fund was set up to 'sustain' the uprising. Every Arab State was exhorted to contribute to it. There was also a sense in the Arab world that pressure could now be intensified for the withdrawal of Israel from the West Bank and Gaza, and for Palestinian self-rule.

On 5 February 1988 the United States Assistant Under-Secretary of State, Richard Murphy, began a six-day visit to the Middle East, bringing with him, to Jerusalem, Cairo and Amman, a new American plan. Under the Murphy proposals, two delegations, one from Israel and one made up of a joint Jordanian–Palestinian delegation, would negotiate interim autonomy for the Palestinians, coupled with an Israeli withdrawal, and municipal elections for the Palestinians. An international conference, though it would have no power to enforce a solution, would meet to guide this outcome. Palestinian autonomy would be in place within three years (that is, by 1991).

Murphy left the Middle East on February 11. His plan offered a way forward beyond the uprising and into the realm of a practical settlement that would tackle the core issue of Palestinian national aspirations. But on the day that Murphy departed, a new organization was established in the Middle East that indicated how hard it would be to secure Palestinian unanimity. The organization was called Movement of the Islamic Resistance (*Harakat al-Muqawama al-Islamiyya*), and quickly became known by its Hebrew acronym, Hamas, the Arabic word for zeal.

Hamas described itself as a branch of the Muslim Brotherhood, an Egyptian Islamic fundamentalist group. It also associated itself with what it called the 'chain of Jihad', extending from the uprising against the British in Palestine in 1936 to the Jihad proclaimed by Islamic leaders after the United Nations partition resolution of 1947. The new force was committed to Islamic rule for the whole of Palestine. It rejected any form of Jewish State or territorial presence. It was also committed to the use of terror against both Israel and Palestinian Arabs who opposed its rejectionist goals.

As the uprising intensified, Hamas operated as an independent organization. It declared its own strike days. It punished Palestinians who did not follow its injunctions. It also set up its own social network to help the poor and the sick. Often it opposed strikes which had been called by Yasser Arafat and the PLO-dominated United National Command, because many traders who were Hamas supporters would have suffered. It also opposed the frequent calls for strikes in schools, for it attached importance to education.

On April 15, five weeks after Richard Murphy left the Middle East, the Israeli government launched a commando raid against Abu Jihad (Khalid al-Wazir), the PLO military chief, who was then in Tunis, and was believed to be the mastermind behind the intifada. Israeli naval vessels took the commandos across the Mediterranean to Tunis, where the Israeli Deputy Chief of Staff, Ehud Barak, ran the operation from a command centre on a navy missile boat off the shore. Mossad agents inside Tunis had located Abu Jihad's house and provided the layout. The commandos were inside the house for less than ten minutes; Abu Jihad, coming out of his bedroom to investigate the commotion, was shot dead.

Ten years later, Ehud Barak was leader of the Israeli Labour Party, and Abu Jihad's widow, Intissar al-Wazir, was Welfare Minister in Arafat's autonomy government on the West Bank and Gaza Strip.

In 1988 Chaim Herzog was elected to a second term as President and was sworn in on May 9. During his address that evening to the Knesset, after expressing his pride at Israel's achievements, Herzog voiced his fears at the way in which the society was developing – what he called 'the clouds and dark shadows which threaten us', and he went on to explain:

> This malaise has many faces, beginning with irresponsible interparty incitement, a poisoned atmosphere and a tendency to demonize one's political opponent. This, in addition to outbreaks of racism, and the encouragement of hatred between Jew and Jew, and Jew and Arab.
>
> We are witness to violence in industrial relations and the conduct of industrial struggle on the back of the ordinary citizen, and even of the sick, wounded, disabled and old. We are witness to an unfair approach as far as the dignity of the individual is concerned, and a lack of sensitivity in regard to the sanctity of a citizen's good name.

'Above all,' Herzog warned, 'we must protect the democratic nature of our country, for without a democratic society based on the will of the people, the State of Israel will have no future.'

A book that was published in 1988 echoed Herzog's sentiments. Its author was Arie Lova Eliav, who, born in Moscow in 1921, had emigrated to Palestine with his parents at the age of three. In the 1960s Eliav had been Israel's Ambassador in Moscow; later he served as a Labour Member of Knesset. In his book, he reflected on what he saw as the four goals of Zionism. The first was the establishment of a Jewish State, which had been achieved forty years earlier. The second was the establishment of a democratic State, which had been the goal from the outset and which, following the 'initial difficult period' of military government, had led to the enfranchisement of Israel's Arabs. The third goal was the establishment of 'a tolerant State', of which Eliav wrote, 'True, from the moment of its establishment, Israel encompassed numerous areas of friction: national – between Jews and Arabs; ethnic – between Jews of European and Oriental origin; and religious, economic, social and class-oriented conflicts. Nevertheless, in general, it succeeded in approaching the goal of "live and let live".'

Of the fourth goal, 'the goal of peace', Eliav wrote that not only was it not achieved but, following the Six Day War, Israel 'became a conqueror and an occupying power', with subsequent evil consequences that had, he argued, lasted for two decades:

> The results of the brilliant military victory achieved in the 1967 war included a horrifying exposure of all the evil impulses hidden within Israel as individual human beings and as a people: arrogance, vanity, indifference to the fate of the defeated, a strong desire to control the conquered territories and to enslave their population to the economy of the victors, and a mystic ritual of 'sanctification' of the conquered lands. And one more thing: the insouciance of those who led the victory.
>
> Drunk and dizzy with glory and fame, gorged on the fruits of victory,

the leaders imagined themselves all-powerful and believed that time was bound to work on their behalf. They began to think, speak, and act in terms of the supremacy of force, of 'might is right,' to which they added a sort of 'The sword shall devour forever' fatalism on one hand, and a belief in divine miracles, which would hasten the coming of the Messiah, on the other.

These patterns of thought, speech, and action have led ever-increasing sections of Israeli society, and primarily Israeli youth, to hate the 'strangers' in their midst, and to increase their hostility toward the Arabs. This, in turn, has engendered the settlements on the West Bank – some of them built upon the tricks of land speculators.

The 'Greater Israel' movement, whose main goal is to incorporate the West Bank and Gaza into the nation's borders, has led Zionism and Israel astray, diverting them from their proper path and deflecting them from the achievement of their fundamental goals.

We must now ask ourselves some painful questions: Is Israel still a Jewish state? Is Israel still a democratic state? Are Israel's economy and society still productive? Is Israel a state – or a society – of tolerance? And is Israel approaching peace?

*   *   *

In August 1988 Hamas published its covenant. The legitimacy of the PLO as the sole leader of the Palestinian people was rejected. Also rejected was the legitimacy of any permanent compromise that might be reached with Israel. The whole of Mandate Palestine, 'from the sea to the river', was declared a Muslim endowment (*Waqf*). Holy war (Jihad) was to be conducted not only against Israel but against 'corrupt and degenerate' elements in Palestinian society. This was not mere rhetoric: as well as Palestinians who had collaborated with Israel, or had worked for better relations with Israel, Palestinain drug dealers and prostitutes were among those punished – and often killed – by Hamas.

As well as Hamas, there was a splinter group from the Egyptian-based Muslim Brotherhood active in the Gaza Strip – Islamic Jihad. Its main objective was 'instant military action' against Israel. The principal support for its terrorist activities came from Iran. One centre of its influence was in Hebron. Other Palestinian factions operated from Jordan and from the Sudan and from southern Lebanon (in conjunction with Hizballah – the Party of God – some 5,000 fundamentalists who had been trained in Iranian-operated camps at Balbeck, 50 miles north of Beirut).

On 1 November 1988 the Israeli electorate went to the polls once more. When the election results were declared, the Likud remained the largest single Party in the Knesset, but with only forty seats. The business of coalition-building began at once, and it was to the right-wing Parties that the Likud turned. The forces which Eliav had feared most were those which emerged considerably strengthened.

With the addition of three smaller right-wing Parties – Tehiyah, Tsomet and Moledet – Likud could command forty-seven seats. The religious Parties gained five seats, reaching a combined total of eighteen. The second largest (and oldest) of these, the National Religious Party, won five seats. A recent break-away group from the National Religious Party, Meimad, which rejected the steady right-wing drift of the National Religious Party, and favoured territorial compromise in return for peace, failed to secure sufficient votes to receive any seats. Its leader, Rabbi Yehuda Amital, was the head of a group of Yeshivot whose students, unlike most of their ultra-Orthodox confrères, accepted the duty of national military service.

The largest of the religious Parties, Shas – Shomrei Torah Sephardim, Sephardi Torah Guardians – won six seats. This was the largest number of seats won by any of the thirteen small Parties represented in the new Knesset. A relative newcomer on the Israeli electoral scene, founded four years earlier, Shas represented the Sephardi Orthodox population. Many of its members and supporters were of Moroccan origin. A former Sephardi Chief Rabbi, Ovadia Yosef, became the spiritual leader of Shas. Born in Baghdad in 1921, and brought to Palestine at the age of three, Ovadia Yosef was already a leader of the Sephardi *teshuva* (religious penitence) movement, which stressed Jewish education dedicated to Torah learning and the observance of the *mitzvot* (commandments).

The origins of Shas lay in the memory of discrimination felt among Sephardi Jews in the ma'abarot shortly after the founding of the State forty years earlier. But Shas had much more than historic discontents to draw on. It also gained considerable support from a network of social welfare activities which it had instituted on behalf of the Sephardi Orthodox community, and from the spiritual leadership of Rabbi Elazar Menachem Shach, a ninety-year-old Ashkenazi sage under whom many Sephardi religious leaders had studied, and by whom they had been inspired. Religion had entered the Israel body politic in an intensified form, representing Orthodox forces committed to an enhancement of religious observance, and laws in which rabbinic interpretations would predominate.

The second largest Party in the 1988 election, Labour, lost five of its seats, leaving it with thirty-nine. It could count on a further ten seats from its three natural supporters on the Left: the Civil Rights Movement (with five seats), Mapam (three seats) and Shinui (two seats). The two predominantly Arab Parties received six seats between them.

Under the mathematics of coalition building, Likud could, by bringing in the religious Parties, have pieced together an adequate majority of sixty-five, sufficient to form a government. From this coalition Labour would be excluded. Labour tried to get the religious Parties to support a Labour-led coalition (such a coalition was also possible mathematically) but the religious Parties refused. Their ideological interests had become over the previous decade far closer to Likud.

Likud chose to form a new National Unity Government. This meant that it could neutralize what it had long denounced as Labour's 'dovish' tendencies. On 22 December 1988 Yitzhak Shamir became Prime Minister for the third time. Shimon Peres was appointed Minister of Finance and Rabin – who had throughout the coalition-building discussions pressed for a National Unity Government – remained Defence Minister. The leading light of the new ultra-Orthodox Shas Party, Arye Deri, was appointed Minister of the Interior. The political leader of Shas, Rabbi Yitzhak Peretz, was appointed Minister of Immigrant Absorption. The Sephardi Orthodox voters had found two powerful patrons in their Shas representatives. The leader of the National Religious Party, Zvulun Hammer, was appointed Minister of Religious Affairs.

The previous National Unity Government had been established on the basis of the rotation of the premiership. This one was established without rotation. Israel was ruled by a cross-Party coalition in which there was no plan to replace Shamir by Peres (or by any other Labour leader) at the end of any fixed or unspecified term.

The Parties to the left of Labour felt that their fragmentation was leading them into a permanent wilderness. Not long after the 1988 elections they decided to form a common list, Democratic Israel, to be known as Meretz – the Hebrew word for energy or vigour – Israel's only secular political constellation. Three Parties came together on the Meretz list: the Civil Rights Movement of Shulamit Aloni (which had won five seats in the 1988 Knesset); Mapam, the veteran left-wing Party in Israeli politics which for the first time had run independently of Labour (and won three seats); and Shinui, the Party headed by Professor Amnon Rubinstein that had been an important element in the Democratic Movement for Change in the 1977 election (and had won two seats in the 1988 election).

Meretz, a potentially influential Party, if it could amalgamate its component seats, and even extend them, considered itself the representative of the Peace Camp, and drew support from those on the left of the political spectrum who found the concept of a National Unity Government frustrating with regard to real progress in the peace negotiations. Also established at the end of 1988 was Betselem (In the Image) a human rights organization that focused its attention on breaches of human rights by the Israeli military authorities in the occupied territories. One of those who was instrumental in setting it up was a Member of the Knesset for the Civil Rights Movement, Dedi Zucker. At first the military authorities on the West Bank and in Gaza tended to ignore the complaints and charges laid by Betselem, but as the foreign press corps in Israel came to rely on its findings more and more, the Israeli authorities took them seriously, and a greater effort than hitherto was made to prevent the abuses, or to punish those that came to light.

The struggle within Israel between the forces of occupation and those of human rights was continuous. In 1988 a new organization, Hotline: Centre for the Defence of the Individual was established, located in East Jerusalem, and inspired by a Holocaust survivor, Lotte Salzberger. The spur to Hotline was the growing number of refusals at Israel's borders to allow Palestinians to leave. Lists at the borders were supposedly of those whom it was judged unsafe for the security of the State to allow to cross into Jordan or to go abroad. Most of those who contacted Hotline did so, one of its reports stated, after going to one of the bridges or another border terminal 'carrying all necessary permits' and being turned back. A typical case was that of a sixty-year-old woman from the Jerusalem area who received a telegram from the Red Cross in Jordan informing her of the death of her brother. 'After taking the appropriate documents,' Hotline reported, 'she left for the bridge for a condolence call to her sister, and was sent back without explanation.'

By dint of raising each individual case, Hotline was able to ensure that 70 per cent of those refused permission to leave were eventually allowed to do so, including the woman turned back from making her condolence call, though even then it took the Ministry of the Interior forty-five days to respond to her written appeal.

Convinced that there was a government-inspired policy of 'intentional delays', Lotte Salzberger and her team were tenacious in their determination to assert the human rights of the Palestinians to leave, as all peoples are by international law allowed to leave, their countries of residence, and to return to them.

On 19 April 1989 Israeli police announced that they had uncovered 'a network of illegal classes held by two West Bank universities at private schools in East Jerusalem'. The classes were immediately closed down. In a public protest, Stanley Cohen, professor of criminology at the Hebrew University, wrote that 'this designation of education as illegal (closing of schools, banning alternative classes) is itself only of the most dubious legality. It goes clearly against local (Jordanian) law, against the Fourth Geneva Convention and against the Universal Declaration of Human Rights. But neither respect for law nor sensitivity to international declarations of human rights have disturbed Israeli governments during the last twenty-one years and certainly not during the intifada. The ubiquitous system of military orders justifies everything in the name of security. Schools and universities have to be closed, we are told, because they are centres of disturbance and unrest.'

This justification, Cohen wrote, 'is extremely weak. While it is true that at some times and in some places schools have been focus points for demonstrations, protest and stone throwing (often in response to provocation and harassment from soldiers, settlers or police), this cannot justify the collective punishment of indefinitely prohibiting the entire education process.' Cohen produced statistics to support his argument. 'In the academic year 1987–1988, pupils in the West Bank lost some 175 out of 210 school days because of forced closures. In the current year, schools have only been open for 40 days. The most recent (January 20) order closed all West Bank schools "until further notice". Effectively, two complete years of schooling have been lost for some 290,000 school-age children. Universities have not functioned at all. Why should they all be punished? And how can kindergartens be a security threat or gathering point for violence? And what about teachers being suspended, arrested and harassed? Schools being used (as they were last year) as military camps?'

The indignation expressed by Cohen reflected the frustration as well as the anger of the Israeli human rights activists. As he went on to explain:

> Teachers in the West Bank have even been prevented by threats of punishments from distributing homework assignments to parents to give their children. Off-campus university classes have been raided by the police or army.
>
> None of this makes sense except as a deliberate attempt to suppress all manifestations of Palestinian self-organization and to increase their ties to dependency on Israel. This has been happening for many years – and is the essence of the occupation. The uprising has simply allowed for more extreme forms of collective punishment – which (as the authorities have correctly judged) will evoke no protest from the Israeli public.
>
> The Ministry of Education, the Teachers' Unions and the Israeli university authorities can all be relied upon to keep quiet. Perhaps they are impressed by the stigmatization of Palestinian educational institutions as 'nationalistic' – a bizarre notion coming from one of the most nationalistic educational systems in the world, where institutions like the Hebrew University (even in its name) were historically associated with a movement for national revival.

The emigration of Jews from the Soviet Union had reached an unprecedented 8,155 in 1987, but of these, only 2,072 went to Israel, and almost all the remainder to the United States, attracted by a vision of economic prosperity far from the Arab–Israeli conflict. This 'dropout' problem caused deep anxiety in Israel, and much anger that the efforts of the Government of Israel to prise open the gates of the Soviet Union were being used by the emigrants to go elsewhere. For Israel the Jewish masses of the Soviet Union had long been regarded as by far the most important future addition to the numerical building up of the State. In 1988 the problem intensified, with 18,961 Jews being allowed to leave, 10,000 more than the previous year, but only 2,173 – scarcely more than in the previous year – went to Israel. When, in 1989, a staggering 71,000 Jews were given exit visas

– the largest ever legal exodus from the Soviet Union – the number who decided to go to Israel was 12,117, with 58,888 going elsewhere.

An anguished debate took place inside Israel, with the government urging the United States not to grant refugee status to the Jews reaching them from Russia, but to let them go to Israel, for which their exit visas had been issued and where they would be received not as refugees but as citizens. On at least two occasions the Prime Minister, Yitzhak Shamir, asked the United States administration to prohibit Soviet Jews with exit visas for Israel to enter the United States. His efforts were in vain. The United States prided itself both on its own policy towards refugees, and on the efforts its people and government had made in putting pressure on the Soviet Union to open the gates, particularly the Jackson–Vanick Amendment, which had specifically linked the increase in American–Soviet trade with the granting of exit permits. The fact that almost the only permits ever granted were for the journey to Israel was seen by the Americans as a Soviet device; if individual Jews chose, as so many tens of thousands clearly did, to leave their Israel journey in Vienna and transfer to a flight to New York, that was their affair and acceptable. A United States Senator, Charles E. Grassley, who tried to explain this to the Immigration and Absorption Committee of the Knesset was given a rough verbal passage.

This was the most strident and long drawn-out debate between Israel and the Diaspora since the founding of the State. Shamir insisted on a 'Zionist' answer to the drop-out phenomenon, arguing that every Jewish emigrant from the Soviet Union with an Israeli visa in his passport should be obliged to go to Israel. If he then wished to leave Israel for good, that was his right. But he should go to Israel first.

The American Jewish establishment countered that every Jew departing from Russia should be granted the right of freedom of choice. This phrase 'freedom of choice' became the code-word of the whole argument, which had its origins when Rabin was Prime Minister in 1975, and which had continued throughout the Begin era and remained an acrimonious one during Shamir's premiership. At the heart of this debate, which raged with great passion on both sides, was the Zionist assertion that ever since the founding of the State of Israel in 1948 there was no such thing as a Jewish refugee, for since then every Jew had a country of his own – Israel.

In 1989 American policy changed, not on moral but on economic grounds. On reaching the United States, each refugee was given a grant of some $3,000, but the Federal budget was not an open-handed exchequer. The American immigration authorities had already begun to turn down the visa applications of about 20 per cent of those Soviet Jews who, already equipped with their visas for Israel, wanted to go on to America instead. On 1 October 1989 an even more stringent United States policy came into force, whereby any Soviet Jew who wanted to go to the United States could not use his Israel exit visa, but would have to apply to the United States Embassy in Moscow, and take the usual route of emigrants to the United States. A quota was put in force, first set at 24,500 a year and then raised by President Bush to 50,000. All other Soviet Jews who left must go elsewhere. With one exception, no other country would take in a tenth that number. Israel became the main option.

Even as the United States closed the door to more than 50,000 Soviet Jews a year, the number of Jews being allowed out was rising. Estimates of those who would leave in the coming years ranged from 100,000 to 500,000 (figures inconceivable five years earlier when fewer than 1,000 Jews had been allowed to leave). Another anguished debate took place in Israel over what scale of efforts should be made to provide facilities, especially housing, for the newcomers. The debate was not only on the statistics of the day, but of projections and aspirations as to what might happen in the future.

The Government of Israel decided to base its budgeting and planning on an estimated 100,000 new immigrants over the coming three years. This led to a strong protest from Avraham Harman, who headed the Israel Public Council for Soviet Jewry, and who urged – in the forceful tones for which he was famous – that a far more comprehensive plan had to be devised. He was particularly emphatic that the Jews of the Soviet Union must be made aware that Israel not only wanted them, but was prepared to take them all in, not 100,000 over three years, but as many millions as presented themselves. It was the duty of the Government of Israel, he argued, and of Jews in the Diaspora, to make sure that housing was available for whatever numbers came. To plan for 100,000 over three years was, he wrote to the head of the Jewish Agency and the Minister of Absorption on 11 November 1989, 'narrow and myopic'.

Avraham Harman was right: the numbers were far greater than the bureaucracy could foresee. In 1990 a total of 185,227 Jews left the Soviet Union for Israel. The first direct flight between the Soviet Union and Israel took place on 1 January 1990. It took the actors of the Tel Aviv Habima Theatre to Moscow, and brought back 125 immigrants to Tel Aviv. But within a month this route was closed. In a speech in the predominantly Russian immigrant neighbourhood of Neve Yaakov, Jerusalem on 23 January 1990, Yitzhak Shamir spoke about the need for 'a large and strong Jewish people in a large and strong State'. His remark was seen by many observers as a call to immigrants to settle in the occupied territories, and as an assertion that Israel must keep the territories in order to absorb the new immigrants. The Soviet Foreign Minister, Eduard Shevardnadze, at once ended the direct flights because, he said, Israel would not guarantee that the new immigrants would not settle in the occupied territories.

Neve Yaakov, although within the Jerusalem municipality, was indeed beyond the Green Line; but it had existed as a Jewish settlement before 1948, when it was overrun by the Arabs.

Flights were soon resumed from Russia through third countries: including

Finland, Hungary, Roumania and Poland. It was not until August 1991 that direct flights from the Soviet Union were resumed. A few months later the Soviet Union disintegrated. Soon afterwards, flights began from all the capital cities of the republics of the Commonwealth of Independent States (CIS), especially from Kiev, Minsk, Tashkent, Baku, Tbilisi – and Moscow. Within two years, almost every former Soviet city with a large Jewish population became a staging point for direct flights: in the summer of 1995 Jews from the war-torn region of Chechnya were flown from the North Caucasian town of Mineralnye Vody direct to Tel Aviv.

Israeli embassies opened in all the capitals of the new republics, and hundreds of Israeli emissaries were despatched to process the accelerated exodus. Teachers from Israel were sent to establish Jewish schools. Within a two-year period, Jewish emigration from Russia exceeded 330,000, creating a new element in Israeli society.

There was another, smaller, immigration that was to affect Israeli society in the 1990s: the arrival of yet more Jews from Ethiopia. In 1990, under Operation Solomon, a total of 4,137 Ethiopian Jews reached Israel, almost the same number in a single year as had arrived in the previous five. Famine and civil war in Ethiopia were a powerful catalyst for their departure. Welfare centres were opened in Addis Ababa by the Jewish Agency and the Israeli Foreign Ministry for Ethiopian Jews who made their way to the capital, often amid great hardship, from their remote villages. An Israeli negotiator, Uri Loubrani, negotiated with the ruler of Ethiopia, Colonel Haile Mariam Mengistu, for the emigration of these remaining Jews to Israel. The Colonel asked for money and a sum of $30 million was agreed on. On 24 May 1991, even as Mengistu's regime was in its final stages, an airlift took place in which 15,000 Ethiopian Jews were flown in Israeli planes to Israel.

The absorption of the Ethiopian Jews created many problems. During their national military service, some of them found themselves taunted because of their colour, and several Ethiopian Jewish soldiers who felt that they had been humiliated by such racial abuse committed suicide. The Chief Rabbinate of Israel added to the newcomers' sense of alienation by refusing to accept them automatically as Jews (they had to undergo a symbolic circumcision ceremony) and refusing to let their spiritual leaders – the Kessim – carry out marriage ceremonies without being retrained first. Many of the Kessim were old; all of them resented the implication that they were not fully or adequately Jewish.

At the end of 1988 the intifada entered its second year. Yitzhak Rabin, the Minister of Defence since 1984, in search of a way to lessen tension, suggested that Palestinian wishes should be ascertained by means of elections in the West Bank and Gaza Strip. This plan was put to the United States administration in April 1989 by Shamir, during a visit to Washington. It was then presented publicly on May 14 by the National Unity Government, as part of a package of four proposals. The first was the strengthening of peace between Egypt and Israel, on the basis of the Camp David Accords, which had languished in what had become known as a 'cold peace' with very little trade or tourism or other contact between the two countries. The second proposal was to seek peace agreements with the other Arab States – which, since the armistice agreements of 1949, were still technically in a state of war with Israel. The third proposal was to try to 'resolve' the problem of the Arab refugees in camps outside Israel: this, too, was unfinished business from the War of Independence and the Six Day War. The fourth proposal was the one relating to elections on the West Bank and in the Gaza Strip. These were to be held 'for nominating a representation that would conduct negotiations for a transitional period of self-rule, and later for a permanent settlement'.

With the peace initiative of 14 May 1989 – an initiative by a National Unity Government led by a Likud Prime Minister – the 'peace process', as it later became known, had begun. But the initiative came with a number of conditions. The first was that Israel would not negotiate with the PLO, only with Palestinians unconnected with the PLO who lived in the occupied areas. The second condition was that there would be 'no change in the status' of the occupied territories: whatever form of self-rule was devised, Israel would remain the sovereign and occupying power. The third condition was that Israel would not agree to 'an additional Palestinian State in the Gaza District and in the area between Israel and Jordan'. These conditions were unacceptable to the PLO, and did nothing to lessen the intensity of the intifada. On May 20 the underground leadership of the uprising issued a leaflet – Leaflet 40 – urging Palestinians to kill a soldier or settler for every 'martyr' killed in a clash with troops. The leaflet explained: 'Stemming from a position of self-defence and the need to make the enemy pay a high price for his crimes, the Unified National Leadership of the Uprising calls on its strike forces to liquidate one soldier or settler for every martyr of our people.'

Leaflet 40 condemned the United States support for Israel's election pro-posal, and expressed 'absolute rejection of Shamir's conspiracy, and holding political elections in the occupied areas, under occupation'. Any political elections, it said, 'can only be held after the end of the occupation and under international supervision as a first step to an international conference'.

An appeal in Leaflet 40 called for fresh attacks on Palestinian 'collaborators', noting that they were not targeted for their divergent political views, but because they were 'traitors, instruments of the occupation's tyranny, and carry weapons to kill and terrorize our people'.

During May 1989, as the intifada continued, thirty-five Palestinians were killed and one Israeli soldier. This was the highest monthly death toll for more than a year. Inside Israel there were growing protests with regard to the nature of military service in the West Bank and Gaza Strip. Omri Frisch, a company commander in the reserves and a teacher from Ashdod, told a meeting called by Concerned Parents of Israeli Soldiers: 'Eighteen year olds ask me if it is frightening to serve in the territories. I tell them the greatest fear is of myself – what I could become, what I could be drawn into. It's a jungle with its own laws.'

At this same meeting, a lawyer, Moshe Negbi, accused the army of military, legal and moral betrayal of its soldiers. It had given soldiers with moral compunctions the choice, he said, of either being passive accomplices or refusing to serve, an alternative which he rejected as likely to undermine the entire basis of Israel as a democratic State. In the discussion that followed, one parent, who described himself as a right-winger, said that the army had never left wounded in the field – 'but this is what we are doing to our children.'

Repeated efforts were made by the Israeli authorities to prevent excesses by the troops sent to the West Bank. In May, a group of Israeli soldiers were reprimanded for desecrating a Koran by using pages from it as toilet paper while occupying a school building. Orders were also given by Israeli military authorities to prevent West Bank settlers from damaging Arab property. When a settler distributed leaflets calling on soldiers to disobey these orders, he was arrested: his name was David Axelrod, the twenty-eight-year-old great-grandson of Leon Trotsky.

In July thirty ultra-Orthodox students at a Yeshiva near Nablus went on the rampage. A thirteen-year-old Palestinian girl was shot dead. During the trial of the students, the head of the Yeshiva, Rabbi Yitzhak Ginzburg, caused consternation when he told the court, 'The people of Israel must rise and declare in public that a Jew and a goy are not, God forbid, the same. Any trial that assumes that Jews and goyim are equal is a travesty of justice.'

Had Rabbi Ginzburg cared to read Judge Haim Cohn's study *Human Rights in the Bible and the Talmud*, he would have seen just how mistaken he was with regard to both Jewish law and ethics.

Throughout the autumn of 1989 the Palestinian kindergartens and schools on the West Bank remained closed. The head of the West Bank Civil Administration, Brigadier General Shaike Erez, insisted that they were 'a focal point of violent disturbances'. In the second week of June a group of Jerusalem parents appealed to the Minister of Education, Yitzhak Navon (a former President of Israel), calling for an immediate end to the school closure. 'This collective punishment is contrary to all the values that you, as education minister, and we, as parents, are trying to teach our children,' the petition reads. 'This action is immoral and ineffective and will cause irreversible damage in the long and short run to Palestinian children and to our own.'

Yvonne Deutsch, one of the initiators of the petition, made the point, 'In the short run, we are hurting the Palestinians by closing the schools. But the kind of society we are creating in the West Bank is affecting us, too. We are becoming more violent, and as a mother I fear for the future of my son.'

The schools remained closed. During August, following the deportation of four leading intifada activists whose petition to remain in the country was rejected by the Israeli Supreme Court, clashes between Palestinian youth and Israeli troops intensified. West Bank Palestinians who wished to open a dialogue with Israel found themselves subjected to growing pressures: a meeting between the Mayor of Bethlehem, Elias Freij, and the Israeli Minister of Health, Yaakov Tsur, had to be kept secret for fear that it might endanger Freij's life.

The number of Palestinian moderates killed by fellow Palestinians grew each month. The climate of violence affected every aspect of the occupation. On September 1 the London *Jewish Chronicle* reported that soldiers on the West Bank were 'now permitted to open fire at masked youths even though they may not be involved in stone throwing incidents. Orders permitting opening fire at masked youths have already been introduced in the Gaza Strip. These orders increase the Army's ability to combat youths who are considered the intifada's chief activists.'

Seeking a way out the impasse a group of concerned Israelis and Palestinians founded the Israel Palestine Centre for Research and Information (IPCRI). Those who came together under its auspices hoped that the Israeli peace initiative of May 14 could still offer a way forward to negotiations and a settlement. One way forward was proposed by the Egyptian President, Hosni Mubarak, who proposed that Israeli–Palestinian talks should take place in Cairo. Israel's refusal to talk with members of the PLO made progress difficult. The United States Secretary of State, James Baker, brought the United States into the discussions as a facilitator. On December 6 he issued a five-point document which contained, in a sixteen-word sentence, what he hoped would prove to be the key to the impasse: 'Israel will attend the dialogue only after a satisfactory list of Palestinians has been worked out.' But for more than two months no such list could be devised to Israel's satisfaction.

The decade ended with the intifada still being fought. In Jerusalem, fewer and fewer Jews went into the Arab bazaars of the Old City, which had so teemed with life and commerce after 1967. 'Israelis no longer shop in Salah ed-Din Street outside the walls,' the British-born journalist Eric Silver wrote in the London *Jewish Chronicle* on June 2. 'They no longer walk in the Arab quarters of the Old City, or take their cars to be resprayed in Wadi Joz. Drivers no longer take the short cut past the American Colony to French Hill and Mount Scopus. The Dolphin fish restaurant, a popular symbol of Jewish–Arab co-ownership opposite the Rockefeller Museum, has closed.'

Reflecting on the Arab reaction that 'the Jews are afraid', Eric Silver commented, 'The Jews say it is not so much fear as prudence. Why risk a knife in the back, a rock through the windscreen? Who needs it?' On June

22, three weeks after Silver wrote those words, Professor Menachem Stern – a Hebrew University scholar and member of the Israel Academy of Arts and Sciences – was stabbed to death by two teenage Arabs. The killing took place in West Jerusalem. Professor Stern, who had come to Jerusalem from Poland as a teenager fifty years earlier, was walking from his home through the valley of the Monastery of the Cross to the National Library on the Givat Ram campus. His attackers later told the police that they had killed him as an 'initiation rite', in order to qualify for membership of Yasser Arafat's Fatah. They had not known who Stern was, nor had they cared.

The site of Stern's killing was a popular park. The murderer could not have chosen 'a better arena for his despicable crime to elicit revulsion and terror among Jerusalemites,' Menachem Shalev wrote in the *Jerusalem Post*. The pleasant valley, in which several Israeli youth movements held their weekly meetings, had been turned 'into an ugly place which, henceforth, will fill parents' hearts with foreboding'.

Repeated Israeli police and army raids into East Jerusalem sought out the perpetrators of these intifada attacks. There was increased nervousness in the streets of West Jerusalem after each incident. Suspicion mounted on each side of the divide, with hatred often breaking to the surface. The Israeli police and army again resorted to tear gas and rubber bullets, and to live ammunition. At night, an incident would often be followed by the release from helicopters of intensely bright magnesium flares, to illuminate an area through which police and soldiers would search for some fugitive who had just killed or maimed.

On July 6, at Telshe Stone (the site of Colonel Marcus's death in 1948) five miles west of Jerusalem, an Arab passenger on the regular 405 bus service from Tel Aviv to Jerusalem seized the wheel from the driver. Crying out 'Radwan, Radwan' – the name of a friend who had been injured in the intifada – he forced the bus off the road and into a ravine, where it burst into flames. Sixteen people were killed. Many of them were burned while still trapped in their seats as a strong wind fanned the flames. Among the dead were Etti and Yitzhak Naim, a married couple on their first holiday alone in twenty-two years. They had planned to make the journey in a friend's car, but it had broken down.

That evening, at a concert in Jerusalem, the conductor Zubin Mehta asked the audience as a mark of respect to stand in silence for two minutes before the concert started, and to refrain from applause during the performance.

The twenty-five-year-old Arab who had forced the bus into the ravine, Abed al-Mufti Gneim, was from the Nusseirat refugee camp in Gaza. His wife, a Negev Bedouin, was expecting their first baby. One of his brothers had been arrested in the first months of the intifada. Gneim survived the downward plunge of the bus and the fire, and was imprisoned for life. He belonged to no Palestinian faction and had acted alone.

In the following three days, twenty-four Jews were arrested by the

Jerusalem police for throwing stones at Arab cars and trying to attack Arabs in the streets. In retaliation, petrol bombs were thrown at Jewish buses, but no one was hurt. A Jew who threw a stone at an Arab truck was sentenced to eight months in prison. On July 13 all public transport stopped for one minute in memory of the dead of the bus attack.

On July 28 in an unprecedented act of retaliation, an élite Israeli military unit crossed into southern Lebanon and abducted Sheikh Abd al-Karim Obeid, who was both a senior Hizballah cleric and the regional military commander of Islamic Jihad.

A terrorist attack, moments of tribute, sudden deaths and the life stories of each of the victims published in every newspaper and read by the whole country, the frightened anticipation of the next bloodshed, the next account of lives cut short: these were the terrible interruptions of daily existence that marked and marred the evolution of the Jewish State in its fifth decade.

CHAPTER TWENTY-EIGHT

Towards Madrid and Oslo

In August 1989 Yitzhak Rabin told the officers at the Israel Defence Forces Staff College that the uprising represented 'the will of small groups to discover their national identity and demand its realization'. This recognition of the national nature of the intifada was followed at the beginning of March 1990 by the American Secretary of State, James Baker, again using his good offices with regard to the future of the West Bank and Gaza by posing the question: 'Will the Government of Israel be ready to agree to sit with Palestinians on a name-by-name basis who are residents of the West Bank and Gaza?'

The Israeli Foreign Minister, Moshe Arens, one of the leading ideologists of Likud, wanted to answer 'yes' to James Baker's question, thus enabling the talks between Israel and the Palestinians to go ahead. But Shamir objected, and as Prime Minister, his authority prevailed. The Labour Party, angry at what it saw as a deliberate attempt to prevent the talks from getting under way at all, decided to withdraw its support from the National Unity Government, and to defeat the government on a vote of no confidence.

Inside the National Unity Government that had been formed in December 1988, Yitzhak Shamir, as Prime Minister, with the fullest support of his Likud followers, planned a substantial increase in Jewish settlements on the West Bank. Many of these lay far from the pre-1967 borders and were in no way connected with the 'security settlements' that the Labour leaders felt were needed in different areas of the West Bank as a protection against Palestinian incursions. At the beginning of March 1990, Shimon Peres decided to use the settlement issue to bring an end to the government, over which, under the terms of its creation, Shamir would remain Prime Minister without any automatic rotation to Labour.

Peres began negotiations with the religious Parties, seeing them as possible partners in a Labour-led coalition. Rabin, as Minister of Defence –

and Peres's long-term rival in the Labour leadership – warned that such an attempt to break away would play into the hands of the Likud. He went so far as to call Peres's activities a 'bad-smelling manoeuvre'. But Peres was determined to break away from the political stranglehold of Likud, and hoped to be able to lead a government that could move matters forward in the direction of a settlement not only with the Palestinians, but with all Israel's Arab neighbours. When Shamir's government asked for the money needed to build eight new settlements on the West Bank, Peres, as Minister of Finance, refused to agree.

On 15 March 1990 the Labour Alignment left the coalition, Shamir was defeated on a vote of no-confidence and the National Unity Government collapsed. The President then called on Peres to form a government. It was the moment Peres had waited for, when Labour would once again reassert its powers as the ruling party. The date, constitutionally, on which he was meant to complete his negotiations was April 12.

Meanwhile, on April 10, in the Lebanese weekly newspaper *Al-Moharer*, published in Paris, Yasser Arafat made a statement that caused consternation inside Israel. 'I want to say clearly,' he declared, 'open fire on the new Jewish immigrants, be they Soviet, Falasha or anything else. It would be disgraceful of us if we were to see herds of immigrants conquering our land and settling our territory and not raise a finger.' Arafat added, 'I want you to shoot, on the ground or in the air, at every immigrant who thinks our land is a play-ground and that immigration to it is a vacation or a picnic. I give you explicit instructions to open fire. Do everything to stop the flow of immigration.'

Arafat's call came at a turning point in the history of Jewish immigration. In 1989 just under 13,000 Jews had been allowed to leave the Soviet Union for Israel, the highest figure for ten years. But in 1990, as part of the general collapse of Communism, the restrictions on Jews leaving, which had hitherto been rigidly enforced, were swept away. In the course of that one year, an unprecedented 185,227 Jews arrived in Israel from the Soviet Union. In the following year a further 147,839 Soviet Jews arrived: making a total in two years of more than 5 per cent of the Israeli population.

In an attempt to form a Labour-led coalition, from which Likud would be excluded, and to move ahead with talks with the Palestinians, despite Arafat's inflammatory remarks, Peres, as Labour Party chairman, was able to win over the former Likud Minister of Tourism, Avraham Sharir, to the Labour cause. But Likud found a Labour member of the Knesset, Efraim Gur, who left Labour and received a Ministerial portfolio. The Knesset did not look kindly on this manoeuvre, and was eventually to pass a law (on 12 February 1991) whereby a Knesset member who changed sides, while still able to serve and vote in the Knesset itself, could not be made a Minister or a deputy minister, and could not be promised a seat in the next Knesset.

While Shamir remained caretaker Prime Minister, Peres sought the support of the Sephardi religious Party, Shas, to give him the majority he needed to form a government. All depended on whether Arye Deri would agree to join a Labour-led coalition, bringing with him six Knesset seats. When Rabin heard from Deri that he would support a Labour approach to him, he agreed that it should go ahead. In the end it was Deri who pulled out. Seven years later, Peres commented wistfully, 'Deri had led his party to topple the Government, but then did not support the creation of an alternative.'

A crucial influence on Deri was Elazar Menachem Shach, the ninety-two-year-old ultra-Orthodox (and anti-Zionist) leader who in a television broadcast castigated the left, the kibbutzim and secular Israelis for having deviated from the true path of religion. Rabbi Shach urged the Sephardi Shas Party, and also the Ashkenazi ultra-religious Degel Hatorah Party, not to join Labour, but to make an alliance with Likud.

Peres still thought that all would be well in his attempts to form a Labour-led coalition. But then the Orthodox Agudat Yisrael Party, which had committed its five seats to Peres, had second thoughts. Without having secured enough seats from the religious or minor Parties to obtain even a majority of one, Peres was forced to ask President Herzog for an extension of the coalition-forming time. The extension was granted, but still Peres (while acting Prime Minister Shamir watched from the wings) could not win over the religious Parties.

At the last moment it looked as if two Agudat Yisrael seats might come Peres's way. That evening Shamir and Peres were at a State banquet in honour of the Czech President, Vaclav Havel. Hardly had the dinner begun than a note was handed to Shamir and Herzog. The Agudat Yisrael had decided to support Likud.

Peres's chances of forming a government were over. Watching him as the State banquet continued was Leah Rabin. 'I must say he remained controlled,' she later wrote, 'and continued to make small talk left and right. He was then called outside to speak to reporters and his mood suddenly turned.' Disappointed and angered by his unexpected failure, Peres asked the reporters: 'What have I done wrong? For forty-nine days I've been fighting to form a new government.'

On June 11 a Likud-led government was formed, with Shamir as Prime Minister. As Shamir was supported by the right-wing and religious Parties, he had no need for Labour participation. Shas and Degel Hatorah both joined the administration. The Shas leader, Arye Deri, remained Minister of the Interior. Another Shas leader, Rabbi Yitzhak Peretz, who had broken away from Shas when the party withdrew its support from the National Unity Government, and had formed his own Party (Moriah), was reappointed as Minister of Immigrant Absorption. He was a strong supporter of the Greater Israel philosophy, and an outspoken opponent of allowing the non-religious kibbutzim to absorb Ethiopian children whose way of life was traditionally Jewish.

*   *   *

On 2 August 1990 Saddam Hussein ordered an Iraqi invasion of Kuwait. In Israel, Abdel Wahab Darawshe, leader of the Arab Democratic Party, supported the Iraqi action. The Government of Israel, against which Saddam issued regular threats – including the threat, first used on August 31, to fire missiles against Israel – supported the coalition, led by the United States, which demanded Saddam's withdrawal from Kuwait. Saddam countered by offering an Iraqi withdrawal from Kuwait in return for an Israeli and Syrian withdrawal from southern Lebanon, and an Israeli withdrawal from the occupied territories. He thus ensured that the Arab–Israeli conflict, which was far removed from his own territorial ambitions in the Gulf, was given a high place on the international agenda.

A summit was held in Helsinki on September 8, at which Presidents George Bush and Mikhail Gorbachev discussed Saddam Hussein's proposal. Gorbachev wanted to accept it. The Americans would not. Three weeks later, on October 1, as Iraqi threats to use missiles against Israel intensified, the Israeli government began to distribute gas masks to the whole population, Jews and Arabs alike.

On 17 January 1991 the Allied coalition, overwhelmingly American but with a large British contingent, advancing from bases in Saudi Arabia, attacked the Iraqi forces in Kuwait. Jordan did not join the coalition, and was seen, ostensibly at least, as an ally of Iraq. Throughout the war the main supply route to Iraq ran from the port of Akaba – within sight of Eilat – to Baghdad.

To maintain the unity of the Allied coalition, in which Saudi Arabia was the crucial Arab participant, the United States and Britain urged Israel not to enter the conflict. It was unthinkable to the Arabs in the coalition that they should be in alliance with a Jewish State that was striking at an Arab State. Despite an acceleration of Iraqi threats, and the launching of Scud missiles against Tel Aviv – the first on January 18 – Shamir's government agreed not to retaliate. On three occasions at least, Israeli fighter planes were on their runways, ready to search out the Scud missile launching sites deep inside Iraq. But they did not take off.

Two influential political figures, and former military commanders, Ariel Sharon and Ezer Weizman, opposed Israel remaining passive in the Gulf War, as did the Minister of Defence, Moshe Arens. Their argument was two-fold: that by remaining passive Israel was negating its whole defensive capacity and making a mockery of its claim to be able to defend its citizens; and that Israeli forces could act more effectively against the Scud launching sites in western Iraq than could the Americans. Unknown to Sharon or Weizman, or the Israeli public, until several years later, the new British Prime Minister, John Major, instructed a special British military unit to operate behind Iraqi lines in seeking out and destroying Scud missile batteries targeting Israel.

The Gulf War was a testing time for Israel. The main cluster of Scud missile launching sites was in and around H3 (the third staging post on the prewar Iraq to Haifa oil pipeline). These missiles carried an explosive warhead, and could have – and were feared to have – chemical warheads. Every Israeli was urged to spend the nights in specially sealed rooms, gas mask on, as a protection against possible gas attack.

The distance from H3 to Tel Aviv was 440 kilometres: a flying time for the missile of between five and seven minutes meant that no effective warning could be given. Forty Scuds fell on Tel Aviv, Ramat Gan and Haifa during the course of the war. Four thousand apartments were totally destroyed. But only one civilian, fifty-one-year-old Eitan Grundland, was killed, when a missile hit his home in Tel Aviv on January 25. Several other civilians died of heart attacks during the missile strikes. Several hundred were wounded.

There was anger in Israel at the reports of West Bank Arabs delighting in the Israeli civilian agony, and even standing on their roof tops to cheer the Scud missiles on their way.

The defeat of Saddam Hussein was a relief for Israel, which had entrusted its security to the Allied coalition. One of the results of Israeli restraint had been that American-manned Patriot anti-missiles and their launchers were brought to Tel Aviv, the first on January 20, to try to shoot down the incoming Scuds. This brought American service personnel on active duty on Israeli soil for the first time in Israel's history. This was much resented by many Israelis who had always insisted with regard to American public opinion that they would never ask for American boys to fight for Israel. The presence of the Americans was also seen to gravely undermine Israel's own much-vaunted independent defence capacity.

Whether the Patriots were effective was much debated. The noise they made certainly imparted to the citizens of Tel Aviv a much-needed sense that counter-action was in progress. The Government of the Netherlands also sent Patriots to Israel to help ward off the missile danger.

The nights spent in gas masks and sealed rooms had a traumatic effect on Israel. Every family was affected. There was a sense of vulnerability; Israel was not at war, her army and air force could not show its prowess, her borders could not keep away the danger.

The intifada had continued throughout 1990. During 1991, stabbings of individual Jews, twelve of them fatal, became more frequent in Jewish urban areas. The murder and maiming of Arabs by their fellow-Arabs, who accused them of 'collaboration with the authorities', was on a far larger scale. In September 1991 the Israeli army issued the figures of known deaths since the uprising had begun. There had been 1,225 Arab deaths: 697 were Arabs killed by Israeli soldiers. Of these, 78 were aged fourteen and younger. The remaining 528 Arab dead were killed by Arabs.

In addition, thirteen Israeli soldiers had been killed by Arabs.

The Likud Government was determined not to recognize the PLO. When Ezer Weizman – the architect of the Likud electoral victory in 1977 – met Arafat in Vienna, he was sacked by Shamir from the government, on the grounds that it was illegal to consort with the enemy. It was seen by many as an act of treachery to meet the PLO. An appeal by a leading member of the Labour Party, Gad Yaacobi, that 'the time has come for Israel to talk with its Palestinian neighbours whoever they may be', was rejected by Likud. During the course of seven months of shuttle diplomacy by James Baker, Shamir insisted that any conference held under international auspices could only serve as a 'preamble' to direct talks between Israel and its Arab neighbours; it was also made clear to the United States that Israel would not agree to any conference with the Palestinians if the PLO were invited to participate.

As a result of James Baker's efforts, a formula was eventually found whereby the Palestinians would be represented by individuals from the occupied territories who would form part of a joint Jordanian–Palestinian delegation. Invitations to a conference were sent out on 18 October 1991. The sponsors were the United States and the Soviet Union. The conference would take place in Spain.

The Madrid Conference opened on 30 October 1991. The first two speakers were Presidents Bush and Gorbachev, the joint sponsors of the conference. Of the parties to the dispute, only Israel was represented by its Prime Minister, Yitzhak Shamir. The Arab States were represented by their Foreign Ministers. Despite protests from Israel, the leading Palestinian delegate, Haider Abd al-Shafi, was given equal time with the other delegations to deliver his opening and closing speeches. Although Israel and its Arab neighbours were sitting at the same table, the Syrian Foreign Minister, Farouk a-Sharaa, spoke so strongly against Israel that it was thought by many of those present that the Israeli delegates might walk out. They stayed, and agreed to participate in separate negotiating teams with the Syrian, Lebanese and Jordanian–Palestinian delegations. These talks also took place in Madrid, and were resumed in Washington on December 10.

This was a historic turning point. The belligerents had met around the same table at Madrid, and their representatives were talking directly to each other for the first time since the War of Independence forty-three years earlier.

The Israelis and Arabs were at last talking, and were doing so face to face, and – in Washington – without American or Russian intermediaries. Indeed, even as the Washington talks began, the Soviet Union was collapsing, and its republics breaking away to form independent States. The collapse of the Soviet Union lost the Arab States the influence and support of their former champion, and knocked out the Soviet Union as a force for disruption in the

Middle East. It also gave the United States, as the one remaining superpower, increased influence as a mediator. The American military action against Iraq had also enhanced her influence. 'The parties had no choice but to go to the table,' Yehuda Avner has commented.

The Washington talks were concerned mostly with the procedures to be followed in future talks. Israel was not prepared to respond to questions of 'territorial concessions' to the Palestinians, but wished to limit the discussion to some form of limited Palestinian autonomy, in keeping with the Camp David commitment. The Likud formula was of autonomy for people, not for land. For the Palestinian Arabs this was a deceptive, disingenuous offer. Jerusalem-born Edward W. Said, a Palestinian writer resident in the United States, expressed the general frustration and anger when he wrote on 10 January 1992:

> A few weeks ago the Israeli Defence Ministry renewed for three months its closure of Bir Zeit University, the leading West Bank institution of higher learning, continuously forbidden to open its doors since early 1988. There has been little outcry among Western intellectuals or academicians, no rights to teach and to be taught for four years by the government of a state that has received $77 billion since 1967 from the United States.
>
> Unlike South Africa, Israel has not been boycotted, although what Israel does on the West Bank and Gaza more than rivals the practices of the South African government during the worst days of *apartheid*.

The negotiations started at Madrid and continued at Washington were not at an end, however. Following the Washington talks, various regional issues were discussed at face-to-face meetings between Israel and the Arab delegations in various cities throughout the world. Five issues were covered in these talks: water, ecology, economic cooperation, refugees, and arms control. On January 28 the first of these sets of talks opened in Moscow – which, less than a month earlier, had ceased to be the capital of the Soviet Union and had become the capital of the Russian Federation.

The Moscow talks focused on water sharing and economic co-operation. In the Canadian capital, Ottawa, there were direct talks on the refugee question. In Tokyo, environmental protection and marine pollution were discussed; in Vienna, water sharing; in Brussels, economic co-operation; and in Washington, arms control. Israeli and Palestinian negotiators were slowly becoming used to sitting down together and going in detail into aspects of life that had to be resolved by agreement if their two societies were to live together without bloodshed, and with political and economic benefits to both. These various bilateral talks were still going on when Israel was plunged into another election.

The election was to be held at a testing time for Israeli society. During the intifada more than 2,000 Arab houses had been demolished or seized, 120,000 trees uprooted in reprisal actions, and on any given day 14,000 Palestinians held in detention (1 per cent of the Arab population). According to the Israeli human rights organization, Betselem – whose members included lawyers, doctors and academics – as many as 4,500 Palestinians had been tortured while in detention. The Labour Party challenged the attitude of mind but did not find such statistics appalling. 'We would not like to dominate another people against their will,' Shimon Peres declared on the eve of the election. 'It is a moral issue. It is a political issue. Throughout our history as a Jewish people, we have never dominated other people, and whoever dominated us disappeared from history. We do not want to copy that.'

With the disintegration of the Soviet Union in the last weeks of 1991, tens of thousands of Jews began to leave for Israel. The pressure on absorption, always heavy, became an economic burden on Israel. Shamir, in an attempt to find the money needed to house the new immigrants, and to start them on their new lives, asked the United States for $10 billion in loan guarantees. The United States, which was already giving Israel loans of more than $3 billion each year for arms purchases and other defence needs, refused to give the loan guarantees, arguing that the settlement building in the occupied territories was illegal, and that many of the Russian Jews would be directed there. Even if, as proved to be the case, relatively few Russians were sent to the West Bank, the money would, the Americans argued, free the Israeli government for further West Bank settlements. It could also be used to subsidize building on the West Bank.

The dispute over the loan guarantees worsened relations between Shamir and the Americans. But his days as Prime Minister were numbered. The elections held on 23 June 1992 ended the Likud ascendancy after an almost unbroken fifteen years. Labour obtained forty-four seats – a gain of five – and became the largest Party in the Knesset. It was therefore able to form a coalition with the left-wing Meretz (whose members, standing for the first time as a single political bloc, won twelve seats) and with the Arab Democratic Party (two seats).

Although Rabbi Shach, who in 1990 had persuaded his ultra-Orthodox Shas (predominantly Sephardi) and Degel Hatorah (Ashkenazi) followers not to support a Labour-led coalition, continued to push this view, Shas, with its six seats, declined to follow his lead a second time. On the instructions of its spiritual mentor, Rabbi Ovadia Yosef, himself of Sephardi origin, who had at one time been Chief Rabbi of Israel's Sephardi community, Shas entered the coalition. Indeed, its participation made the coalition possible. The Shas leader, Arye Deri, who was under police investigation for alleged misappropriation of Party funds, was made Minister of the Interior.

The High Court permitted him to keep his post until he was actually indicted. Likud, with thirty-two seats, went into opposition.

Yitzhak Rabin, who had successfully challenged Peres for the leadership of the Party four months earlier, became Prime Minister of the new Labour-led coalition. His government was committed to continuing the peace process, ending government financing of Jewish settlement in the occupied territories, and making greater efforts to absorb the growing number of Russian immigrants whose disillusionment with Likud absorption policies had been one of the factors in Labour's victory.

In his speech to the Knesset when he was sworn in as Prime Minister on July 13, Rabin made it clear that the Labour Party was finally turning its back on its Socialist legacy. His government would, he pledged, increase economic growth by 'retooling the economy for open management, free of administrative restrictions and superfluous government involvement'. There was, he said, 'too much paperwork, not enough production'.

The advent of the new Labour government led to an immediate change in Israel's attitude to the peace process, and to the outside world. In his first speech as Prime Minister, Rabin set out what he hoped to achieve, telling the Knesset that his government would embark on the pursuit of peace with a 'fresh momentum', determined to turn 'a new page in the annals of the State of Israel'. His government was willing 'to do everything necessary, everything possible, and more, for the sake of national and personal security, to achieve peace and prevent war, to do away with unemployment, for the sake of immigration and absorption, for economic growth, to strengthen the foundations of democracy and the rule of law, to ensure equality for all citizens, and to protect human rights.'

The revolution over which Rabin intended to preside was one in which the Palestinians would become partners, not enemies, and in which Palestinian rights, perceived in the widest national and personal sense, would be restored. In private conversations, Rabin stressed his belief that it would take no more than nine months to reach agreement with the Palestinians. In his speech, he quoted the poet Rachel, who had asked, 'A concerted, stubborn, and eternal effort of a thousand arms, will it succeed in rolling the stone off the mouth of the well?'

Rabin then set out his philosophy and programme of Israel's place in the world at a time when apartheid was ending in South Africa, when there were hopes of negotiations between the adversaries in Northern Ireland, and when the post-Communist world was coming to terms with a new reality in Eastern Europe. 'It is our duty to ourselves and to our children,' he said, 'to see the new world as it is now – to discern its dangers, explore its prospects and do everything possible so that the State of Israel will fit into this world whose face is changing.'

Rabin then spoke of peacemaking with the Palestinians, a credo the revolutionary nature of which was understood not only in Israel but by Israel's friends – and perhaps even by her enemies – around the world. His words marked a revolution in Israel's thinking:

> No longer are we necessarily 'a people that dwells alone', and no longer is it true that 'the whole world is against us'. We must overcome the sense of isolation that has held us in its thrall for almost half a century. We must join the international movement toward peace, reconciliation and cooperation that is spreading over the entire globe these days – lest we be the last to remain, all alone, at the station.
>
> The new Government has accordingly made it a central goal to promote the making of peace and take vigorous steps that will lead to the end of the Arab–Israeli conflict. We shall do so based on the recognition by the Arab countries, and the Palestinians, that Israel is a sovereign state with the right to exist in peace and security.
>
> We believe wholeheartedly that peace is possible, that it is imperative, and that it will ensue. 'I shall believe in the future', wrote the poet Shaul Tchernikovsky. 'Even if it is far off, the day will come when peace and blessings are borne from nation to nation' – and I want to believe that the day is not far off.

Rabin then set out his specific agenda, based on, but going far beyond the talks that had taken place, first at Madrid and then in Washington, between Israel and a joint Palestinian–Jordanian delegation. The Israeli government would propose to the Arab States and the Palestinians, he said, 'the continuation of the peace talks based upon the framework forged at the Madrid Conference'. But it would move swiftly, and towards a more ambitious outcome:

> As a first step toward a permanent solution we shall discuss the institution of autonomy in Judaea, Samaria, and the Gaza District. We do not intend to lose precious time. The Government's first directive to the negotiating teams will be to step up the talks and hold ongoing discussions between the sides.
>
> Within a short time we shall renew the talks in order to diminish the flame of enmity between the Palestinians and the State of Israel.
>
> As a first step, to illustrate our sincerity and good will, I wish to invite the Jordanian–Palestinian delegation to an informal talk, here in Jerusalem, so that we can hear their views, make ours heard, and create an appropriate atmosphere for neighbourly relations.

Rabin had stern yet realistic words, and also words of encouragement, for the Palestinians living in the occupied territories:

I wish to say from this rostrum: We have been fated to live together on the same patch of land, in the same country. We lead our lives with you, beside you and against you.

You have failed in the war against us. One hundred years of your bloodshed and terror against us have brought you only suffering, humiliation, bereavement and pain. You have lost thousands of your sons and daughters, and you are losing ground all the time. For forty-four years you have been living under a delusion. Your leaders have led you through lies and deceit. They have missed every opportunity, rejected all the proposals for a settlement, and have taken you from one tragedy to another.

And you, Palestinians who live in the territories, who live in the wretched poverty of Gaza and Khan Yunis, in the refugee camps of Hebron and Nablus; you who have never known a single day of freedom and joy in your lives – listen to us, if only this once. We offer you the fairest and most viable proposal from our standpoint today – autonomy – self-government – with all its advantages and limitations. You will not get everything you want. Perhaps neither will we.

So once and for all, take your destiny in your hands. Don't lose this opportunity that may never return. Take our proposal seriously – to avoid further suffering and grief; to end the shedding of tears and blood.

The new Government urges the Palestinians in the territories to give peace a chance – and to cease all violent and terrorist activity for the duration of the negotiations on autonomy.

We are well aware that the Palestinians are not all of a single mold, that there are exceptions and differences among them. But we urge the population, which has been suffering for years, and the perpetrators of the riots in the territories, to forswear stones and knives and await the results of the talks that may well bring peace to the Middle East.

If the Palestinians rejected this proposal, Rabin ended, 'we shall go on talking but treat the territories as though there were no dialogue between us. Instead of extending a friendly hand, we will employ every possible means to prevent terror and violence. The choice, in this case, is yours.'

This was a strong statement, looking forward to Palestinian self-government in the West Bank and Gaza Strip – the Camp David formula for a settlement with the Palestinians – to be followed three years after its institution by discussions on a 'permanent solution', which would itself be put in place within five years. Rabin's words launched a new agenda, set a new tone, and were the prelude to a series of new initiatives, whereby Israel embarked on a course that was to transform the Middle East.

The attitudes of forty-four years were being challenged. Rabin, who had been victorious in war, and who understood the terrible cost of war, was determined to be victorious in peace. In his speech, which constituted both a landmark and a turning point in the history of Israel, Rabin declared:

> Members of Knesset, from this moment on the concept of a 'peace process' is no longer relevant. From now on we shall not speak of a 'process' but of making peace.
>
> In that peace-making we wish to call upon the aid of Egypt, whose late leader, President Anwar Sadat, exhibited such courage and was able to bequeath to his people – and to us – the first peace agreement. The Government will seek further ways of improving neighbourly relations and strengthening ties with Egypt and its President, Hosni Mubarak.
>
> I call upon the leaders of the Arab countries to follow the lead of Egypt and its President and take the step that will bring us – and them – peace.
>
> I invite the King of Jordan and the Presidents of Syria and Lebanon to this rostrum in Israel's Knesset, here in Jerusalem, for the purpose of talking peace.
>
> In the service of peace, I am prepared to travel to Amman, Damascus and Beirut today or tomorrow, for there is no greater victory than the victory of peace. Wars have their victors and their vanquished, but everyone is a victor in peace.

Rabin then turned to the role of the United States, which would, he said, be 'sharing with us in the making of peace' – the United States, 'whose friendship and special closeness we prize'. But the pursuit of peace with her neighbours was primarily Israel's task alone:

> We shall spare no effort to strengthen and improve the special relationship we have with the single super-power in the world. Of course we shall avail ourselves of its advice, but the decisions will be ours alone, those of Israel as a sovereign and independent state.
>
> We shall also take care to cultivate and strengthen our ties with the European Community. Even if we have not always seen eye to eye and have had our differences with the Europeans, we have no doubt that the road to peace will pass through Europe as well.
>
> We shall strengthen every possible tie with Russia and the other States of the Commonwealth, with China and with every country that responds to our outstretched hand.

Rabin ended with a powerful statement of the philosophy which under-pinned the efforts he intended to make:

> Mr Speaker, Members of Knesset, security is not only the tank, the plane,

and the missile boat. Security is also, and perhaps above all, the person; the Israeli citizen.

Security is a man's education; it is his home, his school, his street and neighbourhood, the society that has fostered him. Security is also a man's hope. It is the peace of mind and the livelihood of the immigrant from Leningrad, the roof over the head of the immigrant from Gondar in Ethiopia, the factory that employs a demobilized soldier, a young native son. It means merging into our way of life and culture – that, too, is security.

Mr Speaker, distinguished Members of Knesset, this is our declaration of intent, this is our 'identity card', these are the desires that we wish to turn into reality.

Everything I have said on behalf of the Government and myself has been stated in good faith and in an eagerness to set out on a new path, to stimulate, to reawaken, to create and maintain here a State that every Jew, everywhere, will consider his home and the object of his dreams.

Our entire policy can be summarized by a single verse from the Book of Books: 'May the Lord give His people strength, may the Lord bless His people with peace.'

From his very first hours as Prime Minister, Rabin acted with the speed and authority of a man determined to break the mould of stagnation and confrontation that had characterized the premiership of his predecessor. When the army sought permission to enter the Palestinian university campus at Nablus, to search out six armed Palestinians who were believed to be trying to influence student council elections there, Rabin refused. Within a week of his coming to power, he ordered a freeze on all new building, including settlers' houses, in the West Bank and Gaza. Shamir voiced his disapproval. 'Freezing construction, freezing settlements on lands in Judaea and Samaria,' he said, 'the meaning is giving up parts of Greater Israel without negotiations, before negotiations, in return for money.' This was a reference to $10 billion in loan guarantees from the United States, needed to absorb the growing number of Russian immigrants; guarantees which Shamir had been refused because of the settlement building in the occupied territories. The Jewish settlers denounced the freeze as a 'declaration of war'. Rabin was neither intimidated nor deterred.

On July 19, only six days after Rabin's clarion call in the Knesset, the United States Secretary of State, James Baker, returned to the Middle East. Rabin was determined to find a way forward that would make the Palestinians partners rather than adversaries. Two days after Baker's visit, Rabin flew to Cairo. He had believed from the outset that Egypt had an important bridging role to play. A momentum of conciliation had begun. Rabin's Cairo visit was the first by an Israeli Prime Minister for six years.

A way to renewing negotiations for a permanent peace settlement in the Middle East was once more on Israel's agenda.

On August 10, Rabin flew to the United States for a two-day meeting with President Bush at Kennebunkport, the President's summer home in Maine, on the Atlantic Ocean. It had been two years since the last meeting between Shamir and Bush. When the summit ended, Bush announced that the United States would grant Israel the loan guarantees. 'I am committed to assisting Israel with the task of absorbing immigrants,' Bush announced when the talks were over. 'I am delighted that the Prime Minister and I have agreed to an approach which will assist these new Israelis without frustrating the search for peace.'

On August 24 Israel cancelled deportation orders against eleven Palestinians regarded as a disruptive influence. That same day, talks between Israel and the Palestinians were resumed in Washington. Four days later, Israel released 800 of the 7,429 Palestinians whom it was holding in detention.

Rabin's efforts at peacemaking took place side by side with his former rival, Shimon Peres, who became Foreign Minister in the new government. 'It was remarkable how the two men now worked together,' one of those who often saw them in action commented. In September 1992, Peres visited London. In John Major he found a man sympathetic to Israel's position in pushing forward the peace process. Major agreed not only to end the arms embargo which had been imposed by Margaret Thatcher a decade earlier at the time of the Lebanese War, but to end the ban on British oil companies selling their North Sea oil to Israel. Major also agreed to take the initiative with Britain's partners in the European Community (later the European Union) to end the twenty-year-old Arab boycott on British and European companies doing business with Israel.

With the fall of Communism in Eastern Europe, and the disintegration of the Soviet Union in 1991, many Israelis made the journey back to their home towns, or the home towns of their parents and grandparents. Israeli schoolchildren were encouraged to travel to Poland and to visit Auschwitz with their teachers, in what became an annual pilgrimage known as The March of the Living. A sense of roots, which had been absent for so many years – during the years when many Israelis were taught to scorn those who had remained in Europe when Zionist Palestine had been open to them – became strong. The Israel Museum devoted a series of exhibitions to life in the former Jewish heartland where Yiddish had been the language of daily life.

In August 1992 Shimon Peres was invited on an official visit to Belarus, one of the new republics of the former Soviet Union. It was in Belarus – then part of Poland – that he had been born, and spent the first years of his life. The Foreign Minister of Belarus, Piotr Kravchenko, accompanied Peres on his visit to his home of sixty years earlier. Such journeys remind all those who make them of the murder of the six million Jews (Peres's grandparents were among those murdered in the Holocaust), of the destruction of a great heartland of Jewish life, and of the emergence of Israel as a new heartland. As Peres wrote:

> All the way – the village is about a hundred kilometres from Minsk – I compared the actual scenery to what was embedded in my memory. Everything seemed bigger in real life. The trees looked leafier, the river wider. The sky was greyer than I remembered.
>
> In Vishneva, the whole village turned out to welcome us, some of the young girls wearing the national costume. We were greeted with bread and salt. The girls sang local folk-songs in my honour. The tunes were catchy and cheerful, but to me they sounded distant, other-worldly.
>
> There are no Jews left in Vishneva today. By the town hall, a pile of stones stands on the site of the collective grave where the remains of the synagogue victims were buried. Among those remains, under those stones, are my grandparents. The stones seemed silent and uncaring, but in my heart I felt I heard their cry.
>
> The old wooden houses are almost all gone. New ones have been built of bricks and concrete. The streets are also different. I walked slowly along what I thought had been the street on which our house had stood.
>
> Suddenly, I saw it. It had been rebuilt and is now overgrown with ivy, but I recognized it by the well that still stands outside. The fires of war had consumed everything, but not the well. We let down the bucket and brought it up, full of clear water. I put it to my lips – and it tasted the same. I was overwhelmed. Standing beside the house where I was born, I recited Kaddish (the prayer for the dead). Deep inside me, I wept.

Visitors to Israel during November and early December 1992 were aware of a growing tension throughout the West Bank, and in Jerusalem. The revival of the peace process had created an intensification of the activities of Hamas and Islamic Jihad, the two Muslim fundamentalist groups most opposed to any compromise with Israel. Hamas in particular had grown in reputation among the West Bankers at the expense of Arafat's PLO. Largely funded by Iran, whose fundamentalist ideology it shared, Hamas became the dominant force, not only in denouncing Israeli occupation, but in enhancing welfare and health services, as well as educational opportunities, for the Palestinian population. In so doing, it was consolidating Muslim fundamentalism in the very centre of Israel.

Israeli troops, often trapped by gangs of youths in narrow streets and dead end alleyways, responded with live ammunition. On November 10, in Khan

Yunis, Israeli soldiers killed three Palestinians during a demonstration that had been declared illegal. On November 11 in Beit Omar, south of Bethlehem, Israeli troops killed a fifteen-year-old Palestinian youth who was part of a stone-throwing crowd. Two days later, in Hebron, a seventeen-year-old stone thrower who refused an order to halt was shot dead. On November 23 at A-Ram, just north of Jerusalem, a twelve-year-old Palestinian boy was killed during a burst of fire directed against stone throwers. On December 3 in the Balata refugee camp outside Nablus a seventeen-year-old Palestinian was killed while a bomb that he was preparing exploded.

The almost daily battle in the streets was taking its toll on the sanity of both sides. Bitterness intensified. Anger pullulated through both societies. In Jerusalem there were twenty-four stabbings by Palestinians (none of them fatal) and 3,000 stoning incidents in a single month; and 400 cars belonging to Israelis were set on fire in regular night-time attacks throughout the city. It was a time of deep anguish inside Israel at the disintegration of civilized behaviour. Israeli civil rights organizations castigated the Israeli soldiers who had often opened fire against those who were young and unarmed.

On 7 December 1992, the fifth anniversary of the start of the intifada, three Israeli soldiers were murdered in Gaza City, and their deaths proclaimed as an act of heroism by the self-styled 'military wing' of Hamas. Six days later, on December 15, Hamas militants kidnapped an Israeli army sergeant, Nissim Toledano. He was found, stabbed to death and bound, on December 16.

The Israeli government decided to take drastic action against Hamas and Islamic Jihad. There were already 1,600 members of theses two organizations being held in detention, most of them at the Ketziot prison complex south of Beersheba. On December 17, 415 of their leaders were flown to the north of Israel and deported across the border into Lebanon. The Lebanese government refused to allow them in, and the 415 established a tented encampment on the Lebanese side of the border to which the world's journalists quickly came. The sight of the deportees, some of whom were elderly, in their pious devotions created a storm of anti-Israel feeling. Day after day the plight of the deportees was shown on television throughout the world, and considerable pressure was put on Israel to allow them to return to their homes. This pressure culminated in United Nations Security Council resolution 799, demanding their return to Israel. The Israeli government rejected this. To ensure that the spotlight remained on the deportees, to Israel's detriment, the Lebanese government refused to allow the Red Cross to take food or supplies to them through Lebanese territory.

On December 18, the day after the Hamas deportations, Israeli troops shot dead an eighteen-year-old stone thrower in the Askar refugee camp near Nablus, and a seventeen-year-old in the El Arab refugee camp near Hebron. On December 19 Israeli troops killed six Palestinians in Khan Yunis, including a nine-year-old girl and two ten-year-old boys. Three days later, also in Khan Yunis, ten-year-old Iman Abu Amar was shot dead.

On December 30, Israel admitted that a few of the Hamas deportees had been deported 'in error'. Inside Israel, the terror continued. On 15 January 1993 a Palestinian from Gaza drew a knife and stabbed to death four people at the Central Bus Station in Tel Aviv, and at a café. One of the dead was a recent Jewish immigrant from Russia. Another was a Lebanese Arab visiting Tel Aviv. A statement issued in Beirut by Islamic Jihad declared: 'In revenge for the Zionist enemy's heinous crime of deporting the strugglers, and within the context of our nation's continuous struggle against the enemy, one of our strugglers stabbed four invading settlers at the Central Bus Station in Tel Aviv.'

In the midst of these personal and national tragedies, talks had begun, in the strictest secrecy, between Israel and the PLO, with whom, until then, successive Israeli governments had refused to talk. It was a momentous event in the history of Israel. The first of these talks took place on January 20, five days after the Tel Aviv killings, at a secluded villa just outside Oslo, and lasted for three days. The host was a Norwegian research organization, headed by Terje Rod Larsen, who had earlier befriended the Israeli Deputy Foreign Minister, Yossi Beilin. 'The time had come to make good our campaign commitments,' Peres later wrote. 'Rabin had pledged to implement the Palestinian autonomy plan within nine months of our taking office, and that deadline was now approaching. I, too, was impatient to move forward.'

On the Israeli side at that first meeting outside Oslo was a Haifa University political scientist, Dr Yair Hirschfeld, and a senior journalist, Dr Ron Pundak. Neither of them had any official status. On the PLO side was the 'Minister of Finance', Abu Ala'a, and two of his close colleagues. Abu Ala'a outlined the PLO's proposals:

1. A 'Gaza first' option, whereby Israel would commit itself to withdrawing from the Gaza Strip 'within two or three years'. The Strip would then become a trusteeship, under Egypt or under a multinational mandate, for a limited period. Meanwhile, negotiations would continue on an interim autonomy scheme for the West Bank, which, once agreed, would be put in place there.
2. A 'mini-Marshall Plan' for the West Bank and Gaza, in which the international community would undertake, through aid and investment, to invigorate and massively expand the economies of the territories.
3. Intensive economic cooperation between Israel and the Palestinian interim authorities.

Abu Ala'a's reference to the Marshall Plan harked back to the American economic measures immediately after the Second World War, devised by the former United States Chief of Staff, General George C. Marshall, to revive the war torn nations of Europe. On learning what Abu Ala'a proposed, Peres recognized many similarities with his own thinking. The economic aspects of peacemaking were particularly a part of his own outlook and ideas on the way forward. Inside Israel the question of the Hamas deportees remained uppermost in the public mind. On January 23 seventeen of the deportees were allowed back to their homes. But an Israeli offer a week later to take back 101 of the deportees was rejected, amid great protestations of victimization, by the deportees themselves, who continued to invite journalists to witness their daily life and prayers.

The intifada again intensified. On January 27 Israeli troops killed an armed Palestinian in Gaza City. On February 5 they killed a fourteen-year-old boy, Khaled Itawi, in Nusseirat refugee camp in the Gaza Strip. Three other Palestinian stone throwers in Bureij refugee camp were also shot dead. On February 6 Israeli troops, confronted by a crowd of rock-throwers, killed seventeen-year-old Ashraf Da'or in Gaza City.

Inside Israel there was a fierce debate about these killings. Between August 1992 and January 1993 two-thirds of the Palestinian deaths occurred, according to the army's own announcements, during incidents where the lives of soldiers were not in danger. Under such circumstances, army regulations forbade any 'lethal' shooting – that is, aiming above the legs when using live ammunition. Most Palestinian deaths therefore occurred as a result of the army's failure to comply with its own regulations. The number of Palestinian children aged sixteen and under who were killed had also risen in the previous six months. When the *Jerusalem Post* wrote of the children sent by their elders to throw rocks and cinder-blocks: 'It is their senders who are guilty of murder, not the soldiers,' Yuval Ginbar, of the Israel human rights organization Betselem, noted bitterly, 'Our soldiers are never wrong. Who is?'

On February 7 five of the Hamas deportees who were sick were allowed back to Israel and agreed to go. Then, as was perhaps inevitable, the interest of the media declined, and the focus returned to the mounting violence inside Israel. On the West Bank and in Gaza Palestinian stone and rock throwing and taunting of Israeli soldiers reached a crescendo following the deportation of the Hamas and Islamic Jihad leaders.

At its heart, Zionism had striven for a hundred years for the recognition of its legitimacy by the Palestinians. The many conflicts before and after 1948, often marked by harsh and cruel actions, could not hide the basic imperative, that a way had to be found for the Jews and Arabs of the small strip of land running between the Mediterranean Sea and the River Jordan to find a way of accepting each other's right to live and prosper. It was to this imperative that Shimon Peres and the small group of negotiators in Oslo were addressing themselves and investing much time and energy throughout the early months of 1993. 'A close-knit group of top government people made itself available for passing messages between ourselves and the PLO,' Peres later wrote. The group included the Norwegian Foreign Minister, Johann Jurgen Holst, and his wife Marianne. 'When discussions began to intensify they spared no logistical or other effort to keep the momentum going and to shield us from the curious.'

On February 9 the Oslo talks took another step forward, when Peres saw Rabin alone and expressed his 'firm opinion' that Israel should 'try to induce' Arafat to leave Tunis and return to the West Bank and Gaza. Peres then set out for Rabin the advantages in the PLO proposals that Abu Ala'a had presented at the seminar less than three weeks earlier:

1. Israel would announce that it intended to withdraw from Gaza within a fixed period of two or three years.
2. The 'mini-Marshall Plan' for the Gaza Strip would get under way.
3. These developments would not prejudice the ongoing bilateral negotiations in Washington for a full-scale autonomy agreement with the Palestinians.
4. Meanwhile, we could simultaneously draw up plans for long-term economic cooperation between Israel and the Gaza Strip.
5. Desalination plants could be built in the Strip, as well as tourism facilities, a harbour and an oil pipeline terminal, all of which would help boost its economy.
6. The talks on this proposal would proceed discreetly, without the official Palestinian negotiators knowing about them.

On February 11 the Oslo talks were resumed for another two days, again in strictest secrecy. A draft declaration of principles was produced, as well as a paper setting out 'Guidelines for a Regional Marshall Plan'. The urgency of finding a new way forward was underlined a month later, when on March 18 it was announced that, in the three months since December 18 – the day after the deportation of the Islamic Jihad and Hamas leaders – ten Israelis had been killed by Palestinians, and thirty Palestinians by Israelis. Figures were also released for the death toll during 1992. It was clear that the uprising was without precedent in the history of Israel in terms of loss of life. Palestinians had killed eleven Israeli soldiers and eleven Israeli civilians during the year. Israeli troops had killed a hundred Palestinians. And Palestinians, mostly members of Hamas and Islamic Jihad, had killed 220 fellow-Palestinians.

The murder of Jews during March caused distress among Israelis of all political persuasions. On March 1 two civilians in their twenties, Natan Azaria and Gregory Avramov, were stabbed to death in Tel Aviv by a Palestinian from the Gaza Strip. On the following day Yehoshua Weissbrod was stoned and then shot dead in his car in the Gaza Strip town of Rafa. On March 8 a gardener, Uri Magidish, was stabbed to death by two Palestinians while working in a hothouse in the Gaza Strip settlement of Gan Or. On March 12 Yehoshua Freidberg, a twenty-four-year-old immigrant from Canada, was shot dead on the Tel Aviv to Jersualem highway. On the following day, at Khan Yunis, a woman driver, Simha Levy, who worked taking Palestinians from the Gaza Strip to their jobs inside the pre-1967 borders, was axed to death in her van. On March 20 two Israeli soldiers were shot dead: Sergeant Yossi Shabtai, who was patrolling the alley-ways of the Jabalya refugee camp in the Gaza Strip, and Sergeant Gitai Avisar, from the West Bank settlement of Alei Zahav, who was killed when his army jeep was ambushed by three Palestinians near the settlement of Ariel.

On the day of these last two murders (and there were to be five more before the end of the month) a third meeting was held in Oslo. It lasted three days, until March 22. Even as the post-Madrid talks continued to drag on inconclusively in Washington, in the glare of newspaper and television coverage, the secret Oslo talks progressed. An Israel–PLO accord was beginning to emerge. Peres was kept in touch with the discussions at every stage, and on May 14 he returned to see Rabin, to report on what had been discussed, and achieved:

> I stressed that the PLO men in Oslo were more flexible, more imaginative and more authoritative than the West Bank–Gaza team negotiating in Washington.
>
> In Oslo, interesting proposals had been made to define the jurisdiction of the autonomy and to establish Israel's residual status and powers during the interim period – two issues that had long been deadlocked in Washington. Moreover, the 'Gaza first' concept, I stressed to Rabin, was most definitely in the interest of Israel: an overwhelming majority of Israelis wanted to get out of the teeming, terror-ridden Gaza Strip.
>
> In the evolving Oslo agreement, I also pointed out, the five-year interim period was to begin at once. In the 1978 Camp David agreement, by contrast, the five-year clock was to start ticking only after all the details of the autonomy had been agreed and put in place. The Camp David process had an end, but no definite beginning; it set up the target, but failed to provide the bow and arrows to shoot at it.

'In the dramatic weeks that followed,' Peres noted, 'Rabin, by disposition always cautious, moved slowly and warily. He was sceptical about the Oslo talks; sometimes he wholly disbelieved in them. When asked later why he did not share the secret with any of his close aides, he replied frankly that he doubted anything would come of Oslo. None the less he gave me, and the talks, a chance. And ultimately, when the final goal became attainable, he did not draw back.'

In Oslo, the talks were joined by Uri Savir, the Director-General of the Israeli Foreign Office. This raised their status considerably. 'Savir must have surprised the Palestinians,' Peres reflected, 'by indicating that "Gaza first", in the Israeli view, need not necessarily mean a commitment now and implementation only after many months, or even years. Could it not mean, he asked, a handover of Gaza within three or four months of the signing of a Declaration of Principles, even *before* a detailed agreement on interim self-government for the West Bank had been concluded?' The trusteeship of the Gaza Strip, in Savir's scenario, would be run jointly by the Palestinians, the Egyptians and the Jordanians, in full cooperation with Israel.

'Gaza first' had begun to evolve into 'Gaza and Jericho first'. Peres kept a close watch on every development. 'I had hinted to the Egyptians some time before,' he later wrote, 'that we were prepared to consider a "Gaza-plus" proposal, and that Jericho might be the "plus".'

On June 14 another secret meeting was held in Oslo. Uri Savir and Abu Ala'a were joined by Yoel Singer, who had been a member of the Israeli negotiating team at Camp David almost two decades earlier. One point at issue was Israel's insistence on control of security areas in the West Bank. 'My instructions', Abu Ala'a disclosed, 'are that in matters of security I am to be open to your suggestions.' Israel should define its own security locations. 'But please,' Abu Ala'a added with a smile, 'don't declare the entire West Bank a security area!'

It was at this session that the Israeli delegation set aside its earlier suggestion of a Jordanian–Egyptian trusteeship over the areas to be handed over to Palestinian control. 'Is it feasible from your point of view', Savir asked Abu Ala'a, 'to do without it?' 'It's your choice,' Abu Ala'a replied. Savir then told Abu Ala'a that the Palestinians should think in terms of taking over Gaza and Jericho within three or four months of the signing of a Declaration of Principles.

On 20 August 1993 Shimon Peres celebrated his seventieth birthday. That night, his were among the initials put on the document at which the negotiators on both sides had worked so long: the Oslo Accords. 'Here,' he later wrote, 'was a small group of Israelis, Palestinians and Norwegians – partners to one of the best-guarded secrets ever, a secret whose imminent revelation would mark a watershed in the history of the Middle East.' With immense enthusiasm, born of a lifelong quest for the Zionist ideal of peace with her neighbours, Peres told Abu Ala'a: 'The fate of Gaza can be like that of Singapore. From poverty to prosperity in one sustained leap.' How strange, Peres reflected that night, 'that we Israelis are now the ones granting the Palestinians what the British had granted us more than seventy years ago, a "homeland in Palestine", in the words of the Balfour Declaration of November 1917.' Peres added, 'I didn't even try to sleep that night, but lay waiting for the dawn of the new day.'

Peres and his fellow-negotiators had achieved not only a way forward in the Palestinian-Israeli conflict, but a blueprint for wider regional cooperation, prosperity and peace – from the Mediterranean to the Persian Gulf: something that had been envisaged both in the United Nations partition plan in 1947 and in the Israeli Declaration of Independence in 1948. Peres flew with Johann Jurgen Holst to California, to inform the Secretary of State, Warren Christopher, who was on holiday, and Dennis Ross, the State Department coordinator on the Middle East, about the agreement. 'They were generous in their support,' Peres later wrote, 'in spite of the fact that they didn't have the full picture beforehand.'

The Palestine Liberation Organization, hitherto a pariah for most Israelis, became a partner. The opposition at first derided this, and later denounced Rabin and Peres in the crudest, most threatening of language for having shaken the hand of the enemy. But they, in their turn, were to shake that hand, and to promise publicly to honour the agreements that had been signed by their predecessors, and to build on them.

Israel was insistent that Yasser Arafat, hitherto the leader of a group that had carried out many spectacular acts of terror, should specifically renounce terror before any public agreement could be reached between them, and before a Declaration of Principles, embodying the Oslo Accords, could be signed. As President Clinton awaited Rabin and Arafat at the White House for the signing, the turning point in the negotiations came – at the eleventh hour – when Arafat agreed to sign a letter, written to the Norwegian Foreign Minister, which acceded to Israel's demand. The letter read: 'I would like to confirm to you that, upon signing the Declaration of Principles, I will include the following positions in my public statements: In light of the new era marked by the signing of the Declaration of Principles, the PLO encourages and calls upon the Palestinian people in the West Bank and Gaza Strip to take part in the steps leading to normalization of life, rejecting violence and terrorism, contributing to peace and stability, and participating actively in shaping reconstruction, economic development and co-operation.'

An impressive ceremony was being prepared on the White House lawn for the signing of the Declaration of Principles. President Clinton invited a remarkable array of dignitaries, past and present, to witness the signing. As Rabin prepared to fly from Israel to Washington, he too decided to bring with him a number of special guests. One of them was Semadar Haran. Several years earlier, Palestinian terrorists had burst into her home in the coastal town of Nahariya, five miles from the border with Lebanon, and seized her husband and baby daughter. She managed to hide in the attic with her other daughter, listening to the voices of the terrorists, and to the sound of shooting. 'Her young daughter could not contain a muffled cry in that hiding place, just above the terrorists' gunshots', Rabin later wrote, 'and Semadar feared that the cries would reveal them.' Rabin added, 'We associate this kind of horror story with the time of our darkest hours in the Holocaust. Her daughter died in that attic hideaway. Her second daughter's infant head was smashed into a rock and crushed. Her husband was killed by bullets.' Rabin's account (written three months later) continued:

> I asked Semadar Haran if she would join me in Washington, in order to be with me for a special, and difficult moment for the Jewish people, for the State of Israel and for Semadar herself. I know that for her, what I was asking was as difficult as parting the Red Sea. She would be sitting at the ceremony just metres away from the man who had given the orders to her family's murderers.
>
> Semadar Haran agreed to come. She arrived at Ben-Gurion airport just minutes before take-off. I had great respect and tremendous admiration for her ability to respond positively. But Semadar did not accompany us on our flight. At the last moment, her memories overwhelmed her.
>
> She wished us luck, blessed us for our decision to pursue peace with the Palestinians, said that she would pray for us, and for us to return to Israel with peace. She said that she would support us publicly, that she would dream peace, and scream peace.
>
> But she didn't come with us. 'I wouldn't be able to bear it', she told me with her eyes filled with tears, 'I can't shake his hand,' she added, 'but you, the Prime Minister, you are my messenger'.
>
> We parted. Our plane rose into the sky on its way to Washington. Semadar remained in Israel. From the airport, she travelled directly to the cemetery, and laid olive branches on the grave of her husband, on the grave of her first born daughter, and on the grave of her second daughter.

*   *   *

The Declaration of Principles was signed in Washington on 13 September 1993. Yitzhak Rabin and Yasser Arafat were the main signatories. The signing ceremony took place on the lawn of the White House, on the very desk that had been used at Camp David for the signing of the Begin–Sadat agreement in 1978. Immediately after the signing, Rabin's uneasy handshake with Arafat, and his wry smile, were seen by millions of television viewers across the world. That handshake, as much as any other single act, symbolized the revolution and the new reality: Israel had recognized the PLO, was talking to it, and was signing agreements with it; and the PLO had recognized Israel.

Within a month, what had been so secret in Oslo was open to the world's view in Washington. In the months that followed, Israel and the PLO were in almost continuous negotiations, with a view to an Israeli withdrawal from some, and eventually most, of the West Bank and the Gaza Strip. This was a direction inconceivable a few years earlier, breaking the stalemate, the hardships and the violence generated by occupation.

Rabin was often asked in the months ahead what had gone through his mind during that handshake. A private, shy man, he did not easily give an answer. But when his 1979 memoirs were reissued three months after Washington, he set down his feelings. 'I knew,' he wrote, 'that the hand outstretched to me from the far side of the podium was the same hand that held the knife, that held the gun, the hand that gave the order to shoot, to kill. Of all the hands in the world, it was not the hand that I wanted or dreamed of touching. But it was not Yitzhak Rabin on that podium, the private citizen who lives on Rav Ashi Street in Tel Aviv, it was not the father of Dalia, and of Yuval, who both completed their army service, or the grandfather of a soldier today, Yonatan – a grandfather who does not sleep too well at nights, and worries like all parents and grandparents in Israel.' Rabin's reflection continued:

> I would have liked to sign a peace agreement with Holland, or Luxembourg, or New Zealand. But there was no need to. That is why, on that podium, on that world stage, I stood as the representative of a nation, as the emissary of a State that wants peace with the most bitter and odious of its foes, a State that is willing to give peace a chance.
>
> As I have said, one does not make peace with one's friends. One makes peace with one's enemy.
>
> The world is turning upside-down before our eyes: the globes and atlases in your homes have become archeological findings. Your geography books are about to become collectors' items. The most unlikely events are unfolding before our very eyes. Ideologies that moved hundreds of millions vanished without a trace: ideas which brought about the death of millions died themselves overnight.
>
> Borders were erased, or were moved. New States came into being, others fell. Heads of State left centre-stage, while new leaders arose. Almost every day in recent years is more dramatic than the one before it. The great revolution in Moscow, and in Berlin, in Kiev and in Johannesburg, in Bucharest and in Tirana, is reaching Jerusalem, Tel-Aviv, Beersheba and Tiberias.
>
> We are undergoing the revolution of peace.

Rabin knew that Arafat and the PLO were the last vestige of secular Palestinian nationalism with which Israel could deal. The PLO was then at its lowest ebb, he explained at the time to a confidant. On the West Bank and in Gaza, Hamas and Islamic Jihad were pressing for more radical, fundamentalist and essentially violent solutions, and this extremism was swiftly winning over the allegiance of the inhabitants. Rabin knew that if they were to succeed, if the conflict was to be theologized, there never would be peace. For, to theological conflicts, there are no compromises, and therefore no solutions. Hence the handshake, and hence Rabin's resolve to strengthen the PLO arm of Palestinian nationalism as a partner for peace.

In his speech on the White House lawn, Rabin also revealed an attitude of mind that had first been hinted at when he had spoken on Mount Scopus in 1967 after the Six Day War. At the White House, he spoke of how, 'for the many thousands who have defended our lives with their own, and have even sacrificed their lives for our own – this ceremony had come too late'. He had come, with his delegation, 'from a people, a home, a family, that has not known a single year – not a single month – in which mothers have not wept for their sons'.

During his speech, Rabin appealed directly to the Palestinians. 'We are destined to live together, on the same soil in the same land,' he told them. 'We, the soldiers who have returned from battle stained with blood, we who have seen our relatives and friends killed before our eyes, we who have attended their funerals and cannot look into the eyes of parents and orphans, we who have come from a land where parents bury their children, we who have fought against you, the Palestinians – We say to you today in a loud and clear voice: Enough of blood and tears. Enough. We harbour no hatred towards you. We have no desire for revenge. We, like you, are people who want to build a home, plant a tree, love, live side by side with you – in dignity, in empathy, as human beings, as free men. We are today giving peace a chance and saying to you: Enough. Let's pray that a day will come when we all will say: Farewell to arms.'

CHAPTER TWENTY-NINE

The peace process

Following the agreement with the PLO, Rabin and Peres discussed what their next move should be. Rabin felt that Syria should be approached for an agreement on the Golan Heights, based on their return to Syria, in full or almost so, in exchange for 'full peace'. He felt that any approach to Jordan would founder on King Hussein's three conditions: to return all the land, to give back a great deal of the water Israel was taking from the River Jordan, and to allow the return of the Palestinian refugees to their former homes.

Peres suggested that they should tackle the Jordanian option first, and 'storm' it. 'I told Rabin that if he would meet two of the King's requests we can make peace.' The two that should be met were the requests for land and water. They could then refuse the return of the refugees. 'If we can promise the King to keep his status in Jerusalem,' Peres added, 'and help him solve Jordan's foreign debt in America, we can reach agreement.' Rabin agreed that Peres should try his hand. In November, Peres went privately to Amman to see Hussein. They reached a full agreement, which later became the public agreement.

The carrying out of the Oslo Accords became known as the 'peace process' (though Rabin had hoped that the phrase 'making peace' would be used instead). This process offered a way forward for Israel and the Palestinians that had not existed since the United Nations partition resolution in 1947. For many Palestinians, the self-government put in place by Oslo and the withdrawal plans being negotiated after Washington were a first step to statehood – the very status that their leaders had rejected in 1947: the two-State solution for the land between the Mediterranean and the River Jordan, divided as closely as possible on national grounds. Arafat understood that self-governing institutions on the scale offered by Oslo throughout the West Bank and Gaza – not unlike the National Institutions that the Jewish Agency had established between the two world wars – were an essential prelude to statehood. But, just as in 1947, the Palestinian Arabs were encouraged to take up arms against the two-State solution, and encouraged by external as well as internal Arab forces, so, after Oslo, the leaders of Hamas and Islamic Jihad, representing now a minority of Palestinians, turned to terror to try to make the Oslo Accords unworkable.

Israeli extremists also sought to sabotage the accords. In January 1994 a former head of Israeli Military Intelligence, Yehoshafat Harkabi – who was dying of cancer – told two researchers from Ben-Gurion University who were interviewing him about the political situation that he feared that the internal Israeli debate over the withdrawal from the West Bank would have disastrous consequences. 'Rabin will not die a natural death,' he said. 'Not that I wish this on him, God forbid.' Harkabi had become a staunch supporter of the peace process, but he recognized how deeply it was dividing Israeli opinion.

On February 25, a month after Harkabi's warning, an Israeli gunman, Baruch Goldstein, opened fire inside the main mosque in Hebron – the site of the Tomb of the Patriarchs venerated by Jews and Muslims alike – and killed twenty-nine Arab worshippers.

Following the Hebron massacre, Arafat broke off negotiations with Israel on the withdrawal of Israeli troops and administration from Gaza and Jericho. It took some weeks for Rabin to persuade him not to allow violence to deflect him from the path of agreement and self-rule. The talks were resumed. But Palestinian Arab extremism was also at work seeking to destroy the peace process. The new Palestinian terror tactic was the 'suicide bomber', a young man who, in return for the promise of paradise, would blow himself up in a bus or public place in order to kill as many people as possible. On April 6 a suicide bomber, a member of Hamas, killed eight Israelis in Afula. Hamas announced that this was only the first of a series of attacks in revenge for the Hebron killings. Seven days later another Hamas member, with explosives strapped to his body, detonated a bomb on a bus in Hadera. Six people were killed.

In the words of Dr Iyad Sarraj, a Palestinian psychiatrist, 'Suicide bombings are not prohibited for these Islamists, because they consider it a Holy War against Jews, and every Jew is an enemy. They all believe that they are not going to die, that they are moving from one phase of life to another.' Each suicide bomber is told that when the very first drop of his blood is spilled, it cleans all his sins and sends him into paradise.

Rabin would not allow two Arab acts of terror, however shocking to Israeli public opinion, to distract him from the pursuit of an agreement with Arafat that was intended, among other aims, to make terror a thing of the past. He was insistent that to halt the peace process because of terror would give the terrorists their victory. Echoing Ben-Gurion's dictum in 1939 that the Jews of Palestine would fight Hitler as if there were no White Paper, and would fight the White Paper as if there were no war, Rabin declared, 'We will continue the process as if there is no terror. And we will fight the terror as if there is no process.'

The final negotiations took place in secret in Bucharest and Cairo, with Peres as Israel's chief negotiator. When Rabin and Arafat met in Cairo on May 3, it was to finalize an agreement that built on Oslo and Washington, and did so substantially. Signing was planned for May 4. The world's press gathered to witness the historic occasion. But at the very last minute, well after midnight of May 3, Arafat continued to demand tiny territorial changes to the detailed maps that had been drawn up, and previously argued over for so many hours (there were six detailed maps in all). At one point President Mubarak, who was to preside over the public ceremony, went to see Rabin privately to ask him to agree to give Arafat a small area on the Gaza coast that Israel wanted to retain for security reasons, as it was close to some of the Jewish settlements. Rabin agreed.

The signing ceremony on May 4 was shown live on television throughout the world – something that was to become a feature of every subsequent phase of the public Arab–Israel negotiations. It was Mubarak's sixty-sixth birthday, a date deliberately chosen to give him some kudos. The signing ceremony went ahead, and Rabin was seen by the participants, and by millions of television viewers, about to sign the pile of documents that constituted the agreement. He stopped, and went over to the other dignitaries – including Mubarak, Peres, and the United States Secretary of State, Warren Christopher – and pointed out to them that Arafat had not signed the maps on the Jericho area. The assembled dignitaries took it in turns to try to persuade Arafat to sign. He would not do so. Rabin remained impassive, his arms crossed, his expression dour. The leaders then left the carefully prepared stage and argued the matter out away from the cameras. Mubarak was said to have used stern language in insisting that Arafat sign. The leaders then returned to the stage and the cameras and Arafat appended his signature to the maps.

It was a moment of high drama, and near absurdity. For the Israeli public, it was a worrying display of Arafat's volatility. But for the Palestinians, and for many in the Arab world, it showed – as probably Arafat intended – that he would not agree without some sort of protest to what he was being asked to sign – even though it was what he and his senior advisers had agreed to after lengthy negotiations.

As finally signed that day, the Cairo Agreement was comprehensive. Under it, the Palestinian Authority – which would be headed by Arafat – was to be given 'legislative, executive, and judicial powers and responsibilities', including its own armed police force, and full control over internal security, education, health and welfare. The Palestinian Authority would also be empowered to negotiate agreements with foreign powers on economic matters, regional development, and agreements on cultural, educational and scientific matters. Israel would retain control of foreign affairs and defence.

This gave the Palestinian Authority (which the Palestinians called the Palestinian National Authority) even wider powers than those which the Jewish national institutions had obtained from the British between the two world wars.

By the agreement, Arafat assumed the title of Chairman of the Palestinian Authority. He chose, however, for himself, the title of 'President', creating the aura of Head of State.

A few days after the Cairo Agreement was signed, having flown to Johannesburg, Arafat spoke in a mosque of how the Palestinians would 'continue their Jihad until they had liberated Jerusalem'. To speak of Jihad – Holy War – so soon after concluding an agreement aimed at peace and co-existence was a shock to the Israeli public, and to Rabin. 'Any continuance of violence and terror,' Rabin warned, 'contravened the letter . . . that brought about mutual recognition, and called the entire peace process into question.'

Arafat explained that by 'Jihad' he had not meant 'holy war' but 'sacred campaign'. Israeli experts were puzzled by this hitherto unknown concept, but let it pass. At best, all Arafat's pronouncements would have to be taken as rhetoric and exaggeration. At worst, they boded ill for the process that was only just beginning. On May 13, a few days after Arafat's 'Jihad' speech, all Israeli troops and administrative personnel were withdrawn from Jericho. It was only nine days after the signing of the Cairo Agreement. Four days later, they were withdrawn from the Gaza Strip. Israel had been punctilious in carrying out its obligations. The Palestinian flag, which four years earlier had been hauled down by Israeli troops whenever it flew in national defiance – and those who tried to raise it often shot, wounded, and even killed – was raised in pride and ceremony over Jericho and over Gaza City. These were heady moments for Palestinian Arab nationalism, and moments of truth for the Israeli public, which since 1967 had become used to Israel's presence, however disliked, in both these cities.

To calm the situation in Hebron a United Nations observer force took up positions between the Jewish and Arab sections of the city on May 8. The force, known as the Temporary International Presence in Hebron (TIPH) consisted of 160 unarmed soldiers from Norway, Denmark and Italy. A similar force, the Temporary International Presence (TIP) was despatched to Gaza, consisting of 400 unarmed foreign observers.

In Gaza, Palestinian self-rule extended to all the 800,000 Arab inhabitants. Excluded from Palestinian control were the sixteen Israeli settlements in Gaza, with a total population of 5,000.

Following the transfer of Gaza and Jericho to the Palestinian Authority, the way was clear for Rabin and Peres to begin intensive negotiations with

Jordan, with a view to signing a peace treaty. They, and King Hussein, not wanting to lose the momentum of conciliation, were determined if possible to sign the treaty before the end of the year. On May 19, two days after the Israeli withdrawal from Gaza, Rabin and King Hussein met secretly in London to draft the outlines of a peace treaty between their two countries. On this basis, Israeli and Jordanian negotiating teams worked on the details. The Israeli team was headed by the Cabinet Secretary, Elyakim Rubinstein, who sixteen years earlier had been among the drafters of the Camp David Agreements. Many of the details were worked out in the presence of the Jordanian Crown Prince Hassan, in the Royal Palace at Akaba, within sight of the hotels and pleasure beaches of Eilat. These negotiations, like those with the Palestinians, were bilateral: the United States was no longer needed to act as an intermediary.

A substantial breakthrough in the relations between Jordan and Israel – which were still nominally at war – took place on July 20, when Peres flew from Jerusalem to the Dead Sea Spa Hotel, on the Jordanian side of the sea, and negotiated with the Jordanians a plan to develop the two countries' water resources, a canal to link the Red Sea to the Dead Sea, the joining of the electricity grids of Israel and Jordan, and the vision – shared by Peres and Crown Prince Hassan – of turning the dry desert valley of the Arava into a farming and tourist paradise. With water and electricity – and peace – everything could be achieved.

Peres was elated by his visit. 'The flight took only fifteen minutes,' he said, 'but it crossed a gulf of forty-six years of hatred and war.' Four days later, on July 24, Rabin and King Hussein flew to Washington, where, on the following day, they signed a declaration ending the state of war between their two countries which had broken out in May 1948, and had included Jordan's loss of East Jerusalem and the West Bank in 1967.

The momentum of peace appeared to be as powerful as the momentum of war had once been. On August 3 King Hussein flew, at the controls of his royal jet, over the city of Jerusalem, circling the golden Dome of the Rock which had recently been gilded as a result of his own munificence. It was the first time that a Jordanian aircraft had flown over Israeli air space except in war. Jerusalemites looked up in wonderment at the scene, as the King talked over his aircraft radio to Rabin, who was in his office in the city. Five days later, on August 8, the first border crossing was opened between Israel and Jordan in the Arava Valley just north of Eilat and Akaba, for their respective citizens. A ceremony was held, once more before the world's television cameras, as the border crossing was opened and the King entertained Rabin to lunch at the Royal Palace in Akaba. Israelis, denied since the foundation of their State the opportunity to visit Petra – the 'rose red city, half as old as time' – welcomed the new era in which instead of a barbed wire barrier as their eastern border they had a gateway.

*   *   *

Another cruel jolt to the peace process came in the second week of October, when four Hamas kidnappers, three from Jerusalem and one from Gaza, seized an Israeli soldier, Corporal Nahshon Wachsman, and held him captive at an unknown hideaway. They would kill Wachsman, they said, unless Israel released 200 Hamas prisoners, including the Hamas spiritual leader, Sheikh Ahmed Yassin, a quadriplegic whom Israel had arrested and imprisoned four years earlier.

Rabin insisted that there would be 'no negotiations' with Hamas. At the same time seeking to show toughness he suspended Egyptian–Israeli negotiations that were taking place in Cairo, and called back the Israeli negotiating team. On Israel television, Sheikh Yassin appealed to Hamas from his prison cell 'not to harm the soldier and to treat him well, since killing has no point.'

As with the kidnapping of Sergeant Toledano almost two years earlier, there was anguish throughout Israel as to the abducted corporal's fate, and fear that he too would be killed. There was also extreme public distress when a video was released by Hamas to the world's television stations, showing the terrified young soldier in captivity. Rabin decided that if Wachsman was being held in Gaza out of reach of the Israeli security services, he would release Sheikh Yassin in return for the soldier (who, it was learned during the crisis, used to look after his disabled younger brother). But when Wachsman's captors were tracked down to the Arab village of Bir Nabala, north of Jerusalem, an area under Israeli military jurisdiction, it was decided to storm their hiding place. As the Israeli commando troops were about to burst in, Wachsman was killed.

One of the Israeli commandos in the raid, Captain Nir Poraz, was also killed. He had been on demobilization leave when called back to his unit to try to rescue Wachsman. His father, Maoz Poraz, a combat veteran and El Al pilot, had been shot down and killed over the Suez Canal in the October War when his son Nir was two years old.

Following Wachsman's death, there was rejoicing in the streets of Gaza by Hamas. Negotiations for Palestinian autonomy were suspended. But negotiations with Jordan continued, and at a rapid pace. On October 16, two days after Wachsman's death, Rabin followed in Peres's footsteps and visited Jordan. Rabin's destination was the Hashemiyya Palace just outside Amman, a beautiful new royal residence. The visit was not a social one, but an extraordinary working session in which the King and Rabin, sometimes both on their knees poring over aerial photographs, worked out, literally mile by mile, the border modifications which were to be a part of the peace treaty. This complex work began at nine in the evening and ended seven hours later.

The border to be delineated stretched from the confluence of the Yarmuk and Jordan Rivers just south of the Sea of Galilee to the Dead Sea shore at Kaliya, and from the salt pans and mineral works at Sdom to the Gulf of

Akaba. There were points on the map where Rabin agreed that land (which had been taken by Israel at the end of the War of Independence) should be handed over to Jordanian sovereignty – and King Hussein agreed that the Israeli farmers who had cultivated that land should remain in possession of it as lessees. Rabin also agreed that Jordan would receive from Israel 50 million cubic metres of water from the River Jordan, equivalent to 8 per cent of Israel's annual consumption of water from the river. It was also agreed that a desalinization scheme would be begun, financed by both countries, and benefiting both.

The negotiations ended at four in the morning of October 17. The King and Rabin then retired to their rooms and slept while their respective officials wrote up what had been agreed. At eight in the morning the two leaders initialled their handiwork. They then held a joint press conference to announce their achievement. Two days later, on October 19, another suicide bomber blew himself up on a bus in Ramat Gan. Twenty-two people were killed. Among the dead was fifty-seven-year-old Haviv Tishbi, whose parents had immigrated to Israel from Iran. Tishbi had fought in the wars of 1956, 1967 and 1973. He had been on his way by bus to open his flower shop.

Less than an hour after the Ramat Gan explosion, the leader of the opposition, Benjamin Netanyahu, was at the scene. In an interview with journalists, he denounced the government's peace policies. It was 'absurd', he added, 'that we are entrusting our security to Yasser Arafat. Gaza has become a safe haven for terrorists. Gaza has to be closed, and it has to be cleared out.' Rabin was greatly angered by what he denounced as Netanyahu's use of the tragedy to make a political point so soon after the killings. The opposition, Rabin said, was 'dancing on the blood' of the victims: by reacting in such a way, they would encourage more such killings by Hamas in an attempt to destroy the peace process altogether. The opposition, Rabin declared, were 'Hamas collaborators'.

'The Prime Minister must have fallen on his head,' was Netanyahu's reaction. Of greater concern to Rabin was the public comment by President Weizman, whose office might be thought above politics, calling on the government to 'rethink' the peace process. Anger inside Israel was intense. When Weizman and his wife Reuma visited the injured in hospital, a crowd at the entrance shouted out: 'Death to the Arabs' and 'Rabin is to blame for this.' At the funeral of one of the victims, seventy-four-year-old Pnina Rapoport – a survivor of the Holocaust – the Chief Rabbi of Tel Aviv, Chaim David Halevi, declared, 'These deaths are more painful than all of the losses the Jewish people suffered while in exile. Here, they are trying to flush us out of our homeland. But we will stay in this land, despite everything.'

The blast of the bomb had been so fierce that by midnight on October 19 only nine of the dead had been identified. The carnage stunned the nation. 'This is our country,' the Tel Aviv police chief, Commander Gabi Last, commented, 'and although this attack stunned us, we mustn't let these terrorists prevent us from going about our normal lives. There is no need to go into hiding.' That night, Rabin assured television viewers that the man thought to have been behind the manufacture of the bomb, and of several others (including those at Afula and Hadera in May) was being actively sought. Four nights after the Ramat Gan bomb, Israeli police shot and killed an Arab who failed to stop at a road block near Kalkilya. He was not the wanted man.

The wanted man was Yahyia Ayash, a former chemistry student at Bir Zeit university on the West Bank, known because of his bomb-making skills as 'the Engineer'. Like the suicide bomber, Saleh Nazal Souwi, he was from the Kalkilya region. Both men were in their late twenties.

That October Yitzhak Rabin visited Moscow. It was an emotional visit, with the Prime Minister being greeted with enormous enthusiasm by the Jewish community, several thousand of whom were in the process of emigrating to Israel. While in Moscow, Rabin saw something of the efforts being made under the leadership of Chaim Chessler (who had worked for him in the 1992 election) to maintain contact with Jews living in hundreds of towns from the Baltic Sea to the Pacific Ocean, and to encourage their emigration to Israel. Chessler's travels had taken him tens of thousands of miles in search of Jewish communities, often only a few hundred strong, in the remotest geographic regions.

Summer camps had been set up to introduce Russian youth to Israeli and Zionist themes. Jews had been rescued from the fighting in Chechnya, and from the hardships of life in cities like Tbilisi, the Georgian capital, where the disintegration of civic amenities made the Israeli option attractive from the practical as well as the ideological standpoint. There were direct flights to Israel from more than thirty cities in the former Soviet Union. In 1994 more than 60,000 Jews made use of these flights to go to Israel, and to become Israeli citizens.

Public support for the peace process, which had been estimated at more than 60 per cent in the opinion polls at the time of the signing of the Declaration of Principles in Washington on 10 September 1993, fell back towards the 50 per cent mark after the Ramat Gan bomb. It was pointed out by critics of the process that Arafat's letter to the Norwegian Foreign Minister on the eve of the signing of the Declaration of Principles had specifically called upon the Palestinian people to reject terrorism and violence. Indeed, Rabin had been unwilling to sign the Declaration without the letter.

In an effort to prevent suicide bombers from crossing into Israel from Gaza or the West Bank, Rabin instituted a comprehensive closure of the two areas, whereby the Jewish and Arab populations were effectively sealed off from each other. His policy, he declared, was 'a complete separation' between the two peoples. Road blocks were set up on all the roads leading from the

West Bank and Gaza into the rest of Israel, and Palestinian workers turned back. Overnight, 50,000 Palestinians who travelled every day from Gaza to Israel to work lost their livelihood. Inside Gaza, Arafat continued to search for Hamas extremists, and to arrest them. The workers of Gaza, unable to earn their living, blamed Hamas for their plight.

On 26 October 1994, five days after the Ramat Gan bus bomb, President Clinton – the ninth American President to involve himself in the Arab-Israel conflict – embarked on what he called 'a mission inspired by a dream of peace'. He started near Cairo, with a visit to the tomb of President Sadat. Then, after talks with President Mubarak and Yasser Arafat in Cairo, he flew to Akaba for the signing of the Israel–Jordan Treaty of Peace. From Akaba he flew to Amman, and then on to Damascus, before travelling to Israel and addressing the Knesset.

The Israel–Jordan peace treaty, signed amid great festivities on October 26, ended the nominal state of war along Israel's longest border. As agreed by Rabin and Hussein in Amman ten days earlier, Israel gave back certain strips of land in the Arava Valley, and just south of the Sea of Galilee. It was a time 'to make the desert bloom', Rabin told the assembled multitude – of politicians, journalists, citizens of both countries, and well-wishers headed by President Clinton. Hussein pledged that the peace between Israel and Jordan would be a 'warm peace' that would enhance the quality of life in both countries. Israelis had long complained that the peace with Egypt was no more than a 'cold peace'.

Three days after the signing of the Israel–Jordan peace treaty, 2,500 Israeli, Arab, American and European political and business leaders met in Casablanca, at a Middle East/North Africa economic summit hosted by King Hassan of Morocco. It marked a historic widening of the dialogue between Israel and the Arab States, and a start to what was intended to be economic cooperation and joint ventures throughout the Middle East and the Mediterranean. Opening the conference, King Hassan declared, 'We must prove that the Mediterranean can become a region of solidarity and equilibrium, a true Sea of Galilee, around which the three religions and the sons of Abraham, united by historical bonds, will be able to build a magnificent bridge for the century to come.'

Both Rabin and Peres were present at the conference, which Peres addressed on October 30:

> This conference is the first attempt to view the region with economic eyes, with the intention of improving the lot of the people. Marshall Plans, in themselves, cannot salvage our region: the resources required are too vast, and the Middle East has its own resources.
>
> No outsiders can or should be expected to do that which we can and must do ourselves. It is up to us to unleash the potential of our region

and launch our area into a policy of 'Seven Good Years', which was the policy of Joseph of Egypt.

Yet to achieve this we must reduce the unjustifiable expenditure in the arms race. Over the past ten years, 700 billion dollars have been invested in this region for the purchase of metals of hatred, of arms that produce nothing but fear and destruction. Collectively, we can cut this expense by a half or a third and direct it on a constructive course.

Peres went on to point out that 90 per cent of the region was desert. At the same time, given current trends, the population was growing at a rate that would see it doubled in twenty-five years. Water was, and would remain, the key need. 'Water is unimpressed with political boundaries or national sovereignties,' he said, 'and the effort must be regional. Our choice is clear – national deserts or regional bloom.' The Middle East could not achieve prosperity 'hiding behind national boundaries'. Technology ignored borders. Western companies should be encouraged to invest in the Middle East with schemes that would enhance the well-being of the whole region. Peres continued:

There are almost 300 million people in the Middle East. The choice before them and the outside world is whether to be a poor and bitter people, wearing the cloak of protest and the mantle of fundamentalism – or to be 300 million productive consumers, investing in the education of their own posterity and thus able to enter the twenty-first century as equals with the peoples of Europe, the United States and Asia.

Here in Casablanca, we are entrusted with the obligation to take the first step in transforming the Middle East from a hunting ground into a field of creativity.

It is here, in this city rich with history, along the shores of our common Mediterranean, and under the chairmanship of His Majesty King Hassan II, that we are entrusted with an actual mandate – to dream a new dream about the Middle East and to seek concrete ways to implement it.

The dream is to erect a new Middle East. A region with no wars, no front lines, yielding a missile-free and hunger-free Middle East.

This remained Peres's theme, and was always expressed by him without prevarication. Two and a half years after Casablanca, writing about the Negev in a private letter, he commented, 'The Negev is over 50 per cent of the Land of Israel, it is empty, and that instead of fighting the Palestinians, it is better to fight the desert.'

When Rabin spoke to the Casablanca conference he recalled the time in October 1978 when he had made a secret visit to King Hassan: 'I came here; you graciously agreed to receive me. I was disguised. It had to be secret – in search of peace.' In his speech, Rabin urged the international community to 'encourage the PLO and the Palestinians to develop projects in Gaza, in Jericho', and to put their money behind such projects. Israel was interested, he said, 'that the Palestinians in Gaza will entertain better economic and social conditions, because poverty is a fertile ground for the growth of the Hamas and the Islamic Jihad. And it is first and foremost human to see people doing better, earning better, eating better, with better housing.'

Rabin ended by setting out his vision for the future of the whole region. Like Peres, he believed that there were enormous strides that could be taken towards prosperity, once the habit of conflict had been set aside. As Rabin told the Arab, American and European leaders and businessmen assembled there:

> The Casablanca meeting could be a landmark in peace development. The diplomatic agreements, the peace treaties, the political arrangements – someday, that will be translated to the life of every citizen in Gaza and Amman, in Tel Aviv and Tiberias. I don't expect tomorrow, immediately, but the mere fact that this unique, large conference was convened, is the expression of a new opening. It will create, not immediate results, but people will meet one another.
>
> I don't remember any conference where so many representatives of Arab countries, Europeans, Americans, from all religions – the mere fact that they are convened, talk to one another, get to know one another, creates a better basis for whatever resolution, creates new realities in the economic life, more readiness to do it and more likelihood of signing a peace treaty.

Rabin then spoke, as Peres had done, of water. 'For Jordan and Israel, water is vital,' he said. 'We can build two dams on the Jordan and the Yarmuk Rivers. We hope very much, as a result of this conference, that a different approach, better cooperation, will be developed. We will not say that immediately we will meet the needs of Jordan, or those of Israel.' Rabin continued:

> I hope that as a result of this conference, the word 'boycott' will disappear from the relationship between the countries of the region and the countries of the world.
>
> Peace is a sacred goal. It is not so easy to achieve, it is very difficult to maintain. And only by concerted efforts of countries in the region, assistance from abroad and readiness to invest here – not just to pay lip service, to invest in peace – if this will be the result of this conference, there is no doubt in my mind, we will see an entirely different Middle East being developed.

The Casablanca Conference marked a high point in the prospect of a full and fruitful peace. But much remained to be done to create the confidence needed whereby, on the basis of the Palestinian–Israeli reconciliation, the regional economic projects to which Casablanca looked forward could go ahead.

On 22 January 1995 nineteen Israeli soldiers and a civilian were killed at a bus station at Beit Lid, near Netanya, by a suicide bomber who blew himself up close to a snack bar crowded with soldiers. Israel was devastated by the atrocity, burying more young soldiers in a single day – including several women – than at any time in a decade. Islamic Jihad called the victims 'pigs' and 'monkeys'. When Arafat telephoned Rabin to offer his condolences, Rabin pressed upon him the need for immediate and strong action against the extremists, who were endangering the peace process and putting further Palestinian autonomy at risk. Arafat promised strong action, and took it. In a series of sweeps through Gaza by Palestinian police, many Islamic fundamentalist leaders were arrested; military tribunals were set up which, often in the space of a few hours, sentenced those arrested to prison. Weapons were confiscated.

Rabin had been scheduled to appear that day at Yad Vashem in Jerusalem at a ceremony to commemorate the fiftieth anniversary of the liberation of Auschwitz by the Russian army. Because of the Beit Lid bomb, he had cancelled his visit. Awaiting him at Yad Vashem, had he been able to carry out his original schedule, was a Jewish student, Yigal Amir, a vitriolic opponent of the peace process, who had sworn to kill him.

The peace process continued. In March, Rabin – whose rivalry with Peres hitherto had been much commented on in the Israeli press, and had been a striking feature of Rabin's published memoirs – entrusted to Peres the negotiations with the Palestinians for extending their autonomy in Gaza and Jericho to the whole of the West Bank. He, Rabin, would concentrate on the issue of terror: both internal Palestinian terror, and the wider, insidious and dangerous terror sponsored worldwide by outside forces, most notably Iran.

On the morning of 29 March 1995 the British Prime Minister, John Major, flew to Israel. That afternoon, in Jerusalem, Yitzhak Rabin told him that there were two issues facing Israel: how to expand the Declaration of Principles to the West Bank, and 'how to reach agreement with Syria that would also mean Lebanon'.

Rabin also spoke of the darker side of the coin. 'There are two major problems,' he said. 'The main obstacle is terror, the ugly wave of radical Islamic terrorism, in many parts of the Middle East and beyond, what I call Khomeinism without Khomeini. Most terror is extremist. We did not experience suicidal terror before in Israel, only in Lebanon. Then there is overseas terror. Terror today is the enemy of peace. Terror endangers moderate Arab regimes. They look at Sudan. It is now in Algeria. It endangers peace in Egypt. I'm not worried about the military strength of Syria. What worries me is terror. It is paid by Iran. You find Iranians in Lebanon, Sudan, everywhere.'

The second problem, Rabin told Major, was 'the need to give economic assistance. I don't believe peace can become viable unless the man in the street in Gaza and Jericho sees that peace gives new hope to his daily life. Without economic advance, the chances of stable peace will be reduced.'

There were two sides to the Israeli–Palestinian equation. Rabin told Major that when Shimon Peres had met Arafat a few days earlier, Peres told him that 'the Palestinians must act differently on law and order. They have to have only one law-enforcement law. They have to cope with those who instigate, preach and carry out terror; to cope with the threat of the use of Gaza and Jericho as a base and a refuge for terrorists in Israel and the West Bank. Today we found forty kilogrammes of TNT on the road to Gaza that could have been operated electrically, I don't know for sure against whom.'

Rabin went on to explain the need for an economically stable, and prosperous Jordan, seeking British help – which was forthcoming – in helping to reduce Jordan's overseas debt. He then turned further east: 'In the longer term, there is the danger of Iran building its military capacity. The combined effects of terrorism and military capacity are a danger to peace in the whole region. The Casablanca conference could be and should be a major step, but some Arab countries fear the openness it created.'

John Major responded with a sympathetic approach, based on an understanding of the impact of terror on Israel. 'The terror bomb in Tel Aviv,' he said, 'had a profound effect in the United Kingdom. As to peace, third parties cannot negotiate. I don't propose to interfere. But we are ready to help where you want us to.' For its part, Britain hoped that Israel would increase its economic assistance to the Palestinian Authority. 'Not necessarily in the short term, but in the long term,' Major told Rabin, 'improving economic circumstances, I believe, will prove the greatest *de*-motivator for terrorism. But in the international community much of the resources, especially of the United States, compared with a few years ago, are now diverted to Eastern Europe, to Russia; to sustain democracy, mostly in Russia. The Gulf States are not so willing to provide help for the Palestinians as they were a few years ago.' Britain was prepared to help. With Britain's encouragement, there would also be more money from the European Union for the Mediterranean area. Jordan would also be a beneficiary.

Both Rabin and Peres were anxious that the growth of economic activity, and successful economic growth, in the region, would not be characterized as Israeli economic imperialism. As Rabin expressed it later that same evening, 'We don't want to create fear among neighbouring countries that, under the name of a New Middle East, we are going to dominate them economically. We don't want to patronize them. We have to build it gradually.' Speaking of Israeli hopes that King Hussein would join in a joint Israeli–Jordanian project to build an international airport at Akaba, Rabin said, 'It could be like Geneva airport, serving the two countries, with the Jordanian border at one exit and the Israeli border at the other. If we will approach in a pragmatic way such ventures there is much that can be done: a joint electricity grid for example, and the export of Israeli phosphates through the Jordanian road and Akaba port facilities.' Rabin added, 'It is feasible. It takes time. People have to change the disks in their computers.' As for the Palestinian Authority and its attitude to such joint economic ventures, Rabin declared, 'There are many capable Palestinians. I wish that their system would be run by educated people.'

Inside Gaza a struggle was taking place between those who wanted to respond to Israel's willingness to grant a meaningful autonomy and those who rejected any dealings with Israel whatsoever. On April 2, in Gaza City, two members of Hamas accidentally blew themselves up while preparing for a future attack on Israeli targets. Six Palestinians were killed in the explosion, among them a young girl aged five or six.

Also killed in the explosion on April 2 was Kamal Kahil, a leader of the Issadin Kassem Brigades, the armed wing of Hamas. Kahil was believed to have been responsible for the ambush and killing in December 1993 of the most senior Israeli officer to have been killed in Gaza before the autonomy, Lieutenant-Colonel Meir Mintz. Among the weapons and munitions found in the ruined building were several explosive-packed body belts of the sort used by suicide bombers.

The chief of Gaza's civil police, Brigadier-General Ghazi Jabali, appealed to Hamas to 'stop all these acts and come join us – for the mercy of our people.' His appeal was ignored. On April 9, seven Israelis and an American student, Alisa Flatow, were killed in two suicide bombings in Kfar Darom, an Israeli settlement in the Gaza Strip. Two weeks later, on April 22, Yigal Amir, who had sworn to kill Rabin, tried to enter the hall in Jerusalem where Rabin was present as the guest of honour at a Moroccan Jewish folklore festival. He had no invitation, and was unable to get past the security guards.

The Government of Israel was determined to see the Oslo peace process brought to fruition. On July 4 Shimon Peres went to Gaza for talks with Arafat. Both men wanted to conclude what was being called Oslo II, the extension of Palestinian rule to the West Bank. A complex plan had been devised, with a timetable for the establishment of the Palestinian Authority in different areas of the West Bank, and a phased withdrawal of Israeli troops. These troops would be withdrawn at once from the main cities, and at a later stage from most of Hebron – the one West Bank city with a small Jewish presence. Then they would be withdrawn from the Palestinian villages, and finally from the countryside. Israel would remain responsible for the West Bank settlements, and there would be a small Israeli military enclave. But the bulk of the West Bank would be ruled and policed by the Palestinians, whose flag, stamps, symbols and authority would be put in place for the first time in history. Neither Jordanians, nor British, nor Turks, but Palestinian Arabs, would rule the areas in which Palestinians lived.

This was a remarkable agreement, after so many years of hostility and suspicion culminating in the bloodshed, expulsions and imprisonments of the intifada. But as a result of back-breaking negotiations over many months it was almost in place. Peres and Arafat agreed to continue talking that day until they had reached a final agreement. Then at eight o'clock in the evening it was announced over Israeli television and radio that agreement had been reached. In two weeks' time the agreement would be signed on the White House lawn, as Oslo I had been.

That evening the United States Ambassador to Israel, Martin Indyk, was giving a party at his residence in Herzliya to celebrate American Independence Day. Rabin was among a host of Israeli dignitaries attending. The news that a draft agreement had just been reached by Peres and Arafat electrified the gathering. But on a cautionary note, Rabin's close confidant and speechwriter, Eitan Haber, told those who came up to congratulate his master, 'There is no Oslo II accord. The deal has not been done. There'll be no White House signing ceremony this month.'

Eitan Haber was right. Much work remained to transform what Peres and Arafat had agreed that day in Gaza into an agreement that could be signed. The issues were so complex, the map itself so intricate, that Rabin personally, with his particular concerns about the security aspects, would have to devote many hours to checking each aspect, and to seeking amendments wherever he felt they were needed. This task took him twelve weeks. Only then was Oslo II ready for signature.

Many outsiders pressed Rabin, and Israel, for a quick resolution. The Palestinians and the Americans were particularly emphatic. But Rabin had always been methodical and cautious, and in the matter of transferring so much land and so much power to the former enemy, his action was tempered with the knowledge that once Israel handed over responsibilities for fighting terror to the Palestinians, as it proposed to do (and eventually did), there could be many difficult moments ahead.

On July 24, ten days after Rabin began working on the draft of Oslo II, six Israelis were killed in Tel Aviv by a bomb on a bus during the rush hour. Rabin continued to refuse to allow the continuing acts of terror, horrible though they were, to deflect him from his determination to see the process of Palestinian autonomy through to its conclusion. The negotiations continued. Peres and Arafat met six times in the course of as many months to put the details of an agreement into place, and to review the work of the joint Israeli–Palestinian subcommittees which were examining the questions of Israel's military withdrawal, Palestinian policing, the release by Israel of Palestinian prisoners, the delineation of the areas to be transferred to the Palestinian Authority, the timetable of transfer, and – an issue of particular complexity – the protection of the Jewish minority in Hebron.

The most difficult part of the Hebron negotiations centred around the Tomb of the Patriarchs. According to Israel, as explained by Peres to Arafat, this was a Holy Place in the eyes of Jews. It was also, Arafat explained, a Holy Place in the eyes of Muslims. It was eventually agreed that since no agreement could be reached about the Tomb, the status quo would be maintained until a later date. The two negotiators were determined not to allow difficult questions to serve as a road block.

During 1994 the Israeli government had transferred five areas of civilian control to the Palestinians on the West Bank and in the Gaza Strip. These were education, health, taxation, tourism, and social welfare. On 20 August 1995 the Israeli Cabinet approved the transfer of eight more aspects of Palestinian life to the Palestinian Authority: commerce and industry, agriculture, local government, fuel and petrol, postal services, labour, insurance, and statistics. Israeli–Palestinian discussions continued in Eilat for the transfer of yet more areas of daily civilian life; the American Special Middle East Coordinator, Dennis Ross, was present as a facilitator. Of particular difficulty were four questions: the water resources on the West Bank; the Israeli army's redeployment from Hebron; the extent to which Arabs in East Jerusalem would be able to participate in forthcoming Palestinian elections; and the release of more than a thousand Palestinians being held in Israeli prisons.

Inside Israel, opposition to the negotiations with Arafat and the Palestinians was reaching a fever pitch. The Likud claimed that Rabin had no mandate for the negotiations as his parliamentary majority depended on the Arab vote: the two seats of the Arab Democratic Party, and those Israeli Arabs who had voted Labour. Likud arguments and slogans declared that the whole peace process was a delusion, that Arafat was a murderer, and that he should not be Israel's partner. Extremist rhetoric abounded, as did car stickers and street placards denouncing the negotiations, and excoriating Rabin and Peres in particular. Posters appeared showing Rabin wearing Arab headdress, an Arafat-style keffiyah. On some of these posters, it appeared that the Arabized Rabin was in the sights of a sniper's rifle.

On September 11, while Rabin was dedicating a new highway interchange at Kfar Shmaryahu, the Jewish student who had sworn to kill him, Yigal Amir, prepared to make a third attempt but could not get close enough to him because of the crush of people.

On September 22 Peres and Arafat went to the Egyptian resort village of Taba (essentially a single hotel), which a decade earlier had been the subject of a prolonged territorial dispute between Egypt and Israel. For two days they discussed the final details of the extension of Palestinian self-rule throughout the West Bank. The document to which they agreed was 314 pages long. Peres and Rabin had proposed, and Arafat accepted, a staged solution.

The West Bank – other than Jericho, which was already under Palestinian control – was divided into three areas. Area A (3 per cent of the total land area of the West Bank, and containing 29 per cent of its Palestinian population) was the seven large Arab towns which would come under Palestinian control, with only the small Jewish enclave in Hebron excluded and protected by Israel. Area B were the Palestinian villages, which would initially be under joint Israeli and Palestinian control, and from which, at a later date, Israel would withdraw. Area C (74 per cent of the West Bank, but containing only 4 per cent of its Arab population) was the rest of the West Bank, including Jewish settlements in the West Bank, whose 140,000 settlers would remain under Israeli control, linked by specially built Israeli-controlled by-pass roads that would enable them to avoid Arab population centres while travelling. The Arab areas of Jerusalem were excluded from the agreement.

On 28 September 1995 Rabin returned to Washington, and once again signed an agreement with Arafat. Known as Oslo II it provided a timetable and a clear pattern for the extension of Palestinian self-rule to the West Bank, going far beyond the Gaza and Jericho transfers of Oslo I. In his speech Rabin referred to the problem of terrorism, urging Arafat to 'prevent terrorism from triumphing over peace'. If that were not done by the Palestinians, Rabin said, 'we will fight it by ourselves'.

Inside Israel the opposition parties reviled Rabin for Oslo II. Even his wife was subjected to harassment and abuse. Outside their apartment block in Tel Aviv, opponents of the peace process called out whenever Rabin arrived, 'Traitor' and 'Murderer'. Once, when Leah Rabin pulled into the driveway underneath her apartment, someone in the crowd shouted out, 'After the next election, you and your husband will hang from your heels in the town square like Mussolini and his mistress. This is what we are going to do to you. Just you wait.'

On September 30, Shaul Arlosoroff, the son of the Zionist leader who had been assasssinated in 1933, wrote an article in *Ha'Aretz* headed 'The Price of Incitement'. In it he pointed out that in the months before the murder his father had been consistently and wildly attacked in the Revisionist newspapers 'in a way which, like today, created a dangerous atmosphere which eventually brought about the assassination of one of the most outstanding leaders of the State-in-the-making.' The real danger, Shaul Arlosoroff warned, 'is that incitement from the right may again bring about an attack on leaders on the left of the political map in Israel', and he went on: 'The leaders of the right must cease to incite, and must explain to their followers what can happen if incitement continues, otherwise all the blame will fall on them, as it did with the murder of Arlosoroff.'

No such words of restraint were forthcoming. When Netanyahu was present at a rally in which a mock coffin was paraded with Rabin's name on it, he said nothing. The religious leaders were likewise silent. A second article by Shaul Arlosoroff, published in *Davar* on October 1, and headed 'Fatal Incitement', likewise went unheeded. On the extreme right, a recent immigrant from Russia, Avigdor Eskin, denounced Rabin's policies of peace by putting an ancient biblical death threat on him, the *pulsa dinura*: Aramaic for 'lashes of fire'. Standing outside Rabin's house on October 4, the eve of Yom Kippur, Eskin delared: 'And on him, Yitzhak, the son of Rosa, known as Rabin, we have permission to demand from the angels of destruction that they take a sword to this wicked man to kill him, for handing over the Land of Israel to our enemies.'

Eskin believed, as he later told the Jeruslaem Magistrates Court which in July 1997 sentenced him to four months in prison, that the curse 'generally worked' within thirty days. It was thirty-two days before November 4.

The Knesset debated Oslo II on October 5 and 6. The leader of the right-wing Moledet Party, Rehavam Ze'evi, told his fellow-legislators, 'This is an insane government that has decided to commit national suicide.' Ariel Sharon declared that Rabin and his government had 'collaborated with a terror organization'.

In the streets of Jerusalem, throughout October 6, right-wing demonstrators also protested against the agreement. For them, it was a surrender of part of the biblical Land of Israel to murderers. Their denunciations tinged with racism, they declared that with the help of the Arabs, on whose votes the Labour government depended, Rabin was surrendering parts of the Land of Israel. Netanyahu then addressed them, attacking Rabin for basing his coalition on the Arab Knesset members 'whose children did not serve in the army' (but neither do the children of most ultra-Orthodox Jews whose support Netanyahu was courting – and eventually won). The appreciative crowd bayed in response, 'Traitors, traitors'. Netanyahu asked them to stop, but they ignored him. Leaflets were handed out at the demonstration portraying Rabin in the uniform of an SS officer.

The demonstrators marched from Zion Square to the Knesset, where the debate was in progress. They were estimated to number in their tens of thousands. A police cordon kept them away from the Knesset building. Burning torches were thrown at the police, but the line did not break. When the Housing Minister, Binyamin Ben-Eliezer, drove away from the building, his car was pelted with stones. Rabin's car was located by the demonstrators; it was waiting for him to leave, and was empty. The demonstrators broke the windows and damaged the interior.

When Oslo II was voted on in the Knesset, it passed with a narrow margin: 61 to 59. Even this narrow victory for Rabin and Peres was secured because a right-wing member of the Knesset, Alex Goldfarb, of the Tsomet Party, had decided to break away from his former colleagues, and their ideology, and support the government. He was made a deputy minister. Cynics even suggested that he had made his 'yes' vote conditional on being promised a better government car.

After the vote the protests against the extension of Palestinian autonomy continued. President Weizman, who had emerged as a persistent critic, said that the agreement had been negotiated 'too quickly, and without sleep'.

Weizman was critical of the small government majority, and quipped that the passage of the agreement through the Knesset had depended on 'Goldfarb's Mitsubishi'.

Rabin did not allow such taunts to deter him, much though he was angered by them. At the end of October he flew by helicopter to Amman for the Amman Economic Conference, the follow-up to the Casablanca Conference of the year before where Peres had unveiled his plan for a comprehensive economic regional development. Amman, like Casablanca, was the forum at which the leaders of Arab States which hitherto had had no contact with Israel, and whose populations had been repeatedly inflamed in the past against Israel, were to meet and negotiate and plan for a common future throughout the region. Several important steps forward were made. Egypt, which had agreed to come to the conference only at the last moment, announced that it was willing to negotiate a comprehensive free-trade agreement with Israel. Qatar agreed to supply Israel with natural gas, to be pumped in Israeli pipelines, within five years. Jordan announced the ending of trade barriers with Israel.

In his speech to the Amman Economic Conference, Rabin referred to the first efforts at economic cooperation, launched at Casablanca. 'There were cynics who scoffed that after the final handshakes of the Casablanca Conference, the dust would once again settle on the Middle East,' he said. 'But the hundreds of representatives of large corporations and multi-nationals, which have invested in this area in the last year, are present here today, and are the proof that the process which was started in Casablanca was a success.' Since Casablanca, Israel and her Arab neighbours had begun work on joint projects covering water, transportation, agriculture, environment and energy. Those efforts, inconceivable a few years earlier, were weaving a fabric, Rabin said, of coexistence and cooperation, 'a fabric resistant to the pressures, scepticism and outright sabotage which attempt to disrupt and derail the peace process'.

Those pressures were continuous. At a rally in Jerusalem on October 28, at which Netanyahu was the main speaker, Rabin was denounced by several speakers at the rally as a traitor who was abandoning the Land of Israel. During the rally, more leaflets were distributed showing Rabin dressed as a Nazi officer. A video film of the meeting was later shown, in which a woman plunged a knife into a photograph of Rabin.

Rabin had begun to feel that, in terms of public support for the peace process, the government was losing ground. On November 4, a week after the Netanyahu rally in Jerusalem, a mass rally was held in Tel Aviv in support of the government and the peace process. Rabin and Peres were both on the platform. The supporters of Oslo II were determined to show that they were both numerical and vociferous. It was a joyful gathering. Israel – having already left Gaza and Jericho – would be leaving the densely populated areas of the West Bank where so much blood had been shed, and so much hardship endured, in the twenty-eight-year-long occupation of Arab towns, and in confrontation with Arab youth.

To those at the rally, Rabin spoke from the platform about the peace process with emotion and understanding:

> I was a military man for twenty-seven years. I fought as long as there was no chance for peace. I believe that there is now a chance for peace, a great chance. We must take advantage of it for the sake of those standing here, and for those who are not here – and they are many.
>
> I have always believed that the majority of the people want peace and are ready to take risks for peace. In coming here today, you demonstrate, together with many others who did not come, that the people truly desire peace and oppose violence.
>
> Violence erodes the basis of Israeli democracy. It must be condemned and isolated. This is not the way of the State of Israel. In a democracy there can be differences, but the final decision will be taken in democratic elections, as the 1992 elections which gave us the mandate to do what we are doing, continue on this course.
>
> I want to say that I am proud that representatives of the countries with whom we live in peace are present with us here, and will continue to be here: Egypt, Jordan and Morocco, which opened the road to peace for us.

A few moments later Rabin declared:

> This is a course which is fraught with difficulties and pain. For Israel there is no path without pain. But the path of peace is preferable to the path of war. I say this to you as one who was a military man, someone who is today Minister of Defence and sees the pain of the families of the IDF soldiers. For them, for our children, in my case for our grandchildren, I want this government to exhaust every opening, every possibility, to promote and achieve a comprehensive peace. Even with Syria, it will be possible to make peace.

As the rally came to an end Rabin, although normally a shy man, joined in the singing of 'The Song of Peace', which had been composed immediately after the October War. He then left the platform and, as he went to his car, was shot dead by an assassin.

CHAPTER THIRTY

'Shalom, haver'
Peace, my friend

Rabin's murderer was a Jew, Yigal Amir, a religious student at Bar-Ilan University who considered the peace process a betrayal of Jewish values. Rabin's assassination was a watershed in the history of the State of Israel: the killing of a great Jewish leader and visionary by a fellow-Jew filled with hatred and vitriol. The shock of the murder created a stunned numbness throughout the land. Schoolchildren gathered in small groups, lit memorial candles (often spelling out in Hebrew letters either the word 'Yitzhak', or the question 'Why?'), and spoke in whispers of the death of a man who could have been their grandfather, and whose vision of an Israel at peace meant so much for their future well-being.

When he was told the news of Rabin's death, President Clinton came out on to the White House lawn, visibly shaken, and spoke a few words expressing his deep sorrow, and stressing how much Rabin had done to advance the peace process. Then, before returning inside, he said in a quiet, emotion-laden voice the two words *Shalom, haver* (Peace, my friend).

On the evening of November 4, at a meeting of the Israeli Cabinet in the Ministry of Defence in Tel Aviv, Shimon Peres was elected Prime Minister by all his colleagues. On the following morning he worked from dawn in the Prime Minister's office in Jerusalem. In promising Rabin's staff that they would not lose their jobs, he told them, 'I know you've lost a father'. Unwilling to sit in Rabin's chair, he conducted government business from the sofa.

At Rabin's lying in state in the Knesset forecourt, an estimated one million people filed passed his coffin throughout the afternoon and evening of November 4 and the early hours of November 5: old soldiers, young soldiers, parents and children, teenagers and students, Jerusalemites, people from all over Israel, people from abroad. As hundreds of thousands of Jews walked in a daze through the streets of Jerusalem, and hundreds of thousands more gathered in the square in Tel Aviv were Rabin had been shot down, 'The Song of Peace' became an anthem, sung in mournful yet defiant voices, voices shaken by tragedy yet still determined to hope. Its words seemed so eerily apposite in the aftermath of the tragedy – words that had been on Rabin's lips minutes before he was gunned down:

> Let the sun rise, and give the morning light
> The purest prayer will not bring back
> He whose candle was snuffed out and was buried in the dust
> A bitter cry won't wake him, won't bring him back
> Nobody will return us from the dead dark pit
> Here, neither the victory cheers nor songs of prayer will help
> So sing only a song for peace
> Do not whisper a prayer
> Better sing a song for peace
> With a great shout.

Rabin's own copy of 'The Song of Peace', stained with his blood, had been found in his jacket pocket by staff in the hospital where he succumbed to his terrible wounds. At his funeral on Mount Herzl in Jerusalem on November 5, his friend and speechwriter Eitan Haber produced the actual sheet and read the lyrics of the song from it.

Rabin's funeral was attended by an extraordinary gathering of dignitaries. King Hussein came: it was forty-four years since he had seen his grandfather King Abdullah killed by an assassin in Jerusalem, less than two miles from where Rabin was to be buried. President Mubarak of Egypt came, the successor to another assassinated leader, Anwar Sadat. While Rabin was alive, Mubarak had declined several invitations from him to visit Jerusalem. Now that Rabin was dead, Mubarak came. He too had been the target of an assassin, only a few weeks earlier, while on a visit to Ethiopia.

Two Gulf States, Qatar and Oman, were represented by Cabinet Ministers. The Prime Minister of Morocco was present. Representatives of more than eighty countries came, including Queen Beatrix of the Netherlands and the British heir to the throne, Prince Charles. Two Heads of State gave up crucial ceremonies to attend: President Lennart Meri of Estonia was due to swear in a new government the next day, but came to Jerusalem instead, and sent his official seal of approval by fax from President Weizman's office. Eduard Shevardnadze, who had been re-elected President of Georgia on the day before the funeral, decided to forgo his own celebrations and fly to Israel instead.

There were eleven eulogies at Rabin's graveside. Mubarak spoke appreciatively of Rabin's vision. Only by redoubling the efforts for peace, he said, could 'those traitorous hands hostile to our goal' be thwarted. Hussein, looking desperately sad, told the mourners and television viewers across the world:

> I never thought that the moment would come like this, when I would grieve the loss of a brother, a colleague, and a friend; a man, a soldier who met us on the opposite side of a divide, whom we respected as he respected us, a man I came to know because I realized as he did that we had to cross over the divide, establish the dialogue, and strive to leave also for us a legacy that is worthy of him.
>
> And so he did. And so we became brethren and friends.
>
> Never in all my thoughts would it occur to me that my first visit to Jerusalem . . . would be on such an occasion.

Hussein then spoke as if to Rabin in person. 'You lived as a soldier,' he said. 'You died as a soldier for peace. And I believe it is time for all of us to come out openly and speak for peace. Not here today, but for all the times to come. We belong to the camp of peace. We believe that our one God wishes us to live in peace and wishes peace upon us.' Then, in an appeal addressed to all Israelis and all Arabs, an appeal that was heard around the world, he declared, 'Let us not keep silent. Let our voices rise high to speak of our commitment for all times to come, and let us tell those who live in darkness, who are the enemies of light, "This is where we stand. This is our camp."'

In his eulogy, Shimon Peres recalled that last evening at the rally in Tel Aviv. 'I didn't know that those would be our last hours together,' he said. 'But I felt as if a special grace had descended on you, and suddenly you could breathe freely at the sight of a sea of friends who came to support your way and cheer you. You reached the summit, broke through the clouds, and from there you could view the new tomorrow – the view that was promised to the youth of Israel.'

The penultimate speaker on that solemn occasion was Rabin's eighteen-year-old granddaugher, Noa Ben-Artzi. Speaking in Hebrew (the only speaker to do so), she recalled a beloved grandfather. 'I want you to know,' she said, 'that in all I have done, I have always seen you before my eyes. Your esteem and love accompanied us in every step and on every path, and we lived in the light of your values. You never neglected anyone. And now you have been neglected – you, my eternal hero – cold and lonely.'

There was one more speaker, President Clinton. Addressing the people of Israel, he said:

> Even in your hour of darkness, his spirit lives on, and so you must not lose your spirit. Look at what you have accomplished, making a once-barren desert bloom, building a thriving democracy in a hostile terrain, winning battles and wars, and now winning the peace, which is the only enduring victory.

Your Prime Minister was a martyr for peace, but he was a victim of hate. Surely, we must learn from his martyrdom that if people cannot let go of the hatred of their enemies, they risk sowing the seeds of hatred among themselves.

I ask you, the people of Israel, on behalf of my own nation, that knows its long litany of loss from Abraham Lincoln to President Kennedy to Martin Luther King, do not let that happen to you.

In the Knesset, in your homes, in your places of worship, stay the righteous course.

Clinton's speech was ended, and he stepped back from the microphone. Then he came forward again and repeated, as he had done on the White House lawn, almost in a whisper, the words 'Shalom, haver.'

Those two words, 'Shalom, haver', echoed and re-echoed through the hearts and minds of millions of Israelis; they were to be found within days as car stickers on tens of thousand of cars; and for the next few years they were to be a potent visual symbol throughout Israel of Rabin's loss. There was another car sticker that also spoke volumes in a few words. Printed with a photograph of Rabin it read: 'No to violence'.

Arafat had very much wanted to go to Rabin's funeral, but it was felt that his visit at that particular traumatic time might be too difficult for many Israelis, despite the recognition of the PLO by Israel, and despite the peace process. In telephone calls to both Peres and Leah Rabin, Arafat recalled how, after he had made a particularly long speech during the Washington talks on Oslo II, Rabin had said he was 'beginning to wonder if maybe I was close to being Jewish'. Arafat agreed not to go to the funeral, but, four days later, he made a secret night journey to Rabin's apartment in Tel Aviv to offer his personal condolences to Leah Rabin and her family. It was his first ever visit to Israel.

Memorial meetings to Yitzhak Rabin were held in Jewish communities all over the world. 'While the killer uses the name of God in vain,' the Israeli Ambassador in London, Moshe Raviv, told a vast, solemn gathering in the Royal Albert Hall, 'some will ask "Why were we silent?" when violent and inciting rhetoric was used against the government and Yitzhak Rabin personally. As a society we shall have to ponder how to instil tolerance and restraint and how to prevent the nurturing of zealots, for zealots are dangerous and destructive.'

Also in London, members of Mahal, the volunteers who had fought in Israel's War of Independence, wept as they listened to a recording of Rabin's speech at one of their past reunions in which he recalled their fortitude and his affection for them. In Moscow the Israeli emissaries from throughout the former Soviet Union – more than a hundred in all – gathered to remember him, and to sing the Palmach songs of the War of Independence.

*   *   *

Eight days after Rabin's assassination, a memorial meeting was held in Tel Aviv, in the same square in which he had been shot. Turning to Shimon Peres, Leah Rabin urged him 'to lead the people of Israel to peace', and to do so 'in the spirit of Yitzhak'. Shimon Peres continued with the peace process. The Oslo Accords had been his creation: he now had the full authority as Prime Minister to pursue their timetable. Of Rabin, he wrote on November 27, in reply to worldwide letters of sympathy, 'His life was cut short, but not his work. His spirit has left him, but this spirit has not abandoned us, nor the process in which we are engaged.'

That process was about to be struck a severe blow. On 5 January 1996 one of the leading instigators of terrorist actions inside Israel, Yahiya Ayash, the West Bank university graduate known as 'the Engineer' because of his skills with explosives, was killed in Gaza City while talking on a portable telephone that exploded in his face. Israel did not deny responsibility. But Hamas swore to be avenged, and on February 25 a suicide bomber, entering a bus in Jerusalem, killed twenty-five people, most of them Israeli soldiers. A Muslim Arab, Wael Kawasmeh, who was waiting for a bus, was also killed. That same day a suicide bomber in Ashkelon blew himself up at a bus stop. One Israeli was killed, twenty-year-old Hofit Ayash, who had recently chosen a wedding gown for her marriage in four months' time.

Arafat's adviser, Ahmed Tibi, condemned the bus bombs. 'The circle of violence must be broken and stopped,' he said. 'There is no place for revenge attacks.' But on March 13, thirteen more Israelis were killed by a suicide bomber in the heart of Jerusalem on the same bus route, No. 18, as the previous bomb. One of those killed, nineteen-year-old Chaim Amedi, had unintentionally missed the bus that had been destroyed in the last attack. Another of those killed, thirty-eight-year-old George Yonan, was a Christian Arab who had been deaf from birth.

On the following day a suicide bomber struck in Tel Aviv, in a crowded shopping street in the centre of the city, killling eighteen. These were enormous explosions that ripped the buses apart, mutilating many of the dead beyond recognition. The mood inside Israel was one of near despair. It seemed impossible that the peace process could go on while such terrorist killings, on a far larger scale than before, went on.

Immediately after the March 3 bus bomb, Peres had warned Arafat that the future of the peace process 'hangs in the balance' unless the Palestinian Authority took immediate action against Hamas. Israel could not be the only party to the agreements to keep its commitments. 'It cannot be unilateral.' Meanwhile, Peres instituted a separation between Israelis and Palestinians, setting up specific crossing points for vehicles and goods between Israel and the occupied territories, and effectively cutting off most Palestinians from contact with Israel.

Two acts by the Palestinian Authority caused anger throughout Israel: the permission given for mass solidarity meetings in Gaza after the death of

Yahyia Ayash, and the fact that Arafat made a condolence call on Ayash's parents: 'This,' said the Israeli government, 'sent the wrong message to the Palestinians.' The decision of the Palestinian Council to condemn the Israeli closure of the occupied territories without condemning the bus bombing was also a cause of Israeli anger.

The acts of terror by Hamas, Peres commented, 'occurred in spite of the fact that – even after the assassination of Rabin – we had redeployed our army from 450 villages and six cities in the West Bank, in the face of tremendous resentment.' Peres added, with understandable anger: 'Instead of thanks, we got bombs.'

In an attempt to enlist worldwide governmental support against terror, and to find ways of stopping the funding of terrorist groups by Iran and Libya, the Israeli and Egyptian governments sponsored an international conference against terrorism at Sharm el-Sheikh, the Red Sea promontory that had once been the focus of Egyptian–Israeli tension and a cause of war. President Clinton, who was one of the many Heads of State who attended the conference, then flew to Israel to express his solidarity there with the devastated Israeli public. Since the suicide bombings had begun, more than 100 Israeli civilians had been killed in the most horrible circumstances. In addition, more than 500 people were injured in these attacks, many of them severely.

The continuation of the Oslo Accords was under great strain. The Government of Israel, first under Rabin and then under Peres, repeatedly declared that it would not allow terror to derail the peace process, and negotiations with the Palestinians continued on the many issues relating to Palestinian autonomy and Israeli withdrawal. But the opposition Parties denounced the peace process and called for 'peace with security' – a phrase that was interpreted by many as meaning an end to Oslo.

Peres, the architect of Oslo, was himself under enormous public pressure to react to the killings. But he declined to suspend the timetable of the Oslo Accords. Instead, in agreement with the Palestinians, he postponed the redeployment on Hebron, and called an election. In doing so, and thus inviting the Israeli public to express its opinion through the ballot box, he hoped to win an endorsement for continuing the peace negotiations. These included negotiations with Syria, to which Peres, like Rabin before him, was prepared to return most, and even all, of the Golan Heights in return for a full peace between the two countries.

Peres knew that the Hamas suicide terror attacks were shifting Israeli public opinion towards Likud hard-line policies. In an attempt to show that terror would be responded to from whatever direction it came, on the eve of the election, following spasmodic shelling by Islamic fundamentalists from across the Lebanese border, when three Israelis were injured, Peres launched Operation Grapes of Wrath.

For seventeen days, Israeli artillery and aircraft bombarded the fundamentalist positions north of the security zone. In the first forty-eight hours, twenty Lebanese civilians were killed, five when an Israeli helicopter gunship hit an ambulance filled with villagers fleeing from their village north of the security zone. In a terrible tragedy, a civilian shelter was hit by mistake, and 105 Lebanese were killed. Shocked by this accident, Peres offered financial compensation to the Lebanese government.

Following strong international condemnation, and the direct mediation of the United States, the Israeli bombardment ended with an agreement that the rocket attacks on northern Israel would also cease. The Likud were swift to condemn the government for having 'caved in' to outside pressure, and for having given the 'green light' to the fundamentalists to bombard the northern settlements again whenever they chose to do so.

The 1996 election campaign was a fierce one, almost violent, with groups of Yeshiva students and Likud youngsters roaming the streets denouncing Peres and the peace process in terms almost as savage as those which had been used in the weeks leading up to Rabin's assassination. Within the Labour coalition, the left-wing Meretz Party created a backlash of hostility by vaunting its secularism. The message that came from Likud, which had always been a predominantly secular Party, was a strong identification with religious values. For the religious Parties, Likud's strident nationalism, echoing the pre-State extremism of the Revisionists (Netanyahu's father had been a leading Revisionist writer and thinker), was proving attractive. Shas rabbis distributed amulets to the faithful, whereby a vote for Netanyahu would be a blessing.

As a result of a constitutional change made in the previous Knesset, with Rabin's support, Israelis would be asked for the first time to cast two votes, one for Prime Minister, and the other for a political Party. In leading the Likud campaign, Netanyahu, popularly known as Bibi, said he would carry out all of Israel's international commitments, including Oslo. But he also promised to slow the peace process down, and to insist upon total reciprocity. His slogan was: 'Peace with Security'.

Election advertisements on television mocked at a photograph of Peres and Arafat walking up a staircase together. As the photograph was shown, the glass on it was smashed, revealing the two men beneath it. In another televised advertisement, aimed at the Russian immigrant voters, the horrifying scenes that had followed the suicide bus bombings in February and March were shown repeatedly. On the eve of the election, the Lubavich religious movement, Habad, put up thousands of posters with the words: 'Bibi is good for the Jews'.

The election was held on May 29. The results were not as clear cut a rejection of Peres's vision of the peace process as the opposition had hoped. Indeed, Labour emerged from the election with the largest single number of seats in the Knesset: 34 seats as against Likud's 32. But in the separate vote for Prime Minister the former leader of the opposition, Benjamin Netanyahu, won, by the narrowest of margins, 1,501,023 votes as against 1,471,566 for Peres (50.4 per cent as against 49.5 per cent), and automatically became Prime Minister.

The extremist Parties opposed to the Oslo peace process had been considerably strengthened in the election, as had the religious Parties. Netanyahu was able to form a coalition by bringing in these religious Parties (with 19 seats between them), and also the new Russian immigrant Party, Yisrael B'Aliyah, headed by Natan Sharansky, which had secured seven seats, and which was on the right of the political spectrum. Netanyahu also brought in to his government the four members of the centrist Third Way Party, led by a defector from Labour, Avigdor Kahalani, one of the heroes of the fighting on the Golan Heights in 1967. The Third Way had run on a platform against withdrawal from the Golan Heights.

Netanyahu's premiership was to see an intensification of conflict, both between Israel and the Palestinians, and within Israeli society. The divide between left and right, which had first been seen with such intensity in the debates several decades earlier between Labour and Herut, and then between Labour and Likud, was revived, as was the divide between the ultra-Orthodox and the rest of Israel, and the Ashkenazi and the Sephardi.

Even the Supreme Court was attacked, first by the extreme religious Parties, who alleged that it was 'unfair' to the ultra-Orthodox, and then by members of Shas, who alleged that it was unfair to the Sephardi community (a Sephardi Cabinet Minister, Arye Deri, was under indictment). It was 'simply a lie', declared a former Supreme Court Justice, Haim Cohn, to say that the judges were biased against any section of the population. When an anonymous telephone caller told the President of the Supreme Court, Aharon Barak, 'You'll rot next to Rabin's grave, then you will understand', Barak had to be given twenty-four-hour police protection.

Following his defeat in the 1996 election, Shimon Peres had refused to give up his vision. 'We shall continue to dream together,' he wrote, 'of a Middle East of light and hope.' In pursuit of that dream, he continued to advance the cause of economic cross-border activities, and to 'tutor' his successor, Benjamin Netanyahu, in what could be achieved for the region through mutually acceptable agreements with all its neighbours. But the Netanyahu era was a time of increasing confrontation between Israel and the Palestinians. At Netanyahu's insistence, the withdrawal from Hebron – the only Arab city in the West Bank not yet evacuated – and the subsequent rural-area withdrawals agreed upon in Oslo II, were severely delayed. There were repeated complaints that Arafat was not doing enough to combat terrorism that originated in the areas under his control; but he had taken action after the bus bombs to track the killers down, and continued to do this after the election.

Netanyahu was not able to make the sort of commitment of trust that had been the intention and hallmark of the Rabin-Peres approach to the Palestinians, and the evolving relationship of cooperation between them and Israel. In a move that had the effect of exacerbating Palestinian frustration built up since he had come to power almost four months earlier, Netanyahu went ahead with the opening of the exit to the northern end of an ancient tunnel that ran under the Old City, adjacent to the Temple Mount. The tunnel nowhere ran beneath the Mount itself; its new exit was in the populous Muslim Quarter. Although this would have the effect of increasing the tourist trade, and thus benefit the Muslim shopkeepers, the sudden, unannounced opening of the exit (in the middle of the night) created an upsurge of Palestinian anger, stimulated by fiery sermons in the mosques.

At the time of his electoral defeat, Shimon Peres had been negotiating with the Muslim religious authorities on the Mount for a reciprocal agreement, whereby an area of the Mount would be turned into an additional mosque, in return for the opening of the tunnel, which had an entirely tourist use. Netanyahu had decided to go ahead without further negotiations with the Palestianians.

In the violence that followed throughout the occupied territories, fifteen Israeli soldiers and fifty-six Palestinian civilians were killed. Amid intense Palestinian anger, the peace process seemed about to collapse. In an attempt to diffuse the crisis, and to persuade Netanyahu to return to an Oslo-style timetable, President Clinton invited him, Arafat and King Hussein to Washington. The meeting was not a success.

In a television interview on September 25, Shimon Peres criticized Netanyahu for having opened the tunnel when he did. 'I would have tried to talk with the Palestinians before making any sensitive move,' he said. 'The last four years were the best in all of Jewish history – so why stop this?' For the Arab States (with the exception of Jordan) the opening of the tunnel provided an opportunity for extremism of the sort that Peres and Rabin had hoped to relegate to the moods and attitudes of the past. On September 26 the Council of the Arab League, meeting in Cairo, applauded the Palestinian protests against the opening of the tunnel and described the opening itself as 'part of an Israeli-Zionist plot to destroy the al-Aksa mosque, set up the Temple of Solomon and obliterate all Islamic Arab land-marks'.

With no agreement yet reached between Israel and the Palestinian Authority on Hebron, it looked as if Palestinian protests would once again, as during the intifada, turn to unending violence in the streets. On October 7, in a speech in the Knesset, Shimon Peres made a powerful appeal for Netanyahu to conclude the Hebron agreement. 'One hundred and eleven days of talk-talk-talk have passed,' he said scathingly, 'One hundred and eleven days of non-communication', and he continued:

I want to say what real peace is in my experience. True peace is the way of agony. I remember what my comrades and I have gone through over the past year, seeing that man, the great military leader and the courageous statesman Yitzhak Rabin murdered before my eyes.

For what? For what crime? For what sin? Just unbridled incitement at gatherings, in which some of those present here participated. With a keffiyah on his head. We won't photograph anyone with a keffiyah, don't worry.

I know what a shock it was for most of the nation. I know what a shock it was for me, seeing a comrade and a leader, a man who stood at the helm of government and gave 4,000 years of Jewish history four glorious years. The world admired Israel, believed in it. Jews began to come. Investors began to come. What was he murdered for?

And afterwards I saw – I, a man who pursues peace – the terrorist attack in Jerusalem. I know what it is to leave one's office and be told that a bus has exploded. You also showed it on television. Thank you. I went there and I saw the blood and the flesh and the murder and the killing, and I saw the people screaming at me: 'You are guilty'.

Peres continued with a warning that Israel must not delude itself. 'If you want peace,' he said, 'there will be blood and agony and murder,' and he went on to quote a remark by Menachem Begin's son Benjamin, a Likud member of the Knesset, who had said, 'I prefer the agony of peace to the blood of war.' Peres commented, 'If you genuinely go for peace – not a verbal, declarative peace, not a pretend peace but a real peace – we will support you. We are not asking to be members of your government. We will support a real peace if you have the courage to make one.'

Appealing to Netanyahu to continue with the peace process, Peres also spoke of the fundamental values which must sustain the State of Israel as it embarked on its second fifty years. 'War has a heavy price,' he said. 'So does peace. But if we want to give the next generation a world without war, our generation must undergo the agonies of peace and of crucial decisions, and make Israel what it is: not the addition of a kilometre here or there but an Israel of values, an Israel of moral considerations, an Israel which knows how to respect the Arabs as we do ourselves.'

After considerable American pressure that the negotiating process should not be allowed to fail, the Hebron Agreement was concluded on 17 January 1997. The small Jewish presence inside the city was to remain, in an enclave guarded by Israeli soldiers. But 80 per cent of the city, like the whole of every other West Bank city – Jenin, Kalkilya, Nablus, Tulkarm, Ramallah and Bethlehem – came under the jurisdiction of the Palestinian Authority.

By far the largest part of Hebron, designated Area H-1, with 100,000

Palestinian residents, would be transferred to the control of the Palestinian authorities, who would be responsible for both security and civil-related matters, as in the other Palestinian cities in the West Bank. The second area (area H-2), with 20,000 Palestinian and 500 Jewish residents, would enjoy a unique status in that the Israeli authorities would retain responsibility for security and public order, and for all matters related to the Israeli residents, while the Palestinian authorities would be responsible for all civil matters relating to the Palestinian population residing in this area.

The Israelis on behalf of whom this complex argument was reached constituted only forty-five families and 150 Yeshiva students. As a result of their determination not to leave the former Jewish Quarter in the centre of Hebron, and the Netanyahu government's support, a partition plan, inherited from Labour, had been put in place with confrontation lines in the centre of a populous city. But with the signing of the Hebron Agreement, the Palestinian national aspiration to fly the flag, and to legislate and govern Palestinian cities conquered by Israel in 1967, was effectively fulfilled.

As an integral part of the Hebron Agreement, a document entitled 'Note for the Record' was signed. Under this, Israel would begin further withdrawal of its troops from the rural areas of the West Bank in March 1997, a process that was to be completed not later than 'mid-1998'. The term 'mid-1998' was agreed to encompass the three months June, July and August. This extended by a full year the original Rabin–Peres–Arafat timetable whereby the final withdrawal was to take place no later than 7 September 1997. Sceptics doubted whether Netanyahu would adhere to that timetable.

In one of the numerous and often tense meetings which took place during the final days of the negotiations, Yasser Arafat invited the American mediator, Dennis Ross, to join him in that year's Christmas celebrations in Bethlehem. A Palestinian negotiator pointed out to an Israeli colleague the humour in the fact that a Muslim had invited a Jew to celebrate a Christmas holiday.

A year after Rabin's assassination, Chaim Herzog, a former President, spoke at Bar-Ilan University, where Rabin's assassin had been a student. He recalled Rabin's military service in the War of Independence, through which Israel had come into being, and in the subsequent wars, through which she had survived. He recalled the Six Day War – 'the days of terror, the sense of being closed in from every side' – when Rabin had been Commander-in-Chief. Herzog then spoke of how the secular and religious soldiers, all soldiers, 'those who went into battle as brothers with a joint mission, heads covered and uncovered, sons of kibbutzim and moshavim, of cities and village, of suburban neighbourhoods and development towns, loved and admired their commander. They knew his life was sacred to his people and his land. They followed him to war, in fire and flood. They followed him into victory, and they followed him towards peace. They will not forget.'

In his speech, Herzog (a son of a former Chief Rabbi of Israel) spoke angrily of the religious fanatics. 'If the murder of such a man, of a Prime Minister, does not set the very fibres of our national being atremble', he said, 'if it does not shock us to our very foundations; if we have not vomited out the curse, and uprooted the cancer, and not done away with that group of insane zealots – that badge of dishonour for our people – we are, God forbid, in danger of seeing this nightmare recur.'

Pointing to what he regarded as the failure of the public at large to address themselves to what had happened, Herzog said, 'I am afraid; I am afraid and anguished because I cannot see that the shock of this murder has fully impacted on this nation. It seems as though the trauma too quickly gave way to the demands of routine. It is as if the terrible lesson was not learned, as if national introspection simply stopped at a certain point, and people returned to their everyday concerns.'

Then, in words which echoed the passion of the Hebrew prophets, Herzog warned his listeners, 'The fires of destruction are burning at the edge of the camp. If we do not, together, hasten to extinguish them, they will destroy our entire house.'

The continuing Israeli presence in southern Lebanon was a topic that had generated strong debate for many years. The accidental crash of two helicopters at She'ar Yashuv in northern Israel on 4 February 1997, in which seventy-three Israeli soldiers were killed while on their way to military duties in southern Lebanon, a crash which caused anguish through the whole country, also intensified the debate about the future of the northern border. Yoel Marcus, one of the most thoughtful of journalistic commentators, wrote in the newspaper *Ha-Aretz* on February 7, 'Lebanon haunts us like a curse. True, the helicopter crash was the sort of accident that might have occurred anywhere – it could have been two civilian planes, two planes, two yachts – still, the fact that this tragedy happened where it happened, because of our activity, our involvement and the presence of our forces in Lebanon, the seventy-three soldiers who lost their lives in the accident join the ranks of 400 killed soldiers and 1,420 who have been injured in the security zone since the Lebanon War ended.'

Marcus was far from alone in questioning Israel's continuing presence in southern Lebanon. Yet he sought, as did all Israelis at that time of anguish and heart-searching, to suggest a way forward. 'From our point of view,' he wrote, 'when we are out of Lebanon, and no longer a "foreign conqueror" there, we will be more at liberty to strike out against Lebanese targets, and Syrian targets in Lebanon, than we are today. I am not saying we should stand up and flee the place unilaterally – but the time has come to reconsider, together with the United States, which is prepared to take part in an international force, alternatives that would allow us to get out of there soon. Each drop of blood is too much.'

In the aftermath of the helicopter crash, the photographs and stories of the lives – the drastically curtailed lives – of all the victims were given wide press coverage, as happens with every national disaster. In an unprecedented gesture, President Weizman paid a condolence call on every bereaved family. Poems by children were published which combined in their fragile stanzas the personal and national grief. One, by fifteen-year-old Yana Barkun, was addressed as if to her brother, and to all the brothers of Israel's scarred history:

> When you left, you promised me to come back, you gave me a kiss and I couldn't tell that it was going to be the last one.
>
> After a day somebody called and mother started to cry, she explained to me that you won't be coming home any more. And I just couldn't understand, how could you do a thing like that, promise to come and just forget.
>
> Then, they were crying next to your grave and everybody left you something: letters, flowers, candles. I left you something too, a place in my heart, just for you and your friends who have gone with you.
>
> We will never forget how young and strong you were.

On March 13, within six weeks of the helicopter crash, while Israelis still pondered with deep sadness the high death toll and many personal tragedies, another tragedy struck. Seven schoolgirls aged twelve and thirteen, all from the same school at Beit Shemesh, were shot dead by a Jordanian soldier who went berserk on the Jordan border. The girls were visiting The Hill of Peace, at Naharayim, one of the areas on the Jordan River handed back to Jordan by Israel under the 1994 peace treaty, and visited by thousands of tourists as a beauty spot and dividend of peace on a peaceful border which had once been enemy territory.

In a move that made a considerable impact in Israel, King Hussein paid a surprise condolence call on the grieving families. His decision was hailed as a remarkable gesture of concern. But in the Israeli Cabinet, in a move that exacerbated both Palestinian and world opinion – and was hardly likely to please the King – it was decided on the day after the killings to begin work early the following week on a controversial building project at Har Homa in southern Jerusalem.

Situated between kibbutz Ramat Rahel and Bethlehem, Har Homa is a tree covered hillside (known to the Arabs as Gebel Abu-Ghneim). Beyond it are two Arab villages. The land of the hillside was three-quarters Jewish owned, one-quarter Arab, much of the Arab land having been purchased by Jews over a number of years. From the perspective of the Netanyahu government, it was a site on which Jewish homes would constitute an added Jewish presence on the southernmost, and remotest, section of the Jerusalem border as established after the Six Day War thirty years earlier – since when the hill had been afforested.

The decision of Netanyahu to build on Har Homa, and to build homes for as many as 32,000 Jews, provoked Palestinian and left-wing Israeli protests. But neither public protests, nor the condemnation of the United Nations – where 130 nations voted against the proposed construction, and only the United States and the Pacific island group of Micronesia voted in Israel's favour – could deter Netanyahu from his resolve to build. Indeed, the condemnation served only to strengthen the government's determination to push ahead with what was clearly a political statement: that Jewish Jerusalem would be extended without regard to Arab sensitivities. The Oslo Accords had specifically stated that there would be no change on the ground in the status of Jerusalem until agreement was reached at the negotiating table.

As the compromises and the conciliatory attitudes of the Rabin–Peres era were unravelled, acts of terror continued. On March 21 a West Bank Arab, Mahmoud Abdel Khader Ranimat, killed three people in the Café Apropo in Tel Aviv when the bomb he was intending to leave inside the café blew up prematurely. The Palestinian Authority also hardened its stance. On May 2, after a Palestinian Authority Cabinet meeting in Ramallah, presided over by Arafat, the Palestinian Justice Minister, Freih Abu Medein, issued an order reviving an old Jordanian law which imposed the death penalty on any Arab who sold land to a Jew – some of the land of Har Homa had been sold by Arabs to Jews. The first Arab to suffer this penalty was seventy-year-old Farid Bashiti, an East Jerusalem real estate agent who had frequently sold land to Israelis in the past. He was lured from East Jerusalem to Ramallah on May 9 and, after his mouth had been taped and his hands bound, killed by a blow to the head.

During the sermon in the al-Aksa mosque in Jerusalem shortly after Bashiti's body was discovered, the Mufti of Jerusalem, Akrime Sabri, said that the murdered man should be denied a Muslim burial. The Mufti – an appointee of the Palestinian Authority – added that according to Islamic law a Muslim who sells land to a 'non-believer' in Jerusalem is considered an 'infidel'. Within a month, four more Palestinian land dealers had been murdered. The Israeli government protested to the international community at this violation of Palestinian rights by Palestinians. In June four members of the Palestinian Authority's Preventive Security Service (established as a result of the Oslo Agreements) were charged by Israel with Bashiti's murder.

In the early months of 1997 a political scandal threatened to bring the Netanyahu government down. An Israeli television journalist, Ayala Hasson, who worked on Channel One, claimed that she had found evidence that the appointment of the Attorney General, Roni Bar-On (who had resigned after only two days in office) had been made as the result of a clandestine deal. Under the deal, according to Hasson, Bar-On would use his authority as Attorney General to protect the Shas leader, Arye Deri, who was under indictment for corruption. Deri, for his part, would secure a favourable Shas vote with regard to the then imminent Israeli withdrawal from Hebron, to which many Shas supporters, as well as many Likud supporters, were opposed.

A police investigation was opened. Among those questioned was Netanyahu himself, who was cautioned that anything he might say could be used in evidence against him. Had he known of the Deri–Bar-On deal? Had it gone through his appointee as Minister of Justice, and fellow Likud politician, Tsachi Hanegbi? Had the head of the Prime Minister's office, Avigdor Lieberman, brought it to him for decision? These were among the questions to which the police sought answers.

In their report, the police recommended that charges be brought not only against Deri, Hanegbi and Lieberman, but against Netanyahu himself. The new Attorney General, Elyakim Rubinstein, and the State Prosecutor, Edna Arbel, examined the police recommendations. It was April 21, two days before Passover. Arbel read out the report on television, watched by more than a million Israelis. Only Deri would be indicted. The cases against Hanegbi and Lieberman would continue. Netanyahu would not face trial, though the report stated that 'there is in fact cause for suspicion that the Prime Minister proposed to the Government to appoint attorney Roni Bar-On to the post of Attorney General in order to please Arye Deri, acting wilfully, or closing his eyes to the possibility that there was an illegitimate connection between Deri and Bar-On.'

Immediately after it was known that Netanyahu was not to be indicted, he went to the main television studio and made a strong attack on the television itself, and on the Israeli left wing which, he said, had never been prepared to accept the result of the election that brought him to power. 'Some members of the media who are identified with the Left,' he said, 'were happy to adopt all malicious accusations, ones which were totally unwarranted, entirely for the purpose of making me stand in the middle of them.' Netanyahu added, 'A number of persons, especially on Channel One of the television, are still not ready to accept the voters' decision in the last election and almost nightly they try to undermine the legitimacy of the government.'

The Attorney General, Elyakim Rubinstein, took a different view. 'In the end,' he said, in his official statement following that of Edna Arbel, 'the media filled a positive role in disclosing a difficult and painful issue. The public interest was served, even if not all the details were accurate.'

In his own television appearance, Netanyahu was outspoken in his denunciation of his critics. 'They cannot accept the fact that the people voted for us and not for them,' he said. 'They cannot accept the fact that we are building on Har Homa. They refuse to accept our vigorous objection to a Palestinian State. They refuse to be reconciled to the fact that we are guarding the Golan Heights.'

As Israel's forty-ninth anniversary drew near, Netanyahu took up a challenge on behalf of Jewish Orthodoxy in Israel. To the distress of the strongly held opinion of Jews inside Israel who follow less stringent forms of observance, and challenging in particular the concerns of a large number of non-Orthodox but practising Jews in the United States, he supported the first reading of a parliamentary bill that would recognize as true conversions to Judaism inside Israel only those that were carried out by Orthodox rabbis. The bill had been put forward by Shas, on whose seats Netanyahu depended for his Knesset majority, and one of whose senior figures, Eli Suissa, was Netanyahu's Interior Minister. Passage of the bill had been promised by the Likud in the coalition agreements it had signed with the religious Parties when it was forming the government, and had become part of the 'government guidelines'.

The proposed law, Shimon Peres declared, would 'divide the Jewish people'. The bill was passed by the Knesset on April 1. So great, however, had been the protests inside and outside Israel that the government announced it would 'freeze' all further progress on the bill 'to allow time to find a compromise' that would enable it not to go ahead.

On the evening of 11 May 1997, Israel observed the start of Memorial Day. That evening, at the traditional ceremony in Jerusalem, torches were lit by thirteen people, chosen to represent the diversity of the Israeli national experience. Their stories encapsulated many aspects of the extraordinary mixture of people that make up modern Israel. Anat Madmony, a doctoral student at the Hebrew University faculty of agriculture, was a descendant of Jews who had immigrated from Yemen in 1881; her father had fought in the Palmach. Emmanuel Cohen had emigrated from Switzerland in 1993; it was his great-grandfather who had invited Herzl to hold the First Zionist Congress in Basle in 1897. Shulamit Cohen Peretz, a social worker and the director of a community centre, had lost her brother Eran in the war of 1973; his body had only been returned from Egypt in 1995.

Peretz Hochman had escaped from the Warsaw Ghetto; after immigrating, he settled on a kibbutz in northern Israel. The parents of Ora Zar were members of the Persian Jewish community in Meshed that had been forced to convert to Islam in the early nineteenth century but had retained their Judaism; she had been born in Jerusalem, and worked to help new immigrants. Asaf Mordechai's parents were Indian Jews. In 1975, when Palestinian terrorists held him and many of his family hostage in their home, his uncle was killed by the terrorists, as was his father during the army rescue operation. His mother died in hospital from her wounds.

Moshe Benziman was an eighth generation Jerusalemite who had fought in the defence of the Jewish Quarter in 1948 and been taken prisoner by the Arab Legion. Rivka Lankry was a recent immigrant from Morocco. Alona Illarionova, at nineteen the youngest of the torch lighters, had emigrated from the Ukraine in 1994. Elisa Ben-Rafael was from Puerto Rico, brought up as a Catholic, but in fact a Marrano, whose ancestors had been converted to Catholicism at the time of the Inquisition. She had married an Israeli diplomat, David Ben-Rafael, who was later posted to Buenos Aires, where he died in a terrorist bombing of the Israeli Embassy there in 1987 that had killed more than eighty people.

Yakov Elias was an Ethiopian Jew who had been accused of spying for Israel and imprisoned for two years; he later took a leading part in Operation Moses in 1984 and Operation Solomon in 1990, bringing Ethiopian Jews to Israel. Zion Shenkor, a paratroop officer, had come from Ethiopia with his family at the age of twelve, after a month-long trek through the desert to Sudan. Sheikh Zaki Zaher, a Druse notable, was co-founder, in 1977, and President of the Druse Zionist Movement, set up to strengthen the ties between the Druse people and the State of Israel.

Since the foundation of Peace Now in 1979 by those who feared that the initiative for peace following the Sadat visit would be lost, several peace groups had been created, devoted to reconciliation between Jews and Arabs. The International Centre for Peace in the Middle East (ICPME) had been created in response to the Lebanese War of 1982. The Adam Institute had been founded in 1983 in the wake of the murder of Emil Grunzweig. In the aftermath of Rabin's assassination, *Dor Shalem Doresh Shalom* (A Whole Generation Demands Peace) – known in English as the Peace Generation – had been established, with Rabin's son Yuval a leading supporter. In the fourth month of Netanyahu's premiership, Mothers and Women for Peace was created as a reaction to the opening of the tunnel in the Old City.

Much of the work done by the protest movements in the late 1990s was caught up in attempts to lessen the alienation of what were called the 'neighbourhoods' – those areas of cities and towns in which Jews from oriental lands, then often called the *Mizrachim*, or Easterners, were living, frequently in conditions of considerable hardship. One of those who studied the peace movements, Rolly Rosen, wrote in July 1997 of how 'the decision to "promote peace among unconvinced population groups" is a decision to swim against the current, to break down the increasing over-lap of social cleavages around which the basic world views of Israeli society are organized'. Rosen also wrote of 'the deep rifts in Israeli society following the Bar-On/Hebron affair, the "ethnic devil" that threatens to come out of the bottle, the attack on principles that had been regarded as within the consensus, such as the rule of law and the primacy of the Supreme Court'.

As Israel approached its fiftieth anniversary, at least eighteen organizations worked, often with limited resources, but with immense determination, to try to create a more tolerant society and to bridge the widening gap of ethnic divergencies; a few of them focused on the relations between Jews and Arabs, but most of them were active with regard to the even more potentially divisive relations between Jew and Jew.

*   *   *

On the morning of May 11, during the two-minute silence of Memorial Day at ceremonies held at military cemeteries throughout the country, in the newspaper coverage for the day, and at home, Israelis remembered the 18,538 soldiers who had fallen in battle since the start of the War of Independence forty-nine years earlier. Of this number, 327 fell during the year that had just ended, some in southern Lebanon, others in the fighting that broke out after the opening of the tunnel, seventy-three in the helicopter crash in February.

For the first time in its history, following a special law introduced in the Knesset, Israel also commemorated during Memorial Day those Israelis who had fallen victims to terrorist attacks since the foundation of the State.

With the sounding of the sirens, a moment of deep respect and solemnity throughout Israel, an episode took place on the main road running through the ultra-Orthodox section of Jerusalem, and also in the ultra-Orthodox suburb of Bnei Brak on the outskirts of Tel Aviv that shocked most Israelis: the refusal of the ultra-Orthodox to stand for two minutes in silence and to pay their respect to the soldiers who had died in Israel's wars and in the defence of the State. Interviewed by the newspaper *Ma'ariv*, one ultra-Orthodox young man told the newspaper, 'Each Sabbath you hurt our sensitivities and travel on a holy day. Now it is our turn to hurt your feelings. For us, to stand in silence is a sin. It is a goy custom. This is a secular Zionist state with which we have no truck. There is no way we are going to stand to remember them in the middle of the street.'

The Ashkenazi Chief Rabbi, Israel Lau – a survivor of the Holocaust – described the behaviour of the ultra-Orthodox elements as 'an abomination' that was causing 'a schism in the people which will be hard to mend'. But that extremist behaviour continued. In June 1997 a group of a hundred Jews who were praying at the Wailing Wall were attacked by a crowd of ultra-Orthodox boys and men, spat on, and cursed. As soldiers intervened, the ultra-Orthodox cried out at the worshippers and soldiers alike, 'Nazi, Nazi, You killed the six million, piece of shit, Christian . . .' A week later a Jewish-owned pub in Jerusalem popular with teenagers, the Gotham, was attacked by the ultra-Orthodox because it was open on the Sabbath. The owner was threatened with death by stoning, and prayers were said for his death: the same prayers that had been offered up by ultra-Orthodox extremists in the months leading up to Rabin's death.

Memorial Day is followed rapidly by Independence Day, a day of celebration. It was also a day, in 1997, on which Israelis reflected that the population of Israel was higher than it had ever been, 5,716,000 people. Of these, 4,620,000 (80.8 per cent) were Jews; 835,000 (14.6 per cent) were Israeli Muslim Arabs; 166,000 (2.9 per cent) were Israeli Christian Arabs; 95,000 (1.7 per cent) were Druse; and 4,000 (a tiny minority that had dwindled over the centuries) were Armenians. In the West Bank and Gaza Strip, increasingly under the jurisdiction of Yasser Arafat and the Palestinian

Authority, as a result of the Oslo Agreements, but still within the sovereign boundaries of the State of Israel, were a further 1,850,000 Palestinian Arabs, mostly Muslims.

During Independence Day, the majority of Israelis hang the national flag from their windows and cars. A survey conducted for Israel Army radio in 1997 showed that the 72 per cent of the population who flew flags from their homes were divided equally into those who had supported Peres and Netanyahu in the previous election, while of the 62 per cent who put flags on their cars a clear majority were Netanyahu supporters. The ultra-Orthodox did not fly flags, holding as they did in contempt the State which protected them and provided them with the amenities of civilized daily life.

In March 1997 a group of women whose sons were serving in Lebanon had set up a group which they called 'A Voice Calling for Peace'. 'It was the helicopter crash that had opened my eyes,' one of their leaders, Miri Sela, explained in the summer of 1997. 'We came to realize that most of the soldiers killed in the North die because of logistics. It is primarily a war of logistics – how to move troops and how to supply them with food and ammunition. When we moved them in trucks the Hizballah blew them up with suicide bombers. We gave the trucks armour but that didn't help. We started using helicopters and look what happened. Why not take all the resources needed for logistics and invest them in strengthening the border defences – reinforcing the fence, building more watch towers etc.'

Protest movements in favour of peace and withdrawal are common in Israel, a feature of her vigorous democracy: Peace Now was one such group which had earlier made an impact. In the Netanyahu era, however, such protest groups could not expect to be effective, or even to arouse particularly widespread public support. Withdrawal from the Lebanon would require political will at the highest level. During the four Rabin–Peres years it might have been possible to combine political will and public pressure, but as the first Netanyahu year came to its close, such an outcome was unrealistic. The government position was put clearly by the Minister of Defence, Yitzhak Mordechai, in the first week of June: 'I promise the mothers that the moment we find a partner with whom we can reach agreement or any element which can assume responsibility for establishing a serious force in south Lebanon, we will quickly reach an agreement,' he told Channel One television. 'Today, in my opinion, there is no such element. Any change in the currrent situation means bringing terrorism closer to Israeli territory and bringing weapons closer to the northern border and a substantial threat to large portions of the population of northern Israel.'

A member of A Voice Calling for Peace, Shoshana Saban, was not convinced by General Mordechai's attitude. 'Simply sit and think,' she said. 'We have been in the muck of the Lebanon for fifteen years. It doesn't get better. We continue to suffer casualties. That means there has been no change in the conception. Nothing has changed. Let them sit and think. I still believe that maybe, despite everything, it's still possible.'

Saban was the administrator of the Hadassah Hospital neighbourhood branch in Kiryat Ha-Yovel. 'What moved me so much about the mothers is that they are sick and tired of the situation,' she said. 'It's hard for us as mothers to live from one ambush to another, or from the time our sons go up to Lebanon to when they come down again. When my son goes up, it is terrible. Lebanon is a cancer in the souls of all the parents.'

A Voice Calling for Peace was not without supporters in high places. Among them was Yossi Beilin, one of the architects of the Oslo Agreements, and the runner-up in the Labour leadership contest in June 1997. Beilin, a potential successor to the intellectual mantle of Shimon Peres, held views that spanned the whole liberal spectrum, including support for a Palestinian State and the belief that Israel should exchange peace with security guarantees from Syria for a return to the international borders on the Golan Heights. He also advocated a unilateral withdrawal from Lebanon, 'preferably' with a signed agreement, but, failing that, through informal understandings. 'It is easier to guard our country from within rather than without,' Beilin said. 'There is no need for our soldiers to continue dying in Lebanon.'

In the summer of 1997 there was an outcry among many liberal-minded Israelis at the revelations of efforts to reduce the Palestinian population of Jerusalem. 'In the past eighteen months,' wrote Naomi Chazan, a Meretz Member of Knesset and a Deputy Speaker, 'Israeli efforts to stake a claim to all of Jerusalem have taken on a distinctly anti-democratic, anti-humanitarian cast. By refusing to renew the identity cards of Palestinian Jerusalemites, Israeli authorities are forcing thousands to leave their homes, and often their families, on fifteen days' notice. The authorities are denying these residents not only the social benefits associated with residency in Jerusalem, but also their fundamental right to live in the city of their birth.'

Naomi Chazan's protest was published in the *Jerusalem Post* under the heading 'End this oppression'. In it she explained the deep unfairness of the refusal to reissue identity cards:

Many Palestinians who have carefully observed the already stringent standing regulations for years, and whose families have lived in the same place for generations, are now left with nowhere to go. Many more live in danger of expulsion from the city. Permits are being revoked left and right, with no regard for either the policy that existed for nearly thirty years or the real circumstances of peoples' lives.

The Interior Ministry has tried to conceal the very existence of this new policy, as well as its effects. Official numbers are significantly underrated. They also fail to include family members listed on rejected permits, who are forced to leave as well.

Before 1996, permits were revoked if a permanent resident lived in a foreign country for more than seven years, or became a permanent resident or citizen of another country. That itself was problematic. But merely moving out of Jerusalem did not endanger the right to residency.

Now, however, the Supreme Court holds that this right expires automatically if one's 'centre of life' – cryptically defined by Israeli authorities – moves outside of Israel.

What made the new policy doubly offensive to its Israeli critics was that there was one rule for Israelis and another for Palestinians. As Naomi Chazan explained, 'an Israeli Jew who moves to a West Bank settlement is an Israeli citizen entitled to all benefits of citizenship. Yet a registered Jerusalem resident who marries a resident of Ramallah, or even of a village just outside Jerusalem's borders, and moves to join his or her spouse, is treated as though he or she has moved abroad.'

More than a question of residence was at stake. Naomi Chazan – who before entering the Knesset had been a professor at the Hebrew University of Jerusalem – set out in strong tones her objection, and that of the human rights activists who for so many years, since the days of Peace Now, had sought to uphold the rule of a tolerant law throughout Israel for Jew and Arab alike, and not to allow the thirty-year occupation of the West Bank to undermine the values of democracy and tolerance:

> Israel has no right to play games with Palestinian lives in this manner. Particularly in light of our history and essential values, we cannot justify uprooting people from their homes.
>
> We cannot maintain such a policy yet continue to call ourselves a democracy; democracies do not flagrantly violate international laws like the Geneva Convention and the Universal Declaration of Human Rights.
>
> This oppression policy must stop. The rights of those who have been expelled from their homes and separated from their families must be reinstated.
>
> Integrity and justice demand that we develop a new definition of Palestinian residency in Jerusalem, one that does not expire with bureaucratic whim, that protects rather than tramples upon fundamental human rights.
>
> Such a formulation, honouring equality and Jerusalem residents' basic rights, must acknowledge the inherent injustice in Israel's quiet but insidious efforts to depopulate East Jerusalem.

In 1997, the Association for Civil Rights in Israel (ACRI) celebrated its twenty-fifth anniversary. That year, it appointed one of its founders, Professor Ruth Gavison, an internationally renowned jurist, as its President. Reflecting on the conflicts in Israeli society, she said, 'The schism, however, is deeper and more pronounced than it used to be. Initially, at least among Jews in Israel, there was an understanding that the existential needs of the State demanded a united front on ideological issues, and that compromise was necessary. But with growth and strengthening of Israel, many people have come to feel that the existential issues are no longer relevant, that we can 'afford' the cultural war that is now being waged with new intensity.'

Professor Gavison was also concerned with the differences of attitude towards Jews and Arabs. 'When a young Haredi was detained at a demonstration not long ago, everyone was concerned, but no one paid any attention to the fact that young Arab children who were caught doing basically the same thing were given stiff prison sentences, without going through the same legal processes. There is a kind of double standard in human rights thinking in Israel, reflecting the fact that many people feel allegiance to their own groups and demand rights for themselves but don't think that these apply to others as well. One of the major contributions of human rights groups, especially in conflict situations, is to remind us that this is not the case.'

A spate of building on the West Bank signalled the complete departure of the Netanyahu government from its predecessor. On June 9 Netanyahu went to the West Bank settlement of Ariel, where he was photographed pouring concrete during a corner-stone laying ceremony for an agricultural school. In an interview the following day, the first anniversary of his premiership, he told Christopher Walker, the Israel correspondent of *The Times,* that the disputed construction of houses on Har Homa would continue. 'Mr Arafat must tell his people openly and squarely,' Netanyahu said, 'that peace will not be achieved on the 1967 lines. Israel will not reduce itself to a fragile Ghetto State on the Mediterranean shores.'

In December 1996 the settlements had been designated 'national priority areas', making them eligible for considerable financial benefits, which the Shamir government had given them almost a decade earlier, but which Rabin and Peres had taken away from them. The settlements vied with each other in search of new residents. Elkana, for example, offered 'huge villas' in the latest architectural styles, separated by 'well-maintained roads and footpaths . . . surrounded by well-kept green areas'. Barkan offered a 100-factory industrial park, a shopping mall with post office, bank and cafeteria, and 'build your own home plots' for those who wished to settle there. A newspaper advertisement for a promotional visit to the settlements on the Golan Heights noted, 'We'll pick blueberries to our heart's content.'

*   *   *

On June 11 the ultra-Orthodox again showed their displeasure at more liberal forms of Judaism, and their hatred of Arabs. The *Jerusalem Post* report shocked fair-minded Jews thorughout the world:

Hundreds of Haredim rioted yesterday in Jerusalem, attacking Conservative Jews, Palestinian residents and Border Patrol policemen after prayers held by tens of thousands of Haredim at the Western Wall on the occasion of the Shavuot holiday. The riots began in the morning, immediately following prayers. A group of Conservative Jews praying near the entrance gate to the Western Wall was attacked by dozens of Haredim, who cursed, shoved, and spat on them.

Members of the Unit for Protection of the Holy Places formed a barrier between the Haredim and Conservatives and accompanied the latter to beyond the plaza. As they were being escorted out, stones and bags of excrement were thrown on them from one of the Yeshivas. One of the Conservative Jews was hit by a bag of excrement.

Later, as the tens of thousands left the Western Wall plaza after prayers, several hundred Haredim began to attack Palestinian property. Near the Dung Gate, Haredim youths with hammers smashed windshields and mirrors of cars. Eighteen vehicles were damaged. Police arrested three Haredim, aged thirteen to sixteen, who admitted to having caused damage. They explained the destructive acts as being based on 'hatred of Arabs'.

On the way from the Wailing Wall to the Haredi neighbourhoods in West Jerusalem, hundreds of Haredim gathered at the Damascus Gate and attacked Palestinian passers-by. A border patrol unit tried to disperse them. Several of the Haredim sprayed tear gas at the border patrol policemen and on Palestinians, and an officer and several Palestinians were slightly injured.

The Haredim also threw stones at police and Palestinians, but police ultimately moved them toward the Me'a Shearim neighbourhoods by pushing and using clubs.

There were no arrests. Police spokesmen said that their goal was to prevent unrest and fights between Haredim and Palestinians, and that in these cases arrests are secondary.

*   *   *

During the second week of June the Netanyahu government signalled a major shift away from the map envisaged in the Oslo Accords, under which as much as 80 per cent of the West Bank would be transferred to the Palestinian Authority. Four years earlier an Israeli expert on Palestinian affairs (and champion of the Bedouin), Clinton Bailey, had devised a map with three self-governing Palestinian enclaves. These enclaves would cover

50 per cent of the West Bank. Under this plan, there would be an Israeli corridor through Samaria, linking pre-1967 Israel at almost its narrowest point with the Jordan Valley and including within it six major West Bank settlements. The Bailey map was devised in such a way that almost 90 per cent (1,706,000) of the Arabs of the West Bank would be released from Israeli control, while leaving 90 per cent (101,000) of the Jewish West Bank settlers in contiguous Israeli territory. The remaining 10 per cent (12,470) of the settlers would be moved from inside the Arab enclaves. In all, only 124,000 Arabs would be outside the three enclaves, as would, in addition, all 140,000 Arabs of East Jerusalem.

In basing his new ideas on the Bailey map, and elaborating on it, Netanyahu envisaged Israeli annexation of the larger part of the territory captured from Jordan in the 1967 war. Among the lands that would pass from Israeli military rule to outright Israeli sovereignty under Netanyahu's plan were the principal water aquifers of the whole region, several columns of territory along the West Bank's borders with Israel and Jordan, and a corridor from Jerusalem to the Jordan River that would cut the West Bank in half.

Under the broad principles used to describe the plan, the Palestinians would receive three or four enclaves, amounting to roughly 40 per cent of the territory of the West Bank, and drawn, as in the Bailey map, to enclose nearly all of the Palestinian population. The Palestinian entity, while being ruled by Arafat's Palestinian Authority, would be lacking statehood and would possess no common border with Jordan. It would be sandwiched between territories annexed by Israel, and, unlike the Bailey map, sliced by four east–west highways controlled by the Israeli army.

Netanyahu publicly described the new plan as something 'I've offered', but before making it public his government had not discussed it with the Palestinians, and said that it had no plans to do so. Drafters of the proposal, which had been outlined to Cabinet members and a few journalists, said that its main aim was political – to reassure the right-wing dissenters in Netanyahu's governing coalition, and to challenge the opposition Labour Party's newly elected leader, Ehud Barak.

'Reports about the proposal', wrote the *Washington Post* correspondent Barton Gellman, 'came amid the most serious sustained impasse in Israeli–Palestinian talks since the two peoples reached mutual recognition in 1993. Palestinian leaders, who broke off talks when Mr Netanyahu sent bulldozers to East Jerusalem in March to begin work on a new Jewish neighbourhood, said the leaks about the proposal are further evidence that Mr Netanyahu means to impose unacceptable terms.'

The chief Palestinian negotiator, Saeb Erekat, commented, 'We have heard about this plan only in the newspaper. Why bother telling us? The real negotiation is taking place within his own coalition, and with us he feels he can dictate.'

*   *   *

Following the agreements negotiated at Oslo, the Israeli army had left six Palestinian cities and, most recently, much of the seventh, Hebron. These cities constituted 3 per cent of the West Bank's territory and 29 per cent of its Arab population. Just over 24 per cent of the West Bank – containing 67 per cent of the Arab population – remained under a mixture of Israeli military control and Palestinian civil rule. The largest part of the West Bank – more than 72 per cent of the territory but only 4 per cent of the Arab population – remained entirely in Israeli hands.

Under Oslo, Israel was committed to hand three further areas of the West Bank to the Palestinian Authority, but Netanyahu halted those transfers as part of his dispute with Arafat, whom he accused of not fulfilling the Palestinian side of the agreement, including a rigorous searching out of Palestinian terrorists. Peres, the former leader of the opposition, insisted that Arafat was carrying out his obligations with regard to seeking out terrorists.

An army map of Israeli security interests in the West Bank, which Netanyahu's staff described in July 1997 as the basis for the new proposal, showed the strategic Jordan Valley as a permanently Israeli barrier against attack from the east, as in Yigal Allon's plan from the early 1970s. On the western side of the West Bank, at Israel's pre-1967 border, the army has drawn in a north–south 'seam strip' of varying width, as well as enclaves in the northern West Bank to incorporate heavy Jewish settlement there, and to include the aquifer, the underground source of fresh water for the region.

Other Israeli interests on the army map included four east–west 'strategic axes' enabling the movement of heavy equipment between Israel and the Jordan Valley and ensuring that the settlements and their surroundings remained Israeli; a broad 'defence zone' around Jerusalem and nearby Jewish settlements; the retention by Israel of high ground on the Samarian hills for intelligence and air defence emplacements; and additional sovereign Israeli 'lifelines' of traffic, electricity and water-pipes.

Israel is faced with many sensitivities from beyond her borders. With regard to what became known as 'the Armenian genocide' of the First World War, the Turkish government resents any publicity through Israeli television or museum exhibitions. At the same time, joint economic ventures such as the industrial park at Izmir inspired by Stef Wertheimer draw the two countries together. The Israeli government has clashed with Lithuania over the record of individual Lithuanians who participated in the mass murder of Jews in the Second World War. But diplomatic relations with Lithuania are maintained, and joint cultural activities do take place. Switzerland and Israel have also clashed over the gold deposits in Swiss banks that derive from wartime illicit activity – including the melting down by the Nazis of the gold teeth of Jewish victims. But Israel was officially represented in Basle at the hundredth anniversary celebrations of the First Zionist Congress, though not by the Prime Minister.

Another irritant in Israel's international relations concerned the growing number of foreign workers in Israel: more than a quarter of a million by the autumn of 1997. Many came from Roumania, others from as far away as Thailand and the Philippines. The often harsh conditions under which Israeli contractors, mostly in the building trade, put them to work, angered sensitive Israelis, and was of concern to the countries from which the workers came. Each time the Gaza Strip was sealed off, however, for political reasons, there seemed a need for foreign workers to replace the Arabs not allowed to enter.

Palestinian anger was intensified during 1997 by an increase in the number of Arab homes in East Jerusalem that were bulldozed to the ground by the Israeli authorities. The Jerusalem municipality, led by the Mayor Ehud Olmert (a Likud stalwart), based its actions upon the fact that these homes had been built without the necessary building permits. But these were permits which it was increasingly difficult for Jerusalem Arabs to obtain.

Thirty-two houses were demolished in one week alone in the summer of 1997. What a British traveller, David Pryce-Jones, called in 1964 'the intransigence towards the Arabs' had in many ways intensified in the 1990s among a wide spectrum of Israeli society. This intransigence, although constantly challenged by moderate Israelis, is a cause of continual Arab–Jewish friction and mutual suspicion. Yet as early as 1955, Ben-Gurion told a group of army officers, among them Mordechai Bar-On (who recorded his words): 'In both our military preparedness and in the belligerent actions imposed on us we must never lose sight of the fact that our ultimate goal in our relations with our neighbours is to attain peace and coexistence.' In 1997 that coexistence, above all with the Palestinian Arabs, was fragile and continually endangered.

On 29 July 1997 the first Nahal outpost to be created in eleven years – Yatir – was established west of the city of Arad. Its fifty inhabitants came from various youth movements, among them the religious Bnei Akiva. To begin with they decided to live in tents, to remind themselves of the way the first Nahal settlements had been founded half a century earlier. Among the tasks assigned to the settlers was patrolling the former Green Line which lay close to the settlement, and doing community work in Arad. Several more Nahal settlements were being set up elsewhere, among them Eshbal in Galilee and Kerem Ha-Shalom near the Egyptian border.

In the Gaza Strip, 1997 saw the celebration on August 17 of a two-day public festival to celebrate the twenty-fifth anniversary of the Gush Katif settlement bloc. In all, there were 6,000 Jews living in the Gaza Strip settlements: their celebrations included an arts and crafts fair, horseback riding and a song festival. The settlers were confident that they could double their population within the next twenty-five years, and continue to improve their flourishing agriculture.

The autumn of 1997 proved a testing time for Israel, both at home and abroad. On July 30 two suicide bombers blew themselves up in the main Jewish market in Jerusalem, the Mahane Yehuda, killing sixteen shoppers and stallholders. Among the dead was Simha Fremd, a ninety-two-year-old survivor of Auschwitz who had immigrated to Israel from South America eighteen years earlier, and Mohi Othman, a thirty-three-year-old Muslim Arab from Abu Ghosh who worked in the market. Also killed was fifty-two-year-old Shalom Zevulun, who had been seriously wounded in the head during a Syrian shelling attack on the Golan Heights during the October War. He had never recovered from the shock, added the word 'Golan' to his name, and spent his days wandering around the market.

In response to the Mahane Yehuda bombs Israel not only imposed a total closure of the movement of Palestinian workers into Israel (as the Labour government had also done after such bombings) but also decided to suspend all financial transfers from Israel to the Palestinian Authority. These transfers were laid down in the Oslo Accords. Their suspension meant that 60 per cent of all Palestinian funding was cut off, including customs duties and VAT payments. The unilateral cancellation of a revenue-sharing agreement (signed in 1994 and formally known as the Paris Protocol) angered the Palestinian Authority. The Netanyahu government's anger was reserved for what it denounced as the failure of the Palestinian Authority to seek out more actively the perpetrators of the crime. Netanyahu's spokesman, Shai Bazak, issued a formal statement explaining his government's position. 'Agreements must be respected by both sides,' it read. 'Until now Israel fulfilled all its obligations according to the agreements, yet the Palestinian Authority is not doing so and in the main it is not fulfilling its obligations in the most important area – the fight against terror,' Bazak added: 'When the Palestinian Authority fulfils its obligations Israel will transfer moneys. Israel will no longer pay with dead lying in the streets and while the Palestinian Authority does not fight terror or prevent it from being committed.'

A second suicide bomb attack, by three Palestinians, killed five people in Ben Yehuda Street, Jerusalem, on the afternoon of September 4. Three of the dead were fourteen-year-old schoolgirls; another was the son of a Holocaust survivor. Outside Israel, twelve Israeli naval commandos were killed on the morning of September 5 when a seaborne attack which had been carried out in strictest secrecy against a Hizballah target inside Lebanon was intercepted. Among the dead was Lieutenant Zvi Grossman from Tel Aviv, whose friends had planned a surprise twenty-second birthday party for him, but attended his funeral instead. Following the deaths of the twelve naval commandos – which brought the number of Israeli servicemen killed in Lebanon since the beginning of the year to thirty-four – calls to withdraw from southern Lebanon intensified. The Chief of Staff, General Amnon Lipkin-Shahak, went so far as to warn over army radio: 'There are definite signs of fatigue in Israeli society.'

Throughout Israeli society the soul-searching was intense. Nurit Peled-Elchanan, the mother of one of the fourteen-year-old girls killed in the Ben Yehuda Street bombing, blamed the bombing on the government, for its treatment of the Palestinians. 'When you put people under a border closure,' she said, 'when you humiliate, starve and suppress them, when you raze their villages and demolish their houses, when they grow up in garbage and in holding pens, that is what happens.'

The sale of weapons technology from the former Soviet arsenal to Iran raised the spectre of a revived fundamentalist onslaught. In September, Israel asked the new British Prime Minister, Tony Blair, to raise this issue with President Yeltsin on his first visit to Moscow. In the United States, Israel pressed for an end to American aid to Russia while the sale of missile technology from Russia to Iran continued. In the last week of September, Yitzhak Mordechai, the Israeli Defence Minister, warned publicly that Russian aid was making it possible for Iran to complete the development of ballistic missiles that could strike at any part of Israel. Iran was 'on the verge', he said, of achieving such a capability, and would do so 'by 1999'.

Although belittled by many Likud supporters, and given somewhat short shrift by the Netanyahu government, the hopes of many Israelis in the Oslo peace process remained. In an attempt to enhance public awareness of the need to maintain the Oslo process, that autumn Shimon Peres launched the Peres Centre for Peace. Its aim was 'to build on the opportunities established between governments to create tangible improvements in the day-to-day life of all the peoples of the Middle East'. One of its projects was the establishment of eight partnerships, each between an Israeli, a Palestinian and a European city. It also intended to embark upon regional joint projects for agricultural research, information technology and youth dialogue.

The establishment of the Peres Centre for Peace, with its high hopes of regional cooperation throughout the Middle East, was in stark contrast with the establishment of Jewish homes on land purchased in 1990 by Irving Moskowitz – a retired physician, bingo-hall magnate and philanthropist from Florida – at Ras al-Amud, a few hundred yards from the Dome of the Rock in a predominantly Arab area of East Jerusalem, and adjacent to the Jewish cemetery on the Mount of Olives. Both Yitzhak Rabin (when acting Minister of the Interior) and Teddy Kollek had earlier opposed the construction of these homes. The decision to establish them led to a crisis in the autumn of 1997, when three Jewish families moved into houses at Ras al-Amud which Moskowitz had also bought. Ironically, the land had originally been purchased more than a hundred years earlier by Jerusalem's Jews to be used as an extension of the cemetery.

Following the direct intervention of the American Secretary of State, Madeleine Albright, the Netanyahu government agreed to move the families out; but it insisted that a group of religious students should be allowed to use the houses as a place of prayer, and that they should be guarded by Israeli soldiers. Fifteen people moved out, and ten ultra-Orthodox students, together with ten security men to guard them, moved in. Such was the compromise which the Americans accepted, though Peace Now and the Palestinians both derided it. It was, wrote the journalist Danna Harman in the *Jewish Chronicle*, 'the latest act in a drama that has, even more inexorably, pitted Israeli against Palestinian, and Israeli against Israeli'.

Hardly had the much-disputed compromise been reached about Ras al-Amud than Netanyahu announced a further expansion of the Etzion bloc settlements by 300 housing units. In the normal course of events, in a normal country, such a relatively small increase in new homes would have been unremarked and unremarkable. In the climate of suspicion and accusation generated by Israeli and Palestinian actions during 1997, it served to exacerbate hostility. 'We are building in Judaea and Samaria,' Netanyahu told a group of religious teenagers in the last week of September. The American Ambassador to Israel, Martin Indyk, responded immediately. 'We are unhappy with that announcement,' he said. 'The Secretary of State is engaged in an almost full-time effort to help Israel, because there is an effort under way to isolate Israel, and it is in that context that this announcement comes and it undermines her efforts.'

A spokesman for Madeleine Albright said that she did not 'regard this kind of building as consistent with the kind of climate for negotiation that she hoped to create'. Indyk was even blunter, telling an audience in Tel Aviv that the Oslo peace process, to which he and his government were committed, 'some days seems to be turning into a nightmare'. Albright was firm in her own recommendations: the Palestinian Authority should act more vigorously to search out terrorists, and Israel should stop confiscating Palestinian land, stop demolishing Palestinian homes, and stop building on Har Homa in East Jerusalem.

At the end of September a different conflict intensified, the cloak-and-dagger world of secret assignments. After a Jordanian extremist – called in the press reports an 'Islamic gunman' – had wounded two Israeli embassy guards in Amman, an undercover operation was undertaken by two Mossad agents – posing as Canadian tourists – to assassinate the Hamas political representative in Amman, Khaled Mashaal. A toxic substance was injected in Mashaal's ear, but the agents were caught.

Deeply embarrassed by the failed assassination, and under intense pressure from King Hussein, the Israeli government – which was forced to send its own doctor with an antidote to save Mashaal – agreed to release from prison the spiritual founder of Hamas, Sheikh Ahmed Yassin, who was in the eighth year of a life sentence. On October 1 the Sheikh was flown from Israel to Amman in a medical helicopter sent by the King. It was two weeks before the talks between Israel and the Palestinian Authority, talks that had been halted for six months, were due to resume. Yassin returned to Gaza, a Hamas hero.

With each diplomatic development, with each outburst of violence, with each moment of uncertainty and self-doubt, the words of Julian Huxley, a British traveller to Israel in 1954 – when the State was only eight years old – came to mind with growing force: 'Time alone will show whether the result will be merely another jealous nationalistic country in the Middle East, or whether constructive ideas and energies at its core become a stimulus instead of a stumbling block to its neighbours, a stimulus to the real job of recreating a high civilization in the region, and taking a constructive hand in the human adventure as a whole.'

In 1936, during the discussions on the nature and status of the Jewish National Home, Chaim Weizmann was asked by the Peel Commissioners when he would consider the National Home to be 'finished'. 'Will Belgium ever be finished?' was Weizmann's reply. The achievements and problems of the Jewish State are likewise unfinished, and a source of daily wonderment. Remarkable achievements in science and technology, art and agriculture, across five decades, contrast with the tensions and violence of the political and national debate as Israel enters its sixth decade.

In the first fifty years of statehood Israel has made remarkable strides. In the sphere of economy it has gone, in the words of Walter Eytan (who was an eyewitness at the foundation of the State), 'from Jaffa oranges to high tech'. In the cultural sphere it has created an Israeli literature, theatre, art and music that, while deriving from the countries and cultures of origin of its inhabitants, have their distinctive Israeli style. Several Israeli novelists, among them Amos Oz, Aharon Appelfeld, David Grossman and A. B. Yehoshua, have achieved international status – and have also often been highly critical of aspects of intolerance inside the society.

Israeli scientists have been pioneers in every branch of the physical and medical sciences. Israeli industrialists, prominent among them Stef Wertheimer, have pioneered high-tech industries and robotics; and in Wertheimer's case built a model village on a hillside in Galilee next to his factories. Israeli hospitals are among the most advanced in the world, catering for Jews and Arabs alike, and equipped with the most modern operating and after-care facilities, often provided by Jewish overseas philanthropists. Israel suffers, as do all modern States, from road deaths that exact a heavy annual toll in death and injury; in 1995 more than 550 Israelis were killed on the roads, a figure that, having fallen in earlier years, has begun to rise again. This is a higher per capita road death figure than for any European country: higher even per capita than the United States. As well as the considerable personal tragedies and economic hardships which such frequent deaths entail, Israel also remains burdened by the continuing need to face the possibility of war with Syria, its one remaining neighbour which has not yet signed a peace treaty, and by the unremitting conflict inside its military zone in southern Lebanon.

Until full peace is secured with Syria and Lebanon, compulsory national military service remains an additional burden on the national economy, and on the smooth evolution of young men and women from school to university and work, despite the educational aspects of army training. America's enormous economic contribution to the Israeli budget, more than three billion dollars a year – the largest single American overseas commitment – creates a dependence which successive Israeli governments have sought to reduce and which, in an address to Congress shortly after coming to power, Netanyahu said would be reduced. Efforts have also been made by successive governments to persuade Israelis living abroad to return, and to make their contribution to the economic upbuilding of the State. As many as half a million Israelis live and work outside the country, participating in the life – as the Jews in the Diaspora did, and still do – of the countries in which they have chosen to stay.

Life in Israel has been transformed in fifty years. Increasing in numbers from 600,000 to almost 6,000,000, in each decade its fast growing population has put enormous burdens on the State, as well as a strain on the environment, but has at the same time given it a vibrancy and energy that no visitor can fail to notice. Jerusalem, the capital, is a centre of conferences, teaching institutions, fairs (including a bi-annual Book Fair), music and pilgrimage – for Christians, Muslims and Jews. Tel Aviv, which was only founded after the turn of the century, is a vigorous metropolis. Eilat, a mere border post at the time of independence, is a thriving holiday resort, offering coral reefs and desert canyons as well as beaches. Many kibbutzim have guest houses from which the visitor can explore the beauty of their surroundings.

Agriculture, trade and industry have made Israel a dynamic contributor to the world economy. Universities and research institutes flourish. Modern suburbs have grown up around every city. Tourism brings more than a million visitors every year. National Parks help preserve the astounding natural beauty; through the Nature Preservation Society, tens of thousands of people visit these areas every year, marvelling at the rich variety of landscape and wildlife. Museums abound. Shopping malls, restaurants, cafés and night clubs have transformed a once sleepy lifestyle into a pulsating one.

Day-to-day life in Israel has reached a high standard. In the main, the climate is mellifluous. The forests that were planted by the Jewish National Fund during more than half a century provide shade, especially at weekends, for vast numbers of picnickers. Also at weekends, the beaches from Nahariya in the north to Ashkelon in the south are crowded. Family life is prized. The

Jewish festivals provide a focus throughout the year for family, school and community celebration.

Israel is often at the centre of world attention. This is seldom for her achievements, which are considerable, or for the quality of life which she has created, and which is the envy of many nations. It is Israel's wars, her political and social divisions, her conflict with her neighbours and with the Palestinians, and the stark intrusion of acts of terror into daily life, that make its locations and its leaders internationally recognized.

Within Israel, as the State reaches its fiftieth year, a sharp struggle has been taking place between different and often conflicting attitudes. Divisions that had marked the earlier years of statehood, such as that between Ashkenazi and Sephardi Jews, and a recent intensification of the religious–secular divide, have led – especially in the period since the assassination of Yitzhak Rabin in October 1995 – to harsh words and fears of a deeper rift. Yet the vision of a forward-looking Israel remains, inspired by the vision of Ben-Gurion at the time of the declaration of statehood, and made all the more powerful, and urgent, after the assassination of Rabin. In his speech on the first anniversary of the assassination, Shimon Peres told a special session of the Knesset:

> Yitzhak knew that it is impossible to bring up generations of young people only on the tragic memories accumulated in the store house of experiences of our people.
>
> He knew, and we knew, that once we recovered from the Holocaust we overcame the animosity, we conquered the wilderness.
>
> The time of new Israel has arrived; an Israel that will justify – not only compensate for – the long journey that we made.
>
> Yitzhak knew and we knew that the time has come to unveil Israel, to reveal an Israel of moral considerations, with a heritage of deep truth, an Israel who carries inside her the talent to obtain bread from science as much as flour from the fields.
>
> We knew that the moment had come for Israel to turn away from being a compensation for the past, and become a Jewish invitation for the future.

CHAPTER THIRTY-ONE

Conflict and conciliation

As the twenty-first century began, Israel had been in occupation of southern Lebanon since 1982, trying to contain the Iranian-backed, fiercely anti-Israel Hizballah movement there. On 15 February 2000, following the deaths of seven Israeli soldiers in two weeks at the hands of Hizballah, the Labour Prime Minister Ehud Barak faced calls from the Israeli public for an immediate withdrawal of all Israeli troops from Lebanon. He promised to do so by July 2000, by which time he hoped to be in direct negotiations with Syria, with a view to reaching a wider agreement. A unilateral withdrawal was, he warned, 'fraught with risks' of Hizballah taking its fight into northern Israel. It was appropriate for him 'to exhaust all the chances and bring the boys home under an agreement'.

The Israel Defence Forces insisted that the pullback from southern Lebanon should follow a peace agreement with Syria, an agreement that would include a Syrian promise to rein in Hizballah. But the mothers of many soldiers, as well as reserve officers and even soldiers serving with the élite paratroops, were demanding a withdrawal before there were more casualties.

Syria having made no response to Barak's request for direct negotiations with Israel, Barak ordered a unilateral withdrawal. It was completed on 24 May 2000. Israel expressed its hope that the Lebanese Government could restore its sovereignty over the area, and exercise its sovereign rights with regard to Hizballah. Barak's critics, particularly his predecessor as Prime Minister, Benjamin Netanyahu, head of the Likud Party, warned that the withdrawal would be interpreted by Hizballah, and by the Palestinians in the West Bank and Gaza Strip, as a sign of Israeli weakness.

Since conquering the West Bank and Gaza Strip in 1967, Israel had been enmeshed in the burdens of being an occupying power, and facing the demands of Palestinian Arab nationalists for statehood. Under the Interim Agreement (Oslo II) signed on 28 September 1995 between the Israeli

Government and the Palestinian Authority, portions of the West Bank and Gaza had been transferred to the Palestinian Authority, headed by Yasser Arafat. A date had also been set for a Final Status Agreement. That date was 4 May 1999.

Despite intense negotiations, that target was not reached. The Palestinian leadership refused to give up the continuing armed resistance to the occupation. The Israelis refused to give up the two hundred settlements built on the West Bank and in Gaza since 1967, and allocated substantial military resources to protect the 250,000 settlers, who were living, most of them in small enclaves, amid two and a half million Palestinians. Armed Palestinian attacks on Israeli Jews continued. Israeli settlement building continued.

The search for a lasting agreement, and for a secure and recognized Israel side by side with an independent sovereign Palestinian State on the West Bank and in Gaza, was given a lift in July 2000, when President Clinton – as he embarked on the last six months of his eight-year presidency – invited Ehud Barak and Yasser Arafat to the presidential retreat at Camp David in Maryland. The summer meetings made progress, but then came an unexpected obstacle. On 28 September 2000, Netanyahu's successor as leader of the opposition Likud Party, Ariel Sharon, visited the Temple Mount in Jerusalem, escorted by several hundred Israeli policemen. His visit was prompted by allegations made by Israeli archaeologists – accusations that proved well founded – that Palestinian building works on the Temple Mount were destroying ancient artefacts relating to the Jewish past on a site that is sacred to Jews, Christians and Muslims alike. While on the Mount, standing a few yards from the al-Aksa mosque, near which the digging was being done, Sharon announced that the complex would remain under Israeli control for all time.

Sharon's visit to the Temple Mount was met by an immediate Palestinian protest on the Mount itself. During the clash that day with Israeli police, four Palestinians were killed. Within days, sustained and intense violence broke out throughout the West Bank and Gaza Strip. This became the second intifada, known to the Palestinians as the al-Aksa intifada. Palestinians claimed that Sharon's visit was provocative, even though the head of the Palestinian Authority's security service, Jabril Rajoub, had assured Sharon that if he did not enter the al-Aksa mosque itself there would be no problem; nor had Sharon tried to do so. Imad Falouji, the Communications Minister of the Palestinian Authority, later stated that the violence had been planned since Arafat's return from Camp David more than two months before Sharon's visit to the Mount.

One of the first traumas of this second intifada was a scene, shown repeatedly on world television, in which a twelve-year-old Palestinian, Mohammad al-Dura, was killed – his slow death filmed in terrible detail – as his father sought to shelter him during fierce Israeli–Palestinian cross-fire in Gaza. In videos of the young boy's death that were shown repeatedly on Palestinian television during the following three years, a lyric proclaimed: 'How sweet is the fragrance of the martyr.'

As the second intifada continued, it spilled behind the pre-1967 borders, with riots among Israeli Arabs – citizens of Israel – of whom twelve were killed. But the search for an agreement was not abandoned, and on December 19 the Camp David meetings were renewed. This was a month after the Democrats had been defeated in the presidential election by George W. Bush, but a month before Clinton would have to relinquish office. At the renewed Camp David talks, Israel, pressed by Clinton, offered terms that were at the limit of Barak's powers of concession: first and foremost, the establishment of a sovereign, independent Palestinian State. Israel also offered to remove all Israeli settlements in the Gaza Strip and almost all those on the West Bank.

The pre-1967 borders – the Green Line – would be the borders of the Palestinian State, with only minor modifications. The transfer of pre-1967 Israeli land just south of the West Bank to the Palestinian State would be compensation for the minor modifications of the Green Line. A safe transit route, a main road, would be constructed on Israeli land between the Gaza Strip and the West Bank; it would be free of any Israeli control, and possibly under Palestinian sovereignty. Jerusalem, while remaining under Israeli sovereignty, would be divided between its Jewish and Arab inhabitants, with a strong measure of Palestinian control over the Arab sections. Some Palestinian refugees would be allowed to return to Israel. Those who were willing to leave the refugee camps that had been so tenaciously maintained since 1948, with United Nations funding, by Lebanon, Syria, Jordan and even the Palestinian Authority, would be offered homes abroad: Canada offered to take a substantial number.

By December 23, the Camp David terms were almost signed. The Israeli and Palestinian negotiators were in near agreement. A Palestinian State would be established on more than 90 per cent of the West Bank and 100 per cent of the Gaza Strip. Then, at the last minute, Arafat refused to sign, insisting that all Palestinian refugees, including the children and grand-children of the original refugees, numbering several million in all, be allowed to return to their former homes. Most of those homes no longer existed. Some had been bulldozed by Israel in 1948 and 1949. Others had become part of substantial post-1948 Israeli cities and suburbs.

On December 27 the Israeli Cabinet voted to accept the terms negotiated at Camp David. Arafat again rejected them. Although the talks continued from 21 to 27 January 2001 between senior Israeli and Palestinian negotia-tors at Taba, just south of the Israeli resort town of Eilat, and were then renewed in Stockholm, it was too late. On January 19, even as the Taba talks were about to begin, Clinton's presidency came to an end.

On 6 February 2001, elections were held throughout Israel to decide who would be Prime Minister. Ehud Barak was defeated by Ariel Sharon, who had earlier voted against the Oslo Accords. Sharon won the prime ministerial vote with the largest ever margin in Israeli politics: 62.6 per cent of the vote against Barak's 37.2 per cent.

Hitherto, Sharon had seemed to represent the harsh, uncompromising, confrontational face of Israel. He was still regarded by many as the man who had provoked the Palestinians by his visit to the Temple Mount. His actions in Lebanon almost twenty years earlier, actions condemned by an Israeli judicial inquiry, were still a source of hostile recollections. But in his victory speech at Likud Party headquarters on February 6 he struck a note of compromise and hope. 'Today,' he said, 'the State of Israel has embarked on a new path, a path of domestic unity and harmony, of striving for security and a real genuine peace.' Israel was 'the State of all of us, and all of us have a joint future and a single destiny. In the course of the years, divisions, incisions in our people and society have proliferated. Baseless hatred and fury have multiplied. The time now has come to seek for as wide-based unity and agreement as possible.' Sharon added: 'There is a deep felt desire among the people to stand firmly together in dealing with the challenges of the future. I call now for the establishment of as wide a government of national unity as possible. And I here turn to the Labour Party and call upon it to join us in a partnership in pursuing the difficult path towards security and peace.'

Sharon was aware, he stressed, 'of the fact that peace requires painful compromises by both sides. Any and every political arrangement will be based on security for all peoples of the region. I call upon our Palestinian neighbours to cast off the path of violence and to return to the path of dialogue and solving the conflicts between us by peaceful means.' Nor were domestic issues to be neglected. 'In Israeli society, there are many areas of distress with which I am familiar. We will do everything we can to wipe these out. A people cannot exist if there is no social solidarity. We shall act to reduce the social gaps and to achieve equality of opportunity to everybody.'

As he had hoped, Sharon persuaded senior members of the defeated Labour Party to join his administration. The unity government that he formed consisted of the largest number of coalition members in Israel's history, a broad swathe of left, right, centre and religious Parties. On March 7, following the Knesset's approval of his coalition, Sharon was sworn in as Prime Minister.

Despite Sharon's call for dialogue, the second intifada continued. Throughout 2001 Palestinians with guns and knives targeted individual Israelis, most of them living in Jewish settlements on the West Bank and in the Gaza Strip. Some Israelis were killed when roadside bombs, and bombs left inside cars, were detonated. Others were killed while walking in the street, or sitting in restaurants and cafés, or waiting at bus stops.

The Israeli public, including those who were in favour of a Palestinian State and the removal of Israeli settlements from the West Bank and Gaza, were deeply disturbed by these killings, so different from the confrontation between troops and insurgents. Those Israelis murdered included a five-month-old boy killed by a rock, two fourteen-year-old boys stoned to death, and a mother of five who was five months pregnant when she was killed. Ten Jews were killed in the West Bank settlement of Emanuel during a sustained bomb and grenade assault by several attackers against passenger cars and a bus. In sixteen months, eighty-eight Israeli civilians were killed. More than forty Israeli soldiers and border policemen were also killed in ambushes and drive-by shootings, usually when they were on leave.

Suicide bombers added their terror. The willingness of a Palestinian youngster to commit suicide in order to kill as many Israelis as possible cast a pall of fear over the Israeli public. The suicide bombers were stimulated by the belief that in the cause of Jihad – Holy War – they would go directly to paradise. Their actions were applauded in the streets of the West Bank and Gaza. From Iraq, Saddam Hussein sent money to their families once their deed was done.

During 2001 Palestinian suicide bombers killed more than a hundred Israelis, most of them young people out for an evening in a café or disco, or waiting at a bus stop. The largest number of deaths in a single attack was in Tel Aviv, where twenty-one youngsters were killed at the Dolph-Disco near the seashore Dolphinarium on June 1. To combat the upsurge in these random killings and suicide bombings, not only were the homes from which the suicide bombers came blown up, but Israel embarked on an intensified policy of what was called 'targeted assassination'.

Those whom Israel knew, through its intelligence network, to be organizing and preparing terrorist acts, were tracked down and killed. Previously, such killings – aimed at the organizers of terrorism – were carried out abroad, the targets being Palestinians in Malta, Rome, Paris and Beirut. Between July and December 2001, however, the main targeted assassinations were in the Palestinian West Bank towns. In almost every such operation, for each Palestinian who was intentionally killed, at least one Palestinian bystander, often more, was killed.

Inside Israel and beyond, the policy of targeted assassinations was fiercely debated. Yitzhak Rabin's daughter, Dalia Rabin-Pelossof, a member of the Knesset and Deputy Defence Minister, gave the argument in favour of such action: 'It is a policy of self-defence. When we know of a terrorist who is a ticking bomb, it is incumbent on us to prevent it. And that is what we do.' This was the predominant view, but it was not held universally. Naomi Chazan, a Deputy Speaker of the Knesset, disagreed. 'It is an inefficient policy,' she declared, and went on to explain: 'It breeds more hatred and more terrorism instead of eliminating or even reducing it.' From the United States, the Secretary of State, Colin Powell, whose department had already expressed unease at the policy, warned Israel: 'We continue to express our distress and opposition to these kinds of targeted killings and will continue to do so.'

There was a moment of alarm in Israel with regard to the Hizballah forces that had begun to gain the ascendancy in southern Lebanon, when on 4 January 2002 the Israeli navy captured fifty tons of munitions, including several hundred 107 millimetre Katyusha rockets, which were being smuggled by ship to Lebanon. But it was within Israel that immediate danger loomed, as a series of suicide attacks on Israeli civilians culminated in the 'Passover Massacre' on the night of March 27. That night the Park Hotel in Netanya held a communal Passover dinner for its 250 guests, many of them elderly Jews with no relatives. A Palestinian suicide bomber entered the hotel dining room and detonated the explosive device he was carrying in a suit-case. Twenty-eight people were killed and twenty seriously injured. Two of the injured later died from their wounds. Many of the victims were survivors of the Holocaust.

In the following two weeks there were six more suicide bombings. In response, Israel launched a large-scale counter-terrorist offensive, Operation Defensive Shield, reoccupying the towns of the West Bank. One source of attacks on Israel had been the refugee camp in Jenin, known to Palestinians as 'The Martyrs' Capital'. Israel charged that the camp had 'served as a launch site' for twenty-eight suicide bombers. On 3 April 2002, when Israeli troops entered Jenin in search of the masterminds behind the suicide bomb attacks and their arsenal, they were lured into an ambush.

In the battle that followed, Israeli troops were accused of carrying out a massacre. In the words of the chief Palestinian negotiator, Saeb Erekat: 'People were massacred, and we say the number will not be less than five hundred.' Israel denied that any such massacre had taken place, a denial sup-ported by Amnesty International and Human Rights Watch, both of which reported that they had found no evidence of a massacre. Kadoura Moussa, the Fatah director for the Northern West Bank, confirmed this, stating that Fatah investigators put the death toll at fifty-six. The United Nations put the final death toll at fifty-two, more than half of them armed Palestinian fighters, and concluded that no Palestinian civilians were killed deliberately. The accusation of five hundred dead was shown to be false.

Twenty-three Israeli soldiers had been killed in Jenin, the highest Israeli death toll for a single confrontation during the second intifada. The suicide bombings continued, with increased frequency and higher casualties. During 2002, more than two hundred Israeli civilians (211 is the most accurate figure) were killed and 1,448 injured, some of the injuries being of the severest kind, permanently disfiguring and maiming. In addition to the suicide bombings, where the bomb carrier was also killed, 142 Israelis were killed in targeted killings, most by gunfire, or bombs thrown at individuals and small groups at bus stops, in cars and buses, and in their homes. Some were killed while celebrating the Sabbath and Jewish festivals; on one occasion while casting their votes in a polling booth. One Israeli, 71-year-old Avraham Boaz, was kidnapped at a checkpoint controlled by the Palestinian Authority and killed. Among the nine Jews killed when their bus was attacked on the way to the West Bank settlement of Emanuel were an eight-month-old girl, her father and her grandmother, as well as a premature infant delivered after its mother had been seriously injured.

In response, Israel continued with its policy of targeted assassinations. During 2002, thirty-six Palestinians known to be behind the dispatch of suicide bombers – and the preparation of their bombs – were killed. Twenty-seven Palestinian bystanders were also killed.

How could this terrifying spiral of violence be halted? How could the peace process that had collapsed after Camp David in 2000 be renewed? How could the Oslo prospect of a Final Status Agreement be revived? On 24 June 2002 the American President, George W. Bush, called for 'an independent Palestinian State living side by side with Israel in peace'. He also announced a 'Road Map for Peace': a three-year plan to resolve the Israel–Palestinian conflict through a 'Quartet' of the United States, the European Union, Russia and the United Nations. Bush described the Road Map as 'a starting point toward achieving the vision of two States, a secure State of Israel and a viable, peaceful, democratic Palestine'.

The Road Map envisaged the establishment of a Palestinian State and an end to the conflict by the end of 2005. Phase One involved five steps: an immediate halt to Palestinian violence; internal Palestinian political reform; Israeli withdrawal from the West Bank and Gaza Strip, though not necessarily to the pre-1967 borders; a freeze on all Israeli settlement expansion; and Palestinian elections. Phase Two, envisaged for June–December 2003, consisted of four steps: an international conference to support Palestinian economic recovery; the launching of a process leading to establishment of an independent Palestinian State with provisional borders; the revival of multilateral engagement on issues including regional water resources, environment, economic development, refugees, and arms control issues; and the restoration by the Arab States of pre-intifada links to Israel such as trade offices.

Phase Three, the final phase, was to be carried out in 2004 and 2005. Its six steps were: a second international conference; a permanent status agreement; the end of armed conflict; agreement on final borders; clarification of the status of Jerusalem, refugees and settlements; and the agreement of the Arab States to sign peace terms treaties with Israel.

The Road Map was ambitious and optimistic, offering the hope of a final and peaceful settlement of a conflict that had lasted more than seventy-five years. To ensure a Palestinian partner for the Road Map, a Palestinian Prime Minister was appointed, something Arafat had until then refused to contemplate. The man chosen by Arafat was Mahmoud Abbas, also known as Abu Mazen. Born in Safed in 1935, in what was then Mandate Palestine, Abbas had fled with his family to Syria in 1948. After studying in Cairo and Moscow, he became a founder member of Fatah in 1959, and had been a leading nego- tiator of the Oslo Accords in 1993.

Abbas was appointed Prime Minister on 19 March 2003. On April 30 the details of the Road Map were made public. On May 27, in his first public response, Sharon stated that the 'occupation' of Palestinian territories was 'a terrible thing for Israel and for the Palestinians', and that it 'cannot continue endlessly'. On June 4 a political summit was held at the Jordanian Red Sea port of Akaba, near where the Israeli–Jordanian peace treaty had been signed almost a decade earlier. The main participants were President George W. Bush, Palestinian Authority Prime Minister Mahmoud Abbas and Ariel Sharon. But Yasser Arafat – who was seen by Israel as the instigator of the continuing terror, and had for many months been isolated and boycotted by Israel and the Americans, a virtual prisoner in his compound in Ramallah – refused repeatedly to renew the 'presidential' talks he had broken off at Camp David.

It seemed that the Road Map could work. On 2 June 2003, the first day of a three-day visit by President Bush to the Middle East, Israel freed a hundred Palestinian political prisoners. In Egypt on June 3, when Bush met the leaders of Egypt, Saudi Arabia, Jordan and Bahrain, and also Mahmoud Abbas, the Arab leaders announced their support for the Road Map and promised to end funding to 'terrorist groups'.

These events marked the start, it was hoped, of a pattern of progress. But hardly had the American President returned to Washington than violence was renewed. On June 5 the bodies of two Israelis were found near the Hadassah Ein Kerem hospital in Jerusalem; they had been beaten and stabbed to death.

Hamas, which since its foundation in 1988 had been committed to the elimination of Israel, took the lead in seeking to undermine Fatah and to sabotage the Road Map. On June 8 Abdel Aziz al-Rantisi, one of the Hamas leaders, organized an attack that killed four Israeli soldiers at the Erez check- point at the entrance to the Gaza Strip. On June 10, in a failed attempt to assassinate Rantisi, Israeli helicopters fired missiles at a car in Gaza. Two Palestinians were killed. On the following day, a Palestinian suicide bomber killed seventeen passengers and bystanders on an Israeli bus. In response, Israel continued targeting Hamas leaders with helicopter attacks.

The anguish of conflict was briefly but profoundly eclipsed for the Israeli public on 1 February 2003, when Israel's only astronaut, Ilan Ramon, was killed with the six other members of the *Columbia* space flight crew during re-entry into the earth's atmosphere. The destruction of the spacecraft took place over Texas, sixteen minutes before it was due to touch down. It had been sixteen days in orbit.

Ramon was already a hero for young Israelis. As a fighter pilot he had been decorated for bravery in both the 1973 and 1982 wars. In 1981 he had taken part in the successful Israeli attack on the unfinished Iraqi nuclear reactor. Before going to Texas in 1998 to start training as an astronaut, he had been the Colonel in charge of Weapons Development and Acquisition for the Israeli Air Force. He had already spent fifteen days in space before setting off, on January 16, on the fatal flight.

Ramon's mother and grandmother were survivors of Auschwitz. A secular Jew, Ramon sought to follow Jewish observances while in orbit. 'I feel I am representing all Jews and all Israelis,' he said in an interview before the flight. He was the first astronaut to ask for kosher food. He also obtained rabbinic opinions about observing the Jewish Sabbath while in space, since the period between sunrises in orbit is approximately ninety minutes, so that six 'days' on earth pass in nine hours while in space.

On board the spacecraft, Ramon carried a pencil sketch, entitled 'Moon Landscape', which had been drawn in the Theresiensadt Ghetto by fourteen-year-old Petr Ginz, a Czech Jewish boy, who later died in Auschwitz.

On 29 June 2003, as part of the first phase of the Road Map, Islamic Jihad and Hamas announced a joint three-month ceasefire, and Arafat's Fatah movement declared a six-month truce. The maintenance of the truce was made conditional on the release of prisoners from Israeli prisons, a demand that was not part of the Road Map process. At the same time, Israel withdrew troops from the northern Gaza Strip and began drawing up plans for the transfer of territory to Palestinian control. This further progress took place during a visit to the Middle East by the United States National Security Adviser, Condoleezza Rice.

On July 1, in Jerusalem, Ariel Sharon and Mahmoud Abbas held a ceremonial opening to Road Map peace talks, televised live in both Arabic and Hebrew. The two Prime Ministers declared that the violence had gone on too long, and stressed that they were committed to the Road Map. Further action was swift. On July 2 Israeli troops pulled out of Bethlehem and transferred control of the city to the Palestinian Authority security forces. That week, the United States announced a $30 million aid package to the Palestinian Authority, to help rebuild infrastructure destroyed by Israeli incursions. Despite such progress, however, both the random killing of Israeli civilians by Palestinian terrorists, and the targeted killing by the Israeli forces of Palestinian organizers of terror, continued. In 2003, 182 Israelis were killed in suicide bomb attacks. During that same year, twenty-two Palestinians were killed in targeted assassinations. More than twice that number of Palestinian bystanders, forty-nine, were killed, including several children.

Israel had become a country under attack from an enemy that could not be fought on the battlefield. That enemy used individuals who were persuaded to kill others by killing themselves. The Israeli Government decided to strike at the most senior organizers. On 22 March 2004 an Israeli helicopter gunship fired missiles at the spiritual leader of Hamas, Sheikh Ahmed Yassin, a paraplegic who was being wheeled out of an early morning prayer meeting. Yassin and his two bodyguards were killed, together with nine bystanders.

The killing of Sheikh Yassin created a storm of outrage worldwide. The United Nations Secretary-General, Kofi Annan, condemned the killing and called on Israel to halt its policy of targeted assassination. The British Foreign Secretary, Jack Straw, was outspoken in his condemnation. 'All of us understand Israel's need to protect itself,' he said, 'and it is fully entitled to do that against the terrorism which affects it, within international law. But it is not entitled to go in for this kind of unlawful killing and we condemn it. It is unacceptable, it is unjustified, and it is very unlikely to achieve its objectives.' Those Israelis who felt as Jack Straw did were in a minority. With the carnage of each suicide bomb, that minority became even smaller, and the Israeli Government even more determined to strike at the organizers of terror.

On 17 April 2004 Abdel Aziz al-Rantisi, who had replaced Sheikh Yassin as the Hamas leader in the Gaza Strip, was assassinated. The Palestinian response was an intensification of attacks on Israeli civilians. In a sixteen-day period in mid-September there were eighty-two violent deaths and several close confrontations. On September 14 two would-be female suicide bombers surrendered to Israeli troops at a checkpoint near Nablus. They had been instructed to blow themselves up in Tel Aviv. A few hours before their surrender, Israeli soldiers, acting on intelligence information, had killed the man who had planned their mission. On the following day a fifteen-year-old would-be suicide bomber was taken to an Israeli checkpoint by his grandfather and made to give himself up: the grandfather feared that the family home in Nablus would be blown up as a reprisal for his grandson's suicide attack.

On September 24 an Israeli woman was killed at Neve Dekalim, an Israeli settlement in the Gaza Strip, when a Palestinian mortar shell exploded in her living room. In retaliation, Israeli forces destroyed twenty Palestinian homes in the Gaza Strip town of Khan Yunis; one Palestinian was killed during the army incursion. Following the firing of rockets from the Gaza Strip into the nearby Israeli town of Sderot on September 29, when a two-year-old girl and a four-year-old boy were killed, thirty-four Palestinian civilians were killed in an Israeli reprisal air strike.

Also on September 29, in the Gaza Strip settlement of Alei Sinai, an Israeli woman jogger was shot dead. An Israeli army doctor who reached the scene was also killed. Two of the attackers then killed themselves when one of them detonated the explosive belt he was wearing.

Within the Palestinian community, on September 14, a suspected

Palestinian collaborator was kidnapped in Ramallah by fellow-Palestinians and shot dead, the sixth such execution in Ramallah in eighteen months. On September 20 two suspected Palestinian collaborators were executed by Palestinian gunmen in the West Bank town of Tulkarm. On the Israeli side, September 28 saw the indictment of five Israeli border police accused of abusing Palestinians at a road block in Abu Dis, a Palestinian suburb just east of Jerusalem. Such abuse, which was believed to be widespread, was creating bitterness among the Palestinians. The Israeli judicial system did not condone this abuse; the Israeli judge was emphatic in denouncing the acts committed by the border police as 'dreadful deeds that descend to the lowest level of interpersonal behaviour'.

These were harsh times, creating fear, anger and frustration throughout Israeli and Palestinian society. On 11 November 2004 Arafat died in hospital in Paris. He had been ill for some time. He was succeeded as President of the Palestinian Authority by Mahmoud Abbas, who had resigned a year earlier as Palestinian Prime Minister, claiming lack of support from Israel and the United States with regard to the Road Map, as well as 'internal incitement' against his government by fellow-Palestinians. Abbas promised to do everything possible to end the lawlessness created by the rivalry between Fatah and Hamas, and among the several clan armies and competing security services, especially in the Gaza Strip.

Even as Abbas sought to end Palestinian near-anarchy, Israeli politics were in turmoil. In December 2004, after protracted secret negotiations, Sharon formed a coalition with the Labour Party in order to implement his plan for Palestinian–Israeli conciliation and amelioration. It seemed an uphill task. In the four years between September 2000 and September 2004 more than a thousand Israelis had been killed, and twice that number severely injured. During that same four-year period 2,736 Palestinians had been killed. The Palestinian deaths included 466 members of Hamas, 408 members of a Fatah military group, the Tanzim (the Organization) and 334 members of Force 17, part of the Palestinian Authority police force. These 1,208 Palestinians were killed while carrying out or planning acts of terror. The remaining 1,528 Palestinians who were killed were bystanders, five hundred of them under the age of eighteen and several of them babies.

Most of the suicide bombers and targeted killers had come from the West Bank. In order to prevent suicide bombers crossing into Israel, in September 2002 the Israeli Government had decided on a drastic step. It would build a substantial barrier, part barbed-wire fence, part concrete wall, to divide the Jewish and Arab areas of the West Bank, and to cut off the West Bank from pre-1967 Israel. The structure, which would take more than five years to complete, was intended to be a formidable deterrent to terror. Covering 480 miles (720 kilometres), where it was in the form of a wall it would be between 20 and 26 feet (6 to 8 metres) high – as tall as a two-storey house; where it was a barbed-wire fence, it would have electronic monitors, and deep trenches dug on either side of it. To build the wall and the fence – known variously as the Separation Barrier, the Wall, the Fence and the Anti-Terrorist Fence – Israel expropriated seven thousand acres of Palestinian land, one-half of 1 per cent of the land of the West Bank.

Some 263,000 Palestinians lived in enclaves entirely surrounded by the Wall. Where the Wall went through the eastern suburbs of Jerusalem, it left 210,000 Palestinians living on the Israeli side of it. In all, 402,400 Palestinians found themselves living west of the Wall, often cut off from their schools and health clinics and fields and farms. There were few gates in the Wall, and long waits at them. Most Palestinians had to make enormous detours to get from one side to the other. Checkpoints were intensified at and near each crossing.

The Wall not only scarred the landscape, it constituted a formidable barrier. A leading Israeli Labour parliamentarian, Haim Ramon, called on the government to build the wall along the 1967 border – the Green Line. The government refused. For Sharon, and for a majority of the Israeli population, the Wall was an essential protection. Between September 2000 and the completion of the northern and most of the Jerusalem sectors of the Wall, 431 Israelis had been killed in 137 suicide bombings, one hundred of them in March 2002 alone. With the construction of the Wall, the number of suicide attacks decreased dramatically. After the completion of the Wall in the north there was not a single terrorist attack across that section. By December 2004, the number of suicide attacks initiated from the West Bank had fallen by 84 per cent in less than two years.

For Palestinians, the route of the Wall was as distressing as the Wall itself. Following an application from several Israeli human rights organizations, on 30 June 2004 the Israeli High Court of Justice, Israel's Supreme Court, ruled that the Wall north-west of Jerusalem should be rerouted to 'minimize hard-ship to the Palestinians'. This ruling obliged the Israeli Government to make substantial revisions in its proposed route in order to prevent the Wall from placing a broad swathe of Palestinian homes and land on the Israeli side, or cutting off Palestinian villagers from their fields and schools.

Three of Ariel Sharon's Ministers urged him to initiate legislation that would circumvent the High Court's ruling. He refused to do this. Three weeks earlier, on 6 June 2004, he had persuaded his Cabinet to agree to a dramatic departure in Israeli policy: the withdrawal of all Israeli settlements in the Gaza Strip, and of four settlements in the northern West Bank, whose land circumscribed the growth of several neighbouring Palestinian villages.

The proposed Gaza withdrawal meant an end to thirty-seven years of Jewish settlement there, during which small but flourishing communities had been built up, many situated on the sea shore; eventually taking up almost a fifth of the land area of Gaza, their inhabitants together numbered 8,800, surrounded by more than a million Palestinians. Their protection against repeated Palestinian attacks had involved the use of Israeli army units, security zones and special roads. The Cabinet, while agreeing to financial compensation of between $200,000 and $350,000 for each family that left its Gaza Strip home, set penalties – including imprisonment – for settlers who resisted evacuation.

The evacuation was planned for the end of 2005. Before it could be carried out, there was an intensification of Palestinian terrorist acts and of Israeli counter-measures. In one 48-hour period, 6 and 7 October 2004, forty-six people were killed. These included twenty-three Israelis and two Israeli Arabs killed by a truck bomb in the Egyptian resort town of Taba, just south of the Israeli resort town of Eilat; a Palestinian greenhouse worker in an Israeli settlement in the Gaza Strip killed by a Palestinian sniper; a fifty-one-year-old Palestinian kidnapped from hospital and executed for having sold land to Jews on the West Bank; and two Palestinian boys in the Gaza Strip, Raid Abu Zaid, aged thirteen, and Suliman Abu Ful, aged fourteen, killed by a missile fired from an Israeli helicopter after an unmanned Israeli drone aircraft had spotted them, with two others, preparing to fire a Kassam rocket into Israel. That same week, a sixteen-year-old Palestinian girl was arrested on the West Bank on suspicion of having agreed to carry out a suicide bomb attack.

In parallel with this ongoing violence, continuous efforts were being made to bring the Israelis and Palestinians closer together. On 3 February 2005, meeting at the Egyptian port of Sharm el-Sheikh, Abbas and Sharon agreed to a ceasefire between their soldiers and militias. On March 13, despite a suicide bomb two and a half weeks earlier that had killed four Israelis in Tel Aviv, the Israeli Cabinet approved the dismantling of unauthorized settler outposts in the West Bank. Three days later, Israeli forces withdrew from Jericho and on March 23 from Tulkarm. At the same time, a gate in the Separation Barrier near Tulkarm was unlocked, allowing movement of Palestinians from Tulkarm into the West Bank. On April 13, sixteen Israeli and twenty Palestinian mayors and heads of local town councils met in Jericho to discuss a one-year truce. The Mayor of the Israeli town of Sderot, Eli Moyal, told the assembled mayors: 'I would like to convey my apologies for all people who were hurt in this terrible cycle of bloodshed. On behalf of millions of Israelis, I would like to convey my deep regret over what has happened to the Palestinian people.'

The search for reconciliation continued. In January 2007 five Palestinian and five Israeli youngsters, who had just participated in a Seeking Common Ground programme in New York, volunteered in Israel at both a Jewish school and an Arab orphanage. They also started a parents' group for their friends and families to get to know each other. One of the youngsters, seventeen-year-old Saleh Alzjary, from the Arab suburb of Beit Safafa, in Jerusalem, visited the school of his Jewish travel companion. 'When I came into the room,' he said, 'they were totally in shock. It was a challenge to be one of your kind in the middle of a group, which, maybe, hated you. But it was so good, and it benefited me and them to know things they didn't know before.' Commented the Israeli student, Avi Gordis, 'I was afraid of every Palestinian I saw.' Until the New York programme, he had never held a conversation with a Palestinian.

Active in Israel since 1979, the New Israel Fund provided financial and technical support to help community organizations in Israel, in both the Jewish and Arab sectors, to safeguard civil and human rights, promote religious tolerance and pluralism, and close the social and economic gaps in Jewish and Israeli Arab society. Another joint venture, Project COPE (named because its members sought means of 'coping' with their illness), worked to improve support for Palestinian and Israeli breast cancer patients, creating the opportunity for amicable, medically helpful exchanges between Palestinian and Israeli women. Over three years beginning in 2002, COPE applied for ninety permits to enable Palestinian women to attend meetings in Israel. Eighty of these permits were approved by the Israeli authorities, although fifteen could not be used because they were issued too close to the meeting. Eight were not respected at a checkpoint. Several more could not be used because checkpoints were unexpectedly closed due to a deterioration in the security situation.

Both Israelis and Palestinians benefited from the work of COPE. A Palestinian woman remarked: 'At the first meeting I was frightened because it was between Israelis and Arabs, but after fifteen minutes, I realized this disease makes no distinction.' An Israeli woman commented: 'At the first meeting I was very excited, because I had hated Arabs in the past, and here I found myself simply wanting to help, because they didn't have the same access to prostheses, wigs, medicines. I felt compassion.'

Another group seeking to bridge the gulf between Israelis and Palestinians is the Peres Center for Peace. Established in Tel Aviv in 1996, and named after Shimon Peres, its participants believe, in the words of their charter, that 'meaningful peace is only possible between peoples with direct and personal knowledge of each other. One of the greatest barriers to peace in today's environment is the negative images and stereotypes that abound in the region. It is our role to create activities that help dispel these myths, while simultaneously addressing real needs within the community, whether social or economic.'

The Peres Center for Peace initiated the Twinned Peace Sports Schools, a scheme whereby young Palestinians and Israelis from the schools concerned play and compete in soccer tournaments throughout Israel. It was actively involved in supporting Palestinian farm cooperatives in Gaza in high-value, labour-intensive crop production for export. The crops included straw-berries, tomatoes, peppers, sweet potatoes and flowers. Exports went to Britain, the Netherlands, Germany, Belgium, France and several countries of the former Soviet Union.

Joint projects brought Israelis and Palestinians together on a daily basis. Muslim Palestinian and Jewish Israeli schoolgirls participated in joint basketball tournaments. Under the Saving Children Project, Palestinian physicians trained in Israeli hospitals to become independent and to provide parallel medical services in their own communities. Palestinian and Israeli fashion designers launched a 'Peace Women' collection, combining modern Israeli design and traditional Palestinian embroidery. Starting in 2001, the Viewpoints Theatre Project involved a mixed Palestinian–Israeli cast, performing to Palestinian and Israeli teenage audiences. By December 2004 it had played to more than ten thousand youngsters.

A three-year Peace in Sight programme, initiated in 2003, fostered co-operation in the field of ophthalmology between St John's Hospital in Arab East Jerusalem and the Hadassah Hospital in Jewish West Jerusalem. At the Hadassah Hospital, Peace in Sight provided, in one year, more than three hundred clinical consultations for Palestinian patients. At both St John's and Hadassah hospitals, Palestinian babies and young children were brought from the West Bank and Gaza Strip for operations and medical procedures not available in the hospitals there. In November 2004, in Florence, Israeli and Palestinian paediatricians discussed innovations in paediatric medicine.

At Beit Jala, a Palestinian Authority town just south of Jerusalem, Palestinian and Israeli university students met to discuss their different historical narratives. The main focus of discussion was a single event that took place in 1948, known to the Israelis as the War of Independence, associated with victory and statebuilding, and to the Palestinians as the Catastrophe – the Nakhba – associated with the humiliation and pain of defeat, flight, expulsion and expropriation. These discussions were continued in October 2004 in Salzburg.

Much work was being done in the first decade of the twenty-first century to bring Israelis and Palestinians together in the economic sphere. During three months at the start of 2004 the Aix Group, made up of Palestinian, Israeli and international economists, met in Paris to discuss future economic cooperation. At a press conference that year in Arab East Jerusalem the Peres Center launched an Economic Road Map, under which Israeli and Palestinian businesspeople meet regularly to discuss banking, tourism, shipping and international transportation, essential elements in sustaining and enhancing both the Palestinian economy – with its near-catastrophic rates of unemployment – and joint Israeli–Palestinian ventures.

Israeli armed incursions into the West Bank and Gaza Strip in search of Palestinians organizing acts of terror often acted as setbacks to these economic initiatives. Roadblocks cut off factories from their supplies and prevented Palestinian workers from reaching their places of work. Curfews imposed by Israel on Palestinian towns, with the aim of making the hunt for wanted men easier, hindered economic activity, as well as creating general bitterness.

At the centre of Sharon's coalition projects was the withdrawal of all Israeli settlements in the Gaza Strip and, with the withdrawal, an end to Israel's military presence there. It was a plan that many members of Likud, including his Finance Minister, Benjamin Netanyahu, opposed. Sharon was determined that Israel's disengagement from the Gaza Strip should go ahead. He was helped by an informal ceasefire adopted by Hamas in January 2005, whereby nine months followed without a single suicide bomb attack against Israel. But opposition within the Likud party to the Gaza withdrawal remained strong. On 7 August 2005 Netanyahu resigned as Finance Minister, insisting that unilateral withdrawal 'endangered the safety of Israeli citizens'. Subsequent frequent rocket attacks from the Gaza Strip into Israel, in breach of the Hamas ceasefire, were taken by Netanyahu and those who thought as he did as proof that he had been right, even though Israel struck forcefully against the rocket launch sites and the perpetrators of the attacks, killing many dozens.

The evacuation of Israeli settlers – most of them farmers – from Gaza was completed on 12 September 2005. The evacuation of the four settlements in the northern West Bank was completed ten days later. Sharon let it be known that he intended as soon as possible to remove the bulk of the settlements on the West Bank. If the Palestinian leadership was unwilling to be the negotiating partner, Sharon was prepared, as with Gaza, to carry out a unilateral disengagement.

A month after Israel had evacuated all Jewish settlements in the Gaza Strip, the President of Iran, Mahmoud Ahmadinejad, declared that he wanted to see 'the Zionist regime wiped off the map'. This threat, uttered by one member of the United Nations about another, was unprecedented. At that very moment, Iran was confronting the international community with its determination to create nuclear power, and with it the possibility of a nuclear bomb. The distance from the western border of Iran to Israel is less than six hundred miles. Readily available rockets could cover that distance with ease.

Israel took immediate, and at the time secret, defensive action, beginning preparations for a satellite, Techstar, which could be launched from India, with whose government Israel concluded successful negotiations. Techstar had the ability to track Iran's nuclear compounds, as well as Syria's rocket bases, twenty-four hours a day, and to send back the clearest pictures available. These precautions were, and remained, Israel's national defence.

In Damascus Khaled Mashaal, the political leader of Hamas, whom Israel had tried earlier to assassinate when he was in Jordan, called Ahmadinejad's remarks 'courageous'. But Saed Erekat, the chief Palestinian negotiator and member of the Palestinian Legislative Council, spoke out against the Iranian President. 'Palestinians', he said, 'recognize the right of the State of Israel to exist and I reject his comments. What we need to be talking about is adding the State of Palestine to the map, and not wiping Israel from the map.'

This path, the very prospect of a Palestinian State, was deeply disliked by many members of the Likud Party, among whom Netanyahu emerged as the leader of an anti-withdrawal faction. He and many Likud members rejected Sharon's policies of unilateral disengagement – in the absence of a Palestinian negotiating partner – which involved the removal of a large proportion of Jewish settlements from Palestinian territory, and the fixing of Israel's borders with a prospective Palestinian State. Facing up to this challenge from among his own Likud colleagues and co-workers, on 21 November 2005 Sharon resigned as head of Likud, the Party he had helped to establish more than thirty years earlier. That same day he announced the formation of a new political Party, Kadima ('Forward').

On November 28 Kadima announced a 'national agenda'. This constituted a complete break with Likud thinking, and offered a way forward to a peaceful resolution of the Palestinian–Israeli conflict. Its basic tenet was: 'The Jewish majority in Israel will be preserved by territorial concessions to the Palestinians.' Kadima's manifesto was emphatic. 'The Israeli national agenda to end the Israeli–Palestinian conflict and achieve two States for two nations will be the Road Map,' it declared. 'It will be carried out in stages: dismantling terror organizations, collecting firearms, implementing security reforms in the Palestinian Authority, and preventing incitement. At the end of the process, a demilitarized Palestinian State devoid of terror will be established.'

Many Likud and Labour politicians joined Kadima. From Likud came Ehud Olmert, former Mayor of Jerusalem. From Labour came Shimon Peres, a member of the Labour Party for more than sixty years, and its leader until three weeks earlier, when he had been defeated by a senior trade-union leader, Amir Peretz. Eager to assert an independent Labour stance, Peretz called on all Labour Party members of Sharon's coalition to resign. This they did, undermining Sharon's majority and forcing him to call an election. This suited Sharon well. As Prime Minister he was determined to lead Kadima and the nation towards a peace agreement. The election was called for March. But on 4 January 2006, less than forty-four days after launching his dramatic initiative, Sharon suffered a massive stroke and fell into a coma.

CHAPTER THIRTY-TWO

'Permit me to be a dreamer'

As a result of Ariel Sharon's stroke, Ehud Olmert was appointed Acting Prime Minister, until such time as Sharon should recover. Olmert took immediate steps to support what Sharon had begun, telling the annual Herzliya Security Conference on 25 January 2006 that Israel would have to give back the Palestinian-inhabited areas of the West Bank. His words were emphatic. 'In order to ensure the existence of a Jewish national homeland, we will not be able to continue ruling over the territories in which the majority of the Palestinian population lives. We must create a clear boundary as soon as possible, one which will reflect the demographic reality on the ground.' Israel would retain control over the existing security zones, the main settlement blocs and Jerusalem, but the only solution for the national aspirations of the two peoples was two independent States. As for the Palestinian refugees, they would have to be absorbed into the Palestinian State. The key was to pursue the Road Map, which was 'a simple and just idea'.

Olmert then made a radical declaration. 'If the Palestinians abandon the path of terror and stop their war against the citizens of Israel,' he said, 'they can receive national independence in a Palestinian State, with temporary borders, even before all the complicated issues connected to a final agreement are resolved. All these issues will be resolved later, during negotiations between the two countries.' This offer to give the Palestinians statehood even before the final status issues had been resolved meant that the Palestinians would come to the negotiating table as a sovereign entity.

In giving up large parts of the West Bank, Olmert explained, Israel would have to make 'painful concessions'. As a first step, on 1 February 2006 he ordered the demolition of the six houses built – alongside several caravans – on the Israeli hilltop outpost of Amona. Hundreds of Israeli troops and policemen ensured that the order was carried out. But Hamas had set its heart against any acceptance whatsoever of the 'Zionist entity', and was determined to maintain the armed struggle, not only against Israel but against all Fatah attempts at a two-State solution: the statehood that had been on offer in 1937 and again in 1947, and had twice been rejected.

On January 25, the day of Olmert's forward-looking speech at Herzliya, elections were held throughout the Palestinian Authority. Hamas emerged as the winner, securing 76 of the 132 seats in the Legislative Council. Fatah won 43. The seats won by Hamas in East Jerusalem and Bethlehem included all but the Christian Arab seats, even though Israel had prohibited Hamas from campaigning in those two cities. Those committed to Holy War with Israel, and opposed to any negotiated peace settlement, had been democratically elected to rule the West Bank and Gaza Strip.

On February 16 Mahmoud Abbas, as President of the Palestinian Authority, appointed the Hamas leader Ismail Haniyeh as Prime Minister. Hamas made it clear that it would continue to refuse to recognize Israel and that it rejected negotiations with Israel: it wanted nothing to do with the Palestinian efforts of past years in working towards a Road Map, or towards a Final Status Agreement with the object of a Palestinian State existing alongside Israel, at peace, within secure borders.

Alarmed at the triumph of fundamentalism and rejectionism, the Quartet – the United States, the European Union, Russia and the United Nations – set out three conditions for continuing to support the Palestinian economy, and for any negotiations with the new government: it must recognize Israel; it must give up the armed struggle; and it must recognize previously signed agreements, including the Oslo Agreements. The Hamas government refused to accept these conditions. Western aid was cut off and Israel halted the transfer of all tax revenues it had collected on behalf of the Palestinian Authority.

In Gaza, fighting broke out between Hamas militants, loyal to Haniyeh, and the Palestinian Presidential Guard, loyal to Abbas. At least twenty-five Palestinians were killed in this internecine struggle. With some difficulty, Abbas achieved a compromise whereby Hamas agreed to a joint Fatah–Hamas 'unity' government that would accept the three conditions, even if Hamas would not. Israel's Foreign Minister, Tzipi Livni, found this compromise hypocritical, capable only of leading to 'further stagnation'. The European Union, Britain and the United States continued to withhold funds, except for emergency humanitarian needs.

On 28 March 2006, with Sharon still in a coma, elections were held for the Seventeenth Knesset. Kadima, which had not previously fought an election, won 29 seats, making it the largest single Party, as against 19 seats for Labour, its nearest rival. Even together they did not hold a majority in the 120-member Knesset. Voter turnout was the lowest ever in an Israeli national election, with only 63.6 per cent of eligible voters casting their ballots. As Acting Prime Minister, Olmert formed a coalition government with his Party's chief electoral rival, the Labour Party, whose leader, Amir Peretz, became Minister of Defence.

On April 14 it was announced that Sharon would never recover. Olmert became Prime Minister, at the head of a predominantly Kadima administration, with Shimon Peres – Labour's former stalwart and elder statesman – as his Foreign Minister. Also in the coalition, to make it numerically less vulnerable, were representatives of the Shas religious party, with its 12 seats, and of Gil – the newly formed and unexpectedly successful Pensioners' Party – that had received 7 seats. Likud, headed by Benjamin Netanyahu and reduced to 12 seats, a mere one-tenth of the Knesset membership, was not invited to join.

Crisis soon confronted Olmert's administration: not from within but from the outside. On May 25, Mahmoud Abbas gave Hamas a ten-day deadline to accept the pre-1967 borders of Israel. Hamas contemptuously rejected this. Abbas, as President, consolidated his position on the West Bank, while trying in vain to regain some measure of control in the Gaza Strip – where Hamas militia were by then fighting in the streets to gain mastery. On June 25, twenty-four hours after Israeli troops had crossed into Gaza and seized two Hamas militants, Hamas gunmen, crossing into southern Israel from the Gaza Strip, killed two Israeli soldiers and kidnapped nineteen-year-old Corporal Gilad Shalit. He was the first Israeli soldier to be abducted since Corporal Wachsman had been kidnapped and killed in 1994.

Shalit's kidnappers were Izz ad-Din al-Kassam, the military wing of Hamas. As a condition of his release, they demanded that Israel release all female Palestinian prisoners, and all Palestinian prisoners under the age of eighteen: several hundred in all. Israel rejected this demand, and on June 28 Israeli forces crossed into the Gaza Strip, entering Khan Yunis. After Abbas failed to persuade Hamas to release Shalit, Israeli air force jets attacked a Hamas training camp on the outskirts of Gaza, while Israeli troops and tanks took up positions inside the Gaza Strip. On July 2 a helicopter-launched missile hit Ismail Haniyeh's office in Gaza City. Three days later, Hamas fired Kassam rockets into Israel, hitting Ashkelon, where eight residents were treated for shock. On the following day, Israeli troops entered northern Gaza, creating a buffer zone between Gaza and southern Israel. One Israeli soldier and twenty Palestinian fighters were killed. Mahmoud Abbas sent two Palestinian emissaries to Syria, to urge the Hamas leader Khaled Mashaal to order the release of Corporal Shalit. The emissaries were rebuffed.

The Kassam rockets – free-flying, with a small explosive charge but no guidance system – continued to fall on southern Israel, mainly on the town of Sderot. In the Gaza Strip, thirty-eight more armed Palestinians were killed by Israeli troops in the second week of July. Then, on July 11, Hizballah forces in southern Lebanon, under the command of the Iranian-backed Sheikh Hassan Nasrullah, began the intensive firing of Katyusha rockets – inexpensive, easy to produce, with low accuracy and a high explosive charge – across Israel's northern border. Israel responded on the following day by imposing a naval blockade on Lebanon, bombing Beirut International airport, striking at Hizballah missile launch sites, and sending tanks and troops into southern Lebanon. On July 12 a small Hizballah force, crossing into northern Israel, kidnapped two Israeli soldiers, Ohad Goldwasser and Elad Regev. While attempting to rescue the kidnapped men, four Israeli soldiers were killed when their tank was blown up as it drove over a large explosive device.

Within forty-eight hours, eight Israeli soldiers had been killed in southern Lebanon. Rockets continued to be fired from inside Lebanon, hitting Haifa and Nahariya and killing two Israeli civilians. Another Israeli civilian was killed that day in Safed and twenty wounded. Israel intensified its air strikes on Lebanon. In a southern suburb of Beirut, Hizballah's main television studios were hit. In an air strike on a civic centre attached to a Shiite mosque in central Lebanon, seventeen people – two families – were killed. In Israel, eight civilians were killed when a rocket fired from southern Lebanon hit the Haifa railway depot. Hundreds of thousands of Israelis spent their days and nights in deep shelters. Tens of thousands sought safety further south.

On July 17 Israeli aircraft located and destroyed a rocket launcher holding an Iranian-made Zelzal missile, capable of reaching Tel Aviv, a hundred miles south of the Lebanese border. On the battlefield, hundreds of Hizballah fighters were killed. But the Katyusha rockets continued to fall in northern Israel. More than a hundred launch sites were hit, as well as Hizballah camps, radar stations and road bridges. Hundreds of Lebanese civilians were killed in these air attacks. The Israeli Defence Minister, Amir Peretz, the leader of the Labour Party, directed the Israeli forces not to be deterred by Hizballah's use of 'human shields' – Lebanese civilians placed by Hizballah at their rocket launch sites.

On July 19 two Israeli Arab boys, nine-year-old Rabiya Taluzi and his three-year-old brother Muhammad, were killed when a Hizballah Katyusha rocket hit the courtyard of their home in the Israeli Arab town of Nazareth. Also that day two Israeli soldiers were killed in southern Lebanon. In Gaza, where the Israeli army was still in action against armed Hamas fighters, twelve Palestinians were killed. In three weeks Israel had fired a thousand artillery shells into the Gaza Strip and carried out 168 bombing raids. More than a hundred Palestinians were killed, the vast majority of them armed Hamas fighters.

Israel's response to the rocket attacks from Lebanon had attracted international criticism. The United Nations Secretary-General, Kofi Annan, denounced Israel's response as 'disproportionate'. Annan called for a cease-fire, but Israel, with the rockets still falling, rejected this. In Nablus, four thousand Palestinians demonstrated in the streets in support of Hizballah, calling on its leader: 'Nasrallah, our dearest, strike, strike Tel Aviv.'

Several hundred of Sheikh Nasrallah's troops had been killed, but he continued to order the rockets to be fired. On July 25 a sixteen-year-old Palestinian Arab girl, Doa Abbas, was killed when a rocket hit her home in the Israeli Arab village of Mughar, near Tiberias. On the following day, in a fierce struggle for the Hizballah stronghold of Bint Jbail, eight Israeli soldiers and forty Hizballah gunmen were killed. That day in the Gaza Strip twenty-three Palestinians were killed, including five civilians and three young girls.

On July 27 the Chief of the Israeli General Staff, Lieutenant-General Dan Halutz, asked the Cabinet for permission to send extra troops into southern Lebanon 'to sweep through the Hizballah strongholds', including several hundred bunkers near the Israeli border that were still intact. The Cabinet rejected his request. The main assault weapon continued to be the air force. On July 30, in an Israeli air strike on a building in the southern Lebanese village of Kafr Kana, from which rockets had earlier been fired into Israel, fifty-seven Lebanese – old people, women and twenty-seven children – were killed. They had taken refuge in the building to be safe from Israeli air attacks.

The deaths at Kafr Kana led to intensified international condemnation of Israel and calls for a ceasefire. Ten years earlier, Israel had been forced to suspend its operation against Hizballah in southern Lebanon after Israeli artillery shells killed more than a hundred civilians seeking refuge in a United Nations building in this same village, Kafr Kana. But as Hizballah continued to fire its rockets into northern Israel, the ground and air war continued. On August 3 eight Israeli civilians, most of them in Acre, were killed by rockets, the largest Israeli death toll in a single day. On the following day six civilians were killed, all of them Israeli Arabs. On August 6, twelve Israeli army reservists were killed when a rocket exploded at their unit's staging point in northern Israel.

Prime Minister Olmert still held back against authorizing a more sub-stantial military force. On August 9, a day on which 190 rockets fell on northern Israel, fifteen Israeli soldiers were killed in southern Lebanon, the highest number in a single day since the start of the operation. On August 12, after Olmert had approved a massive troop advance to the Litani River, twelve miles inside Lebanon, twenty-four Israeli soldiers were killed. Ten more were killed in the following two days, the last on August 14, when – under Security Council Resolution 1701 – a United Nations brokered cease-fire came into effect.

Within a week of the ceasefire Israeli troops had withdrawn from southern Lebanon, and Hizballah was being disarmed by the Lebanese army, which had hitherto treated southern Lebanon as a no-go zone. An international force, to consist of 15,000 men, then began to take up positions on the Lebanese side of the border. During the thirty-four days of fighting, Hizballah had fired four thousand rockets into Israel, forcing a million residents into shelters. One hundred and fifty-eight Israelis had been killed – 117 soldiers and 41 civilians – along with 270 Hizballah fighters.

The highest casualties were among Lebanese civilians, more than a thousand of whom were killed in Israeli air strikes on Hizballah positions. In the months following the war several dozen more Lebanese civilians were killed by the remnants of the 1,800 American-made cluster bombs that Israel had dropped, releasing more than a million bomblets. As a result of Israel's use of cluster bombs, a resolution was put forward in the United States Senate by several leading Democrats, including the two Jewish senators from California, requiring recipients of American-made cluster bombs not to use them in or near civilian centres. The resolution was defeated, on 6 September 2006, by seventy votes to thirty.

In the British Parliament, the Prime Minister, Tony Blair, had been criticized by members of his own Labour Party for refusing to condemn Israel's 'disproportionate' response to the Hizballah rocket attacks. But a year later, Human Rights Watch, which had initially condemned the Israeli reaction, issued a report describing Hizballah's shelling of Israeli civilians as 'indiscriminate' and called it a 'war crime'.

In Israel there were calls for a public inquiry into why Hizballah had been able to resist Israel's attacks – despite suffering serious setbacks – and into the unexpectedly high Israeli military death toll. Olmert at first resisted an inquiry, then conceded, establishing a commission under Judge Eliyahu Winograd. Its report took five months to complete.

On 13 November 2006, amid these calls for an inquiry, President Ahmadinejad of Iran announced: 'Israel is destined for destruction and it will disappear soon.' Along the seven-mile border between the Gaza Strip and the Philadelphi Road – which Israel had evacuated when it left the Gaza Strip – Hamas controlled a series of deep tunnels along which a vast quantity of weapons and munitions was being smuggled each day from Egypt into Gaza.

The Lebanon war intensified problems in the Israeli Arab sector. In December 2006 a group of Israeli Arab academics published a paper entitled *The Future Vision of the Palestinian Arabs in Israel*. In the words of one of its authors, Yousef Jabareen, Director of the Arab Centre for Law and Policy in Nazareth, Israel's Arabs seek 'full equality and inclusion in all sectors of Israeli society', including equal housing and employment opportunities, and 'equal status' for the Arab language, the language of 20 per cent of Israel's population – a population that, in Jabareen's words, is 'among the lowest echelons of nearly every aspect of Israeli society'.

Jabareen's concerns were echoed by Mohammed Saif-Alden Wattad, an Israeli Arab member of the Israeli bar, and former law clerk to an Israeli Supreme Court justice. Wattad warned, in the spring of 2007 while a visiting scholar at the University of Toronto, that Arab municipalities and schools within Israel did not receive their fair share of State funds. He also pressed for the return of land and property expropriated by Israel in the past, or proper compensation.

Efforts were made to bridge the gap between Israeli Jews and Israeli Arabs. In 2004 a group of Jewish and Arab parents had established a bi-national, bilingual school in the mainly Israeli Arab region of the Wadi Ara. Called the Bridge over the Wadi School, it started with fifty Jewish and fifty Arab students. In a film made about the school, an Arab mother explained why she enrolled her daughter there: 'I did not want my daughter to hate Jews as I did,' she said. A Jewish father was more sceptical, telling his interviewer: 'We are training children who will one day come to kill us.' He was referring to a particular concern of a majority of Israelis, namely the strong anti-Jewish as well as anti-Zionist content of many Palestinian school textbooks.

Inside Israel, a succession of scandals rocked society. When the Attorney General announced plans to indict the President, Moshe Katsav, on rape charges, the President suspended himself from office. Investigations were taking place into alleged financial improprieties by Olmert himself, and by Sharon's son Omri. In August 2006 Olmert's Justice Minister, Haim Ramon, resigned after being indicted for allegedly having kissed a female soldier against her will. The Finance Minister, Avraham Hirschson, who was under investigation for financial irregularities, resigned at the end of June 2007.

The rule of law was to be upheld. In January 2007 Amir Peretz, the Defence Minister, denounced the abuse inflicted by a Jewish settler in Hebron on her Palestinian neighbours. Her insulting behaviour had been caught on film. The settler's conduct, Peretz told the annual Herzliya Conference, was 'a disgrace to our moral codes as a State'. Peretz added that although he had earlier ordered 'less strict searches at roadblocks' and the opening of additional terminals at the crossing points, such efforts to improve Israeli–Palestinian cooperation were 'dwarfed' by comparison with the 'footage of a settler who curses and humiliates a Palestinian family'.

Conditions inside Israeli prisons were also being investigated. In the summer of 2006 the Public Defender's Office, in its Annual Report, had revealed that sanitary conditions in many of the twelve prisons it had inspected were so poor as to endanger the health of the prisoners: Israeli criminals convicted in civil courts. 'We are confronted with a national problem,' the report warned, 'which leads to daily violation of the basic rights of the detainees and prisoners, and injury to their human rights.' In eight of the twelve prisons, inmates were held for long hours in 'hard and intolerable conditions'. Prisoners were chained to their beds for hours at a time as a punitive measure. Cells for minors were one square metre in size – with no windows or furniture. The Israeli newspapers gave prominence to the report. The Israeli police and the Prison Service promised that the abuses would be corrected. Decent human behaviour would be upheld.

It was, however, turmoil inside the Palestinian Authority that dominated Israeli concerns. Israelis watched with alarm as the power of Hamas increased in the Gaza Strip, and in certain towns in the West Bank. In the autumn of 2006 the Interior Minister of the Palestinian Authority, Said Siam, a member of Hamas, had established an Executive Force of several thousand armed men, to assert the authority of Hamas against Fatah in what was proving an increasingly bloody struggle for control. In November, a Hamas suicide bomber blew herself up in an unsuccessful attempt to kill Israeli troops in the northern Gaza town of Beit Hanoun. Aged fifty-seven, she was honoured in Gaza for her status as the oldest female suicide bomber.

The spectre of fundamentalism created fear not only among all Israelis, but among hundreds of thousands of moderate Palestinians. On December 15, Ismail Haniyeh survived an assassination attempt by Fatah gunmen, during which one of his bodyguards was killed. A few hours later, at a mass rally in Gaza City, Hamas supporters called for the death of their Fatah rivals. That same day, Fatah gunmen attacked a mosque in Ramallah where Hamas gunmen were praying; thirty-two people were injured. On 5 January 2007 the offices of Hamas legislators in Ramallah were set on fire by Fatah vigilantes. That same day in Gaza a prominent Islamic scholar, Sheikh Adel Nasser, who was affiliated with Fatah, was assassinated by Hamas gunmen shortly after he had preached a sermon appealing for calm and expressing regret over the continuing fighting between Hamas and Fatah.

On the day after Sheikh Nasser was killed, Abbas, as President of the Palestinian Authority, outlawed the Hamas-controlled armed Executive Force in the Gaza Strip. Hamas responded by announcing that the force's personnel would be doubled from 6,000 to 12,000. Fatah denounced the expansion of the force, whose members, it said, 'are operating as death squads'. During that day three Palestinians were killed in the Gaza Strip, as Hamas and Fatah forces clashed yet again. In Nablus, Fatah gunmen kidnapped the Deputy Mayor, Mahdi al-Hanbali, who was affiliated with Hamas. In Ramallah, Fatah gunmen kidnapped Ihab Suleiman, the director of Said Siam's Interior Ministry office. Suleiman was released an hour later, having been shot in both legs.

On 24 April 2007, after further Kassam rockets had been fired at Sderot, Israeli forces killed nine Palestinian militants on the West Bank. Hamas responded by firing more rockets into Sderot, timing the first barrage, on May 14, with the celebrations inside Israel for the fifty-ninth anniversary of statehood. But the main fury of Hamas was reserved for its Fatah rivals. In May 2007 Hamas and Fatah forces clashed again in the Gaza Strip. Fifteen gunmen had been killed on May 12 in an intensification of fighting; on the following day, eleven gunmen were killed.

In one week, the Hamas–Fatah death toll reached forty-one, including five Hamas fighters who were killed when their jeep was ambushed in error by other Hamas fighters. In a Hamas attack on the home of Rashid Abu Shbak, the former head of the Palestinian Preventive Security Service, Shbak survived, but six of his bodyguards were killed. In a Hamas attack on a Fatah police training camp, eight Fatah men were killed – some of them executed at close range after they had been wounded. One Palestinian policeman, fleeing towards the Israeli border, was shot dead by Israeli troops who thought he was part of an attacking force.

On May 16, as a result of the third Egyptian-brokered ceasefire in a week, the Hamas–Fatah fighting died down. That same day, Hamas again fired rockets into Sderot. On May 17 a resident of Tel Aviv, Oshri Oz, was killed in Sderot when a rocket fell near his car. Eight rockets fell on the following day, damaging several buildings. On May 20, in an attempt to destroy the rocket-making facilities, Israel carried out a series of air attacks into northern Gaza. A Hamas military position in southern Gaza was also hit. Eleven Palestinians were killed. On May 21 an Israeli woman was killed when a rocket hit her car in the centre of Sderot. Israeli air attacks later that day killed five Palestinians. The rocket fire continued: more than 250 rockets within a week. On May 22 Ismail Haniyeh announced: 'We will keep to the same path until we win one of two goals, victory or martyrdom.'

Israel's fifth decade ended with many imponderables. One thing was certain: that its Jewish population, having reached 5,309,000 in 2007, exceeded for the first time the Jewish population of the United States. The margin was only 35,000, but that margin was growing by the day.

One source of public anger in Israel was the high number of people killed in road accidents. The figure of 158 Israelis killed in the struggle with Hizballah contrasted with the 414 killed on the roads in 2006. This figure was highlighted by Metuna, the Organization for Road Safety, in a public campaign for greater care on the roads and more active government involvement, drawing attention to the appalling fact that between 2000 and 2006 more than 4,320 Israelis had been killed in road accidents. During that same period, less than a third of that number of Israelis, 1,134 in all, had been killed by Palestinian acts of violence. The roads were the greater danger to life.

The search for Palestinian–Israeli links and bridges was continuous. In October 2007 a leading Israeli soccer team, Hapoel Tel Aviv, launched a scheme for soccer schools in five Palestinian villages near Nablus, whereby 2,500 Palestinian children would be trained by Israeli coaches over the following eighteen months. Hapoel was working with the United Kibbutz Movement in this endeavour. 'We will try to give the children hope of a better future,' commented the kibbutz leader Yoel Marshak. 'Who knows? Maybe this activity will sow seeds of peace where now there is only enmity.'

Ehud Olmert's future had seemed in doubt at the end of April 2007, when the Israeli government commission set up under Judge Eliyahu Winograd after the war against Hizballah in Lebanon concluded that Olmert had 'made up his mind hastily' to launch the air, sea and land campaign the previous July, and accused him of 'a serious failure in exercising judgment, responsibility and prudence'. Olmert's declared aims in going to war, to free the two soldiers seized by Hizballah, and to crush Hizballah itself, were criticized as 'overly ambitious and impossible to achieve'. Paul Michaels, the communications director for the Canada–Israel Committee, commented: 'Only a nation strong and self-confident in its democratic institutions, and capable of self-correction, could have publicly released such a critical report, to the delectation of its sworn enemies.'

The 232-page report also criticized the Defence Minister, Amir Peretz, and the Chief of Staff, Dan Halutz. Peretz stayed at his post. Halutz resigned. Olmert carried on, confident that he could rebuild his popularity and be an effective national leader. He particularly wanted to tackle, as Sharon had done, the rapidly widening gap between Israel's rich and poor – in August 2006 an Israeli government report had revealed that one million, six hundred thousand Israelis, more than one-quarter of the population, were living 'below the poverty line' – and also to negotiate with the Palestinians.

Amir Peretz's performance as Minister of Defence during the war was considered so poor by the public that when early elections were held for the Labour Party leadership, he was defeated by the former Labour Party Chairman and former Prime Minister Ehud Barak. Peretz then resigned from the Defence Ministry, and Barak was invited by Olmert to become Minister of Defence, and accepted.

One outcome of the Winograd Report was the discovery that, in a country proud of its citizen army, some 25 per cent of Israeli men did not want to serve and did not turn up to enlist; 17.5 per cent of those who did enlist left the army during their compulsory military service; and 42 per cent of women of military age did not enlist. The Ministry of Defence announced in the first week of August 2007 that those who did not serve would be considered second-class citizens, and that legislation would be put forward to deprive them of State benefits.

To try to improve the Palestinian economy and create constructive links between Israeli and Palestinian businessmen and women, the Portland Trust, a British-based group with offices in Tel Aviv and Ramallah, helped provide finance for new Palestinian businesses on the West Bank. This included support for construction, to ease the severe housing shortage and to provide employment. The Trust also encouraged agricultural projects, and the land purchase needed for such schemes, as well as the creation of an Israeli–Palestinian Chamber of Commerce. The object of this initiative was to facilitate a prospering Palestinian business environment and an expanding Palestinian economy. It was also intended to serve as a stimulus to normalization, with a view to creating goodwill and positive working partnerships between Israelis and Palestinians.

The Portland Trust also embarked on microcredit loans, both in Galilee – for Jews and Arabs alike – and on the West Bank. 'Getting the partners into the same room, that is the great challenge,' commented Sir Ronald Cohen,

Chairman of the Trust, at a meeting in April 2007. At the same meeting, a Palestinian businessman noted that the economic component 'is to make sure that extremism will not grow'. Working with the Israeli authorities, the Portland Trust makes every effort to help Palestinian goods, including perishable food exports, move as speedily as possible through the crossings and checkpoints.

As Hamas sought full control of Gaza, Fatah fought back. Ismail Haniyeh's home and the Hamas-run Ministry of Culture were both attacked. On June 10, Hamas gunmen killed Mohammed Sweirki, a member of Abbas's Presidential Guard. Later that day Fatah gunmen abducted and killed a Hamas preacher, Mohammed al-Rifati. Fearful that Hamas would impose a rigid Islamic regime in Gaza, drive out all Palestinian Muslim moderates, many of whom were being harassed and even killed, and seek to spread its military power and fundamentalist ideology into the West Bank, on June 14 Abbas dissolved the Hamas-led unity government and declared a State of Emergency. The Hamas takeover in Gaza, he declared, was 'armed insurrection', and its government there 'illegal'.

On June 17, in Ramallah, Abbas swore in a new emergency Palestinian Authority government, with no Hamas members. On the following day, the European Union agreed to unfreeze the money it had withheld from the Palestinian Authority since the day Hamas had formed a government. The United States likewise ended its fifteen-month embargo on the Palestinian Authority and resumed aid. Israel restored the tax revenues it had collected on behalf of the Palestinian Authority.

Abbas appointed a former Palestinian Finance Minister, Salam Fayyad, as Prime Minister. Fayyad, a former World Bank Vice-President, was highly regarded in the West, and by Palestinians, for his integrity. Ismail Haniyeh refused to acknowledge his dismissal, however, and continued to exercise executive authority in the Gaza Strip. On June 19 the Fatah Central Committee, based in Ramallah, cut off all ties and dialogue with Hamas. Israel renewed the talks with the Palestinian Authority that it had suspended when Hamas had come to power.

Slowly, areas of agreement were found between Olmert and Abbas. On 25 June 2007, when they met at the Egyptian resort of Sharm el-Sheikh with the Egyptian President, Hosni Mubarak, and King Abdullah of Jordan, Olmert agreed to release 250 Fatah prisoners from Israeli jails, and to transfer more than $500 million in tax revenue to the Palestinians, provided none of it reached terrorist groups. A new outside initiative was also in the offing. In the aftermath of his resignation as British Prime Minister, Tony Blair was appointed the Middle East envoy of the Quartet. Blair's remit was to help the Palestinians to create a viable administrative and economic structure: the essential basis for statehood. During a meeting of the Quartet in Lisbon on July 19, Blair declared: 'There is a sense that we can regain momentum. That is the crucial thing. If we are able to regain that momentum, then a whole lot of things become possible, not least the fact that those people of peace can then feel that the force is with them, and not with those who want conflict.' This was something, Blair confided, that he felt 'passionate about'.

On July 25 the Jordanian Foreign Minister, Abdel Ilah al-Khatib, and the Egyptian Foreign Minister, Ahmed Abdoul Gheit, came to Jerusalem for talks with Olmert and the newly elected Israeli President, Shimon Peres. Their aim, they said, was to extend 'a hand of peace' by facilitating Israeli–Palestinian negotiations. On August 2 the United States Secretary of State, Condoleezza Rice, expressed her hopes that a new round of talks between Olmert and Abbas could go beyond the 'modest trust-building measures' of the previous few months. At a joint conference with Abbas in Ramallah, she told him that Olmert 'will support new discussions with you' and that he was 'ready to discuss the fundamental issues that will lead to negotiations soon for the creation of a Palestinian State'. Four days later, on August 6, Olmert went to Jericho for discussions with Abbas – the first time an Israeli Prime Minister had negotiated with the Palestinian Authority on Palestinian territory. After the meeting, Olmert stressed that the object of the meeting was 'to move forward towards the establishment of a Palestinian State'.

Economic initiatives made steady progress. A conference of fifty Israeli and fifty Palestinian business leaders set itself the task of establishing joint Palestinian–Israeli enterprises in Palestinian Authority territory and along the effective border between the Palestinian Authority and Israel. Political initiatives were not far behind. On August 6, the day of Olmert's Jericho meeting with Abbas, twelve Israeli soldiers who refused to obey orders that would enable their colleagues to remove some two hundred Jewish settlers who had barricaded themselves into two houses in Hebron were court-martialled, sentenced to between two and four weeks in the stockade, and banned from further combat service. The army wanted it made clear that illegal Jewish settlements would continue to be dismantled. The path to an agreement with the Palestinian Authority was not to be unnecessarily encumbered.

On August 10 the *Jerusalem Post* reported that Olmert was said to have 'promised Palestinian Authority Chairman Mahmoud Abbas to release in the coming days a list of major West Bank roadblocks that will be removed'. At the same time Israel's Defence Minister and the country's most highly decorated soldier, Ehud Barak, emerged as the voice of scepticism. Also on August 10 an Israeli newspaper reported Barak's view that suggestions of a forthcoming peace deal were 'fantasies', and that Israel would not withdraw from the West Bank for at least five years because of the threat of rockets and missiles. Barak also opposed the removal of any of the several hundred checkpoints and barriers across the West Bank.

Two opposing points of view were in conflict. On the one hand, Olmert as Prime Minister, with the full support of his former colleague in Kadima, three times Prime Minister Shimon Peres, who at the age of eighty-three had been elected Israel's ninth President, was seeking a means of reconciliation and constructive dialogue with the Palestinians. On the other hand, the confrontational nature of occupation involved harshness and death on both sides. In the week ending August 10, an eight-year-old Palestinian boy and his six-year-old sister were killed in the Gaza Strip when a Kassam rocket, fired at Israel, fell short and hit their home. That same week, an Israeli commander was dismissed from the army over the shooting of a teenage Palestinian girl in the Gaza Strip: his soldiers had mistaken her for a terrorist and opened fire, not knowing that the girl was being used as a 'human shield' by the Israel Defence Forces to warn militants to leave their homes.

Small incidents can betoken deeper changes. On 27 August 2007 Palestinian police rescued an Israeli soldier who had taken a wrong turn and driven by mistake into Jenin, where he had been surrounded by a mob that later burned his car. Seven years earlier, two Israeli army reservists who had likewise strayed – into Ramallah – had been beaten to death, their murderers displaying the mutilated bodies for all to see, and raising their blood-covered hands in triumph. The Jenin rescue by Palestinian police gave hope in Israel that Abbas would uphold the rule of law and prevent such ugly incidents as in Ramallah. Two days later, in talks between Abbas and Olmert in Jerusalem, the prospect of negotiations for Palestinian statehood was moved gently yet perceptibly forward. The message that came out of the meeting was that both men were seeking to find ways forward for the creation of 'two States for two peoples'.

Gaza, under Hamas control, saw no such amelioration. An Israeli army inquiry at the end of August discovered that three Palestinian cousins – ten-year-old Mahmoud and Sara Ghazal, and twelve-year-old Yehiya Ghazal – who had been playing near rocket launchers in northern Gaza when they were killed, were not connected with terrorists. Israeli troops had spotted them moving near the rocket launchers. 'We already do all we can to prevent harming innocent people,' an Israeli officer told the press. 'Palestinians need to know that it is dangerous to be near or play near Kassam rocket launchers.' The Israeli army expressed sorrow at the children's death, but said that Palestinian extremists bore responsibility because they put rocket launchers in civilian areas. A week earlier, a fifteen-year-old Palestinian boy was caught on his way to perpetrate a suicide attack against Israeli troops.

On August 30, Human Rights Watch called on both Israelis and Palestinians to stop endangering children in the Gaza Strip. In the words of Joe Stork, deputy director of Human Right Watch's Middle East division: 'Rocket launchers constitute legitimate military objectives, and Palestinian armed groups endanger civilians when they place them near residential areas. But Israeli forces must take all feasible precautions when conducting attacks to avoid unnecessary loss of civilian life, especially of children.'

Alongside the uncertainties and pain of the Israeli–Palestinian situation, Israel has created a vigorous economy. It attracts global investment, and is in the forefront of the worldwide computer technology industry. Exports of agricultural and industrial products flourish. Theatre and film, art and music, dance and song, sport and recreation, learning and leisure, all thrive. Open-air cafés rival those of Paris and Rome. Museums are a focal point of study and enjoyment. Israel's national football team competes at the highest level of world soccer. Schools and universities are havens of work, play and scholarship. Medical, scientific and technological research benefits hundreds of millions around the world. In October 2006, an initiative launched at the Weizmann Institute by David Cahen and other scientists began intensive research for alternatives to non-sustainable fossil fuel. At Tel Aviv University, research was being undertaken by Dr Amir Sharon for the conversion of hitherto unusable parts of plants into bio-ethanol, a chemical used for biofuel, as an alternative to fuel oil. In another initiative, Lucien Bronicki and his wife Ormat work, through their company Ormat, to move biodiesel fuel into the mainstream, phasing out the polluting traditional diesel.

In the medical sphere an Israeli company, BioControl Medical, developed the first electrical stimulator to treat urinary incontinence, and in 2007 was working on an implanted electrical nerve stimulator to treat congestive heart failure. An Israeli doctor, Uri Kramer, has developed the Epilert, a device to detect and recognize epileptic seizures within twenty seconds, and to alert caregivers or parents that an epileptic episode is occurring.

In the social sphere, the Israel Asper Community Action Centre was established in four Israeli cities to fulfil the vision of a Canadian Jew, Israel Asper, to enable disadvantaged Israeli youth, including new immigrants from Ethiopia, to develop basic computer skills and gain access to the job market.

Overseas, Israel and Israelis have been at the forefront of global ameliorative action. Israeli rescue teams were put at the disposal of disaster areas, including those devastated by the Armenian earthquake in 1999, the 2004 tsunami in Sri Lanka, the 2007 earthquake in Peru and the 2007 forest fires in Greece, where Israeli firefighters made up the largest of the eight-nation contingent. An Israeli woman from Haifa, Aya Shneerson, heads the United Nations World Food Programme's activities in Congo. Other Israelis are working in Chad as part of the IsraAid refugee programme to help refugees from Darfur.

The Israeli Supreme Court, situated in Jerusalem, has established, under the rubric of human dignity, the rights to equality of all Israeli Arabs, women, gays and lesbians. It has repeatedly asserted freedom of expression, conscience and religious pluralism. When a leading Israeli Arab member of the Knesset, Azmi Bishara, was accused of sympathizing with Hizballah by visiting Lebanon and Syria, the Israeli Supreme Court dismissed the criminal charges against him. Under the leadership for more than a decade of Judge Aharon Barak, who stepped down in September 2006, the court challenged the exemption from military conscription given to Jewish religious students, outlawed spanking, ruled against a Knesset law forbidding the sale of pork, required (as we have seen) that the army alter the route of the security fence

– the Wall – in order to minimize harm to Palestinian communities, and insisted that military interrogations should not use 'physical pressure' – that is, torture – when interrogating Palestinian prisoners, even when they might know of a terror act in prospect. Above all, Aharon Barak asserted the Supreme Court's responsibility to review all Knesset legislation to ensure that it abided by the Knesset's own 'Basic Laws' protecting human rights. In the words of a leading Israeli jurist, Frances Raday, 'democracy is not only the will of the majority but also the guardian of the human rights of the minority and the individual'. Despite Israel's sense of siege and isolation, its Supreme Court is fearless in asserting what it considers essential restraints on executive and military policy.

Such fearlessness is not always recognized for what it is. In March 2007, knowing little or nothing of the efforts being made in Israel to create a just society for all its inhabitants and to live in peace with its Palestinian neighbours, 130 British doctors proposed boycotting the Israel Medical Association. In April the British National Union of Journalists voted to boycott Israeli products. In May a British academic union, the University and College Union, voted to recommend its members to boycott Israeli academics.

Such attitudes create anger and bewilderment in Israel, and a sense of being demonized by outsiders. Yet internally the search for constructive compromise is continuous: in May, the month of the recommended boycott, King Abdullah II of Jordan, through his Abdullah Fund for Development, and Elie Wiesel, through his Foundation for Humanity, gave their support to the Arava Institute in southern Israel, where forty students, including Israelis, Israeli Arabs, West Bank Palestinians and Jordanians, examine local problems regarding air and water pollution, and ways of turning waste into energy.

In August 2007 the Palestinians of the Jenin district of the West Bank, headed by the Palestinian Authority governor, Kadura Musa, and the Israelis of the adjacent Gilboa region, headed by the leader of the Gilboa Regional Council, Danny Attar, began work on the Palestinian side of the security fence to create a 250-acre free-trade industrial zone. Its aim is to benefit the economies on both sides of the border, generate 10,000 Palestinian jobs, attract foreign investment, and serve as a model for future free-trade zones elsewhere along the border. To bring such initiatives under a single roof, and to link them to the peace process, Tony Blair was making plans in the autumn of 2007 for an Israeli–Palestinian Economic Council that would take the existing economic needs and hardships and transform them into ventures of opportunity across the Palestinian–Israeli divide. Blair is determined to see sanity prevail, and two nations with so much to gain in productive partnership live in harmony.

Alongside economic opportunity, political progress continued to be made. Within a week of his meeting with Abbas in Jerusalem in August 2007, Olmert agreed to the dismantling of twenty-six illegal Israeli outposts on the West Bank. On 2 September 2007 the Israeli army chiefs predicted that the evacuation of these outposts would be 'difficult and violent'. But Olmert was determined that it should go ahead, in the cause of Israeli–Palestinian reconciliation. The Israeli army was then in the final stages of drawing up an operational plan that could be implemented within a matter of weeks after the government gave the green light. The outposts had all been established since 2001. Sharon had promised the United States that he would evacuate them. Olmert had destroyed the permanent structures of one of them, Amona, on becoming Prime Minister a year and a half earlier.

Steps to ameliorate the situation were continuous. On 4 September 2007 the Israeli Supreme Court, under its new President, Dorit Beinish, the first woman to hold this high position, accepted an appeal by the Palestinian residents of the West Bank village of Bil'in to reroute the line of the security fence, which had been built in such a way that it cut off the villagers' access to parts of the agricultural land they cultivated. This decision reasserted the 20 June 2004 Supreme Court ruling that justice for the Palestinians was an integral part of Israel's justice system.

Israel is a country animated by resilience and hope. Shimon Peres, on becoming President, reflected in his acceptance speech on 16 July 2007:

'My years place me at an observation point from which the scene of our life as a reviving nation is seen, spread out in all its glory. It is true that in the picture stains also appear. It is true that we are flawed and have erred – but please believe me – there is no room for melancholy. The outstanding achievements of Israel in its sixty years, together with the courage, wisdom and creativity of our young generation, give birth to one clear conclusion: Israel has the strength to reach great prosperity and to become an exemplary State as commanded us by our prophets. Permit me to remain an optimist. Permit me to be a dreamer of his people. Permit me to present the sunny side of our State. And also, if sometimes the atmosphere is autumnal, and also if today, the day seems suddenly grey, the President whom you have chosen will never tire of encouraging, awakening and reminding – because spring is waiting for us at the threshold. The spring will definitely come!'

The alternative, already often glimpsed, is too painful. But with renewed efforts, both Israeli and Palestinian leaders – and those who support them – seek to move forward to that hitherto elusive Final Status Agreement. A Palestinian State living in peace with its Israeli neighbour is essential if the region is to have the calm, prosperity and harmony that, after so many decades of conflict, all of its people so deserve. The State of Israel will have an even brighter future when there is peace across the whole of that ninety-mile-wide sliver of land between the Mediterranean Sea and the Jordan River: home to two separate national aspirations, and – when sanity and wisdom prevail – home also to two flourishing, vibrant, contented peoples.

Epilogue

After continuous discussions throughout the summer and autumn of 2007 between Israel's Prime Minister, Ehud Olmert, and the Palestinian Authority President, Mahmoud Abbas, the way was set for a new attempt to launch the Israeli–Palestinian peace process. These meetings had been facilitated by repeated visits to the region by the United States Secretary of State Condoleezza Rice. The breakthrough came on 27 November 2007, when, under the auspices of the United States, a conference was held in Annapolis, Maryland. Its purpose was to revive the Road Map for the creation of a Palestinian State.

The Muslim world was represented at Annapolis by fifteen Arab States. These were Israel's four Arab neighbours, Lebanon, Syria, Jordan and Egypt, and, from the wider Arab world, Algeria, Bahrain, Morocco, Qatar, Oman, Saudi Arabia, Sudan, Tunisia, Yemen and the United Arab Emirates, most of which had never recognized Israel and had no diplomatic relations with Israel.

The Arab League was represented at Annapolis by its Secretary-General, Amre Moussa. Four months earlier the Arab League had sent a mission to Israel, headed by the Jordanian and Egyptian foreign ministers, Abdel Ilah Al-Khatib and Ahmed Abdoul Gheit, to promote the 2002 Saudi Arabian initiative, based on Israel's withdrawal from all territories occupied since 1967, a solution to the Palestinian refugee problem, and the establishment of an independent Palestinian State. The idea of a Palestinian State, the ultimate aim of Annapolis, had earlier been endorsed by President Bush, and by Ehud Olmert and his predecessor as Israeli Prime Minister, Ariel Sharon.

The twenty-one participants in the Annapolis talks from outside the Middle East included three Muslim nations, Indonesia, Malaysia and Pakistan, as well as Norway, the host of the 1993 Olso talks, and Canada. At the time of the Camp David talks in 2000, Canada was one of the countries that had offered to take in Palestinian refugees. Also present at Annapolis, as observers, were the World Bank and the International Monetary Fund.

The talks at Annapolis were devoted to creating a plan for future direct discussions between Israel and the Palestinians about the crucial disputed elements relating to any future Palestinian State: Jerusalem as the joint capital of Israel and the Palestinians, the Palestinian refugees, Israeli settlements on the West Bank, mutual security, and the borders of the two States. These issues were not discussed at Annapolis, but a timetable was agreed within which they were to be the subject of continuous negotiations between Israel and the Palestinian Authority.

During the Annapolis talks, which lasted for two days, Ehud Olmert and Mahmoud Abbas both expressed the desire to create a meaningful dialogue and to reach an agreement whereby a Palestinian State would come into being. Olmert told the conference: 'I have no doubt that the reality created in our region in 1967 will change significantly. While this will be an extremely difficult process for many of us, it is nevertheless inevitable. I know it. Many of my people know it. We are ready for it.' In Abbas's words: 'Neither we nor you must beg for peace from the other. It is a joint interest . . . Peace and freedom is a right for us, just as peace and security is a right for you and us . . . War and terrorism belong to the past.'

For Israel, this last statement was an essential pledge that dialogue and not terror would be the way forward. As the meeting came to an end, President Bush announced on behalf of those present a 'Joint Understanding', which read: 'We agreed to launch immediately, in good faith, bilateral negotiations in order to conclude a peace treaty resolving all outstanding issues, including core issues, without exception.'

The obstacles in the way of making Annapolis work, especially within the timetable proposed by President Bush, which set the target date for the conclusion of a treaty at the end of 2008, were formidable. Hardly had the conference ended when, on December 2, it was revealed that three Palestinian policemen had been responsible for the shooting attack that had killed Ido Zoldan, a 29-year-old father of two from the West Bank settlement of Shavei Shomron. He had been killed – the day before the Annapolis summit began – when shots were fired at his car as he drove past the Palestinian village of al-Punduk. The three killers were members of the Fatah-controlled Palestinian National Security Force. Distressingly for many of the Israeli public, both Israel and the United States had been supporting this force financially and with weapons, as part of an international effort to strengthen Abbas and his Fatah-led government.

Two of the killers, both 22-year-old Palestinian policemen and members of Fatah, were arrested by Israeli forces. During their interrogation, the two confessed their involvement in the attack and handed over the weapon used in the shooting. They said that they had parked their car on the side of the road and waited for an Israeli car to pass by. When Zoldan's car appeared, they emerged onto the road, drove past him and opened fire. They told their interrogators that they decided to carry out the attack to 'scare settlers'. The third killer, also a policeman, was taken into Palestinian police custody.

Israelis watched with apprehension as the Fatah connection was made public. In the same week, from Hamas-controlled territory in the Gaza Strip, yet more rockets were fired across the border into Israel, hitting the town of Sderot. In an Israeli air attack on December 3 on a Hamas training camp in the Gaza Strip, three Palestinians were killed. But neither the West Bank killing nor the rocket firing and retaliation were allowed to derail the ongoing talks agreed upon at Annapolis.

In order to achieve a final agreement and Palestinian statehood by the end of 2008, Olmert and Abbas agreed that they would meet every two weeks, while at the same time Israeli and Palestinian negotiators would be in continuous session on each of the issues that needed to be resolved. As a confidence-building gesture, Olmert announced on December 2 that Israel would release 400 of the 8,000 Palestinians held in Israeli prisons, most of whom had been convicted of terrorist offences.

The issues that would need to be resolved for the Annapolis timetable to be fulfilled, and for a Palestinian State to come into being by the end of 2008, were to be discussed in the year ahead at a series of international meetings. On the day after Annapolis, Bush, Olmert and Abbas met at the White House. At this meeting, held on November 29 – sixty years to the day after the United Nations voted to set up a Jewish and a Palestinian State in a partitioned Palestine – Israel, the Palestinian Authority and the United States agreed to begin work on three parallel tracks.

The first track, that of Israeli–Palestinian bilateral negotiations, started with a Steering Committee meeting on December 12 between Israeli and Palestinian representatives. No American representatives were present. The Steering Committee was to meet on a regular basis to establish the framework for the negotiations that would lead to an agreement by the end of 2008. It was in part to oversee the work of the Steering Committee that Olmert and Abbas would meet every other week.

The second track agreed upon by Olmert, Abbas and Bush at the White House was implementation of the Road Map, with the United States in charge of a monitoring mechanism. Israel having declared that it would not implement any agreement until the Palestinians fulfilled their Road Map requirements, the monitoring mechanism, which would be the judge of when this happened, was a crucial element in furthering Palestinian statehood. A retired United States general and former NATO Supreme Allied Commander in Europe, James Jones, was appointed head of the monitoring mechanism.

The third post-Annapolis track was to enlist the practical support of the Arab countries present at Annapolis, and of the international community, to produce 'a favourable regional environment' – essentially, Palestinian economic security and growth – to advance the two-State solution. To this end, the donor countries met in Paris on December 17 to put in place practical steps to help create workable Palestinian governing institutions and a flourishing Palestinian economy. The host of the Paris talks, Tony Blair, had been at Annapolis as the representative of the Quartet, and was able to enlist in his endeavours the British-based Portland Trust, an initiator of projects designed to help Palestinian businesses and stimulate housing development in the West Bank. From the United States, Condoleezza Rice gave her strong support to the American-based Aspen Institute's US–Palestinian Public–Private Partnership, committed to attracting international investment, supporting existing Palestinian businesses, creating jobs and training for Palestinians, and otherwise fostering economic opportunities in the Palestinian Authority through economic and social projects.

One of the goals of the Annapolis process was to delegitimize Hamas in the eyes of Palestinians and the Arab world by showing that the Palestinian Authority is a viable alternative to it. To do this, the economic track was considered essential.

Under the agreement at Annapolis, negotiations aimed at Palestinian statehood were to take place in the six months leading up to the sixtieth anniversary of the creation of Israel, and were to continue for seven months beyond. Such an outcome – an independent, sovereign Israel with recognized, secure and peaceful borders living alongside the equally independent, secure and peaceful Palestine – would offer calm, productive, harmonious, peaceful lives to two peoples caught up in what has seemed at times, and to many on both sides of the Israeli–Palestinian divide still seems, an intractable conflict.

MAPS

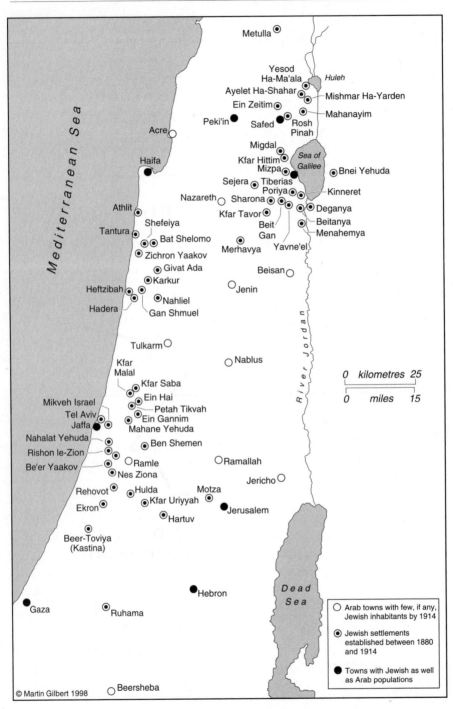

1 Jewish settlements in Ottoman Palestine by 1914

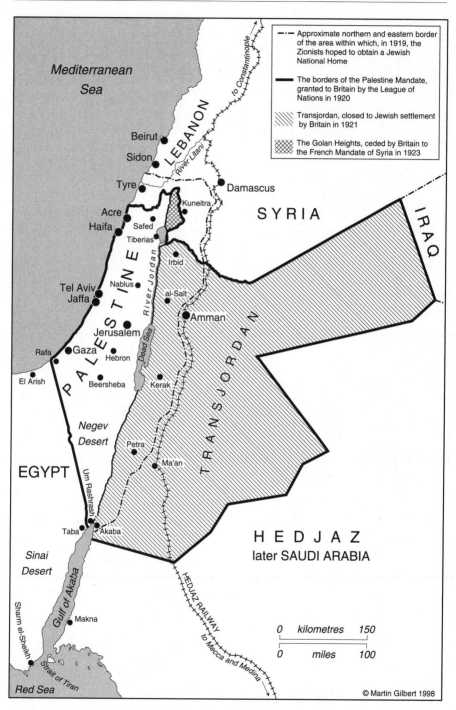

Key (top right):

- - · - · - Approximate northern and eastern border of the area within which, in 1919, the Zionists hoped to obtain a Jewish National Home

——— The borders of the Palestine Mandate, granted to Britain by the League of Nations in 1920

///// Transjordan, closed to Jewish settlement by Britain in 1921

The Golan Heights, ceded by Britain to the French Mandate of Syria in 1923

Mediterranean Sea

to Constantinople

LEBANON

Beirut

Sidon

River Litani

Tyre

Damascus

Kuneitra

Acre

SYRIA

Haifa

Safed

Tiberias

IRAQ

PALESTINE

River Jordan

Irbid

Tel Aviv

Nablus

al-Salt

Jaffa

Amman

Dead Sea

Jerusalem

TRANSJORDAN

Rafa

Gaza

Hebron

El Arish

Beersheba

Kerak

Negev Desert

EGYPT

Petra

Ma'an

Um Rashrash

HEDJAZ
later SAUDI ARABIA

Taba

Akaba

Sinai Desert

Gulf of Akaba

HEDJAZ RAILWAY

to Mecca and Medina

Sharm el-Sheikh

Makna

Strait of Tiran

| 0 | kilometres | 150 |
| 0 | miles | 100 |

Red Sea

© Martin Gilbert 1998

2 *The British Mandate for Palestine, 1920–48*

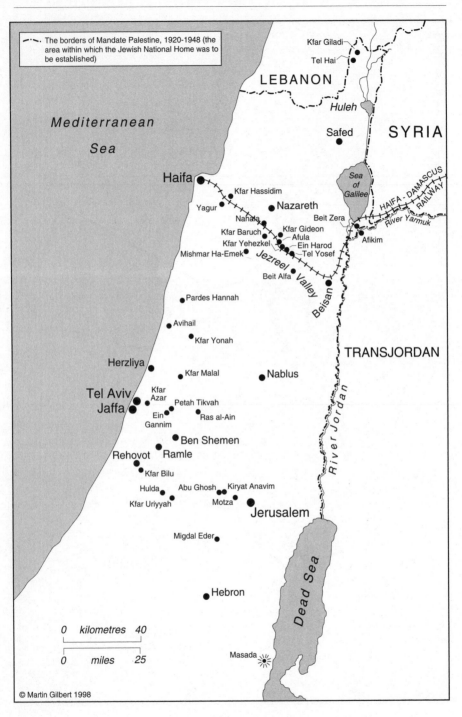

The borders of Mandate Palestine, 1920-1948 (the area within which the Jewish National Home was to be established)

Kfar Giladi
Tel Hai
LEBANON
Huleh
Safed
SYRIA
Mediterranean
Sea
Sea of Galilee
HAIFA - DAMASCUS RAILWAY
Haifa
Kfar Hassidim
Yagur
Nazareth
Beit Zera
River Yarmuk
Nahala
Kfar Gideon
Kfar Baruch
Afula
Afikim
Kfar Yehezkel
Ein Harod
Mishmar Ha-Emek
Tel Yosef
Jezreel Valley
Beit Alfa
Beisan
Pardes Hannah
Avihail
Kfar Yonah
TRANSJORDAN
Herzliya
Kfar Malal
Nablus
Tel Aviv
Kfar Azar
Petah Tikvah
Jaffa
Ras al-Ain
Ein Gannim
River Jordan
Ben Shemen
Rehovot **Ramle**
Kfar Bilu
Hulda
Abu Ghosh
Kiryat Anavim
Kfar Uriyyah
Motza
Jerusalem
Migdal Eder
Dead Sea
Hebron

0   kilometres   40
0   miles   25

Masada

© Martin Gilbert 1998

3 Mandate Palestine, 1921–32

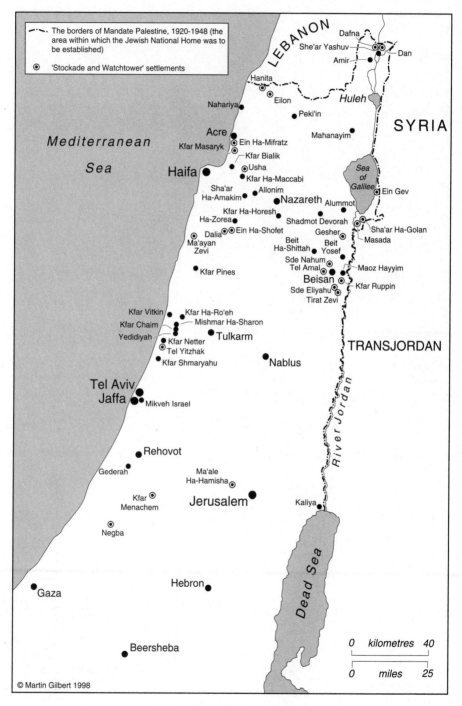

The borders of Mandate Palestine, 1920-1948 (the area within which the Jewish National Home was to be established)

'Stockade and Watchtower' settlements

LEBANON

Dafna

She'ar Yashuv

Dan

Amir

Hanita

Eilon

Huleh

Nahariya

Peki'in

SYRIA

Acre

Mahanayim

Kfar Masaryk

Ein Ha-Mifratz

Kfar Bialik

Mediterranean Sea

Usha

Haifa

Kfar Ha-Maccabi

Sea of Galilee

Ein Gev

Sha'ar Ha-Amakim

Allonim

Nazareth

Alummot

Kfar Ha-Horesh

Ha-Zorea

Shadmot Devorah

Dalia

Ein Ha-Shofet

Gesher

Sha'ar Ha-Golan

Ma'ayan Zevi

Beit Ha-Shittah

Beit Yosef

Masada

Kfar Pines

Sde Nahum

Tel Amal

Maoz Hayyim

Beisan

Sde Eliyahu

Kfar Ruppin

Tirat Zevi

Kfar Vitkin

Kfar Ha-Ro'eh

Kfar Chaim

Mishmar Ha-Sharon

Yedidiyah

Kfar Netter

Tulkarm

Tel Yitzhak

Kfar Shmaryahu

Nablus

TRANSJORDAN

Tel Aviv

Jaffa

Mikveh Israel

Rehovot

Gederah

Ma'ale Ha-Hamisha

Kfar Menachem

Jerusalem

Kaliya

Negba

River Jordan

Gaza

Hebron

Dead Sea

Beersheba

0 kilometres 40

0 miles 25

© Martin Gilbert 1998

4 Mandate Palestine, 1933–39

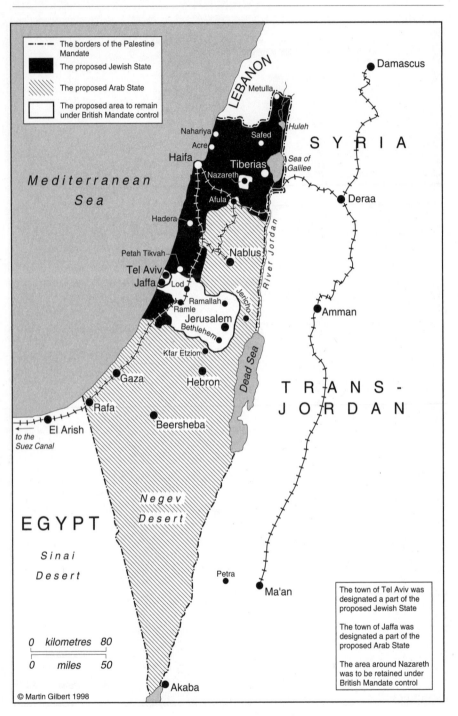

Legend:
- –·–·– The borders of the Palestine Mandate
- ■ The proposed Jewish State
- ▨ The proposed Arab State
- □ The proposed area to remain under British Mandate control

Damascus

LEBANON

Metulla

Huleh

Nahariya

Safed

Acre

S Y R I A

Haifa

Tiberias

Sea of Galilee

Nazareth

Mediterranean

Sea

Afula

Deraa

Hadera

River Jordan

Petah Tikvah

Nablus

Tel Aviv

Jaffa

Lod

Jericho

Ramallah

Ramle

Jerusalem

Amman

Bethlehem

Kfar Etzion

Dead Sea

Gaza

Hebron

T R A N S -

Rafa

J O R D A N

El Arish

to the Suez Canal

Beersheba

EGYPT

Negev

Desert

Sinai

Desert

Petra

Ma'an

0   kilometres   80

0   miles   50

Akaba

© Martin Gilbert 1998

The town of Tel Aviv was designated a part of the proposed Jewish State

The town of Jaffa was designated a part of the proposed Arab State

The area around Nazareth was to be retained under British Mandate control

5   *The Peel Commission proposals, 1937*

- - - The borders of Mandate Palestine, 1920-1948 (the area within which the Jewish National Home was to be established)

⊙ The eleven kibbutzim founded on 6 October 1946

LEBANON

Sde Nehemyah
Kfar Blum

Huleh

Mediterranean Sea

Yehiam

SYRIA

Acre

Ammiad

Haifa

Tiberias

Sea of Galilee

Yagur
Athlit Beit Oren Mount Carmel

Nazareth

Kfar Kisch

Deganya

River Yarmuk

Ein Harod

Naharayim

Caesarea
Sdot Yam

Narbata (Ma'anit)

Jezreel Valley

Beisan

Kfar Monash
Netanya Beit Yitzhak

TRANSJORDAN

Nablus

Tel Aviv
Jaffa

Petah Tikvah

Sarafand Be'erot Yitzhak

Yibnah Rishon le-Zion

Rehovot

Yavneh

Latrun

Jericho

River Jordan

Jerusalem

Beit Ha-Aravah

Kaliya

Kedma ⊙

Beit Jalla

Bethlehem

Massuot Yitzhak Kfar Etzion

Yad Mordechai

Gat ⊙ Gal-On

Ein Zurim Revadim

Erez

Dead Sea

Dorot

Hebron

Be'eri ⊙ ⊙ Tekuma

⊙ Kfar Darom

⊙ Shoval

⊙ Nirim

⊙ Mishmar Ha-Negev

Urim ⊙

Hazerim ⊙ Beersheba

0 kilometres 40

0 miles 25

© Martin Gilbert 1998

⊙ Nevatim

6 Mandate Palestine, 1940–47

7 *The United Nations Partition Plan, 1947*

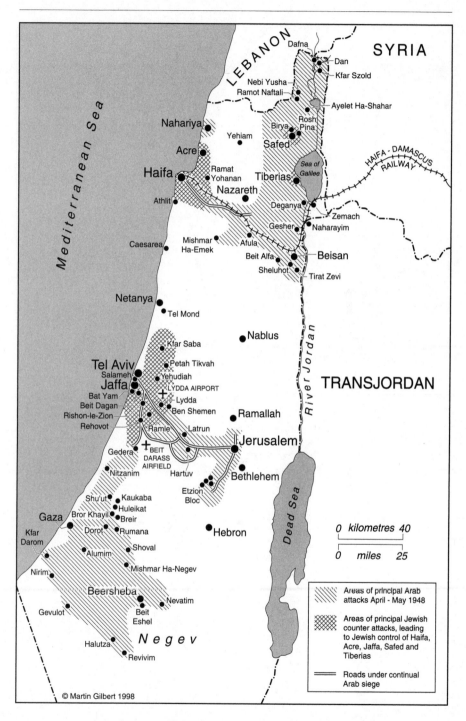

8 Prelude to independence, November 1947 to May 1948

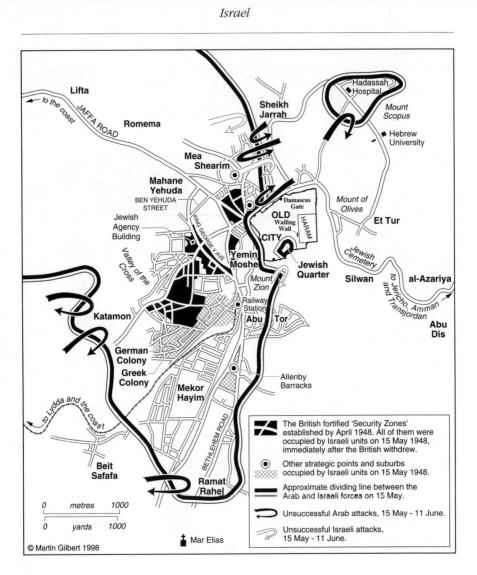

Lifta

to the coast

JAFFA ROAD

Romema

Sheikh
Jarrah

Hadassah
Hospital

Mount
Scopus

Hebrew
University

Mea
Shearim

Mahane
Yehuda

BEN YEHUDA
STREET

Damascus
Gate

Mount of
Olives

Et Tur

OLD
Wailing
Wall

HARAM

Jewish
Agency
Building

Valley of the Cross

KING GEORGE V AVE

CITY

Yemin
Moshe

Jewish
Quarter

Jewish
Cemetery

Silwan

to Jericho, Amman
and Transjordan

al-Azariya

Mount
Zion

Railway
Station

Katamon

Abu Tor

Abu
Dis

German
Colony

Greek
Colony

Mekor
Hayim

BETHLEHEM ROAD

Allenby
Barracks

to Lydda and the coast

Beit
Safafa

Ramat
Rahel

The British fortified 'Security Zones'
established by April 1948. All of them were
occupied by Israeli units on 15 May 1948,
immediately after the British withdrew.

Other strategic points and suburbs
occupied by Israeli units on 15 May 1948.

Approximate dividing line between the
Arab and Israeli forces on 15 May.

Unsuccessful Arab attacks, 15 May - 11 June.

Unsuccessful Israeli attacks,
15 May - 11 June.

0   metres   1000

0   yards   1000

© Martin Gilbert 1998

Mar Elias

9 Jerusalem besieged, 1948

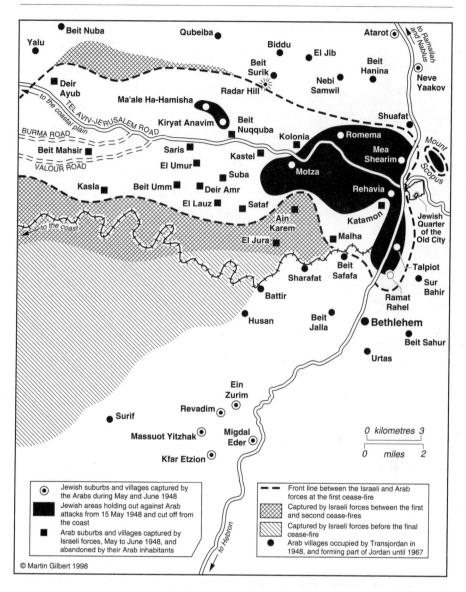

10 The Jerusalem Corridor

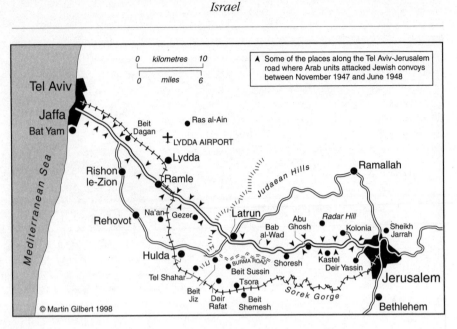

11 The Tel Aviv to Jerusalem Road, 1947–48

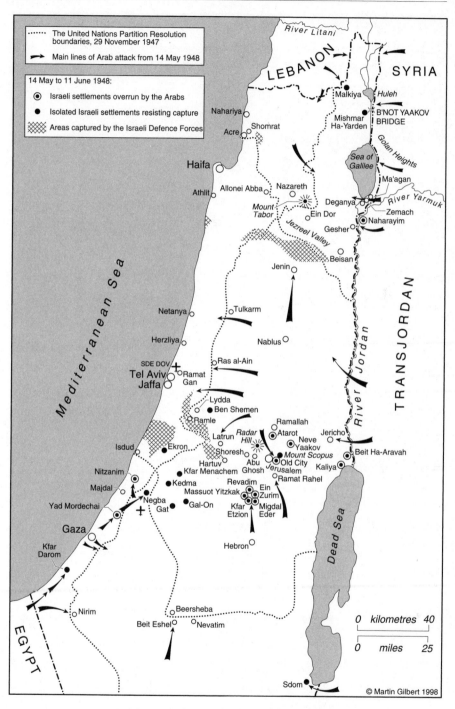

12 The War of Independence, May–June 1948

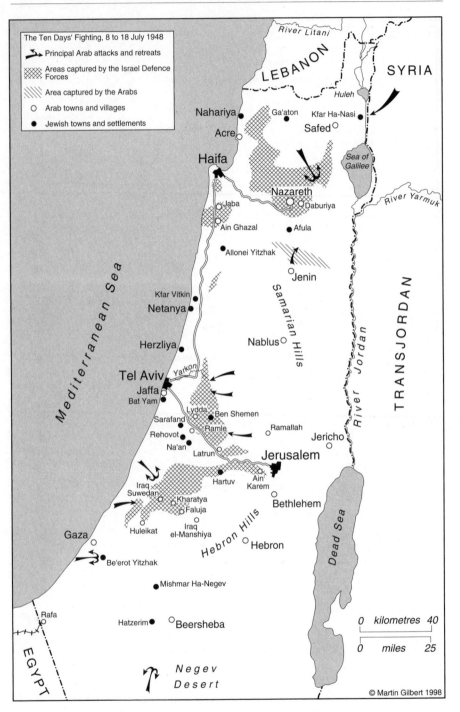

The Ten Days' Fighting, 8 to 18 July 1948

➤ Principal Arab attacks and retreats

▨ Areas captured by the Israel Defence Forces

▧ Area captured by the Arabs

○ Arab towns and villages

● Jewish towns and settlements

River Litani

LEBANON

SYRIA

Huleh

Nahariya

Ga'aton

Kfar Ha-Nasi

Acre

Safed

Haifa

Sea of Galilee

Nazareth

River Yarmuk

Jaba

Daburiya

Ain Ghazal

Afula

Allonei Yitzhak

Jenin

Kfar Vitkin

Netanya

Samarian Hills

Herzliya

Nablus

TRANSJORDAN

Tel Aviv

Yarkon

Jaffa

Bat Yam

Lydda

Ben Shemen

Sarafand

Ramle

Ramallah

Rehovot

Jericho

Na'an

Latrun

Jerusalem

Hartuv

Ain Karem

Iraq Suwedan

Kharatya

Bethlehem

Faluja

Dead Sea

Huleikat

Iraq el-Manshiya

Gaza

Hebron Hills

Hebron

Be'erot Yitzhak

Mishmar Ha-Negev

Mediterranean Sea

Rafa

Hatzerim

Beersheba

0 kilometres 40

0 miles 25

EGYPT

Negev Desert

© Martin Gilbert 1998

13 *The War of Independence, June–October 1948*

670

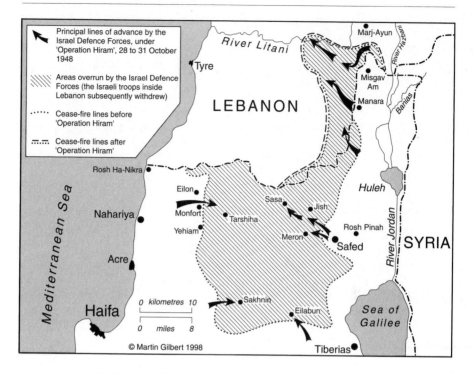

Principal lines of advance by the Israel Defence Forces, under 'Operation Hiram', 28 to 31 October 1948

Areas overrun by the Israel Defence Forces (the Israeli troops inside Lebanon subsequently withdrew)

Cease-fire lines before 'Operation Hiram'

Cease-fire lines after 'Operation Hiram'

River Litani

Tyre

Marj-Ayun

Misgav Am

Manara

LEBANON

River Hazbani

Banias

Rosh Ha-Nikra

Mediterranean Sea

Eilon

Monfort

Tarshiha

Yehiam

Nahariya

Acre

Haifa

0   kilometres   10

0   miles   8

© Martin Gilbert 1998

Sasa

Jish

Meron

Safed

Rosh Pinah

Huleh

River Jordan

SYRIA

Sakhnin

Eilabun

Sea of Galilee

Tiberias

14 The War of Independence, the final phase in the north,
October 1948

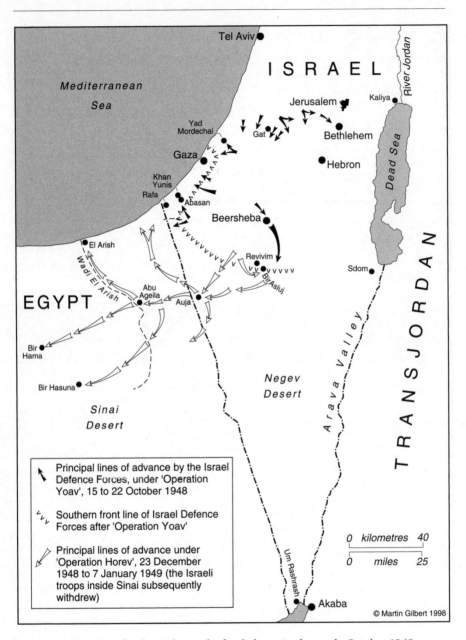

15 *The War of Independence, the final phases in the south, October 1948
to January 1949*

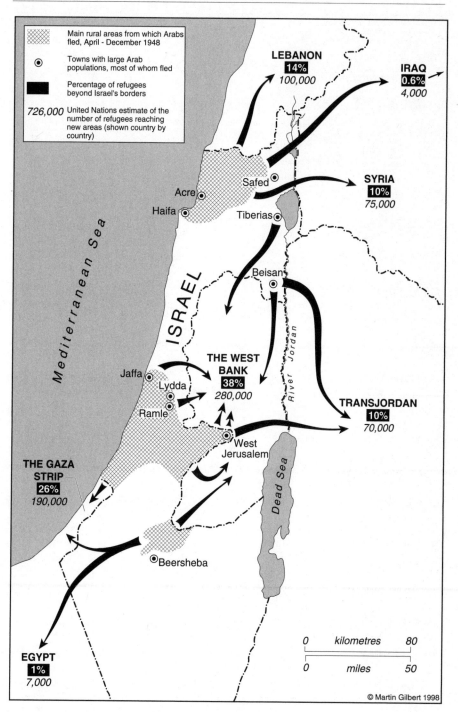

Main rural areas from which Arabs
fled, April - December 1948

Towns with large Arab
populations, most of whom fled

Percentage of refugees
beyond Israel's borders

726,000 United Nations estimate of the
number of refugees reaching
new areas (shown country by
country)

LEBANON
14%
100,000

IRAQ
0.6%
4,000

SYRIA
10%
75,000

Acre

Haifa

Safed

Tiberias

Beisan

ISRAEL

Mediterranean Sea

River Jordan

Jaffa

Lydda

Ramle

**THE WEST
BANK**
38%
280,000

West
Jerusalem

TRANSJORDAN
10%
70,000

**THE GAZA
STRIP**
26%
190,000

Dead Sea

Beersheba

EGYPT
1%
7,000

0   kilometres   80

0   miles   50

© Martin Gilbert 1998

16 Arab refugees, 1948

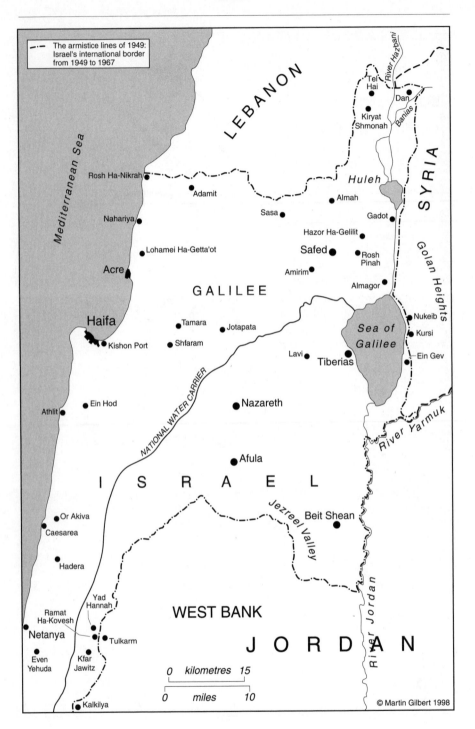

The armistice lines of 1949:
Israel's international border
from 1949 to 1967

River Hazbani

L E B A N O N

Tel Hai

Dan

Banias

Kiryat Shmonah

S Y R I A

Mediterranean Sea

Rosh Ha-Nikrah

Huleh

Adamit

Almah

Sasa

Gadot

Nahariya

Hazor Ha-Gelilit

Golan Heights

Lohamei Ha-Getta'ot

Safed

Rosh Pinah

Acre

Amirim

Almagor

G A L I L E E

Nukeib

Haifa

Tamara

Jotapata

Sea of Galilee

Kursi

Kishon Port

Shfaram

Lavi

Ein Gev

Tiberias

Ein Hod

NATIONAL WATER CARRIER

Nazareth

Athlit

River Yarmuk

Afula

I   S   R   A   E   L

Or Akiva

Jezreel Valley

Beit Shean

Caesarea

Hadera

River Jordan

Yad Hannah

WEST BANK

Ramat Ha-Kovesh

Netanya

Tulkarm

J   O   R   D   A   N

Even Yehuda

Kfar Jawitz

| 0 | kilometres | 15 |

| 0 | miles | 10 |

Kalkilya

© Martin Gilbert 1998

17 Israel, 1949–67: the north

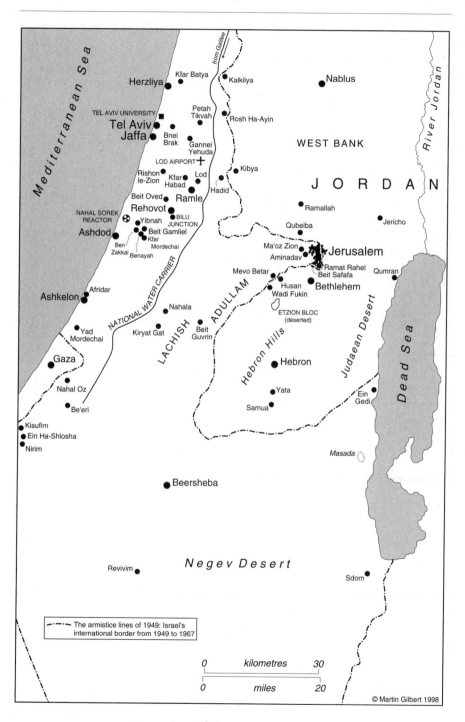

18 Israel, 1949–67: the centre and south

19 *The Negev*

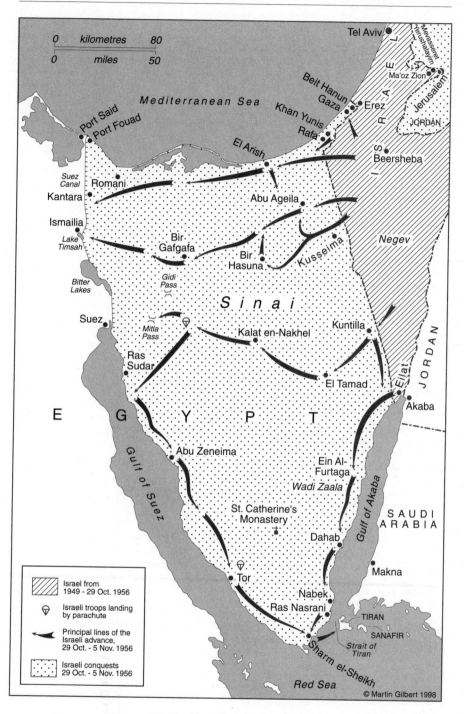

20 The Sinai Campaign, 1956

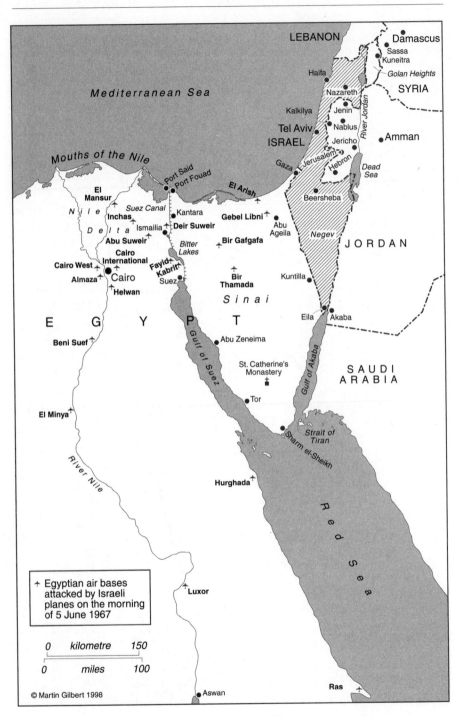

LEBANON

Damascus
Sassa
Kuneitra
Haifa
Golan Heights
Nazareth
SYRIA
Kalkilya
Jenin
Tel Aviv
Nablus
ISRAEL
Jericho
Amman
Mediterranean Sea
Gaza
Jerusalem
Hebron
Dead Sea

Mouths of the Nile
Port Said
Port Fouad
El Arish

El
Mansur
Kantara
Beersheba
Nile
Suez Canal
Inchas
Gebel Libni
Abu
Ageila
Delta
Ismailia
Deir Suweir
Abu Suweir
Bir Gafgafa
Negev
*Bitter
Lakes*
Cairo
International
JORDAN
Cairo West
Fayid
Kabrit
Almaza
Cairo
Suez
Bir
Thamada
Kuntilla
Helwan
S i n a i

E   G   Y   P   T
Eila
Akaba

Beni Suef
Abu Zeneima

St. Catherine's
Monastery
SAUDI
ARABIA

El Minya
Tor

River Nile
Gulf of Suez
Gulf of Akaba
*Strait of
Tiran*
Sharm el-Sheikh

Hurghada

R e d   S e a

> ✈ Egyptian air bases
> attacked by Israeli
> planes on the morning
> of 5 June 1967

Luxor

| 0 | kilometre | 150 |
| 0 | miles | 100 |

© Martin Gilbert 1998

Aswan
Ras

21 The Six Day War, 1967: the preliminary air attack

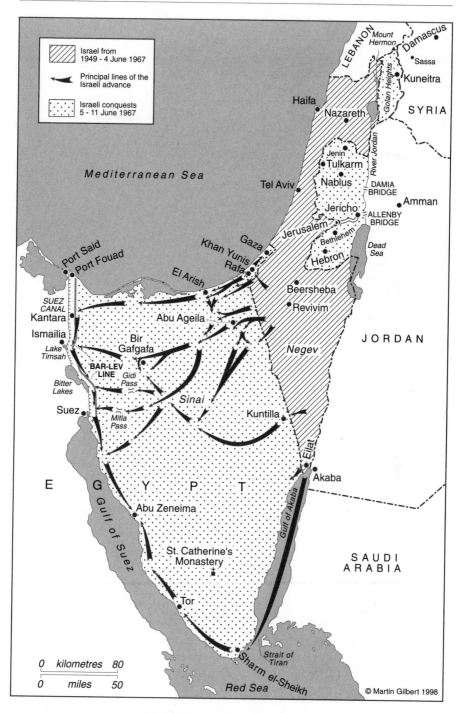

22   The Six Day War, 1967: Sinai

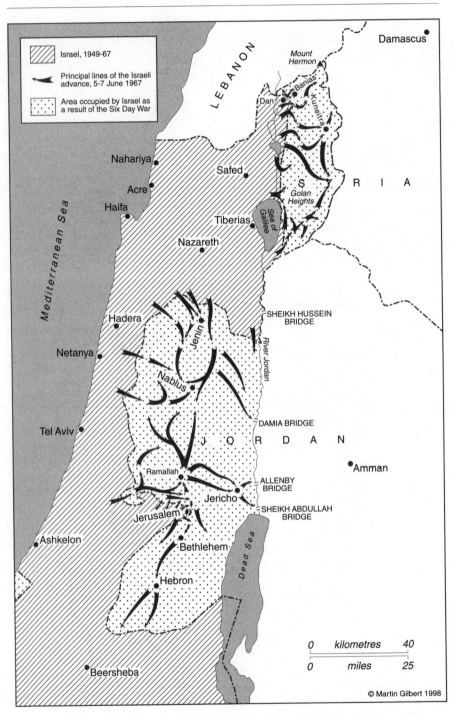

23 The Six Day War, 1967: the West Bank and the Golan Heights

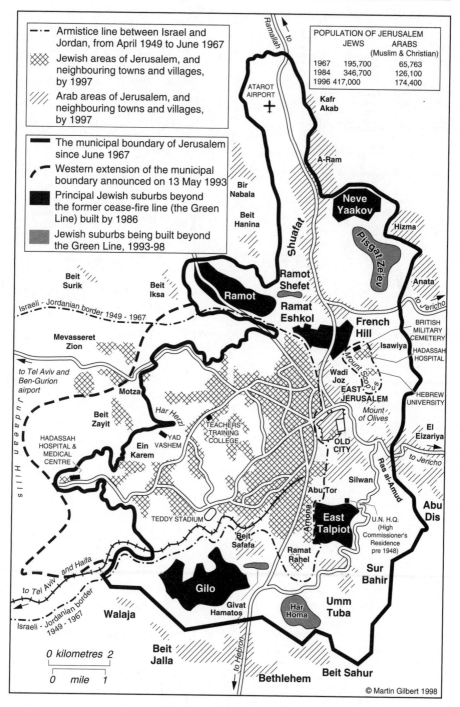

POPULATION OF JERUSALEM

	JEWS	ARABS (Muslim & Christian)
1967	195,700	65,763
1984	346,700	126,100
1996	417,000	174,400

Legend:

- –·– Armistice line between Israel and Jordan, from April 1949 to June 1967
- ▨ Jewish areas of Jerusalem, and neighbouring towns and villages, by 1997
- ⫽ Arab areas of Jerusalem, and neighbouring towns and villages, by 1997
- ▬ The municipal boundary of Jerusalem since June 1967
- ⊏⊐ Western extension of the municipal boundary announced on 13 May 1993
- ■ Principal Jewish suburbs beyond the former cease-fire line (the Green Line) built by 1986
- ▨ Jewish suburbs being built beyond the Green Line, 1993-98

© Martin Gilbert 1998

24 Jerusalem since 1967

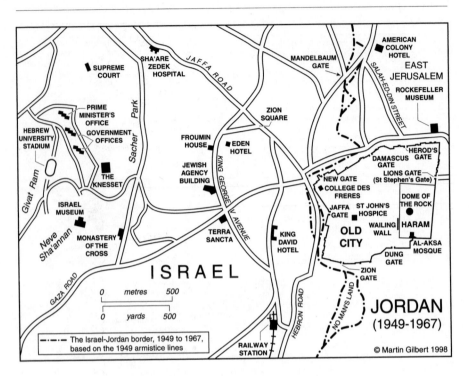

ISRAEL

Givat Ram

Neve Sha'annan

GAZA ROAD

HEBRON ROAD

SUPREME COURT

SHA'ARE ZEDEK HOSPITAL

JAFFA ROAD

MANDELBAUM GATE

AMERICAN COLONY HOTEL

EAST JERUSALEM

SALAH-ED-DIN STREET

ROCKEFELLER MUSEUM

PRIME MINISTER'S OFFICE

GOVERNMENT OFFICES

Sacher Park

ZION SQUARE

HEBREW UNIVERSITY STADIUM

FROUMIN HOUSE

EDEN HOTEL

JEWISH AGENCY BUILDING

THE KNESSET

KING GEORGE V AVENUE

ISRAEL MUSEUM

MONASTERY OF THE CROSS

TERRA SANCTA

KING DAVID HOTEL

HEROD'S GATE

DAMASCUS GATE

LIONS GATE (St Stephen's Gate)

NEW GATE

COLLEGE DES FRERES

JAFFA GATE

ST JOHN'S HOSPICE

DOME OF THE ROCK

WAILING WALL

HARAM

OLD CITY

AL-AKSA MOSQUE

DUNG GATE

ZION GATE

NO MAN'S LAND

JORDAN
(1949-1967)

0   metres   500
0   yards   500

— · — · — The Israel-Jordan border, 1949 to 1967, based on the 1949 armistice lines

RAILWAY STATION

© Martin Gilbert 1998

25 Central Jerusalem divided, and reunited, 1949–97

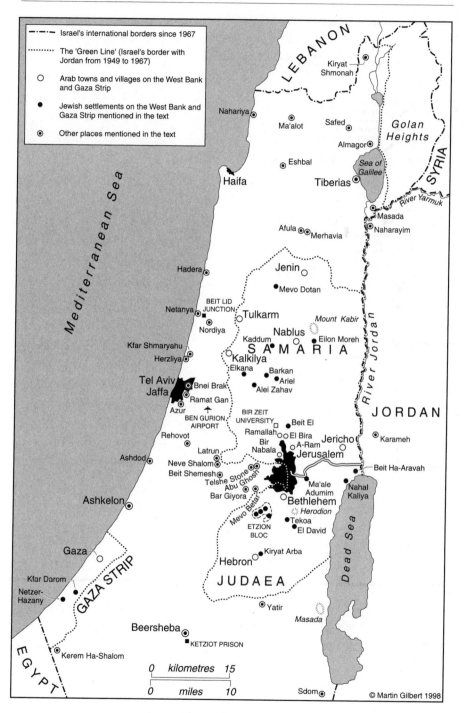

26 Israel since 1967

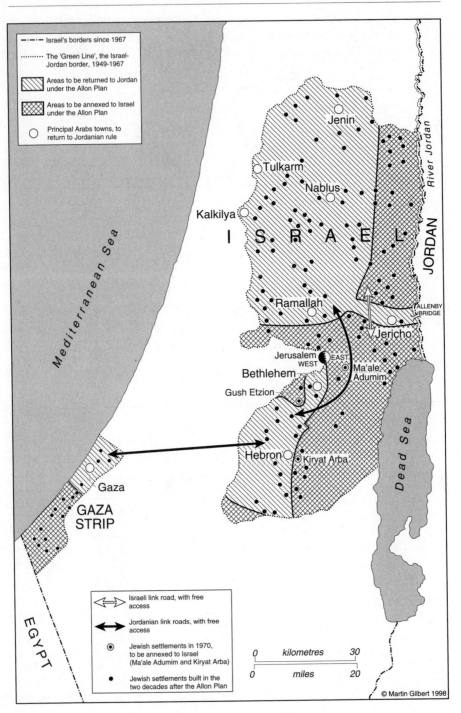

27 *The Allon Plan, 1970*

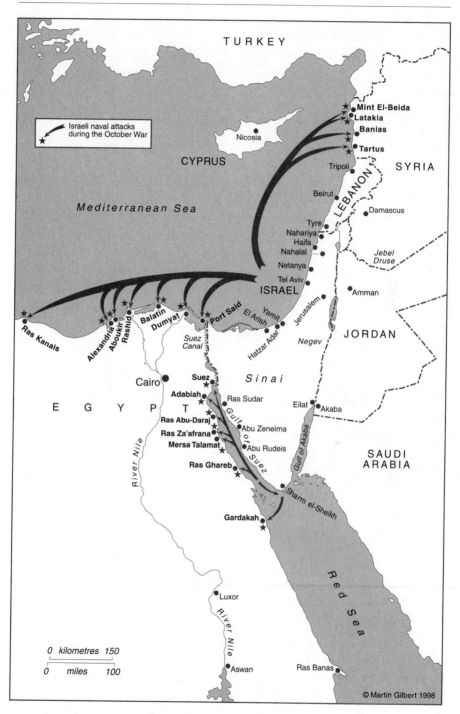

TURKEY

Mint El-Beida
Latakia
Banias
Tartus

Nicosia

Tripoli

CYPRUS

SYRIA

Beirut

Mediterranean Sea

Damascus

Tyre
Nahariya
Haifa
Nahalal

Jebel Druse

Netanya

Tel Aviv

ISRAEL

Amman

Port Said
El Arish
Yamit

Jerusalem

Ras Kanais

Balatin
Dumyat

Suez
Canal

Alexandria
Aboukir
Rashid

Hatzar Adar

Negev

JORDAN

Sinai

Cairo

Suez

Ras Sudar

Eilat
Akaba

Adabiah

E G Y P T

Ras Abu-Daraj

Abu Zeneima

Ras Za'afrana

Mersa Talamat

Abu Rudeis

SAUDI
ARABIA

River Nile

Ras Ghareb

Sharm el-Sheikh

Gardakah

Red Sea

Luxor

River Nile

Israeli naval attacks
during the October War

Gulf of Akaba

Gulf of Suez

0   kilometres 150

0   miles   100

Aswan

Ras Banas

© Martin Gilbert 1998

28 The October War, 1973: regional map

685

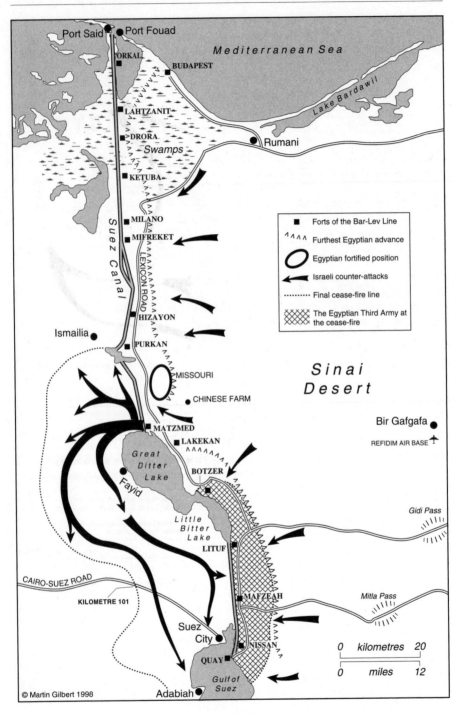

29 The October War, 1973: Sinai

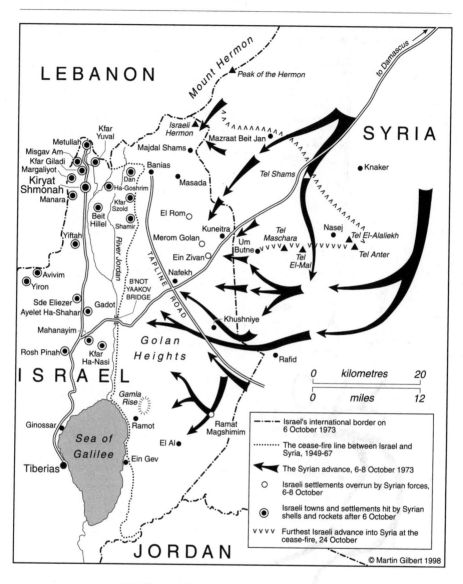

30 *The October War, 1973: Golan Heights*

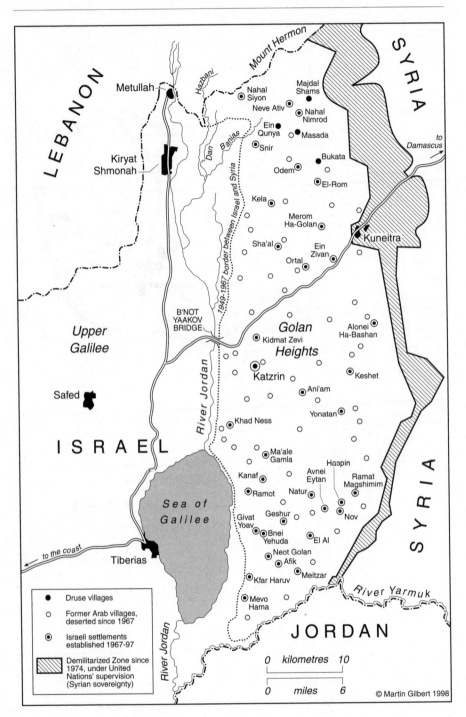

31 The Golan Heights settlements, 1967–97

Legend:
- ◆ Palestinian Arab towns in the Gaza Strip
- ⊙ Principal Palestinian Arab refugee camps
- ● Palestinian Arab villages
- ▨ Areas transferred to Palestinian Authority, 1994
- ○ Israeli settlements near the Gaza Strip
- □ Israeli settlements inside the Gaza Strip
- ▦ Areas retained by Israel, 1994
- ╍╍╍ Access roads to Israeli settlements

Mediterranean Sea

GAZA STRIP

Zikkim · Karmiya · Elei Sinai · to Tel Aviv
Gush Erez · Dugit · Netiv Ha-Asara · Yad Mordechai
Nisanit · Erez
Beit Lahiya · Jabalya · Nazla · Beit Hanun
Nir Am
Gaza City
Netzarim · Melfasim · Kfar Azar · Sa'ad
Nahal Oz · Alumim
Kfar Maimon
Nusseirat · Bureij · El Mauzi · Be'eri · Shokeda
Deir el-Balah
Kfar Darom · Re'im
Netzer Hazani · Kisufim
Katif · Ein Ha-Shlosha
GUSH KATIF · Ganei Tal · Nirim
Neve Dekalim · **ISRAEL**
Pe'at Sade Slav · Gadid · Khan Yunis · Nir Oz
Gan Or · Abasan el-Kabir · Magen
Bnei Atsmon · Morag · Khirbet Ikhza'a
Rafiah Yam
Rafa · *Negev*
EGYPT · Yesha · Ami Oz · Mivtahim
Nir Yitzhak · Sufa · Peri Gan
Holit · 0 kilometres 6
Kerem Shalom · Yated · Yesud Ha-Darom · Talme Yosef · 0 miles 4
Sinai · Yevul · Dekel
to Cairo
© Martin Gilbert 1998

32 The Gaza Strip, 1967–97

689

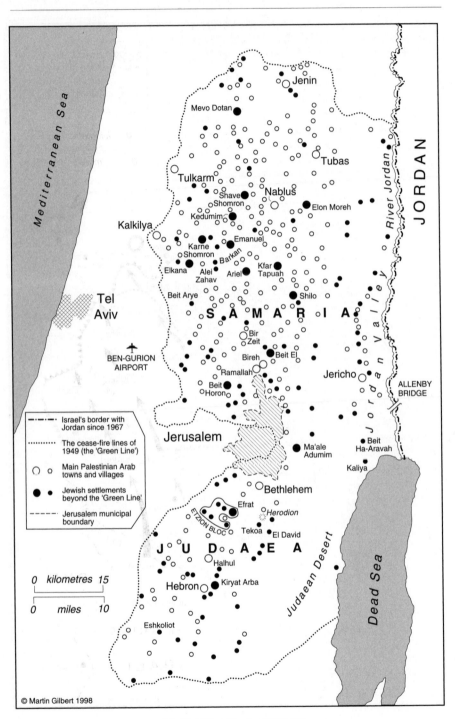

Mediterranean Sea

JORDAN

River Jordan

Jenin

Mevo Dotan

Tubas

Tulkarm

Nablus

Shave Shomron

Elon Moreh

Kalkilya

Kedumim

Emanuel

Karne Shomron

Barkan

Elkana

Alei Zahav

Ariel

Kfar Tapuah

Beit Arye

Shilo

Tel Aviv

S A M A R I A

Jordan Valley

BEN-GURION AIRPORT

Bir Zeit

Bireh

Beit El

Ramallah

Jericho

ALLENBY BRIDGE

Beit Horon

Israel's border with Jordan since 1967

The cease-fire lines of 1949 (the 'Green Line')

Main Palestinian Arab towns and villages

Jewish settlements beyond the 'Green Line'

Jerusalem municipal boundary

Jerusalem

Ma'ale Adumim

Beit Ha-Aravah

Kaliya

Bethlehem

Efrat

Herodion

ETZION BLOC

Tekoa

El David

0 kilometres 15

0 miles 10

J U D A E A

Halhul

Dead Sea

Hebron

Kiryat Arba

Judaean Desert

Eshkoliot

© Martin Gilbert 1998

33 The West Bank, 1967–97

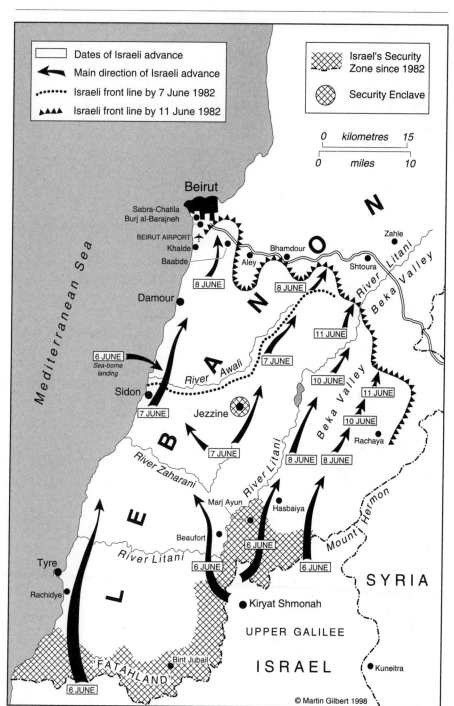

34 The Lebanon War, 1982

THE CAIRO AGREEMENT, 4 MAY 1994

The Gaza Strip (800,000 inhabitants): to be under the jurisdiction of the Palestinian Authority

Areas of Jewish settlement within the Gaza Strip, excluded from the Palestinian Authority (16 settlements, total population 5,000)

Jericho: approximate area of the Palestinian Authority (area not finalized during the Cairo negotiations)

THE OSLO II AGREEMENT, 28 SEPTEMBER 1995

Cities to which Palestinian autonomy was extended

City (Hebron) in which Palestinian autonomy was extended to 80% of the city

THE GREEN LINE (1949-1967 BORDER)

Jenin

Tulkarm

Nablus

Kalkilya

JORDAN

River Jordan

SAMARIA

Tel Aviv

ISRAEL

WEST BANK

El Auja

ALLENBY BRIDGE

Ramallah

Jericho

Mediterranean Sea

Jerusalem

Nebi Musa

Bethlehem

THE GREEN LINE (1949-1967 BORDER)

Dead Sea

Jibalya

Gaza

Hebron

GUSH KATIF

THE GAZA STRIP

JUDAEA

Khan Yunis

Rafa

EGYPT

0   kilometres   30

0   miles   20

© Martin Gilbert 1998

35 The Cairo and Oslo II Agreements, 1994, 1995

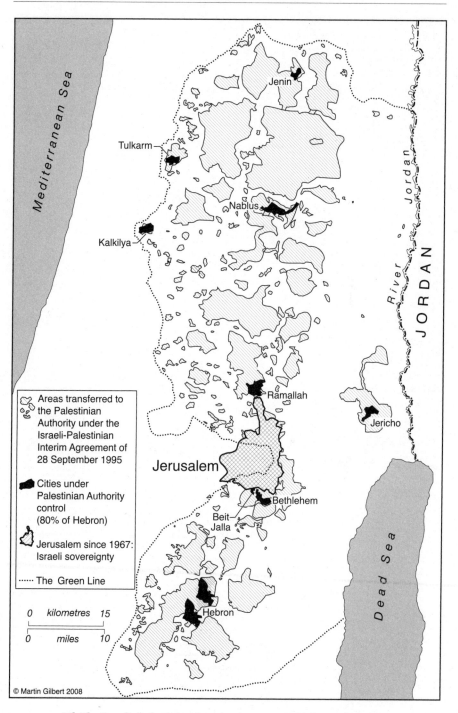

36 The Israeli–Palestinian Interim Agreement, 28 September 1995

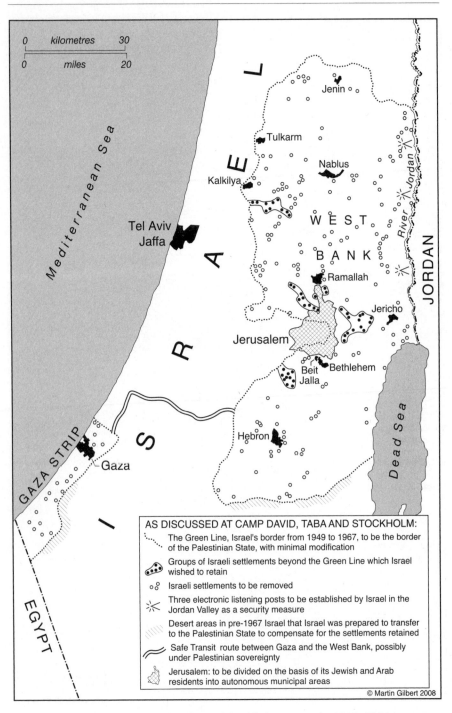

AS DISCUSSED AT CAMP DAVID, TABA AND STOCKHOLM:

⋯⋯ The Green Line, Israel's border from 1949 to 1967, to be the border of the Palestinian State, with minimal modification

Groups of Israeli settlements beyond the Green Line which Israel wished to retain

∘₀° Israeli settlements to be removed

⛌ Three electronic listening posts to be established by Israel in the Jordan Valley as a security measure

Desert areas in pre-1967 Israel that Israel was prepared to transfer to the Palestinian State to compensate for the settlements retained

Safe Transit route between Gaza and the West Bank, possibly under Palestinian sovereignty

Jerusalem: to be divided on the basis of its Jewish and Arab residents into autonomous municipal areas

© Martin Gilbert 2008

37 The Camp David, Taba and Stockholm proposals, 2000–2001

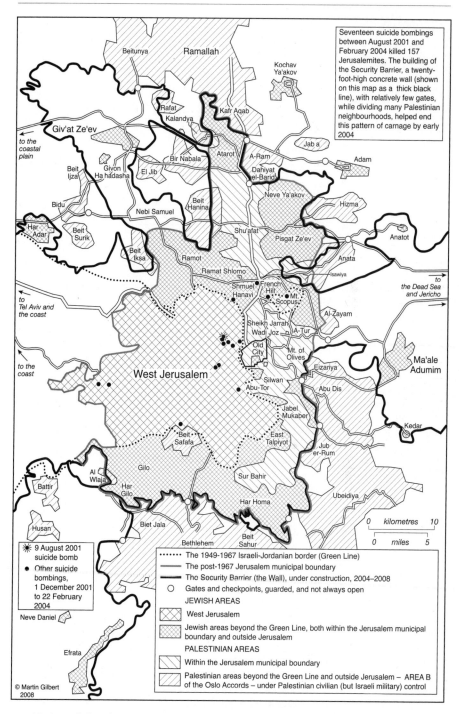

Seventeen suicide bombings between August 2001 and February 2004 killed 157 Jerusalemites. The building of the Security Barrier, a twenty-foot-high concrete wall (shown on this map as a thick black line), with relatively few gates, while dividing many Palestinian neighbourhoods, helped end this pattern of carnage by early 2004

Beitunya

Ramallah

Kochav Ya'akov

Rafat

Kalandya

Kafr Aqab

Giv'at Ze'ev

to the coastal plain

Beit Ijza

Givon Ha hadasha

El Jib

Bir Nabala

Atarot

A-Ram

Jab a

Adam

Dahiyat el-Barid

Bidu

Nebi Samuel

Beit Hanina

Neve Ya'akov

Hizma

Beit Surik

Shu'afat

Pisgat Ze'ev

Anatot

to Tel Aviv and the coast

Beit Iksa

Ramot

Ramat Shlomo

Anata

Isawiya

to the Dead Sea and Jericho

Shmuel Hanavi

French Hill

Mt. Scopus

Al-Zayam

to the coast

Sheikh Jarrah

Wadi Joz

A-Tur

West Jerusalem

Old City

Mt. of Olives

Eizariya

Ma'ale Adumim

Silwan

Abu-Tor

Abu Dis

Kedar

Jabel Mukaber

Beit Safafa

East Talpiyot

Jub er-Rum

Gilo

Sur Bahir

Al Wlaja

Har Gilo

Ubeidiya

Battir

Har Homa

0   kilometres   10

0   miles   5

Husan

Biet Jala

Beit Sahur

Bethlehem

* 9 August 2001 suicide bomb

• Other suicide bombings, 1 December 2001 to 22 February 2004

Neve Daniel

Efrata

The 1949–1967 Israeli-Jordanian border (Green Line)

The post-1967 Jerusalem municipal boundary

The Security Barrier (the Wall), under construction, 2004–2008

○ Gates and checkpoints, guarded, and not always open

JEWISH AREAS

West Jerusalem

Jewish areas beyond the Green Line, both within the Jerusalem municipal boundary and outside Jerusalem

PALESTINIAN AREAS

Within the Jerusalem municipal boundary

Palestinian areas beyond the Green Line and outside Jerusalem – AREA B of the Oslo Accords – under Palestinian civilian (but Israeli military) control

© Martin Gilbert 2008

38 Suicide bombings and the security barrier ('The Wall') in the Jerusalem area,
2001–2008

Glossary

Agudat Yisrael (the Agudah): ultra-religious Jewish political Party with an anti-Zionist orientation, established in Germany in 1912.

Alignment (Ma'arach): Hama'arach Le'achdut Po'alei Yisrael – The alignment for the unity of Israeli workers: left of centre Party formed in 1965 by Mapai and Ahdut Ha'avodah-Po'alei Zion.

Aliyah (literally, Ascent): immigration to the Land of Israel.

Alliance Israelite Universelle: school system founded in Paris in 1860, operating in Palestine since 1890s.

America Palestine Fund: founded 1939. Renamed America-Israel Cultural Foundation in 1948.

American Jewish Joint Distribution Committee (The Joint): charitable organization providing help for Jewish communities worldwide since 1914.

Amidar: national immigrant housing company set up in 1949.

Arab Democratic Party: founded in 1987 by Abdel Wahab Darawshe; first purely Arab Party to enter the Knesset.

Arab Legion: British-led Transjordanian military force active in the 1948 war.

Arab Liberation Army: Arab military force led by Fawzi el-Kaukji.

Ateret Cohanim (Crown of the Priests): founded 1979; considered it imperative for Jews to settle throughout the Land of Israel.

Banai (Brit Ne'emanei Eretz Yisrael): Alliance of the Land of Israel Faithful; formed in 1977 by Geula Cohen and Moshe Shamir in opposition to Begin's commitments at Camp David.

Betselem (In the Image): Israeli human rights organization established in 1988.

Bilu: (Beth Jacob Lechu Venelcha – House of Jacob, Come ye and let us go, Isaiah 2:5); secular and Socialist movement founded in Russia in the nineteenth century to encourage emigration of Jews to Palestine.

Brit Shalom (Peace Covenant): organization established by Palestinian Jews in 1925 to foster good Arab–Jewish relations.

Brit Trumpeldor: Revisionist youth movement founded in 1925; Known as Betar.

Bund: Jewish Socialist Workers' Party founded in Vilna in 1897; dedicated to the coming of a Russian Socialist government.

Dash: Democratic Movement for Change (DMC); political Party formed in 1976 under the leadership of Yigael Yadin.

Degel Hatorah Party: Ashkenazi ultra-Orthodox religious Party founded on eve of 1988 Knesset elections.

Elisha: organization set up in 1948 for treatment and rehabilitation of disabled soldiers.
Etzel: see Irgun.

Fatah (Victory): Movement for the National Liberation of Palestine set up in 1965, led by Abu Ammar (Yasser Arafat).

Gahal: immigrants who became soldiers on emigrating to Israel in 1948.
Galuth (Diaspora): Jews living outside Palestine/Israel.
Gedud Ha-Avodah: The Labour Legion; kibbutz movement.
Gush Emunim (Bloc of the Faithful): extra-parliamentary religious Zionist movement established in 1975; advocating extensive Jewish settlement throughout the West Bank and Gaza Strip.

Hadassah Woman's Zionist Organization of America: established in America in 1912; active in provision of medical care and health services in Palestine/Israel.
Haganah (Defence): Jewish defence force; absorbed into Israel Defence Forces in 1948.
Hakibbutz Hadati: cooperative organization founded in 1935.
halutzim (pioneers): Jews trained in Europe and the Americas and elsewhere, mostly in agriculture, to prepare them for emigration to Palestine.
Hamas (Harakat al-Muqawama al-Islamiyya): the Movement of the Islamic Resistance established in 1988, a branch of the Muslim Brotherhood; committed to Islamic rule for the whole of Palestine.
Hamashbir: national consumers' cooperative.
Haredi (plural Haredim): ultra-Orthodox, non-Zionist or anti-Zionist Jews.
Ha-Shomer (The Watchman): first self-defence association in Palestine; formed in 1909.
Ha-Shomer Hatzair (The Young Watchman): A Zionist youth movement founded in Vienna during the First World War.
Herut (Freedom): political Party founded in 1948 by Menachem Begin; part of the Likud bloc after 1973.
Hever Hakvutzot: collective association, non-Marxist, non-Socialist.
Histadrut: General Federation of Jewish Labour incorporating Solel Boneh, Hamashbir and Tnuva.
Hizballah (Party of God): Iranian-backed Palestinian movement active in South Lebanon.

IDF Israel Defence Forces (Zva Haganah le-Israel): Israel's army, navy and air force; known as both Zahal and IDF; formed from Haganah, the Palmach, the Irgun and the Stern Gang in 1948.
Irgun: (Irgun Zvai Leumi, IZL; National Military Organization) founded in 1931, known as both Irgun and, from its Hebrew initials, Etzel.
Islamic Jihad: splinter group from the Muslim Brotherhood.

Jewish Agency: the Jewish public body established in 1922, as a result of the League of Nations Mandate for Palestine, to advance the interests of the Jewish population in Palestine and to take part in the development of the country; independent of the World Zionist Organization from 1929.

Jewish Colonial Trust (Juedische Colonialbank): instrument for Jewish economic development and land purchase in Palestine. Established prior to the First World War.

Jewish National Council of Palestine Jews (Va'ad Leumi): the main administrative body of Palestinian Jewry under the Mandate; also known as the General Council.

Jewish National Fund (Karen Kayemet Le'Yisrael): fund to buy land in Palestine; established 1901.

Jewish Resistance Movement: brief working together of the Irgun, the Stern Gang and Haganah during 1946. Broke down following the bombing of the King David Hotel, 22 July 1946.

Jewish Settlement Police: established by the British working with the Jewish Agency in 1937.

kibbutz (literally, ingathering): collective settlement.

Kibbutz Artzi (My Land): Marxist kibbutz organization founded in 1935.

Kibbutz Hameuhad: The United Kibbutz movement established 1927; Marxist orientated.

Knesset (Assembly): Israel's parliament.

Land of Israel Movement: political lobby founded August 1967 advocating retention of all land captured by Israel on the West Bank and Gaza Strip.

Lebanese Front: coalition of Christian leaders formed in 1976, supported by Israel.

Lehi (Lohamei Herut Yisrael): Fighters for the Freedom of Israel, founded in 1940 by a breakaway group of Irgun members; cooperated with Haganah 1946; disbanded 1948; also known as the Stern Gang after its founder Avraham Stern.

Likud (union): Parliamentary bloc of right-wing Parties brought together in 1973 under General Ariel Sharon.

Lovers of Zion (Hovevei Zion): European Jewish movement established predominantly in Russia in 1884 to encourage emigration to Palestine.

ma'abara: immigrant transit camp (plural, ma'abarot).

Ma'arach: see Alignment.

Magen David Adom: Red Shield of David; Jewish health organization, equivalent to Red Cross and Red Crescent.

Mahal (Mitnavdei Hutz La-aretz): Foreign volunteers in the Israeli War of Independence.

Maki: Israeli Communist Party founded in 1948, split 1949, and again in 1965 and then in 1975.

Mapai (Mifleget Po'alei Eretz Yisrael): the Party of the Workers of the Land of Israel, founded in 1930.

Mapam (Mifleget Poa'lim Meuhedet): United Workers' Party, founded in 1948.

Meretz (Democratic Israel): political grouping founded in 1988, incorporating Mapam, Shinui and the Civil Rights Movement of Shulamit Aloni.

Mizrachi: Religious Zionist Party founded in Vilna in 1902.

Moledet (Homeland): political Party founded on the eve of the 1988 elections which favoured 'voluntary transfer' of Arabs unwilling to accept Jewish dominance in the Land of Israel.

moshav: village of smallholders combining cooperative and private farming; first established before the First World War.

Mossad (Institution for Intelligence and Special Tasks): founded in 1951; central organization in Israel for intelligence and security, including activities beyond the borders of the State.

Mossad le-Aliyah Bet: organization for 'Immigration B' (illegal immigration) set up by Haganah in 1938 and active until 1948.

Moses Montefiore Testimonial Fund: nineteenth-century, Russian-based charitable fund to support Jewish agricultural work in Palestine; chaired by Judah Leib Pinsker.

Muslim Brotherhood: Islamic national movement founded in Egypt in 1928 aimed at giving 'new soul in the heart of this nation to give it life by means of the Koran'; wanted Egypt to become an Islamic kingdom.

Nahal: Pioneering and Fighting Youth; set up by the Government of Israel in 1948 to provide a combination of military and agricultural training.

National Religious Party (NRP): Zionist religious Party established 1956 with the aim of preserving 'the religious character of the State'.

Nili (Nezah Yisrael Lo Yeshakker – The Eternity of Israel Will Not Lie): Jewish spy ring founded in 1915 in Palestine to assist the British war effort behind enemy lines.

Oded: Moroccan immigrant youth movement founded in Israel in 1962.

Palestine Foundation Fund (Keren Hayesod): founded in 1920, the financial arm of the World Zionist Organization; known in the US since 1939 as the United Jewish Appeal, and later as the United Israel Appeal.

Palestine Jewish Colonization Association (PICA): founded in 1891, a philanthropic organization to assist Jews worldwide and to encourage Jewish emigration from Russia and Eastern Europe to the Americas and Palestine.

Palestine Land Development Corporation: founded in 1908; purchased land with money raised by the Jewish National Fund.

Palestine Liberation Organization (PLO): established in 1964, an umbrella organization for Palestinian groups inside Israel and Jordan; since 1967 advocating a 'secular democratic State' in the West Bank and Gaza Strip as a possible prelude to wider statehood.

Palestine National Covenant: also known as the Palestine Charter; adopted by the PLO in 1964, redrafted in 1968, declares Palestine 'the homeland of the Arab Palestine people'.

Palmach (Plugot Mahatz): specially trained 'striking force' of the Haganah established in 1941.

Peace Now: extra-parliamentary, non-Party peace movement established in 1978 to keep the peace process 'at the forefront of the public agenda'.

Pioneers (Hehalutz): American Jewish organization set up by David Ben-Gurion and David Ben-Zvi to encourage large numbers of would-be immigrants to Palestine at the end of the First World War.

Provisional Council of State: Legislative body of the Jews of Palestine replacing the Jewish National Council; established March 1948.

Rafi (Reshimat Po'alei Yisrael U'bilti Miflagtiyim): the List of Israel Workers and Non-Partisans; a political Party formed by David Ben-Gurion in 1965.

Rakah Party (Reshimah Kommonistit Hadashah): New Communist List; predominantly Arab, established in 1965; regarded as 'bourgeois, chauvinist' and 'serving imperialism'.

Sabras: Palestinian-born Jewish youth; from the cactus fruit that is tough and prickly outside, sweet and succulent inside.

Shas (Shomrei Torah Sephardim): Sephardi Torah Guardians; ultra-religious, non-Zionist, Sephardi political Party established before 1984 elections.

She'erit Ha-Peletah (Surviving Remnant/The Saving Remnant): organization of Jewish displaced persons in Europe after 1945.

Shinui (Change): a centrist liberal, political protest movement founded in 1974, led by Amnon Rubinstein; later joined Democratic Movement for Change (Dash), and later still, Meretz.

Shlomzion: political Party founded by Ariel Sharon before 1977 election.

Small Zionist Actions Committee: Zionist movement's highest parliamentary forum before 1948.

Stern Gang: see Lehi.

Tehiyah (Revival): nationalist Party established in 1977 by the Loyalist Circle in Herut, the Land of Israel Movement and Gush Emunim in opposition to the Camp David Accords.

UNEF: United Nations Emergency Force; created after Suez War of 1956 to replace Israeli troops in Sinai.

UNIFIL: United Nations Interim Force in Lebanon established in 1982.

UNRRA: United Nations Relief and Rehabilitation Administration formed in 1943 to assist European refugees.

UNRWA: United Nations Relief and Works Agency established to help Palestinian refugees after 1948.

UNSCOP: United Nations Special Committee on Palestine, 1947.

UNTSO: United Nations Truce Supervision Organization.

United Religious Front: grouping of all the Israeli religious Parties formed before the first Knesset and subsequently disbanded.

World Zionist Organization: founded by Theodor Herzl in 1897; active as the Jewish Agency for Palestine until 1929. After 1952 recognized by the Knesset, together with the Jewish Agency, as 'the authorized agencies which continue to operate in Israel for the development and settlement of the country, absorption of immigration and the coordination of activities in Israel of Jewish institutes active in those fields'.

Yad Vashem: The Holocaust Martyrs' and Heroes' Remembrance Authority established in 1953 in Jerusalem.

Yeshiva: Jewish religious seminary or Talmudic college for men.

Yishuv (literally: to settle, settlement): the Jewish community in Palestine before 1948.

Yisrael B'Aliyah: Russian immigrant Party founded before the 1996 elections, headed by Natan Sharansky.

Youth Aliyah: established in 1932 to help youngsters from prewar Germany. From 1949 it undertook responsibility for youth immigration and care from every land of the Diaspora.

Zahal: see IDF.

Bibliography of works consulted

Reference Books

Yehoshua Freundlich (editor), *Documents on the Foreign Policy of Israel*, Volume 1, *14 May –30 September 1948*, Israel State Archives, Jerusalem, 1981.

Yehoshua Freundlich (editor), *Documents on the Foreign Policy of Israel*, Volume 5, *1950*, Israel State Archives, Jerusalem, 1988.

Ruth Lapidoth and Moshe Hirsch (editors), *The Arab–Israeli Conflict and its Resolution: Selected Documents*, Martinius Nijhoff, Dordrecht, The Netherlands, 1992.

Meron Medzini (editor), *Israel's Foreign Relations, Selected Documents, 1947–1974*, Ministry for Foreign Affairs, Jerusalem, 1976.

Susan Hattis Rolef (editor), *Political Dictionary of the State of Israel*, The Jeruslem Publishing House, second edition, Jerusalem, 1993.

Yemima Rosenthal (editor), *Documents on the Foreign Policy of Israel*, Volume 2, *October 1948–April 1949*, Israel State Archives, Jerusalem, 1983.

Yemima Rosenthal (editor), *Documents on the Foreign Policy of Israel*, Volume 3, *Armistice Negotiations with the Arab States*, December 1948–July 1949, Israel State Archives, Jerusalem, 1983.

Yemima Rosenthal (editor), *Documents on the Foreign Policy of Israel*, Volume 4, *May–December 1949*, Israel State Archives, Jerusalem, 1986.

Yaacov Shimoni (editor), *Biographical Dictionary of the Middle East*, Facts on File, The Jerusalem Publishing House, New York, Oxford, Sydney, 1991.

Geoffrey Wigoder (editor-in-chief), *New Encyclopedia of Zionism and Israel*, 2 volumes, Associated University Presses, London and Toronto, 1994.

Gedalia Yogev (general editor), Yehoshua Freundlich, Michael Engel and Yemima Rosenthal, *Political and Diplomatic Documents, December 1947–May 1948*, Israel State Archives, Central Zionist Archives, Jerusalem, 1979.

Encyclopaedia Judaica, Keter Publishing House, Jerusalem, 1972.

Guide Books

Joan Comay, *Introducing Israel*, Methuen, London, 1963.

Elian-J Finbert, *Israel*, Librairie Hachette, Paris, 1956.

Eugene Fodor (editor), *Fodor's Israel 1971*, Hodder and Stoughton, London, 1971.

Zev Vilnay, *The Guide to Israel*, Ahiever, Jerusalem, 1955 (16th edition, 1973).

General Histories

Yossi Beilin, *Israel: A Concise Political History*, St Martin's Press, New York, 1992.

Abba Eban, *Heritage: Civilization and the Jews*, Weidenfeld and Nicolson, London, 1984.

Yuval Elizur and Eliahu Salpeter, *Who Rules Israel?*, Harper and Row, New York, 1973.

Walter Eytan, *The First Ten Years, A Diplomatic History of Israel*, Simon and Schuster, New York, 1958.

William Frankel, *Israel Observed, An Anatomy of the State*, Thames and Hudson, London, 1980.

Gerald de Gaury, *The New State of Israel*, Derek Verschoyle, London, 1952.

David J. Goldberg, *To The Promised Land, A History of Zionist Thought from Its Origins to the Modern State of Israel*, Penguin Books, London, 1996.

Noah Lucas, *The Modern History of Israel*, Weidenfeld and Nicolson, London, 1974.

George Mikes, *The Prophet Motive: Israel Today and Tomorrow*, André Deutsch, London, 1969.

Terence Prittie, *Israel, Miracle in the Desert*, Frederick A. Praeger, New York, 1967.

Abraham Rabinovich, *Israel*, Flint River Press, London, 1989.

L. F. Rushbrook Williams, *The State of Israel*, Faber and Faber, London, 1957.

Howard M. Sachar, *A History of Israel from the Rise of Zionism to Our Time*, Alfred A. Knopf, New York, 1976.

Autobiographical Works

Menachem Begin, *The Revolt*, Nash, New York, 1951.

Alex Bein (editor), Arthur Ruppin, *Memoirs, Diaries, Letters*, Weidenfeld and Nicolson, London, 1971.

David Ben-Gurion, *Israel: Years of Challenge*, Massadah-P.E.C. Press, Tel Aviv, 1963.

David Ben-Gurion, *Israel, A Personal History*, New English Library, London, 1972.

Izhak Ben-Zvi, *The Hebrew Battalions, Letters*, Yad Izhak Ben-Zvi, Jerusalem, 1969.

Lieutenant-General E. L. M. Burns, *Between Arab and Israeli*, George G. Harrap, London, 1962.

Moshe Dayan, *Story of My Life*, Morrow, New York, 1976.

Yaël Dayan, *My Father, His Daughter*, Weidenfeld and Nicolson, London, 1985.

Placido Domingo, *My First Forty Years*, Weidenfeld and Nicolson, London, 1983.

Abba Eban, *Personal Witness, Israel through My Eyes*, Jonathan Cape, London, 1993.

Chaim Herzog, *Living History, The Memoirs of a Great Israeli Freedom-Fighter, Soldier, Diplomat and Statesman*, Weidenfeld and Nicolson, London, 1997.

Theodor Herzl, Zionist Writings, *Essays and Addresses*, 2 volumes, Herzl Press, New York, 1975.

Ben-Zion Ilan, *An American Soldier/Pioneer in Israel*, Labour Zionist Letters, New York, 1979.

Doris Katz, *The Lady was a Terrorist, During Israel's War of Liberation*, Futuro Press, New York, 1953.

Shmuel Katz, *Days of Fire*, W. H. Allen, London, 1968.

Teddy Kollek (with Amos Kollek), *For Jerusalem, A Life*, Random House, New York, 1978.

Golda Meir, *My Life*, Weidenfeld and Nicolson, London, 1975.

Yaakov Meridor, *Long is the Road to Freedom*, Newzo Press, Johannesburg, South Africa, 1955.

Mordechai Nurock (editor), *David Ben-Gurion, Rebirth and Destiny of Israel* (speeches), Philosophical Library, New York,1954.

Shimon Peres, *Battling for Peace, Memoirs*, Weidenfeld and Nicolson, London, 1995.

Leah Rabin, *Rabin, Our Life, His Legacy*, G. P. Putnam's Sons, New York, 1997.

Yitzhak Rabin, *The Rabin Memoirs*, second edition, Steimatzky, Bnei Brak, 1994.

Gideon Rafael, *Destination Peace, Three Decades of Israeli Foreign Policy*, Weidenfeld and Nicolson, London, 1981.

Ariel Sharon (with D. Charnoff), *Warrior*, Simon and Schuster, New York, 1989.

Naomi Shepherd, *Alarms and Excursions, Thirty Years in Israel*, Collins, London, 1990.

Chaim Weizmann, *Trial and Error*, Harper and Brothers, New York, 1949.

Ezer Weizman, *The Battle for Peace*, Bantam Books, London, 1981.

Biographies

David Aberbach, *Bialik*, Peter Halban, London, 1988.

Shlomo Avineri, *Arlosoroff*, Peter Halban, London, 1989.

Michael Bar-Zohar, *Ben-Gurion*, Weidenfeld and Nicolson, London, 1978.

Humphrey Burton, *Leonard Bernstein*, Doubleday, New York, 1994.

Matti Golan, *Shimon Peres, A Biography*, Weidenfeld and Nicolson, London, 1982.

David Horovitz (editor), *Yitzhak Rabin, Soldier of Peace*, Peter Halban, London, 1996.

Shmuel Katz, *A Biography of Vladimir (Ze'ev) Jabotinsky*, 2 volumes, Barricade Books, New York, 1996.

Simcha Kling, *Joseph Klausner*, Thomas Yoseloff, New York, 1970.

Dr Joseph Klausner, *Menachem Ussishkin, His Life and Work*, Joint Zionist Publication Committee, London, no date.

Naomi Layish, *Wilfred Israel, German Jewry's Secret Ambassador*, Weidenfeld and Nicolson, 1984.

Maurice Pearlman, *Collective Adventure, An Informal Account of the Communal Settlements of Palestine*, William Heinemann, London, 1938.

Jehuda Reinharz, *Chaim Weizmann, The Making of a Statesman*, Oxford University Press, New York and Oxford, 1993.

Neil Asher Silberman, *A Prophet from Amongst You, The Life of Yigael Yadin: Soldier, Scholar, and Mythmaker of Modern Israel*, Addison-Wesley, Reading, Massachusetts, 1993.

Leon Simon, *Ahad Ha-Am, Asher Ginzberg, A Biography*, East and West Library, London, 1960.

Robert Slater, *Warrior Statesman, The Life of Moshe Dayan*, St Martin's Press, New York, 1991.

Robert Slater, *Rabin of Israel, A Biography*, revised edition, Robson Books, London, 1993.

Shabtai Teveth, *Moshe Dayan, The Soldier, the Man, the Legend*, Weidenfeld and Nicolson, London, 1972.

Bernard Wasserstein, *Herbert Samuel, A Political Life*, Clarendon Press, Oxford, 1992.

Ronald W. Zweig (editor), *David Ben-Gurion, Politics and Leadership in Israel*, Frank Cass, London, and Yad Izhak Ben-Zvi, Jerusalem, 1991.

Specific Periods and Episodes

Donna E. Arzt, *Refugees Into Citizens, Palestinians and the End of the Arab-Israeli Conflict*, Council on Foreign Relations, New York, 1997.

Ehud Avriel, *Open the Gates! A Personal Story of 'Illegal' Immigration to Israel*, Weidenfeld and Nicolson, London, 1975.

Clinton Bailey, *Bedouin Poetry from Sinai and the Negev, Mirror of a Culture*, Clarendon Press, Oxford, 1991.

Mordechai Bar-On, *The Gates of Gaza, Israel's Road to Suez and Back, 1955–1957*, St Martin's Press, New York, 1994.

Mordechai Bar-On, *In Pursuit of Peace, A History of the Israeli Peace Movement*, United States Institute of Peace Press, Washington DC, 1996.

N. Bar-Yaacov, *The Israeli–Syrian Armistice, Problems of Implementation, 1949–1966*, Magnes Press, Jerusalem, 1967.

Arnold Behr, *Israel 1948, Photographs and Recollections*, Gordon Fraser, London, 1988.

Alex Bein, *The History of Jewish Agricultural Settlement in Palestine*, Rubin Mass, Jerusalem, 1939.

David Ben-Gurion, *My Talks with Arab Leaders*, Keter Books, Jerusalem, 1997.

Eli Ben-Hanan, *Elie Cohn, Our Man in Damascus*, A. D. M. Publishing House, Tel Aviv, 1967.

Yeshayahu Ben-Porat, Eitan Haber and Zeev Schiff, *Entebbe Rescue*, Delacorte Press, New York, 1976.

Michael Berkowitz, *Zionist Culture and West European Jewry before the First World War*, Cambridge University Press, Cambridge, 1993.

Uri Bialer, *Between East and West: Israel's Foreign Policy Orientation, 1948–1956*, Cambridge University Press, Cambridge, 1990.

Ian Black and Benny Morris, *Israel's Secret Wars, A History of Israel's Intelligence Services*, Grove Weidenfeld, New York, 1991.

Michael Brecher, *The Foreign Policy System of Israel, Settings, Images, Process*, Yale University Press, New Haven, 1972.

Yigal Bronner and Assaf Oron (editors), *Restrictions on Travel Abroad for East Jerusalem and West Bank Palestinians*, Hotline: Centre for the Defence of the Individual, Jerusalem, 1992.

Daniel Carpi and Gedalia Yogev (editors), *Zionism, Studies in the History of the Zionist Movement and of the Jewish Community in Palestine*, Masada, Tel Aviv, 1975.

Randolph S. Churchill and Winston S. Churchill, *The Six Day War*, Heinemann, London, 1967.

Israel Cohen, *The Zionist Movement*, Frederick Muller, London, 1945.

Haim H. Cohn, *Human Rights in the Bible and Talmud*, Ministry of Defence Books, Tel Aviv, 1989.

Richard Crossman, *Palestine Mission, A Personal Record*, Hamish Hamilton, London, 1946.

Moshe Dayan, *Diary of the Sinai Campaign*, Weidenfeld and Nicolson, London, 1966.

Yaël Dayan, *Israel Journal: June, 1967*, McGraw-Hill, New York, 1967.

Abba Eban, *Voice of Israel*, Horizons Press, New York, 1957.

Abba Eban, *The New Diplomacy, International Affairs in the Modern Age*, Random House, New York, 1983.

Arie Lova Eliav, *New Heart, New Spirit, Biblical Humanism for Modern Israel*, Jewish Publication Society, Philadelphia, 1988.

Amos Elon, *The Israelis, Founders and Sons*, Weidenfeld and Nicolson, London, 1971.

Amos Elon and Sana Hassan, *Between Enemies, An Arab–Israeli Dialogue*, Andre Deutsch, London, 1974.

D. Jason Fenton, *Volunteers in the War of Independence*, privately printed, 1995.

William Frankel, *Israel Observed, An Anatomy of the State*, Thames and Hudson, London, 1980.

Isaiah Friedman, *The Question of Palestine, British–Jewish–Arab Relations: 1914–1918*, second expanded edition, Transaction Publishers, New Brunswick, USA, 1992.

Moshe Gat, *The Jewish Exodus from Iraq, 1948–1951*, Frank Cass, London, 1997.

Mordechai Gazit, *President Kennedy's Policy Toward the Arab States and Israel, Analysis and Documents*, Tel Aviv University, Tel Aviv, 1983.

Eliyahu Golomb, *The History of Jewish Self-Defence in Palestine (1878–1921)*, The Zionist Library, Tel Aviv, 1946.

Anthony Gorst and Lewis Johnman, *The Suez Crisis*, Routledge, London, 1997.

Stephen Green, *Living by the Sword, America and Israel in the Middle East, 1968–87*, Faber and Faber, London, 1988.

Ruth Gruber, *Rescue, The Exodus of the Ethiopian Jews*, Atheneum, New York, 1987.

Baruch Gur-Gurevitz, *Open Gates, The Story Behind the Mass Immigration to Israel from the Soviet Union and its Successor States*, Jewish Agency for Israel, Jerusalem, 1996.

Ze'ev Hadari and Ze'ev Tsahor, *Voyage to Freedom, An Episode in the Illegal Immigration to Palestine*, Vallentine, Mitchell, London, 1984.

Y. Harkabi, *Arab Attitudes To Israel*, Vallentine, Mitchell, London, 1972.

Eric Hammel, *Six Days in June, How Israel Won the 1967 Arab–Israeli War*, Charles Scribner's Sons, New York, 1992.

Joseph Heller, *The Stern Gang, Ideology, Politics and Terror, 1940–1949*, Frank Cass, London, 1995.

Theodor Herzl, *Altneuland, Old-New Land*, illustrated edition, W. Turnowsky and Son, Haifa, 1960.

Brigadier General C. Herzog, *Israel's Finest Hour*, Ma'ariv, Tel Aviv, 1967.

Chaim Herzog, *The Arab–Israeli Wars, War and Peace in the Middle East,* Arms and Armour Press, London, 1982.

Yaacov Herzog, *Israel in the Middle East, An Introduction*, Jerusalem Papers on Peace Problems, Hebrew University, Jerusalem, February 1975.

Dilip Hiro, *Sharing the Promised Land, An Interwoven Tale of Israelis and Palestinians*, Hodder and Stoughton, London, 1996.

F. Robert Hunter, *The Palestinian Uprising, A War by Other Means*, I.B. Tauris, London, 1991.

Julian Huxley, *From an Antique Land*, Crown Publishers, New York, 1954.

Raphael Israeli, *Muslim Fundamentalism in Israel*, Brassey's (UK), London, 1993.

Dominique-D Junod, *The Imperilled Red Cross and the Palestine-Eretz Yisrael Conflict, 1945–1952, The Influence of Institutional Concerns on a Humanitarian Operation*, Kegan Paul International, London, 1996.

Samuel Katz, *Days of Fire, the Secret Story of the Making of Israel*, W. H. Allen, London, 1968.

Shmuel Katz, *Battletruth, The World and Israel*, Dvir, Tel Aviv, 1983.

Zachariah Kay, *Canada and Palestine, The Politics of Non-Commitment*, Israel Universities Press, Jerusalem, 1978.

Elie Kedourie, *The Chatham House Version and other Middle-Eastern Studies*, Weidenfeld and Nicolson, London, 1970.

Aharon Kellerman, *Society and Settlement, Jewish Land of Israel in the Twentieth Century*, State University of New York Press, New York, 1993.

Walid Khalidi (editor), *All that Remains: The Palestinian Villages Occupied and Depopulated by Israel in 1948*, Institute for Palestine Studies, Washington D.C., 1992.

David Kimche and Dan Bawly, *The Sandstorm, The Arab–Israeli War of June 1967: Prelude and Aftermath*, Secker and Warburg, London, 1968.

Jon and David Kimche, *The Secret Roads, The 'Illegal' Migration of a People 1938–1948*, Secker and Warburg, London, 1954.

Jon and David Kimche, *Both Sides of the Hill, Britain and the Palestine War*, Secker and Warburg, London, 1960.

I. J. Kliger and I. Weitzmann, *Malaria Control Demonstrations in Palestine*, Hadassah Medical Organization, Jerusalem, 1922.

Arthur Koestler, *Promise and Fulfilment, Palestine 1917–1949*, Macmillan, London, 1949.

Keith Kyle, *Suez*, Weidenfeld and Nicolson, London, 1991.

Karl Lenk, *The Mauritius Affair, The Boat People of 1940/41*, R. S. Lenk, Brighton, 1993.

Gordon Levett, *Flying Under two Flags, An ex-RAF Pilot in Israel's War of Independence*, Frank Cass, London, 1994.

Aryeh Levin, *Envoy to Moscow, Memoirs of an Israeli Ambassador 1988–1992*, Frank Cass, London, 1996.

Harry Levin, *Jerusalem Embattled, A Diary of the City under Siege, March 25th, 1948 to July 18th, 1948*, Victor Gollancz, London, 1950.

Irene Lewitt (editor), *The Israel Museum*, Laurence King, London, 1995.

Berl Locker, *A Stiff-necked People, Palestine in Jewish History*, Victor Gollancz, London, 1946.

Lt. Col. Netanel Lorch, *The Edge of the Sword, Israel's War of Independence, 1947–1949*, 2nd revised edition, Masada Press, Jerusalem, 1968.

Misha Louvish (editor), *My Talks with Arab Leaders, David Ben-Gurion*, Keter Books, Jerusalem, 1972.

Ehud Luz, *Parallels Meet, Religion and Nationalism in the Early Zionist Movement, 1882–1904*, Jewish Publication Society, Philadelphia, 1988.

James G. McDonald, *My Mission in Israel, 1948–1951*, Simon and Schuster, New York, 1951.

Ian McIntyre, *The Proud Doers, Israel after Twenty Years*, British Broadcasting Corporation, London, 1968.

Bruce Maddy-Weitzman and Efraim Inbar (editors), *Religious Radicalism in the Greater Middle East*, Frank Cass, London,1997.

David Makovsky, *Making Peace with the PLO, The Rabin Government's Road to the Oslo Accord*, Westview Press, Boulder, Colorado, 1996.

Neville J. Mandel, *The Arabs and Zionism before World War I*, University of California Press, Berkeley and Los Angeles, 1976.

Benny Morris, *The Birth of the Palestinian Refugee Problem, 1947–1949*, Cambridge University Press, Cambridge, 1987.

Benny Morris, *Israel's Border Wars, 1949–1956, Arab Infiltration, Israeli Retaliation, and the Countdown to the Suez War*, Clarendon Press, Oxford, 1993.

Lon Nordeen, *Fighters Over Israel, The Story of the Israeli Air Force from the War of Independence to the Bekaa Valley*, Orion Books, London, 1990.

Yehuda Ofter, *Operation Thunder, The Entebbe Raid, The Israelis' Own Story*, Penguin, London, 1976.

Ritchie Ovendale, *The Origins of the Arab–Israeli Wars*, second edition, Longman, London, 1992.

Amos Oz, *Israel, Palestine and Peace, Essays*, Vintage, 1994.

Palestine Land Transfer Regulations, Letter to the Secretary-General of the League of Nations, His Majesty's Stationery Office, London, 1940.

Palestine: Statement of Information Relating to Acts of Violence, His Majesty's Stationery Office, London, 1946.

Palestine: Termination of the Mandate, 15th May, 1948, His Majesty's Stationery Office, London, 1948.

Lt. Colonel J. H. Patterson, *With the Zionists in Gallipoli*, Hutchinson, London, 1916.

Moshe Pearlman, *The Army of Israel*, Philosophical Library, New York, 1950.

Moshe Pearlman, *The Capture of Adolf Eichmann*, Weidenfeld and Nicolson, London, 1961.

Shimon Peres, *David's Sling, The Arming of Israel*, Weidenfeld and Nicolson, London, 1970.

Shimon Peres, *From These Men, Seven Portraits*, Weidenfeld and Nicolson, London, 1979.

Shimon Peres (with Arye Naor), *The New Middle East*, Henry Holt, New York, 1993.

Don Peretz, *Israel and the Palestine Arabs*, Middle East Institute, Washington DC, 1958.

Joel Peters, *Israel and Africa, The Problematic Friendship*, British Academic Press, London, 1992.

Y. Porath, *The Emergence of the Palestinian–Arab National Movement 1918–1929*, Frank Cass, London, 1974.

Zipporah Porath, *Letters from Jerusalem, 1947–1948*, Association of Americans and Canadians in Israel, Jerusalem, 1987.

David Pryce-Jones, *Next Generation, Travels in Israel*, Weidenfeld and Nicolson, London, 1964.

William B. Quandt, *Camp David, Peacemaking and Politics*, The Brookings Institution, Washington DC, 1986.

Magnus Ranstorp, *Hizb'allah in Lebanon, The Politics of the Western Hostage Crisis*, Macmillan Press, London, 1997.

Louis Rapoport, *Confrontations, Israeli Life in the Year of the Uprising*, Quinlan Press, Boston, 1988.

Jehuda Reinharz and Anita Shapira (editors), *Essential Papers on Zionism*, Cassell, London, 1996.

Rolly Rosen, *Peace in the Neighbourhoods? The Activity of Peace Organizations to Widen Their Circles of Support*, Shatil, Jerusalem, 1997

Barry Rubin, Joseph Ginat and Moshe Ma'oz (editors), *From War to Peace: Arab–Israeli Relations, 1973–1993*, Sussex Academic Press, Brighton, 1994.

H. Sacher (editor), *Zionism and the Jewish Future By Various Writers*, John Murray, London, 1916.

Harry Sacher, *Israel, The Establishment of a State*, Weidenfeld and Nicolson, London, 1952.

Harry Sacher, *Zionist Portraits and Other Essays*, Anthony Blond, London, 1959.

Edward W. Said, *The Question of Palestine*, Vintage, New York, 1992 (first published 1979).

Edward W. Said, *The Politics of Dispossession, The Struggle for Palestinian Self-Determination 1969–1994*, Chatto and Windus, London, 1994.

Edwin Samuel, *Handbook of the Jewish Communal Villages in Palestine*, Zionist Organization Youth Department, Jerusalem, 1938.

Ze'ev Schiff, *October Earthquake, Yom Kippur, 1973*, University Publishing Projects, Tel Aviv, 1974.

Ze'ev Schiff, *A History of the Israeli Army, 1874 to the Present*, Macmillan, New York, 1985.

Avraham Shapira (principal editor), *The Seventh Day, Soldiers' Talk About the Six-Day War*, Andre Deutsch, London, 1970.

Zeev Sharef, *Three Days*, W. H. Allen, London, 1962.

Naomi Shepherd, *The Russians in Israel, The Ordeal of Freedom*, Simon and Schuster, London, 1993.

Colin Shindler, *Ploughshares Into Swords? Israelis and Jews in the Shadow of the Intifada*, I. B. Tauris, London, 1991.

Nir Shohet, *The Story of an Exile, A Short History of the Jews of Iraq*, Association for the Promotion of Research, Literature and Art, Tel Aviv, 1982.

Nahum Sokolow, *Hibbath Zion (The Love for Zion)*, Rubin Mass, Jerusalem, 1934.

Jacob Sonntag (editor), *New Writing from Israel 1976, Stories, Poems, Essays*, Corgi, London, 1976.

Stephen Spender, *Learning Laughter* (Youth Aliyah in Israel), Weidenfeld and Nicholson, London, 1952.

Yael Stein, *The Quiet Deportation, Revocation of Residency of East Jerusalem Palestinians*, Hamoked, Centre for the Defence of the Individual, Jerusalem, 1997.

Eliezer Tauber, *The Arab Movements in World War I*, Frank Cass, London, 1993.

Shabtai Teveth, *Ben-Gurion's Spy, The Story of the Political Scandal That Shaped Modern Israel*, Columbia University Press, New York, 1996.

S. Ilan Troen and Noah Lucas (editors), *Israel, The First Decade of Independence*, State University of New York Press, Albany, NY, 1995.

Dov Vardi, *New Hebrew Poetry*, Sefer Press, Tel Aviv, 1947.

Evelyn Wilcock, *Pacifism and the Jews*, Hawthorn Press, Stroud, Gloucestershire, 1994.

Mary C. Wilson, *King Abdullah, Britain and the Making of Jordan*, Cambridge University Press, Cambridge, 1987.

Gad Yaacobi, *Wanted: A New Course*, Steimatzky, Bnei Brak, 1991.

Gad Yaacobi, *Breakthrough, Israel in a Changing World*, Cornwall Books, New York, 1996.

Yigael Yadin, *Bar-Kokhba, The Rediscovery of the Legendary Hero of the Last Jewish Revolt against Imperial Rome*, Weidenfeld and Nicolson, London, 1971.

Yigael Yadin, *Masada*, Weidenfeld and Nicolson, London, 1966.

Yigael Yadin, *Hazor, The Rediscovery of a Great Citadel of the Bible*, Weidenfeld and Nicolson, London, 1975.

Yigael Yadin, *The Temple Scroll, The Hidden Law of the Dead Sea Sect*, Weidenfeld and Nicolson, London, 1985.

Shmuel Yilma, *From Falasha to Freedom, An Ethiopian Jew's Journey to Jerusalem*, Gefen, Jerusalem, 1996.

Itzhak Zamir and Allen Zysblat, *Public Law in Israel,* Clarendon Press, Oxford, 1996.

Shlomo Zemach, *An Introduction to the History of Labour Settlement in Palestine*, Zionist Library, Tel Aviv, 1945.

I have also drawn on material in the *Palestine Post* and the *Jerusalem Post*; and on the following of my own works: *Sir Horace Rumbold, Portrait of a Diplomat* (Heinemann, London, 1974); *Exile and Return, the Emergence of Jewish Statehood* (Weidenfeld and Nicolson, London, 1977); *Winston S. Churchill*, volumes 4 and 8 (Heinemann, London, 1977, 1988); *The Routledge Atlas of the Arab–Israeli Conflict*, sixth edition, (Routledge, London, 1996); and *Jerusalem in the Twentieth Century*, (Chatto and Windus, London, 1996).

Index

Compiled by the author

assassination *(continued)*
of the Emir Abdullah, 274; of Yitzhak
Rabin, 274, 587, 588–92, 618; of
Nokrashy Pasha, 242; of Israel Kastner,
304, 349; of President Kennedy, 348;
of President Sadat, 499; and
'liberation', 504; of President-elect
Bashir Jemayel, 509; of Yahiya Ayash,
592; attempted, against Max Nordau,
22; attempted, against Khaled Mashaal,
616
Assassins: recalled, 122
Association for Civil Rights in Israel
(A.C.R.I.): 608
Association of Jews from Arab Countries:
271
Association of Wine Growers: 26
Assyria: and the road to war in 1967,
369
Ateret Cohanim: advocates widespread
settlement in the Occupied Territories,
524
Athens Airport: a terrorist attack on,
404; a hijacking after take-off from,
471
Athlit: an experimental station at, 25;
'illegal' immigrants detained at, 101,
107; a raid on, 130–1; a leading Jew
murdered while on his way from, 157
Atlantic (refugee ship): 105, 106, 108
Atlantic Ocean: 75
Atomic Reactor (Dimona): 522
Attar, Danny: and a cross-community
cooperative venture, 651
Attlee, Clement: and 'pure hypocrisy',
128; and the future of the 100,000, 135
Auerbach, Elias: opens a hospital, 28
Auschwitz: and a trial for slander, 303;
visits to, 556; the liberation of,
commemorated, 579; a survivor of,
killed by a suicide bomb, 613; son of a
survivor of, in space, 628
Austin, Warren: and American
opposition to partition, 165
Australia: 137, 138, 150, 185, 210, 268;
volunteers from (1973), 437
Australian forces (in Syria): 111
Austria: 79, 85, 93, 95, 97, 124, 125, 126,
128, 134, 136, 336, 352, 426
Austro-Hungarian Empire: 11, 19, 30,
90, 259
Auxiliary Territorial Service (ATS): 103
Avenue of the Righteous (Yad Vashem,
Jerusalem): 289

Avidar, Yosef: and military training, 85
Avigur, Shaul: see Meirov, Shaul
Avihail: founded, 69
Aviner, Rabbi Shlomo: 'the Arabs are
squatters', 524
Avisar, Sergeant Gitai: shot dead, 562
Avner, Yehuda: xii; and the Eichmann
Trial, 337; and the 'retreat of the
pioneer ethos', 404; and the
immigration of Ethiopian Jews, 497;
and the rotation experiment, 524; and
the Washington talks, 549
Avrahami, District Police Superintendent
Levi: and the use of tear-gas, 281
Avramov, Gregory: stabbed to death,
561
Avriel, Ehud: in Istanbul, 115; in
Czechoslovakia, 168, 199, 286
Awali River (Lebanon): Israeli troops
reach, 505; Israel withdraws south of,
512
Axelrod, David: incitement by, 539
Ayalon, Assistant Superintendent Moshe:
injured, 281
Ayash, Hofit: killed, 592
Ayash, Yahiya: a wanted man, 575;
assassinated, 592
Ayelet Ha-Shahar: safety of, 175
Azar, Samuel: hanged, 296
Azaria, Natan: stabbed to death, 561
Azma'ut (steamship): brings refugees,
247
Azur: an accident at, 404
Azzadin, Amin Bey: leaves, 172
Azzam Pasha: and partition, 141

Baabde (Lebanon): Israeli and Syrian
tanks battle near (1982), 506
Bab al-Wad: ambushes at, 157, 167, 173
Babylon: and Jerusalem 99; and the
Jews, 113; and Jewish terrorism, 139;
exiles from, remembered, 231; and the
road to war in 1967, 369
Baghdad: 32, 74, 257, 258, 303; 'savage
visions' in, 369; and the Gulf War, 546
Baghdad Pact (1955): 304
Bahir, Arieh: at Basle (1946), 139
Bahrain: and Palestine, 141; supports
Road Map, 627; represented at
Annapolis (2007), 653
Bailey, Clinton: champions the Bedouin,
360; and a plan for the West Bank,
610–11

Churchill, Winston *(continued)*
and the Jewish Brigade Group, 117;
and a Stern Gang promise, 119; and a
'squalid war', 145; in 1940, 202;
precedents by, 214, 346
Chuvakhin, Dmitri: protests, 366
Circassians: in Palestine, 26; in Israel,
344
'City Line' (Jerusalem): established, 241
Civil Rights Movement: protests, 509; its
electoral success, 531; and the
formation of a new political Party, 532
Claims Conference: 279, 284
Clinton, President William J.: rejects an
appeal for clemency, 523; and the
Oslo Accords, 564; and 'a mission
inspired by a dream of peace', 576;
and the assassination of Rabin, 588;
his eulogy at Rabin's funeral, 590–1;
attends an anti-terrorism conference,
593; seeks to diffuse a crisis, 596; hosts
talks at Camp David, 621, 622
Cohen, Eddie: killed in action, 200
Cohen Eli: hanged, 355–6
Cohen, Emmanuel: his ancestors, 603
Cohen family (of Ma'alot): and a
terrorist attack, 466
Cohen, Geula: forms a new political
Party, 494
Cohen, Nehama: killed, 155
Cohen, Peretz: killed in action (1973),
603
Cohen, Sir Ronald: and 'the great
challenge', 646–7
Cohen, Samuel: and Hatikvah, 7
Cohen, Professor Stanley: and human
rights on the West Bank, 533–4
Cohen, Yehoshua: an assassin, 228
Cohn, Haim: and the 'Lavon Affair', 338;
and human rights in the Bible, 539;
and a 'lie', 595
Cold War: and the October War, 449
Collège des Frères (Jerusalem): 221
Cologne (Germany): 18
Columbia space flight: destruction of,
627–8
Comay, Joan: witnesses opera and
gunfire, 170; describes Jaffa, 350
Commercial Centre (Jerusalem):
attacked, 158
Committee of Seven: and the 'Lavon
Affair', 338–9
Commonwealth of Independent States
(CIS): and Soviet Jewish emigration,

537, 575; and Rabin's second
premiership, 554
Communism: in the Ukraine, 49; among
Zionist pioneers in Palestine, 62;
opposed, 63; and the Popular Front
(France), 114; and Czechoslovakia
(after 1948), 165, 224–5, 285, 286; and
Roumania, 259, 482; and China, 288;
and Yugoslavia, 345–6; and the Soviet
Union, 347, 444; and the Communist
bloc, 347, 460, 468; and the post-
Communist era, 551, 556
Communist Party (of Israel): 189, 250,
275, 286, 399–400, 471, 479
Conference on Jewish Material Claims
against Germany: see Claims
Conference
Concerned Parents of Israeli Soldiers:
and the intifada, 539
Conservative Party (Britain): 121
Constansa (Roumania): 115
Constantinople: 12, 17, 30, 31, 270; for
subsequent index entries see Istanbul
Constituent Assembly (1949): 250
The Conventional Lies of our Civilization
(Max Nordau): 10
COPE: an Israeli-Palestinian initiative
(2002-5), 633
Cossacks (in Russia): 41
Cotler, Irwin: and Soviet Jewry, 520
Council of Europe (Strasbourg): 426
Council of State: and the Altalena, 212;
Ben-Gurion's warning to, 213–4
Creech Jone, Arthur: and 'law and
order', 154–5
Crescent (French destroyer): in action,
321
Crete: 119
Crimean War: 5
Criminal Investigation Department (CID):
103
Croats: and national self-determination,
11
Crusader ruins: and a golf course, 343
Crossman, Richard: and the Jewish
Displaced Persons, 128; protests,
132–3
Cuba: votes against, 150; tanks of, in
Syria, 466
Cukierman, Yitzhak: reaches Israel, 264
Cunliffe-Lister, Sir Philip: warns, 77
Custodian of Absentee Property: 256
Cyprus: 13, 135, 136, 142, 145, 151, 157,
159, 165, 176, 191, 197, 209, 247, 249,

Ha-Shimshoni, Zion: leads the Scouts, 68
Ha-Shomer (self-defence organization): 23, 27, 30, 32, 47
Ha-Shomer Ha'tzair: its programme, 62
Hasidism: after the Holocaust, 289; a centre of, in Israel, 305; music of, 403
Hasmonaeans: recalled, 122
Hasneh (insurance company):   52
Hasson, Ayala: launches a political scandal, 601
Hassan, Crown Prince (of Jordan):   and the Jordanian peace treaty with Israel, 572
Hassan, King (of Morocco):   and the October War, 437–8; and the road to an Israeli-Egyptian peace, 482; and the Casablanca economic conference, 576, 577; and a secret visit from Rabin, recalled, 577
Hatikvah: 7, 100, 186, 281; at the White House, 416
Hatikvah Quarter (Tel Aviv):   155
Hatzar Adar: settlers in, removed by force, 499
Hatzofeh (newspaper):   and a political crisis, 474
Havel, Vaclav: in Jerusalem, 545
Haviv, Avshalom: executed, 148
Haviv, Moshe: and a new political Party, 358; his death in action, 358–9
Hawaii: arms purchased in, 225
Hayarkon Street (Tel Aviv):   a battle in sight of, 203
Hazan, Yaakov: and reparations, 280
Hazbani River: and a war crisis, 353
Hazit Ha-Am (newspaper):   attacks the Labour movement, 72
Hazor: an immigrant village founded near, 287–8
Hazor Ha-Gelilit: founded, 287–8
Ha-Zorea: founded, 79; a museum in, 80
Heath, Edward: and the October War, 448
Heaven: and Hell, 81; and the Jezreel Valley, 143
Hebrew Gymnasium (Jaffa):   39
Hebrew language: to be a spoken language, 8–9, 19, 23–4, 72, 187, 287; the primacy of, in Palestine, 27, 29, 40; at the Paris Peace Conference, 41–2; recognized by the Mandate, 48; and Bialik, 50–1; on Cyprus, 165; and the opera Thais, 170; and statehood, 187;

and the battle of Latrun, 197; and the death of Colonel Marcus, 209; and new immigrants, 275; and a Language Academy, 299–300; the 'universalist associations' of, 335; in Jaffa, 350; at the opera, 352; and a Nobel Prize, 360; and music, 403; and the 'Oriental' Jews, 420; a unifying element, 520; and Soviet Jewry, 521
Hebrew National Opera: see National Opera (Tel Aviv)
Hebrew Patriarchs: 397, 422
Hebrew Prophets: their 'universalistic nationalism', 335; the descendants of, 421; Chaim Herzog echoes, 598–9
'Hebrew Spirit':   40
Hebrew Teachers' Federation in Palestine: 24
Hebrew University: in prospect, 20, 29, 35, 85; foundation stones of, laid, 39–40; its early years, 53–4, 57, 79; isolated, 249; opens doors to Israeli Arab students, 292; and a military tattoo (1967), 366; after the Six Day War, 394, 407; a professor of, and human rights, 533; and the tradition of 'national revival', 534; a Professor of, murdered, 541; a doctoral student at, on Memorial Day, 603; a professor of, and Palestinian rights in Jerusalem, 608
Hebrew Writers' Association: 79
Hebron Agreement (17 January 1997): 597–8
Hebron: violence in (1929), 60; and the Etzion bloc settlements (1948), 161–2; the Mayor of, and Transjordan, 241; and an Israeli reprisal action near, 363; the Patriarchs buried at, 397; return of Jews to (after 1967), 404–5, 495; and Dayan's 'view of the map', 405–6; tourism in, 420; and the 'Palestinian people', 492; and a tear-gas attack near, 503; and the Kach political Party's racism, 516; and Islamic Jihad, 530; and an appeal by Rabin, 553; a stone thrower shot dead in, 558; another shooting near, 558; the killing of twenty-nine Arabs in, 569; a United Nations presence in, 571; and Oslo II, 581, 582–3, 583–4; and the impact of terror, 593; withdrawal from, delayed, 595, 596; withdrawal from, agreed, 597–8, 611; and an Israeli political scandal, 601; and a court martial, 648

Jordan River *(continued)*
and a water crisis, 290–1, 352–3; raids
and reprisals across, 306, 343, 356;
trade across, 403, 412; Israel's
presence along, 406; Golda Meir's
'dream' for, 417; settlements along,
419; and the October War, 433, 434;
and negotiations with Jordan, 568,
573; Rabin's hopes for, 578
'Jordan is Palestine' (political slogan):
480
Jordan–Israel Agreement (1955): 299
Jordan Valley: 37, 193, 386; and the
Allon Plan, 406, 419, 469, 612; and the
Bailey map, 610–11
Jordanian–Egyptian Trusteeship: an
option (1993), 563
'Jordanian-Palestinian Delegation':
proposed, 527; created, 548; invited,
552
'Jordanian–Palestinian State': acceptable
to Israel, 476
Joseph: his policy in Pharaonic Egypt
recalled, 577
Joseph, Dov: and Count Bernadotte,
228; enters Cabinet, 252
Josephus Flavius: his account
confirmed, 351
Joshua: his spies recalled, 83
Joshua, Book of: and the name of a
new settlement, 470
Jotapata: founded, 343
Judaeo-Tat (dialect): speakers of, reach
Palestine, 55
Judah, Tribe of: and the name of a new
settlement, 268, 470
Judas Maccabeus: recalled, 12
'Judaea': a name proposed, 187
Judaea: the concept of, revived (1967),
397; 'our homeland', 422; the name,
revived (1977), 480; plans for,
broached by Begin, 481; autonomy
for, broached by Rabin, 552; Jewish
settlements in, frozen, 555; Jewish
settlements in, renewed, 615
Judaean Hills: reclamation in, 269; a
distant view of, 278
Judin (crusader fort): a settlement near,
137

Ka'anan, Hamdi: to become an
administrator, 398
Kabri: founded, 249

Kach (political Party): its electoral
success, 516
Kaddum: a new settlement at, 470
Kadima (political Party): founded
(2005), 636; its first election (2006),
638–9
Kafr Kana (Lebanon): civilians killed in
(2006), 641
Kagan, Helena: and a source of
precious milk, 33
Kahalani, Avigdor: joins coalition
(1996), 595
Kahan Commission Report: its criticisms,
510; a protest following, 511
Kahan, Yitzhak: heads Sabra and Chatila
enquiry, 510
Kahana, Baruch: his generosity
remembered, 55
Kahane, Meir: enters Knesset, 516
Kalischer, Rabbi Zevi Hirsh:
remembered, 84–5
Kaliya (Dead Sea): 85, 153, 249, 573
Kalkilya (West Bank): reprisals against,
315, 354; under Israeli military
occupation, 397; a suicide bomber,
and 'The Engineer', from, 575; under
the Palestinian Authority, 597
Kamchatka (Siberia): 147
Kamel, Mohamed Ibrahim: negotiates,
490
Kantara (Suez Canal): and the War of
Attrition, 410
Kaplan, Eliezer: and a political crisis,
139–40; his welcome message, 191
Kaplan Hospital (Rehovot): wounded
brought to, 298
Karameh (Jordan): a reprisal against,
404
Karun Lake (Lebanon): Israeli and
Syrian tanks clash at (1982), 505
Kashani, Eliezer: executed, 143
Kastel: the struggle for (1948), 168–9,
173; Jews from Kurdistan settle at,
272–3
Kastner, Israel: and a trial for slander,
303–4, 349
Katamon (Jerusalem): battle for, 183;
Irgun base at, 229
Katsav, President Moshe: charges
against, 643
Kattowitz (Upper Silesia): a pioneering
conference at, 5
Katz, Doris: and the Irgun, 229
Katz, Israel: and a new political Party,

Rice, Condoleezza: visits Middle East
(2000), 628; and the renewed peace
process (2007), 648, 653, 656
Richter, Glenn: and Soviet Jewry, 520
al-Rifati, Mohammed: killed, 647
Riga (Latvia): Jews from 9; Betar
founded in, 43; an incident in, 93
Rimalt, Elimeleh: and reparations, 280
Rishon le-Zion: founded, 6–7; Ben-
Gurion at, 24, 25; an Association of
Wine Growers in, 26; protected, 27; a
settlement founded near, 29; and
Jewish postwar immigration, 144
Rivlin, David: and the premiership
succession (1969), 408–9
roadblocks and checkpoints (West Bank):
631, 633, 634; searches at, to be 'less
strict', 643; some to be removed, 648
road deaths (Israel): 645
Road Map (for Peace): launched (2002),
626, 627; proceeds, 628, 630; Kadima
supports, 636; Olmert pursues, 637;
Hamas rejects, 638; and the renewed
Israeli-Palestinian peace process
(2007), 653, 655
Rodolfo (in La Bohème): a Tel Aviv
debut as, 351
Rogers, William P.: brokers a cease-fire,
414, 415; his plan recalled, 477
Roitberg, Roi: killed by infiltrators,
311–2
Rohlings, August: denounces Jews, 18
Romans: recalled, 41, 113, 122, 163, 235;
an old road of, re-used, 242; and a
new kibbutz, 264; and an immigrant
camp, 266–7, 273; and Masada, 267,
351; and Bar Kochba's revolt, 341; and
an ancient town, 343; and a 'historical
catastrophe', 360–1; and the Diaspora,
519
Rome: an ambassador in, and Jewish
emigration, 124; terrorism at, 624;
cafes of, rivalled, 650
Rome and Jerusalem (Moses Hess): and
Jewish nationality, 4
Romema (Jerusalem): Arabs leave, 163
Ron, Chaim: and an expedition, 116
Roosevelt, President Franklin D.: his
rural retreat, and the Middle East
peace process, 491
Roscher-Lund, Colonel: in Jerusalem,
164–5; his warning, 178
Rosen, Pinhas: enters Cabinet, 252–3;
and the 'Lavon Affair', 338, 339

Rosen, Rolly: and the 'ethnic devil', 604
Rosenblatt, Zevi: charged with murder,
72
Rosh Ha-Ayin: and an immigrant camp,
261–2; becomes a farming village, 272
Rosh Ha-Nikrah: founded, 264
Rosh Pinah: founded, 4–5, 6; a
settlement founded near, 137; and the
withdrawal of British troops from
Palestine, 175; and the Arabs near,
177; an immigrant camp near, 288
Ross, Dennis: and the Oslo Accords,
564; and the extension of Palestinian
autonomy, 583; a mediator during the
Hebron talks, 598
Rotblat, Professor Joseph: his appeal,
522–3
Rothschild family: and Palestine, 5, 10,
18
Rothschild, Baron Edmond de: supports
early Jewish settlements, 4, 6, 9, 12,
13, 26
Rothschild, Dorothy de: a benefactress,
97–8
Rothschild, Hannah: remembered, 62
Rothschild, Lord (Lionel Walter
Rothschild): and the Balfour
Declaration, 34
Roumania: Jews from, in Palestine, 4,
22, 27, 30, 44, 55, 76, 96; and the
Jewish (later Israeli) anthem, 7; anti-
Semitic policies in, 65, 95; British
anti-refugee pressure on, 98, 119, 136;
survivors from, reach Palestine, 196,
198, 199; immigrants from, reach
Israel, 233, 259, 266, 352; mediates,
482; Soviet Jews in transit through,
536–7; foreign workers from, 613
Rousso, Claire: killed, 135
Royal Air Force: and a parachutist, 102;
attacks an Arab stronghold, 159; small
aircraft purchased from, 189; an Israeli
clash with, 247
Royal Albert Hall (London): Balfour's
speech in, 119; a memorial meeting to
Rabin in, 591
Royal Commission on Alien Immigration:
Herzl's evidence to, 21
Royal Fusiliers: Jewish battalions of, 36
Royal Navy: intercepts would-be
immigrants, 105, 145; and the Israeli
navy, 269
Royal Palace (Akaba): negotiations at,
572

St Jérome (Marseille): refugees at, 254
St John's Hospice (Jerusalem): purchased by Jews, 524
St John's Hospital (East Jerusalem): 634
St Ottilien (near Munich): Displaced Persons at, 123
St Petersburg (Russia): Herzl visits, 22; see also Leningrad
Sakaria (refugee ship): hostile reaction to, 101
Salah ed-Din Street (Jerusalem): and the intifada, 540
Salameh (near Tel Aviv) a target, 162
Salameh, Hassan: his headquarters attacked, 168
Salonica (Greece): Jewish dockworkers of, 50
al-Salt (Transjordan): Turks driven from, 37
Salzberger, Lotte: and human rights, 533
Salzburg: an Israeli-Palestinian initiative in, 634
Samaria: a kibbutz set up in, 113; Arab refugees seek safety of, 218; an ancient Jewish kingdom in, 397; carpenters of, 403; 'our homeland', 422; the name of, revived, 480; Begin's proposal for, 481; Rabin's proposal for, 552; Jewish settlements in, frozen, 555; a proposed Israeli corridor through, 610; parts of, to be retained by Israel, 612; Jewish settlements in, to be expanded, 616
Samson: his birthplace, 221; and the Philistine sea god, 290
Samua (West Bank): Israeli reprisal raid on, 363–4
Samuel, Book of: and a spy ring, 33
Samuel, Sir Herbert: in Palestine, 48, 49
San'a (Yemen): Jews from, 263
Sandström, Emil: and Palestine, 145, 148
Sapir, Pinhas: and the emergence of Golda Meir as Prime Minister, 407–8, 409; opposes West Bank settlements, 423, 424; provides funds for settlements, 425
Sarafand: Detention Camp at, 100, 101, 103; army base at, 212
Sarajevo (Yugoslavia): a Jew from, 183
Saris (Arab village): occupied, 168; levelled to the ground, 196
Sarraj, Dr Iyad: and the psychology of the suicide bomber, 569
Sa'sa' (Arab village): abandoned, 264
Sasa: founded, 264

Sassa (Syria): and the October War, 454
Sasson, Eliahu: meets Abdullah, 150
Sassoon, Yehezkel: remembered, 44
Saudi Arabia: and the Arab nation, 76; opposes Jewish immigration, 97; in the Middle East, 120; and oil, 141; and Israel's War of Independence, 193, 232; opposes recognition, 254; and the United States, 273; a threat from, 309; an uncompromising stance by, 332; and 'a cheque for a billion dollars', 451; and 'petro-dollar diplomacy', 460–1; troops of, in Syria, 466; supports 'Zionism is racism' resolution, 468; refuses to join Camp David talks, 491; and the Gulf War, 546; supports Road Map, 627; represented at Annapolis (2007), 653
Saul: his encampment, 44; a witch visited by, 205
Saving Children Project: 634
Savir, Uri: and the Oslo talks, 563
Scandinavia: Jews 'comfortable' in, 147
Schapira, Hermann: and a Jewish university, 8, 20
Schatz, Boris: founds an art school, 27
Schiff, Ze'ev: and the status of Israeli Arabs, 292; his reflections on the Lebanon War (1982), 512
Schneersohn, Rabbi Joseph Isaac: and a Hasidic centre in Israel, 305
Schoenau (Austria): refugee camp at, 426
Schultz, George: and the London Agreement, 523
Schwimmer, Al: acquires aircraft, 200; builds aircraft, 273–4
Scorpion Ascent (Negev): a terrorist attack at, 294
Scud missiles: strike Israel, 546–7
Sde Boker: founded, 277–8; Ben-Gurion retires to, 293, 294; a visit to, 297; Ben-Gurion emerges from, 356; Ben-Gurion returns to, 358
Sde Dov airfield (Tel Aviv): an attack on, 189; aircraft prepared at, 189–90
Sde Eliyahu: founded, 98
Sde Nahum: founded, 85
Sde Nehemyah: founded, 102
Sderot: rocket attacks on, from Gaza, 629, 639, 644, 645, 654; an initiatve by the mayor of, 632
Sdom (the biblical Sodom): relieved, 240, 358; an evacuation to, 249;

Solomon, King *(continued)*
  Israel at the time of, 253; and the
  location of an immigrant camp, 287–8
Song of Deborah (symphonic poem):
  403
'Song of Peace': commissioned, 400;
  sung, 587; becomes an anthem, 589
Sonnenschein, Rosa: an early Zionist, 14
Sorek Gorge: Israel gains control of,
  234; an immigrant camp near, 268
South America: immigrants from 205; an
  immigrant from, killed, 614
South Africa: Russian Jews emigrate to,
  5; 'closed', 65; and a new settlement,
  137; a volunteer from, killed in action,
  200; volunteers from, in action, 201,
  222; immigrants from, in Israel, 268;
  and a model township, 290; a Jew
  from, at the White House, 480; and a
  comparison, 549, 551
Southern Lebanon: Israel withdraws
  from (1949), 253; shelling from (1982),
  503; Israel advances through (1982),
  503–5; PLO supporters under arms in,
  513; Israeli military control of, 515–6,
  518; Hizballah active in, 530; an
  abduction from, 542; and the Helsinki
  Summit (1990), 546; shelling from
  (1996), 593; the bombardment of
  (1996), 594; the growing debate over a
  withdrawal from, 599–600, 606–7; the
  continuing conflict in, 617
Southern Lebanese Army (SLA): Israel's
  support for, 518
Souwi, Saleh Nazal: a suicide bomber,
  575
Soviet Communism: disillusionment
  with, 273
Soviet Cultural Centre (Damascus): and
  the October War, 439
Soviet Union: 50, 112, 150, 165, 189;
  Jews of, 226, 257, 286–7, 304–5, 414,
  420, 426–7, 497, 520–1; Israel's
  relations with, 251, 253, 279, 304–5,
  326, 347; and Egypt, 300, 309, 317; a
  warning from (1956), 326; and Syria,
  333, 365; and a precedent, 346; and
  the road to war (in 1967), 365–6,
  373; and the Six Day War, 387, 388,
  390, 391, 393; and the occupied
  territories, 402; and Arab arsenals, 407;
  and the War of Attrition, 410; and
  'Black September', 417; and the
  October War, 427, 431, 433, 439, 440,
  441, 443–4, 445, 448, 449–51, 452,
  455–6; and the aftermath of the
  October War, 463; seen as contributing
  to a 'mortal danger' to Israel, 481; to
  be excluded, 483; Jews from, reach
  Israel, 519; and the London
  Agreement, 523; mass Jewish
  emigration from, 535–6, 544, 550; and
  the Madrid Conference, 548;
  disintegration of (1991), 548–9, 550
Spain: an expulsion from, remembered,
  352
Special Night Squads (Palestine):
  established, 93
Spragg, Flight Lieutenant Brian: and a
  'Yiddish' Spitfire, 247
Sprinzak, David: killed in action, 203
Sprinzak, Yosef: and a political crisis,
  139
Sri Lanka tsunami (2004): Israel sends
  aid for, 650
Stalin, Joseph: curbs Jewish emigration,
  68; and the Nazi–Soviet pact, 99; anti-
  Jewish policy of, 273
Stalingrad, Siege of (1942–3): recalled,
  443–4
Stanley, Oliver: defends land purchase
  restrictions, 104
Star of David: soldiers wear, 36; aircraft
  display, 200; and an Arab symbol, 236;
  flies at the Gulf of Akaba, 248
Stars and Stripes: in Jerusalem, 164
State Department (Washington DC):
  245, 273, 327, 483
Stavsky, Avraham: charged with murder,
  72; killed, 212
Steiger, Yitzhak: remembered, 91
Stern, Avraham: breaks away, 111–2
Stern, Barbara: and Soviet Jewry, 520
Stern Gang: formed, 111; and an
  assassination, 118; and a promise, 119;
  and continuing action by, 121–2; and
  the Haganah, 132, 135; and the Jewish
  Agency, 142; and two suicides, 143;
  killings by, 157, 158, 159, 164; and
  Deir Yassin, 169; and the struggle for
  Palestine (1947–8), 178; and the Israel
  Defence Forces, 202; and the
  assassination of Bernadotte, 228; a
  fighter of, killed in action, 244; and
  Israel's first general election, 250; and
  the Land of Israel Movement, 400
Stern, Professor Menachem: murdered,
  541

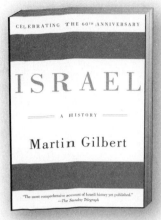
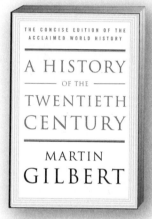
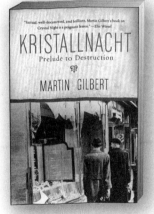